Springer Monographs in Mathematics

This series publishes advanced monographs giving well-written presentations of the "state-of-the-art" in fields of mathematical research that have acquired the maturity needed for such a treatment. They are sufficiently self-contained to be accessible to more than just the intimate specialists of the subject, and sufficiently comprehensive to remain valuable references for many years. Besides the current state of knowledge in its field, an SMM volume should ideally describe its relevance to and interaction with neighbouring fields of mathematics, and give pointers to future directions of research.

More information about this series at https://link.springer.com/bookseries/3733

Marco Manetti

Lie Methods in Deformation Theory

Springer

Marco Manetti
Department of Mathematics
Sapienza University
Rome, Italy

ISSN 1439-7382 ISSN 2196-9922 (electronic)
Springer Monographs in Mathematics
ISBN 978-981-19-1184-2 ISBN 978-981-19-1185-9 (eBook)
https://doi.org/10.1007/978-981-19-1185-9

Mathematics Subject Classification: 18G55, 13D10, 53D17

This Springer imprint is published by the registered company Springer Nature Singapore Pte Ltd.
The registered company address is: 152 Beach Road, #21-01/04 Gateway East, Singapore 189721, Singapore

Preface

Deformation theory is at the same time a classical subject in mathematics and a very active field of contemporary research, especially in algebra and geometry. During the years a variety of approaches and methods have been proposed to understand deformations, and this book aims to give an accessible and self-contained treatment of a particular class of these techniques.

Although the first ideas of deformation theory date back to Riemann, the modern approach to deformation theory started with the seminal works by Kodaira, Spencer and Kuranishi in deformation theory of compact complex manifolds [147–150, 162] and by Gerstenhaber in deformation theory of associative algebras [88]. A few years later, in a series of papers by Nijenhuis and Richardson [208, 209] the authors recognized that in both Gerstenhaber's theory and Kodaira—Spencer—Kuranishi's theory, a basic role is played by a certain equation among the elements of degree one in a suitable differential graded Lie algebra.

The contribution by Nijenhuis and Richardson was then developed by several people, including Schlessinger and Stasheff [238], Deligne [54], Goldman and Millson [93–95], and Drinfeld [60]. Later the same ideas were developed and clarified in the nineties by Kontsevich, Hinich, Schechtman and others of the Russian school. From then on, this approach has grown exponentially in the scientific literature and it's impossible to give here a comprehensive treatment of this new chapter of mathematics.

Very roughly, the underlying philosophy can be described by Deligne's claim: "*in characteristic 0, a deformation problem is controlled by a differential graded Lie algebra, with quasi-isomorphic DG-Lie algebras giving the same deformation theory*" [54]. Here a deformation problem is intended in an imprecise and heuristic way that can be extracted from all the concrete examples arising in algebra, algebraic geometry, etc. It is also possible to transform the above philosophy into a theorem, whenever a formal axiomatic definition of deformation problem is given; this line of research, carried out by several people, has culminated in a celebrated theorem obtained independently by Pridham [214] and Lurie [174].

In this book we do not follow this axiomatic approach and we continue to consider deformation problems as suggested by concrete examples, in particular by deformations of complex manifolds, vector bundles and holomorphic maps. Needless to say that the above philosophy has received a lot of additional evidence, and the construction of the DG-Lie algebra controlling a given deformation problem is nowadays easier and more conceptually clear than 30 years ago.

It is important to point out that, with few exceptions, there isn't any canonical way to associate an isomorphism class of DG-Lie algebras to a given deformation problem; in fact, the controlling DG-Lie algebra is usually defined only up to quasi-isomorphisms. Since the general quasi-isomorphism class of DG-Lie algebras does not admit any canonical or minimal representative, it is very useful to include the category of DG-Lie algebras into the bigger category of L_∞-algebras; every DG-Lie algebra is an L_∞-algebra, two DG-Lie algebras are quasi-isomorphic if and only if they are homotopy equivalent as L_∞-algebras and every homotopy class of L_∞-algebras is represented by a unique (up to isomorphism) minimal model.

A large part of the second half of this book is devoted to the definition and homotopy classification of L_∞-algebras. These algebraic structures are an essential and extremely important tool in deformation theory, not only for the unicity of minimal models but also because they frequently encode deformation problems in a clearer way than DG-Lie algebras.

According to the standard point of view of algebraic and complex geometry, every problem in (infinitesimal) deformation theory is formally described by a functor of Artin rings, namely, by a covariant functor from the category of local Artin rings to the category of sets. Every differential graded Lie algebra pro-represents, in a suitable way, a functor of Artin rings; the advantage of this kind of pro-representability is that we can describe all the classical aspects of deformation theory (tangent space, obstruction maps, Massey powers, etc.) whenever we know the quasi-isomorphism class of the controlling DG-Lie structure. One must be aware that the functor of Artin rings does not determine uniquely the controlling differential graded Lie algebra, even up to quasi-isomorphism; in fact this has been one of the main motivations for the study of derived deformation theory, where functors of Artin rings are replaced by functors from the category of differential graded Artin rings to the category of Kan complexes.

In this book, which is introductory in its nature, we concentrate on the basic theory, carried out over an arbitrary field of characteristic 0, and applications to deformations of complex manifolds, vector bundles and holomorphic maps. We do not explicitly consider derived deformation theory, although its general principles and methods are guiding our approach, and the homotopical point of view is kept to the minimum. Our hope is that, while the reader with a wide mathematical background can find this book useful for several deformation problems, the limitations to complex manifolds and holomorphic maps should make this book accessible also to graduate students with a basic knowledge in algebraic topology, homological algebra and complex algebraic geometry.

The table of contents is planned according to the well-known principle that "good general theory does not search for the maximum generality, but the right generality";

in particular the involved notions are presented in order of increasing generality. As a by-product of this fact some theorems are proved several times, at different levels of generality and methods of proof. As an example, the Bogomolov–Tian–Todorov theorem is proved twice, the first time for compact Kähler manifolds and the second time, with a greater number of preliminary results, for manifolds for which the Hodge to de Rham spectral sequence degenerates at E_1.

The book is ideally divided into four parts. The Part I (Chaps. 1–4) is devoted to some classical aspects of deformation theory; in particular, we develop here Schlessinger's theory and obstruction calculus for functors of Artin rings. In the Part II (Chaps. 5–9) we deal with the use of differential graded Lie algebras in deformation theory. In the Part III (Chaps. 10–14) we enlarge the category of differential graded Lie algebras to the category of L_∞-algebras. The last part consists of three appendices, each one containing either mathematical background or complementary topics.

As the title suggests, this book covers only a small part of what is nowadays called deformation theory and it is intended as a complement to, and not as a replacement, of any of the excellent introductory books on deformation theory already in literature, for instance [7, 39, 110, 145, 242]. Almost all the results contained here have already appeared in other books and research papers and are related to the state of the art reached around the year 2015; however, we always tried to look for simplified or more elementary proofs and a better exposition. For these reasons, and also to ensure a reasonable number of pages, lots of equally interesting topics are missing. In particular, cotangent complex and deformations of (singular) schemes, deformations of algebras over operads and derived deformation theory are not covered. It is worth to saying that most of the methods described in this book equally apply to the above-mentioned topics; however, for a satisfying theory of cotangent complex and deformations of (possibly singular) schemes, DG-Lie methods alone are not sufficient and a significant background of homotopical algebra is also required.

The material contained in this book grew out of some graduate courses and some short advanced lecture series given in the period 1996–2016 at several locations, mainly Sapienza University, Rome, and Scuola Normale Superiore, Pisa. During the preparation of this book, the following people have greatly contributed with ideas, comments, remarks and corrections: Enrico Arbarello, Ruggero Bandiera, Alberto Canonaco, Francesca Carocci, Carmelo Di Natale, Barbara Fantechi, Domenico Fiorenza, Donatella Iacono, Emma Lepri, Luigi Lunardon, Francesco Meazzini, Kieran O'Grady, Giulia Ricciardi, Luca Simi, Bruno Vallette, Angelo Vistoli; to all of them my sincere sense of gratitude.

I'm grateful to Francesca Bonadei, Marina Reizakis and Masayuki Nakamura of Springer-Verlag for their guidance from the first draft to the published version. I would like to thank also the anonymous reviewers for sharing some very useful comments, and the institutions who hosted the author during the preparation of the first draft of this book, viz., Scuola Normale Superiore (January 2012), University of Milan (March 2012) and Max–Planck Institute for Mathematics (May 2012).

Rome, Italy
February 2022

Marco Manetti

Contents

Chapter 1
An Overview of Deformation Theory of Complex Manifolds

The aim of this chapter is to give a partial and informal introduction to classical deformation theory of complex manifolds and moves around the notion of a **family of compact complex manifolds**, defined as a proper holomorphic submersion $f: M \to B$ of complex manifolds. The origin of the problem lies in the fact that, while the differential structure of the fibres $M_t := f^{-1}(t)$ remains unchanged when B is connected, several examples show that in general M_t and M_s have different complex structures for $t \neq s$.

According to Kodaira, Nirenberg and Spencer we define a deformation of a compact complex manifold X as a triple composed of a family $f: M \to B$, a base point $0 \in B$ and a biholomorphic map $X \to f^{-1}(0)$: there exists a natural notion of isomorphism of deformations that involves only the structure of f in a neighbourhood of the base point. The Kodaira–Spencer map of such a deformation is a linear map $\rho: T_0B \to H^1(X, \Theta_X)$ which is an obstruction to analytic triviality of the family in a neighbourhood of the point 0, and plays an important role in the notion of semiuniversal deformation and in the Kuranishi existence theorem.

In this first chapter we implicitly assume that the reader has had some practical experience in algebraic geometry and familiarity with the notions of a complex manifold and a complex space; many results are only sketched and most proofs are replaced by suitable references. However, the complete understanding of this chapter is certainly useful but not strictly necessary for the comprehension of the remaining part of the book.

1.1 Proper Smooth Families of Complex Manifolds

For every complex manifold M, the sheaf of holomorphic functions is denoted by $\mathcal{O}_M$. Every choice of local holomorphic coordinates $z_1, \ldots, z_n$ in a neighbourhood of $p \in M$ gives a local isomorphism of the stalk $\mathcal{O}_{M,p}$ with the ring $\mathbb{C}\{z_1, \ldots, z_n\}$

M. Manetti, *Lie Methods in Deformation Theory*, Springer Monographs in Mathematics,
https://doi.org/10.1007/978-981-19-1185-9_1

of convergent power series. We recall that $\mathbb{C}\{z_1, \ldots, z_n\}$ is a local Noetherian ring with residue field $\mathbb{C}$ and maximal ideal generated by $z_1, \ldots, z_n$.

The (complex) tangent space at a point $p \in M$ is denoted by T_pM. If $Z \subset M$ is a complex submanifold we denote by $N_{Z|M}$ the normal bundle of Z in X.

Definition 1.1.1 A **proper smooth family** of complex manifolds is a proper holomorphic map $f: M \to B$ of complex manifolds such that:

1. M is nonempty and B is connected;
2. the differential $df: T_pM \to T_{f(p)}B$ is surjective at every point $p \in M$.

Two families $f: M \to B$, $g: N \to B$ over the same base are isomorphic if there exists a holomorphic isomorphism $N \to M$ commuting with f and g.

From now on, when there is no risk of confusion, we shall simply say proper smooth family instead of proper smooth family of complex manifolds.

As an example, for every compact complex manifold X and every connected complex manifold B, the projection map $f: X \times B \to B$ is a proper smooth family.

If $f: M \to B$ is a proper smooth family, then f is open, closed and surjective. For every $b \in B$ we denote by $M_b = f^{-1}(b)$ the fibre of f over the point b; by the implicit function theorem, M_b is a compact complex submanifold of M and the normal bundle of M_b in M is the trivial vector bundle of rank equal to the dimension of B. For every connected open subset $V \subset B$, the restriction $f: f^{-1}(V) \to V$ is a proper smooth family and, more generally, for every holomorphic map of connected complex manifolds $C \to B$, the pull-back $M \times_B C \to C$ is a proper smooth family.

Definition 1.1.2 A proper smooth family $f: M \to B$ is called **trivial** if it is isomorphic to $M_b \times B \to B$ for some (and hence all) $b \in B$. It is called **locally trivial** if there exists an open cover $B = \cup U_a$ such that every restriction $f: f^{-1}(U_a) \to U_a$ is trivial.

The following examples of families also show that, in general, if $a, b \in B, a \neq b$, then M_a is not biholomorphic to M_b and therefore that not every family is locally trivial.

Example 1.1.3 (*Plane cubics*) Consider $B = \mathbb{C} - \{0, 1\}$,

$$M = \{([x_0, x_1, x_2], \lambda) \in \mathbb{P}^2 \times B \mid x_2^2 x_0 = x_1(x_1 - x_0)(x_1 - \lambda x_0)\},$$

and $f: M \to B$ the projection. Then f is a proper smooth family and the fibre M_λ is the smooth plane cubic with j-invariant

$$j(M_\lambda) = 2^8 \frac{(\lambda^2 - \lambda + 1)^3}{\lambda^2(\lambda - 1)^2}.$$

In particular, the function $\lambda \mapsto j(M_\lambda)$ is not locally constant and it is sufficient to recall that two elliptic curves are biholomorphic if and only if they have the same j-invariant.

Example 1.1.4 (*Projective hypersurfaces*) For a fixed pair of integers $n, d > 0$, consider a set of indeterminates $a_{i_0,\dots,i_n}$, indexed by the set of multi-indices

$$I = \{(i_0, \dots, i_n) \in \mathbb{N}^d \mid i_0 + \cdots + i_n = d\}, \qquad |I| = \binom{d+n}{n},$$

and the bihomogeneous polynomial of bidegree $(1, d)$

$$F(a, x) = \sum_{(i_0,\dots,i_n)\in I} a_{i_0,\dots,i_n} x_0^{i_0} \cdots x_n^{i_n}.$$

For every $[a] \in \mathbb{P}^N$, $N = \binom{d+n}{n} - 1$, the closed subset

$$X_a = \{[x] \in \mathbb{P}^n \mid F(a, x) = 0\}$$

is a projective hypersurface of degree d; conversely, every hypersurface of degree d in $\mathbb{P}^n$ is equal to X_a for some $a \in \mathbb{P}^N$. Defining

$$X = \{([x], [a]) \in \mathbb{P}^n \times \mathbb{P}^N \mid F(a, x) = 0\},$$
$$Z = \left\{([x], [a]) \in X \;\middle|\; \frac{\partial F(a, x)}{\partial x_i} = 0 \;\forall i\right\},$$

it is well known, and in any case easy to prove, that X is a smooth hypersurface of $\mathbb{P}^n \times \mathbb{P}^N$ and the differential of the projection $f : X \to \mathbb{P}^N$ is not surjective at a point $([x], [a])$ if and only if $([x], [a]) \in Z$, or equivalently if and only if $[x]$ is a singular point of X_a. Therefore, since $f(Z)$ is a Zariski closed subset of $\mathbb{P}^N$, the subset

$$B = \mathbb{P}^N - f(Z) = \{ [a] \in \mathbb{P}^N \mid X_a \text{ is smooth}\}$$

is a Zariski open subset of a projective space and hence a connected complex manifold. Denoting $M = f^{-1}(B)$ we obtain that $f : M \to B$ is a proper smooth family of complex manifolds and every smooth hypersurface of degree d of $\mathbb{P}^n$ is isomorphic to a fibre of f.

Example 1.1.5 (*Hopf surfaces*) The set of matrices

$$Z = \left\{ \begin{pmatrix} a & b \\ 0 & c \end{pmatrix} \in M_{2,2}(\mathbb{C}) \;\middle|\; |a| > 1,\ |c| > 1 \right\}$$

carries a natural structure of a connected complex manifold of dimension 3.

For every matrix $A \in Z$, let $\langle A \rangle \subset GL_2(\mathbb{C})$ be the infinite cyclic subgroup generated by A. It is easy to see that the action of $\langle A \rangle$ on $U := \mathbb{C}^2 - \{0\}$ is free and properly discontinuous and the quotient $S_A = U/\langle A \rangle$ is a compact complex manifold called a **Hopf surface**. The (holomorphic) projection map $U \to S_A$ is the universal covering space and then, for every point $x \in S_A$ there exists a natural isomorphism $\pi_1(S_A, x) \simeq \langle A \rangle$. When $A \in Z$ is a real multiple of the identity we can immediately see that S_A is diffeomorphic to $S^1 \times S^3$.

It is easy to prove that two matrices $A, B \in Z$ are similar if and only if S_A is biholomorphic to S_B. The "only if" part is clear; in fact every linear isomorphism $C: \mathbb{C}^2 \to \mathbb{C}^2$ changes the action of $\langle A \rangle$ into the action of $\langle CAC^{-1} \rangle$ and this immediately implies that C induces an invertible holomorphic map $C: S_A \to S_{CAC^{-1}}$. Conversely, if $f: S_A \to S_B$ is a biholomorphism, then f lifts to a biholomorphism $g: U \to U$ such that $gA = B^k g$ for some integer k; since f induces an isomorphism of fundamental groups we must have $k = \pm 1$. By Hartogs' theorem, g extends to a biholomorphism $g: \mathbb{C}^2 \to \mathbb{C}^2$ such that $g(0) = 0$, and differentiating at 0 the equality $gA = B^k g$ we obtain that A and B^k are similar matrices; in particular $\det(A) = \det(B)^k$. Finally, since $A, B \in Z$ we have $k > 0$ and therefore A is similar to B.

Define now $X = Z \times U$ and let $G \subset \mathrm{Aut}(X)$ be the infinite cyclic subgroup generated by the holomorphic isomorphism:

$$(A, u) \mapsto (A, Au), \qquad A \in Z, \ u \in U.$$

The action of G on X is free and properly discontinuous; denote by $M = X/G$ its quotient and by $f: M \to Z$ the projection on the first factor. Then f is a proper holomorphic family whose fibres are Hopf surfaces.

From the point of view of differential topology, every smooth proper family of complex manifolds is locally trivial.

Theorem 1.1.6 (Ehresmann's theorem) *Let $M \xrightarrow{f} B$ be a proper submersive differentiable map of differentiable manifolds. Then for every point $0 \in B$ there exist an open neighbourhood $0 \in U \subset B$ and a diffeomorphism $\phi: M_0 \times U \to f^{-1}(U)$ such that $\phi(x, 0) = x$ and $f\phi(x, t) = t$ for every $x \in M_0$, $t \in U$.*

In particular, if B is connected, then the diffeomorphism type of the fibre M_b is independent of the point $b \in B$.

Proof For the proof we refer either to the original paper [63] or to the more recent books [145, Theorem 2.3], [264, Proposition 9.3]. □

We recall that submersive means that the differential is surjective at every point and then every proper smooth family of compact complex manifolds is locally trivial from the differentiable point of view. For instance, Ehresmann's theorem applied to Example 1.1.5 above gives that any two Hopf surfaces are diffeomorphic to each other and then diffeomorphic to $S^1 \times S^3$.

In order to avoid possible mistakes, we point out that Theorem 1.1.6 becomes false if the properness assumption is replaced by the compactness of every fibre. As an example consider a locally finite open cover $\{U_a\}$ of B, take $M = \coprod_a U_a$ the disjoint union and define $f: M \to B$ as the morphism induced by all the inclusion maps $U_a \hookrightarrow B$.

1.2 A Short Review of Čech and Dolbeault Cohomology

We assume that the reader is familiar with cohomology theory of sheaves and with Dolbeault cohomology of vector bundles on complex manifolds; the goal of this section is to fix some notation used throughout the book. For a sheaf $\mathcal{F}$ on a topological space we shall use both the symbols $\mathcal{F}(U)$ and $\Gamma(U,\mathcal{F})$ to denote the space of its sections over an open subset U. The stalk of a sheaf $\mathcal{F}$ at a point x will be denoted by $\mathcal{F}_x$. For any complex manifold X we shall denote:

- by $\mathcal{A}_X^{p,q}$ and $A_X^{p,q}=\Gamma(X,\mathcal{A}_X^{p,q})$ the sheaf and the space of differential forms of type (p,q) respectively;
- by $\mathcal{O}_X$ and Θ_X the sheaves of holomorphic functions and holomorphic vector fields;
- by Ω_X^p the sheaf of holomorphic differential p-forms.

Similarly, for every locally free sheaf of $\mathcal{O}_X$-modules $\mathcal{E}$ we shall denote by $\mathcal{A}_X^{p,q}(\mathcal{E})$ and $A_X^{p,q}(\mathcal{E})=\Gamma(X,\mathcal{A}_X^{p,q}(\mathcal{E}))$ the sheaf and the space of differential forms of type (p,q) with values in $\mathcal{E}$, respectively. Notice that for every p,q we have a natural isomorphism of sheaves $\mathcal{A}_X^{p,q}(\mathcal{E})\simeq\mathcal{A}_X^{0,q}(\Omega_X^p\otimes_{\mathcal{O}_X}\mathcal{E})$. The qth Dolbeault cohomology group of $\mathcal{E}$ is denoted by $H_{\overline{\partial}}^q(X,\mathcal{E})$; by definition it is the qth cohomology group of the Dolbeault complex

$$0\to A_X^{0,0}(\mathcal{E})\xrightarrow{\overline{\partial}}A_X^{0,1}(\mathcal{E})\xrightarrow{\overline{\partial}}A_X^{0,2}(\mathcal{E})\xrightarrow{\overline{\partial}}\cdots.$$

Let $\mathcal{U}=\{U_a\}$ be an open cover of a complex manifold X and denote by $U_{a_0\cdots a_k}=U_{a_0}\cap\cdots\cap U_{a_k}$. For every sheaf of abelian groups $\mathcal{F}$ on X we denote by

$$C(\mathcal{U},\mathcal{F}):\qquad C^0(\mathcal{U},\mathcal{F})\xrightarrow{\delta}C^1(\mathcal{U},\mathcal{F})\xrightarrow{\delta}C^2(\mathcal{U},\mathcal{F})\xrightarrow{\delta}\cdots$$

the corresponding complex of Čech cochains; recall that

$$C^k(\mathcal{U},\mathcal{F})=\prod_{a_0,\ldots,a_k}\mathcal{F}(U_{a_0\cdots a_k})$$

is the abelian group of all the parametrized collections $\{f_{a_0a_1\cdots a_k}\}$, where $a_0,\ldots,a_k$ is an ordered sequence of indices and $f_{a_0a_1\cdots a_k}\in\mathcal{F}(U_{a_0\cdots a_k})$. The Čech differential δ is defined by the formula

$$(\delta f)_{a_0\cdots a_n}=\sum_{i=0}^{n}(-1)^i f_{a_0\cdots\widehat{a_i}\cdots a_n}{}_{|U_{a_0\cdots a_n}}\tag{1.1}$$

and the cohomology of the above complex is denoted by $\check{H}^*(\mathcal{U},\mathcal{F})$.

Remark 1.2.1 It is often useful to consider the subcomplex $C_a(\mathcal{U}, \mathcal{F}) \subset C(\mathcal{U}, \mathcal{F})$ of alternating Čech cochains. By definition a Čech cochain $\{f_{a_0a_1\cdots a_k}\}$ is alternating if:

1. $f_{a_{\sigma(0)}a_{\sigma(1)}\cdots a_{\sigma(k)}} = (-1)^\sigma f_{a_0a_1\cdots a_k}$ for every permutation σ of the set $\{0, \ldots, k\}$, where $(-1)^\sigma$ denotes the signature of σ;
2. $f_{a_0a_1\cdots a_k} = 0$ whenever $a_i = a_j$ for some $i \neq j$.

Similarly, for every total order in the set of indices, we can consider the complex $C_o(\mathcal{U}, \mathcal{F})$ of ordered Čech cochains, where

$$C_o^k(\mathcal{U}, \mathcal{F}) = \prod_{i_0<\cdots<i_k} \mathcal{F}(U_{i_0\cdots i_k}), \qquad k \geq 0,$$

and the differential δ is defined by the same formula (1.1) of the usual complex. The natural projection $C(\mathcal{U}, \mathcal{F}) \to C_o(\mathcal{U}, \mathcal{F})$ is a morphism of complexes whose restriction to the subcomplex $C_a(\mathcal{U}, \mathcal{F})$ is an isomorphism. By well known and standard results in homology theory, the two morphisms of complexes

$$C_a(\mathcal{U}, \mathcal{F}) \xrightarrow{\text{inclusion}} C(\mathcal{U}, \mathcal{F}) \xrightarrow{\text{projection}} C_o(\mathcal{U}, \mathcal{F})$$

induce isomorphisms in cohomology. A morphism of cochain complexes that induces an isomorphism in cohomology is called a quasi-isomorphism.

Proposition 1.2.2 *For every locally free sheaf $\mathcal{E}$ of $\mathcal{O}_X$-modules on a complex manifold X and every open cover $\mathcal{U} = \{U_a\}$, $a \in \mathcal{I}$, there exists a natural morphism of vector spaces*

$$\varphi\colon \check{H}^k(\mathcal{U}, \mathcal{E}) \to H^k_{\bar{\partial}}(X, \mathcal{E}).$$

Proof Let $t_a\colon X \to \mathbb{C}$, $a \in \mathcal{I}$, be a fixed differentiable partition of unity subordinate to the open cover $\{U_a\}$. For every Čech cochain $f \in C^k(\mathcal{U}, \mathcal{E})$ we consider the differential forms:

$$\phi_a(f) = \sum_{c_1,\ldots,c_k} f_{ac_1\cdots c_k}\bar{\partial}t_{c_1} \wedge \cdots \wedge \bar{\partial}t_{c_k} \in \Gamma(U_a, \mathcal{A}^{0,k}_X(\mathcal{E})), \qquad a \in \mathcal{I},$$

$$\phi(f) = \sum_a t_a\phi_a(f) \in A^{0,k}_X(\mathcal{E}) = \Gamma(X, \mathcal{A}^{0,k}_X(\mathcal{E})).$$

Then $\phi\colon C(\mathcal{U}, \mathcal{E}) \to A^{0,*}_X(\mathcal{E})$ is a morphism of complexes; given $f \in C^k(\mathcal{U}, \mathcal{E})$, since $\sum_{c_i} \bar{\partial}t_{c_i} = 0$ and $\bar{\partial}\phi_a(f) = 0$ for every a we have

$$\bar{\partial}\phi(f) = \sum_a \bar{\partial}t_a \wedge \phi_a(f) = \sum_{c_0,\ldots,c_k} f_{c_0\cdots c_k}\bar{\partial}t_{c_0} \wedge \cdots \wedge \bar{\partial}t_{c_k},$$

$$\begin{aligned}
\phi(\delta f) &= \sum_a t_a \sum_{c_0,\dots,c_k} (\delta f)_{ac_0\cdots c_k}\overline{\partial} t_{c_0}\wedge\cdots\wedge\overline{\partial} t_{c_k}\\
&= \sum_a t_a \overline{\partial}\phi(f)\\
&\quad -\sum_a t_a \sum_{i=0}^{k}\sum_{c_i}\overline{\partial} t_{c_i}\wedge\sum_{c_0,\dots,\widehat{c_i},\dots,c_k} f_{ac_0\cdots\widehat{c_i}\cdots c_k}\overline{\partial} t_{c_0}\wedge\cdots\wedge\widehat{\overline{\partial} t_{c_i}}\wedge\cdots\wedge\overline{\partial} t_{c_k}\\
&= \sum_a t_a\overline{\partial}\phi(f) = \overline{\partial}\phi(f).
\end{aligned}$$

Denoting by φ the morphism induced by ϕ in cohomology, we want to prove that φ is independent of the choice of the partition of unity. Let v_a be another partition of unity subordinate to the cover $\{U_a\}$ and define, for every $f \in C^k(\mathcal{U},\mathcal{E})$, by

$$\tilde{\phi}_a(f) = \sum_{c_1,\dots,c_k} f_{ac_1\cdots c_k}\overline{\partial} v_{c_1}\wedge\cdots\wedge\overline{\partial} v_{c_k},$$

$$\psi_a^j(f) = \sum_{c_1,\dots,c_k} f_{ac_1\cdots c_k}\overline{\partial} t_{c_1}\wedge\cdots\wedge\overline{\partial} t_{c_{j-1}}\wedge v_{c_j}\overline{\partial} v_{c_{j+1}}\wedge\cdots\wedge\overline{\partial} v_{c_k},\quad j=1,\dots,k.$$

We first note that if $\delta f = 0$, then in the open subset $U_a\cap U_b$ we have:

$$\begin{aligned}
\phi_a(f)-\phi_b(f) &= \sum_{c_1,\dots,c_k}(f_{ac_1\cdots c_k}-f_{bc_1\cdots c_k})\overline{\partial} t_{c_1}\wedge\cdots\wedge\overline{\partial} t_{c_k}\\
&= \sum_{c_1,\dots,c_k}\sum_{i=1}^{k}(-1)^{i-1} f_{abc_1\cdots\widehat{c_i}\cdots c_k}\overline{\partial} t_{c_1}\wedge\cdots\wedge\overline{\partial} t_{c_k}\\
&= \sum_{i=1}^{k}(-1)^{i-1}\sum_{c_1,\dots,c_k} f_{abc_1\cdots\widehat{c_i}\cdots c_k}\overline{\partial} t_{c_1}\wedge\cdots\wedge\overline{\partial} t_{c_k}\\
&= \sum_{i=1}^{k}\sum_{c_i}\overline{\partial} t_{c_i}\wedge\sum_{c_1,\dots,\widehat{c_i},\dots,c_k} f_{abc_1\cdots\widehat{c_i}\cdots c_k}\overline{\partial} t_{c_1}\wedge\cdots\wedge\widehat{\overline{\partial} t_{c_i}}\wedge\cdots\wedge\overline{\partial} t_{c_k}\\
&= 0.
\end{aligned}$$

The same argument as above shows that, if $\delta f = 0$ then $\tilde{\phi}_a(f) = \tilde{\phi}_b(f)$ and $\psi_a^j(f) = \psi_b^j(f)$ for every a, b, j. Therefore the elements $\psi_a^j(f)$ are the restrictions of a global section $\psi^j(f)\in A_X^{0,k-1}(\mathcal{E})$. Moreover, $\phi(f)-\tilde{\phi}(f)=\sum_j(-1)^{j-1}\overline{\partial}\psi^j(f)$ and then $\phi(f)$, $\tilde{\phi}(f)$ determine the same cohomology class. □

Remark 1.2.3 The morphism $\varphi\colon \check{H}^1(\mathcal{U},\mathcal{E})\to H^1_{\overline{\partial}}(X,\mathcal{E})$ of Proposition 1.2.2 is always injective. In fact, if $f=\{f_{ab}\}\in C^1(\mathcal{U},\mathcal{E})$ is a cocycle, we have proved $\phi_a(f)=\phi_b(f)$ on $U_a\cap U_b$ and then $\phi(f)=\phi_a(f)$ on U_a, for every a, b. Thus, if $\phi(f)=\overline{\partial}\psi$ for some $\psi\in\Gamma(X,\mathcal{A}_X^{0,0}(\mathcal{E}))$, and defining

$$g_a = \psi_{|U_a} - \sum_c t_c f_{ac} \in \Gamma(U_a, \mathcal{A}_X^{0,0}(\mathcal{E}))$$

we have $\overline{\partial} g_a = 0$ and

$$g_a - g_b = \sum_c t_c(f_{ac} - f_{bc}) = \sum_c t_c f_{ab} = f_{ab}$$

for every a, b.

Remark 1.2.4 In the notation of Proposition 1.2.2, it is easy to prove directly that the morphism $\varphi\colon \check{H}^1(\mathcal{U}, \mathcal{E}) \to H^1_{\overline{\partial}}(X, \mathcal{E})$ is bijective whenever $H^1_{\overline{\partial}}(U_a, \mathcal{E}) = 0$ for every index a. Under this assumption, an explicit description of φ^{-1} can be obtained in the following way: given a $\overline{\partial}$-closed form $\omega \in A_X^{0,1}(\mathcal{E})$, for every a there exists $\psi_a \in \Gamma(U_a, \mathcal{A}_X^{0,0}(\mathcal{E}))$ such that $\overline{\partial}\psi_a = \omega$.

Setting $f_{ab} = \psi_b - \psi_a$ we have $f = \{f_{ab}\} \in C^1(\mathcal{U}, \mathcal{E})$ and $\delta f = 0$. Then the cohomology class of f in $\check{H}^1(\mathcal{U}, \mathcal{E})$ is well defined and we leave to the reader the easy verification that the induced map

$$H^1_{\overline{\partial}}(X, \mathcal{E}) \to \check{H}^1(\mathcal{U}, \mathcal{E}), \qquad \omega \mapsto [f],$$

is the inverse of φ.

By an open Stein subset of a complex manifold we intend an open subset which is a Stein manifold. We refer to the existing literature (e.g. [80, 97, 107]) for the notion and the main properties of Stein manifolds. However, for the comprehension of this book it is sufficient to recall that Stein manifolds are the analogue of affine varieties in algebraic geometry and satisfy the following properties:

1. every complex manifold biholomorphic to a convex open subset of $\mathbb{C}^n$ is Stein;
2. every closed submanifold of a Stein manifold is a Stein manifold;
3. every finite product of Stein manifolds is a Stein manifold;
4. if U, V are open Stein subsets of a complex manifold, then $U \cap V$ is also a Stein open subset;
5. every compact submanifold of a Stein manifold is finite;
6. every locally free sheaf on a Stein manifold is generated by global sections;
7. every higher cohomology group of a locally free sheaf on a Stein manifold vanishes.

By definition, a Stein covering of a complex manifold X is a locally finite open covering $\mathcal{U} = \{U_i\}$ such that every U_i is a Stein mainfold. The above properties easily imply that: every open covering admits a refinement that is a Stein covering; if $\{U_i\}$ is a Stein covering, then $U_{i_0 \cdots i_n}$ is a Stein open subset for every $i_0, \ldots, i_n$.

As an immediate consequence of the above items and of Leray's theorem on acyclic covers, we have that if $\mathcal{U}$ is a Stein covering of a complex manifold X, then for every locally free sheaf $\mathcal{E}$ we have that $\check{H}^k(\mathcal{U}, \mathcal{E}) = H^k(X, \mathcal{E})$ and the morphism φ of Proposition 1.2.2 is an isomorphism [107, VI.D.4].

Example 1.2.5 Let's compute the cohomology groups of the holomorphic tangent sheaf $\Theta_{\mathbb{P}^1}$ on the projective line. Fixing a system of homogeneous coordinates x_0, x_1 we define $U_i = \{x_i \neq 0\}$. Since every $U_i \cong \mathbb{C}$ is a Stein open subset we have $H^1(U_i, \Theta_{\mathbb{P}^1}) = 0$ and by Leray's theorem $H^i(\mathbb{P}^1, \Theta_{\mathbb{P}^1}) = \check{H}^i(\{U_0, U_1\}, \Theta_{\mathbb{P}^1})$, $i = 0, 1$; according to Remark 1.2.1 we can compute Čech cohomology by using the complex of alternating cochains.

Consider the affine coordinates $s = x_1/x_0$, $t = x_0/x_1$, then the holomorphic sections of $\Theta_{\mathbb{P}^1}$ over U_0, U_1 and $U_{01} = U_0 \cap U_1$ are given respectively by convergent power series

$$\sum_{i=0}^{+\infty} a_i s^i \frac{\partial}{\partial s}, \qquad \sum_{i=0}^{+\infty} b_i t^i \frac{\partial}{\partial t}, \qquad \sum_{i=-\infty}^{+\infty} c_i s^i \frac{\partial}{\partial s} .$$

In the open set U_{01} we have $t = s^{-1}$, $\dfrac{\partial}{\partial t} = -s^2 \dfrac{\partial}{\partial s}$ and then the Čech differential is given by

$$\delta\left(\sum_{i=0}^{+\infty} a_i s^i \frac{\partial}{\partial s}, \sum_{i=0}^{+\infty} b_i t^i \frac{\partial}{\partial t}\right) = \sum_{i=0}^{+\infty} a_i s^i \frac{\partial}{\partial s} + \sum_{i=-\infty}^{2} b_{2-i} s^i \frac{\partial}{\partial s} .$$

It is therefore immediate to observe that $\check{H}^1(\{U_0, U_1\}, \Theta_{\mathbb{P}^1}) = 0$ and

$$H^0(\{U_0, U_1\}, \Theta_{\mathbb{P}^1}) = \operatorname{Span}\left(\left(\frac{\partial}{\partial s}, -t^2 \frac{\partial}{\partial t}\right), \left(s\frac{\partial}{\partial s}, -t\frac{\partial}{\partial t}\right), \left(s^2 \frac{\partial}{\partial s}, -\frac{\partial}{\partial t}\right)\right).$$

Example 1.2.6 If $X = \mathbb{P}^1 \times \mathbb{C}^n$ then $H^1_{\bar{\partial}}(X, \Theta_X) = 0$. If $t_1, \ldots, t_n$ are linear coordinates on $\mathbb{C}^n$ and s is an affine coordinate in $\mathbb{P}^1$, then $H^0(X, \Theta_X)$ is the free $\mathcal{O}(\mathbb{C}^n)$-module generated by

$$\frac{\partial}{\partial t_1}, \ldots, \frac{\partial}{\partial t_n}, \frac{\partial}{\partial s}, s\frac{\partial}{\partial s}, s^2\frac{\partial}{\partial s}.$$

The proof is essentially the same (replacing the constant terms a_i, b_i, c_i with holomorphic functions over $\mathbb{C}^n$) as that of Example 1.2.5.

1.3 Locally Trivial Families

Every holomorphic map of complex manifolds $f : X \to Y$ induces a morphism of locally free sheaves $f_* : \Theta_X \to f^*\Theta_Y$. In fact $f^*\Theta_Y$ can be interpreted as the sheaf of sections of the pull-back of the holomorphic tangent bundle of Y and f_* is induced by the usual differential $df : T_x X \to T_{f(x)} Y$, $x \in X$. In general, the morphism f_* is neither injective nor surjective; its kernel and cokernel are called respectively the **sheaf of relative holomorphic vector fields**, and the **normal sheaf** of f:

$$0 \to \Theta_{X/Y} \to \Theta_X \xrightarrow{f_*} f^*\Theta_Y \to \mathcal{N}_f \to 0.$$

For instance, if X is a submanifold of Y and f is the inclusion, then the normal sheaf $\mathcal{N}_f$ is the sheaf of sections of the usual normal bundle $N_{X|Y}$. If f is a submersion, then the sheaf of relative holomorphic vector fields $\Theta_{X/Y}$ is the subsheaf of holomorphic vector fields on X that are tangent to the fibres of f at every point.

The goal of this section is to define, for every proper smooth family of complex manifolds $f: M \to B$, a morphism of sheaves of $\mathcal{O}_B$-modules $\rho: \Theta_B \to R^1 f_*\Theta_{M/B}$ with the remarkable property that $\rho = 0$ if and only if the family is locally trivial.

The derived image sheaf $R^1 f_*\Theta_{M/B}$ may be defined as the sheaf of $\mathcal{O}_B$-modules associated to the presheaf $V \mapsto H^1(f^{-1}(V), \Theta_{M/B})$. Since the sheaf $\Theta_{M/B}$ is locally free, it becomes convenient to define the map ρ in terms of Dolbeault cohomology.

The map $f_*: \Theta_X \to f^*\Theta_Y$ is a morphism of locally free $\mathcal{O}_X$-modules and therefore extends to a morphism of Dolbeault complexes $f_*: A_X^{0,*}(\Theta_X) \to A_X^{0,*}(f^*\Theta_Y)$. Similarly, the pull-back of differential forms $f^*: A_Y^{*,*} \to A_X^{*,*}$ commutes with the de Rham differentials and extends naturally to a morphism of Dolbeault complexes $f^*: A_Y^{0,*}(\Theta_Y) \to A_X^{0,*}(f^*\Theta_Y)$. For later use we point out that a pair of vector fields $\eta \in A_X^{0,0}(\Theta_X)$ and $\gamma \in A_Y^{0,0}(\Theta_Y)$ are f-related, i.e., $df(\eta(x)) = \gamma(f(x)) \in T_{f(x)}Y$ for every $x \in X$, if and only if $f^*(\gamma) = f_*(\eta)$.

Lemma 1.3.1 *Let $f: M \to B$ be a proper smooth family of complex manifolds. Then for every $\gamma \in A_B^{0,0}(\Theta_B)$ there exists $\eta \in A_M^{0,0}(\Theta_M)$ such that $f_*(\eta) = f^*(\gamma)$.*

Proof Let n be the dimension of B and $n+m$ the dimension of M. Given any point $x \in M$, let $(v_1, \ldots, v_n)$ be a local system of holomorphic coordinates in an open neighbourhood $V \subset B$ of $f(x)$. By assumption the differential $df: T_xM \to T_{f(x)}B$ is surjective and then, by the implicit function theorem there exists a local system of holomorphic coordinates $(z_1, \ldots, z_m, t_1, \ldots, t_n)$ in an open neighbourhood $U \subset M$ of x such that $f(U) \subset V$ and $t_i = f^*(v_i) = v_i \circ f$. If $\gamma_{|V} = \sum a_i(v_1, \ldots, v_n)\dfrac{\partial}{\partial v_i}$, then

$$f_{|U}^*(\gamma) = (f_{|U})_*\left(\sum a_i(t_1, \ldots, t_n)\frac{\partial}{\partial t_i}\right).$$

Therefore it is possible to find an open cover $M = \cup U_a$ and a family of vector fields $\eta_a \in \Gamma(U_a, \mathcal{A}_M^{0,0}(\Theta_M))$ such that $f_*(\eta_a) = f^*\gamma$ for every a. Finally, we take $\eta = \sum_a \varrho_a \eta_a$, where $\varrho_a: U_a \to \mathbb{C}$ is a partition of unity subordinate to the open cover $\{U_a\}$.

□

Given a smooth family of compact complex manifolds $f: M \to B$, if $\dim B = n$ and $\dim M = n+m$, we shall say that a holomorphic coordinate chart

$$(z_1, \ldots, z_m, t_1, \ldots, t_n): U \hookrightarrow \mathbb{C}^{m+n}, \qquad U \subset M \text{ open subset},$$

is **admissible** if $f(U)$ is contained in a coordinate chart

$$(v_1,\dots,v_n)\colon V\hookrightarrow \mathbb{C}^n,\qquad V\subset B \text{ open subset},$$

such that $t_i = v_i \circ f$ for every $i = 1,\dots,n$. The same argument, as used in the proof of Lemma 1.3.1 shows that M admits an open cover of admissible coordinate charts. If $z_1,\dots,z_m,t_1,\dots,t_n$ is an admissible system of local holomorphic coordinates for a smooth family $f\colon M\to B$, then $\dfrac{\partial}{\partial z_1},\dots,\dfrac{\partial}{\partial z_m}$ is a local basis of the locally free sheaf $\Theta_{M/B}$ of relative holomorphic vector fields; in particular the restriction of $\Theta_{M/B}$ to any fibre M_b is equal to Θ_{M_b}.

We are now ready to define the morphism $\rho\colon \Theta_B\to R^1f_*\Theta_{M/B}$ for a fixed smooth family $f\colon M\to B$. Let $V\subset B$ be an open subset and let $\gamma\in\Gamma(V,\Theta_B)$ be a holomorphic vector field. By Lemma 1.3.1 there exists $\eta\in\Gamma(f^{-1}(V),\mathcal{A}^{0,0}_M(\Theta_M))$ such that $f_*(\eta)=f^*(\gamma)$. Then $f_*(\overline{\partial}\eta)=\overline{\partial}(df(\eta))=\overline{\partial}f^*(\gamma)=f^*(\overline{\partial}\gamma)=0$ and therefore $\overline{\partial}\eta\in\Gamma(f^{-1}(V),\mathcal{A}^{0,1}_M(\Theta_{M/B}))$. The form $\overline{\partial}\eta$ is obviously $\overline{\partial}$-closed and its cohomology class

$$[\overline{\partial}\eta]\in H^1(f^{-1}(V),\Theta_{M/B})$$

depends only on γ. In fact, if $\tilde\eta\in\Gamma(f^{-1}(V),\mathcal{A}^{0,0}_M(\Theta_M))$ satisfies $f_*(\tilde\eta)=f^*(\gamma)$, then $\eta-\tilde\eta\in\Gamma(f^{-1}(V),\mathcal{A}^{0,0}_M(\Theta_{M/B}))$ and therefore $[\overline{\partial}\tilde\eta]=[\overline{\partial}\eta]\in H^1(f^{-1}(V),\Theta_{M/B})$. Notice that for every $g\in\mathcal{O}_B(V)$ we have $f_*(g\eta)=f^*(g\gamma)$ and $\overline{\partial}(g\eta)=g\overline{\partial}\eta$; in other words the map

$$\Gamma(V,\Theta_B)\ni\gamma\mapsto[\overline{\partial}\eta]\in H^1(f^{-1}(V),\Theta_{M/B}),\qquad f_*(\eta)=f^*(\gamma),\tag{1.2}$$

is a morphism of presheaves of $\mathcal{O}_B$-modules.

Definition 1.3.2 The (sheafified) **Kodaira–Spencer** map of a proper smooth family of complex manifolds $f\colon M\to B$ is the morphism of sheaves of $\mathcal{O}_B$-modules $\rho\colon\Theta_B\to R^1f_*\Theta_{M/B}$, induced by the morphism of presheaves (1.2).

Keep in mind the fact that in general the sheaf $R^1f_*\Theta_{M/B}$ is not locally free. We have anticipated that $\rho=0$ if and only if our family is locally trivial. One implication is easy and follows immediately from the following lemma.

Lemma 1.3.3 *Let $f\colon M\to B$ be a proper smooth family of complex manifolds and let $\gamma\in\Gamma(B,\Theta_B)$ be a holomorphic vector field. Then, in the above notation, we have $\rho(\gamma)=0$ if and only if for every $b\in B$ there exists an open neighbourhood $b\in V$ and a section $\eta\in\Gamma(f^{-1}(V),\Theta_M)$ such that $f_*(\eta)=f^*(\gamma)$. In particular, if $f\colon M\to B$ is locally trivial, then $\rho=0$.*

Proof Since $R^1f_*\Theta_{M/B}$ is a sheaf, in order to prove the "if" part it is sufficient to prove that if there exists an open subset $V\subset B$ and a section $\eta\in\Gamma(f^{-1}(V),\Theta_M)$ such that $f_*(\eta)=f^*(\gamma)$, then $\rho(\gamma)$ vanishes in $\Gamma(V,R^1f_*\Theta_{M/B})$. This is clear since $\overline{\partial}\eta=0$ and $\rho(\gamma)$ is the image of $[\overline{\partial}\eta]$ under the natural map

$$H^1(f^{-1}(V),\Theta_{M/B})\to\Gamma(V,R^1f_*\Theta_{M/B}).$$

Conversely, choosing $\mu \in \Gamma(M, \mathcal{A}_M^{0,0}(\Theta_M))$ such that $f_*(\mu) = f^*(\gamma)$, the assumption $\rho(\gamma) = 0$ means that for every $b \in B$ there exists an open neighbourhood $b \in V$ such that the cohomology class $[\overline{\partial}\mu]$ is trivial in $H^1(f^{-1}(V), \Theta_{M/B})$. Thus there exists $\tau \in \Gamma(f^{-1}(V), \mathcal{A}_M^{0,0}(\Theta_{M/B}))$ such that $\overline{\partial}\eta = \overline{\partial}\tau$. In order to conclude, it is sufficient to observe that $f_*(\tau) = 0$, $\eta = \mu - \tau \in \Gamma(f^{-1}(V), \Theta_M)$ and $f_*(\eta) = f^*(\gamma)$.

□

For every $0 < R \leq +\infty$, the (open) polydisc of dimension n and radius R is defined as

$$\Delta_R^n = \{(z_1, \dots, z_n) \in \mathbb{C}^n \mid |z_i| < R,\ i = 1, \dots, n\}.$$

Lemma 1.3.4 *Let $f: M \to \Delta_R^n$ be a proper smooth family of compact complex manifolds and let $t_1, \dots, t_n$ be a set of linear coordinates in the polydisc Δ_R^n. Suppose that there exist holomorphic vector fields $\chi_1, \dots, \chi_n$ on M such that $f_*(\chi_h) = f^*\left(\dfrac{\partial}{\partial t_h}\right)$ for every $i = 1, \dots, n$. Then there exists a smaller radius $0 < r \leq R$ such that the family $f: f^{-1}(\Delta_r^n) \to \Delta_r^n$ is trivial.*

Proof For every $0 < r \leq R$, $0 \leq h \leq n$, define:

$$\Delta_r^h = \{(z_1, \dots, z_n) \in \mathbb{C}^n \mid |z_1| < r, \dots, |z_h| < r, z_{h+1} = 0, \dots, z_n = 0\} \subset \Delta_R^n.$$

We prove by induction on $h = 0, \dots, n$ that there exists $0 < r_h \leq R$ such that the restriction of the family f over $\Delta_{r_h}^h$ is trivial; this is obvious for $h = 0$ and $r_0 = R$. Assume now that the family is trivial over $\Delta_{r_h}^h$, for some $h < n$. The integration of the vector field χ_{h+1} gives an open neighbourhood $M \times \{0\} \subset U \subset M \times \mathbb{C}$ and a holomorphic map $H: U \to M$ with the following properties (see e.g. [36, Chap. VII]):

1. For every $x \in M$, $\{x\} \times \mathbb{C} \cap U = \{x\} \times \Delta(x)$ with $\Delta(x)$ an open disc.
2. For every $x \in M$ the map $H_x = H(x, -): \Delta(x) \to M$ is the solution of the Cauchy problem

$$\begin{cases} \dfrac{dH_x}{dt}(t) = \chi_{h+1}(H_x(t)), \\ H_x(0) = x\,. \end{cases}$$

   In particular, if $H(x, t)$ is defined then $f(H(x, t)) = f(x) + (0, \dots, t, \dots, 0)$ (t in the $(h + 1)$th coordinate).
3. If $V \subset M$ is open and $V \times \Delta \subset U$ for some polydisc Δ, then the map $H(-, t)$: $V \to M$ is an open embedding for every $t \in \Delta$.

Since f is proper there exists $0 < r_{h+1} \leq r_h$ such that $f^{-1}(\Delta_{r_{h+1}}^h) \times \Delta_{r_{h+1}}^1 \subset U$; then the holomorphic map $H: f^{-1}(\Delta_{r_{h+1}}^h) \times \Delta_{r_{h+1}}^1 \to f^{-1}(\Delta_{r_{h+1}}^{h+1})$ is a biholomorphism giving a trivialization of the family over $\Delta_{r_{h+1}}^{h+1}$.

□

Example 1.3.5 Lemma 1.3.4 is generally false if the map f is not proper. Consider for instance an irreducible polynomial $Q \in \mathbb{C}[x_1, \dots, x_n, t]$ and define

$$V = \left\{(x,t) \in \mathbb{C}^n \times \mathbb{C} \;\middle|\; Q(x,t) = \frac{\partial Q}{\partial x_i}(x,t) = 0, \quad \forall\, i = 1, \dots, n\right\}.$$

Let $f\colon \mathbb{C}^n \times \mathbb{C} \to \mathbb{C}$ be the projection on the second factor and assume that $f(V)$ is a finite set of points. Setting

$$B = \mathbb{C} - f(V), \qquad X = \{(x,t) \in \mathbb{C}^n \times B \mid Q(x,t) = 0\},$$

we have that X is a smooth manifold of dimension n, the restriction $f\colon X \to B$ is surjective and its differential is surjective everywhere. Since X is a closed affine subvariety of $\mathbb{C}^n \times B$, by the Hilbert Nullstellensatz there exist a sequence of regular functions $g_1, \dots, g_n \in \mathcal{O}(\mathbb{C}^n \times B)$ such that

$$g := \sum_{i=1}^{n} g_i \frac{\partial Q}{\partial x_i} \equiv 1 \qquad (\text{mod } Q);$$

in particular X is contained in the open subset $U = \{g \neq 0\} \subset \mathbb{C}^n \times B$. The algebraic vector field

$$\chi = \sum_{i=1}^{n} \frac{g_i}{g}\left(\frac{\partial Q}{\partial x_i}\frac{\partial}{\partial t} - \frac{\partial Q}{\partial t}\frac{\partial}{\partial x_i}\right) = \frac{\partial}{\partial t} - \sum_{i=1}^{n} \frac{g_i}{g}\frac{\partial Q}{\partial t}\frac{\partial}{\partial x_i}$$

is holomorphic on U, it is tangent to X and lifts $\dfrac{\partial}{\partial t}$.

In general the fibres of $f\colon X \to B$ are not biholomorphic to each other; consider for example the case $Q(x, y, \lambda) = y^2 - x(x-1)(x-\lambda)$. Then $B = \mathbb{C} - \{0, 1\}$ and $f\colon X \to B$ is the restriction to the affine subspace $x_0 \neq 0$ of the family $M \to B$ of Example 1.1.3. The fibre $X_\lambda = f^{-1}(\lambda)$ is $M_\lambda - \{\text{point}\}$, where M_λ is an elliptic curve with j-invariant $j(\lambda) = 2^8(\lambda^2 - \lambda + 1)^3\lambda^{-2}(\lambda-1)^{-2}$. If X_a is biholomorphic to X_b then, by Riemann extension theorem, also M_a is biholomorphic to M_b and then $j(a) = j(b)$.

Theorem 1.3.6 *A proper smooth family $f\colon M \to B$ of complex manifolds is locally trivial if and only if the sheafified Kodaira–Spencer map*

$$\rho\colon \Theta_B \to R^1 f_* \Theta_{M/B}$$

is trivial.

Proof If the family is locally trivial, then $\rho = 0$ by Lemma 1.3.3. Conversely, if $\rho = 0$, then for every point $b \in B$ it is not restrictive to assume that B is a polydisc centred at b with coordinates $t_1, \dots, t_n$ and $\rho\left(\frac{\partial}{\partial t_i}\right) = 0$ for every $i = 1, \dots, n$. According to Lemma 1.3.3, after a possible shrinking of B, there exist holomorphic vector fields $\xi_1, \dots, \xi_n$ such that $f_*(\xi_i) = f^*\frac{\partial}{\partial t_i}$, and by Lemma 1.3.4 the family is trivial over an open polydisc $\Delta \subset B$ centered at b. □

1.4 The Kodaira–Spencer Map

In the previous section, for a proper smooth family of complex manifolds $f : M \to B$ we have defined a morphism of sheaves of $\mathcal{O}_B$-modules $\rho : \Theta_B \to R^1 f_*\Theta_{M/B}$ called a sheafified Kodaira–Spencer map; for every $b \in B$ the map ρ gives a commutative diagram

$$\begin{array}{ccc} \Theta_B & \xrightarrow{\rho} & R^1 f_*\Theta_{M/B} \\ \downarrow & & \downarrow \\ T_bB & \xrightarrow{\rho_b} & H^1(M_b, \Theta_{M_b}), \end{array} \tag{1.3}$$

where the vertical arrows are the natural restriction maps.

Definition 1.4.1 The morphism of vector spaces ρ_b defined in (1.3) is called the **Kodaira–Spencer map** of the family f at the point b.

We have seen that if a proper smooth family $f : M \to B$ is locally trivial, then for every $b \in B$ the Kodaira–Spencer map $\rho_b : T_bB \to H^1(M_b, \Theta_{M_b})$ is trivial. The converse is generally false but becomes true under the additional assumptions described in the following theorem.

Theorem 1.4.2 *Let $f : M \to B$ be a proper smooth family of complex manifolds such that the dimension of $H^1(M_b, \Theta_{M_b})$ is independent of the point b and the map $\rho_b : T_bB \to H^1(M_b, \Theta_{M_b})$ vanishes at every point $b \in B$. Then the family is locally trivial.*

Proof According to the continuity theorem for direct image sheaves, also known as the *exactness criterion of Grauert*, see e.g. [15, Chap. 3, Theorem 4.12], [109, Corollary III.12.9], [145, Theorem 7.4], [150, I, Theorem 2.2], if the dimension of $H^1(M_b, \Theta_{M_b})$ is constant then the sheaf $R^1 f_*\Theta_{M/B}$ is locally free and for every $b \in B$ the natural map

$$R^1 f_*\Theta_{M/B} \otimes_{\mathcal{O}_B} \mathbb{C}_b \to H^1(M_b, \Theta_{M_b})$$

is an isomorphism. Thus the sheafified Kodaira–Spencer map ρ is trivial if and only if $\rho_b = 0$ for every $b \in B$ and the conclusion follows from Theorem 1.3.6. □

Corollary 1.4.3 *Let $f\colon M \to B$ a proper smooth family of complex manifolds. Suppose that there exists a point $b \in B$ such that $H^1(M_b, \Theta_{M_b}) = 0$. Then the family is locally trivial in a neighbourhood of b.*

Proof By the semicontinuity theorem we have $H^1(M_c, \Theta_{M_c}) = 0$ for every c in an open neighbourhood of b, and the proof follows from Theorem 1.4.2. □

Definition 1.4.4 A compact complex manifold X is called **infinitesimally rigid** if $H^1(X, \Theta_X) = 0$.

Thus every proper smooth family of infinitesimally rigid manifolds is locally trivial.

Example 1.4.5 We have already proved that $H^1(\mathbb{P}^1, \Theta_{\mathbb{P}^1}) = 0$. More generally, the Euler exact sequence (see e.g. [242, p. 35])

$$0 \to \mathcal{O}_{\mathbb{P}^n} \to \mathcal{O}_{\mathbb{P}^n}(1)^{\oplus n+1} \to \Theta_{\mathbb{P}^n} \to 0$$

gives $H^i(\mathbb{P}^n, \Theta_{\mathbb{P}^n}) = 0$ for every n and every $i > 0$; in particular, every projective space is a infinitesimally rigid manifold. Similarly, every finite product of projective spaces is infinitesimally rigid.

It is useful to describe the Kodaira–Spencer map of a family $f\colon M \to B$ at a point $0 \in B$ in terms of Čech cocycles; to this end, assume that $B \subset \mathbb{C}^n$ is a polydisc centred at 0 with holomorphic coordinates $t_1, \dots, t_n$ and it is given a locally finite collection $\{U_a\}$ of admissible charts such that $\mathcal{U} = \{U_a \cap M_0\}$ is an open cover of the fibre M_0. By assumption every chart U_a has a set of holomorphic coordinates

$$z_1^a, \dots, z_m^a, t_1^a, \dots, t_n^a\colon U_a \to \mathbb{C}$$

such that $t_a^i = f^*(t_i)$ for every i. Then, on the double intersections $U_{ab} = U_a \cap U_b$ we have the transition functions:

$$\begin{cases} z_i^b = g_{i,a}^b(z^a, t^a) & i = 1, \dots, m, \\ t_i^b = t_i^a & i = 1, \dots, n. \end{cases}$$

Consider a fixed integer $h = 1, \dots, n$ and choose a section η of $\mathcal{A}_M^{0,0}(\Theta_M)$ such that $f_*(\eta) = f^*\left(\dfrac{\partial}{\partial t_h}\right)$. The restriction of η to U_{ab} is written in the local coordinates of U_a and U_b respectively as:

$$\eta = \sum_i \eta_i^a(z^a, t^a)\frac{\partial}{\partial z_i^a} + \frac{\partial}{\partial t_h^a}, \qquad \eta = \sum_i \eta_i^b(z^b, t^b)\frac{\partial}{\partial z_i^b} + \frac{\partial}{\partial t_h^b}.$$

Since $\eta - \dfrac{\partial}{\partial t_h^a} \in \Gamma(U_a, \mathcal{A}_M^{0,0}(\Theta_{M/B}))$ and $\bar\partial\left(\eta - \dfrac{\partial}{\partial t_h^a}\right) = \bar\partial\eta$, according to Remark 1.2.4, the Kodaira–Spencer class $\rho_0\left(\dfrac{\partial}{\partial t_h}\right) \in \check{H}^1(\mathcal{U}, \Theta_{M_0})$ is represented by the cocycle:

$$\rho\left(\frac{\partial}{\partial t_h}\right)_{a,b} = \left(\eta - \frac{\partial}{\partial t_h^b}\right) - \left(\eta - \frac{\partial}{\partial t_h^a}\right) = \frac{\partial}{\partial t_h^a} - \frac{\partial}{\partial t_h^b} = \sum_{i=1}^{m} \frac{\partial g_{i,a}^b}{\partial t_h^a}\frac{\partial}{\partial z_i^b}\,. \tag{1.4}$$

Using formula (1.4) it is easy to prove the invariance of the Kodaira–Spencer map under base change. More precisely, let $f\colon M \to B$ be a smooth family, $\phi\colon C \to B$ a holomorphic map, and $\hat\phi$, $\hat f$ the pull-backs of ϕ and f respectively:

$$\begin{array}{ccc} M \times_B C & \xrightarrow{\hat\phi} & M \\ \downarrow{\scriptstyle \hat f} & & \downarrow{\scriptstyle f} \\ C & \xrightarrow{\phi} & B. \end{array}$$

If $0 \in C$ is a point, then via the natural isomorphism $M_{\phi(0)} = \hat f^{-1}(0)$, the Kodaira–Spencer map $\rho_0\colon T_0C \to H^1(M_{\phi(0)}, \Theta_{M_{\phi(0)}})$ is equal to the composition

$$T_0C \xrightarrow{d\phi} T_{\phi(0)}B \xrightarrow{\rho_{\phi(0)}} H^1(M_{\phi(0)}, \Theta_{M_{\phi(0)}}). \tag{1.5}$$

In fact, it is not restrictive to assume $B \subset \mathbb{C}^n$, $C \subset \mathbb{C}^s$ polydiscs centred at 0 and $\phi(0) = 0$; denoting by $t_1, \ldots, t_n$ the coordinates on B, by $u_1, \ldots, u_s$ the coordinates on C we have $t_i = \phi_i(u)$ for a sequence of holomorphic functions $\phi_1, \ldots, \phi_n$ such that $\phi_i(0) = 0$ for every i.

If $z^a, t^a\colon U_a \to \mathbb{C}$, $z^b, t^b\colon U_b \to \mathbb{C}$ are admissible local coordinate charts on M with transition functions $z_i^b = g_{i,a}^b(z^a, t^a)$, then

$$z^a, u^a\colon U_a \times_B C \to \mathbb{C}, \qquad z^b, t^b\colon U_b \times_B C \to \mathbb{C},$$

are admissible charts with transition functions $z_i^b = g_{i,a}^b(z^a, \phi(u^a))$. Therefore for every a, b

$$\rho_0\left(\frac{\partial}{\partial u_h}\right)_{a,b} = \sum_i \frac{\partial g_{i,a}^b}{\partial u_h^a}\frac{\partial}{\partial z_i^b} = \sum_{i,j} \frac{\partial g_{i,a}^b}{\partial t_j^a}\frac{\partial \phi_j}{\partial u_h^a}\frac{\partial}{\partial z_i^b} = \rho_{\phi(0)}\left(d\phi\frac{\partial}{\partial u_h}\right)_{a,b}.$$

The Example 1.4.6 below shows that the conclusion of Theorem 1.4.2 may fail if the dimension of $H^1(M_b, \Theta_{M_b})$ depends on b.

Example 1.4.6 Consider the family of Hopf surfaces $f\colon M \to B$, where $B = \mathbb{C}$, $M = X/G$, $X = B \times (\mathbb{C}^2 - \{0\})$ and $G \simeq \mathbb{Z}$ is the subgroup of holomorphic automorphisms generated by $(b, z_1, z_2) \mapsto (b, 2z_1 + b^2 z_2, 2z_2)$. In the notation of Example 1.1.5, the fibre M_b is the Hopf surface $S_{A(b)}$, where $A(b) = \begin{pmatrix} 2 & b^2 \\ 0 & 2 \end{pmatrix}$ and then M_0 is not biholomorphic to M_b for every $b \neq 0$. This family is isomorphic to $N \times_{\mathbb{C}} B$, where $B \to \mathbb{C}$ is the map $b \mapsto b^2$ and N is the quotient of $\mathbb{C} \times (\mathbb{C}^2 - \{0\})$ by the group generated by $(s, z_1, z_2) \mapsto (s, 2z_1 + sz_2, 2z_2)$. By base-change property (1.5), the Kodaira–Spencer map $\rho_0 : T_0 B \to H^1(M_0, \Theta_{M_0})$ is trivial.

On the other hand, the family is trivial over $B - \{0\}$, since the map

$$(B - \{0\}) \times (\mathbb{C}^2 - \{0\}) \to (B - \{0\}) \times (\mathbb{C}^2 - \{0\}), \quad (b, z_1, z_2) \mapsto (b, z_1, b^2 z_2)$$

induces an isomorphism $(B - \{0\}) \times M_1 \simeq (M - f^{-1}(0))$. Therefore the Kodaira–Spencer map $\rho_b : T_b B \to H^1(M_b, \Theta_{M_b})$ is trivial for every b, while the family is not locally trivial at 0.

According to Theorem 1.4.2 the dimension of $H^1(M_b, \Theta_{M_b})$ cannot be constant; in fact it is proved in [150, II.15.12] that $\dim H^1(M_0, \Theta_{M_0}) = 4$ and $\dim H^1(M_b, \Theta_{M_b}) = 2$ for $b \neq 0$.

Example 1.4.7 Let $M \subset \mathbb{C}_b \times \mathbb{P}^3_x \times \mathbb{P}^1_u$ be the submanifold defined by the equations

$$u_0 x_1 = u_1(x_2 - b x_0), \qquad u_0 x_2 = u_1 x_3.$$

We leave to the reader the easy verification that the projection $f : M \to \mathbb{C}$ onto the first factor is a smooth family of compact complex manifolds. We claim that:

1. the restriction $f : M - f^{-1}(0) \to \mathbb{C} - \{0\}$ is locally trivial;
2. f is not locally trivial at 0.

After the linear change of coordinates $x_2 - b x_0 \mapsto x_0$ the equations of

$$M - f^{-1}(0) \subset (\mathbb{C} - \{0\}) \times \mathbb{P}^3 \times \mathbb{P}^1$$

become $u_0 x_1 = u_1 x_0$, $u_0 x_2 = u_1 x_3$, and there exists an isomorphism of families $(\mathbb{C} - \{0\}) \times \mathbb{P}^1_t \times \mathbb{P}^1_u \to M - f^{-1}(0)$, given by

$$(b, [t_0, t_1], [u_0, u_1]) \mapsto (b, [t_0 u_1, t_0 u_0, t_1 u_1, t_1 u_0], [u_0, u_1]).$$

Let $Y \subset M_0$ be the subvariety of equation $b = x_1 = x_2 = x_3 = 0$; we can immediately see that Y is a projective line and therefore a compact manifold of positive dimension. If f is locally trivial at 0, then there exist an open subset $0 \in U \subset \mathbb{C}$ and a commutative diagram of holomorphic maps

$$\begin{array}{ccc} Y \times U & \xrightarrow{\;j\;} & M \\ \Big\downarrow{\scriptstyle pr} & & \Big\downarrow{\scriptstyle f} \\ U & \xrightarrow{\;i\;} & \mathbb{C}, \end{array}$$

where i is the inclusion, j is injective and extends the identity $Y \times \{0\} \to Y \subset M_0$. Since Y is a compact subset of the open set $\{x_0 \neq 0\}$, possibly shrinking U we may assume that the image of j is contained in the open subset $\{x_0 \neq 0\}$. For every $b \neq 0$ the holomorphic map

$$\sigma \colon \{x_0 \neq 0\} \cap M_b \to \mathbb{C}^3, \qquad \sigma(b, [x_0, x_1, x_2, x_3], [u_0, u_1]) = \left(\frac{x_1}{x_0}, \frac{x_2}{x_0}, \frac{x_3}{x_0}\right),$$

is injective. Therefore for $b \in U, b \neq 0$, the holomorphic map $\sigma j(-, b) \colon Y \simeq \mathbb{P}^1 \to \mathbb{C}^3$ is injective, in contradiction with the maximum principle of holomorphic functions. Therefore f cannot be locally trivial at 0.

Applying the base change $\mathbb{C} \to \mathbb{C}, b \mapsto b^2$, to the family $M \to \mathbb{C}$ we get a family with trivial Kodaira–Spencer map at every point which is not locally trivial at 0. It is not difficult to prove that M_0 is the Segre–Hirzebruch surface $\mathbb{F}_2$ and $H^1(\mathbb{F}_2, \Theta_{\mathbb{F}_2}) = \mathbb{C}$, while $\mathbb{P}^1 \times \mathbb{P}^1$ is infinitesimally rigid.

1.5 Deformations over Smooth Bases

A germ of a complex manifold is a pair (X, x) where X is a complex manifold and x is a point of X. A system of holomorphic coordinates in a germ (X, x) is a system of holomorphic coordinates in an open neighbourhood of x in X. For every pair of pointed complex manifolds (X, x), (Y, y) we denote by $\mathrm{Mor}((X, x), (Y, y))$ the set of germs of holomorphic maps $f \colon (X, x) \to (Y, y)$; every element of $\mathrm{Mor}((X, x), (Y, y))$ is an equivalence class of pairs (U, f), where $x \in U \subset X$ is an open neighbourhood of x, $f \colon U \to N$ is a holomorphic map such that $f(x) = y$ and $(U, f) \sim (V, g)$ if and only if there exists an open subset $x \in W \subset U \cap V$ such that $f_{|W} = g_{|W}$.

Remark 1.5.1 Every germ of holomorphic map $f \colon (X, x) \to (Y, y)$ between germs of complex manifolds induces by composition a local morphism of $\mathbb{C}$-algebras $f^* \colon \mathcal{O}_{Y,y} \to \mathcal{O}_{X,x}$. Conversely, for a local morphism of $\mathbb{C}$-algebras $\phi \colon \mathcal{O}_{Y,y} \to \mathcal{O}_{X,x}$ there exists a unique germ of holomorphic map $f \colon (X, x) \to (Y, y)$ such that $f^* = \phi$. In fact, fixing two systems z_i and w_j of holomorphic coordinates in (X, x) and (Y, y) respectively, we have $\mathcal{O}_{X,x} = \mathbb{C}\{z_1, \dots, z_n\}, \mathcal{O}_{Y,y} = \mathbb{C}\{w_1, \dots, w_m\}$; then the germ of holomorphic map f defined as

$$f(z_1, \dots, z_n) = (f_1(z_1, \dots, z_n), \dots, f_m(z_1, \dots, z_n)), \quad \text{where} \quad f_i = \phi(w_i),$$

is the unique morphism satisfying the condition $f^* = \phi$: this is well known and a proof is reproduced for completness' sake in Lemma A.1.5.

Definition 1.5.2 A **deformation** of a compact complex manifold X over a germ of complex manifold $(B, 0)$ is represented by a commutative diagram of complex manifolds and holomorphic maps

$$\begin{array}{ccc} X & \xrightarrow{\ i\ } & \mathcal{X} \\ \big\downarrow & & \big\downarrow{\scriptstyle f} \\ 0 & \longrightarrow & B \end{array} \tag{1.6}$$

such that:

1. there exists an open neighbourhood $0 \in U \subset B$ such that $f\colon f^{-1}(U) \to U$ is a proper smooth family. In particular, the fibre $f^{-1}(0)$ is a smooth compact complex manifold containing the image of i;
2. the morphism $i\colon X \to f^{-1}(0)$ is a holomorphic isomorphism.

Two diagrams

$$\begin{array}{ccc} X & \xrightarrow{\ i\ } & \mathcal{X} \\ \big\downarrow & & \big\downarrow{\scriptstyle f} \\ 0 & \longrightarrow & B, \end{array} \qquad\qquad \begin{array}{ccc} X & \xrightarrow{\ j\ } & \widetilde{\mathcal{X}} \\ \big\downarrow & & \big\downarrow{\scriptstyle g} \\ 0 & \longrightarrow & B, \end{array}$$

as above define isomorphic deformations if there exist an open subset $0 \in U \subset B$ and a commutative diagram

$$\begin{array}{ccc} X & \xrightarrow{\ \ i\ \ } & f^{-1}(U) \\ {\scriptstyle j}\big\downarrow & \swarrow & \big\downarrow{\scriptstyle f} \\ g^{-1}(U) & \xrightarrow[g]{} & U, \end{array}$$

with the diagonal arrow a holomorphic isomorphism.

The manifold $\mathcal{X}$ is called the **total space** of the deformation and $(B, 0)$ the **base germ** space.

For notational simplicity, from now on we shall refer to the representative diagram (1.6) with the one-line notation $X \xrightarrow{i} \mathcal{X} \xrightarrow{f} (B, 0)$.

Definition 1.5.3 For every deformation of a complex manifold

$$X \xrightarrow{i} \mathcal{X} \xrightarrow{f} (B, 0) \qquad = \qquad \begin{array}{ccc} X & \xrightarrow{\ i\ } & \mathcal{X} \\ \big\downarrow & & \big\downarrow{\scriptstyle f} \\ 0 & \longrightarrow & B, \end{array}$$

the **Kodaira–Spencer map** $\rho_0\colon T_0B \to H^1(X, \Theta_X)$ is defined as the composition of the map $\rho_0\colon T_0B \to H^1(f^{-1}(0), \Theta_{f^{-1}(0)})$ introduced in the Definition 1.4.1, and the isomorphism $i^*\colon H^1(f^{-1}(0), \Theta_{f^{-1}(0)}) \to H^1(X, \Theta_X)$.

It is immediate from the definition that the Kodaira–Spencer map is well defined on isomorphism classes of deformations, i.e., that it is independent of the representative diagram (1.6) defining the deformation. On the other hand, the Kodaira–Spencer map depends on the choice of the isomorphism i between the complex manifold X and the fibre $f^{-1}(0)$.

Definition 1.5.4 A deformation $X \to \mathcal{X} \to (B, 0)$ is called **trivial** if it is isomorphic to

$$X \xrightarrow{\mathrm{Id}\times\{0\}} X \times B \xrightarrow{\text{projection}} (B, 0).$$

The following result is an immediate consequence of Corollary 1.4.3.

Corollary 1.5.5 *Let X be a compact complex manifold such that $H^1(X, \Theta_X) = 0$. Then every deformation of X is trivial.*

According to Ehresmann's Theorem 1.1.6, every deformation of a complex manifold X over a germ $(B, 0)$ may be represented by a diagram (1.6) in which $\mathcal{X}$ is diffeomorphic to the product $X \times U$, for a certain open connected neighbourhood $0 \in U \subset B$. Thus every isomorphism class of deformation can be also represented by a suitable pair (U, J), where U is an open subset of B and J is a complex structure on the differentiable manifold $X \times U$.

For every germ $(B, 0)$ of a complex manifold we denote by $\mathrm{Def}_X(B, 0)$ the set[1] of isomorphism classes of deformations of X over $(B, 0)$. Every morphism of germs $g\colon (C, 0) \to (B, 0)$ induces a natural map $g^*\colon \mathrm{Def}_X(B, 0) \to \mathrm{Def}_X(C, 0)$. In fact, given a deformation $\xi \in \mathrm{Def}_X(B, 0)$, after a possible shrinking of B and C, we may assume that ξ is represented by a smooth proper holomorphic map $f : \mathcal{X} \to B$ (together with a holomorphic isomorphism $X \simeq f^{-1}(0)$) and g by a holomorphic map $g\colon C \to B$, $g(0) = 0$. Then $g^*\xi \in \mathrm{Def}_X(C, 0)$ is defined as the deformation associated to the usual fibred product:

$$\mathcal{X} \times_B C \xrightarrow{g^* f} C.$$

We can reinterpret the above facts by saying that Def_X is a functor from the opposite of the category of germs of complex manifolds to the category of sets. The natural question whether this functor is representable leads to the notion of universal deformation.

Definition 1.5.6 A deformation $\xi \in \mathrm{Def}_X(B, 0)$ of a compact complex manifold X is called **universal** if for every germ $(C, 0)$ and every deformation $\eta \in \mathrm{Def}_X(C, 0)$ there exists a unique germ of holomorphic map $g\colon (C, 0) \to (B, 0)$ such that $\eta = g^*\xi$.

[1] This is a set and not a proper class; in fact every complex structure can be interpreted as an almost complex structure and therefore as an element of the set of splittings of the complexified tangent bundle. The same fact also follows immediately from Corollary 4.3.9. We inform the reader that this footnote should be considered an exception and, for simplicity of exposition, from now on we shall frequently ignore any kind of set theoretic difficulties.

If the universal deformation exists, then it is necessarily unique up to canonical isomorphism. In fact if $\xi \in \mathrm{Def}_X(B, 0)$ and $\xi' \in \mathrm{Def}_X(B', 0)$ are both universal, then there exists a unique morphism $g\colon (B', 0) \to (B, 0)$ such that $g^*\xi = \xi'$ and a unique morphism $h\colon (B, 0) \to (B', 0)$ such that $h^*\xi' = \xi$. Clearly, h and g are one the inverse of the other.

In many interesting cases the universal deformations exist. On the other hand, some of the simplest examples of compact complex manifolds do not have universal deformations.

Example 1.5.7 The algebraic surface $X \subset \mathbb{P}^3_x \times \mathbb{P}^1_u$ defined by the equations

$$u_0x_1 = u_1x_2, \qquad u_0x_2 = u_1x_3,$$

does not have a universal deformation. Let $M \subset \mathbb{P}^3_x \times \mathbb{P}^1_u \times \mathbb{C}_b$ be the algebraic submanifold defined by equations

$$u_0x_1 = u_1(x_2 - bx_0), \qquad u_0x_2 = u_1x_3.$$

We have proved in Example 1.4.7 that the projection $f\colon M \to (\mathbb{C}, 0)$ on the third factor is a nontrivial deformation of X. Given the germ of holomorphic map

$$g\colon (\mathbb{C}, 0) \to (\mathbb{C}, 0), \qquad g(b) = b + b^2,$$

we claim that f and g^*f are isomorphic deformations of X over the germ $(\mathbb{C}, 0)$. In fact, the equation of g^*M in $\mathbb{P}^3_x \times \mathbb{P}^1_u \times \mathbb{C}_b$ is

$$u_0x_1 = u_1(x_2 - (b + b^2)x_0), \qquad u_0x_2 = u_1x_3,$$

and an isomorphism of deformations $g^*M \simeq M$ is induced by the map

$$(b, [x_0, x_1, x_2, x_3], [u_0, u_1]) \mapsto (b, [(1 + b)x_0, x_1, x_2, x_3], [u_0, u_1]).$$

Assume now that $X \to \mathcal{X} \xrightarrow{h} (B, 0)$ is a universal deformation of the manifold X; then $X \to M \xrightarrow{f} (\mathbb{C}, 0)$ is the pull-back of h via a unique germ of holomorphic map $\phi\colon (\mathbb{C}, 0) \to (B, 0)$. The unicity of ϕ implies $\phi(b + b^2) = \phi(b)$. This is possible if and only if ϕ is a constant map and this gives a contradiction since the deformation f is not trivial.

If we drop the unicity from Definition 1.5.6 and we require only the existence of g we get the notion of **complete** deformation. A very useful criterion for completeness was proved very early on by Kodaira and Spencer in their pioneering works on deformation theory.

Theorem 1.5.8 (Completeness theorem [148]) *Let $(B, 0)$ be a germ of a smooth complex manifold and let $\xi \in \mathrm{Def}_X(B, 0)$ be a deformation of a compact complex* *manifold X such that the Kodaira–Spencer map* $\rho_0 \colon T_0 B \to H^1(X, \Theta_X)$ *is surjective. Then for every germ* $(C, 0)$ *and every deformation* $\eta \in \mathrm{Def}_X(C, 0)$ *there exists a germ of holomorphic map* $g \colon (C, 0) \to (B, 0)$ *such that* $\eta = g^*\xi$.

For instance, one can prove that the Kodaira–Spencer map of the deformation $f \colon M \to (\mathbb{C}, 0)$ of Example 1.5.7 is bijective and therefore that such a deformation is complete. In the same paper Kodaira and Spencer showed that not every compact complex manifold X has a deformation ξ satisfying the assumption of Theorem 1.5.8; see Example 1.7.6.

It is an instructive exercise in algebraic geometry to prove that the deformation of a smooth hypersurface of degree d in $\mathbb{P}^n$ described by the family of Example 1.1.4 satisfies the condition of Theorem 1.5.8 except for the two special cases $n = 2, d \geq 4$ and $n = 3, d = 4$, cf. [150, Theorem 14.1].

1.6 Deformations over Singular Bases

Historically, a giant step forward in deformation theory was the introduction of deformations of complex manifolds over (possibly unreduced) analytic singularities; in other words, it is useful and important to extend Definition 1.5.2 to the case when $(B, 0)$ is a germ of a complex space.

Here we briefly recall the basic definitions and some properties of analytic singularities, referring to [80, 96, 97, 171, 204] for proofs and more details.

We assume that the reader is familiar with the notion of a ringed space [109, p. 72]. Following [80], a $\mathbb{C}$-ringed space is a ringed space $(X, \mathcal{O}_X)$ such that, for every $x \in X$, the stalk $\mathcal{O}_{X,x}$ is a local $\mathbb{C}$-algebra with residue field $\mathbb{C}$; a morphism of $\mathbb{C}$-ringed spaces is a morphism of ringed spaces $f \colon (X, \mathcal{O}_X) \to (Y, \mathcal{O}_Y)$ such that for every $x \in X$ the map $f^* \colon \mathcal{O}_{Y, f(x)} \to \mathcal{O}_{X,x}$ is a morphism of local $\mathbb{C}$-algebras.

Let $U \subset \mathbb{C}^n$ be an open subset and let $\mathcal{I} \subset \mathcal{O}_U$ be the ideal sheaf generated by a finite set of holomorphic functions $g_1, \ldots, g_m \in \Gamma(U, \mathcal{O}_U)$; in other words, $\mathcal{I}$ is the image of the morphism of sheaves of $\mathcal{O}_U$-modules

$$\mathcal{O}_U^m \xrightarrow{(g_1, \ldots, g_m)} \mathcal{O}_U$$

and then, at every point $u \in U$, the stalk $\mathcal{I}_u$ is the ideal of $\mathcal{O}_{U,u}$ generated by $g_1, \ldots, g_m$. Setting

$$Z = \{u \in U \mid g_1(u) = \cdots = g_m(u) = 0\}, \qquad \mathcal{O}_Z = \frac{\mathcal{O}_U}{\mathcal{I}},$$

we have that the pair $(Z, \mathcal{O}_Z)$ is a locally $\mathbb{C}$-ringed space called a *local model* of a complex space.

Definition 1.6.1 A **complex space** is a Hausdorff $\mathbb{C}$-ringed space locally isomorphic to a local model defined as above.

In particular, if $(X, \mathcal{O}_X)$ is a complex space, then for every $x \in X$ the stalk $\mathcal{O}_{X,x}$ is isomorphic to a quotient $\mathbb{C}\{z_1, \dots, z_n\}/I$ for some $n \geq 0$ and some ideal I. Conversely, every local $\mathbb{C}$-algebra $R = \mathbb{C}\{z_1, \dots, z_n\}/I$ is isomorphic to the stalk of the structure sheaf of a complex space; in fact since $\mathbb{C}\{z_1, \dots, z_n\}$ is a Noetherian ring, the ideal I is generated by a finite number of convergent power series $f_1, \dots, f_m$. Then we have $R = \mathcal{O}_{X,0}$, where X is the complex space defined by the equations $f_1 = \cdots = f_m = 0$ in a sufficiently small open neighbourhood of $0 \in \mathbb{C}^n$.

Definition 1.6.2 An **analytic singularity** is a germ (X, x) of a complex space at a point. A morphism of analytic singularities is a germ of a morphism of complex spaces.

An **analytic algebra** is a local $\mathbb{C}$-algebra isomorphic to $\mathbb{C}\{z_1, \dots, z_n\}/I$, for some $n \geq 0$ and some ideal $I \subset (z_1, \dots, z_n)$. A morphism of analytic algebras is a local morphism of $\mathbb{C}$-algebras.

Every analytic algebra is a local Noetherian $\mathbb{C}$-algebra with residue field $\mathbb{C}$; other properties of analytic algebras will be studied in Appendix A. Throughout this book we shall denote by $\mathbf{C}^{op}$ the opposite of a category $\mathbf{C}$ and, unless otherwise specified, every functor is intended covariant. Every contravariant functor from a category $\mathbf{C}$ to a category $\mathbf{D}$ will be expressed as a functor $\mathbf{C}^{op} \to \mathbf{D}$.

Recall that a functor $F: \mathbf{C} \to \mathbf{D}$ is called an equivalence of categories if it satisfies the following three conditions:

1. F is *surjective on isomorphism classes:* for every $d \in \mathbf{D}$ there exists $c \in \mathbf{C}$ and an isomorphism $\gamma \in \operatorname{Mor}_{\mathbf{D}}(F(c), d)$;
2. F is *full:* for every $a, b \in \mathbf{C}$ the map $F: \operatorname{Mor}_{\mathbf{C}}(a, b) \to \operatorname{Mor}_{\mathbf{D}}(F(a), F(b))$ is surjective;
3. F is *faithful:* for every $a, b \in \mathbf{C}$ the map $F: \operatorname{Mor}_{\mathbf{C}}(a, b) \to \operatorname{Mor}_{\mathbf{D}}(F(a), F(b))$ is injective.

Theorem 1.6.3 (Grothendieck) *The functor*

$$\textbf{\textit{Analytic singularities}}^{op} \to \textbf{\textit{Analytic algebras}}, \qquad (X, x) \mapsto \mathcal{O}_{X,x},$$

is an equivalence of categories.

Proof For the proof we refer to [80, Proposition 0.21], [100, Remark 1.47.1] or [104]. □

For instance, an analytic singularity (X, x) is isomorphic to a germ of a complex manifold if and only if the analytic algebra $\mathcal{O}_{X,x}$ is isomorphic to $\mathbb{C}\{z_1, \dots, z_n\}$ for some n.

Corollary 1.6.4 *Let $\mathfrak{m}_R$ be the maximal ideal of an analytic algebra R. Then for every finite sequence $f_1, \dots, f_n \in \mathfrak{m}_R$ there exists a unique morphism of analytic algebras $f^*: \mathbb{C}\{z_1, \dots, z_n\} \to R$ such that $f^*(z_i) = f_i$.*

Proof Immediate consequence of Theorem 1.6.3 since, for a fixed analytic singularity (X, x), every morphism $f: (X, x) \to (\mathbb{C}^n, 0)$ is determined by its components $f_1, \dots, f_n \in \mathcal{O}_{X,x}$; the condition $f(x) = 0$ implies that every f_i belongs to the maximal ideal. In Appendix A we shall give a direct elementary proof of the same result (Lemma A.1.5). □

The dimension of an analytic singularity (X, x) is defined as the dimension of the Noetherian local ring $\mathcal{O}_{X,x}$. Recall that for a local Noetherian ring R with maximal ideal $\mathfrak{m}_R$, the dimension of R is the minimum integer d such that there exists a sequence $a_1, \dots, a_d \in R$ with the property that $\sqrt{(a_1, \dots, a_d)} = \mathfrak{m}_R$ [10, Theorem 11.14]. In particular, if the maximal ideal $\mathfrak{m}_R$ is generated by n elements then $\dim R \le n$. Recall also that a local Noetherian ring has dimension 0 if and only if its maximal ideal is nilpotent.

Let R be an analytic algebra of dimension 0, then its maximal ideal $\mathfrak{m}_R$ is nilpotent and R is a local Artin $\mathbb{C}$-algebra with residue field $\mathbb{C}$. Conversely, every local Artin $\mathbb{C}$-algebra A with residue field $\mathbb{C}$ is an analytic algebra of dimension 0; choosing a finite set of generators $a_1, \dots, a_n \in \mathfrak{m}_A$ of its maximal ideal, for every positive integer d the ideal $\mathfrak{m}_A^d$ is generated by the monomials of degree d in $a_1, \dots, a_n$. Since $\mathfrak{m}_A^d = 0$ for some $d > 0$ there are two well defined surjective morphisms of $\mathbb{C}$-algebras

$$\mathbb{C}[z_1, \dots, z_n] \to A, \qquad \mathbb{C}\{z_1, \dots, z_n\} \to A, \qquad z_i \mapsto a_i.$$

In other words, the full subcategory of analytic algebras of dimension 0 is equivalent to the category of local Artin $\mathbb{C}$-algebra with residue field $\mathbb{C}$. For every local Artin $\mathbb{C}$-algebra A with residue field $\mathbb{C}$ we shall denote by $(\operatorname{Spec} A, 0)$ the analytic singularity such that $\mathcal{O}_{\operatorname{Spec} A,0} = A$.

Definition 1.6.5 The **Zariski tangent space** of an analytic singularity (X, x) is the space

$$T_xX = \{\text{morphisms } \phi\colon (\operatorname{Spec} \mathbb{C}[\epsilon]/(\epsilon^2), 0) \to (X, x) \text{ of analytic singularities}\}.$$

The differential of a morphism $f: (X, x) \to (Y, y)$ of analytic singularities is the linear map $df: T_xX \to T_yY$ defined by the formula $df(\phi) = f \circ \phi$.

By Theorem 1.6.3 we also have

$$T_xX = \{\text{morphisms } \alpha\colon \mathcal{O}_{X,x} \to \mathbb{C}[\epsilon]/(\epsilon^2) \text{ of analytic algebras}\}.$$

If $\mathfrak{m}_{X,x}$ is the maximal ideal of $\mathcal{O}_{X,x}$, then $\alpha(\mathfrak{m}_{X,x}) \subset \mathbb{C}\epsilon$ and $\alpha(\mathfrak{m}_{X,x}^2) = 0$ for every $\alpha \in T_xX$. Therefore we can also write:

$$
\begin{aligned}
T_xX &= \left\{\text{morphisms } \alpha\colon \frac{\mathcal{O}_{X,x}}{\mathfrak{m}_{X,x}^2} \to \frac{\mathbb{C}[\epsilon]}{(\epsilon^2)} \text{ of analytic algebras}\right\} \\
&= \left\{\text{morphisms } \alpha\colon \frac{\mathfrak{m}_{X,x}}{\mathfrak{m}_{X,x}^2} \to \mathbb{C} \text{ of complex vector spaces}\right\}.
\end{aligned}
$$

The above identification of T_xX with the dual of $\mathfrak{m}_{X,x}/\mathfrak{m}_{X,x}^2$ induces a vector space structure on T_xX, of dimension equal to the minimum number of generators of $\mathfrak{m}_{X,x}$, cf. [10, Proposition 2.8]. We leave to the reader the easy verification that the differential of a morphism is a linear map.

Remark 1.6.6 The inverse function theorem does not hold for morphism of analytic singularities. For instance, if $X = \{z_1^2 + z_2^2 = 0\} \subset \mathbb{C}^2$, then the inclusion of singularities $(X, 0) \to (\mathbb{C}^2, 0)$ is not an isomorphism, while its differential $T_0X \to T_0\mathbb{C}^2$ is an isomorphism, cf. Exercise 1.9.3.

However, it is possible to prove, see Proposition A.3.2, that the inverse function theorem holds for analytic endomorphisms; in other words, if $f\colon (X, 0) \to (X, 0)$ is a morphism from an analytic singularity into itself, then f is an isomorphism if and only if $df\colon T_xX \to T_xX$ is an isomorphism.

The **basic dimension upper bound** for analytic singularities is given by the inequality $\dim(X, x) \le \dim_{\mathbb{C}} T_xX$, cf. [10, Corollary 11.15]. In fact, if $\dim_{\mathbb{C}} T_xX$ has dimension n, then the maximal ideal $\mathfrak{m}_{X,x}$ can be generated by n elements and therefore $\dim(X, x) = \dim \mathcal{O}_{X,x} \le n$.

Definition 1.6.7 A morphism of analytic singularities $f\colon (X, x) \to (Y, y)$ is said to be **smooth** if there exists an isomorphism of analytic singularities

$$\phi\colon (X, x) \xrightarrow{\simeq} (Y \times \mathbb{C}^n, (y, 0))$$

such that f is the composition of ϕ and the projection on the first factor.

A morphism of complex spaces $f\colon X \to Y$ is smooth if for every $x \in X$ the morphism of analytic singularities $f\colon (X, x) \to (Y, f(x))$ is smooth.

In the above definition the product $(Y \times \mathbb{C}^n, (y, 0))$ is defined in the obvious way:

$$\text{if } \mathcal{O}_{Y,y} = \frac{\mathbb{C}\{z_1, \dots, z_m\}}{(f_1(z), \dots, f_n(z))}, \quad \text{then } \mathcal{O}_{(Y\times\mathbb{C}^n,(y,0))} = \frac{\mathbb{C}\{z_1, \dots, z_m, u_1, \dots, u_n\}}{(f_1(z), \dots, f_n(z))}.$$

In particular, an analytic singularity (X, x) is a germ of a complex manifold if and only if the constant map $(X, x) \to (0, 0) = \operatorname{Spec}\mathbb{C}$ is smooth. Notice that, by the implicit function theorem, a morphism $f\colon (X, x) \to (Y, y)$ of germs of complex manifolds is smooth if and only if its differential $df\colon T_xX \to T_yY$ is surjective. Be careful of the fact that if (X, x) is not a germ of a complex manifold, then the surjectivity of $df\colon T_xX \to T_yY$ does not imply the smoothness of f.

We are now ready to extend the notion of smooth family (Definition 1.1.1) over any complex space.

Definition 1.6.8 A **smooth family** of compact complex manifolds is a proper smooth morphism $f\colon M \to B$ of complex spaces such that M is nonempty and B is connected.

Two families $f\colon M \to B$, $g\colon N \to B$ over the same base are isomorphic if there exists an isomorphism of complex spaces $N \to M$ commuting with f and g.

Since fibred products exist in the category of complex spaces [80, Corollary 0.32], for every smooth family $f\colon M \to B$ and every morphism of complex spaces $C \to B$, the pull-back $M \times_B C \to C$ is again a smooth family.

As above, a deformation is a "framed germ" of a family. More precisely, a deformation of a compact complex manifold X over an analytic singularity $(B, 0)$ is given by a commutative diagram of complex spaces

$$\begin{array}{ccc} X & \xrightarrow{i} & \mathcal{X} \\ \downarrow & & \downarrow{\scriptstyle f} \\ 0 & \longrightarrow & B \end{array}$$

such that:

1. the morphism $i\colon X \to f^{-1}(0)$ is an isomorphism;
2. there exists an open neighbourhood $0 \in U \subset B$ such that $f\colon f^{-1}(U) \to U$ is a proper smooth family.

The diagrams

$$\begin{array}{ccc} X & \xrightarrow{i} & \mathcal{X} \\ \downarrow & & \downarrow{\scriptstyle f} \\ 0 & \longrightarrow & B, \end{array} \qquad \begin{array}{ccc} X & \xrightarrow{j} & \widetilde{\mathcal{X}} \\ \downarrow & & \downarrow{\scriptstyle g} \\ 0 & \longrightarrow & B, \end{array}$$

give the same deformation if there exist an open neighbourhood $0 \in U \subset B$, and a commutative diagram

$$\begin{array}{ccc} X & \xrightarrow{i} & f^{-1}(U) \\ {\scriptstyle j}\downarrow & \swarrow & \downarrow{\scriptstyle f} \\ g^{-1}(U) & \xrightarrow[g]{} & U, \end{array}$$

with the diagonal arrow an isomorphism. As before, we shall denote by $\mathrm{Def}_X(B, 0)$ the set of (isomorphism classes) of deformations of a compact complex manifold X over the analytic singularity $(B, 0)$.

Every morphism of analytic singularities $g\colon (C, 0) \to (B, 0)$ induces, by taking fibre products, a natural map $g^*\colon \mathrm{Def}_X(B, 0) \to \mathrm{Def}_X(C, 0)$. The definitions of universal and complete deformations extend in the obvious way to deformations over singular bases and the same argument used in Example 1.5.7 proves that, in general, universal deformations do not exist.

The extension of the Kodaira–Spencer map is less obvious and requires some additional work. The key result is that, for every compact complex manifold X, there exists a natural isomorphism

$$\mathrm{Def}_X(\mathrm{Spec}\,\mathbb{C}[\epsilon]/(\epsilon^2)) \cong H^1(X, \Theta_X). \tag{1.7}$$

A complete and detailed proof of this fact will be given in Corollary 4.3.9 and we give here only a very rough idea of such a bijection. We first note that every cohomology class $\eta \in H^1(X, \Theta_X) = \mathrm{Ext}^1_X(\Omega^1_X, \mathcal{O}_X)$ determines an extension of locally free sheaves

$$0 \to \mathcal{O}_X \xrightarrow{\epsilon} \mathcal{E} \xrightarrow{p} \Omega^1_X \to 0.$$

Taking the pull-back of p under the de Rham differential $d\colon \mathcal{O}_X \to \Omega^1_X$ we obtain a sheaf of $\mathbb{C}[\epsilon]/(\epsilon^2)$-algebras

$$\mathcal{O}_{\mathcal{X}} := \mathcal{O}_X \times_{\Omega^1_X} \mathcal{E} = \{(f, s) \mid f \in \mathcal{O}_X,\ s \in \mathcal{E},\ df = p(s)\},$$

where the product in $\mathcal{O}_{\mathcal{X}}$ is defined by the formula $(f, s)(g, r) = (fg, fr + gs)$, and the $\mathbb{C}[\epsilon]/(\epsilon^2)$-algebras structure is given by the morphism of rings

$$\frac{\mathbb{C}[\epsilon]}{(\epsilon^2)} \to H^0(X, \mathcal{O}_{\mathcal{X}}), \qquad (a + b\epsilon) \mapsto (a, \epsilon(b)).$$

Finally, it is not difficult to prove that the ringed space $(X, \mathcal{O}_{\mathcal{X}})$ is a complex space which is smooth over $\mathrm{Spec}\,\mathbb{C}[\epsilon]/(\epsilon^2)$.

The Kodaira–Spencer map $\rho\colon T_0B \to H^1(X, \Theta_X) = \mathrm{Def}_X(\mathrm{Spec}\,\mathbb{C}[\epsilon]/(\epsilon^2))$ of a deformation $\xi \in \mathrm{Def}_X(B, 0)$ is defined in terms of the pull-back $\rho(g) = g^*(\xi)$, where every $g \in T_0B$ is interpreted as in Definition 1.6.5, i.e., as a morphism of analytic singularities $\mathrm{Spec}\,\mathbb{C}[\epsilon]/(\epsilon^2) \to (B, 0)$.

1.7 The Completeness Theorem of Kuranishi

Probably, the most important result in deformation theory of complex manifolds is that, whenever deformations over analytic singularities are taken into account, then every compact complex manifold admits a complete deformation. For a more precise statement we need to introduce the notion of semiuniversal deformation.

Definition 1.7.1 Let X be a compact complex manifold and $(B, 0)$ an analytic singularity. A deformation $\xi \in \mathrm{Def}_X(B, 0)$ is called **semiuniversal** if for every analytic singularity $(C, 0)$ and every deformation $\eta \in \mathrm{Def}_X(C, 0)$ there exists a morphism

$g: (C, 0) \to (B, 0)$ such that $\eta = g^*\xi$. Moreover, the differential $dg: T_0C \to T_0B$ is unique.

The unicity of differential means that if $h: (C, 0) \to (B, 0)$ is another morphism such that $h^*\xi = \eta$, then $dg = dh: T_0C \to T_0B$. The semiuniversal deformation, if it exists, is unique up to a *non-canonical* isomorphism. In fact if two deformations $\xi \in \mathrm{Def}_X(B, 0)$ and $\xi' \in \mathrm{Def}_X(B', 0)$ are both semiuniversal, then there exist a morphism $g: (B', 0) \to (B, 0)$ such that $g^*\xi = \xi'$ and a morphism $h: (B, 0) \to (B', 0)$ such that $h^*\xi' = \xi$. Since $(gh)^*\xi = \xi$ the differential of gh must be the identity and then, by Remark 1.6.6, the morphism gh is an analytic isomorphism of $(B, 0)$; similarly, hg is also an analytic isomorphism of $(B', 0)$ and the conclusion follows from the 2 out of 6 property of isomorphisms (Exercise 1.9.1).

It is clear from the definition that the Kodaira–Spencer map of a semiuniversal deformation is bijective.

Theorem 1.7.2 (Kuranishi [162]) *Every compact complex manifold X admits a semiuniversal deformation over an analytic singularity $(B, 0)$. The analytic singularity $(B, 0)$ is isomorphic to the fibre of a germ of holomorphic map*

$$q: (H^1(X, \Theta_X), 0) \to (H^2(X, \Theta_X), 0)$$

with trivial differential at the origin: $dq(0) = 0$. Moreover, the quadratic part of the map q is equal to the Lie-cup product in cohomology.

In view of the above theorem, very often a semiuniversal deformation of a compact complex manifold is also called a **Kuranishi family**.

As an immediate consequence of Theorem 1.7.2 we get the following results: the first was proved earlier by Kodaira, Nirenberg and Spencer, while the second is nothing else than basic dimension theory.

Theorem 1.7.3 (Kodaira–Spencer existence theorem [146]) *Let X be a compact complex manifold such that $H^2(X, \Theta_X) = 0$. Then X admits a semiuniversal deformation over the germ $(H^1(X, \Theta_X), 0)$.*

Corollary 1.7.4 *Let $X \to \mathcal{X} \xrightarrow{f} (B, 0)$ be the semiuniversal deformation of a compact complex manifold X. Then*

$$\dim H^1(X, \Theta_X) \geq \dim(B, 0) \geq \dim H^1(X, \Theta_X) - \dim H^2(X, \Theta_X).$$

Moreover, $(B, 0)$ is smooth if and only if $\dim H^1(X, \Theta_X) = \dim(B, 0)$.

One can ask whether the semiuniversal deformation $\xi \in \mathrm{Def}_X(B, 0)$ of a compact complex manifold X is universal, i.e., if it has the property that for every analytic singularity $(C, 0)$ and every deformation $\eta \in \mathrm{Def}_X(C, 0)$ there exists a unique morphism $g: (C, 0) \to (B, 0)$ such that $\eta = g^*\xi$, cf. Definition 1.5.6.

The answer to the above question is generally difficult; a sufficient condition for universality, which is quite useful and applies in several cases, is the following.

Theorem 1.7.5 (Schlessinger [237], Wawrik [268]) *Let $X \to \mathcal{X} \xrightarrow{f} (B, 0)$ be the semiuniversal deformation of a compact complex manifold X. If the direct image sheaf $f_*\Theta_{\mathcal{X}/B}$ is locally free at 0, then f is the universal deformation of X.*

A compact complex manifold is called **unobstructed** if has a semiuniversal deformation over a smooth basis; otherwise it is called **obstructed**. By Theorem 1.7.3 every compact complex manifold X with $H^2(X, \Theta_X) = 0$ is unobstructed. In particular; every compact complex manifold of dimension 1 is unobstructed.

Conversely, looking at the quadratic part of the Kuranishi map q, a necessary condition for a compact complex manifold X to be unobstructed is that the Lie-cup product

$$H^1(X, \Theta_X) \to H^2(X, \Theta_X), \qquad x \mapsto [x, x],$$

vanishes identically. It should be noted that the vanishing of the Lie-cup product in cohomology is *not sufficient* to ensure unobstructedness, see Example 1.7.11; in fact, from the author's personal point of view, this is one of the main motivations for the study of Lie methods in deformation theory.

We conclude this section by listing (without proofs) a very small selection of classical examples of obstructed manifolds.

Example 1.7.6 (*Kodaira and Spencer* [150]) Let $\mathbb{C}^q/\Gamma$ be a complex torus of dimension $q \geq 2$; for every $n \geq 1$ the manifold $X = \mathbb{C}^q/\Gamma \times \mathbb{P}^n$ does not have a complete deformation over a smooth basis. In effect, it is possible to prove (see Sect. 8.4) that the base space of its semiuniversal deformation is $(\mathbb{C}^{q^2} \times Y, 0)$, where

$$Y = \{(A_1, \ldots, A_q) \in \mathfrak{sl}_{n+1}(\mathbb{C})^q \mid [A_i, A_j] = 0 \quad \forall\, i, j\}.$$

These examples also show that for every integer $n \geq 3$ there exists an obstructed compact complex manifold of dimension n. The first example of an obstructed surface was found later by Kas [138]. In the paper [183] it is proved that every smooth and sufficiently ample divisor $S \subset \mathbb{C}^2/\Gamma \times \mathbb{P}^1$ is obstructed.

Example 1.7.7 (*Burns and Wahl* [33]) Let $L_1, \ldots, L_n$, $n \geq 5$, be general linear forms in the variables x_1, x_2, x_3 and let S be the minimal resolution of singularities of the surface $X \subset \mathbb{P}^3$ defined by the equation $L_1 \cdots L_n = x_0^n$. Then S is obstructed.

The above example follows from a general study of deformations of smooth minimal surfaces of general type S for which the canonical bundle is not ample [33, 40]. If X is the canonical model of S, then $\mathrm{Aut}(X) = \mathrm{Aut}(S)$ is a finite group and S, X have universal deformations, with base spaces $\mathrm{Def}(S)$, $\mathrm{Def}(X)$. Then Burns and Wahl proved that there exists a natural finite surjective map $\mathrm{Def}(S) \to \mathrm{Def}(X)$.

It is interesting to point out that the above result does not hold for equivariant deformations; if $G \subset \mathrm{Aut}(X) = \mathrm{Aut}(S)$ is a subgroup, then G acts on both $\mathrm{Def}(S)$ and $\mathrm{Def}(X)$ but the map between the G-invariant parts $\mathrm{Def}(S, G) \to \mathrm{Def}(X, G)$ is not surjective in general, even in the case where G acts trivially on $\mathrm{Def}(X)$; we refer to [17] for more details and some specific examples.

Example 1.7.8 (*Mumford and Horikawa* [119, 200]) Let $S \subset \mathbb{P}^3$ be a smooth cubic surface and let $E \subset S$ be one of its 27 lines. Denoting by $H \subset S$ the hyperplane divisor, the complete linear system $|4H + 2E|$ is base point free and then contains a smooth curve $C \subset S$ (cf. [39, 9.11], [110, Theorem 13.1], [242, 4.6.2]). Let $X \to \mathbb{P}^3$ be the blow-up along C, then the base space of the semiuniversal deformation of X is unreduced (Mumford). More precisely, Curtin proved [50] that it is isomorphic to $(\mathbb{C}^{56} \times \operatorname{Spec} \mathbb{C}[\epsilon]/(\epsilon^2), 0)$. Finally, every sufficiently ample and generic divisor of X is obstructed (Horikawa).

Example 1.7.9 (*Catanese* [40]) In the notation of Corollary 1.7.4, for an algebraic surface X, the ratio $\dim H^1(X, \Theta_X)/\dim(B, 0)$ can be arbitrarily large. The examples considered in [40] include the "generalized Kas surfaces", namely the minimal resolutions of certain quotient surfaces $C_1 \times C_2/\mu_n$, where μ_n in the group of nth roots of unity and C_1, C_2 are both simple cyclic covers of degree n of smooth curves.

Example 1.7.10 (*Fantechi and Pardini* [71]) For every pair of integers $n, m \geq 2$ there exists a projective manifold of dimension n whose base space of its semiuniversal deformations has at least m irreducible components.

Example 1.7.11 (*Murphy's law in deformation theory*) Let $(B, 0)$ be any analytic singularity defined by a finite number of polynomials with rational coefficients. Then, by a remarkable result by Vakil [259], for every $n \geq 2$ there exists a projective manifold of dimension n whose base space of its semiuniversal deformation is isomorphic to $(B \times \mathbb{C}^m, 0)$ for some $m \geq 0$.

In particular, for every $n, r \geq 2$ there exists a projective manifold of dimension n whose base space of its semiuniversal deformation is isomorphic to $(\mathbb{C}^m \times \operatorname{Spec}(\mathbb{C}[t]/(t^r)), 0)$ for some $m \geq 0$.

Example 1.7.12 (*Rigid but not infinitesimally rigid compact complex manifolds*) A compact complex manifold is called rigid if the base space of their semiuniversal deformation is zero-dimensional. According to Corollary 1.4.3 every infinitesimally rigid compact manifold is also rigid.

A long-standing question, posed by Morrow and Kodaira [198, p. 45], about the existence of rigid but not infinitesimally rigid compact complex manifolds has been recently solved by Bauer and Pignatelli [19]. They proved that for each dimension $d \geq 2$ there exists an infinite series of compact complex manifolds that are rigid but not infinitesimally rigid. A different set of examples, based on a different approach, is described in [26].

1.8 Infinitesimal Deformations

In Sect. 1.6 we have used deformations over $\operatorname{Spec} \mathbb{C}[\epsilon]/(\epsilon^2)$ in order to give a nice and useful description of the Kodaira–Spencer map for any deformation of a compact complex manifold.

By definition, a deformation over Spec $\mathbb{C}[\epsilon]/(\epsilon^2)$ is called a **first order deformation**, while a deformation over an analytic singularity of dimension 0 is called an **infinitesimal deformation**.

Clearly, every first order deformation is infinitesimal; more generally one can define the set of nth order deformations as the set of deformations over Spec $\mathbb{C}[t]/(t^{n+1})$.

Although it is not possible to give a complete explanation at this point of the book, the theory of infinitesimal deformations is the core of deformation theory and it is extremely useful also in order to prove results concerning deformations over higher-dimensional bases.

For instance, consider the following problem: given a compact complex manifold X together with a cohomology class $\xi_1 \in H^1(X, \Theta_X)$, does there exist a deformation $X \to \mathcal{X} \to (\mathbb{C}, 0)$ such that $\rho_0(d/dt) = \xi_1$?

Interpreting the class ξ_1 as a first order deformation, it is possible to prove that the above problem has a positive answer if and only if ξ_1 can be lifted to an nth order deformation, for every $n > 1$.

As an additional motivation, Kuranishi's Theorem 1.7.2 is an existence result and its proof gives very little information about the geometry of the semiuniversal deformation $X \to \mathcal{X} \xrightarrow{f} (B, 0)$. In particular, and with a few exceptions, Kuranishi's theorem leaves unanswered the question whether $(B, 0)$ is smooth. In fact, the condition $H^2(X, \Theta_X) = 0$ is sufficient but it is very far from being necessary; the "majority" of unobstructed manifolds have the cohomology group $H^2(X, \Theta_X)$ different from 0.

Now, the smoothness of $(B, 0)$ depends only on the properties of the formal completion of the analytic algebra $\mathcal{O}_{B,0}$, and therefore depends only on the collection of morphisms $\mathcal{O}_{B,0} \to A$, where A ranges over Artin local $\mathbb{C}$-algebras. On the other hand, every morphism $\mathcal{O}_{B,0} \to A$ gives a morphism of analytic singularities Spec $A \to (B, 0)$; taking the pull-back of the semiuniversal deformation f we obtain a deformation of X over Spec A. Conversely, the properties of the semiuniversal deformation imply that every deformation of X over Spec A is obtained in this way.

Therefore we can affirm that: *the answer to the question about the smoothness of the semiuniversal deformation depends only on the family of all infinitesimal deformations of X.*

The advantage of this point of view is that infinitesimal deformations are usually much easier to study in view of the following facts. Very roughly, if A is a local Artin $\mathbb{C}$-algebra with residue field $\mathbb{C}$, then:

- A morphism of complex spaces $f : \mathcal{X} \to$ Spec A is a smooth family if and only if the fibre $X = f^{-1}(\text{Spec}\,\mathbb{C})$ is a smooth manifold and $\mathcal{O}_{\mathcal{X}}$ is a sheaf of A-algebras which is locally isomorphic to $A \otimes_{\mathbb{C}} \mathcal{O}_X$.
- The notion of infinitesimal deformation makes sense also for non compact manifolds. In other words the topological triviality of every infinitesimal deformation makes inessential the properness assumption.
- Every infinitesimal deformation of a Stein manifold is trivial, cf. Example 1.3.5.

- By the above items, every set $\mathrm{Def}_X(\mathrm{Spec}\, A)$ may be interpreted as a "first cohomology set" of a nonabelian sheaf, which can be conveniently described (see Chap. 4) by using Lie algebras (Chap. 2), differential graded Lie algebras (Chap. 6) and functors of Artin rings (Chap. 3).

1.9 Exercises

1.9.1 (*The 2 out of 6 property of isomorphisms*) Let

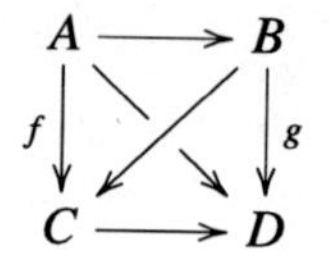

be a commutative diagram of 4 objects and 6 arrows in a category $\mathbf{C}$. Prove that if f and g are isomorphisms, then every arrow in the diagram is an isomorphism.

1.9.2 Let $\tau = a + ib$, $b < 0$, be a complex number with negative imaginary part and consider the diagonal matrix $A = e^{2\pi i\tau} I \in GL_2(\mathbb{C})$ together with the elliptic curve $T = \mathbb{C}/(\mathbb{Z} + \tau\mathbb{Z})$; notice that $|e^{2\pi i\tau}| = e^{-2\pi b} > 1$. Prove that, in the notation of Example 1.1.5, the Hopf surface S_A is the total space of a holomorphic T-principal bundle $S_A \to \mathbb{P}^1$.

1.9.3 Let $f\colon (X, x) \to (Y, y)$ be a morphism of analytic singularities and assume that $f^*\colon \mathcal{O}_{Y,y} \to \mathcal{O}_{X,x}$ is surjective. Prove that $df\colon T_xX \to T_yY$ is injective and df is a linear isomorphism if and only if the kernel of f^* is contained in $\mathfrak{m}_{Y,y}^2$.

1.9.4 Let X be a compact complex manifold and $(B, 0)$ an analytic singularity. There exists a natural action of the group $\mathrm{Aut}(X)$ of holomorphic isomorphisms of X on the set of deformations $\mathrm{Def}_X(B, 0)$; if $g \in \mathrm{Aut}(X)$ and $\xi\colon X \xrightarrow{i} \mathcal{X} \xrightarrow{f} (B, 0)$ is a deformation we have

$$\xi^g\colon X \xrightarrow{ig^{-1}} \mathcal{X} \xrightarrow{f} (B, 0).$$

Prove that $\xi^g = \xi$ if and only if $igi^{-1}\colon f^{-1}(0) \to f^{-1}(0)$ can be extended to an isomorphism of complex spaces $\hat{g}\colon f^{-1}(V) \to f^{-1}(V)$, $0 \in V$ open neighbourhood, such that $f\hat{g} = f$.

1.9.5 Prove that a universal deformation $\xi : X \xrightarrow{i} \mathcal{X} \xrightarrow{f} (B, 0)$ of a compact complex manifold X induces a homomorphism of groups:

$$\theta\colon \mathrm{Aut}(X) \to \mathrm{Aut}(B, 0), \qquad \theta(g)^*\xi = \xi^g, \quad g \in \mathrm{Aut}(X).$$

Moreover, for every other universal deformation over the same germ $(B, 0)$ the new action θ differs from the old one by an inner automorphism.

References

Following Bourbaki, a dangerous bend is a passage in the text that is designed to forewarn the reader against possible errors. The "dangerous bend" symbol ☡ used here is borrowed from the TeXbook.

The material of this chapter is completely standard and can be found in several places, for instance in the books [123, 145, 199, 203, 242, 262] and in the papers by Kodaira and Spencer [146–150] devoted to the subjects. It is worth mentioning also the first chapter of [248], the papers [211, 212, 245], and the more technical contributions [39, 59, 95].

The notation at the beginning of Sect. 1.2 are taken from Horikawa's paper [117]. Example 1.3.5 is taken from [156, Exercise 1.4]. The quasi-isomorphism between alternating and standard Čech cochains is a standard consequence of some general fact of simplicial homology theory, see e.g. [65, VI.6], [91], [243, p. 214].

In addition to projective spaces, there exist many examples of infinitesimally rigid manifolds; a classification of of infinitesimally rigid surfaces is given in [18].

The completeness theorem of Kodaira and Spencer (Theorem 1.5.8) finds large application for the description of the semiuniversal deformation of several complex manifolds; for instance Segre–Hirzebruch surfaces (e.g. [39, 180]) and abelian covers of manifolds (e.g. [38, 71, 177, 213, 259]).

In the footnote on page 20 we mentioned the possibility to represent deformations as small perturbations of the complex structure in the space of integrable almost complex structures on the underlying differentiable manifold M. One can also globalize this construction by defining $C(M)$ as the set of integrable almost complex structures on M and $\mathcal{M}(M)$ as the quotient of $C(M)$ by the group of orientation-preserving diffeomorphisms [44]. In analogy with the usual Teichmüller space one can also consider the space $\mathcal{T}(M)$ as the quotient of $C(M)$ by the group of diffeomorphisms that are isotopic to the identity. The study of the spaces $\mathcal{M}(M)$, $\mathcal{T}(M)$ and their relation with Kuranishi's theory is a new and interesting research direction, although almost completely skew with the content of this book.

Chapter 2
Lie Algebras

In this chapter, after a brief review of Lie algebras and descending central series, we study free Lie algebras over fields of characteristic 0 and the Baker–Campbell–Hausdorff (BCH) product. For simplicity of exposition, the BCH product is defined here only for nilpotent Lie algebras, since this restriction is sufficient for our purposes. At the end of the chapter we introduce the notion of a semicosimplicial Lie algebra, which plays a central role in the algebraic theory of infinitesimal deformations of complex manifolds and vector bundles.

2.1 Magmatic Algebras

In this book the term *algebra* is intended widely as the data of a (possibly graded) vector space equipped with a (possibly infinite) set of multilinear operations satisfying a certain set of relations. One of the simplest cases is when the algebra is defined by one binary operation without any further relation.

Definition 2.1.1 A **magmatic algebra** over a field is a vector space M equipped with a bilinear map

$$M \times M \to M, \qquad (x, y) \mapsto x \cdot y.$$

The binary operation in a magmatic algebras is usually called a *product*; sometimes it is called a *bracket*, while in few cases it is called a *brace*. The usual notions of morphism, unity, ideal and derivation extend naturally to the framework of magmatic algebras over a fixed field $\mathbb{K}$. More precisely:

- A **morphism** $f\colon (M, \cdot) \to (N, *)$ of magmatic algebras is a linear map $M \to N$ commuting with the binary operations, i.e., $f(x \cdot y) = f(x) * f(y)$ for every $x, y \in M$.
- A magmatic algebra $(M, \cdot)$ is called **unitary** if there exists an element $1 \in M$, called unity, such that $1 \cdot m = m \cdot 1 = m$ for every $m \in M$.

M. Manetti, *Lie Methods in Deformation Theory*, Springer Monographs in Mathematics,
https://doi.org/10.1007/978-981-19-1185-9_2

- A **subalgebra** of a magmatic algebra $(M, \cdot)$ is a vector subspace $N \subset M$ such that $x \cdot y \in N$ for every $x, y \in N$. In particular, every subalgebra is an algebra in its own right with respect to the inherited binary operation.
- A **left ideal** (resp.: **right ideal**) of a magmatic algebra $(M, \cdot)$ is a vector subspace $I \subset M$ such that $m \cdot i \in I$ (resp.: $i \cdot m \in I$) for every $m \in M$ and $i \in I$. A subspace is called an **ideal** if it is both a left and right ideal.
- A $\mathbb{K}$-**derivation** of a magmatic algebra $(M, \cdot)$ is a linear map $\alpha \in \operatorname{Hom}_{\mathbb{K}}(M, M)$ which satisfies the **Leibniz rule**:

$$\alpha(r \cdot s) = \alpha(r) \cdot s + r \cdot \alpha(s), \qquad \text{for every} \quad r, s \in M.$$

A nilpotent derivation is a derivation that is nilpotent as a linear endomorphism.

Lemma 2.1.2 *Let $(M, \cdot)$ be a magmatic algebra over a field of characteristic 0 and let $\alpha \colon M \to M$ be a nilpotent derivation. Then the linear map*

$$e^{\alpha} = \sum_{n \geq 0} \frac{\alpha^n}{n!} \colon M \to M$$

is an invertible morphism of magmatic algebras with inverse $e^{-\alpha}$. Moreover, e^{α} is the identity if and only if $\alpha = 0$.

Proof An easy inductive application of the Leibniz rule gives the formulas

$$\alpha^n(x \cdot y) = \sum_{i=0}^{n} \binom{n}{i} \alpha^i(x) \cdot \alpha^{n-i}(y), \qquad n \geq 0,$$

and therefore

$$\begin{aligned} e^{\alpha}(x \cdot y) &= \sum_{n \geq 0} \frac{1}{n!} \sum_{i=0}^{n} \binom{n}{i} \alpha^i(x) \cdot \alpha^{n-i}(y) \\ &= \sum_{i,j \geq 0} \frac{1}{i!\, j!} \alpha^i(x) \cdot \alpha^{n-i}(y) = e^{\alpha}(x) \cdot e^{\alpha}(y). \end{aligned}$$

Finally,

$$e^{-\alpha} e^{\alpha} = \sum_{n,m \geq 0} \frac{(-1)^n}{n!\, m!} \alpha^{n+m} = \sum_{k \geq 0} \frac{\alpha^k}{k!} \sum_{i=0}^{k} (-1)^i \binom{k}{i} = \sum_{k \geq 0} (1-1)^k \frac{\alpha^k}{k!} = \mathrm{Id}_M.$$

If e^{α} is the identity, then

$$0 = e^{\alpha} - \mathrm{Id}_M = \sum_{n>0} \frac{\alpha^n}{n!} = \alpha \circ \left(\mathrm{Id}_M + \sum_{n>0} \frac{\alpha^n}{(n+1)!} \right)$$

and this implies $\alpha = 0$, since $\sum_{n>0} \frac{\alpha^n}{(n+1)!}$ is a nilpotent endomorphism of M and therefore $\mathrm{Id}_M + \sum_{n>0} \frac{\alpha^n}{(n+1)!}$ is an invertible linear map. □

In the above proof we denoted by Id_M the identity map; throughout this book the identity map on any set X will be denoted either by Id_X or by Id when the set X is clear from the context.

The reader is certainly already familiar with the notion of associative and commutative algebras:

- An **associative** algebra $(S, \cdot)$ is a magmatic algebra such that $(r \cdot s) \cdot t = r \cdot (s \cdot t)$ for every $r, s, t \in S$.
- A **commutative** algebra $(R, \cdot)$ is an associative algebra such that $r \cdot s = s \cdot r$ for every $r, s \in R$.

Notice that a commutative algebra is also associative by definition, while a magmatic algebra $(M, \cdot)$ such that $x \cdot y = y \cdot x$ for every $x, y \in M$, without assuming associativity, is called commutative magmatic.

As an example, for every vector space V, the space $M = \mathrm{Hom}_{\mathbb{K}}(V, V)$ of linear endomorphisms is an associative unitary algebra, when equipped with the composition product and the identity map Id_V as unity.

2.2 Lie and Pre-Lie Algebras

Two other important subclasses of magmatic algebras are given by Lie and pre-Lie algebras. The notion of a Lie algebra is well known, and we briefly recall its definition for completeness sake and notational setup.

Definition 2.2.1 A **Lie algebra** is a vector space L, equipped with a bilinear binary operator

$$L \times L \to L, \qquad (x, y) \mapsto [x, y],$$

called a **bracket**, satisfying the following conditions:

1. $[x, x] = 0$ for every $x \in L$, and then $[x, y] = -[y, x]$ for every x, y in L;
2. (Jacobi identity) $[[x, y], z]] = [x, [y, z]] - [y, [x, z]]$ for every $x, y, z \in L$.

A Lie algebra L is said to be **abelian** if its bracket is trivial, i.e., if $[x, y] = 0$ for every $x, y \in L$.

According to the general magmatic definition, a linear subspace $H \subset L$ is a **Lie subalgebra** if $[x, y] \in H$ whenever $x, y \in H$. Notice that any nonzero element $x \in L$ defines a one-dimensional abelian subalgebra $H = \mathbb{K}x$.

Similarly, a **morphism of Lie algebras** $f : L \to M$ is a linear map commuting with brackets: $f([x, y]) = [f(x), f(y)]$. An isomorphism of Lie algebras is a morphism of Lie algebras which is also an isomorphism of vector spaces.

Example 2.2.2 The space $\mathrm{End}_{\mathbb{K}}(V) = \mathrm{Hom}_{\mathbb{K}}(V, V)$ of all linear endomorphisms of a vector space V is a Lie algebra with bracket $[f, g] = fg - gf$. If V is finite-dimensional, then the subspace $\mathfrak{sl}(V) \subset \mathrm{End}(V)$ of endomorphisms with trace equal to 0 is a Lie subalgebra. For any $n > 0$, we write $\mathfrak{sl}_n(\mathbb{K}) = \mathfrak{sl}(\mathbb{K}^n)$.

Example 2.2.3 (*Wronskian bracket*) The vector space of C^∞ functions on the real line admits a Lie algebra structure with bracket $[f, g] = fg' - f'g$. More generally, for any derivation $\alpha\colon A \to A$ of a commutative algebra, the pairing $[x, y] = x\alpha(y) - \alpha(x)y$ is a Lie bracket.

Example 2.2.4 For every Lie algebra L, the linear map

$$\mathrm{ad}\colon L \to \mathrm{End}_{\mathbb{K}}(L), \qquad \mathrm{ad}\, x = [x, -], \qquad \mathrm{ad}\, x(y) = [x, y],$$

is a morphism of Lie algebras. In fact, by the Jacobi identity we have:

$$\begin{aligned}[\mathrm{ad}\, x, \mathrm{ad}\, y](z) &= \mathrm{ad}\, x(\mathrm{ad}\, y(z)) - \mathrm{ad}\, y(\mathrm{ad}\, x(z)) = \mathrm{ad}\, x([y, z]) - \mathrm{ad}\, y([x, z]) \\ &= [x, [y, z]] - [y, [x, z]] = [[x, y], z] = \mathrm{ad}[x, y](z).\end{aligned}$$

For every associative $\mathbb{K}$-algebra R we denote by R_L the associated Lie algebra, with bracket equal to the commutator: $[a, b] = ab - ba$. The verification of the following properties is immediate:

1. if $I \subset R$ is an ideal of R, then I is also an ideal of R_L, i.e., $[a, b] \in I$ for every $a \in I, b \in R_L$;
2. if $f\colon R \to R$ is a derivation, then also $f\colon R_L \to R_L$ is a derivation.

Example 2.2.5 As expected, not every Lie bracket is the commutator of an associative product. For instance, if $\mathbb{K}$ is a field of characteristic $\neq 2$, then there does not exist any associative product in $\mathfrak{sl}_2(\mathbb{K})$ such that $[x, y] = xy - yx$. Here we give a completely elementary proof; a slightly more conceptual proof, based on Engel's theorem and valid for every $\mathfrak{sl}_n(\mathbb{K})$, $n \geq 2$, is proposed in Exercise 2.7.2.

The canonical basis of the Lie algebra $\mathfrak{sl}_2(\mathbb{K})$ is given by the matrices

$$A = \begin{pmatrix} 0 & 1 \\ 0 & 0 \end{pmatrix}, \qquad B = \begin{pmatrix} 0 & 0 \\ 1 & 0 \end{pmatrix}, \qquad H = \begin{pmatrix} 1 & 0 \\ 0 & -1 \end{pmatrix},$$

and we have

$$[A, B] = H, \qquad [H, A] = 2A, \qquad [H, B] = -2B.$$

We have to prove that if there exists an associative product on $\mathfrak{sl}_2(\mathbb{K})$ such that

$$AB - BA = H, \qquad HA - AH = 2A, \qquad HB - BH = -2B,$$

then we get a contradiction. Writing $H^2 = \lambda_1 A + \lambda_2 B + \lambda H$ we have

$$0 = H^3 - H^3 = [H^2, H] = \lambda_1[A, H] + \lambda_2[B, H]$$

and therefore $\lambda_1 = \lambda_2 = 0$, $H^2 = \lambda H$. If $\lambda = 0$ a direct computation shows that $[H, -]^3 = 0$, hence $[H, -]$ is a nilpotent operator in contradiction with the equality $[H, A] = 2A$. Hence $\lambda \neq 0$ and the operators of left and right multiplication by H

$$L_H, R_H : \mathfrak{sl}_2(\mathbb{K}) \to \mathfrak{sl}_2(\mathbb{K}), \qquad L_H(X) = HX, \quad R_H(X) = XH,$$

are both annihilated by the polynomial $t(t - \lambda)$; in particular, L_H, R_H are both diagonalizable and, since the adjoint operator $[H, -]$ is not trivial, at least one of L_H, R_H is different from λI. Assume, for fixing the ideas, that $L_H \neq \lambda I$, then $0, \lambda$ are the eigenvalues of L_H and therefore its trace is either equal to λ or equal to 2λ. On the other hand, since $L_H = [L_A, L_B]$ the trace of L_H is trivial and this implies $\lambda = 0$. Notice that in characteristic 2 an associative product inducing the bracket is given by setting $AB = H$ and $A^2 = AH = BA = B^2 = BH = H^2 = HA = HB = 0$.

Example 2.2.6 Let $(M, \cdot)$ be a magmatic algebra over $\mathbb{K}$. Then the vector space

$$\mathrm{Der}_{\mathbb{K}}(M, M) = \{d \in \mathrm{Hom}_{\mathbb{K}}(M, M) \mid d(x \cdot y) = (dx) \cdot y + x \cdot (dy),\ \forall\, x, y \in M\}$$

of $\mathbb{K}$-derivations of M is a Lie subalgebra of $\mathrm{End}_{\mathbb{K}}(M) = \mathrm{Hom}_{\mathbb{K}}(M, M)$.

The above example applies in particular to Lie algebras; if L is a Lie algebra over a field $\mathbb{K}$, then the subspace

$$\mathrm{Der}_{\mathbb{K}}(L, L) = \{d \in \mathrm{Hom}_{\mathbb{K}}(L, L) \mid d[x, y] = [dx, y] + [x, dy],\ \forall\, x, y \in L\}$$

is a Lie subalgebra of $\mathrm{Hom}_{\mathbb{K}}(L, L)$ and the Jacobi identity in L is equivalent to the fact that $\mathrm{ad}\, x \in \mathrm{Der}_{\mathbb{K}}(L, L)$ for every $x \in L$.

Example 2.2.7 (*Scalar extension*) Let L be a Lie algebra over a field $\mathbb{K}$ and A a commutative $\mathbb{K}$-algebra. Then the tensor product $L \otimes_{\mathbb{K}} A$ is a Lie algebra with bracket equal to the bilinear extension of

$$[u \otimes a, v \otimes b] = [u, v] \otimes ab.$$

If A and B are linear subspaces of a Lie algebra H, then it is standard to write $[A, B] = \mathrm{Span}\{[a, b] \mid a \in A,\ b \in B\}$. Equivalently, $[A, B]$ is the image of the linear map $A \otimes B \xrightarrow{[-,-]} H$.

Definition 2.2.8 The **descending central series** $H^{[n]}$, $n \geq 1$, of a Lie algebra H is defined as $H^{[1]} = H$ and

$$H^{[n]} = \mathrm{Span}\{\, [a_1, [\ \ldots\ [a_{n-1}, a_n]..]] \mid a_1, \ldots, a_n \in H\}, \qquad n \geq 2.$$

Equivalently, it is defined by the recursive formulas $H^{[1]} = H$ and $H^{[n]} = [H, H^{[n-1]}]$.

It is easy to prove that for every pair of positive integers n, m we have:

$$H^{[n+1]} \subset H^{[n]}, \qquad [H^{[n]}, H^{[m]}] \subset H^{[n+m]}.$$

When $n = 1$ the above properties follow immediately from the definition of the descending central series. If $n > 1$ we may assume by induction that $H^{[n]} \subset H^{[n-1]}$; then $H^{[n+1]} = [H, H^{[n]}] \subset [H, H^{[n-1]}] = H^{[n]}$. Again by induction on n we may assume that $[H^{[a]}, H^{[m]}] \subset H^{[a+m]}$ for every $1 \le a < n$. The vector space $[H^{[n]}, H^{[m]}]$ is generated by the vectors

$$[[x, y], z], \qquad x \in H,\ y \in H^{[n-1]},\ z \in H^{[m]},$$

and since $[[x, y], z] = [x, [y, z]] - [y, [x, z]]$ by the Jacobi identity, we have:

$$[H^{[n]}, H^{[m]}] = [[H, H^{[n-1]}], H^{[m]}] \subset [H, [H^{[n-1]}, H^{[m]}]] + [H^{[n-1]}, [H, H^{[m]}]].$$

By the inductive assumption

$$[H^{[n-1]}, H^{[m]}] \subset H^{[n+m-1]}, \qquad [H^{[n-1]}, H^{[m+1]}] \subset H^{[n+m]}$$

and then

$$\begin{aligned}[H^{[n]}, H^{[m]}] &\subset [H, [H^{[n-1]}, H^{[m]}]] + [H^{[n-1]}, [H, H^{[m]}]]\\ &\subset [H, H^{[n+m-1]}] + [H^{[n-1]}, H^{[m+1]}] \subset H^{[n+m]}.\end{aligned}$$

Notice that the inclusion $[H, H^{[n]}] = H^{[n+1]} \subset H^{[n]}$ means exactly that $H^{[n]}$ is a Lie ideal of H.

Definition 2.2.9 A Lie algebra H is called **nilpotent** if $H^{[n]} = 0$ for some positive integer n.

Example 2.2.10 Every abelian Lie algebra L is nilpotent, since $L^{[2]} = 0$. The Lie subalgebra $N \subset \mathfrak{sl}_n(\mathbb{K})$ of strictly upper triangular matrices is nilpotent, since $N^{[n]} = 0$.

Example 2.2.11 Let L be a Lie algebra over a field $\mathbb{K}$ and let $\mathfrak{m}$ be the maximal ideal of an Artin local $\mathbb{K}$-algebra. Then $L \otimes_{\mathbb{K}} \mathfrak{m}$ is a nilpotent Lie algebra.

If H is a nilpotent Lie algebra, then every adjoint operator $\operatorname{ad} x$, $x \in H$, is nilpotent in $\operatorname{Hom}_{\mathbb{K}}(H, H)$. According to Engel's theorem, see e.g. [122, 136], the converse is true under the additional assumption that H is finite-dimensional.

Pre-Lie Algebras

Given any magmatic algebra $(R, \triangleleft)$ over a field $\mathbb{K}$, the **associator** of R is the trilinear map

$$A\colon R \times R \times R \to R, \qquad A(x, y, z) = (x \triangleleft y) \triangleleft z - x \triangleleft (y \triangleleft z).$$

Clearly, the algebra is associative if and only if its associator vanishes identically.

Definition 2.2.12 In the above notation, an algebra R is called a **right pre-Lie algebra** if $A(x, y, z) = A(x, z, y)$ for every $x, y, z \in R$, i.e., if the associator is symmetric in the last two variables.

A simple example of non-associative right pre-Lie algebra is given by the real vector space $R = C^\infty(\mathbb{R})$ of differentiable functions on the real line, equipped with the product $f \triangleleft g = f'g$.

Lemma 2.2.13 *If $(R, \triangleleft)$ is a right pre-Lie algebra, then the commutator bracket*

$$[-, -]\colon R \times R \to R, \qquad [x, y] = x \triangleleft y - y \triangleleft x,$$

satisfies the Jacobi identity and therefore $(R, [-, -])$ is a Lie algebra.

Proof Straightforward and left to the reader. □

Similarly, one can define a **left pre-Lie algebra** as an algebra $(R, \triangleright)$ such that the associator is symmetric in the first two variables, i.e., if $A(x, y, z) = A(y, x, z)$ for every $x, y, z \in R$. The notions of left and right pre-Lie algebras are perfectly symmetric; if $(R, \triangleright)$ is a left pre-Lie algebra, then the algebra $(R, \triangleleft)$, with product $x \triangleleft y = -y \triangleright x$ is a right pre-Lie algebra with the same commutator bracket, and conversely. In the literature, left (resp.: right) pre-Lie algebras are also called left (resp.: right) symmetric algebras.

Although perfectly equivalent from the algebraic point of view, it is useful to keep separate the notions of left and right pre-Lie algebras, since both of them appear in several algebraic structures. For the goal of this book the most important examples of right pre-Lie algebras are given by Gerstenhaber and Nijenhuis–Richardson products, introduced in Sect. 5.8.

Example 2.2.14 These are two concrete examples of left pre-Lie algebras which are of a certain interest in mathematics:

1. Let A be a commutative $\mathbb{K}$-algebra and let $V \subseteq \mathrm{Der}_{\mathbb{K}}(A, A)$ be an A-submodule freely generated by a basis $\{D_1, D_2, \ldots\}$ of commuting derivations; this means that $D_iD_j = D_jD_i$ for every i, j. Then one can define a left pre-Lie product $\triangleright$ in V by setting
$$(aD_i) \triangleright (bD_j) = a\, D_i(b)\, D_j, \qquad a, b \in A.$$
2. Let $\nabla\colon \chi(M) \times \chi(M) \to \chi(M)$ be the Levi–Civita connection on a Riemannian manifold. Then the operators of left multiplication
$$\nabla_x\colon \chi(M) \to \chi(M), \qquad \nabla_x(y) = \nabla(x, y),$$
are the covariant derivatives, $[-, -]$ is the usual bracket on vector fields and the curvature tensor is given by the formula $R(x, y) = [\nabla_x, \nabla_y] - \nabla_{[x,y]}$. A simple computation shows that $(\chi(M), \nabla)$ is a left pre-Lie algebra if and only if the metric is flat, see Exercise 2.7.4.

Poisson Algebras

In our terminology, an algebra may be defined by more than one single binary operation, as in the case of Poisson algebras.

Definition 2.2.15 A **Poisson algebra** is a commutative algebra A equipped with a Lie bracket $[-,-]$ such that

$$[ab, c] = a[b, c] + [a, c]b, \qquad [a, bc] = b[a, c] + [a, b]c \tag{2.1}$$

for every $a, b, c \in A$.

Thus any Poisson algebra is in particular a commutative algebra and also a Lie algebra; the two equalities (2.1) are called **Poisson identities** and, since the bracket is skew-symmetric, the first Poisson identity holds if and only if the second identity is satisfied.

It should be noted that the Poisson identities can be expressed by saying that for every $a \in A$ the linear map $b \mapsto [a, b]$ is a derivation of the commutative algebra A.

Example 2.2.16 Any commutative algebra is a Poisson algebra with the trivial bracket, and any Lie algebra is a Poisson algebra with the trivial product.

Example 2.2.17 The most classical example of Poisson algebra is the algebra of differentiable functions on $\mathbb{R}^{2n}$, where the Poisson bracket is

$$\{f, g\} = \sum_{i=1}^{n} \frac{\partial f}{\partial x_i}\frac{\partial g}{\partial y_i} - \frac{\partial g}{\partial x_i}\frac{\partial f}{\partial y_i},$$

with $x_1, \ldots, x_n, y_1, \ldots, y_n$ a system of linear coordinates on $\mathbb{R}^{2n}$.

Example 2.2.18 Let $B = C^\infty(\mathbb{R})$ be the algebra of smooth functions on the real line and denote by $A = B[t]$ be the algebra of polynomials with coefficients in B. Then A is a Poisson algebra with the bracket defined by

$$[ft^n, gt^m] = (nfg' - mgf')t^{n+m-1}, \qquad n, m \geq 0, \qquad f, g \in B.$$

2.3 Differential Operators and Derivations of Pairs

Together with derivations of algebras, another important class of Lie algebras is provided by differential operators. Let $\mathbb{K}$ be a fixed field and let R be a *commutative unitary* $\mathbb{K}$-algebra. Thus R is a commutative unitary ring, and every R-module is also a $\mathbb{K}$-vector space.

For every pair M, N of R-modules, the vector space $\operatorname{Hom}_{\mathbb{K}}(M, N)$ of $\mathbb{K}$-linear maps $M \to N$ is also an R-module, with the scalar product defined by the obvious formula $af(m) = a(f(m))$, $a \in R$, $f \in \operatorname{Hom}_{\mathbb{K}}(M, N)$ and $m \in M$. We may define a bilinear map of vector spaces

$$\operatorname{Hom}_{\mathbb{K}}(M, N) \times R \xrightarrow{[-,-]} \operatorname{Hom}_{\mathbb{K}}(M, N)$$

by setting:

$$[f, a](m) = f(am) - af(m), \qquad f \in \operatorname{Hom}_{\mathbb{K}}(M, N), \quad a \in R, \quad m \in M.$$

By definition, a linear map $f \in \operatorname{Hom}_{\mathbb{K}}(M, N)$ is a morphism of R-modules if and only if $[f, a] = 0$ for every $a \in R$. We can immediately verify that for every $f \in \operatorname{Hom}_{\mathbb{K}}(M, N)$ and every $a, b \in R$ we have:

$$[af, b] = a[f, b], \tag{2.2}$$

$$[[f, a], b] = [[f, b], a] = [f, ab] - a[f, b] - b[f, a]. \tag{2.3}$$

Following Grothendieck we define recursively a sequence of subspaces

$$\operatorname{Diff}^n_{R/\mathbb{K}}(M, N) \subset \operatorname{Hom}_{\mathbb{K}}(M, N), \qquad n \geq 0,$$

by setting

$$\operatorname{Diff}^0_{R/\mathbb{K}}(M, N) = \{f \in \operatorname{Hom}_{\mathbb{K}}(M, N) \mid [f, a] = 0 \quad \forall a \in R\} = \operatorname{Hom}_R(M, N),$$

and for every $n > 0$,

$$\operatorname{Diff}^n_{R/\mathbb{K}}(M, N) = \{f \in \operatorname{Hom}_{\mathbb{K}}(M, N) \mid [f, a] \in \operatorname{Diff}^{n-1}_{R/\mathbb{K}}(M, N) \quad \forall a \in R\}.$$

By using formula (2.2), a simple induction on n shows that $\operatorname{Diff}^n_{R/\mathbb{K}}(M, N)$ is an R-submodule of $\operatorname{Hom}_{\mathbb{K}}(M, N)$ and $\operatorname{Diff}^n_{R/\mathbb{K}}(M, N) \subset \operatorname{Diff}^{n+1}_{R/\mathbb{K}}(M, N)$ for every $n \geq 0$.

Definition 2.3.1 In the above notation, the R-submodule

$$\operatorname{Diff}^n_{R/\mathbb{K}}(M, N) \subset \operatorname{Hom}_{\mathbb{K}}(M, N)$$

is called the module of **differential operators of order** $\leq n$.

Thus we have an increasing filtration of R-submodules

$$\operatorname{Diff}^0_{R/\mathbb{K}}(M, N) \subset \operatorname{Diff}^1_{R/\mathbb{K}}(M, N) \subset \operatorname{Diff}^2_{R/\mathbb{K}}(M, N) \subset \cdots$$

and the union $\operatorname{Diff}_{R/S\mathbb{K}}(M, N) := \bigcup_{n>0} \operatorname{Diff}^n_{R/\mathbb{K}}(M, N)$ is called the module of differential operators (of any order).

The name differential operator is derived from the case $\mathbb{K} = \mathbb{R}$ and $R = M = N$, the ring of smooth functions on an open subset of $\mathbb{R}^n$. In this case it is not difficult to prove that the above algebraic notion is equivalent to the usual analytic notion of a differential operator, cf. Exercise 2.7.11.

Lemma 2.3.2 *In the above notation, if $f \in \mathrm{Diff}^n_{R/\mathbb{K}}(M, N)$ and $g \in \mathrm{Diff}^m_{R/\mathbb{K}}(P, M)$, then $fg \in \mathrm{Diff}^{n+m}_{R/\mathbb{K}}(P, N)$. In particular, $\mathrm{Diff}_{R/\mathbb{K}}(M, M)$ is an associative subalgebra of $\mathrm{Hom}_{\mathbb{K}}(M, M)$.*

Proof This is clear if $n = m = 0$; if $n + m > 0$ by induction on $n + m$ we may assume

$$f[g, a], [f, a]g \in \mathrm{Diff}^{n+m-1}_{R/\mathbb{K}}(P, N) \qquad \text{for every } a \in R\,.$$

The conclusion follows from the identity $[fg, a] = f[g, a] + [f, a]g$. □

Lemma 2.3.3 *If $n + m > 0$, $f \in \mathrm{Diff}^n_{R/\mathbb{K}}(M, M)$ and $g \in \mathrm{Diff}^m_{R/\mathbb{K}}(M, M)$, then*

$$[f, g] = fg - gf \in \mathrm{Diff}^{n+m-1}_{R/\mathbb{K}}(M, M).$$

In particular $\mathrm{Diff}^1_{R/\mathbb{K}}(M,M)$ and $\mathrm{Diff}_{R/\mathbb{K}}(M,M)$ are Lie subalgebras of $\mathrm{Hom}_{\mathbb{K}}(M, M)$.

Proof The result is true by definition whenever $\min(n, m) = 0$. For $n, m > 0$ it is sufficient to prove that $[[f, g], a] \in \mathrm{Diff}^{n+m-2}_{R/\mathbb{K}}(M, M)$ for every $a \in R$. We have

$$\begin{aligned}[][[f, g], a] &= [fg, a] - [gf, a] = f[g, a] + [f, a]g - g[f, a] - [g, a]f\\ &= [f, [g, a]] - [g, [f, a]]\end{aligned}$$

and, since by assumption $[f, a] \in \mathrm{Diff}^{n-1}_{R/\mathbb{K}}(M, M)$ and $[g, a] \in \mathrm{Diff}^{m-1}_{R/\mathbb{K}}(M, M)$ the conclusion follows immediately from an induction on $n + m$. □

In the remaining part of this section we restrict our attention to the differential operator of first order.

Definition 2.3.4 The **symbol** of a differential operator $f \in \mathrm{Diff}^1_{R/\mathbb{K}}(M, N)$ of order ≤ 1 is the $\mathbb{K}$-linear map

$$\sigma(f)\colon R \to \mathrm{Hom}_R(M, N), \qquad \sigma(f)(a) = [f, a].$$

Whenever $N = M$, a differential operator $f \in \mathrm{Diff}^1_{R/S}(M, M)$ is said to have a **principal symbol** if $[f, a] \in \mathrm{Hom}_R(M, M)$ is a scalar multiple of the identity for every $a \in R$.

Thus we have an exact sequence of R-modules

$$0 \to \mathrm{Hom}_R(M, N) \to \mathrm{Diff}^1_{R/\mathbb{K}}(M, N) \xrightarrow{\sigma} \mathrm{Hom}_{\mathbb{K}}(R, \mathrm{Hom}_R(M, N)).$$

If $f \in \mathrm{Diff}^1_{R/\mathbb{K}}(M, N)$ then $[[f, a], b] = 0$ for every $a, b \in R$ and formula (2.3) implies the equality

$$\sigma(f)(ab) = (\sigma(f)a)b + a(\sigma(f)b).$$

If $N = M$ is free of rank 1 then every differential operator has a principal symbol. The importance of differential operators with principal symbols in deformation theory relies on their interpretation as derivations of pairs (ring, module).

Definition 2.3.5 Let R be a unitary commutative $\mathbb{K}$-algebra and let M be an R-module. The R-module of **$\mathbb{K}$-derivations of the pair** (R, M) is defined as the subset

$$D^1_{\mathbb{K}}(R, M) \subset \mathrm{Der}_{\mathbb{K}}(R, R) \times \mathrm{Hom}_{\mathbb{K}}(M, M)$$

of all pairs (h, u) such that $u(am) - au(m) = h(a)m$ for every $a \in R$ $m \in M$.

Equivalently, $(h, u) \in \mathrm{Hom}_{\mathbb{K}}(R, R) \times \mathrm{Hom}_{\mathbb{K}}(M, M)$ is a derivation of the pair if

$$h(ab) = h(a)b + ah(b), \quad u(am) - au(m) = h(a)m, \quad \forall a, b \in R,\ m \in M,$$

and it is clear that $D^1_{\mathbb{K}}(R, M)$ is an R-submodule of $\mathrm{Der}_{\mathbb{K}}(R, R) \times \mathrm{Hom}_{\mathbb{K}}(M, M)$.

Lemma 2.3.6 *In the above notation, the R-submodule $D^1_{\mathbb{K}}(R, M)$ is a Lie subalgebra of $\mathrm{Der}_{\mathbb{K}}(R, R) \times \mathrm{Hom}_{\mathbb{K}}(M, M)$. Denoting by $p\colon D^1_{\mathbb{K}}(R, M) \to \mathrm{Der}_{\mathbb{K}}(R, R)$ and by $q\colon D^1_{\mathbb{K}}(R, M) \to \mathrm{Hom}_{\mathbb{K}}(M, M)$ the projection maps, we have:*

1. *the image of $q\colon D^1_{\mathbb{K}}(R, M) \to \mathrm{Hom}_{\mathbb{K}}(M, M)$ is the submodule of differential operators of first order with principal symbol;*
2. *there exists an exact sequence of R-modules*

$$0 \to \mathrm{Hom}_R(M, M) \to D^1_{\mathbb{K}}(R, M) \xrightarrow{p} \mathrm{Der}_{\mathbb{K}}(R, R);$$

3. *if M is faithful, i.e., if $\mathrm{ann}(M) = \{a \in R \mid aM = 0\} = 0$, then q is injective; if M is free, then p is split surjective.*

Proof The only nontrivial part of the proof is the surjectivity of p whenever M is a free module. Assuming M free with basis e_i, for every $h \in \mathrm{Der}_{\mathbb{K}}(R, R)$ we may consider

$$u \in \mathrm{Hom}_{\mathbb{K}}(M, M), \qquad u\left(\sum a_i e_i\right) = \sum h(a_i)e_i.$$

We have

$$u\left(a\sum a_i e_i\right) - au\left(\sum a_i e_i\right) = \sum (h(aa_i) - ah(a_i))e_i = h(a)\sum a_i e_i$$

and thus $(h, u) \in D^1_S(R, M)$. It should be noted that the R-linear splitting $h \mapsto (h, u)$ is not canonical since it depends on the choice of the basis. □

2.4 Free Lie Algebras

For a vector space V over a field $\mathbb{K}$ we denote by $V^{\otimes n}$ its nth tensor power: $V^{\otimes 0} = \mathbb{K}$, $V^{\otimes 1} = V$, $V^{\otimes 2} = V \otimes V$ etc. The **tensor algebra** generated by V is the vector space

$$T(V) = \bigoplus_{n \geq 0} V^{\otimes n},$$

equipped with the concatenation product, defined as the bilinear extension of the obvious map

$$(v_1 \otimes \cdots \otimes v_a)(w_1 \otimes \cdots \otimes w_b) = v_1 \otimes \cdots \otimes v_a \otimes w_1 \otimes \cdots \otimes w_b .$$

The algebra $T(V)$ is associative and unitary, with unit $1 \in \mathbb{K} = V^{\otimes 0}$; it is commutative if and only if $\dim_{\mathbb{K}} V = 1$, since for every pair of linearly independent vectors $u, v \in V$ we have $u \otimes v \neq v \otimes u$. Whenever $x_1, \ldots, x_n$ is a basis of V, the elements of $T(V)$ are called non-commutative polynomial in $x_1, \ldots, x_n$; sometimes it is written $\mathbb{K}\langle x_1, \ldots, x_n\rangle = T(V)$. The following lemma is a precise formulation of the fact that $T(V)$ is the free unitary associative algebra generated by V.

Lemma 2.4.1 *Let V be a $\mathbb{K}$-vector space and let $\imath\colon V \to T(V)$ be the natural inclusion. For every associative unitary $\mathbb{K}$-algebra R and every linear map $f\colon V \to R$ there exists a unique homomorphism of $\mathbb{K}$-algebras $F\colon T(V) \to R$ such that $F(1) = 1$ and $f = F\imath$.*

Proof Define $F(1) = 1$ and $F(v_1 \otimes \cdots \otimes v_n) = f(v_1)f(v_2)\cdots f(v_n)$ for every $n > 0$, $v_1, \ldots, v_n \in V$. □

In particular, every linear map of vector spaces $f\colon V \to W$ gives by composition a linear map $V \to W \to T(W)$ and therefore induces a morphism of unitary algebras $T(f)\colon T(V) \to T(W)$: more explicitly, $T(f)(v_1 \otimes \cdots \otimes v_n) = f(v_1) \otimes \cdots \otimes f(v_n)$.

Definition 2.4.2 Let V be a vector space. The **free Lie algebra** generated by V is the smallest Lie subalgebra $L(V) \subseteq T(V)_L$ that contains V.

Whenever $x_1, \ldots, x_n$ is a basis of V, the elements of $L(V)$ are called **Lie polynomials in** $x_1, \ldots, x_n$.

Equivalently, $L(V)$ is the intersection of all the Lie subalgebras of $T(V)_L$ containing V.

Lemma 2.4.3 *For every integer $n > 0$ let $L(V)_n \subset V^{\otimes n}$ be the linear subspace generated by all the elements $[v_1, [v_2, [\ \ldots\ , [v_{n-1}, v_n]..]]]$, with $v_1, \ldots, v_n \in V$. Then*

$$L(V) = \bigoplus_{n \geq 1} L(V)_n .$$

Proof Since $L(V)$ is a Lie subalgebra of $T(V)_L$ containing V, we have $L(V)_n \subset L(V)$ for every n and therefore $\oplus_{n>0}L(V)_n \subset L(V)$. On the other hand, by definition we have $L(V)_1 = V$ and $L(V)_n = [V, L(V)_{n-1}]$. The Jacobi identity gives

$$\begin{aligned}[L(V)_n, L(V)_m] &= [[V, L(V)_{n-1}], L(V)_m] \\ &\subset [V, [L(V)_{n-1}, L(V)_m]] + [L(V)_{n-1}, [V, L(V)_m]]\end{aligned}$$

and therefore, by induction on n, we obtain $[L(V)_n, L(V)_m] \subset L(V)_{n+m}$. This implies that the direct sum $\oplus_{n>0}L(V)_n$ is a Lie subalgebra of $L(V)$ containing V and then $L(V) \subset \oplus_{n>0}L(V)_n$. □

The construction $V \mapsto L(V)$ is a functor from the category of vector spaces to the category of Lie algebras: every morphism of vector spaces $V \to W$ induces a morphism of associative algebras $T(V) \to T(W)$ which restricts to a morphism of Lie algebras $L(V) \to L(W)$.

It is possible to prove, see e.g. [224, 244], that $L(V)$ has the universal property of free objects, i.e., that for every Lie algebra H and every linear map $f: V \to H$ there exists a unique morphism of Lie algebras $F: L(V) \to H$ extending f. The proof is nontrivial and relies on general properties of universal enveloping algebras (including the Poincaré–Birkhoff–Witt theorem). Fortunately, in this book we only need free Lie algebras over fields of characteristic 0, and then we can apply the following theorem in order to prove the universal property.

Theorem 2.4.4 (Dynkin–Sprecht–Wever) *Assume that V is a vector space and H is a Lie algebra over a field of characteristic 0. Let $f: V \to H$ be a linear map and define a linear map $F: T(V) \to H$ by setting*

$$F(1) = 0, \quad F(v) = f(v), \quad F(v_1 \otimes \cdots \otimes v_n) = \frac{1}{n}[f(v_1), [\ldots[f(v_{n-1}), f(v_n)]..]].$$

Then the restriction $F: L(V) \to H$ is the unique morphism of Lie algebras extending f.

Proof Since $L(V)$ is generated by V as a Lie algebra we only need to prove that $F: L(V) \to H$ is a morphism of Lie algebras. The linear map

$$\theta: V \to \mathrm{Hom}_{\mathbb{K}}(H, H), \qquad \theta(v)x = [f(v), x],$$

extends to a morphism of associative algebras $\theta: T(V) \to \mathrm{Hom}_{\mathbb{K}}(H, H)$; notice that

$$\theta(v_1 \otimes \cdots \otimes v_s)x = \theta(v_1)\theta(v_2)\cdots\theta(v_s)x = [f(v_1), [f(v_2), [\ldots[f(v_n), x]..]]]$$

and therefore

$$F(v_1 \otimes \cdots \otimes v_n) = \frac{m}{n}\theta(v_1 \otimes \cdots \otimes v_{n-m})F(v_{n-m+1} \otimes \cdots \otimes v_n),$$

for every $0 < m \leq n$, $v_1, \ldots, v_n \in V$.

In order to prove that F is a morphism of Lie algebras we shall prove by induction on n the following propositions:

A_n: if $0 < n$, $y \in L(V)_n$ and $h \in H$ then $\theta(y)h = [F(y), h]$;

B_n: if $0 < m \leq n$, $x \in L(V)_m$ and $y \in L(V)_n$ then $F([x, y]) = [F(x), F(y)]$.

In fact, for proving that $F([x, y]) = [F(x), F(y)]$ for $x, y \in L(V)$ it is not restrictive to assume that $x \in L(V)_m$ and $y \in L(V)_n$ for some positive integers n, m; possibly exchanging the role of x and y we may assume $m \leq n$ and the conclusion follows from B_n. The initial step A_1 is true by definition of F, while for the inductive implication $A_1 + \cdots + A_n \Rightarrow B_n$ we have:

$$\begin{aligned} F([x, y]) = F(x \otimes y) - F(y \otimes x) &= \frac{n}{n+m}\theta(x)F(y) - \frac{m}{n+m}\theta(y)F(x) \\ &= \frac{n}{n+m}[F(x), F(y)] - \frac{m}{n+m}[F(y), F(x)] = [F(x), F(y)]. \end{aligned}$$

Then we prove that $A_1 + A_{n-1} + B_{n-1} \Rightarrow A_n$ for every $n > 1$. By linearity it is not restrictive to assume $y = [v, x]$, with $v \in V$ and $x \in L(V)_{n-1}$. Then

$$\theta(y)h = \theta(v \otimes x)h - \theta(x \otimes v)h = \theta(v)(\theta(x)h) - \theta(x)(\theta(v)h),$$

and by properties A_1 and A_{n-1} we have

$$\theta(y)h = [F(v), [F(x), h]] - [F(x), [F(v), h]] = [[F(v), F(x)], h].$$

Finally, $[[F(v), F(x)], h] = [F(y), h]$ by property B_{n-1}. □

Corollary 2.4.5 (Dynkin projector) *For every vector space V over a field of characteristic 0 the linear map $\rho\colon T(V) \to L(V)$,*

$$\rho(V^{\otimes 0}) = 0, \quad \rho(v_1 \otimes \cdots \otimes v_n) = \frac{1}{n}[v_1, [v_2, [\ \ldots\ [v_{n-1}, v_n]..]]],$$

is a projection. In particular, for every $n > 0$

$$L(V)_n = \{x \in V^{\otimes n} \mid \rho(x) = x\}.$$

Proof The identity on $L(V)$ is the unique morphism of Lie algebras extending the natural inclusion $V \to L(V)$. □

2.5 The Baker–Campbell–Hausdorff Product

It is well known that, given two $n \times n$ real matrices A, B, the formula $e^{A+B} = e^A e^B$ does not hold in general if $AB \neq BA$ and then it makes sense to introduce the binary operator

$$A \bullet B = \log(e^A e^B) , \tag{2.4}$$

at least for A, B in a sufficiently small neighbourhood of 0. It was discovered by Campbell that $A \bullet B$ can be described as an infinite sum of Lie polynomials in A, B, which we are now able to describe explicitly by a universal formula, usually referred to as the Baker–Campbell–Hausdorff (BCH) formula, sometimes also called the Campbell–Hausdorff formula.

For simplicity of exposition we restrict our attention to the case of nilpotent Lie algebras over a field of characteristic 0, which is completely sufficient for the application of this book. This restriction imposes an inversion of the standard approach to BCH formula; we start with a meaningless explicit functorial definition of $\bullet$, via the Baker–Hausdorff recursive formula, and then we relate it to the more significant Eq. (2.4).

From now to the end of this chapter, $\mathbb{K}$ will be a field of characteristic 0. Given two elements a, b in a nilpotent Lie algebra H over $\mathbb{K}$, we define a sequence of elements $Z_n(a, b) \in H, n \geq 0$, by the recursive equations:

$$Z_0(a,b) = b, \quad Z_1(a,b) = a + B_1[b,a] + \frac{B_2}{2!}[b,[b,a]] + \frac{B_3}{3!}[b,[b,[b,a]]] + \cdots$$

$$Z_{r+1}(a,b) = \frac{1}{r+1}\sum_{m\geq 0}\frac{B_m}{m!}\sum_{i_1+\cdots+i_m=r}[Z_{i_1}(a,b),[Z_{i_2}(a,b),[\ldots[Z_{i_m}(a,b),a]..]]], \tag{2.5}$$

where $B_0, B_1, B_2, \ldots$ are the Bernoulli numbers:

$$\sum_{n=0}^{\infty}\frac{B_n}{n!}x^n = \frac{x}{e^x-1} = 1 - \frac{x}{2} + \frac{x^2}{12} - \frac{x^4}{720} + \frac{x^6}{30240} - \frac{x^8}{1209600} + \cdots .$$

We can immediately see that, since H is nilpotent, for every $n > 0$ the element $Z_n(a, b)$ is properly defined and belongs to the nth ideal $H^{[n]}$ of the descending central series.

Definition 2.5.1 For a nilpotent Lie algebra H over a field of characteristic 0 the map

$$H \times H \xrightarrow{\bullet} H, \qquad a \bullet b = \sum_{n\geq 0} Z_n(a,b),$$

is called the **Baker–Campbell–Hausdorff** (BCH) product.

The first terms, up to order 4, of the BCH product are:

$$a \bullet b = a + b + \frac{1}{2}[a, b] + \frac{1}{12}[a, [a, b]] + \frac{1}{12}[b, [b, a]] + \frac{1}{24}[a, [b, [b, a]]] + \cdots .$$

It is plain that $\bullet$ commutes with morphisms of (nilpotent) Lie algebras and that $a \bullet b - a - b$ belongs to the Lie ideal generated by $[a, b]$.

Exponential and Logarithm

Let R be a unitary associative $\mathbb{K}$-algebra and let $I \subset R$ be a nilpotent ideal. Writing $E = \{1 + a \mid a \in I\} \subset R$, we may define the **exponential**:

$$e \colon I \to E \subset R, \qquad e^a = \sum_{n \geq 0} \frac{a^n}{n!},$$

and the **logarithm**:

$$\log \colon E \to I, \qquad \log(1 + a) = \sum_{n=1}^{\infty} (-1)^{n-1} \frac{a^n}{n}.$$

Lemma 2.5.2 *Exponential and logarithm are the inverse of each other; for every* $a, b \in I$ *we have*

$$\log(e^a) = a, \qquad e^{\log(1+b)} = 1 + b.$$

Proof Using the morphism of associative algebras $\mathbb{Q}[[t]] \to R$, $p(t) \mapsto p(a)$, and the embedding $\mathbb{Q}[[t]] \subset \mathbb{R}[[t]]$ the proof is reduced to well known facts of calculus. $\square$

Proposition 2.5.3 *In the above setup, for every* $a \in R$ *denote by* $\operatorname{ad} a \in \operatorname{End}(R)$ *the adjoint operator* $(\operatorname{ad} a)b = [a, b] = ab - ba$.

1. *For every* $a, b \in R$ *and* $n \geq 0$ *we have:*

$$(\operatorname{ad} a)^n b = \sum_{i=0}^{n} (-1)^i \binom{n}{i} a^{n-i} b a^i = \sum_{i=0}^{n} \binom{n}{i} a^{n-i} b (-a)^i .$$

2. *If* a *is nilpotent in* R, *then* $\operatorname{ad} a$ *is nilpotent in* $\operatorname{End}(R)$ *and therefore the following invertible operator is well defined:*

$$e^{\operatorname{ad} a} = \sum_{n \geq 0} \frac{(\operatorname{ad} a)^n}{n!} \in \operatorname{End}(R).$$

For every $a, b \in I$ *and every* $r \in R$ *we have:*

$$e^{\operatorname{ad} a}(r) = \sum_{n\ge 0} \frac{(\operatorname{ad} a)^n}{n!} r = e^a r e^{-a},$$

$$e^{\operatorname{ad} a}(e^b) = e^a e^b e^{-a} = e^c, \qquad \textit{where} \quad c = e^{\operatorname{ad} a}(b).$$

3. *For every $a, b \in I$ we have $ab = ba$ if and only if $e^a e^b = e^b e^a$.*
4. *Given $a, b \in I$, if $ab = ba$, then*

$$e^{a+b} = e^a e^b = e^b e^a, \qquad \log((1+a)(1+b)) = \log(1+a) + \log(1+b).$$

Proof (1) We have $(\operatorname{ad} a)^n b = a(\operatorname{ad} a)^{n-1}(b) - (\operatorname{ad} a)^{n-1}(b)a$ and by induction

$$(\operatorname{ad} a)^n b = \sum_{i=0}^{n-1} (-1)^i \binom{n-1}{i} a^{n-i} b a^i - \sum_{j=0}^{n-1} (-1)^j \binom{n-1}{i} a^{n-1-j} b a^{j+1}.$$

Setting $j = i - 1$ on the second summand we get

$$\begin{aligned}(\operatorname{ad} a)^n b &= \sum_{i=0}^{n-1} (-1)^i \binom{n-1}{i} a^{n-i} b a^i + \sum_{i=1}^{n} (-1)^i \binom{n-1}{i-1} a^{n-i} b a^i \\ &= \sum_{i=0}^{n} (-1)^i \left(\binom{n-1}{i} + \binom{n-1}{i-1}\right) a^{n-i} b a^i = \sum_{i=0}^{n} (-1)^i \binom{n}{i} a^{n-i} b a^i.\end{aligned}$$

(2) According to item (1), if $a^n = 0$ for some $n > 0$, then $(\operatorname{ad} a)^{2n-1} = 0$ and

$$e^{\operatorname{ad} a}(r) = \sum_{n\ge 0} \frac{(\operatorname{ad} a)^n}{n!} r = \sum_{n\ge 0} \sum_{i=0}^{n} \frac{1}{n!} \binom{n}{i} a^{n-i} r (-a)^i.$$

Setting $j = n - i$ we obtain

$$e^{\operatorname{ad} a}(r) = \sum_{i,j\ge 0} \frac{1}{i!j!} a^j r (-a)^i = e^a r e^{-a}.$$

Since $[a, xy] = [a, x]y + x[a, y]$, the operator $\operatorname{ad} a$ is a nilpotent derivation and then, according to Lemma 2.1.2, the operator $e^{\operatorname{ad} a}\colon R \to R$ is an isomorphism of algebras. In particular,

$$e^{\operatorname{ad} a}(e^b) = \sum_{n\ge 0} \frac{(e^{\operatorname{ad} a}(b))^n}{n!} = e^c, \qquad c = e^{\operatorname{ad} a}(b).$$

(3) For $a \in I$ and $r \in R$ we claim that $e^a r = r e^a$ if and only if $ar = ra$. In fact we have $e^a r = r e^a$ if and only if $e^a r e^{-a} - r = 0$ and by the above formula

$$e^a r e^{-a} - r = e^{\operatorname{ad} a}(r) - r = \sum_{n\geq 0} \frac{(\operatorname{ad} a)^n}{(n+1)!}([a,r]) = \frac{e^{\operatorname{ad} a} - 1}{\operatorname{ad} a}([a,r]).$$

Since $\operatorname{ad} a$ is a nilpotent operator, the operator $\frac{e^{\operatorname{ad} a}-1}{\operatorname{ad} a}$ is invertible and therefore $e^a r = r e^a$ if and only if $[a, r] = 0$. Given $a, b \in I$, the above claim applied twice gives

$$e^a e^b = e^b e^a \iff b e^a = e^a b \iff ab = ba\,.$$

(4) Since $ab = ba$, we have for every $n \geq 0$

$$(a+b)^n = \sum_{i=0}^{n} \binom{n}{i} a^i b^{n-i},$$

$$e^{a+b} = \sum_{n\geq 0} \frac{(a+b)^n}{n!} = \sum_{n\geq 0}\sum_{i=0}^{n} \frac{1}{n!}\binom{n}{i} a^i b^{n-i} = \sum_{n\geq 0}\sum_{i=0}^{n} \frac{1}{i!(n-i)!} a^i b^{n-i}.$$

Setting $j = n - i$ we get

$$e^{a+b} = \sum_{i,j\geq 0} \frac{1}{i!j!} a^i b^j = e^a e^b.$$

Setting $x = \log(1+a)$, $y = \log(1+b)$ we have $e^x e^y = e^y e^x$; therefore $xy = yx$ and

$$\log((1+a)(1+b)) = \log(e^x e^y) = \log(e^{x+y}) = x + y = \log(1+a) + \log(1+b).$$

□

The Exponential Derivation Formula

As above, let R be a unitary associative algebra over a field $\mathbb{K}$ of characteristic 0, and let $I \subset R$ be a nilpotent ideal. Let t be an indeterminate and denote by $R[t]$ the algebra of polynomials in the central variable t; more explicitly, $R[t] = \oplus_{n\geq 0} R t^n$ and $(at^i)(bt^j) = abt^{i+j}$ for $a, b \in R$ and $i, j \geq 0$. Denote by

$$\phi\colon R[t] \to \operatorname{Hom}_{\mathbb{K}}(R[t], R[t]), \qquad \phi(p)q = pq,$$

the operator of left multiplication; we can immediately see that ϕ is an injective morphism of unitary associative algebras. The Leibniz formula for the usual derivation operator

$$\frac{d}{dt}\colon R[t] \to R[t], \qquad \frac{d}{dt}\left(\sum a_i t^i\right) = \sum i a_i t^{i-1},$$

may be written as:

$$\phi\left(\frac{dp}{dt}\right)=\left[\frac{d}{dt},\phi(p)\right],\qquad p\in R[t].$$

The subspace $I[t]\subset R[t]$ of polynomials with coefficients in I is a nilpotent ideal of $R[t]$. Therefore for every $p\in I[t]$ we have $e^p\in R[t]$ and, since ϕ is a morphism of associative algebras, we have $\phi(e^p)=e^{\phi(p)}$. By the same argument used in the proof of item 3 of Proposition 2.5.3 we have

$$\begin{aligned}\phi\left(\frac{de^p}{dt}e^{-p}\right)&=\phi\left(\frac{de^p}{dt}\right)e^{-\phi(p)}=\left[\frac{d}{dt},e^{\phi(p)}\right]e^{-\phi(p)}\\&=\frac{d}{dt}-e^{\phi(p)}\frac{d}{dt}e^{-\phi(p)}=\frac{e^{\operatorname{ad}\phi(p)}-1}{\operatorname{ad}\phi(p)}\left[\frac{d}{dt},\phi(p)\right]\\&=\frac{e^{\operatorname{ad}\phi(p)}-1}{\operatorname{ad}\phi(p)}\phi\left(\frac{dp}{dt}\right).\end{aligned}$$

Since ϕ is injective, the above computation gives the **exponential derivation formula**:

$$\frac{de^p}{dt}e^{-p}=\frac{e^{\operatorname{ad}p}-1}{\operatorname{ad}p}\left(\frac{dp}{dt}\right),\qquad p\in I[t].\tag{2.6}$$

Remark 2.5.4 We have already pointed out that if α is a nilpotent endomorphism of some vector space, then the operator

$$\frac{e^\alpha-1}{\alpha}=\sum_{n\geq 0}\frac{\alpha^n}{(n+1)!}$$

is invertible. Its inverse can be explicitly described by the formula:

$$\frac{\alpha}{e^\alpha-1}=\sum_{n\geq 0}\frac{B_n}{n!}\alpha^n,$$

where the B_n's are the Bernoulli numbers. In particular, (2.6) is equivalent to

$$\frac{dp}{dt}=\sum_{n\geq 0}\frac{B_n}{n!}(\operatorname{ad}p)^n\left(\frac{de^p}{dt}e^{-p}\right).$$

Associativity of BCH Product

Our definition of the BCH product applies in particular to every nilpotent ideal I of a unitary associative algebra, considered as a nilpotent Lie algebra equipped with the commutator bracket. In this special case, the interpretation of the BCH product is given by the following theorem.

Theorem 2.5.5 *Let I be a nilpotent ideal of an associative unitary algebra. Then for every $a, b \in I$ we have*

$$e^a e^b = e^{a \bullet b},$$

where $\bullet$ is the Baker–Campbell–Hausdorff product as in Definition 2.5.1.

Proof The proof is an easy consequence of the exponential derivation formula (2.6). Denoting by $Z = \log(e^{ta}e^b) \in I[t]$, we have

$$e^Z = e^{ta}e^b = \sum_n \frac{t^n}{n!}a^n e^b, \qquad \frac{de^Z}{dt} = \sum_n \frac{t^{n-1}}{(n-1)!}a^n e^b = ae^Z.$$

By the exponential derivation formula we get

$$\frac{e^{\operatorname{ad} Z}-1}{\operatorname{ad} Z}\left(\frac{dZ}{dt}\right) = \frac{de^Z}{dt}e^{-Z} = ae^Z e^{-Z} = a,$$

and therefore Z is the solution of the Cauchy problem

$$\frac{dZ}{dt} = \frac{\operatorname{ad} Z}{e^{\operatorname{ad} Z}-1}(a) = \sum_{n\geq 0} \frac{B_n}{n!}(\operatorname{ad} Z)^n(a), \qquad Z(0) = b.$$

Writing $Z = Z_0 + tZ_1 + \cdots + t^n Z_n + \cdots$, the coefficients Z_n can be computed recursively by the equations:

$$Z_0 = b, \quad Z_{r+1} = \frac{1}{r+1}\sum_{m\geq 0} \frac{B_m}{m!} \sum_{i_1+\cdots+i_m=r} (\operatorname{ad} Z_{i_1})(\operatorname{ad} Z_{i_2})\cdots(\operatorname{ad} Z_{i_m})a.$$

Thus $a \bullet b = \sum_n Z_n$ and the conclusion follows from the equality $e^a e^b = e^{Z(1)}$. □

Corollary 2.5.6 *Let H be a nilpotent Lie algebra over a field of characteristic 0. Then the Baker–Campbell–Hausdorff product $\bullet\colon H \times H \to H$ is associative.*

Proof Consider first the case where H is a Lie subalgebra of a nilpotent ideal I of a unitary associative algebra R; here we have $e^{a\bullet b} = e^a e^b \in R$, and then

$$a \bullet (b \bullet c) = \log(e^a e^{a\bullet b}) = \log(e^a e^b e^c) = \log(e^{a\bullet b}e^c) = (a \bullet b) \bullet c$$

for every $a, b, c \in H$.

In general, let n be a positive integer such that $H^{[n]} = 0$; by Theorem 2.4.4 the identity on H extends to a surjective morphism of Lie algebras $F\colon L(H) \to H$ such that $F(L(H)_m) = 0$ for every $m \geq n$.

The subspace $J = \oplus_{m\geq n} H^{\otimes m} \subset T(H)$ is an ideal, the intersection $L(H) \cap J$ is a Lie ideal in $L(H)$ and the quotient $L(H)/(L(H) \cap J)$ is contained in a nilpotent ideal of the algebra $T(H)/J$. Thus the BCH product is associative in $L(H)/(L(H) \cap J)$

and since F factors through a surjective morphism $L(H)/(L(H) \cap J) \to H$, also the Baker–Campbell–Hausdorff product is associative in H. □

Definition 2.5.7 (*Exponential of a nilpotent Lie algebra*) For a nilpotent Lie algebra H we denote by $\exp(H) = \{e^a \mid a \in H\}$ the set of formal exponents of elements of H. It is a group with product

$$\exp(H) \times \exp(H) \to \exp(H), \qquad e^a e^b = e^{a \bullet b},$$

with unit $1 = e^0$ and inverse $(e^a)^{-1} = e^{-a}$.

Remark 2.5.8 The construction of the exponential of nilpotent Lie algebras has the following functorial properties:

1. If $f : L \to M$ is a morphism of nilpotent Lie algebras, then the map

$$f : \exp(L) \to \exp(M), \qquad f(e^a) = e^{f(a)},$$

   is a homomorphism of groups.
2. Let V be a vector space and $f : H \to \mathrm{End}(V)$ a morphism of Lie algebras. If H is a nilpotent Lie algebra and the image of $f(H)$ is contained in a nilpotent ideal of the associative algebra $\mathrm{End}(V)$, then the maps

$$\exp(H) \times V \to V, \qquad (e^a, v) \mapsto e^{f(a)} v,$$

$$\exp(H) \times \mathrm{End}(V) \to \mathrm{End}(V), \qquad (e^a, g) \mapsto e^{f(a)} g e^{-f(a)} = e^{\mathrm{ad} f}(g),$$

   are right group actions.

If L is a non-nilpotent Lie algebra then the Baker–Campbell–Hausdorff product is not defined as a map $L \times L \to L$ and the usual strategy to overcome (partially) this difficulty is to define $\bullet$ in some other algebraic structures depending functorially by L. Three possible constructions are:

1. For every $n > 0$ the quotient Lie algebra $\frac{L}{L^{[n]}}$ is nilpotent and the BCH product commutes with the natural projection maps $\frac{L}{L^{[n+1]}} \to \frac{L}{L^{[n]}}$. Passing to the limit[1] we get a product

$$\lim_n \frac{L}{L^{[n]}} \times \lim_n \frac{L}{L^{[n]}} \xrightarrow{\ \bullet\ } \lim_n \frac{L}{L^{[n]}}.$$

2. Denote by $L[[t]]$ the Lie algebra of all formal power series with coefficients in L in a central variable t (central means that $[t, t] = [t, x] = 0$ for every $x \in L$) and by $tL[[t]] \subset L[[t]]$ the subalgebra of all power series with positive multiplicity. Then we may define the Baker–Campbell–Hausdorff product $tL[[t]] \times tL[[t]] \xrightarrow{\bullet} tL[[t]]$; details are left to the reader.

[1] We adopt the notion of a limit and colimit in the sense usually adopted in category theory, see e.g. [175]. In particular, we shall take limit to mean every inverse or projective limit, and we shall take colimit to mean every direct or inductive limit.

3. Let **nCom** (resp.: **nLie**) be the category of nilpotent commutative (resp.: Lie) $\mathbb{K}$-algebras. Then L defines a functor

$$\widehat{L}\otimes\colon \mathbf{nCom} \to \mathbf{nLie}, \qquad A \mapsto L \otimes_{\mathbb{K}} A,$$

and $\bullet$ gives in the obvious way a natural transformation of functors

$$\widehat{L} \times \widehat{L} \xrightarrow{\bullet} \widehat{L}.$$

2.6 Semicosimplicial Lie Algebras

For later application in deformation theory we introduce here the notion of semicosimplicial objects, which will be treated in a deeper way in Chap. 7.

For every integer $n \geq 0$, consider the finite set $[n] = \{0, 1, \dots, n\}$, equipped with the usual order relation. Let $\overrightarrow{\Delta}$ be the category whose objects are $[0] = \{0\}$, $[1] = \{0, 1\}$, $[2] = \{0, 1, 2\}$ etc. and whose morphisms are the strictly monotone maps. For instance, $\mathrm{Mor}_{\overrightarrow{\Delta}}([n-1], [n])$ contains exactly $n + 1$ morphisms called **face maps**, namely:

$$\delta_k\colon [n-1] \to [n], \qquad \delta_k(p) = \begin{cases} p & \text{if } p < k \\ p+1 & \text{if } p \geq k \end{cases}, \qquad k = 0, \dots, n,$$

Equivalently, δ_k is the unique strictly monotone map whose image avoids k.

It is clear that every morphism in $\overrightarrow{\Delta}$, different from the identity, is a finite composition of face maps, and a straightforward verification shows that the face maps satisfy the **semicosimplicial identities**:

$$\delta_l \delta_k = \delta_{k+1}\delta_l, \qquad \text{for every } l \leq k. \tag{2.7}$$

Every strictly monotone map $f\colon [n] \to [n+k]$, $k > 0$, admits a unique factorization as

$$f = \delta_{i_1} \cdots \delta_{i_k}, \qquad n + k \geq i_1 > i_2 > \cdots > i_k \geq 0. \tag{2.8}$$

The unicity follows from the fact that the sequence $i_1 > i_2 > \cdots > i_k$ can be recovered from f, since

$$i_1 = \max([n+k] - f([n])), \quad i_2 = \max([n+k] - f([n]) - \{i_1\}), \quad \text{etc.}$$

for every f as in (2.8). For the existence it is sufficient to observe that every composition of face maps can be transformed into a composition as in (2.8) by a finite number of cosimplicial identities (2.7).

Definition 2.6.1 Let **C** be a category. A **semicosimplicial** object in **C** is a covariant functor $A\colon \overrightarrow{\Delta} \to \mathbf{C}$.

Equivalently, a semicosimplicial object A is a diagram

$$A_0 \rightrightarrows A_1 \rightrightarrows A_2 \rightrightarrows \cdots,$$

where each A_i is an object of **C**, and, for each $n > 0$, there are $n + 1$ face operators

$$\delta_k\colon A_{n-1} \to A_n, \qquad k = 0, \ldots, n,$$

which are morphisms in the category **C** and satisfy the semicosimplicial identities (2.7).

A morphism of semicosimplicial objects is a natural transformation of functors; equivalently, a morphism $f\colon A \to B$ of semicosimplicial objects is a sequence of morphisms $f_n\colon A_n \to B_n$ such that $\delta_k f_{n-1} = f_n \delta_k$ for every k, n.

Definition 2.6.2 Given a semicosimplicial abelian group

$$A\colon \quad A_0 \rightrightarrows A_1 \rightrightarrows A_2 \rightrightarrows \cdots,$$

its **cochain complex** is the complex of abelian groups

$$C(A)\colon \quad A_0 \xrightarrow{\delta} A_1 \xrightarrow{\delta} A_2 \to \cdots, \qquad \delta = \sum_{i=0}^{n} (-1)^i \delta_i\colon A_{n-1} \to A_n.$$

The proof that $\delta^2 = 0$ is an immediate consequence of the identities (2.7); as usual we write $Z^i(C(A)) = \{x \in A_i \mid \delta x = 0\}$ and $H^i(C(A)) = Z^i(C(A))/\delta A_{i-1}$.

Example 2.6.3 Let $\mathcal{L}$ be a sheaf of abelian groups on a topological space X and let $\mathcal{U} = \{U_i\}$ be an open covering of X; the semicosimplicial abelian group of Čech cochains is naturally defined:

$$\mathcal{L}(\mathcal{U})\colon \qquad \prod_i \mathcal{L}(U_i) \rightrightarrows \prod_{i,j} \mathcal{L}(U_{ij}) \rightrightarrows \prod_{i,j,k} \mathcal{L}(U_{ijk}) \rightrightarrows \cdots,$$

where the face operators $\delta_h\colon \prod_{i_0,\ldots,i_{k-1}} \mathcal{L}(U_{i_0\cdots i_{k-1}}) \to \prod_{i_0,\ldots,i_k} \mathcal{L}(U_{i_0\cdots i_k})$ are given by

$$(\delta_h x)_{i_0\cdots i_k} = x_{i_0\cdots \widehat{i_h} \cdots i_k | U_{i_0\cdots i_k}}, \qquad \text{for } h = 0, \ldots, k.$$

More generally, the operator $f\colon \prod_{i_0,\ldots,i_n} \mathcal{L}(U_{i_0\cdots i_n}) \to \prod_{i_0,\ldots,i_m} \mathcal{L}(U_{i_0\cdots i_m})$ induced by a strictly monotone map $f\colon [n] \to [m]$ is determined by the formula:

$$(fx)_{i_0\cdots i_m} = x_{i_{f(0)}\cdots i_{f(n)} | U_{i_0\cdots i_m}}.$$

Here the cochain complex $C(\mathcal{L}(\mathcal{U}))$ is exactly the Čech complex $C(\mathcal{U}, \mathcal{L})$ of the sheaf $\mathcal{L}$ with respect to the covering $\mathcal{U}$.

Remark 2.6.4 In the same situation as Example 2.6.3, one can choose a total ordering in the set of indices of the covering and consider the semicosimplicial abelian group

$$\mathcal{L}_0(\mathcal{U}): \quad \prod_i \mathcal{L}(U_i) \rightrightarrows \prod_{i<j} \mathcal{L}(U_{ij}) \rightrightarrows\!\!\!\to \prod_{i<j<k} \mathcal{L}(U_{ijk}) \rightrightarrows\!\!\!\rightrightarrows \cdots .$$

The corresponding cochain complex $C(\mathcal{L}_0(\mathcal{U}))$ is clearly isomorphic to the complex $C_0(\mathcal{U}, \mathcal{L})$ (see Remark 1.2.1); thus the natural projection $C(\mathcal{L}(\mathcal{U})) \to C(\mathcal{L}_0(\mathcal{U}))$ induces an isomorphism of cohomology groups.

Keep in mind the fact that the complex of alternating Čech cochains does not have, in general, any canonical semicosimplicial structure.

A semicosimplicial Lie algebra over a field $\mathbb{K}$ is a semicosimplicial object in the category of Lie algebras over $\mathbb{K}$; it is represented by a diagram of Lie algebras

$$\mathfrak{g}: \qquad \mathfrak{g}_0 \rightrightarrows \mathfrak{g}_1 \rightrightarrows\!\!\!\to \mathfrak{g}_2 \rightrightarrows\!\!\!\rightrightarrows \cdots ,$$

in which every face operator δ_i is a morphism of Lie algebras.

It is plain that the 0th cohomology group $H^0(C(\mathfrak{g})) = \{x \in \mathfrak{g}_0 \mid \delta_0(x) = \delta_1(x)\}$ is a Lie subalgebra of $\mathfrak{g}_0$.

Definition 2.6.5 (*Lie-cup product*) For every $a, b \geq 0$ consider the strictly monotone maps:

$$i^b\colon [a] \to [a+b], \qquad i^b(h) = h, \quad h = 0, \dots, a,$$
$$s^a\colon [b] \to [a+b], \qquad s^a(k) = k + a, \quad k = 0, \dots, b.$$

Given a semicosimplicial Lie algebra $\mathfrak{g}$, for every pair of indices $a, b \geq 0$ the **Lie-cup** product is the bilinear map

$$\cup\colon H^a(C(\mathfrak{g})) \times H^b(C(\mathfrak{g})) \to H^{a+b}(C(\mathfrak{g})),$$

induced by the bilinear map

$$q\colon \mathfrak{g}_a \times \mathfrak{g}_b \to \mathfrak{g}_{a+b}, \qquad q(x, y) = [i^b(x), s^a(y)].$$

The first thing to verify is that the Lie-cup product is properly definite, and this follows from the equality

$$\delta q(x, y) = q(\delta x, y) + (-1)^a q(x, \delta y), \qquad x \in \mathfrak{g}_a, \; y \in \mathfrak{g}_b. \tag{2.9}$$

In fact, if $\delta x = \delta y = 0$, then $\delta q(x, y) = 0$, and if one of them is also δ-exact, then $q(x, y)$ is δ-exact, too.

In order to prove (2.9) we notice that for every $j = 0, \ldots, a + b + 1$ we have the following equalities:

$$\delta_j i^b = \begin{cases} i^b \delta_j & \text{for } 0 \le j \le a \\ i^{b+1} & \text{for } a < j \le a + b + 1 \end{cases} : [a] \to [a + b + 1],$$

$$\delta_j s^a = \begin{cases} s^{a+1} & \text{for } 0 \le j \le a \\ s^a \delta_{j-a} & \text{for } a < j \le a + b + 1 \end{cases} : [b] \to [a + b + 1],$$

and then, for every x, y as in (2.9), we have:

$$\begin{aligned} \delta q(x, y) &= \delta[i^b(x), s^a(y)] \\ &= \sum_{j=0}^{a+b+1} (-1)^j \delta_j [i^b(x), s^a(y)] = \sum_{j=0}^{a+b+1} (-1)^j [\delta_j i^b(x), \delta_j s^a(y)] \\ &= \sum_{j=0}^{a} (-1)^j [i^b \delta_j(x), s^{a+1}(y)] + \sum_{j=a+1}^{a+b+1} (-1)^j [i^{b+1}(x), s^a \delta_{j-a}(y)] \\ &= [i^b \delta(x) - (-1)^{a+1} i^b \delta_{a+1}(x), s^{a+1}(y)] + (-1)^a [i^{b+1}(x), s^a \delta(y) - s^a \delta_0(y)] \\ &= [i^b \delta(x), s^{a+1}(y)] + (-1)^a [i^{b+1}(x), s^a \delta(y)] = q(\delta x, y) + (-1)^a q(x, \delta y). \end{aligned}$$

In the particular case $a = 0$ the Lie-cup product is induced by the map

$$q \colon \mathfrak{g}_0 \times \mathfrak{g}_b \to \mathfrak{g}_b, \qquad q(x, y) = [i^b(x), y].$$

Since i^b is a morphism of Lie algebras we have

$$\begin{aligned} q([u, v], y) &= [i^b([u, v]), y] = [[i^b(u), i^b(v)], y] \\ &= [i^b(u), [i^b(v), y]] - [i^b(v), [i^b(u), y]] = q(u, q(v, y)) - q(v, q(u, y)), \end{aligned}$$

and this implies that $H^i(C(\mathfrak{g}))$ is a Lie module over $H^0(C(\mathfrak{g}))$, see Exercise 2.7.1.

It is also possible to prove that $\cup$ is graded skew-symmetric; this means that for $x \in H^a(C(\mathfrak{g}))$ and $y \in H^b(C(\mathfrak{g}))$ we have

$$x \cup y + (-1)^{ab} y \cup x = 0.$$

For simplicity we consider here only the case $a, b \le 1$, which is completely sufficient for our first purposes; the case $a = b = 0$ is clear from the definition. If $x \in \mathfrak{g}_0$ and $y \in \mathfrak{g}_1$ we have

$$x \cup y = [\delta_1 x, y] = [\delta_0 x + \delta x, y] = [\delta_0 x, y] = -[y, \delta_0 x] = -y \cup x.$$

The case $x, y \in Z^1(C(\mathfrak{g}))$ is done in the following computation, in which equalities are intended in the quotient vector space $\mathfrak{g}_2/\delta\mathfrak{g}_1$.

$$\begin{aligned} x \cup y &= [\delta_2 x, \delta_0 y] = [\delta_1 x, \delta_0 y] - [\delta_0 x, \delta_0 y] + \delta[x, y] \\ &= [\delta_1 x, \delta_0 y] - [\delta_1 x, \delta_1 y] + [\delta_2 x, \delta_2 y] = [\delta_1 x, \delta_0 y - \delta_1 y] + [\delta_2 x, \delta_2 y] \\ &= [\delta_1 x, -\delta_2 y] + [\delta_2 x, \delta_2 y] = [\delta_2 x - \delta_1 x, \delta_2 y] = [-\delta_0 x, \delta_2 y] = [\delta_2 y, \delta_0 x] \\ &= y \cup x. \end{aligned} \tag{2.10}$$

2.7 Exercises

2.7.1 Let L be a Lie algebra over a field $\mathbb{K}$. A Lie module over L is the data of a vector space V over $\mathbb{K}$ and a bilinear map $q: L \times V \to V$ which satisfies the Jacobi identity

$$q([x, y], v) = q(x, q(y, v)) - q(y, q(x, v)), \qquad x, y \in L, \ v \in V.$$

Given a Lie algebra L, a vector space V and two bilinear maps

$$q: L \times V \to V, \qquad p: V \times L \to V,$$

consider the magmatic algebra $M = L \oplus V$ with product

$$(x + u) \cdot (y + v) = [x, y] + q(x, v) + p(u, y), \qquad x, y \in L, \quad u, v \in V.$$

Prove that M is a Lie algebra if and only if $p(v, x) = -q(x, v)$ and V is a Lie module over L.

2.7.2 Let A be a finite-dimensional associative algebra over a field of characteristic 0 and denote by $[-, -]$ the commutator bracket. Complete the following proof that the equality $[A, A] = A$ holds if and only if $A = 0$. For every $a \in A$ write

$$R_a, L_a: A \to A, \qquad R_a(x) = xa, \quad L_a(x) = ax.$$

Assume $[A, A] = A$ and use the relations $L_{[a,b]} = [L_a, L_b]$, $R_{[a,b]} = [R_b, R_a]$ in order to prove that the trace of R_a, L_a is trivial for every a. Now use the equalities $L_a^n = L_{a^n}$, $R_a^n = R_{a^n}$ and the assumption about the characteristic of the field for proving that $L_a, R_a, [a, -] = L_a - R_a$ are nilpotent endomorphisms for every a. Finally, use Engel's theorem to deduce that the Lie algebra A_L is nilpotent and therefore $[A, A] \neq A$ unless $A = 0$.

2.7.3 Let L be a Lie algebra over a field $\mathbb{K}$. Prove that there exists a unique structure of a Lie algebra on $\mathrm{Der}_{\mathbb{K}}(L, L) \oplus L$ such that both L and $\mathrm{Der}_{\mathbb{K}}(L, L)$ are Lie subalgebras and $[\eta, x] = \eta(x)$ for every $\eta \in \mathrm{Der}_{\mathbb{K}}(L, L)$, $x \in L$.

2.7.4 For every magmatic algebra $(M, \cdot)$ the operators of left and right multiplications L_x and R_x are defined for every $x \in M$ by the formulas:

$$L_x, R_x \colon M \to M, \qquad L_x(y) = x \cdot y, \qquad R_x(y) = y \cdot x.$$

Denote as usual by $[x, y] = x \cdot y - y \cdot x$ the commutator. Prove that M is a left (resp.: right) pre-Lie algebra if and only if $L_{[x,y]} = [L_x, L_y]$ (resp.: $R_{[x,y]} = [R_y, R_x]$) for every $x, y \in M$.

2.7.5 Prove that a right pre-Lie algebra $(R, \triangleleft)$ is associative if and only if for every $x \in R$ the operator

$$f_x \colon R \to R, \qquad f_x(a) = x \triangleleft a - a \triangleleft x,$$

is a derivation.

2.7.6 Prove the Jacobi and Poisson identities for the bracket of Example 2.2.18; cf. Lemma 9.1.4.

2.7.7 A Lie algebra H is called **locally nilpotent** if every finite subset of H is contained in a nilpotent Lie subalgebra. Prove that:

1. every locally nilpotent finite-dimensional Lie algebra is nilpotent;
2. if H is locally nilpotent, then $\cup_{n>0} \ker(\mathrm{ad}\, x)^n = H$ for every $x \in H$.

2.7.8 Prove that, if R is a unitary commutative $\mathbb{K}$-algebra, then

$$\mathrm{Der}_{\mathbb{K}}(R, R) = \{f \in \mathrm{Diff}^1_{R/\mathbb{K}}(R, R) \mid f(1) = 0\},$$
$$\mathrm{Diff}^1_{R/\mathbb{K}}(R, R) = \mathrm{Hom}_R(R, R) \oplus \mathrm{Der}_{\mathbb{K}}(R, R).$$

2.7.9 Prove that, for $R = \mathbb{C}[t^2, t^3] \subset \mathbb{C}[t]$ the map

$$\mathrm{Diff}^1_{R/\mathbb{C}}(R, R) \otimes \mathrm{Diff}^1_{R/\mathbb{C}}(R, R) \to \mathrm{Diff}^2_{R/\mathbb{C}}(R, R), \qquad f \otimes g \to fg$$

is not surjective.

2.7.10 Consider the commutative algebra $R = \mathbb{K}[x, y]/(y^2)$; every element of R is represented by a unique polynomial in $\mathbb{K}[x, y]$ of the form $f(x) + g(x)y$. Prove that the $\mathbb{K}$-linear map

$$h \colon R \to R, \qquad h(f(x) + g(x)y) = \frac{dg}{dx},$$

is a differential operator of second order, but not of the first.

2.7.11 Setting $R = \mathbb{K}[t]$, prove that every $f \in \mathrm{Diff}^n_{R/\mathbb{K}}(R, R)$ is uniquely determined by $f(1)$ and $[f, t] \in \mathrm{Diff}^{n-1}_{R/\mathbb{K}}(R, R)$. Deduce that f is uniquely determined by $f(1), f(t), \ldots, f(t^n)$ and there exist $p_0, \ldots, p_n \in \mathbb{K}[t]$ such that

$$f = p_0 + p_1 \frac{d}{dt} + \cdots + p_n \frac{d^n}{dt^n}.$$

2.7.12 (The symbol for *n*th order differential operators) Prove that for every $f \in \mathrm{Diff}^n_{R/S}(M, N)$ the $\mathbb{K}$-multilinear map in n variables

$$\sigma(f) \colon R \times \cdots \times R \to \mathrm{Hom}_R(M, N),$$

$$\sigma(f)(a_1, \ldots, a_n) = [\cdots[[f, a_1], a_2], \ldots], a_n],$$

is symmetric, is a derivation in each variable and is represented by a well defined element

$$\sigma(f) \in \mathrm{Hom}_R\left(\mathrm{Sym}^n(\Omega_{R/\mathbb{K}}), \mathrm{Hom}_R(M, N)\right) = \mathrm{Hom}_R\left(\mathrm{Sym}^n(\Omega_{R/\mathbb{K}}) \otimes_R M, N\right),$$

where $\Omega_{R/\mathbb{K}}$ is the module of Kähler differentials and Sym^n denotes the symmetric power.

2.7.13 (Coproduct of Lie algebras) Let H, M be two Lie algebras over the same field. Denote by $F = L(H \oplus M)$ the free Lie algebras generated by the vector space $H \oplus M$ and by $[-,-]_H$, $[-,-]_M$ and $[-,-]_F$ the brackets in H, M and F, respectively. Define $H \amalg M = F/I$, where I is the Lie ideal generated by the elements $[x, y]_H - [x, y]_F$ and $[u, v]_M - [u, v]_F$, with $x, y \in H$ and $u, v \in M$. Prove:

1. the natural linear maps $i \colon H \to H \coprod M$ and $j \colon M \to L \coprod M$ are morphisms of Lie algebras;
2. given two morphisms of Lie algebras $f \colon H \to N$ and $g \colon M \to N$, there exists an unique morphism of Lie algebras $h \colon H \coprod M \to N$ such that $f = hi$ and $g = hj$.

2.7.14 (The Chevalley–Eilenberg complex [47]) Let L be a Lie algebra over a field $\mathbb{K}$; the Chevalley–Eilenberg homology complex of L is

$$\cdots \bigwedge^n L \xrightarrow{d} \bigwedge^{n-1} L \xrightarrow{d} \cdots \xrightarrow{d} \bigwedge^0 L = \mathbb{K},$$

where $d(x) = 0$ for every $x \in \bigwedge^1 L$ and

$$d(x_1 \wedge \cdots \wedge x_n) = \sum_{k<l} (-1)^{k+l+1} [x_k, x_l] \wedge x_1 \wedge \cdots \widehat{x_k} \cdots \widehat{x_l} \cdots \wedge x_n$$

for every $n \geq 2$ and $x_1, \ldots, x_n \in L$. Prove that the operator d is properly defined and $d^2 = 0$. Notice that the first homology group of the above complex is the algebraic is the quotient $L/[L, L]$.

Its dual complex

$$\cdots \xrightarrow{d^*} \operatorname{Hom}(\bigwedge^n L, \mathbb{K}) \xrightarrow{d^*} \operatorname{Hom}(\bigwedge^{n+1} L, \mathbb{K}) \xrightarrow{d^*} \cdots$$

is called the Chevalley–Eilenberg cohomology complex.

2.7.15 Let L be a Lie algebra over a field of characteristic 0 and let $H \subset L$ be a nilpotent Lie ideal. Prove that for every $a \in H$ the operator $e^{\operatorname{ad} a}\colon L \to L$ makes sense. For every fixed element $u \in L$ prove that the map

$$*\colon \exp(H) \times L \to L, \quad e^a * x = e^{\operatorname{ad} a}(x+u) - u = x + \sum_{n\geq 0} \frac{(\operatorname{ad} a)^n}{(n+1)!}([a, x+u]),$$

is a left action of the group $\exp(H)$ on the set L.

2.7.16 Let $\mathbb{K}$ be a field of characteristic 0 and consider a derivation

$$\xi = a(t)\frac{d}{dt} \in \operatorname{Der}_{\mathbb{K}}(\mathbb{K}[[t]], \mathbb{K}[[t]]),$$

where $a(t) \in \mathbb{K}[[t]]$ is a formal power series such that $a(0) = \dfrac{da}{dt}(0) = 0$. Prove that:

1. the exponential $e^{\xi} = \sum_{n\geq 0} \dfrac{\xi^n}{n!}\colon \mathbb{K}[[t]] \to \mathbb{K}[[t]]$ is a well defined isomorphism of algebras;
2. (Julia's equation) if $g(t) = e^{\xi}(t)$, then $a(g(t)) = a(t)g'(t)$. (Hint: $e^{\xi}\xi(t) = \xi e^{\xi}(t)$.)

References

The name magmatic has a clear origin from [170], where the term *magma* is used to define a set equipped with a binary operation without any further relation. The notion of a pre-Lie algebra was introduced by Gerstenhaber [87] in the framework of cohomology theory of associative algebras and, independently, by Vinberg in the setting of differential geometry [263]. However, some early examples of pre-Lie products, together with an operadic vision of them, can be found in a remarkable paper by Cayley [46]. It is possible to prove that, if the base field $\mathbb{K}$ has characteristic 0, then the Lie algebra $\mathfrak{sl}_n(\mathbb{K})$, $n \geq 2$, does not admit any compatible pre-Lie structure. More generally, by a theorem of Helmstetter [111], for every finite dimensional pre-Lie algebra R, the commutator bracket $[-,-]\colon R \otimes R \to R$ is not surjective, unless $R = 0$.

The standard reference for Grothendieck's algebraic definition of differential operators is [106, Sect. 16.8]; another useful reference is [49]. It is easy to see that the map p in Lemma 2.3.6 is not surjective for general R-modules. On the other hand, it is possible to prove that p is surjective whenever M is either a projective or an injective R-module; for a proof of this fact and a more detailed study of derivations of pairs we refer to [133].

Our proof of Corollary 2.4.5 is taken from [136]; for a different proof see [217]. We also refer to Jacobson's book [136] for the proof that the Chevalley–Eilenberg complex (Exercise 2.7.14) computes homology and cohomology of Lie algebras.

The most famous explicit expression of the Baker–Campbell–Hausdorff product in terms of nested commutators is probably the one due to Dynkin [62]:

$$x \bullet y = \sum_{n>0} \frac{(-1)^{n-1}}{n} \sum \frac{1}{p_1!q_1!\dots p_n!q_n!} ad(x)^{p_1} ad(y)^{q_1} \cdots ad(x)^{p_n} ad(y)^{q_n-1} y,$$

where $ad(z) = [z, -]$ is the adjoint operator, and the second sum is over all possible combinations of $p_1, q_1, \dots, p_k, q_k \in \mathbb{N}$ such that $p_i + q_i > 0$, for $i = 1, \dots, k$, and $\sum_{i=1}^{k}(p_i + q_i) = n$.

For the definition of a semicosimplicial object we have followed [66, 269]. We refer to [91] for a deep and detailed treatment of cup products in the cochain complex of semicosimplicial algebras.

Chapter 3
Functors of Artin Rings

In the previous chapters we introduced the notion of infinitesimal deformations of a complex manifold together with an idea of its role in deformation theory. The notion of infinitesimal deformations extends to a wide class of algebro-geometric structures; for instance if R is an associative $\mathbb{C}$-algebra, one can define a deformation of R over an Artin local $\mathbb{C}$-algebra A as an isomorphism class of structures of associative A-algebra on the A-module $R \otimes_{\mathbb{C}} A$, such that the natural projection $R \otimes_{\mathbb{C}} A \to R$ is a morphism of algebras.

One of the most classical and common ways to formalize infinitesimal deformations and formal local moduli spaces is by introducing the notion of functors of Artin rings. Among them a crucial role is played by pro-representable and deformation functors, i.e., by functors satisfying certain conditions which are at the same time sufficiently general to include almost all the concrete examples and sufficiently strong to ensure the existence of the formal analogue of the Kodaira–Spencer map and of formal semiuniversal deformations.

3.1 Artin Rings and Small Extensions

We shall use henceforth the following notation: unless otherwise stated, every ring is considered commutative and unitary; for every local ring R we denote by $\mathfrak{m}_R$ its maximal ideal.

Given a local ring R we denote by $\mathcal{C}_R$ the category of local R-algebras with residue field $R/\mathfrak{m}_R$; more precisely, every object in $\mathcal{C}_R$ is a local morphism of local rings $R \to S$ inducing an isomorphism of fields

$$\frac{R}{\mathfrak{m}_R} \cong \frac{S}{\mathfrak{m}_S}.$$

M. Manetti, *Lie Methods in Deformation Theory*, Springer Monographs in Mathematics,
https://doi.org/10.1007/978-981-19-1185-9_3

A morphism in $\mathcal{C}_R$ is simply a local morphism of R-algebras. For later use we point out that the set theoretic fibre product is also a fibre product in the category $\mathcal{C}_R$; given a diagram

$$\begin{array}{ccc} & & S'' \\ & & \downarrow{\scriptstyle f''} \\ S' & \xrightarrow{f'} & S \end{array}$$

in $\mathcal{C}_R$, the R-algebra

$$S' \times_S S'' = \{(x, y) \in S' \times S'' \mid f'(x) = f''(y)\}$$

is a local ring with maximal ideal $\mathfrak{m}_{S'} \times_S \mathfrak{m}_{S''}$. In fact, denoting by $\mathbb{K} = R/\mathfrak{m}_R$ the residue field and by $p\colon S \to \mathbb{K}$, $p'\colon S' \to \mathbb{K}$, $p''\colon S'' \to \mathbb{K}$ the projections, we have $p'(x) = p(f'(x))$, $p''(x) = p(f''(x))$ and then for any $(x, y) \in S' \times_S S''$ we have

$$p'(x) = 0 \iff p''(y) = 0 \iff (x, y) \in \mathfrak{m}_{S'} \times_S \mathfrak{m}_{S''}.$$

Therefore, for every $(x, y) \notin \mathfrak{m}_{S'} \times_S \mathfrak{m}_{S''}$ there exist $x^{-1} \in S'$, $y^{-1} \in S''$, and since $f'(x^{-1}) = (f'(x))^{-1} = (f''(y))^{-1} = f''(y^{-1})$, we get $(x^{-1}, y^{-1}) = (x, y)^{-1}$.

Given a field $\mathbb{K}$, by Hilbert's Nullstellensatz, for an Artin local $\mathbb{K}$-algebra A the following conditions are equivalent:

1. A is a finitely generated $\mathbb{K}$-algebra;
2. the residue field $A/\mathfrak{m}_A$ is a finite extension of $\mathbb{K}$;
3. A is a finite-dimensional vector space over $\mathbb{K}$.

Now let $\mathbb{K}$ be a fixed field and let $\mathbf{Art}_{\mathbb{K}}$ be the full subcategory of $\mathcal{C}_{\mathbb{K}}$ of all Artin local $\mathbb{K}$-algebras with residue field $\mathbb{K}$: an element $S \in \mathcal{C}_{\mathbb{K}}$ belongs to $\mathbf{Art}_{\mathbb{K}}$ if and only if it is finite-dimensional as a $\mathbb{K}$-vector space. The above considerations about fibre products in $\mathcal{C}_{\mathbb{K}}$ imply also that the category $\mathbf{Art}_{\mathbb{K}}$ is closed under fiber products.

Recall that a **small category** is a category whose class of objects is a set; for instance the category $\overrightarrow{\Delta}$ introduced in Sect. 2.6 is small, while the category of sets is not small.

It should be noted that $\mathbf{Art}_{\mathbb{K}}$ is equivalent to a small category; for example we can fix a countable sequence of indeterminates $x_1, x_2, \ldots$ and denote by $\mathbf{A} \subset \mathbf{Art}_{\mathbb{K}}$ the full subcategory whose objects are the Artin quotients of $\mathbb{K}[[x_1, \ldots, x_n]]$ for some n. It is clear that the inclusion map $\mathbf{A} \to \mathbf{Art}_{\mathbb{K}}$ is an equivalence of categories.

Notice that $\mathbf{Art}_{\mathbb{K}}$ is also a full subcategory of the category of local complete Noetherian $\mathbb{K}$-algebras with residue field $\mathbb{K}$ and, for $\mathbb{K} = \mathbb{C}$ it is also a full subcategory of the category of analytic algebras. For simplicity of notation, when the field $\mathbb{K}$ is clear from the context, we shall write $\mathbf{Art}$ in place of $\mathbf{Art}_{\mathbb{K}}$.

We shall say that a morphism $\alpha\colon B \to A$ in $\mathbf{Art}_{\mathbb{K}}$ is a small surjection if it is surjective and its kernel is annihilated by the maximal ideal $\mathfrak{m}_B$. If $\alpha\colon B \to A$ is surjective with kernel $I = \ker(\alpha)$ there exists an integer $s > 0$ such that $\mathfrak{m}_B^s I = 0$, and then α factors as the composition of the finite sequence of small surjections

$$B \to \frac{B}{\mathfrak{m}_B^{s-1} I} \to \cdots \to \frac{B}{\mathfrak{m}_B I} \to A.$$

Similarly, if $\alpha\colon B \to A$ is a small surjection and $V_1 \subset \cdots \subset V_n \subset \ker(\alpha)$ is a filtration of $\mathbb{K}$-vector subspaces, then every V_i is an ideal of B and α factors as the composition of

$$B \to \frac{B}{V_1} \to \frac{B}{V_2} \cdots \to \frac{B}{V_n} \to A.$$

A **small extension** is a small surjection together with a framing of its kernel. More precisely a small extension e in $\mathbf{Art}_{\mathbb{K}}$ is an exact sequence of abelian groups

$$e\colon \quad 0 \to M \xrightarrow{\varphi} B \xrightarrow{\alpha} A \to 0,$$

such that α is a morphism in the category $\mathbf{Art}_{\mathbb{K}}$, φ is a morphism of B-modules and the ideal $\varphi(M)$ is annihilated by the maximal ideal $\mathfrak{m}_B$. In particular, M is a finite-dimensional vector space over $B/\mathfrak{m}_B = \mathbb{K}$. A small extension as above is called principal if $M = \mathbb{K}$. The above considerations implies in particular that every surjective morphism in $\mathbf{Art}_{\mathbb{K}}$ is a finite composition of small surjections arising from principal small extensions.

The **trivial extension** of an Artin local ring $A \in \mathbf{Art}_{\mathbb{K}}$ by a finite-dimensional $\mathbb{K}$-vector space M is defined as

$$0 \to M \xrightarrow{i} A \oplus M \xrightarrow{p} A \to 0,$$

where i is the inclusion, p is the projection and the product on $A \oplus M$ is defined by the formula

$$(a, m)(b, n) = (ab, an + bm).$$

The small extensions are the objects of a category where the morphisms are the commutative diagrams

$$\begin{array}{llllllllll} e_1\colon & 0 & \longrightarrow & M_1 & \longrightarrow & B_1 & \longrightarrow & A_1 & \longrightarrow & 0 \\ & & & \downarrow{\scriptstyle \alpha_M} & & \downarrow{\scriptstyle \alpha_B} & & \downarrow{\scriptstyle \alpha_A} & & \\ e_2\colon & 0 & \longrightarrow & M_2 & \longrightarrow & B_2 & \longrightarrow & A_2 & \longrightarrow & 0 \end{array} \tag{3.1}$$

with α_A and α_B morphisms in $\mathbf{Art}_{\mathbb{K}}$. The pull-back of a small extension

$$e\colon \; 0 \to M \xrightarrow{\varphi} B \xrightarrow{\alpha} A \to 0,$$

by a morphism $f\colon C \to A$ in $\mathbf{Art}_{\mathbb{K}}$ is the small extension

$$f^*e\colon\ 0 \to M \xrightarrow{(\varphi,0)} B \times_A C \xrightarrow{f^*\alpha} C \to 0, \qquad f^*\alpha(b,c) = c,$$

while the push-out by a morphism of finite dimensional $\mathbb{K}$-vector spaces $g\colon M \to N$ is the small extension

$$g_*e\colon\ 0 \to N \xrightarrow{(0,\mathrm{Id})} \frac{B \oplus N}{\{(\varphi(m), -g(m)) \mid m \in M\}} \xrightarrow{g_*\alpha} A \to 0, \qquad g_*\alpha(b,n) = \alpha(b).$$

Remark 3.1.1 In the above setup it is easy to see that f^*g_*e is isomorphic to g_*f^*e and there exists a commutative diagram of small extensions

$$\begin{array}{ccc} f^*e & \longrightarrow & f^*g_*e \simeq g_*f^*e \\ \downarrow & & \downarrow \\ e & \longrightarrow & g_*e. \end{array}$$

Moreover, in the notation of (3.1), the push-out $\alpha_{M*}(e_1)$ is isomorphic to the pull-back $\alpha_A^*(e_2)$.

3.2 Deformation Functors

We start here a general theory which includes, as a special case, a certain level of categorification of infinitesimal deformations of a fixed object. Although it may seem rather technical at first, the study of deformation functors is important because it gives powerful general results which are immediately applicable to various deformation problems.

As a general setup we denote by **Set** the category of sets and, by a slight and harmless abuse of notation, we shall denote by $0 \in \mathbf{Set}$ the singleton, which is the terminal object in the category of sets.

From now on let $\mathbb{K}$ be a fixed field and as above we denote by $\mathbf{Art} = \mathbf{Art}_{\mathbb{K}}$ the category of Artin local $\mathbb{K}$-algebras with residue field $\mathbb{K}$. Unless otherwise specified every tensor product is intended over the field $\mathbb{K}$.

Definition 3.2.1 A **functor of Artin rings** is a covariant functor $F\colon \mathbf{Art} \to \mathbf{Set}$ such that $F(\mathbb{K}) = 0$.

Since **Art** is equivalent to a small category, the functors of Artin rings are the objects of a category whose morphisms are the natural transformations of functors. For simplicity of notation, if $\phi\colon F \to G$ is a natural transformation, we also denote by $\phi\colon F(A) \to G(A)$ the corresponding morphism of sets, for every $A \in \mathbf{Art}$.

Example 3.2.2 Although the theory of functors of Artin rings is developed for studying the functors of infinitesimal deformations, it is preferable to begin with some more elementary examples:

(1) The **trivial** functor 0, defined by $0(A) = 0$ for every $A \in \mathbf{Art}$, is a functor of Artin rings.

(2) Let V be a $\mathbb{K}$-vector space. Then the functor $\widehat{V}\colon \mathbf{Art} \to \mathbf{Set}$, defined by

$$\widehat{V}(A) = V \otimes \mathfrak{m}_A \quad (= V \otimes_{\mathbb{K}} \mathfrak{m}_A)$$

is a functor of Artin rings. In Chap. 10 we shall call $\widehat{V}$ the formal neighbourhood of V.

(3) Let V be a $\mathbb{K}$-vector space. Then the functor $G\colon \mathbf{Art} \to \mathbf{Set}$, defined by

$$G(A) = \mathrm{Hom}_{\mathbb{K}}(V, V \otimes \mathfrak{m}_A)$$

is a functor of Artin rings. Notice that $G(A)$ is the kernel of the morphism

$$\mathrm{Hom}_A(V \otimes A, V \otimes A) = \mathrm{Hom}_{\mathbb{K}}(V, V \otimes A) \to \mathrm{Hom}_{\mathbb{K}}(V, V \otimes \mathbb{K}) = \mathrm{Hom}_{\mathbb{K}}(V, V)$$

and then $G(A)$ is the set of all the A-linear endomorphisms of $V \otimes A$ whose image is contained in $V \otimes \mathfrak{m}_A$.

Example 3.2.3 For every $B \in \mathbf{Art}$ the functor

$$h_B\colon \mathbf{Art} \to \mathbf{Set}, \qquad h_B(A) = \mathrm{Mor}_{\mathbf{Art}}(B, A),$$

is a functor of Artin rings. By Yoneda's lemma, for every functor of Artin rings F and every $B \in \mathbf{Art}$ there exists a natural bijection between the set $F(B)$ and the set of natural transformations $h_B \to F$: every element $b \in F(B)$ corresponds to the unique natural transformation such that $\mathrm{Id}_B \mapsto b$.

Example 3.2.4 Let $R \in \mathcal{C}_{\mathbb{K}}$ be a local $\mathbb{K}$-algebra with residue field $\mathbb{K}$. Then

$$h_R\colon \mathbf{Art} \to \mathbf{Set}, \qquad h_R(A) = \mathrm{Mor}_{\mathcal{C}_{\mathbb{K}}}(R, A),$$

is a functor of Artin rings. We have seen that the category **Art** is closed under fibre products and then for every pair of morphisms $C \to A$, $B \to A$ in **Art** the natural map

$$h_R(B \times_A C) \to h_R(B) \times_{h_R(A)} h_R(C)$$

is bijective. Since every morphism $\alpha \in h_R(A)$ factors to a morphism $R/\mathfrak{m}_R^n \to A$ for some $n > 0$, there exists a natural isomorphism $h_{\widehat{R}} \xrightarrow{\simeq} h_R$ induced by the completion in the $\mathfrak{m}_R$-adic topology $R \to \widehat{R}$.

Definition 3.2.5 Let $F\colon \mathbf{Art} \to \mathbf{Set}$ be a functor of Artin rings. For every fibre product

$$\begin{array}{ccc} B \times_A C & \longrightarrow & C \\ \downarrow & & \downarrow \\ B & \xrightarrow{\beta} & A \end{array}$$

in the category **Art** consider the induced map

$$\eta\colon F(B \times_A C) \to F(B) \times_{F(A)} F(C).$$

1. The functor F is called a **deformation functor** if:
   a. η is surjective, whenever β is surjective;
   b. η is bijective, whenever $A = \mathbb{K}$.
2. The functor F is said to be **homogeneous** if η is bijective whenever β is surjective.

For instance, for every $R \in \mathcal{C}_{\mathbb{K}}$ the functor h_R is a homogeneous deformation functor. Every homogenous functor of Artin rings is a fortiori a deformation functor; we shall see very soon that there exist deformation functors that are not homogeneous.

Remark 3.2.6 Our definition of deformation functors involves conditions that are slightly more restrictive than the classical Schlessinger's conditions H1, H2 of [237] and the semi-homogeneity condition of [226]. The main motivations of this change are:

1. almost all the examples of functors of infinitesimal deformations of algebro-geometric structures are deformation functors as defined in 3.2.5;
2. functors of Artin rings satisfying the classical Schlessinger's conditions H1, H2 and H3 do not necessarily have a "good" obstruction theory (see [69, Example 6.8]). This is not only an aesthetic question but involves either the validity of many interesting results or their simplicity of proof;
3. the notion of deformation functors extends naturally to the framework of derived deformation theory and extended moduli spaces, cf. [178].

The construction of limits in the category of sets extends immediately to the category of functors of Artin rings. For instance, the product of two functors of Artin rings $F, G\colon \mathbf{Art} \to \mathbf{Set}$ is defined in the obvious way: $(F \times G)(A) = F(A) \times G(A)$. We can immediately see that if F, G are (homogeneous) deformation functors, then $F \times G$ is a (homogeneous) deformation functor; by Example 3.2.3 we also have that $F \times G$ is isomorphic to the categorical product in the full subcategory of (homogeneous) deformation functors.

The category of deformation functors does not have fibre products, see Exercise 3.9.4. Since most functors describing concrete deformation problems are not homogeneous, one must be very careful in performing fibre products in deformation theory. Similarly, the image of a morphism of deformation functors is not a deformation functor in general; an example will be given in Exercise 6.8.4.

The following definition is motivated from the fact that the formal smoothness of $\mathrm{Spec}(R)$ is equivalent to the property that $B \to A$ surjective implies $h_R(B) \to h_R(A)$ surjective, cf. next Lemma 3.3.4.

Definition 3.2.7 A natural transformation $\phi\colon F \to G$ of functors of Artin rings is called **smooth** if for every surjective morphism $B \to A$ in the category **Art**, the natural map

$$F(B) \to G(B) \times_{G(A)} F(A)$$

is surjective. A functor of Artin rings F is called **smooth** or **unobstructed** if the natural transformation $F \to 0$ is smooth; equivalently F is smooth if $F(B) \to F(A)$ is surjective for every surjective morphism $B \to A$ in **Art**.

For instance, for every vector space V the functor $A \mapsto \widehat{V}(A) := V \otimes \mathfrak{m}_A$ is smooth. If $f\colon V \to W$ is a linear map, then the induced natural transformation $\widehat{V} \to \widehat{W}$ is smooth if and only if f is surjective. If $\phi\colon F \to G$ is a smooth natural transformation, then $\phi\colon F(A) \to G(A)$ is surjective for every Artin local ring A; to see this it is sufficient to take $B = \mathbb{K}$ in Definition 3.2.7. We leave to the reader the proof that composition of smooth natural transformations is smooth, and that if G is a smooth functor, then the projection $F \times G \to F$ is smooth for every F.

The smoothness assumption is involved in some useful criteria for the existence of colimits in the category of deformation functors. The following simple coequalizer criterion will be sufficient for most of our applications.

Theorem 3.2.8 *Let*

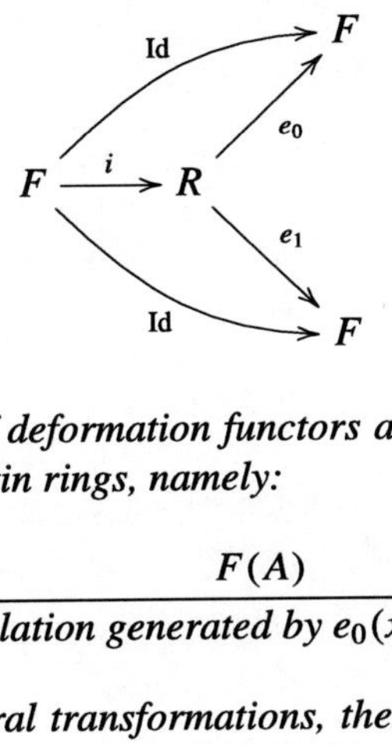

be a commutative diagram of deformation functors and denote by Q its colimit in the category of functors of Artin rings, namely:

$$Q(A) = \frac{F(A)}{\text{equivalence relation generated by } e_0(x) \sim e_1(x),\ x \in R(A)}\,.$$

If e_0 and e_1 are smooth natural transformations, then Q is a deformation functor and the natural projection $\pi\colon F \to Q$ is smooth.

Proof Given $A \in$ **Art** and two elements $x, y \in F(A)$ such that $\pi(x) = \pi(y)$ we denote by $d(x, y)$ the length of the shortest zigzag of pairs $(e_i a, e_{1-i} a)$, $i = 0, 1$, $a \in R(A)$, giving the equivalence $x \sim y$; for instance, we have $d(x, y) = 0$ if and only if $x = y$ and $d(x, y) = 1$ if and only if $x \neq y$ and there exists $a \in R(A)$ such that either $e_0(a) = x$, $e_1(a) = y$, or $e_0(a) = y$, $e_1(a) = x$.

We first prove that the projection $\pi\colon F \to Q$ is smooth, i.e., that for every surjective morphism $\beta\colon B \to A$ in **Art** and every pair $(x, y) \in F(B) \times F(A)$ such that $(\pi(x), y) \in Q(B) \times_{Q(A)} F(A)$ there exists $z \in F(B)$ such that $\pi(x) = \pi(z)$ and $\beta(z) = y$. If $\beta(x) = y$ then there is nothing to prove; otherwise, by induction on $d(\beta(x), y)$ it is sufficient to prove that there exists $z \in F(B)$ such that $\pi(x) = \pi(z)$ and $d(\beta(z), y) < d(\beta(x), y)$. Let $u \in F(A)$ be such that $d(\beta(x), u) = 1$ and $d(u, y) < d(\beta(x), y)$, then there exists $a \in R(A)$ such that $e_i(a) = \beta(x), e_{1-i}(a) = u$. Since e_i is smooth the element a lifts to some $b \in R(B)$ such that $e_i(b) = x$. Now $\beta(e_{1-i}(b)) = e_{1-i}(\beta(b)) = e_{1-i}(b) = u$ and we can take $z = e_{1-i}(b)$.

In order to prove that Q is a deformation functor, consider a pull-back diagram

$$\begin{array}{ccc} B \times_A C & \longrightarrow & C \\ \downarrow & & \downarrow{\scriptstyle \gamma} \\ B & \xrightarrow{\ \beta\ } & A \end{array}$$

in **Art** with β surjective. An element of $Q(B) \times_{Q(A)} Q(C)$ is represented by a pair (x, y) with $x \in F(B), y \in F(C)$ and $\pi(\beta(x)) = \pi(\gamma(y)) \in Q(A)$. Since we already know that π is smooth, there exists $z \in F(B)$ such that $\pi(z) = \pi(x)$ and $\beta(z) = \gamma(y)$. Now it is sufficient to use the fact that F is a deformation functor and there exists a lifting of $(z, y) \in F(B) \times_{F(A)} F(C)$ to an element of $F(B \times_A C)$.

Finally, since $R(B \times_{\mathbb{K}} C) = R(B) \times R(C)$ the existence of the section $i\colon F \to R$ implies that two pairs $(x_1, y_1), (x_2, y_2) \in F(B \times_{\mathbb{K}} C) = F(B) \times F(C)$ are equivalent if and only if x_1 is equivalent to x_2 and y_1 is equivalent to y_2, and this immediately implies that $Q(B \times_{\mathbb{K}} C) = Q(B) \times Q(C)$. □

Theorem 3.2.8 above applies in particular to the case of quotients by actions of smooth group functors. Given a functor of Artin rings $F\colon$ **Art** $\to$ **Set** and a group functor of Artin rings $G\colon$ **Art** $\to$ **Grp**, by a G-action on F we shall mean a natural transformation $G \times F \to F$ such that

$$G(A) \times F(A) \to F(A)$$

is a $G(A)$-action on $F(A)$ in the usual sense for every $A \in$ **Art**. This can be interpreted as the diagram of Theorem 3.2.8 by defining

$$R = G \times F, \quad e_0(g, x) = x, \quad e_1(g, x) = gx, \qquad i(x) = (1, x),$$

and the corresponding colimit is the quotient functor F/G.

If G is smooth then e_0, e_1 are smooth; this is obvious for e_0, while if $\beta\colon B \to A$ is a surjective morphism and $(g, x) \in G(A) \times F(A), y \in F(B)$ are elements such that $\beta(y) = e_1(g, x) = gx$, then $y = e_1(\hat{g}, \hat{g}^{-1}y)$ for every $\hat{g} \in G(B)$ such that $\beta(\hat{g}) = g$. Thus we have proved the following result.

Corollary 3.2.9 *Assume there are given a functor of Artin rings* $F: \mathbf{Art} \to \mathbf{Set}$ *and a group functor of Artin rings* $G: \mathbf{Art} \to \mathbf{Grp}$ *acting on* F. *Suppose that* F *and* G *are both deformation functors and* G *is smooth. Then* F/G *is a deformation functor and the natural projection* $F \to F/G$ *is smooth.*

It is easy to give examples where F and G are homogeneous deformation functors and F/G is not homogeneous (see e.g. Example 3.5.6 and Exercise 3.9.10). Moreover, it is possible to prove that over a field of characteristic 0 every group deformation functor is smooth (Theorem B.3.12).

3.3 Pro-representable Functors

According to general categorical language, a functor of Artin rings is called **representable** if it is isomorphic to h_R, for some $R \in \mathbf{Art}$. In practice, this notion is too restrictive because the category **Art** contains few objects. Nonetheless, representable functors become very important whenever we enlarge **Art** to a sufficiently bigger category. The first, and almost automatic, enlargement is to the category of local complete noetherian $\mathbb{K}$-algebra with residue field $\mathbb{K}$; this leads to the notion of pro-representable functors.

Definition 3.3.1 A functor $F: \mathbf{Art} \to \mathbf{Set}$ is called **pro-representable** if it is isomorphic to h_R (Example 3.2.4), for some local complete noetherian $\mathbb{K}$-algebra R with residue field $\mathbb{K}$.

Recall that a local ring A is complete if the natural morphism

$$A \to \lim_n \frac{A}{\mathfrak{m}_A^n}$$

is an isomorphism. If A is complete and noetherian, then every finite set of generators $m_1, \ldots, m_s$ of $\mathfrak{m}_A$ gives a surjective local morphism

$$\mathbb{K}[[t_1, \ldots, t_s]] \to A, \qquad t_i \mapsto m_i.$$

Therefore a functor of Artin rings is pro-representable if and only if it is isomorphic to h_R, for some quotient R of a formal power series algebra $\mathbb{K}[[t_1, \ldots, t_s]]$.

Example 3.3.2 Let V be a finite-dimensional vector space with basis $(e_1, \ldots, e_n)$, so that for every $A \in \mathbf{Art}$ the vectors of $V \otimes \mathfrak{m}_A$ are precisely the linear combinations $a_1e_1 + \cdots + a_ne_n$, with $a_i \in \mathfrak{m}_A$. Every formal power series $f \in \mathbb{K}[[t_1, \ldots, t_n]]$ induces a map

$$f_A: V \otimes \mathfrak{m}_A \to A, \qquad f_A(a_1e_1 + \cdots + a_ne_n) = f(a_1, \ldots, a_n).$$

Consider a fixed proper ideal $I \subset \mathbb{K}[[t_1, \ldots, t_n]]$; its formal zero locus is the functor of Artin rings

$$Z_I : \mathbf{Art} \to \mathbf{Set}, \qquad Z_I(A) = \{x \in V \otimes \mathfrak{m}_A \mid f_A(x) = 0 \quad \forall f \in I\}.$$

We claim that the functor Z_I is pro-representable, and more precisely that it is canonically isomorphic to the functor h_R, where $R = \mathbb{K}[[t_1, \ldots, t_n]]/I$.

Denoting by $\pi : \mathbb{K}[[t_1, \ldots, t_n]] \to R$ the projection to the quotient, the canonical isomorphism $u : h_R \to Z_I$ is given by

$$h_R(A) \ni \alpha \mapsto u(\alpha) = \sum_{i=1}^{n} \alpha(\pi(t_i)))e_i \in V \otimes \mathfrak{m}_A.$$

The natural transformation u is properly defined because for every $f \in I$ and every $\alpha \in h_R(A)$ as above we have

$$f(\alpha(\pi(t_1)), \ldots, \alpha(\pi(t_n))) = \alpha\pi(f(t_1, \ldots, t_n)) = \alpha(0) = 0.$$

In order to conclude the proof it is sufficient to observe that u is an isomorphism whenever $I = 0$, and a morphism $\beta : \mathbb{K}[[t_1, \ldots, t_n]] \to A$ factors as $\beta = \alpha\pi$ for some $\alpha \in h_R(A)$ if and only if $\beta(f) = f(\beta(t_1), \ldots, \beta(t_n)) = 0$ for every $f \in I$.

If F is a functor of Artin rings, then every natural transformation $\xi : h_R \to F$ is uniquely determined by a sequence of elements $\xi_n \in F(R/\mathfrak{m}_R^n)$, $n > 0$, such that $\pi(\xi_{n+1}) = \xi_n$, where $\pi : R/\mathfrak{m}_R^{n+1} \to R/\mathfrak{m}_R^n$ is the projection. Indeed, given $\xi : h_R \to F$, we denote by $\pi_n : R \to R/\mathfrak{m}_R^n$ the projection and by $\xi_n = \xi(\pi_n)$; conversely, given a coherent sequence $\{\xi_n\}$ as above, for every $A \in \mathbf{Art}$ and every morphism $\alpha : R \to A$ we have

$$\alpha : R \xrightarrow{\pi_n} \frac{R}{\mathfrak{m}_R^n} \xrightarrow{\alpha_n} A$$

for every n sufficiently large and such that $\mathfrak{m}_A^n = 0$, and we may define $\xi(\alpha) = \alpha_n(\xi_n)$.

Lemma 3.3.3 *Let $\eta : F \to G$ and $\xi : H \to G$ be two natural transformations of deformation functors. If η is smooth and H is pro-representable, then there exists a natural transformation $\mu : H \to F$ such that $\eta\mu = \xi$.*

Proof It is not restrictive to assume $H = h_R$; denoting by

$$\pi_n : R \to R_n = \frac{R}{\mathfrak{m}_R^n}, \qquad p_n : R_{n+1} \to R_n,$$

the natural projections, we have $\pi_n \in h_R(R_n)$ and $\xi_n := \xi(\pi_n) \in G(R_n)$. Since the natural transformation η is smooth, we can lift the coherent sequence ξ_n to a coherent sequence $\mu_n \in F(R_n)$ such that $\eta(\mu_n) = \xi_n$, $p_n(\mu_{n+1}) = \mu_n$. By the above remark there exists $\mu \colon h_R \to F$ such that $\mu(\pi_n) = \mu_n$, $\eta\mu(\pi_n) = \eta(\mu_n) = \xi_n$ and then $\eta\mu = \xi$. □

We conclude this section with an useful smoothness criterion for pro-representable functors.

Lemma 3.3.4 *Let $\mathbb{K}$ be an infinite field and let R be a local complete noetherian $\mathbb{K}$-algebra with residue field $\mathbb{K}$. Then the following conditions are equivalent:*

1. *R is isomorphic to a power series ring $\mathbb{K}[[x_1, \dots, x_n]]$;*
2. *the functor h_R is smooth;*
3. *for every $s \geq 2$ the morphism*

$$h_R\left(\frac{\mathbb{K}[t]}{(t^{s+1})}\right) \to h_R\left(\frac{\mathbb{K}[t]}{(t^s)}\right)$$

   is surjective;
4. *for every $s \geq 2$ the morphism*

$$h_R\left(\frac{\mathbb{K}[t]}{(t^{s+1})}\right) \to h_R\left(\frac{\mathbb{K}[t]}{(t^2)}\right)$$

   is surjective.

Proof The only nontrivial implication is $(4 \Rightarrow 1)$. Let n be the minimum number of generators of the maximal ideal $\mathfrak{m}_R$, then we have $R = \mathbb{K}[[x_1, \dots, x_n]]/I$ for some ideal $I \subset (x_1, \dots, x_n)^2$ (cf. Corollary A.1.4), and we want to prove that $I = 0$. To this end, we prove that if $I \neq 0$ and $s \geq 2$ is the greatest integer such that $I \subset (x_1, \dots, x_n)^s$, then the map

$$h_R\left(\frac{\mathbb{K}[t]}{(t^{s+1})}\right) \to h_R\left(\frac{\mathbb{K}[t]}{(t^2)}\right)$$

is not surjective. Choosing $f \in I - (x_1, \dots, x_n)^{s+1}$, after a possible generic linear change of coordinates of the form $x_i \mapsto x_i + a_i x_1$, with $a_2, \dots, a_k \in \mathbb{K}$, we may assume that f contains the monomial x_1^s with a nonzero coefficient, say $f = cx_1^s + \cdots$; let $\alpha \colon R \to \mathbb{K}[t]/(t^2)$ be the morphism defined by $\alpha(x_1) = t$, $\alpha(x_i) = 0$ for $i > 1$.

Assume that there exists a morphism of algebras $\tilde{\alpha} \colon R \to \mathbb{K}[t]/(t^{s+1})$ that lifts α and denote by $\beta \colon \mathbb{K}[[x_1, \dots, x_n]] \to \mathbb{K}[t]/(t^{s+1})$ the composition of $\tilde{\alpha}$ with the projection $\pi \colon \mathbb{K}[[x_1, \dots, x_n]] \to R$. Then $\beta(x_1) - t, \beta(x_2), \dots, \beta(x_n) \in (t^2)$ and then $\beta(f) \equiv ct^s \not\equiv 0 \pmod{t^{s+1}}$, in contradiction with $\pi(f) = 0$. □

Remark 3.3.5 If $\mathbb{K}$ is a finite field, then the equivalence of the first 3 items of Lemma 3.3.4 is still true (Exercise 3.9.3), but the implication $(4 \Rightarrow 1)$ is generally false; consider for instance $R = \mathbb{F}_p[[x, y]]/(x^p y - xy^p)$, for every prime $p > 0$.

3.4 Automorphisms and Exponential Functors

The main theme of this section is to define, for every morphism $S \to R$ of commutative unitary $\mathbb{K}$-algebras, a group functor $\mathrm{Aut}_{R/S}$ and prove that, in characteristic 0, it is obtained by formal integration of the Lie algebra of S-linear derivations of R. The reason for this investigation lies in the fact that many geometrically interesting deformation functors arise from the construction of Corollary 3.2.9 in which the involved group functors are, up to some easy and innocuous generalization, of type $\mathrm{Aut}_{R/S}$.

Let $S \xrightarrow{\imath} R$ be a morphism of commutative unitary $\mathbb{K}$-algebras, then for every $A \in \mathbf{Art}$ we have two natural morphisms of $\mathbb{K}$-algebras

$$S \otimes A \xrightarrow{\imath \otimes \mathrm{Id}_A} R \otimes A, \qquad R \otimes A \xrightarrow{\mathrm{Id}_R \otimes \pi} R \otimes \mathbb{K} = R,$$

where $\pi : A \to \mathbb{K}$ is the projection onto the residue field. We denote by $\mathrm{Aut}_{R/S}(A)$ the subgroup of isomorphisms of $\mathbb{K}$-algebras $f : R \otimes A \to R \otimes A$ making the diagram

$$\begin{array}{ccc} S \otimes A & \xrightarrow{\imath \otimes \mathrm{Id}_A} & R \otimes A \\ {\scriptstyle \imath \otimes \mathrm{Id}_A} \downarrow & \overset{f}{\nearrow} & \downarrow {\scriptstyle \mathrm{Id}_R \otimes \pi} \\ R \otimes A & \xrightarrow[\mathrm{Id}_R \otimes \pi]{} & R \end{array} \qquad (3.2)$$

commutative. Whenever $S = \mathbb{K}$ we simply write $\mathrm{Aut}_R(A) = \mathrm{Aut}_{R/\mathbb{K}}(A)$.

Since S and R are unitary, the commutativity of (3.2) implies that f is also a morphism of $S \otimes A$-algebras and a fortiori also a morphism of A-algebras. In particular, for every morphism $A \to B$ in $\mathbf{Art}$, the morphism

$$f \otimes_A \mathrm{Id}_B : (R \otimes A) \otimes_A B = R \otimes B \to R \otimes B$$

is well defined and belongs to $\mathrm{Aut}_{R/S}(B)$.

Definition 3.4.1 The **automorphisms functor** of a morphism of unitary commutative $\mathbb{K}$-algebras $S \to R$ is the group functor $\mathrm{Aut}_{R/S} : \mathbf{Art} \to \mathbf{Grp}$.

For every $A \in \mathbf{Art}$ the kernel of the surjective morphism $R \otimes A \xrightarrow{\mathrm{Id}_R \otimes \pi} R$ is the nilpotent ideal $R \otimes \mathfrak{m}_A$ and therefore, if $f : R \otimes A \to R \otimes A$ is a morphism of $\mathbb{K}$-algebras such that $(\mathrm{Id}_R \otimes \pi) f = \mathrm{Id}_R \otimes \pi$, then f is an isomorphism. The universal property of the fibre product implies immediately that the functor $\mathrm{Aut}_{R/S}$ is homogeneous.

For a morphism $S \xrightarrow{\imath} R$ as above, we denote by $\mathrm{Der}_S(R, R)$ the R-module of S-linear derivations of R:

$$\begin{aligned} \mathrm{Der}_S(R, R) &= \{d \in \mathrm{Der}_{\mathbb{K}}(R, R) \mid d \circ \imath = 0\} \\ &= \{d \in \mathrm{Hom}_S(R, R) \mid d(xy) = (dx)y + x(dy),\ \forall\, x, y \in R\}. \end{aligned}$$

For every unitary commutative $\mathbb{K}$-algebra A there exists a natural injective morphism of $R \otimes A$-modules

$$u\colon \mathrm{Der}_S(R, R) \otimes A \to \mathrm{Der}_{S\otimes A}(R \otimes A, R \otimes A)$$
$$u(d \otimes a)(x \otimes b) = dx \otimes ab, \qquad d \in \mathrm{Der}_S(R, R),\ x \in R,\ a, b \in A. \tag{3.3}$$

Every element $\phi \in \mathrm{Der}_S(R, R) \otimes A$ can be written as a finite sum $\phi = \sum_{i=1}^n d_i \otimes a_i$ with every $d_i \in \mathrm{Der}_S(R, R)$ and $a_1, \ldots, a_n \in A$ linearly independent over the field $\mathbb{K}$. If $u(\phi) = 0$, then for every $r \in R$ we have

$$0 = u(\phi)(r \otimes 1) = \sum_i d_i(r) \otimes a_i$$

and then $d_1 = \cdots = d_n = 0$. If A is a finite-dimensional $\mathbb{K}$-vector space, for instance if $A \in \mathbf{Art}$, then u is an isomorphism. In fact, if $a_1, \ldots, a_n \in A$ is any linear basis, then every morphism $\psi \in \mathrm{Der}_{S\otimes A}(R \otimes A, R \otimes A)$ determines a sequence of derivations $\psi_1, \ldots, \psi_n \in \mathrm{Der}_S(R, R)$ by the formula

$$\psi(r \otimes 1) = \sum_i \psi_i(r) \otimes a_i,$$

and it is clear that $u(\sum \psi_i \otimes a_i) = \psi$.

Lemma 3.4.2 *Let $S \to R$ be a morphism of unitary commutative $\mathbb{K}$-algebras and $A \in$ **Art**. Then for every ideal $J \subset A$ the morphism u defined in* (3.3) *gives an isomorphism of $R \otimes A$-modules*

$$u\colon Der_S(R, R) \otimes J \xrightarrow{\simeq} \{\phi \in Der_{S\otimes A}(R \otimes A, R \otimes A) \mid \phi(R \otimes A) \subset R \otimes J\}.$$

Proof It is sufficient to choose a linear basis $a_1, \ldots, a_n \in A$ such that $a_1, \ldots, a_m$ is a basis of J and repeat the previous proof of the surjectivity of u. □

Assume now that $\mathbb{K}$ is a field of characteristic 0, then for every $A \in$ **Art** and every $\phi \in \mathrm{Der}_S(R, R) \otimes \mathfrak{m}_A$ the derivation $u(\phi)$ is nilpotent and then, according to Lemma 2.1.2, its exponential is an isomorphism of $S \otimes A$-algebras:

$$e^{u(\phi)} := \sum_{n\geq 0} \frac{u(\phi)^n}{n!} \in \mathrm{Aut}_{R/S}(A), \qquad (e^{u(\phi)})^{-1} = e^{-u(\phi)}.$$

Moreover, since $\{\phi \in \mathrm{Der}_{S\otimes A}(R \otimes A, R \otimes A) \mid \phi(R \otimes A) \subset R \otimes \mathfrak{m}_A\}$ is contained in the nilpotent ideal $I = \{\phi \in \mathrm{Hom}_A(R \otimes A, R \otimes A) \mid \phi(R \otimes A) \subset R \otimes \mathfrak{m}_A\}$ of the associative algebra of A-linear endomorphisms of $R \otimes A$, by the results of Sect. 2.5 we have

$$e^{u(\phi)}e^{u(\psi)} = e^{u(\phi\bullet\psi)},$$

where $\bullet$ is the Baker–Campbell–Hausdorff product on the nilpotent Lie algebra $\mathrm{Der}_S(R, R) \otimes \mathfrak{m}_A$. In other words, the map

$$\exp(\mathrm{Der}_S(R, R) \otimes \mathfrak{m}_A) \to \mathrm{Aut}_{R/S}(A), \qquad e^{\phi} \mapsto e^{u(\phi)},$$

is a morphism of groups.

Proposition 3.4.3 *In the above setup, and with the field $\mathbb{K}$ of characteristic 0, for every $A \in$ **Art** the exponential*

$$\exp\colon Der_S(R, R) \otimes \mathfrak{m}_A \to \mathrm{Aut}_{R/S}(A), \qquad \exp(\phi) := e^{u(\phi)},$$

is bijective. In particular, the functor $\mathrm{Aut}_{R/S}$ *is smooth.*

Proof This is obvious if $A = \mathbb{K}$; by induction on the length of A we may assume that there exists a small extension

$$0 \to J \to A \xrightarrow{\alpha} B \to 0$$

such that $\exp\colon \mathrm{Der}_S(R, R) \otimes \mathfrak{m}_B \to \mathrm{Aut}_{R/S}(B)$ is bijective.

We first note that if $d \in u(\mathrm{Der}_S(R, R) \otimes \mathfrak{m}_A)$ and $h \in u(\mathrm{Der}_S(R, R) \otimes J)$ then $d^i h^j = h^j d^i = 0$ whenever $j > 0$, $j + i \geq 2$. Therefore $e^{d+h} = e^d + h$ and this implies that exp is injective.

Conversely, given $f \in \mathrm{Aut}_{R/S}(A)$, by the inductive assumption there exists a derivation $d \in u(\mathrm{Der}_S(R, R) \otimes \mathfrak{m}_A)$ such that $\alpha(f) = \alpha(e^d) \in \mathrm{Aut}_{R/S}(B)$; denote $h = f - e^d\colon R \otimes A \to R \otimes J$. Since

$$\begin{aligned} h(ab) &= f(a)f(b) - e^d(a)e^d(b) \\ &= f(a)f(b) - e^d(a)f(b) + e^d(a)f(b) - e^d(a)e^d(b) \\ &= h(a)f(b) + e^d(a)h(b) \end{aligned}$$

and J is annihilated by $\mathfrak{m}_A$, we have $h(a)f(b) = h(a)b$ and $e^d(a)h(b) = ah(b)$. Therefore $h \in u(\mathrm{Der}_S(R, R) \otimes J)$ and then $f = e^{d+h}$. □

Definition 3.4.4 The **exponential functor** of a Lie algebra L, over a field $\mathbb{K}$ of characteristic 0, is the functor of Artin rings

$$\exp_L\colon \mathbf{Art} \to \mathbf{Grp}, \qquad \exp_L(A) = \exp(L \otimes \mathfrak{m}_A),$$

where the group structure in $\exp(L \otimes \mathfrak{m}_A)$ is given by the Baker–Campbell–Hausdorff product, as in Definition 2.5.7.

It is plain that, as an abstract functor, the exponential $\exp_L$ is isomorphic to the functor $A \mapsto \widehat{L}(A) = L \otimes \mathfrak{m}_A$ of Example 3.2.2, and therefore $\exp_L$ is a smooth homogeneous deformation functor. On the other hand, every interpretation of a group functor as the exponential functor of a Lie algebra gives an additional geometrical information, and therefore a better understanding of the corresponding deformation problem.

We can therefore rewrite Proposition 3.4.3 in the following form: in characteristic 0 the automorphisms functor $\mathrm{Aut}_{R/S}\colon \mathbf{Art} \to \mathbf{Grp}$ is isomorphic to the exponential functor $\exp_{\mathrm{Der}_S(R,R)}\colon \mathbf{Art} \to \mathbf{Grp}$.

The same argument works, mutatis mutandis, also if $\mathcal{S} \to \mathcal{R}$ is a morphism of sheaves of commutative unitary $\mathbb{K}$-algebras over a topological space. In this case the exponential map gives, for every $A \in \mathbf{Art}$ a bijective map

$$\exp\colon \mathrm{Der}_{\mathcal{S}}(\mathcal{R}, \mathcal{R}) \otimes \mathfrak{m}_A \to \mathrm{Aut}_{\mathcal{R}/\mathcal{S}}(A),$$

where $\mathrm{Der}_{\mathcal{S}}(\mathcal{R}, \mathcal{R})$ is the vector space of $\mathcal{S}$-derivations of $\mathcal{R}$ and $\mathrm{Aut}_{\mathcal{R}/\mathcal{S}}(A)$ is the group of $\mathcal{S} \otimes A$-algebra automorphisms of $\mathcal{R} \otimes A$ lifting the identity on $\mathcal{R}$. For reference purposes we state as a separate lemma a particular case of this situation.

Lemma 3.4.5 *Assume $\mathbb{K} = \mathbb{C}$ and let X be a complex manifold with holomorphic tangent sheaf Θ_X. Then for every $A \in$ **Art** the exponential*

$$\exp\colon H^0(X, \Theta_X) \otimes \mathfrak{m}_A \to \mathrm{Aut}_{\mathcal{O}_X}(A)$$

is a bijection, where $\mathrm{Aut}_{\mathcal{O}_X}(A)$ is the set of automorphisms of sheaves of A-algebras $f\colon \mathcal{O}_X \otimes A \to \mathcal{O}_X \otimes A$ commuting with the projection $\mathcal{O}_X \otimes A \to \mathcal{O}_X \otimes \mathbb{C} = \mathcal{O}_X$.

Proof In every smooth manifold the vector fields are the derivations of the structure sheaf. □

Definition 3.4.1 and Proposition 3.4.3 generalize immediately and without difficulties whenever R is replaced by a diagram of S-algebras. For instance, given a morphism $\phi\colon R \to Q$ of S-algebras, we may consider the space of compatible derivations

$$D = \{(d, \delta) \in \mathrm{Der}_S(R, R) \times \mathrm{Der}_S(Q, Q) \mid \phi d = \delta\phi\}$$

and the group of compatible automorphisms

$$G(A) = \{(f, g) \in \mathrm{Aut}_{R/S}(A) \times \mathrm{Aut}_{Q/S}(A) \mid \phi f = g\phi\}.$$

Then the exponential map $\exp\colon D \otimes \mathfrak{m}_A \to G(A)$ is a bijection. Again, for reference purposes we state a particular case of this fact as a separate lemma.

Lemma 3.4.6 *Assume $\mathbb{K} = \mathbb{C}$ and let X be a complex manifold with holomorphic tangent sheaf Θ_X. Let $Z \subset X$ be a complex submanifold; denote by $\mathcal{I} \subset \mathcal{O}_X$ the ideal sheaf of Z and by $\Theta_X(-\log Z)$ the subsheaf of vector fields in X that are tangent to Z. Then for every $A \in$ **Art** the exponential*

$$\exp\colon H^0(X, \Theta_X(-\log Z)) \otimes \mathfrak{m}_A \to \{f \in \mathrm{Aut}_{\mathcal{O}_X}(A) \mid f(\mathcal{I} \otimes A) \subset \mathcal{I} \otimes A\}$$

is a bijection.

Proof The sections of $\Theta_X(-\log Z)$ are the derivations of $\mathcal{O}_X$ preserving the ideal $\mathcal{I}$. Then we can apply the above argument to the morphism of sheaves of algebras $\mathcal{O}_X \to \mathcal{O}_Z = \mathcal{O}_X/\mathcal{I}$. □

3.5 Tangent Space and Schlessinger's Theorem

The Zariski tangent space of an analytic singularity $(X, 0)$ is defined as the set of morphisms $(\mathrm{Spec}(\mathbb{C}[\varepsilon]/(\varepsilon^2)), 0) \to (X, 0)$ and then it can be recovered from the functor of Artin rings $\mathbf{Art}_{\mathbb{C}} \to \mathbf{Set}$,

$$A \mapsto \{\text{morphisms of analytic singularities } \phi\colon (\mathrm{Spec}(A), 0) \to (X, 0)\}.$$

Consequently, the Zariski tangent space of the base space of the semiuniversal deformation of a complex manifold X can be computed as the set of isomorphism classes of deformations of X over $\mathrm{Spec}(\mathbb{C}[\epsilon]/(\epsilon^2))$. This gives enough motivation for the following definition.

Definition 3.5.1 Let $F\colon \mathbf{Art} \to \mathbf{Set}$ be a deformation functor. The set

$$T^1F = F\left(\frac{\mathbb{K}[\varepsilon]}{(\varepsilon^2)}\right)$$

is called the **tangent space** of F.

Although the definition of T^1F makes sense for every functor of Artin rings, it happens that it is a useful notion only under suitable additional assumptions, that are satisfied in particular by every deformation functor.

Proposition 3.5.2 *The tangent space of a deformation functor has a natural structure of a vector space. For every natural transformation of deformation functors $F \to G$, the induced map $T^1F \to T^1G$ is linear.*

Proof Since $F(\mathbb{K})$ is just one point, by Condition (1) of Definition 3.2.5, there exists a natural bijection

$$F\left(\frac{\mathbb{K}[\varepsilon]}{(\varepsilon^2)} \times_{\mathbb{K}} \frac{\mathbb{K}[\varepsilon]}{(\varepsilon^2)}\right) \xrightarrow{\simeq} F\left(\frac{\mathbb{K}[\varepsilon]}{(\varepsilon^2)}\right) \times_{F(\mathbb{K})} F\left(\frac{\mathbb{K}[\varepsilon]}{(\varepsilon^2)}\right) = T^1F \times T^1F. \qquad (3.4)$$

The morphism of Artin rings

$$+\colon \frac{\mathbb{K}[\varepsilon]}{(\varepsilon^2)} \times_{\mathbb{K}} \frac{\mathbb{K}[\varepsilon]}{(\varepsilon^2)} \to \frac{\mathbb{K}[\varepsilon]}{(\varepsilon^2)}, \qquad (a + b\varepsilon, a + b'\varepsilon) \longmapsto a + (b + b')\varepsilon,$$

induces a map

$$+\colon F\left(\frac{\mathbb{K}[\varepsilon]}{(\varepsilon^2)}\times_{\mathbb{K}}\frac{\mathbb{K}[\varepsilon]}{(\varepsilon^2)}\right)\to F\left(\frac{\mathbb{K}[\varepsilon]}{(\varepsilon^2)}\right)=T^1F,$$

and its composition with the inverse of (3.4) defines the addition on the tangent space:

$$T^1F\times T^1F\xrightarrow{\cong}F\left(\frac{\mathbb{K}[\varepsilon]}{(\varepsilon^2)}\times_{\mathbb{K}}\frac{\mathbb{K}[\varepsilon]}{(\varepsilon^2)}\right)\xrightarrow{+}T^1F.$$

Analogously, the multiplication by a scalar $k\in\mathbb{K}$ is induced by the morphism of Artin rings

$$k\colon\frac{\mathbb{K}[\varepsilon]}{(\varepsilon^2)}\to\frac{\mathbb{K}[\varepsilon]}{(\varepsilon^2)},\qquad a+b\varepsilon\longmapsto a+(kb)\varepsilon.$$

It is easy to see that the axioms of vector space are satisfied. The trivial vector $0\in T^1F$ is the image of the singleton under the map $F(\mathbb{K})\to T^1F$ induced by the unique morphism of unitary algebras $\mathbb{K}\to\mathbb{K}[\varepsilon]/(\varepsilon^2)$. The linearity of the map $T^1F\to T^1G$ induced by a natural transformation of deformation functors $F\to G$ follows from the definition of the $\mathbb{K}$-vector space structure on T^1F and T^1G. □

In the above proposition we have adopted the usual notational convention that *ε is very small and different from* 0; in the framework of deformation theory this means that the symbols $\varepsilon,\varepsilon_1,\varepsilon_2,\ldots$ are used to denote square zero elements of Artin rings annihilated by the maximal ideals; using this convention we may simply write $T^1F=F(\mathbb{K}[\varepsilon])$.

Example 3.5.3 The tangent space of the functor h_R, defined in Example 3.2.4, is

$$T^1h_R=\operatorname{Mor}_{\mathcal{C}_{\mathbb{K}}}(R,\mathbb{K}[\varepsilon])\simeq\operatorname{Hom}_{\mathbb{K}}\left(\frac{\mathfrak{m}_R}{\mathfrak{m}_R^2},\mathbb{K}\right).$$

Therefore T^1h_R is isomorphic to the Zariski tangent space of $\operatorname{Spec}(R)$ at its closed point.

Lemma 3.5.4 *Let $\eta\colon F\to G$ be a natural transformation of deformation functors:*

1. *if G is homogeneous and $\eta\colon T^1F\to T^1G$ is injective, then also F is homogeneous and $\eta\colon F(A)\to G(A)$ is injective for every A;*
2. *if F is smooth and $\eta\colon T^1F\to T^1G$ is surjective, then also G is smooth and η is a smooth natural transformation;*
3. *if $\eta\colon T^1F\to T^1G$ is surjective and $\mu\colon G\to H$ is a natural transformation of deformation functors such that the composition $\mu\eta$ is smooth, then η and μ are smooth.*

Proof For notational simplicity, for every $A\in\mathbf{Art}$ and $a\in A$ we shall denote by $\overline{a}\in\mathbb{K}$ the image of a under the projection onto the residue field. For every small principal extension $0 \to \mathbb{K} \xrightarrow{\alpha} B \xrightarrow{\beta} A \to 0$ there exists an isomorphism of Artin rings

$$B \times_{\mathbb{K}} \mathbb{K}[\varepsilon] \xrightarrow{\simeq} B \times_A B, \qquad (b, \overline{b} + k\varepsilon) \mapsto (b, b + \alpha(k))) \tag{3.5}$$

that, for every deformation functor G, gives a surjective map

$$\theta \colon G(B) \times T^1G = G(B \times_{\mathbb{K}} \mathbb{K}[\epsilon]) \to G(B) \times_{G(A)} G(B)$$

commuting with the projections on the first factors; if G is homogeneous, then θ is bijective. Moreover, $\theta(x, 0) = (x, x)$ since the composition of the isomorphism (3.5) with the map $B \to B \times_{\mathbb{K}} \mathbb{K}[\varepsilon]$, $b \mapsto (b, \overline{b})$, is the diagonal map.

Assume now that G is homogeneous and that $\eta \colon T^1F \to T^1G$ is injective. We shall prove by induction on the length of $B \in$ **Art** that $\eta \colon F(B) \to G(B)$ is injective. Let $x, y \in F(B)$ such that $\eta(x) = \eta(y) \in G(B)$ and let

$$0 \to \mathbb{K} \xrightarrow{\alpha} B \xrightarrow{\beta} A \to 0 \tag{3.6}$$

be a principal small extension. By induction $\beta(x) = \beta(y) \in F(A)$ and then there exists $v \in T^1F$ such that $\theta(x, v) = (x, y)$. Thus $\theta(\eta(x), \eta(v)) = (\eta(x), \eta(y))$ belongs to the diagonal and, since G is homogeneous this implies $\eta(v) = 0$ and therefore $v = 0$, $x = y$. This proves that η is always injective; the homogeneity of F is clear.

Assume now $\eta \colon T^1F \to T^1G$ surjective and $\mu\eta$ smooth for some natural transformation $\mu \colon G \to H$. We first prove that η is smooth, i.e., that for every principal small extension as in (3.6), the map

$$(\beta, \eta) \colon F(B) \to F(A) \times_{G(A)} G(B)$$

is surjective. Let $(x, y) \in F(A) \times_{G(A)} G(B)$; since $\mu\eta \colon F \to H$ is smooth the pair $(x, \mu(y)) \in F(A) \times_{H(A)} H(B)$ lifts to an element $z \in F(B)$; in particular, $\beta(z) = x$. Writing $w = \eta(z)$ we have $(w, y) \in G(B) \times_{G(A)} G(B)$ and then there exists $v \in T^1G$ such that $\theta(w, v) = (w, y)$. Now $\eta \colon T^1F \to T^1G$ is surjective and then $v = \eta(u)$ for some $u \in T^1F$. If $\theta(z, u) = (z, r)$, then the element $r \in F(A)$ satisfies $\beta(r) = \beta(z) = x$ and $\eta(r) = y$; this proves the smoothness of η. The smoothness of $\mu \colon G \to H$ is a simple formal consequence of the smoothness of $\mu\eta \colon F \to H$ and the surjectivity of η, cf. Exercise 3.9.2.

Finally, the second item is exactly the third in the particular case $H = 0$. □

Definition 3.5.5 We shall say that a natural transformation $\eta \colon F \to G$ of deformation functors is a **weak equivalence** if it is smooth and bijective on tangent spaces.

Every isomorphism is a weak equivalence; by Lemma 3.5.4, if G is homogeneous then every weak equivalence $F \to G$ of deformation functors is an isomorphism.

Example 3.5.6 Corollary 3.2.9 gives an easy way to produce examples of weak equivalences that are not isomorphisms. Assume there are given a functor of Artin rings $F\colon \mathbf{Art} \to \mathbf{Set}$ and a group functor of Artin rings $G\colon \mathbf{Art} \to \mathbf{Grp}$ acting on F. We have already proved that if F and G are both deformation functors and G is smooth, then F/G is a deformation functor and the natural projection $p\colon F \to F/G$ is smooth. Clearly, p is an isomorphism if and only if G acts trivially on F, while p is a weak equivalence if and only if T^1G acts trivially on T^1F. Consider for instance the following smooth homogeneous functors:

$$F(A) = \left\{ \begin{pmatrix} a & b \\ 0 & 0 \end{pmatrix} \;\middle|\; a, b \in \mathfrak{m}_A \right\}, \quad G(A) = \left\{ \begin{pmatrix} 1 & t \\ 0 & 1 \end{pmatrix} \;\middle|\; t \in \mathfrak{m}_A \right\}, \qquad A \in \mathbf{Art},$$

The smooth group functor G acts by conjugation on F, so that $F/G = F/\sim$, where

$$\begin{pmatrix} a & b \\ 0 & 0 \end{pmatrix} \sim \begin{pmatrix} 1 & t \\ 0 & 1 \end{pmatrix} \begin{pmatrix} a & b \\ 0 & 0 \end{pmatrix} \begin{pmatrix} 1 & -t \\ 0 & 1 \end{pmatrix} = \begin{pmatrix} a & b - ta \\ 0 & 0 \end{pmatrix}, \quad t \in \mathfrak{m}_A.$$

Thus $G(A)$ acts trivially on $F(A)$ if and only if $\mathfrak{m}_A^2 = 0$ and then the projection $p\colon F \to F/G$ is a weak equivalence but not an isomorphism. According to Lemma 3.5.4 the functor F/G is not homogeneous.

Although weak equivalences are not isomorphisms in general, they satisfy the weaker version of invertibility described in the next proposition.

Proposition 3.5.7 (2 out of 6 property for weak equivalences) *Let*

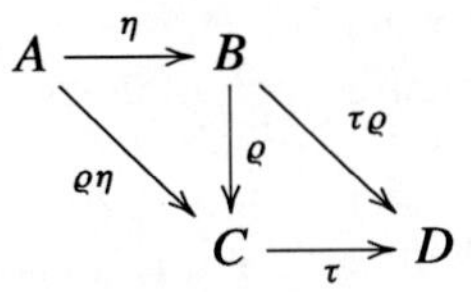

be a commutative diagram of deformation functors. If $\varrho\eta$ and $\tau\varrho$ are weak equivalences then also then also η, ϱ, τ and $\tau\varrho\eta$ are weak equivalences.

Proof It is clear that ϱ is injective and surjective on tangent spaces, and therefore also η and τ are isomorphisms of tangent spaces. Since $\varrho\eta$ is smooth, by Lemma 3.5.4 also the morphisms η and ϱ are smooth. The natural transformation $\tau\varrho\eta$ is smooth because it is the composition of the smooth morphism $\tau\varrho$ and η. Finally, the smoothness of τ is a formal consequence of the smoothness of $\tau\varrho$ and ϱ. □

Definition 3.5.8 A **hull** of a deformation functor F is the data of a pro-representable functor h_R and a weak equivalence $h_R \to F$.

Lemma 3.5.9 *The hull of a deformation functor (if it exists) is unique up to (non-canonical) isomorphisms.*

Proof We need to prove that if F is a deformation functor, then for every pair of hulls $\xi\colon h_R \to F$, $\eta\colon h_S \to F$ there exists an isomorphism of functors $\mu\colon h_R \to h_S$

such that $\eta\mu = \xi$. According to Lemma 3.3.3, there exists a natural transformation $\mu: h_R \to h_S$ such that $\eta\mu = \xi$. The morphism μ is bijective on tangent spaces and h_S is homogenous; according to Lemma 3.5.4 the morphism μ is smooth and injective, hence an isomorphism. □

Unfortunately, the majority of deformation functors arising in concrete situations are not pro-representable. However, a celebrated theorem by Schlessinger asserts that every deformation functor admits a hull, provided that its tangent space is finite-dimensional.

Theorem 3.5.10 (Schlessinger) *Let F be a deformation functor with finite dimensional tangent vector space T^1F. Then, there exists a local complete noetherian $\mathbb{K}$-algebra R with residue field $\mathbb{K}$ and a smooth natural transformation $h_R \to F$ inducing an isomorphism on tangent spaces $T^1h_R = T^1F$.*

A proof of Schlessinger's theorem will be given in Sect. B.1 as a simple consequence of a result regarding obstructions (the approximation theorem), while in Sect. 13.2 we shall prove a slightly stronger version, called the *formal Kuranishi theorem*, under the additional assumption that F is the deformation functor associated to a differential graded Lie algebra, or more generally to an L_∞-algebra.

A Homotopical Interpretation of Schlessinger's Theorem

In order to become familiar with some of the ideas developed later on this book, it may be useful to give an equivalent formulation of Schlessinger's theorem by using the language of abstract homotopy theory. However, this subsection is not necessary for the comprehension of the remaining part of the book and its reading can be safely omitted.

Definition 3.5.11 A **homotopical category** is a category equipped with a distinguished class $\mathcal{W}$ of morphisms, called **weak equivalences**, such that:

1. for every object, the identity on it is a weak equivalence;
2. (2 out of 6 property) given a commutative diagram

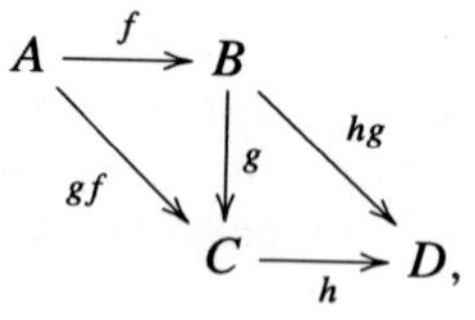

if $gf, hg \in \mathcal{W}$, then also $f, g, h, hgf \in \mathcal{W}$.

Two objects in a homotopical category are said to be **weakly (homotopy) equivalent** if they are equivalent under the equivalence relation generated by weak equivalences.

In other words, two objects a and b are weakly equivalent if and only if there exists a zigzag of finite length

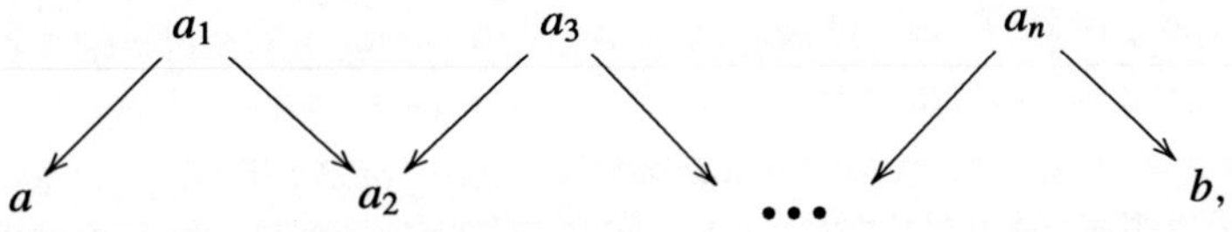

where every arrow is a weak equivalence.

Lemma 3.5.12 *Let $\mathcal{W}$ be the class of weak equivalences in a homotopy category. Then:*

1. *every isomorphism belongs to $\mathcal{W}$;*
2. *(2 out of 3 property) given a diagram $A \xrightarrow{f} B \xrightarrow{g} C$, if two of f, g, gf belong to $\mathcal{W}$, then also the third belongs to $\mathcal{W}$.*

Proof If $f\colon A \to B$ is an isomorphism, the 2 out of 6 property applied to the diagram

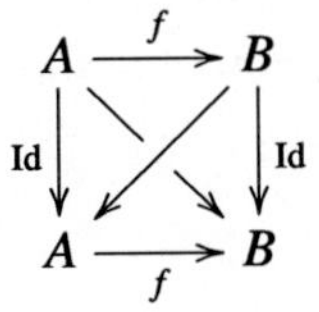

gives $f, f^{-1} \in \mathcal{W}$. Given $A \xrightarrow{f} B \xrightarrow{g} C$, the 2 out of 6 property applied to the diagrams

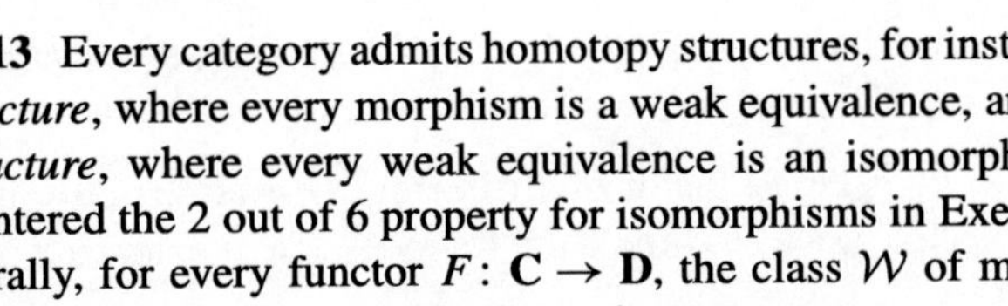

gives the 2 out of 3 property. □

Example 3.5.13 Every category admits homotopy structures, for instance the *trivial homotopy structure*, where every morphism is a weak equivalence, and the *minimal homotopy structure*, where every weak equivalence is an isomorphism. We have already encountered the 2 out of 6 property for isomorphisms in Exercise 1.9.1.

More generally, for every functor $F\colon \mathbf{C} \to \mathbf{D}$, the class $\mathcal{W}$ of morphisms in $\mathbf{C}$ such that $F(f)$ is an isomorphism in $\mathbf{D}$ gives a homotopy structure on $\mathbf{C}$.

The most classical example of nontrivial and nonminimal homotopy structure is given by the class of homotopy equivalences in the category of topological spaces. Another classical example is given by the category of simply connected topological spaces and weak homotopy equivalences; a weak homotopy equivalence is defined as a continuous map inducing isomorphisms between homotopy groups.

The weak equivalences introduced in Definition 3.5.5 give a homotopy structure in the category of deformation functors, where the 2 out of 6 property is proved in Proposition 3.5.7.

Theorem 3.5.14 *Consider the homotopical category of deformation functors, with weak equivalences as in Definition 3.5.5. Then every deformation functors with finite-dimensional tangent space is weakly equivalent to a pro-representable functor. Two pro-representable functors are weakly equivalent if and only if are isomorphic.*

Proof The first part is a direct consequence of Theorem 3.5.10. Assume now that the two functors h_R and h_S are homotopy equivalent, with the equivalence relation given by a zigzag

$$G_0 = h_R \leftarrow F_0 \rightarrow G_1 \leftarrow F_2 \rightarrow \cdots \rightarrow G_n \leftarrow F_n \rightarrow h_S = G_{n+1}$$

of weak equivalences. Since, for every i, the functors F_i, G_i have finite-dimensional tangent space, by Schlessinger's theorem there exist weak equivalences $F_i' \to F_i$, $G_i' \to G_i$, with F_i', G_i' pro-representable functors. By Lemma 3.3.3 every diagram $G_i \leftarrow F_i \rightarrow G_{i+1}$ extends to a diagram of weak equivalences

$$\begin{array}{ccccc} G_i' & \longleftarrow & F_i' & \longrightarrow & G_{i+1}' \\ \downarrow & & \downarrow & & \downarrow \\ G_i & \longleftarrow & F_i & \longrightarrow & G_{i+1} \end{array}$$

and the conclusion follows from the fact that every weak equivalence between pro-representable functors is an isomorphism. □

3.6 Obstruction Theory

We are now ready to talk about obstructions; in the framework of functors of Artin rings, by the term obstructions we intend *obstructions to the smoothness of a deformation functor.*

Definition 3.6.1 Let F be a functor of Artin rings. An **obstruction theory** for F with values in a vector space V is the data, for every small extension in **Art**

$$e\colon \quad 0 \to M \to B \to A \to 0,$$

of an **obstruction map** $v_e\colon F(A) \to V \otimes M$. The obstruction maps should have the *base change property* with respect to morphisms of small extensions: this means that for every morphism of small extensions

$$\begin{array}{llllllllll}
e_1\colon & 0 & \longrightarrow & M_1 & \longrightarrow & B_1 & \longrightarrow & A_1 & \longrightarrow & 0\\
 & & & \downarrow{\scriptstyle \mu} & & \downarrow{\scriptstyle \beta} & & \downarrow{\scriptstyle \alpha} & & \\
e_2\colon & 0 & \longrightarrow & M_2 & \longrightarrow & B_2 & \longrightarrow & A_2 & \longrightarrow & 0
\end{array}$$

we have $v_{e_2}(\alpha(a)) = (\mathrm{Id}_V \otimes \mu)(v_{e_1}(a))$ for every $a \in F(A_1)$.

The name obstruction theory is derived from the following result.

Lemma 3.6.2 *Let (V, v_e) be an obstruction theory for a functor of Artin rings F, and let*

$$e\colon \quad 0 \to M \xrightarrow{\alpha} B \xrightarrow{\beta} A \to 0$$

be a small extension. If an element $a \in F(A)$ lifts to $F(B)$ then $v_e(a) = 0$. In particular, we have $v_e = 0$ whenever there exists a morphism $s\colon A \to B$ in **Art** *such that $\beta s = \mathrm{Id}_A$.*

Proof The vanishing of $v_e(\beta(b))$ for every $b \in F(B)$ is an immediate consequence of the base change property applied to the morphism of small extensions

$$\begin{array}{llllllllll}
 & 0 & \longrightarrow & 0 & \longrightarrow & B & \xrightarrow{\mathrm{Id}} & B & \longrightarrow & 0\\
 & & & \downarrow & & \downarrow{\scriptstyle \mathrm{Id}} & & \downarrow{\scriptstyle \beta} & & \\
e\colon & 0 & \longrightarrow & M & \xrightarrow{\alpha} & B & \xrightarrow{\beta} & A & \longrightarrow & 0.
\end{array}$$

□

Remark 3.6.3 Let $e\colon\ 0 \to M \to B \to A \to 0$ be a small extension and $a \in F(A)$; the obstruction $v_e(a) \in V \otimes M$ is uniquely determined by the values $(\mathrm{Id}_V \otimes f)v_e(a) \in V$, where f varies along a basis of $\mathrm{Hom}_{\mathbb{K}}(M, \mathbb{K})$. On the other hand, by base change we have $(\mathrm{Id}_V \otimes f)v_e(a) = v_{f_*e}(a)$, where f_*e is the push-out extension

$$f_*e\colon \quad 0 \to \mathbb{K} \to \frac{B \oplus \mathbb{K}}{\{(m, -f(m)) \mid m \in M\}} \to A \to 0.$$

This implies that every obstruction theory is uniquely determined by its behaviour on principal small extensions.

Definition 3.6.4 An obstruction theory (V, v_e) for F is called **complete** if the converse of Lemma 3.6.2 holds; i.e., the lifting exists if and only if the obstruction vanishes.

Thus, a functor of Artin rings F is smooth if and only if the trivial obstruction theory $(0, 0)$ is complete. Complete obstruction theories play an essential role when we want to check the smoothness of a natural transformation.

Theorem 3.6.5 (Standard smoothness criterion) *Let $\eta\colon F \to G$ be a natural transformation of deformation functors and let (V, v_e) and (W, w_e) be complete obstruction theories for F and G respectively. Assume that:*

1. *the map $\eta\colon T^1F \to T^1G$ is* surjective*;*
2. *there exists an* injective *linear map $f\colon V \to W$ such that $f \circ v_e = w_e \circ \eta$ for every principal small extension e.*

Then η is smooth.

Proof We need to prove that for every principal small extension

$$e\colon 0 \to \mathbb{K} \xrightarrow{\alpha} B \to A \to 0$$

the map $F(B) \to F(A) \times_{G(A)} G(B)$ is surjective. Fix $(x, y') \in F(A) \times_{G(A)} G(B)$ and denote by $y \in G(A)$ the common image of x and y'. Then $w_e(y) = 0$ because y lifts to $G(B)$, hence $v_e(x) = 0$ by the injectivity of f. By assumption the obstruction theory (V, v_e) is complete and then x lifts to some $x'' \in F(B)$. In general $y'' = \eta(x'')$ is not equal to y'. However, $(y'', y') \in G(B) \times_{G(A)} G(B)$ and therefore there exists a tangent vector $v \in T^1G$ such that $\theta(y'', v) = (y'', y')$ where, as in the proof of Lemma 3.5.4, the map

$$\theta\colon G(B) \times T^1G = G(B \times_{\mathbb{K}} \mathbb{K}[\epsilon]) \to G(B) \times_{G(A)} G(B)$$

is induced by the isomorphism

$$B \times_{\mathbb{K}} \mathbb{K}[\epsilon] \to B \times_A B, \qquad (b, \overline{b} + k\epsilon) \mapsto (b, b + \alpha(k)).$$

By assumption $\eta\colon T^1F \to T^1G$ is surjective and there exists $w \in T^1F$ such that $\eta(w) = v$; setting $\theta(x'', w) = (x'', x')$ we have that x' is a lifting of x that maps to y', as required. □

If a functor of Artin rings F admits a complete obstruction theory (V, v_e) and we embed V in a bigger vector space, then we obtain a "bigger" complete obstruction theory. This leads to the natural problem of looking for minimal complete obstruction theories.

Definition 3.6.6 A **morphism of obstruction theories** $(V, v_e) \to (W, w_e)$ is a morphism of vector spaces $f\colon V \to W$ such that $w_e = (f \otimes \mathrm{Id}_M)v_e$, for every small extension $e\colon 0 \to M \to B \to A \to 0$. An obstruction theory (O_F, ob_e) for F is called **universal** if, for every obstruction theory (V, v_e), there exists a unique morphism of obstruction theories $(O_F, ob_e) \to (V, v_e)$.

It is clear that the universal obstruction theory (O_F, ob_e), if it exists, is unique up to isomorphism and it is uniquely determined by the functor F; the vector space O_F is called the **obstruction space** of F.

Theorem 3.6.7 *Let F be a deformation functor. Then:*

1. *there exists a universal obstruction theory (O_F, ob_e) for F that is complete;*
2. *every element of the obstruction space O_F is of the form $ob_e(a)$, for some principal small extension*

$$e\colon\quad 0 \to \mathbb{K} \to B \to A \to 0$$

and some $a \in F(A)$.

Proof The proof is quite long and for the clarity of exposition it is postponed to Sect. 3.8. □

The first consequence of the above theorem is that every deformation functor admits complete obstruction theories. The second consequence is that a deformation functor F is smooth if and only if the trivial obstruction theory $(0, 0)$ is universal.

Corollary 3.6.8 *Let (V, v_e) be a complete obstruction theory for a deformation functor F. Then the obstruction space O_F is isomorphic to the vector subspace of V generated by all the obstructions arising from principal extensions.*

Proof Denote by $\theta\colon O_F \to V$ the linear map giving the morphism of obstruction theories ensured by the universal property. Then every principal obstruction is contained in the image of θ and, since V is complete, the morphism θ is injective. □

Example 3.6.9 Let R be a local complete noetherian $\mathbb{K}$-algebra with residue field $\mathbb{K}$ and embedding dimension $n = \dim T^1 h_R = \dim \mathfrak{m}_R/\mathfrak{m}_R^2$. Then, we can write $R = P/I$, where $P = \mathbb{K}[[x_1, \ldots, x_n]]$ and $I \subset \mathfrak{m}_P^2$. We claim that

$$T^2 h_R := \operatorname{Hom}_P(I, \mathbb{K}) = \operatorname{Hom}_{\mathbb{K}}\left(\frac{I}{\mathfrak{m}_P I}, \mathbb{K}\right)$$

is the obstruction space of h_R. In fact, for every small extension

$$e\colon\quad 0 \to M \to B \xrightarrow{u} A \to 0$$

and every $\alpha \in h_R(A)$, we can lift the composite morphism $P \to R \xrightarrow{\alpha} A$ to a local morphism of $\mathbb{K}$-algebras $\beta\colon P \to B$, giving the commutative diagram

$$\begin{array}{ccccccccc} 0 & \longrightarrow & I & \longrightarrow & P & \xrightarrow{\pi} & R & \longrightarrow & 0 \\ & & \downarrow{\scriptstyle \beta_{|I}} & & \downarrow{\scriptstyle \beta} & & \downarrow{\scriptstyle \alpha} & & \\ 0 & \longrightarrow & M & \longrightarrow & B & \xrightarrow{u} & A & \longrightarrow & 0. \end{array}$$

Since M is annihilated by the maximal ideal of B we have $\beta(\mathfrak{m}_P I) = 0$ and, since $I \subset \mathfrak{m}_P^2$, the induced map

$$v_e(\alpha) = \beta_{|I} \in \mathrm{Hom}_{\mathbb{K}}\left(\frac{I}{\mathfrak{m}_P I}, M\right) = T^2 h_R \otimes M$$

is independent of the choice of β. Therefore the data $(T^2 h_R, v_e)$ is an obstruction theory for the deformation functor h_R. If $v_e(\alpha) = 0$, then $\beta(I) = 0$ and therefore there exists a morphism $\gamma\colon R \to B$ such that $\beta = \gamma\pi$, $u\beta = u\gamma\pi = \alpha\pi$. The surjectivity of π gives $u\gamma = \alpha$ and this proves that the obstruction theory is complete.

It remains to prove that every element of $T^2 h_R$ is the obstruction of a small principal extension. According to the lemma of Artin–Rees [10, Theorem 10.11] we have $\mathfrak{m}_P^N \cap I \subset \mathfrak{m}_P I$ for some sufficiently large integer N and then there exists a small extension

$$u\colon \quad 0 \to \frac{I}{\mathfrak{m}_P I} \to \frac{P}{\mathfrak{m}_P I + \mathfrak{m}_P^N} \xrightarrow{\pi} \frac{R}{\mathfrak{m}_R^N} \to 0, \qquad N >> 0.$$

It is now clear that every $f\colon \dfrac{I}{\mathfrak{m}_P I} \to \mathbb{K}$ is the obstruction to lifting the projection $R \to R/\mathfrak{m}_R^N$ along the push-out principal small extension $f_* u$.

Let $\phi\colon F \to G$ be a natural transformation of deformation functors. Then, $(O_G, ob_e \circ \phi)$ is an obstruction theory for F; therefore, there exists a unique linear map $ob_\phi\colon O_F \to O_G$ that is compatible with ϕ in the obvious sense.

Corollary 3.6.10 *Let $\phi\colon F \to G$ be a morphism of deformation functors. Then the following conditions are equivalent:*

1. *ϕ is smooth,*
2. *$T^1\phi\colon T^1F \to T^1G$ is surjective and $ob_\phi\colon O_F \to O_G$ is bijective,*
3. *$T^1\phi\colon T^1F \to T^1G$ is surjective and $ob_\phi\colon O_F \to O_G$ is injective.*

Proof In order to avoid confusion we denote by ob_e^F and ob_e^G the obstruction maps for F and G, respectively.

As regards the implication (1) ⇒ (2), since every smooth morphism is surjective, if ϕ is smooth then the induced morphisms $T^1F \to T^1G$ and $O_F \to O_G$ are both surjective. Assume that $ob_\phi(\xi) = 0$ and write $\xi = ob_e^F(x)$, for some $x \in F(A)$ and some small extension $e\colon 0 \to \mathbb{K} \to B \to A \to 0$. Since $ob_e^G(\phi(x)) = 0$, the element x lifts to a pair $(x, y') \in F(A) \times_{G(A)} G(B)$ and then the smoothness of ϕ implies that x lifts to $F(B)$.

The implication (2) ⇒ (3) is clear and the implication (3) ⇒ (1) is a particular case of Theorem 3.6.5. □

Corollary 3.6.11 *Let $\phi\colon F \to G$ be a morphism of deformation functors. Then the following conditions are equivalent:*

1. *ϕ is a weak equivalence,*
2. *the linear maps $T^1\phi\colon T^1F \to T^1G$ and $ob_\phi\colon O_F \to O_G$ are bijective.*

Proof Immediate from Corollary 3.6.10. □

Corollary 3.6.12 *Let F be a deformation functor and let $h_R \to F$ be a smooth natural transformation. Then the dimension of O_F is equal to the minimum number of generators of an ideal I defining R as a quotient of a power series ring, i.e., $R = \mathbb{K}[[x_1, \ldots, x_n]]/I$.*

Proof According to Corollary 3.6.10 the functors F and h_R have the same universal obstruction theory. By the minimality assumption on the number of generators we have $I \subset (x_1, \ldots, x_n)^2$ and the conclusion follows from Example 3.6.9 and Nakayama's lemma applied to the $\mathbb{K}[[x_1, \ldots, x_n]]$-module I. □

3.7 Deformation Functors Associated to Semicosimplicial Lie Algebras

The notion of semicosimplicial Lie algebras, introduced in Sect. 2.6, is a good source of examples of deformation functors; the passage from semicosimplicial Lie algebras to deformation functors is an abstraction of the usual classification of fibre bundles by means of first cohomology sets of sheaves of non-abelian groups.

Consider a semicosimplicial Lie algebra over a field $\mathbb{K}$ of characteristic 0:

$$\mathfrak{g}: \qquad \mathfrak{g}_0 \rightrightarrows \mathfrak{g}_1 \Rrightarrow \mathfrak{g}_2 \;\substack{\longrightarrow\\[-1ex]\longrightarrow\\[-1ex]\longrightarrow\\[-1ex]\longrightarrow}\; \cdots.$$

For every index $i \geq 0$ and every $A \in \mathbf{Art}$, the tensor product $\mathfrak{g}_i \otimes \mathfrak{m}_A$ is a nilpotent Lie algebra. The functor of non-abelian 1-cocycles

$$Z^1_{\mathfrak{g}} \colon \mathbf{Art} \to \mathbf{Set},$$

is defined in a number of equivalent ways:

$$\begin{aligned}
Z^1_{\mathfrak{g}}(A) &= \{e^x \in \exp(\mathfrak{g}_1 \otimes \mathfrak{m}_A) \mid e^{\delta_1(x)} = e^{\delta_2(x)} e^{\delta_0(x)}\}\\
&= \{e^x \in \exp(\mathfrak{g}_1 \otimes \mathfrak{m}_A) \mid e^{\delta_0(x)} e^{-\delta_1(x)} e^{\delta_2(x)} = 1\}\\
&= \exp\{x \in \mathfrak{g}_1 \otimes \mathfrak{m}_A \mid \delta_2(x) \bullet \delta_0(x) = \delta_1(x)\},\\
&= \exp\{x \in \mathfrak{g}_1 \otimes \mathfrak{m}_A \mid \delta_0(x) \bullet (-\delta_1(x)) \bullet \delta_2(x) = 0\},
\end{aligned}$$

where $\bullet$ is the Baker–Campbell–Hausdorff product (Definition 2.5.1). It is clear that $Z^1_{\mathfrak{g}}$ commutes with fibre products in **Art** and then it is a homogeneous deformation functor.

If $\varepsilon^2 = 0$, then the Lie algebras $\mathfrak{g}_i \otimes \mathbb{K}\varepsilon$ are abelian and then the tangent space $T^1 Z^1_{\mathfrak{g}}$ can be described in terms of the differential $\delta = \sum(-1)^i \delta_i$ on the cochain complex $C(\mathfrak{g})$ (Definition 2.6.2) by the formula:

$$\begin{aligned} T^1 Z^1_{\mathfrak{g}} = Z^1_{\mathfrak{g}}(\mathbb{K}[\varepsilon]) &= \{e^x \in \exp(\mathfrak{g}_1 \otimes \mathbb{K}\varepsilon) \mid e^{\delta_0(x)} e^{-\delta_1(x)} e^{\delta_2(x)} = 1\} \\ &= \{x \in \mathfrak{g}_1 \mid \delta(x) = \delta_0(x) - \delta_1(x) + \delta_2(x) = 0\} \otimes \mathbb{K}\varepsilon \\ &= Z^1(C(\mathfrak{g})) \otimes \mathbb{K}\varepsilon \cong Z^1(C(\mathfrak{g})). \end{aligned}$$

Example 3.7.1 (Primary obstruction of $Z^1_{\mathfrak{g}}$) Before going further in the general study of the functor $Z^1_{\mathfrak{g}}$ it is instructive to give an explicit description of $Z^1_{\mathfrak{g}}(\mathbb{K}[t]/(t^3))$. For a generic element $x = x_1 t + x_2 t^2 \in \mathfrak{g}_1 \otimes \mathfrak{m}_{\mathbb{K}[t]/(t^3)}$, the BCH formula gives

$$\delta_2(x) \bullet \delta_0(x) = \delta_2 x_1 t + \delta_2 x_2 t^2 + \delta_0 x_1 t + \delta_0 x_2 t^2 + \frac{1}{2}[\delta_2 x_1, \delta_0 x_1] t^2$$

and then the equation $\delta_2(x) \bullet \delta_0(x) - \delta_1(x) = 0$ is equivalent to the system of two equations

$$\delta x_1 = 0, \qquad \delta x_2 + \frac{1}{2}[\delta_2 x_1, \delta_0 x_1] = 0.$$

In particular, given an element $x_1 \in Z^1(C(\mathfrak{g}))$, there exists $x_2 \in \mathfrak{g}_1$ such that $e^{x_1 t + x_2 t^2} \in Z^1_{\mathfrak{g}}(\mathbb{K}[t]/(t^3))$ if and only if the Lie-cup product $x_1 \cup x_1$ (Definition 2.6.5) vanishes in $H^2(C(\mathfrak{g}))$.

Our next goal is to determine a complete obstruction theory for the functor $Z^1_{\mathfrak{g}}$. Let

$$0 \to J \to B \xrightarrow{\alpha} A \to 0$$

be a small extension in **Art** and $e^y \in Z^1_{\mathfrak{g}}(A)$. Let $x \in \mathfrak{g}_1 \otimes \mathfrak{m}_B$ be any element such that $\alpha(x) = y$, then

$$e^{\delta_0(x)} e^{-\delta_1(x)} e^{\delta_2(x)} = e^r, \qquad \text{with} \quad r \in \mathfrak{g}_2 \otimes J.$$

We claim that the above defined element r is a cocycle in $C(\mathfrak{g}) \otimes J$; this means that $\delta(r) = 0$, or equivalently that

$$\delta_0(r) - \delta_1(r) + \delta_2(r) = \delta_3(r). \tag{3.7}$$

The proof of equality (3.7) is tedious but completely elementary. First, since J is annihilated by the maximal ideal of B, the element e^r belongs to the centre of the group $\exp(\mathfrak{g}_2 \otimes \mathfrak{m}_B)$, it is invariant under inner automorphisms and therefore

$$e^r = e^{\delta_0(x)} e^{-\delta_1(x)} e^{\delta_2(x)} = e^{\delta_2(x)} e^{\delta_0(x)} e^{-\delta_1(x)} = e^{-\delta_1(x)} e^{\delta_2(x)} e^{\delta_0(x)}. \tag{3.8}$$

Similarly, every element $e^{\delta_i(r)}$ belongs to the centre of $\exp(\mathfrak{g}_3 \otimes \mathfrak{m}_B)$ and then for every i, j, k we have $e^{\delta_i(r)} e^{\pm\delta_j\delta_k(x)} = e^{\pm\delta_j\delta_k(x)} e^{\delta_i(r)}$. Since $\mathfrak{g}_3 \otimes J$ is an abelian Lie algebra, for every $i, j = 0, \ldots, 3$ we have

$$e^{\delta_i(r)} e^{\pm\delta_j(r)} = e^{\delta_i(r) \pm \delta_j(r)},$$

and therefore (3.7) is equivalent to

$$e^{\delta_0(r)}(e^{\delta_1(r)})^{-1}e^{\delta_2(r)} = e^{\delta_3(r)}.$$

By (3.8) and the semicosimplicial identities (2.7) we have:

1. $e^{\delta_0(r)} = e^{\delta_0\delta_2(x)}e^{\delta_0\delta_0(x)}e^{-\delta_0\delta_1(x)}$;
2. $e^{\delta_1(r)} = e^{-\delta_1\delta_1(x)}e^{\delta_1\delta_2(x)}e^{\delta_1\delta_0(x)} = e^{-\delta_1\delta_1(x)}e^{\delta_1\delta_2(x)}e^{\delta_0\delta_0(x)}$;
3. $e^{\delta_2(r)} = e^{\delta_2\delta_0(x)}e^{-\delta_2\delta_1(x)}e^{\delta_2\delta_2(x)} = e^{\delta_0\delta_1(x)}e^{-\delta_1\delta_1(x)}e^{\delta_2\delta_2(x)}$;
4. $e^{\delta_3(r)} = e^{\delta_3\delta_0(x)}e^{-\delta_3\delta_1(x)}e^{\delta_3\delta_2(x)} = e^{\delta_0\delta_2(x)}e^{-\delta_1\delta_2(x)}e^{\delta_2\delta_2(x)}$;

and therefore

$$\begin{aligned}
&e^{\delta_0(r)}(e^{\delta_1(r)})^{-1}e^{\delta_2(r)}\\
&= e^{\delta_0\delta_2(x)}e^{\delta_0\delta_0(x)}e^{-\delta_0\delta_1(x)}(e^{\delta_1(r)})^{-1}e^{\delta_0\delta_1(x)}e^{-\delta_1\delta_1(x)}e^{\delta_2\delta_2(x)}\\
&= e^{\delta_0\delta_2(x)}e^{\delta_0\delta_0(x)}(e^{\delta_1(r)})^{-1}e^{-\delta_1\delta_1(x)}e^{\delta_2\delta_2(x)}\\
&= e^{\delta_0\delta_2(x)}e^{\delta_0\delta_0(x)}e^{-\delta_0\delta_0(x)}e^{-\delta_1\delta_2(x)}e^{\delta_1\delta_1(x)}e^{-\delta_1\delta_1(x)}e^{\delta_2\delta_2(x)}\\
&= e^{\delta_0\delta_2(x)}e^{-\delta_1\delta_2(x)}e^{\delta_2\delta_2(x)}\\
&= e^{\delta_3(r)}.
\end{aligned}$$

Thus we have proved that $\delta(r) = 0$ and then the element r defines a cohomology class $[r] \in H^2(C(\mathfrak{g})) \otimes J$ that depends only on y and is independent of the choice of $x \in \mathfrak{g}_1 \otimes \mathfrak{m}_B$. In fact, any other lifting of y is equal to $x + h$, with $h \in \mathfrak{g}_1 \otimes J$, and since every $e^{\delta_j(h)}$ belongs to the centre of $\exp(\mathfrak{g}_2 \otimes \mathfrak{m}_A)$ we have

$$e^{\delta_0(x+h)}e^{-\delta_1(x+h)}e^{\delta_2(x+h)} = e^{\delta_0(x)}e^{-\delta_1(x)}e^{\delta_2(x)}e^{\delta_0(h)-\delta_1(h)+\delta_2(h)} = e^{r+\delta(h)}$$

and the cocycle $r + \delta(h)$ is cohomologous to r. The following lemma summarizes the above computation.

Lemma 3.7.2 *For every semicosimplicial Lie algebra*

$$\mathfrak{g}: \qquad \mathfrak{g}_0 \rightrightarrows \mathfrak{g}_1 \rightrightarrows \mathfrak{g}_2 \cdots$$

over a field of characteristic 0, the homogeneous deformation functor $Z^1_{\mathfrak{g}}$ has a natural complete obstruction theory $(H^2(C(\mathfrak{g})), v_e)$, where for every small extension

$$u: \qquad 0 \to J \to B \xrightarrow{\alpha} A \to 0$$

the obstruction map $v_u \colon F(A) \to H^2(C(\mathfrak{g})) \otimes J = H^2(C(\mathfrak{g} \otimes J))$ is defined by:

$$v_u(e^y) = [\log(e^{\delta_0(x)}e^{-\delta_1(x)}e^{\delta_2(x)})], \qquad x \in \mathfrak{g}_1 \otimes \mathfrak{m}_B, \quad \alpha(x) = y.$$

Proof We have already proved that $(H^2(C(\mathfrak{g})), v_e)$ defined as above is an obstruction theory; it remains to check the completeness. Let u and e^y be as in the lemma and assume that $v_u(e^y) = 0$; then for any chosen lifting $x \in \mathfrak{g}_1 \otimes \mathfrak{m}_B$ there exists $h \in \mathfrak{g}_1 \otimes J$ such that $e^{\delta_0(x)}e^{-\delta_1(x)}e^{\delta_2(x)} = e^{\delta(h)}$. Setting $z = x - h$ we have $\alpha(z) = y$ and $e^{\delta_0(z)}e^{-\delta_1(z)}e^{\delta_2(z)} = 1$. □

There exists a natural left action of the exponential functor $\exp_{\mathfrak{g}_0}$ on $Z^1_{\mathfrak{g}}$, defined for every $A \in \mathbf{Art}$ by the formula:

$$\exp(\mathfrak{g}_0 \otimes \mathfrak{m}_A) \times Z^1_{\mathfrak{g}}(A) \to Z^1_{\mathfrak{g}}(A), \quad (e^a, e^x) \mapsto e^a * e^x := e^{\delta_1(a)}e^x e^{-\delta_0(a)}.$$

In order to verify that the above formula makes sense, i.e., that $Z^1_{\mathfrak{g}}$ is stable under the action $*$, consider two elements $e^x \in \exp(\mathfrak{g}_1 \otimes \mathfrak{m}_A)$ and $e^a \in \exp(\mathfrak{g}_0 \otimes \mathfrak{m}_A)$. Writing $e^y = e^{\delta_1(a)}e^x e^{-\delta_0(a)}$ we have

$$e^{\delta_0(y)} = e^{\delta_0\delta_1(a)}e^{\delta_0(x)}e^{-\delta_0\delta_0(a)}, \quad e^{-\delta_1(y)} = e^{\delta_1\delta_0(a)}e^{-\delta_1(x)}e^{-\delta_1\delta_1(a)},$$

$$e^{\delta_2(y)} = e^{\delta_2\delta_1(a)}e^{\delta_2(x)}e^{-\delta_2\delta_0(a)},$$

and the semicosimplicial identities give

$$e^{\delta_0(y)}e^{-\delta_1(y)}e^{\delta_2(y)} = e^{\delta_0\delta_1(a)}(e^{\delta_0(x)}e^{-\delta_1(x)}e^{\delta_2(x)})e^{-\delta_0\delta_1(a)}. \tag{3.9}$$

Therefore, $e^x \in Z^1_{\mathfrak{g}}(A)$ if and only if $e^y \in Z^1_{\mathfrak{g}}(A)$ and we can define the quotient functor

$$H^1_{\mathfrak{g}} \colon \mathbf{Art} \to \mathbf{Set}, \qquad H^1_{\mathfrak{g}}(A) = \frac{Z^1_{\mathfrak{g}}(A)}{\exp(\mathfrak{g}_0 \otimes \mathfrak{m}_A)}.$$

Notice that when $\mathfrak{g}_0 = \mathfrak{g}_2 = 0$ we recover in this way the exponential functor of $\mathfrak{g}_1$.

Theorem 3.7.3 *The projection $Z^1_{\mathfrak{g}} \to H^1_{\mathfrak{g}}$ is a smooth morphism of deformation functors. The tangent space of $H^1_{\mathfrak{g}}$ is isomorphic to $H^1(C(\mathfrak{g}))$ and there exists a natural complete obstruction theory with values in $H^2(C(\mathfrak{g}))$.*

Proof The proof that the projection $Z^1_{\mathfrak{g}} \to H^1_{\mathfrak{g}}$ is a smooth morphism of deformation functors follows immediately from Corollary 3.2.9. We have already noticed that

$$T^1Z^1_{\mathfrak{g}} \cong \{x \in \mathfrak{g}_1 \mid \delta_0(x) - \delta_1(x) + \delta_2(x) = 0\} \otimes \mathbb{K}\varepsilon$$

and therefore $T^1H^1_{\mathfrak{g}}$ is naturally isomorphic to

$$H^1_{\mathfrak{g}}(\mathbb{K}[\varepsilon]) = \frac{Z^1_{\mathfrak{g}}(\mathbb{K}[\varepsilon])}{\exp(\mathfrak{g}_0 \otimes \mathbb{K}\varepsilon)} \cong \frac{Z^1(C(\mathfrak{g}))}{\{-\delta_1(a) + \delta_0(a) \mid a \in \mathfrak{g}_0\}} = H^1(C(\mathfrak{g})).$$

Given a small extension

$$u: \qquad 0 \to J \to B \xrightarrow{\alpha} A \to 0$$

and two elements $e^y \in Z^1_{\mathfrak{g}}(A)$, $a \in \mathfrak{g}_0 \otimes \mathfrak{m}_A$, in the notation of Lemma 3.7.2 we have $v_u(e^y) = v_u(e^a * e^y)$. In fact, if $x \in \mathfrak{g}_1 \otimes \mathfrak{m}_B$ lifts y and $b \in \mathfrak{g}_0 \otimes \mathfrak{m}_B$ lifts a, then $e^b * e^x$ lifts $e^a * e^y$ and Eq. (3.9) immediately implies the equality $v_u(e^y) = v_u(e^a * e^y)$. Therefore the obstruction map v_u factors through to a map $v_u : H^1_{\mathfrak{g}}(A) \to H^2(C(\mathfrak{g})) \otimes J$. The completeness of this obstruction theory for the functor $H^1_{\mathfrak{g}}$ is an easy formal consequence of the completeness for the functor $Z^1_{\mathfrak{g}}$ of the same obstruction theory. □

Remark 3.7.4 It is clear from the definition that the functor $H^1_{\mathfrak{g}}$ depends only on the truncation

$$\mathfrak{g}_{[0,2]}: \qquad \mathfrak{g}_0 \rightrightarrows \mathfrak{g}_1 \rightrightarrows\!\!\!\!\to \mathfrak{g}_2,$$

while the obstruction theory depends on the truncation

$$\mathfrak{g}_{[0,3]}: \qquad \mathfrak{g}_0 \rightrightarrows \mathfrak{g}_1 \rightrightarrows\!\!\!\!\to \mathfrak{g}_2 \rightrightarrows\!\!\!\!\rightrightarrows \mathfrak{g}_3.$$

This is explained by the fact that, in general, for a deformation functor a (non-universal) obstruction theory is usually determined by some additional data. The theory of extended deformation functors [178], and more generally derived deformation theory, show that the situation changes when considering functors on the category of differential graded Artin local rings; in this extended setting every deformation functor will have a canonical obstruction theory and every element in the obstruction space is a first order deformation over a suitable graded ring.

Every morphism of semicosimplicial Lie algebras $f : \mathfrak{g} \to \mathfrak{h}$ induces in the obvious way a natural transformation of deformation functors $f : H^1_{\mathfrak{g}} \to H^1_{\mathfrak{h}}$. By the explicit description of the obstruction maps given in Lemma 3.7.2 and Theorem 3.7.3 we get that the natural transformation f commutes with obstruction maps and with the morphism in cohomology $H^2(C(\mathfrak{g})) \to H^2(C(\mathfrak{h}))$.

Corollary 3.7.5 *Let $\mathfrak{g}$ be a semicosimplicial Lie algebra. If $H^1(C(\mathfrak{g})) = 0$ then $H^1_{\mathfrak{g}}$ is trivial. If either $\mathfrak{g}_2$ is abelian or $H^2(C(\mathfrak{g})) = 0$ then $H^1_{\mathfrak{g}}$ is smooth.*

Proof The first statement is a consequence of the standard smoothness criterion applied to the morphism of semicosimplicial Lie algebras $0 \to \mathfrak{g}$. If $\mathfrak{g}_2$ is abelian, then

$$(\delta_0(x)) \bullet (-\delta_1(x)) \bullet (\delta_2(x)) = \delta_0(x) - \delta_1(x) + \delta_2(x),$$

$$Z^1_{\mathfrak{g}}(A) = \{x \in \mathfrak{g}_1 \otimes \mathfrak{m}_A \mid \delta_0(x) - \delta_1(x) + \delta_2(x) = 0\} = Z^1(C(\mathfrak{g})) \otimes \mathfrak{m}_A,$$

and therefore $Z^1_{\mathfrak{g}}$ is a smooth functor. If $H^2(C(\mathfrak{g})) = 0$ then $Z^1_{\mathfrak{g}}$ is smooth because $H^2(C(\mathfrak{g}))$ is a complete obstruction space. □

3.8 Existence of Universal Obstruction Theories

This section is entirely devoted to the proof of Theorem 3.6.7, concerning the existence of complete and universal obstruction theories for deformation functors. As usual, let $\mathbb{K}$ be a fixed field and let $\mathbf{Art} = \mathbf{Art}_{\mathbb{K}}$ be the category of local Artin $\mathbb{K}$-algebras with residue field $\mathbb{K}$.

Definition 3.8.1 For every $A \in \mathbf{Art}$ and every finite-dimensional vector space M let $\mathrm{Ex}(A, M)$ be the set of isomorphism classes of small extensions in $\mathbf{Art}$ of type

$$e\colon\quad 0 \to M \to B \to A \to 0.$$

We shall denote by $0 \in \mathrm{Ex}(A, M)$ the trivial extension $0 \to M \to A \oplus M \to A \to 0$.

Recall that if $f : M \to N$ is a linear map of finite-dimensional vector spaces and $\pi : C \to A$ is a morphism in $\mathbf{Art}$ we have two maps

$$f_*\colon \mathrm{Ex}(A, M) \to \mathrm{Ex}(A, N), \qquad \pi^*\colon \mathrm{Ex}(A, M) \to \mathrm{Ex}(C, M) \tag{3.10}$$

defined as follows: given an extension $e\colon\ 0 \to M \to B \to A \to 0$ in $\mathrm{Ex}(A, M)$, its push-out f_*e is the extension

$$0 \to N \to \frac{B \oplus N}{\{(m, -f(m)) \mid m \in M\}} \to A \to 0,$$

and its pull-back π^*e is the extension

$$0 \to M \to B \times_A C \to C \to 0.$$

Given two small extensions $e_1, e_2 \in \mathrm{Ex}(A, M)$:

$$e_1\colon\ 0 \to M \xrightarrow{i} B \to A \to 0, \quad e_2\colon\ 0 \to M \xrightarrow{j} C \to A \to 0, \tag{3.11}$$

we define their sum $e_1 + e_2 \in \mathrm{Ex}(A, M)$ as the push-out of the extension

$$0 \to M \times M \to B \times_A C \to A \to 0$$

under the linear map $+\colon M \times M \to M, (m, n) \mapsto m + n$. By using the isomorphism

$$\phi\colon \frac{(B \times_A C) \oplus M}{\{(i(n), j(m), -n - m) \mid n, m \in M\}} \to \frac{B \times_A C}{\{(i(m), -j(m)) \mid m \in M\}},$$

$$\phi(b, c, x) = (b + i(x), c),$$

we may write

$$e_1 + e_2 : \quad 0 \to M \xrightarrow{(i,0)} \frac{B \times_A C}{\{(i(m), -j(m)) \mid m \in M\}} \to A \to 0.$$

We leave to the reader the easy verification that there exists a natural structure of a vector space on $\operatorname{Ex}(A, M)$, where the sum $e_1 + e_2$ is defined as above and the scalar multiplication by $t \in \mathbb{K}$ is induced by the push-out under the multiplication by t in M;

$$te_1 = (t\mathrm{Id}_M)_* e_1 : \quad 0 \to M \xrightarrow{(0,\mathrm{Id}_M)} \frac{B \oplus M}{\{(i(m), -tm) \mid m \in M\}} \to A \to 0.$$

Moreover, the maps f_* and π^* of (3.10) are linear.

Definition 3.8.2 Let F be a deformation functor, $A \in \mathbf{Art}$, $a \in F(A)$ and M a finite-dimensional vector space. Define $F(A, M, a) \subset \operatorname{Ex}(A, M)$ as the subset of small extensions $0 \to M \to B \to A \to 0$ such that a lifts to $F(B)$.

As a purely formal consequence of the properties of deformation functors we have that the subsets $F(A, M, a)$ are stable by pull-back and push-out. More precisely:

1. if $A \in \mathbf{Art}$, $a \in F(A)$ and $f : M \to N$ is a morphism of finite-dimensional vector spaces, then $f_* F(A, M, a) \subset F(A, N, a)$;
2. if $\pi : C \to A$ is a morphism in $\mathbf{Art}$, M is a finite-dimensional vector space and $c \in F(C)$, then $\pi^* F(A, M, \pi(c)) = \pi^* \operatorname{Ex}(A, M) \cap F(C, M, c)$.

In particular, the subset $F(A, M, a)$ is a vector subspace of $\operatorname{Ex}(A, M)$. In fact, if $e \in F(A, M, a)$ then $te = (t\mathrm{Id}_M)_* e \in F(A, M, a)$ for every $t \in \mathbb{K}$. If e_1, e_2 as in (3.11) belong to $F(A, M, a)$, the surjectivity of

$$F(B \times_A C) \to F(B) \times_{F(A)} F(C)$$

implies that the small extension $0 \to M \times M \to B \times_A C \to A \to 0$ belongs to $F(A, M \times M, a)$ and therefore also $e_1 + e_2 \in F(A, M, a)$.

Lemma 3.8.3 *Let F be a deformation functor, $A \in \mathbf{Art}$, M a finite-dimensional vector space, $e \in \operatorname{Ex}(A, M)$ and $a \in F(A)$. Then $e \in F(A, M, a)$ if and only if $f_* e \in F(A, \mathbb{K}, a)$ for every linear map $f : M \to \mathbb{K}$.*

Proof Let e be the small extension $0 \to M \to B \xrightarrow{\pi} A \to 0$ and assume that $f_* e \in F(A, \mathbb{K}, a)$ for every linear map $f : M \to \mathbb{K}$; we prove that a lifts to $F(B)$ by induction on $\dim_{\mathbb{K}} M$.

If $M = 0$ there is nothing to prove. If $M \neq 0$ there exists a short exact sequence of vector spaces $0 \to N \xrightarrow{i} M \xrightarrow{f} \mathbb{K} \to 0$ and two morphisms of small extensions

$$\begin{array}{lccccccccc}
e' & 0 & \longrightarrow & N & \longrightarrow & B & \xrightarrow{\pi'} & A' & \longrightarrow & 0 \\
 & & & \downarrow{\scriptstyle i} & & \| & & \downarrow{\scriptstyle \delta} & & \\
e & 0 & \longrightarrow & M & \longrightarrow & B & \xrightarrow{\pi} & A & \longrightarrow & 0 \\
 & & & \downarrow{\scriptstyle f} & & \downarrow{\scriptstyle \pi'} & & \| & & \\
f_*e & 0 & \longrightarrow & \mathbb{K} & \longrightarrow & A' & \xrightarrow{\delta} & A & \longrightarrow & 0.
\end{array}$$

The assumption $f_*e \in F(A, \mathbb{K}, a)$ means that a lifts to some $a' \in F(A')$. By the inductive assumption, in order to conclude the proof it is sufficient to prove that $g_*e' \in F(A', \mathbb{K}, a')$ for every linear map $g\colon N \to \mathbb{K}$. Writing $g = hi$ for some linear map $h\colon M \to \mathbb{K}$ we have $h_*e \in F(A, \mathbb{K}, a)$ and then $\delta^*h_*e \in F(A', \mathbb{K}, a')$. By Remark 3.1.1 we have $i_*e' = \delta^*e$, and therefore

$$g_*e' = h_*i_*e' = h_*\delta^*e = \delta^*h_*e \in F(A', \mathbb{K}, a').$$

□

Lemma 3.8.4 *Let F be a deformation functor and let $f, g : B \to A$ be two morphisms in* **Art**. *Let $b \in F(B)$ be such that $f(b) = g(b) = a \in F(A)$. Then for every finite dimensional vector space M we have*

$$f^* = g^* \colon \frac{\mathrm{Ex}(A, M)}{F(A, M, a)} \hookrightarrow \frac{\mathrm{Ex}(B, M)}{F(B, M, b)}.$$

Proof The injectivity of f^* follows immediately from the equality $f^*F(A, M, a) = f^*\mathrm{Ex}(A, M) \cap F(B, M, b)$. Next, we want to prove that $f^*e - g^*e \in F(B, M, b)$ for every small extension

$$e\colon \quad 0 \to M \xrightarrow{i} C \xrightarrow{p} A \to 0.$$

To this end, consider the small extension

$$u\colon \quad 0 \to M \xrightarrow{(i,0)=(0,-i)} D = \frac{C \times_{\mathbb{K}} C}{\{(m, m) \mid m \in M\}} \xrightarrow{(p,p)} A \times_{\mathbb{K}} A \to 0$$

and the morphism $\phi\colon B \to A \times_{\mathbb{K}} A$, $\phi(x) = (f(x), g(x))$. Then $f^*e - g^*e = \phi^*u$ and it is sufficient to prove that $u \in F(A \times_{\mathbb{K}} A, M, \phi(b))$, or equivalently that $\phi(b)$ lifts to $F(D)$. Since $F(A \times_{\mathbb{K}} A) \to F(A) \times F(A)$ is bijective,[1] we must have $\phi(b) = \Delta(a)$, where $\Delta\colon A \to A \times_{\mathbb{K}} A$ is the diagonal. It is now sufficient to observe that Δ lifts to a morphism $A \to D$. □

[1] This is exactly the technical point that explains why functors satisfying the classical Schlessinger's conditions do not have in general a complete obstruction theory.

Definition 3.8.5 Let F be a deformation functor. For every $A \in \mathbf{Art}$ and $a \in F(A)$ define

$$E_F(A, a) = \frac{\mathrm{Ex}(A, \mathbb{K})}{F(A, \mathbb{K}, a)}.$$

We may restate Lemma 3.8.3 by saying that a small extension $e \in \mathrm{Ex}(A, M)$ belongs to $F(A, M, a)$ if and only if $f_* e$ vanishes in $E_F(A, a)$ for every linear map $f : M \to \mathbb{K}$, while Lemma 3.8.4 says that for every morphism $f : B \to A$ and every $b \in F(B)$ such that $f(b) = a$, the linear map $f^* : E_F(A, a) \to E_F(B, b)$ is injective and independent of f.

The construction of the vector spaces $E_F(A, a)$ gives a functor

$$E_F : (\mathbf{Art}, F)^{op} \to \mathbf{Vect},$$

where **Vect** is the category of vector spaces over $\mathbb{K}$, and $(\mathbf{Art}, F)$ is the category of pairs (A, a) with $A \in \mathbf{Art}$ and $a \in F(A)$. A morphism $f : (A, a) \to (B, b)$ in $(\mathbf{Art}, F)$ is a morphism $f : A \to B$ in **Art** such that $f(a) = b$.

Lemma 3.8.6 *The functor* $E_F : (\mathbf{Art}, F)^{op} \to \mathbf{Vect}$ *has the following properties:*

1. *given two objects* (B, b) *and* (C, c) *in* $(\mathbf{Art}, F)$, *there exists a third object* (S, s) *and two morphisms* $(S, s) \to (B, b)$, $(S, s) \to (C, c)$;
2. *given two morphisms* $f, g : (B, b) \to (A, a)$ *in* $(\mathbf{Art}, F)$, *there exists a third morphism* $h : (C, c) \to (B, b)$ *such that* $E_F(hf) = E_F(hg) : E_F(A, a) \to E_F(C, c)$.

Proof Lemma 3.8.4 says that item (2) holds for every h. As regards item (1), given B, C in **Art** and elements $b \in F(B)$, $c \in F(C)$, we can take $S = B \times_{\mathbb{K}} C$; since F is a deformation functor there exists $s \in F(S)$ mapping to $b \in F(B)$ and to $c \in F(C)$. □

It is now natural to define the universal obstruction theory as the colimit of the vector spaces $E_F(A, a)$; to this end we first need to avoid set theoretic difficulties by choosing a small category **A** that is equivalent to **Art**. Then define **C** as the full subcategory of $(\mathbf{Art}, F)$ having as objects all the pairs (A, a) with $A \in \mathbf{A}$.

Since **C** is a small category and **Vect** is cocomplete, it makes sense to define the universal obstruction space O_F as the colimit of the functor E_F. It is easy to see that Lemma 3.8.6 implies that

$$O_F := \operatorname*{colim}_{\mathbf{C}^{op}} E_F(A, a) = \frac{\coprod_{(A,a)\in\mathbf{C}} E_F(A, a)}{\sim},$$

where $x \in E_F(A, a)$ is equivalent to $y \in E_F(B, b)$ if and only if there exist two morphisms $f : (C, c) \to (A, a)$, $g : (C, c) \to (B, b)$ such that $E_F(f)x = E_F(g)y$; by Lemma 3.8.4 the natural maps $E_F(A, a) \to O_F$ are injective morphisms of vector spaces.

Given a small extension $e : 0 \to M \to B \to A \to 0$, the obstruction map $ob_e : F(A) \to O_F \otimes M$ is defined in the following way: take a basis $v_1, \ldots, v_m$ of

M and let $f_1, \ldots, f_m \colon M \to \mathbb{K}$ be the corresponding system of linear coordinates. For every $a \in F(A)$ define $ob_e(a)$ as the image of

$$\tilde{e} := \sum_i (f_i)_* e \otimes v_i \in E_F(A, a) \otimes M$$

under the natural injective map $E_F(A, a) \to O_F$. The fact that (O_F, ob_e) is a complete obstruction theory follows immediately from the previous results regarding the functor E_F, from the fact that the construction of $\tilde{e}$ commutes with push-outs and from Lemma 3.8.3.

Finally, the universal property of (O_F, ob_e) follows almost immediately from the universal property of colimits. Every obstruction theory (V, v_e) for a deformation functor F induces a set of linear morphisms

$$\theta_{(A,a)} \colon \operatorname{Ex}(A, \mathbb{K}) \to V, \quad \theta_{(A,a)}(e) = v_e(a); \qquad A \in \mathbf{A},\ a \in F(A),$$

commuting with pull-backs. If $e \in F(A, \mathbb{K}, a)$ then $v_e(a) = 0$ and therefore the above map factors to $\theta_{(A,a)} \colon E_F(A, a) \to V$. The colimit $\theta = \operatorname{colim} \theta_{(A,a)} \colon O_F \to V$ is the required morphism of obstruction theories.

3.9 Exercises

3.9.1 Prove Remark 3.1.1.

3.9.2 Prove the following properties about natural transformations of functors of Artin rings:

1. If $F \to G$ and $G \to H$ are smooth, then the composition $F \to H$ is smooth.
2. If $u \colon F \to G$ and $v \colon G \to H$ are natural transformations such that u is surjective and vu is smooth, then v is smooth.

3.9.3 Let $\mathbb{K}$ be a field and let $f \in \mathbb{K}[[x_1, \ldots, x_n]]$ be a nontrivial formal power series. Prove by induction on n that there exists a sequence of integers $1 = a_1 \leq a_2 \leq \cdots \leq a_n$ such that $f(t, t^{a_2}, \ldots, t^{a_n}) \neq 0$. Deduce that the implication $(3 \Rightarrow 1)$ of Lemma 3.3.4 holds for every fields.

3.9.4 We have already pointed out that fibre products exist in the category of functors of Artin rings; given two natural transformations $F \to G$ and $H \to G$ of functors of Artin rings we have

$$F \times_G H \colon \mathbf{Art} \to \mathbf{Set}, \qquad F \times_G H(A) = F(A) \times_{G(A)} H(A).$$

Prove that:

1. If $F \to G$ is smooth, then $F \times_G H \to H$ is smooth.
2. Assume that $F \to G$ and $H \to G$ are morphisms of deformation functors and that there exists their categorical fibre product Q in the category of deformation functors. Then the natural morphism $Q \to F \times_G H$ is an isomorphism.
3. Assume that F, H are deformation functors and that G is a homogeneous deformation functor. Then $F \times_G H$ is a deformation functor.
4. In the situation of Example 3.5.6, prove that the diagonal map $F \to F \times_{F/G} F$ is not surjective. Deduce that $F \times_{F/G} F$ is not a deformation functor and that the category of deformation functors does not have fibre products.

3.9.5 Prove that a deformation functor is trivial if and only if its tangent space is trivial.

3.9.6 Consider a fixed unitary commutative ring R as the set of morphisms of a category with one object, and the composition product equal to the multiplication in R. Prove that a non-empty subset $S \subset R$ has the 2 out of 6 property if and only if S is a saturated multiplicative part; recall that a multiplicative subset $S \subset R$ is called saturated if $1 \in S$ and if for every $s, t \in R$ we have $st \in S \iff s, t \in S$.

3.9.7 Prove that there are exactly two structures of homotopical category in the additive monoid $(\mathbb{N}, +)$, considered as a category with one object. Give an example where the 2 out of 3 property does not imply the 2 out of 6 property.

3.9.8 Let $f\colon \mathfrak{g} \to \mathfrak{h}$ be a morphism of semicosimplicial Lie algebras and assume that

1. $f\colon H^1(C(\mathfrak{g})) \to H^1(C(\mathfrak{h}))$ is surjective,
2. $f\colon H^2(C(\mathfrak{g})) \to H^2(C(\mathfrak{h}))$ is injective.

Prove that the morphism $f\colon H^1_{\mathfrak{g}} \to H^1_{\mathfrak{h}}$ is smooth.

3.9.9 (Schlessinger's theorem for semicosimplicial Lie algebras) Let

$$\mathfrak{g}: \qquad \mathfrak{g}_0 \rightrightarrows \mathfrak{g}_1 \,\substack{\longrightarrow\\[-1ex]\longrightarrow\\[-1ex]\longrightarrow}\, \mathfrak{g}_2 \,\substack{\longrightarrow\\[-1ex]\longrightarrow\\[-1ex]\longrightarrow\\[-1ex]\longrightarrow}\, \cdots,$$

be a semicosimplicial Lie algebra over a field of characteristic 0 such that $H^1(C(\mathfrak{g}))$ is finite-dimensional. Let $\mathcal{H} \subset Z^1(C(\mathfrak{g})) \subset \mathfrak{g}_1$ be a vector subspace such that the morphism $\mathcal{H} \to H^1(C(\mathfrak{g}))$ is an isomorphism. Prove that the functor

$$F\colon \mathbf{Art} \to \mathbf{Set}, \qquad F(A) = \{x \in \mathcal{H} \otimes \mathfrak{m}_A \mid e^x \in Z^1_{\mathfrak{g}}(A)\},$$

is pro-representable and the natural transformation $F \to H^1_{\mathfrak{g}}$, $x \mapsto e^x$, is a hull.

3.9.10 Semicosimplicial Lie algebras are a good source of non-pro-representable deformation functors; typically pro-representability fails when the Lie algebras are not abelian and the face operators are "more generic" than the differential of the associated cochain complex. As an example, consider the semicosimplicial Lie algebra

$$\mathfrak{g}: \qquad \mathbb{K} \rightrightarrows \mathfrak{sl}_2(\mathbb{K}) \substack{\rightarrow\\\rightarrow\\\rightarrow} 0 \substack{\rightarrow\\\rightarrow\\\rightarrow\\\rightarrow} \cdots,$$

where $\delta_0(t) = \delta_1(t) = \begin{pmatrix} 0 & t \\ 0 & 0 \end{pmatrix}$ for $t \in \mathbb{K}$, and its semicosimplicial Lie subalgebra

$$\mathfrak{h}: \qquad 0 \rightrightarrows \mathfrak{sl}_2(\mathbb{K}) \substack{\rightarrow\\\rightarrow\\\rightarrow} 0 \substack{\rightarrow\\\rightarrow\\\rightarrow\\\rightarrow} \cdots.$$

Prove that the functor $H^1_{\mathfrak{g}}$ is not homogeneous and that the inclusion $\mathfrak{h} \hookrightarrow \mathfrak{g}$ induces a weak equivalence $H^1_{\mathfrak{h}} \to H^1_{\mathfrak{g}}$ that is not an isomorphism.

References

Together with the original paper by Schlessinger [237], good general references for functors of Artin rings are [69, 110, 226, 242]. The standard reference for cohomology theory of sheaves of non-abelian groups is [103]. The notion of deformation functors comes from [182], in which condition (1) of Definition 3.2.5 replaces the classical Schlessinger's condition:

(H3) *η is bijective, whenever $A = \mathbb{K}$ and $B = \mathbb{K}[\varepsilon]$.*

Most of the results of this chapter hold also under the weak assumption (H3), while (1) is used in the proof of Lemma 3.8.4 and therefore also in the proof of the existence of complete obstruction theory. Functors satisfying the original Schlessinger's conditions do not have in general a complete obstruction theory [69, Example 6.8], at least in the sense described in this book.

Pro-representable functors were introduced by Grothendieck as functors isomorphic to small colimits of representable functors. Here we adopt the notion of pro-representable functors in the more restrictive terms proposed by Schlessinger [237]. We also refer to [69, Lemma 5.6] for the equivalence of the first three items of Lemma 3.3.4 over any field. For the proof of Schlessinger's theorem 3.5.10 we also refer to the original paper [237] and to the books [7, 110, 242].

The definition of obstruction theory together with Theorems 3.6.7 and 3.6.5 are taken from [69], while the notion of a homotopical category is taken from [114, 225]. A deeper study of the functor $H^1_{\mathfrak{g}}$ associated to a (semi)cosimplicial Lie algebra $\mathfrak{g}$ is done in [78].

According to [139, Definition 1.11.2], we can restate Lemma 3.8.6 by saying that the image of the functor E_F is a filtrant subcategory of **Vect**. The relevance of this notion lies in the fact that colimits of small filtrant categories commute with finite limits, cf. [8, Example I, Proposition 2.8]; this explains the fact that the underlying set of the colimit vector space O_F is precisely the colimit of the underlying sets of $E_F(A, a)$.

In several papers in deformation theory, functors of Artin rings are replaced by categories fibred in groupoids, for instance in [6, 82, 226, 227]. Two nice introductions to this language are the paper by Talpo and Vistoli [254] and the (preliminary) manuscript by Buchweitz and Flenner [31].

Chapter 4
Infinitesimal Deformations of Complex Manifolds and Vector Bundles

In this chapter we study deformations, over Artin local rings, of manifolds and vector bundles. We follow a classical approach, similar to every standard introductory book in deformation theory. However, for later applications, we interpret all the involved deformation functors either as the exponential functor of a Lie algebra or as the functor associated with a semicosimplicial Lie algebra.

4.1 Flat Modules over Artin Local Rings

Throughout this section we shall denote by A a Artin local ring with residue field $\mathbb{K} = A/\mathfrak{m}_A$. We assume that the reader is already familiar with the notion of a flat module over a unitary commutative ring.

Theorem 4.1.1 *Let A be an Artin local ring with residue field* $\mathbb{K}$, *let* $f\colon C^* \to D^*$ *be a morphism of cochain complexes of flat A-modules and let n be an integer. Assume that:*

1. $f\colon H^n(C^* \otimes_A \mathbb{K}) \to H^n(D^* \otimes_A \mathbb{K})$ *is surjective;*
2. $f\colon H^{n+1}(C^* \otimes_A \mathbb{K}) \to H^{n+1}(D^* \otimes_A \mathbb{K})$ *is injective.*

Then $f\colon H^n(C^*) \to H^n(D^*)$ *is surjective and* $f\colon H^{n+1}(C^*) \to H^{n+1}(D^*)$ *is injective.*

Proof We prove the lemma by induction on the length of A, defined as the positive integer

$$l(A) = \sum_{n\geq 0} \dim_{\mathbb{K}} \frac{\mathfrak{m}_A^n}{\mathfrak{m}_A^{n+1}}\,,$$

or equivalently as the length of a chain of ideals $0 = I_0 \subset I_1 \subset \cdots \subset I_{l(A)} = A$ such that $\mathfrak{m}_A I_i \subset I_{i+1}$ and $I_{i+1}/I_i \simeq \mathbb{K}$ for every $0 \leq i < l(A)$.

M. Manetti, *Lie Methods in Deformation Theory*, Springer Monographs in Mathematics,
https://doi.org/10.1007/978-981-19-1185-9_4

If $A = \mathbb{K}$ there is nothing to prove. If $\mathfrak{m}_A \neq 0$, let $\varepsilon \in \mathfrak{m}_A$ be a nontrivial element such that $\varepsilon\mathfrak{m}_A = 0$ and consider the short exact sequence of A-modules:

$$0 \to \mathbb{K} \xrightarrow{\varepsilon} A \to B = A/(\varepsilon) \to 0.$$

Since the length of B is smaller than the length of A and $(M \otimes_A B) \otimes_B \mathbb{K} = M \otimes_A \mathbb{K}$ for every A-module M, by induction the map $f\colon H^n(C^* \otimes_A B) \to H^n(D^* \otimes_A B)$ is surjective and $f\colon H^{n+1}(C^* \otimes_A B) \to H^{n+1}(D^* \otimes_A B)$ is injective.

Since C^i and D^i are flat for every i we have a morphism of short exact sequences of complexes

$$\begin{array}{ccccccccc}
0 & \longrightarrow & C^* \otimes_A \mathbb{K} & \longrightarrow & C^* & \longrightarrow & C^* \otimes_A B & \longrightarrow & 0 \\
 & & \Big\downarrow f & & \Big\downarrow f & & \Big\downarrow f & & \\
0 & \longrightarrow & D^* \otimes_A \mathbb{K} & \longrightarrow & D^* & \longrightarrow & D^* \otimes_A B & \longrightarrow & 0
\end{array}$$

inducing a morphism between the corresponding cohomology long exact sequences. Then the surjectivity of $f\colon H^n(C^*) \to H^n(D^*)$ follows from the commutative diagram

$$\begin{array}{ccccccc}
H^n(C^* \otimes_A \mathbb{K}) & \longrightarrow & H^n(C^*) & \longrightarrow & H^n(C^* \otimes_A B) & \longrightarrow & H^{n+1}(C^* \otimes_A \mathbb{K}) \\
\Big\downarrow f & & \Big\downarrow & & \Big\downarrow & & \Big\downarrow f \\
H^n(D^* \otimes_A \mathbb{K}) & \longrightarrow & H^n(D^*) & \longrightarrow & H^0(D^* \otimes_A B) & \longrightarrow & H^{n+1}(D^* \otimes_A \mathbb{K}),
\end{array}$$

while the injectivity of $f\colon H^{n+1}(C^*) \to H^{n+1}(D^*)$ follows from the commutative diagram

$$\begin{array}{ccccccc}
H^n(C^* \otimes_A B) & \longrightarrow & H^{n+1}(C^* \otimes_A \mathbb{K}) & \longrightarrow & H^{n+1}(C^*) & \longrightarrow & H^1(C^* \otimes_A B) \\
\Big\downarrow & & \Big\downarrow f & & \Big\downarrow & & \Big\downarrow \\
H^n(D^* \otimes_A B) & \longrightarrow & H^{n+1}(D^* \otimes_A \mathbb{K}) & \longrightarrow & H^{n+1}(D^*) & \longrightarrow & H^1(D^* \otimes_A B).
\end{array}$$

□

Corollary 4.1.2 *Let $f\colon C^* \to D^*$ be a morphism of cochain complexes of flat modules over an Artin local ring A with residue field $\mathbb{K}$. If the induced map $f\colon H^n(C^* \otimes_A \mathbb{K}) \to H^n(D^* \otimes_A \mathbb{K})$ is an isomorphism for every integer n, then $f\colon H^n(C^*) \to H^n(D^*)$ is an isomorphism for every n.*

Proof Immediate consequence of Theorem 4.1.1. □

Corollary 4.1.3 *Let A be an Artin local ring. Then every flat A-module is free.*

Proof We first observe that for every A-module C there exists a free module F together with a morphism $g: F \to C$ such that the induced map $F \otimes_A \mathbb{K} \to C \otimes_A \mathbb{K}$ is bijective; in fact, since $C \to C \otimes_A \mathbb{K}$ is surjective, we may consider as F the free module generated by a subset of C that lifts a basis of the $\mathbb{K}$-vector space $C \otimes_A \mathbb{K}$.

Let M be a flat A-module and choose a morphism $F \to M$ such that F is free and $F \otimes_A \mathbb{K} \to M \otimes_A \mathbb{K}$ is an isomorphism. Considering F and M as cochain complexes concentrated in degree 0, by Corollary 4.1.2 the map $F \to M$ is an isomorphism. □

Lemma 4.1.4 (Nakayama's lemma for Artin local rings) *Let A be an Artin local ring with residue field $\mathbb{K}$ and let M be any A-module. Then $M = 0$ if and only if*
$$M \otimes_A \mathbb{K} = \frac{M}{\mathfrak{m}_A M} = 0.$$

Proof Assume $M \otimes_A \mathbb{K} = 0$ and consider an exact sequence $C \xrightarrow{f} D \to M \to 0$ with C, D free A-modules. Since the functor—$\otimes_A \mathbb{K}$ is right-exact the induced map $f \otimes_A \mathrm{Id}_{\mathbb{K}}: C \otimes_A \mathbb{K} \to D \otimes_A \mathbb{K}$ is surjective and then, by Corollary 4.1.2 also the map $C \xrightarrow{f} D$ is surjective, hence $M = 0$. □

Proposition 4.1.5 *Let A be an Artin local ring with residue field $\mathbb{K}$ and let $C \xrightarrow{f} D$ be a morphism of A-modules:*

1. *if $f \otimes_A \mathrm{Id}_{\mathbb{K}}: C \otimes_A \mathbb{K} \to D \otimes_A \mathbb{K}$ is surjective, then also f is surjective;*
2. *if D is flat and $f \otimes_A \mathrm{Id}_{\mathbb{K}}: C \otimes_A \mathbb{K} \to D \otimes_A \mathbb{K}$ is injective, then C is flat and f is injective.*

In particular, if D is flat over A, then f is an isomorphism if and only if $f \otimes_A \mathrm{Id}_{\mathbb{K}}$ is an isomorphism.

Proof The first item follows from Lemma 4.1.4 applied to the cokernel of f. As in the proof of Corollary 4.1.3 we can find a free module F together with a morphism $g: F \to C$ such that the induced map $F \otimes_A \mathbb{K} \to C \otimes_A \mathbb{K}$ is bijective. By Lemma 4.1.4 g is a surjective morphism.

Considering the composite map $fg: F \to D$ as a morphism of cochain complexes of flat modules concentrated in degree 0, by Theorem 4.1.1, applied with the integer $n = -1$, the morphism $fg: F \to D$ is injective and therefore g is bijective and f is injective. □

Corollary 4.1.6 *Let $0 \to M \to N \to P \to 0$ be an exact sequence of modules over an Artin local ring A with residue field $\mathbb{K}$, with N flat. Then the following conditions are equivalent:*

1. *P is flat;*
2. *the short exact sequence $0 \to M \to N \to P \to 0$ splits;*
3. *M is flat and $M \otimes_A \mathbb{K} \to N \otimes_A \mathbb{K}$ is injective;*
4. *$M \otimes_A \mathbb{K} \to N \otimes_A \mathbb{K}$ is injective.*

Proof If P is flat then it is free and therefore $0 \to M \to N \to P \to 0$ is a split exact sequence. Split exact means in particular $N \cong M \oplus P$ and this implies immediately that M is flat and $M \otimes_A \mathbb{K} \to N \otimes_A \mathbb{K}$ is injective.

If $M \otimes_A \mathbb{K} \to N \otimes_A \mathbb{K}$ is injective, consider two surjective morphisms $F \xrightarrow{f} P$, $G \xrightarrow{g} M$, such that F, G are free modules and the maps $F \otimes_A \mathbb{K} \to P \otimes_A \mathbb{K}$, $G \otimes_A \mathbb{K} \to M \otimes_A \mathbb{K}$ are isomorphisms. Since F is free there exists a morphism $h \colon F \to N$ that lifts f and we have a morphism of complexes of flat modules

$$\begin{array}{ccccccc} 0 & \longrightarrow & 0 & \longrightarrow & F & \longrightarrow & 0 \\ \downarrow & & \downarrow & & \downarrow{\scriptstyle h} & & \downarrow \\ 0 & \longrightarrow & G & \longrightarrow & N & \longrightarrow & 0. \end{array}$$

Since $G \to N \to P \to 0$ and $0 \to G \otimes_A \mathbb{K} \to N \otimes_A \mathbb{K} \to P \otimes_A \mathbb{K} \to 0$ are exact sequences, by Corollary 4.1.2 the induced map

$$F = \operatorname{coker}(0 \to F) \to \operatorname{coker}(G \to N) = P$$

is an isomorphism. Therefore P is free and the exact sequance splits. □

Example 4.1.7 Let $A = \mathbb{K}[\varepsilon]$, $\varepsilon^2 = 0$, and consider the ideal

$$I = (xy, x^3, y^2 - \varepsilon x^2) \subset A[x, y].$$

Then the A-module $A[x, y]/I$ is freely generated by $1, x, x^2, y$, and therefore the short exact sequence $0 \to I \to A[x, y] \to A[x, y]/I \to 0$ satisfies the equivalent conditions of Corollary 4.1.6. On the other hand, the A-module

$$\frac{A[x, y]}{I^2} = \frac{A[x, y]}{(x^2y^2, x^6, x^4y, xy^3 - \varepsilon x^3y, y^2x^3 - \varepsilon x^5, y^4 - 2\varepsilon x^2y^2)}$$

is not free since $x^5 \notin I^2$ and $\varepsilon x^5 = x(x^2y^2) - (y^2x^3 - \varepsilon x^5) \in I^2$. Notice that εx^5 is not trivial in $I^2 \otimes_A \mathbb{K} = I^2/\varepsilon I^2$, and then the morphism

$$I^2 \otimes_A \mathbb{K} \xrightarrow{\varepsilon \mapsto 0} (I \otimes_A \mathbb{K})^2 \subset A[x, y] \otimes_A \mathbb{K} = \mathbb{K}[x, y]$$

is not injective, in accordance with Corollary 4.1.6.

4.2 Infinitesimal Deformations of Vector Bundles

From here to the end of this chapter we shall work over the field of complex numbers. For simplicity of notation we shall write $\otimes$ to denote the tensor product $\otimes_{\mathbb{C}}$.

Let X be a complex manifold with sheaf of holomorphic functions $\mathcal{O}_X$. For every Artin local $\mathbb{C}$-algebra $A \in \mathbf{Art}_{\mathbb{C}}$, the complex space $X \times \operatorname{Spec} A$ can be interpreted as the topological space X equipped with the structure sheaf $\mathcal{O}_X \otimes A$. Notice that, since A is a finite-dimensional $\mathbb{C}$-vector space, for every open subset $U \subset X$ we have $\mathcal{O}_X \otimes A(U) = \mathcal{O}_X(U) \otimes A$, and the natural projection from $X \times \operatorname{Spec} A$ onto the affine scheme $\operatorname{Spec} A$ is induced by the morphisms of rings $A \to \mathcal{O}_X(X) \otimes A$, $a \mapsto 1 \otimes a$. Similarly, if $p\colon X \times \operatorname{Spec} A \to X$ is the projection and $\mathcal{F}$ is a sheaf of $\mathcal{O}_X$-modules, then for every open subset $U \subset X$ we have

$$p^*\mathcal{F}(U \times \operatorname{Spec} A) = (\mathcal{F} \otimes A)(U) = \mathcal{F}(U) \otimes A.$$

Definition 4.2.1 Let X be a complex manifold and let $\mathcal{F}$ be a sheaf of $\mathcal{O}_X$-modules. A **deformation** of $\mathcal{F}$ over $A \in \mathbf{Art}_{\mathbb{C}}$ is the data of a sheaf $\mathcal{F}_A$ of $\mathcal{O}_X \otimes A$-modules, that is flat as a sheaf of A-modules, and of an isomorphism of sheaves of $\mathcal{O}_X$-modules $\phi\colon \mathcal{F}_A \otimes_A \mathbb{C} \to \mathcal{F}$.

An isomorphism of deformations $\eta\colon (\mathcal{F}_A, \phi) \to (\mathcal{F}'_A, \phi')$ is an isomorphism of sheaves of $\mathcal{O}_X \otimes A$-modules $\eta\colon \mathcal{F}_A \to \mathcal{F}'_A$ such that $\phi = \phi' \circ (\eta \otimes_A \mathrm{Id}_{\mathbb{C}})$.

The reader should keep in mind the fact that the isomorphism ϕ is part of the deformation data; if $g\colon \mathcal{F} \to \mathcal{F}$ is a nontrivial automorphism and $(\mathcal{F}_A, \phi)$ is a deformation, then also $(\mathcal{F}_A, g\phi)$ is a deformation which is isomorphic to $(\mathcal{F}_A, \phi)$ if and only if g lifts to an automorphism of $\mathcal{F}_A$.

The above definition is clearly functorial in A and then it makes sense to consider the functor $\operatorname{Def}_{\mathcal{F}}\colon \mathbf{Art}_{\mathbb{C}} \to \mathbf{Set}$:

$$\operatorname{Def}_{\mathcal{F}}(A) = \{\text{isomorphism classes of deformations of } \mathcal{F} \text{ over } A\}.$$

For every $A \in \mathbf{Art}_{\mathbb{C}}$ the set $\operatorname{Def}_{\mathcal{F}}(A)$ is not empty since it contains the trivial deformation $(\mathcal{F} \otimes A, \mathrm{Id}_{\mathcal{F}})$, where $\mathrm{Id}_{\mathcal{F}}$ is induced by the chain of natural isomorphisms

$$(\mathcal{F} \otimes A) \otimes_A \mathbb{C} \simeq \mathcal{F} \otimes (A \otimes_A \mathbb{C}) \simeq \mathcal{F} \otimes \mathbb{C} \simeq \mathcal{F}.$$

Lemma 4.2.2 *In the above setup, there exists a natural isomorphism of groups*

$$\exp(\operatorname{Hom}_{\mathcal{O}_X}(\mathcal{F}, \mathcal{F}) \otimes \mathfrak{m}_A) \simeq \{\textit{automorphisms of the trivial deformation } \mathcal{F} \otimes A\}.$$

Proof By definition, an automorphism of the trivial deformation is an invertible element $g \in \operatorname{Hom}_{\mathcal{O}_X \otimes A}(\mathcal{F} \otimes A, \mathcal{F} \otimes A)$ such that its reduction

$$g \otimes_A \mathrm{Id}_{\mathbb{C}}\colon \mathcal{F} \otimes A \otimes_A \mathbb{C} \to \mathcal{F} \otimes A \otimes_A \mathbb{C}$$

is the identity, or equivalently such that $\operatorname{Im}(g - \mathrm{Id}) \subset \mathcal{F} \otimes \mathfrak{m}_A$. Since $\mathfrak{m}_A$ is a finite-dimensional vector space we have

$$\operatorname{Hom}_{\mathcal{O}_X}(\mathcal{F}, \mathcal{F}) \otimes \mathfrak{m}_A = \operatorname{Hom}_{\mathcal{O}_X}(\mathcal{F}, \mathcal{F} \otimes \mathfrak{m}_A) = \operatorname{Hom}_{\mathcal{O}_X \otimes A}(\mathcal{F} \otimes A, \mathcal{F} \otimes \mathfrak{m}_A)$$

and then we have an isomorphism of nilpotent Lie algebras

$$\mathrm{Hom}_{\mathcal{O}_X}(\mathcal{F},\mathcal{F})\otimes\mathfrak{m}_A\simeq\{f\in\mathrm{Hom}_{\mathcal{O}_X\otimes A}(\mathcal{F}\otimes A,\mathcal{F}\otimes A)\mid \mathrm{Im}(f)\subset\mathcal{F}\otimes\mathfrak{m}_A\}.$$

If $f\in\mathrm{Hom}_{\mathcal{O}_X\otimes A}(\mathcal{F}\otimes A,\mathcal{F}\otimes A)$ is such that $\mathrm{Im}(f)\subset\mathcal{F}\otimes\mathfrak{m}_A$, the A-linearity of f implies that f is a nilpotent endomorphism, its exponential e^f is defined and $\mathrm{Im}(e^f-\mathrm{Id})\subset\mathcal{F}\otimes\mathfrak{m}_A$. Conversely, if $g\in\mathrm{Hom}_{\mathcal{O}_X\otimes A}(\mathcal{F}\otimes A,\mathcal{F}\otimes A)$ is such that $\mathrm{Im}(g-\mathrm{Id})\subset\mathcal{F}\otimes\mathfrak{m}_A$, then $g-\mathrm{Id}$ is nilpotent and therefore $g=e^f$, where f is the logarithm of $g=\mathrm{Id}+(g-\mathrm{Id})$. □

Theorem 4.2.3 *Let X be a complex manifold, $\mathcal{F}$ a sheaf of $\mathcal{O}_X$-modules and let $\mathcal{U}=\{U_i\}$ be an open covering of X such that every infinitesimal deformation of $\mathcal{F}_{|U_i}$ is isomorphic to the trivial one. Consider the sheaf of Lie algebras $\mathcal{E}=\mathcal{H}om_{\mathcal{O}_X}(\mathcal{F},\mathcal{F})$ and the semicosimplicial Lie algebra of Čech cochains (see Example 2.6.3)*

$$\mathcal{E}(\mathcal{U}):\qquad \prod_i\mathcal{E}(U_i)\rightrightarrows\prod_{i,j}\mathcal{E}(U_{ij})\substack{\longrightarrow\\\longrightarrow\\\longrightarrow}\prod_{i,j,k}\mathcal{E}(U_{ijk})\substack{\longrightarrow\\\longrightarrow\\\longrightarrow\\\longrightarrow}\cdots.$$

Then there exists a natural isomorphism of functors of Artin rings

$$H^1_{\mathcal{E}(\mathcal{U})}\xrightarrow{\simeq}\mathrm{Def}_{\mathcal{F}}.$$

Proof The proof is a slight variation of a classical construction (see e.g. [3, Theorem 7.3.3] and [115, Theorem 3.2.1]). Given $A\in\mathbf{Art}_{\mathbb{C}}$, by Lemma 4.2.2 an element of $Z^1_{\mathcal{E}(\mathcal{U})}(A)$ is the data, for every i,j, of an isomorphism of sheaves of $\mathcal{O}_X\otimes A$-modules

$$g_{ij}\colon\mathcal{F}_{|U_{ij}}\otimes A\to\mathcal{F}_{|U_{ij}}\otimes A$$

such that $g_{ij}\otimes_A\mathrm{Id}_{\mathbb{C}}$ is the identity and, for every i,j,k we have $g_{jk}g_{ik}^{-1}g_{ij}=\mathrm{Id}$ in $\mathcal{F}_{|U_{ijk}}\otimes A$. In particular, taking $i=j=k$ we obtain $g_{ii}=\mathrm{Id}$, while taking $k=i$ we obtain $g_{ji}g_{ij}=\mathrm{Id}$. Finally, we have

$$g_{jk}g_{ik}^{-1}g_{ij}=\mathrm{Id}\quad\iff\quad g_{ij}g_{jk}=g_{ik}$$

and then the usual glueing procedure gives a sheaf $\mathcal{F}_A$ and a set of isomorphisms

$$g_i\colon(\mathcal{F}_A)_{|U_i}\xrightarrow{\simeq}\mathcal{F}_{|U_i}\otimes A$$

such that $g_{ij}=g_ig_j^{-1}$ on U_{ij}. In particular, the induced isomorphisms

$$g_i\otimes_A\mathrm{Id}_{\mathbb{C}}\colon(\mathcal{F}_A)_{|U_i}\otimes_A\mathbb{C}=(\mathcal{F}_A\otimes_A\mathbb{C})_{|U_i}\xrightarrow{\simeq}\mathcal{F}_{|U_i}\otimes A\otimes_A\mathbb{C}$$

coincide on U_{ij} and define an isomorphism of sheaves $\mathcal{F}_A\otimes_A\mathbb{C}\to\mathcal{F}$.

Again by Lemma 4.2.2, two cocycles $g_{ij},h_{ij}\in Z^1_{\mathcal{E}(\mathcal{U})}(A)$ give the same element in $H^1_{\mathcal{E}(\mathcal{U})}(A)$ if and only if there exists a family of isomorphisms

$$q_i\colon \mathcal{F}_{|U_i}\otimes A\to \mathcal{F}_{|U_i}\otimes A$$

lifting the identity on $\mathcal{F}_{|U_i}$ and such that $q_i g_{ij} q_j^{-1}=h_{ij}$ on U_{ij}. In terms of the corresponding glued sheaves

$$(\mathcal{F}_A)_{|U_i}\xrightarrow{g_i}\mathcal{F}_{|U_i}\otimes A\xleftarrow{h_i}(\mathcal{F}'_A)_{|U_i}$$

the condition becomes $h_i^{-1}q_i g_i=h_j^{-1}q_j g_j$ and this is precisely the condition that the diagrams

$$\begin{array}{ccc}(\mathcal{F}_A)_{|U_i} & \xrightarrow{g_i} & \mathcal{F}_{|U_i}\otimes A\\ \downarrow{\scriptstyle q} & & \downarrow{\scriptstyle q_i}\\ (\mathcal{F}'_A)_{|U_i} & \xrightarrow{h_i} & \mathcal{F}_{|U_i}\otimes A\end{array}$$

give a well defined morphism $q\colon \mathcal{F}_A\to\mathcal{F}'_A$. Thus we have defined a natural transformation $H^1_{\mathcal{E}(\mathcal{U})}\to \mathrm{Def}_{\mathcal{F}}$; it is clear from the construction that it is injective and the elements in the image are exactly the isomorphism classes of deformations that are trivial on every open subset U_i. □

According to Lemma 4.2.4 below the above theorem applies in particular to any locally free sheaf of finite rank and every Stein covering trivializing it.

Lemma 4.2.4 *Let X be a complex manifold such that $H^1(X,\mathcal{O}_X)=0$ and let $\mathcal{F}\simeq\mathcal{O}_X^r$ be a free sheaf of rank r. Then every deformation of $\mathcal{F}$ over $A\in\mathbf{Art}_{\mathbb{C}}$ is isomorphic to the trivial deformation.*

Proof Let $(\mathcal{F}_A,\phi)$ be a deformation over A and let $0\to\mathbb{C}\to A\xrightarrow{\alpha}B\to 0$ be a small extension in $\mathbf{Art}_{\mathbb{C}}$. Applying the exact functor $\mathcal{F}_A\otimes_A$— we get an exact sequence

$$0\to\mathcal{F}\to\mathcal{F}_A\xrightarrow{\alpha}\mathcal{F}_A\otimes_A B\to 0$$

and by induction on the length of the Artin local ring we may assume that there exists an isomorphism of deformations $\gamma\colon\mathcal{F}\otimes B\to\mathcal{F}_A\otimes_A B$. Denote by $e_1,\ldots,e_r$ a basis of the free $\mathcal{O}_X$-module $\mathcal{F}$, since $H^1(X,\mathcal{F})=0$, there exist $\varepsilon_1,\ldots,\varepsilon_r\in\mathcal{F}_A$ such that $\alpha(\varepsilon_i)=\gamma(e_i\otimes 1)$. Then the morphism of $\mathcal{O}_X\otimes A$-modules

$$\tilde{\gamma}\colon\mathcal{F}\otimes A\to\mathcal{F}_A,\qquad e_i\otimes 1\mapsto\varepsilon_i,$$

lifts γ and Proposition 4.1.5 implies that $\tilde{\gamma}$ is an isomorphism of deformations. □

Corollary 4.2.5 *Let $\mathcal{F}$ be a locally free sheaf of finite rank on a complex manifold X. Then $\mathrm{Def}_{\mathcal{F}}$ is a deformation functor,*

$$T^1\mathrm{Def}_{\mathcal{F}}=H^1(X,\mathcal{H}om_{\mathcal{O}_X}(\mathcal{F},\mathcal{F}))=\mathrm{Ext}^1_{\mathcal{O}_X}(\mathcal{F},\mathcal{F}),$$

and there exists a complete obstruction theory for $\mathrm{Def}_{\mathcal{F}}$ *with values in the kernel of the trace map*

$$\mathrm{Tr}\colon H^2(X, \mathcal{H}om_{\mathcal{O}_X}(\mathcal{F}, \mathcal{F})) = \mathrm{Ext}^2_{\mathcal{O}_X}(\mathcal{F}, \mathcal{F}) \to H^2(X, \mathcal{O}_X).$$

Proof Consider an open cover $\mathcal{U} = \{U_i\}$ such that U_i is a Stein manifold and $\mathcal{F}_{|U_i}$ is a free sheaf for every i. According to Theorem 4.2.3 and Lemma 4.2.4 there exists an isomorphism of functors $H^1_{\mathcal{E}(\mathcal{U})} = \mathrm{Def}_{\mathcal{F}}$, where $\mathcal{E} = \mathcal{H}om_{\mathcal{O}_X}(\mathcal{F}, \mathcal{F})$, and then $\mathrm{Def}_{\mathcal{F}}$ is a deformation functor. By Theorem 3.7.3 we have

$$T^1\mathrm{Def}_{\mathcal{F}} = H^1(C(\mathcal{E}(\mathcal{U}))) = \check{H}^1(\mathcal{U}, \mathcal{E}) = H^1(X, \mathcal{H}om_{\mathcal{O}_X}(\mathcal{F}, \mathcal{F})),$$

where the last equality is a consequence of Leray's theorem on acyclic covers. Again by Theorem 3.7.3 the functor $H^1_{\mathcal{E}(\mathcal{U})}$ has a complete obstruction theory with values in $H^2(C(\mathcal{E}(\mathcal{U}))) = H^2(X, \mathcal{H}om_{\mathcal{O}_X}(\mathcal{F}, \mathcal{F}))$. Considering $\mathcal{O}_X$ as a sheaf of abelian Lie algebras, the trace morphism $\mathrm{Tr}\colon \mathcal{E} \to \mathcal{O}_X$ commutes with brackets, giving a morphism of semicosimplicial Lie algebras $\mathrm{Tr}\colon \mathcal{E}(\mathcal{U}) \to \mathcal{O}_X(\mathcal{U})$ that induces a natural transformation of functors $\mathrm{Tr}\colon H^1_{\mathcal{E}(\mathcal{U})} \to H^1_{\mathcal{O}_X(\mathcal{U})}$. By Corollary 3.7.5 the functor $H^1_{\mathcal{O}_X(\mathcal{U})}$ is smooth and then every obstruction of $H^1_{\mathcal{E}(\mathcal{U})}$ is annihilated by the trace map. □

4.3 Infinitesimal Deformations of Complex Manifolds

We start here our investigation of infinitesimal deformations of complex manifolds. The first and preliminary step is the proof that every such a deformation is locally trivial; according to Theorem 1.6.3 this is equivalent to the fact that every infinitesimal deformation of a smooth analytic algebra is trivial.

Definition 4.3.1 Let R be an analytic algebra. A **deformation** of R over an Artin local algebra $A \in \mathbf{Art}_{\mathbb{C}}$ is the data of two morphisms of analytic algebras

$$A \xrightarrow{f} S \xrightarrow{\phi} R$$

such that: f is flat, $\phi f(\mathfrak{m}_A) = 0$ and ϕ factors to an isomorphism

$$\overline{\phi}\colon S \otimes_A \mathbb{C} = \frac{S}{f(\mathfrak{m}_A)S} \xrightarrow{\simeq} R.$$

An **isomorphism of deformations** is a commutative diagram of analytic algebras

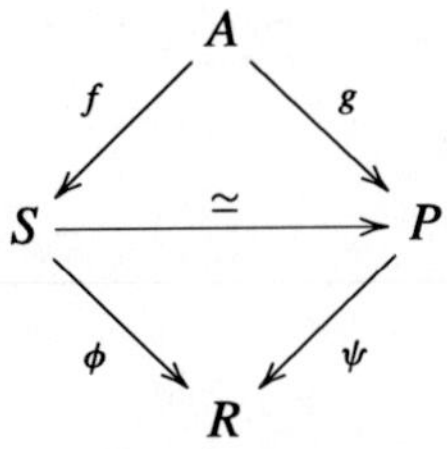

with the horizontal arrow an isomorphism.

Given a morphism $\alpha\colon A \to B$ in the category $\mathbf{Art}_{\mathbb{C}}$ of Artin local $\mathbb{C}$-algebras and a deformation $A \xrightarrow{f} S \xrightarrow{\phi} R$ of the analytic algebra R, the pair of morphisms

$$B \xrightarrow{g} S \otimes_A B \xrightarrow{\psi} R, \qquad g(b) = 1 \otimes b, \quad \psi(s \otimes b) = \phi(s)\pi(b), \tag{4.1}$$

where $\pi\colon B \to B/\mathfrak{m}_B = \mathbb{C}$ is the projection, is again a deformation of R. The only nontrivial verification is the fact that $S \otimes_A B$ is an analytic algebra, since in general the tensor product of two analytic algebras is not a local algebra (cf. Lemma A.2.8 and Exercise A.7.1). However, in our case everything works well in view of the following lemma.

Lemma 4.3.2 *Let $A \to R$ and $A \to B$ be two morphisms of analytic algebras, with $A, B \in \mathbf{Art}_{\mathbb{C}}$. Then $R \otimes_A B$ is an analytic algebra.*

Proof We shall give a complete proof in Corollary A.2.11. We also refer to [100, Lemma 1.89] for a different proof. □

Example 4.3.3 Given an analytic algebra R and $A \in \mathbf{Art}_{\mathbb{C}}$, the **trivial deformation** of R over A is the deformation

$$A \xrightarrow{1\otimes \mathrm{Id}} R \otimes A \xrightarrow{\mathrm{Id}\otimes\pi} R \otimes \mathbb{C} = R,$$

where $\pi\colon A \to \mathbb{C}$ is the projection onto the residue field.

Lemma 4.3.4 *Every deformation of $\mathbb{C}\{z_1, \ldots, z_n\}$ over an Artin local $\mathbb{C}$-algebra is isomorphic to the trivial deformation.*

Proof Let $A \xrightarrow{f} S \xrightarrow{\phi} \mathbb{C}\{z_1, \ldots, z_n\}$ be a deformation over $A \in \mathbf{Art}_{\mathbb{C}}$ and denote by $\overline{\phi}$ the induced isomorphism of analytic algebras $\overline{\phi}\colon S \otimes_A \mathbb{C} \to \mathbb{C}\{z_1, \ldots, z_n\}$. Since ϕ is surjective we may choose $s_1, \ldots, s_n \in \mathfrak{m}_S$ such that $\phi(s_i) = z_i$ and then, by Corollary 1.6.4, there exists a morphism of analytic algebras $h\colon \mathbb{C}\{z_1, \ldots, z_n\} \to S$ such that $\phi h = \mathrm{Id}$, and the universal property of tensor products of algebras gives the commutative diagram of analytic algebras:

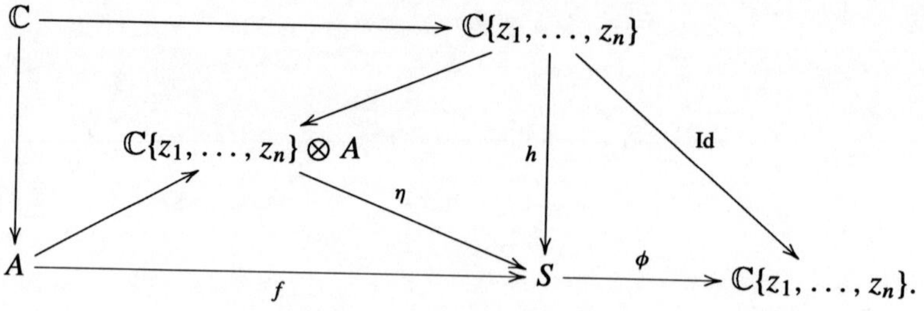

Since S is flat over A and $\eta \otimes_A \mathrm{Id}_{\mathbb{C}} = (\overline{\phi})^{-1}$ is an isomorphism, by Proposition 4.1.5 the morphism η is an isomorphism. □

The following lemma concerns the local triviality of deformations of pairs (manifold, submanifold) and will be used in the next section.

Lemma 4.3.5 *Let $f\colon \mathbb{C}\{z_1, \dots, z_n\} \to R$ be a surjective morphism of analytic algebras, $A \in$ **Art**$_{\mathbb{C}}$ and $F\colon \mathbb{C}\{z_1, \dots, z_n\} \otimes A \to R \otimes A$ an A-linear morphism of analytic algebras such that*

$$f = F \otimes_A \mathrm{Id}_{\mathbb{C}}\colon \mathbb{C}\{z_1, \dots, z_n\} \to R.$$

Then there exists a commutative diagram of analytic algebras

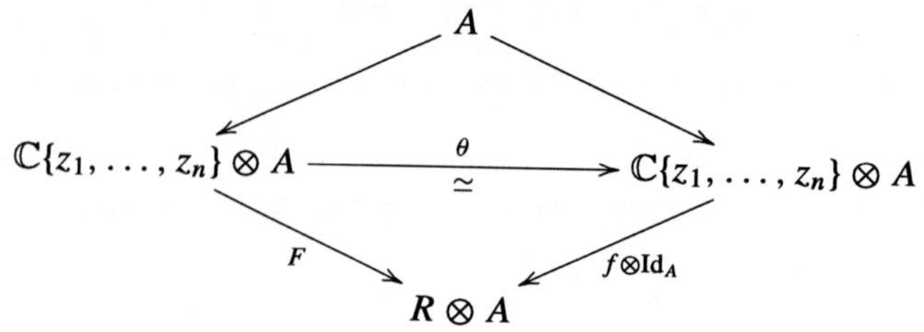

with $\theta \otimes_A \mathrm{Id}_{\mathbb{C}} = \mathrm{Id}$.

Proof Let $m_1, \dots, m_k$ be a basis of $\mathfrak{m}_A$ as a vector space over $\mathbb{C}$. Since the map f is surjective, for every $i = 1, \dots, n$ there exists a sequence $v_{i1}, \dots, v_{ik} \in R$ such that $F(z_i) = f(z_i) \otimes 1 + \sum_j f(v_{ij}) \otimes m_j$. It is now sufficient to define θ as the unique A-linear morphism of analytic algebras such that $\theta(z_i) = z_i \otimes 1 + \sum_j v_{ij} \otimes m_j$. Since $\theta \otimes_A \mathrm{Id}_{\mathbb{C}} = \mathrm{Id}$ by construction, the invertibility of θ follows from Proposition 4.1.5. □

We are now ready to study the global deformations of complex manifolds.

Definition 4.3.6 A **deformation** of a complex manifold X over $A \in$ **Art**$_{\mathbb{C}}$ is the data of a pair of morphisms of complex spaces $X \xrightarrow{i} X_A \xrightarrow{f} \operatorname{Spec} A$ such that f is flat and $i\colon X \xrightarrow{\simeq} f^{-1}(0)$, where $f^{-1}(0)$ is the schematic fibre of f over the closed point of $\operatorname{Spec} A$. An **isomorphism of deformations** is a commutative diagram

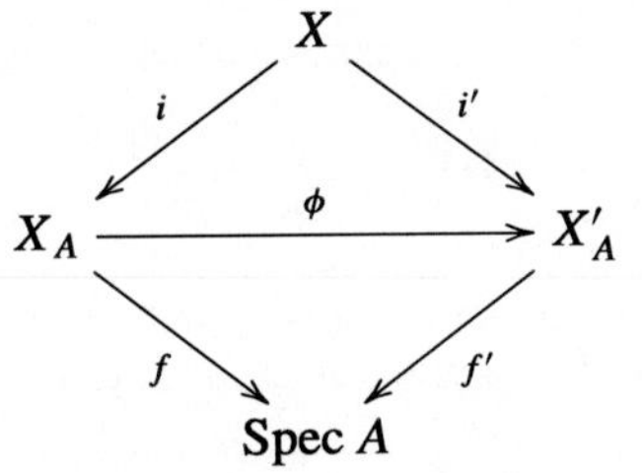

of complex spaces, with ϕ an isomorphism. The collection of isomorphism classes of deformations of X over an Artin local $\mathbb{C}$-algebra A will be denoted by $\operatorname{Def}_X(A)$.

The **trivial deformation** of the complex manifold X is given by $X \times \operatorname{Spec} A$, with $f: X \times \operatorname{Spec} A \to \operatorname{Spec} A$ the projection and $i: X \to X \times \operatorname{Spec} A$ the natural embedding. By a common abuse of notation we shall say that a deformation is trivial if it is isomorphic to the trivial deformation.

Lemma 4.3.7 *For a complex manifold X and an Artin local $\mathbb{C}$-algebra $A \in \mathbf{Art}_{\mathbb{C}}$ we have:*

$$\operatorname{Def}_X(A) = \frac{\left\{\begin{array}{c}\textit{Morphisms of sheaves of unitary } A\textit{-algebras}\\ \mathcal{O}_{X_A} \to \mathcal{O}_X \textit{ over } X \textit{ that are locally isomorphic}\\ \textit{to the projection } \mathcal{O}_X \otimes_{\mathbb{C}} A \to \mathcal{O}_X\end{array}\right\}}{\begin{array}{c}\textit{Isomorphisms of sheaves of } A\textit{-algebras}\\ \textit{lifting the identity on } \mathcal{O}_X\end{array}}.$$

Proof Given a deformation $X \xrightarrow{i} X_A \xrightarrow{f} \operatorname{Spec} A$, since A is an Artin local ring, the complex space X_A is supported on X and then, as a ringed space we have $X_A = (X, \mathcal{O}_{X_A})$, where $\mathcal{O}_{X_A}$ is a sheaf of unitary flat A-algebras and $i^*: \mathcal{O}_{X_A} \to \mathcal{O}_X$ is a morphism of sheaves of algebras inducing an isomorphism $\mathcal{O}_{X_A} \otimes_A \mathbb{C} = \mathcal{O}_X$. Denote by n the dimension of X, then by Lemma 4.3.4 every stalk of $\mathcal{O}_{X_A}$ is isomorphic to $\mathbb{C}\{z_1, \ldots, z_n\} \otimes A$. Then, by Theorem 1.6.3, for every $x \in X$ the germ (X_A, x) is isomorphic to $(X \times \operatorname{Spec} A, x)$ and the sheaf $\mathcal{O}_{X_A}$ is locally isomorphic to $\mathcal{O}_X \otimes_{\mathbb{C}} A$.

Conversely, every sheaf of A-algebras on X locally isomorphic to $\mathcal{O}_X \otimes_{\mathbb{C}} A$ gives a deformation of X over $\operatorname{Spec} A$. □

According to the interpretation of deformations in terms of sheaves given in the above lemma, it makes sense to restrict an infinitesimal deformation of X to any open subset $U \subset X$.

Theorem 4.3.8 *Let Θ_X be the holomorphic tangent sheaf of a complex manifold X and let $\mathcal{U} = \{U_i\}$ be an open covering of X. Consider the semicosimplicial Lie algebra of Čech cochains*

$$\Theta_X(\mathcal{U}): \quad \prod_i \Theta_X(U_i) \rightrightarrows \prod_{i,j} \Theta_X(U_{ij}) \substack{\longrightarrow\\ \longrightarrow\\ \longrightarrow} \prod_{i,j,k} \Theta_X(U_{ijk}) \substack{\longrightarrow\\ \longrightarrow\\ \longrightarrow\\ \longrightarrow} \cdots.$$

Then for $A \in \mathbf{Art}_{\mathbb{C}}$ there exists an injective natural transformation

$$H^1_{\Theta_X(\mathcal{U})} \xrightarrow{\kappa} \mathrm{Def}_X$$

such that its image is the collection of isomorphism classes of deformations that are trivial on every open subset U_i. Moreover, if $H^1(U_i, \Theta_X) = 0$ for every i, then κ is bijective.

Proof According to Lemma 3.4.5, for every open subset $U \subset X$ the group $\exp(\Theta_X(U) \otimes \mathfrak{m}_A)$ is naturally isomorphic to the group of A-linear isomorphisms of $\mathcal{O}_U \otimes A$ lifting the identity on $\mathcal{O}_U$. Thus the first part of the theorem follows from a glueing argument that is the same as the one used in the proof of Theorem 4.2.3.

For the second part it is sufficient to prove that if $H^1(X, \Theta_X) = 0$, then every infinitesimal deformation of X is trivial. Let $X \xrightarrow{i} X_A \xrightarrow{f} \operatorname{Spec} A$ be an infinitesimal deformation; by Lemma 4.3.4 we can choose an open cover $\mathcal{U} = \{U_i\}$ such that X_A is trivial on every open subset U_i and then the isomorphism class of X_A belongs to the image of $\kappa\colon H^1_{\Theta_X(\mathcal{U})}(A) \to \mathrm{Def}_X(A)$. Since the map $\check{H}^1(\mathcal{U}, \Theta_X) \to H^1(X, \Theta_X)$ is injective by Remark 1.2.3, we have $T^1 H^1_{\Theta_X(\mathcal{U})} = \check{H}^1(\mathcal{U}, \Theta_X) = 0$ and therefore the deformation functor $H^1_{\Theta_X(\mathcal{U})}$ is trivial. □

Corollary 4.3.9 *Let X be a complex manifold. Then* Def_X *is a deformation functor, $T^1\mathrm{Def}_X = H^1(X, \Theta_X)$ and there exists a complete obstruction theory for* Def_X *with values in $H^2(X, \Theta_X)$.*

Proof In the notation of Theorem 4.3.8, if $\mathcal{U}$ is a Stein open covering of X, then $\mathrm{Def}_X \cong H^1_{\Theta_X(\mathcal{U})}$ and $H^i(C(\Theta_X(\mathcal{U}))) = \check{H}^i(\mathcal{U}, \Theta_X) = H^i(X, \Theta_X)$ for every i. □

Example 4.3.10 We have already observed in Example 1.4.5 that $H^i(\mathbb{P}^n, \Theta_{\mathbb{P}^n}) = 0$ for every pair of positive integers i, n, while the Lie algebra $H^0(\mathbb{P}^n, \Theta_{\mathbb{P}^n})$ is isomorphic to $\mathfrak{sl}_{n+1}(\mathbb{K})$. In particular, every infinitesimal deformation of a projective space is trivial.

Example 4.3.11 Let X be a compact Riemann surface of genus g, then the functor Def_X is unobstructed and $T^1\mathrm{Def}_X$ has dimension g for $g = 0, 1$ and $3g - 3$ for $g \geq 2$.

In fact, $H^2(X, \Theta_X) = 0$ since X has dimension 1, and by Serre duality the dimension of $T^1\mathrm{Def}_X = H^1(X, \Theta_X)$ is the same as the dimension of $H^0(X, 2K_X)$, where K_X is the canonical divisor. Recall now that $2K_X$ has degree $4g - 4$, K_X is trivial whenever $g = 1$ and the dimension of $H^0(X, 2K_X)$ is computed easily by the Riemann–Roch formula.

Example 4.3.12 Let X be a regular smooth projective surface with an anticanonical divisor; this means that $H^0(X, \Omega^1_X) = 0$ (regularity) and $H^0(X, -K_X) \neq 0$. Then $H^2(X, \Theta_X) = 0$ and therefore X has unobstructed deformations. The vanishing of $H^2(X, \Theta_X)$ is an easy consequence of Serre duality $H^2(X, \Theta_X)^\vee \cong$

$H^0(X, \Omega^1_X(K_X))$ together with the fact that every anticanonical divisor gives an injective map $H^0(X, \Omega^1_X(K_X)) \to H^0(X, \Omega^1_X)$.

Notice that, if the anticanonical divisor is not empty then $H^0(X, K_X^{\otimes 2}) = 0$ and X is a rational surface by Castelnuovo's theorem; then one can use the fact that $H^0(X, \Omega^1_X(K_X))$ is a birational invariant and $H^0(\mathbb{P}^2, \Omega^1_{\mathbb{P}^2}(-3)) = 0$.

The next example shows that, while $H^2(X, \Theta_X) = 0$ is a sufficient condition for the smoothness of Def_X, the same condition is not necessary and therefore, for a general complex manifold X, the space $H^2(X, \Theta_X)$ is not the universal obstruction space for Def_X.

Example 4.3.13 Every complex torus has unobstructed deformations. In fact, considering a complex torus X as a compact complex (abelian) Lie group, we may define the subsheaf $\mathcal{L} \subset \Theta_X$ of locally invariant holomorphic vector fields; notice that for every $U \subset X$ the vector space $\mathcal{L}(U)$ is an abelian Lie subalgebra of $\Theta_X(U)$, although not an $\mathcal{O}_X(U)$-module. By a general and well known result, see e.g. [101, p. 301], the sheaves $\mathcal{L}$ and Θ_X have the same cohomology on X and then for every contractible Stein covering $\mathcal{U}$ the morphism of semicosimplicial Lie algebras $\mathcal{L}(\mathcal{U}) \to \Theta_X(\mathcal{U})$ induces a smooth morphism of deformation functors $H^1_{\mathcal{L}(\mathcal{U})} \to H^1_{\Theta(\mathcal{U})} = \operatorname{Def}_X$. By Corollary 3.7.5 the functor $H^1_{\mathcal{L}(\mathcal{U})}$ is smooth and then also Def_X is smooth.

There exist many other examples of manifolds X with Def_X smooth and $H^2(X, \Theta_X) \neq 0$; for instance, every smooth surface in $\mathbb{P}^3$ of degree at least 6 has this property, see Example 4.4.6 below.

4.4 Deformations of Pairs (Manifold, Submanifold)

The notion of deformations of complex manifolds extends naturally to holomorphic maps between complex manifolds. In this section we study the particular case of regular embeddings of complex manifolds, while the general case will be considered in Section 8.2.

Definition 4.4.1 Let Z be a closed submanifold of a complex manifold X and let $A \in \mathbf{Art}_{\mathbb{C}}$ be a local Artin ring. A **deformation of the pair** (X, Z) **over** $\operatorname{Spec} A$ is a commutative diagram of complex spaces

$$\begin{array}{ccccc} Z & \xrightarrow{i} & Z_A & & \\ \downarrow & & \downarrow & \overset{p}{\searrow} & \\ X & \xrightarrow{j} & X_A & \xrightarrow{q} & \operatorname{Spec} A \end{array}$$

such that $Z \xrightarrow{i} Z_A \xrightarrow{p} \operatorname{Spec} A$ is a deformation of Z, $X \xrightarrow{j} X_A \xrightarrow{q} \operatorname{Spec} A$ is a deformation of X and $Z_A \subset X_A$ is a closed embedding.

An isomorphism of two deformations $Z_A \hookrightarrow X_A$ and $Z'_A \hookrightarrow X'_A$ over A is an isomorphism $\phi\colon X_A \xrightarrow{\simeq} X'_A$ of deformations of X that restricts to an isomorphism $\phi\colon Z_A \xrightarrow{\simeq} Z'_A$.

The **functor of infinitesimal deformations** of the pair (X, Z), will be denoted by

$$\mathrm{Def}_{(X,Z)}\colon \mathbf{Art} \to \mathbf{Set},$$

where $\mathrm{Def}_{(X,Z)}(A)$ is the set of isomorphism classes of deformations of the closed embedding $Z \subset X$ over A. There exist two obvious natural transformations of functors:

$$\mathrm{Def}_{(X,Z)} \to \mathrm{Def}_X, \qquad \mathrm{Def}_{(X,Z)} \to \mathrm{Def}_Z.$$

The first just forgets Z and takes care only of the deformations of X; the second forgets X. For every $A \in \mathbf{Art}_{\mathbb{C}}$ we have that $\mathrm{Def}_{(X,Z)}(A) \subset \{(X_A, Z_A, i_A)\}/\sim$, where X_A is a deformation of X, Z_A a deformation of Z and $i_A\colon Z_A \to X_A$ a closed embedding.

Our goal is to extend Theorem 4.3.8 to deformations of pairs (X, Z) and compute tangent and obstruction spaces in terms of Čech cohomology of the sheaf of Lie algebras $\Theta_X(-\log Z)$. By definition, $\Theta_X(-\log Z)$ is the subsheaf of Θ_X of vector fields that are tangent to Z; equivalently, if $\mathcal{I}_Z \subset \mathcal{O}_X$ is the ideal sheaf defining Z, then $\Theta_X(-\log Z)$ is the subsheaf of $\Theta_X = \mathcal{D}er_{\mathbb{C}}(\mathcal{O}_X, \mathcal{O}_X)$ whose elements are the derivations α such that $\alpha(\mathcal{I}_Z) \subset \mathcal{I}_Z$. It is easy to verify that the inclusion $\Theta_X(-\log Z) \xrightarrow{\chi} \Theta_X$ is a morphism of sheaves of Lie algebras.

Remark 4.4.2 Denoting by $i\colon Z \to X$ the inclusion and by $\mathcal{N}_{Z|X} = \mathcal{H}om_{\mathcal{O}_X}(\mathcal{I}_Z, \mathcal{O}_Z)$ the normal sheaf of Z in X, there exist two short exact sequences of sheaves on X:

$$0 \to \Theta_X(-\log Z) \xrightarrow{\chi} \Theta_X \xrightarrow{\pi} \mathcal{N}_{Z|X} \to 0,$$

$$0 \to \mathcal{I}_Z\Theta_X \to \Theta_X(-\log Z) \to \Theta_Z \to 0,$$

where, by a slight and common abuse of notation we identify a sheaf $\mathcal{F}$ on Z with its direct image $i_*\mathcal{F}$. If the codimension of Z is at least 2, then the sheaf $\Theta_X(-\log Z)$ is not locally free, while if Z has codimension 1 then $\Theta_X(-\log Z)$ is the dual of the locally free sheaf of logarithmic differentials $\Omega^1_X(\log Z)$. Notice that the definition of $\Theta_X(-\log Z)$ makes sense also for Z singular; however, in this case the map π is not surjective in general (cf. [7, 167] and [39, Remark 9.16]).

Theorem 4.4.3 *Let $\Theta_X(-\log Z)$ be the sheaf of holomorphic vector fields on a complex manifold X that are tangent to a closed submanifold Z. Given an open cover $\mathcal{U} = \{U_i\}$ of X, consider the semicosimplicial Lie algebra of Čech cochains*

$$\Theta_X(-\log Z, \mathcal{U})\colon \quad \prod_i \Theta_X(-\log Z)(U_i) \rightrightarrows \prod_{i,j} \Theta_X(-\log Z)(U_{ij}) \rightrightarrows\!\!\!\!\!\rightarrow \cdots .$$

Then there exists an injective natural transformation of functors of Artin rings

$$H^1_{\Theta_X(-\log Z,\mathcal{U})} \xrightarrow{\kappa} \mathrm{Def}_{(X,Z)}.$$

For every $A \in \mathbf{Art}_{\mathbb{C}}$ the image of $\kappa\colon H^1_{\Theta_X(-\log Z,\mathcal{U})}(A) \to \mathrm{Def}_X(A)$ is the set of isomorphism classes of deformations inducing the trivial deformation of the pair $(U_i, Z \cap U_i)$ for every i. Moreover, if $H^1(U_i, \Theta_X(-\log Z)) = 0$ for every i, then κ is an isomorphism of functors.

Proof According to Lemma 3.4.6, for every open subset $U \subset X$ the exponential group $\exp(H^0(U, \Theta_X(-\log Z) \otimes \mathfrak{m}_A))$ is naturally isomorphic to the group of A-linear isomorphisms of $\mathcal{O}_U \otimes A$ lifting the identity on $\mathcal{O}_U$ and preserving the ideal $\mathcal{I}_Z \otimes A$. Moreover, according to Lemmas 4.3.4 and 4.3.5 every deformation of (X, Z) is locally trivial and then the same proof of Theorem 4.3.8 works well also in this case. □

Corollary 4.4.4 *Let X be a complex manifold and let $Z \subset X$ be a closed smooth submanifold. Then* $\mathrm{Def}_{(X,Z)}$ *is a deformation functor, in the sense of Definition 3.2.5, with $T^1\mathrm{Def}_{(X,Z)} = H^1(X, \Theta_X(-\log Z))$, and admitting a complete obstruction theory with values in $H^2(X, \Theta_X(-\log Z))$.*

Proof The proof is the same as Corollary 4.3.9, since $H^i(U, \Theta_X(-\log Z)) = 0$ for every open Stein subset $U \subset X$ and every $i > 0$. □

Example 4.4.5 [Pointed Riemann surfaces] An n-pointed Riemann surface of genus g is the data $(X, p_1, \ldots, p_n)$ of a compact Riemann surface X of genus g and an ordered sequence of n distinct points $p_1, \ldots, p_n \in X$. Then $Z = \{p_1, \ldots, p_n\}$ is a closed submanifold and we define a deformation of $(X, p_1, \ldots, p_n)$ as a deformation of the pair (X, Z). Since every deformation of Z is trivial, giving a deformation of the pair (X, Z) over Spec A is the same as giving a deformation $X_A \to \operatorname{Spec} A$ of X together with n sections $s_1, \ldots, s_n\colon \operatorname{Spec} A \to X_A$ such that $s_i(0) = p_i$, where 0 is the closed point of Spec A.

In this case the sheaf $\Theta_X(-\log Z)$ is nothing else than the sheaf $\Theta_X(-p_1 \cdots - p_n)$ of vector fields vanishing at $p_1, \ldots, p_n$ and then

$$T^1\mathrm{Def}_{(X,p_1,\ldots,p_n)} = H^1(X, \Theta_X(-p_1 - \cdots - p_n)) = H^0(X, 2K_X + p_1 + \cdots + p_n)^\vee,$$

with the functor $\mathrm{Def}_{(X,p_1,\ldots,p_n)}$ unobstructed since $H^2(X, \Theta_X(-p_1 - \cdots - p_n)) = 0$.

When $2g - 2 + n > 0$ the pointed Riemann surface is called *stable* and the vector space $T^1\mathrm{Def}_{(X,p_1,\ldots,p_n)}$ has dimension $3g - 3 + n$.

Example 4.4.6 Here we sketch a proof that every smooth surface $Z \subset X = \mathbb{P}^3$ has unobstructed deformations, although $H^2(Z, \Theta_Z) \neq 0$ whenever the degree of Z is at least 6.

Denote by d the degree of Z. We have seen in Example 4.3.12 that $H^2(Z, \Theta_Z) = 0$ whenever $d \leq 4$, and then it is not restrictive to assume $d \geq 5$. Since $\mathcal{N}_{Z|X} = \mathcal{O}_Z(d)$, the exact sequences

$$0 \to \mathcal{O}_X \to \mathcal{O}_X(d) \to \mathcal{O}_Z(d) \to 0, \quad 0 \to \Theta_X(-\log Z) \to \Theta_X \to \mathcal{N}_{Z|X} \to 0,$$

give $H^i(Z, \mathcal{N}_{Z|X}) = H^{i+1}(X, \Theta_X(-\log Z)) = 0$ for every $i > 0$. The vanishing of $H^2(X, \Theta_X(-\log Z))$ implies in particular that the deformation functor $\mathrm{Def}_{(X,Z)}$ is unobstructed and the forgetful natural transformation $\mathrm{Def}_{(X,Z)} \to \mathrm{Def}_Z$ is smooth if and only if it is surjective on tangent spaces. By the short exact sequence

$$0 \to \Theta_X(-d) \to \Theta_X(-\log Z) \to \Theta_Z \to 0,$$

we get $H^2(Z, \Theta_Z) = H^3(X, \Theta_X(-d))$ and $T^1\mathrm{Def}_{(X,Z)} \to T^1\mathrm{Def}_Z$ is surjective if and only if $H^2(X, \Theta_X(-d)) = 0$. Finally, it is an easy exercise in algebraic geometry to see that $H^2(X, \Theta_X(-d)) = 0$ for every $d \geq 5$ and $H^2(Z, \Theta_Z) \neq 0$ for every $d \geq 6$.

Example 4.4.7 Let $C \subset X = \mathbb{P}^n$ be a smooth curve such that $H^1(C, \mathcal{O}_C(1)) = 0$. Then the functor $\mathrm{Def}_{(C,X)}$ is unobstructed and the forgetful natural transformation $\mathrm{Def}_{(C,X)} \to \mathrm{Def}_C$ is smooth. In fact, the Euler sequence restricted to C gives an exact sequence

$$0 \to \mathcal{O}_C \to \mathcal{O}_C(1)^{\oplus n+1} \to \Theta_X \otimes_{\mathcal{O}_X} \mathcal{O}_C \to 0$$

and the assumption $H^1(\mathcal{O}_C(1)) = 0$ gives $H^1(\Theta_X \otimes_{\mathcal{O}_X} \mathcal{O}_C) = 0$. Since $H^2(\Theta_X) = 0$, the exact sequence

$$0 \to \mathcal{I}_C\Theta_X \to \Theta_X \to \Theta_X \otimes_{\mathcal{O}_X} \mathcal{O}_C \to 0$$

gives $H^2(\mathcal{I}_C\Theta_X) = 0$ and therefore the sequence

$$H^1(\Theta_X(-\log C)) \to H^1(\Theta_C) \to 0 \to H^2(\Theta_X(-\log C)) \to H^2(\Theta_C) = 0$$

is exact. The conclusion follows from the standard smoothness criterion.

Corollary 4.4.8 (Kodaira stability theorem) *Let X be a complex manifold and let $Z \subset X$ be a closed smooth submanifold such that $H^1(Z, \mathcal{N}_{Z|X}) = 0$. Then the natural morphism $\mathrm{Def}_{(X,Z)} \to \mathrm{Def}_X$ is smooth.*

Proof Looking at the long cohomology exact sequence of

$$0 \to \Theta_X(-\log Z) \xrightarrow{\chi} \Theta_X \xrightarrow{\pi} \mathcal{N}_{Z|X} \to 0,$$

we have that $H^1(Z, \mathcal{N}_{Z|X}) = 0$ if and only if $H^1(X, \Theta_X(-\log Z)) \to H^1(X, \Theta_X)$ is surjective and $H^2(X, \Theta_X(-\log Z)) \to H^2(X, \Theta_X)$ is injective. It is now sufficient to apply Corollaries 4.3.9, 4.4.4, and the standard smoothness criterion Theorem 3.6.5. □

Example 4.4.9 Let $\sigma\colon X \to Y$ be the blow-up of a complex manifold at a point $y \in Y$ and let $E = \sigma^{-1}(y)$ be the exceptional divisor. Then $(E, \mathcal{N}_{E|X}) \simeq (\mathbb{P}^n, \mathcal{O}(-1))$, with $n = \dim Y - 1$, and then $H^1(E, \mathcal{N}_{E|X}) = 0$. By the Kodaira stability theorem the forgetful natural transformation $\mathrm{Def}_{(X,E)} \to \mathrm{Def}_X$ is smooth and the exceptional divisor is stable under any deformation of X.

Example 1.4.7 shows that without the assumption $H^1(Z, \mathcal{N}_{Z|X}) = 0$ the above corollary is generally false.

4.5 Deformations of Pairs (Manifold, Vector Bundle)

In Definition 2.3.5 we introduced the module of derivations of pairs

$$D^1_{\mathbb{C}}(R, M) \subset \mathrm{Der}_{\mathbb{C}}(R, R) \times \mathrm{Hom}_{\mathbb{C}}(M, M),$$
$$D^1_{\mathbb{C}}(R, M) = \{(h, u) \mid u(am) - au(m) = h(a)m \text{ for every } a \in R,\ m \in M\},$$

for a commutative unitary algebra R over the field of complex numbers and an R-module M. In a similar way we may introduce the group of automorphisms of the pair.

Definition 4.5.1 Let $A \to S$ be a morphism of commutative unitary rings and let N be an S-module. The group $\mathrm{Aut}_A(S, N)$ of A-linear automorphisms of the pair (S, N) is defined as the group of pairs $(\theta, \phi) \in \mathrm{Hom}_A(S, S) \times \mathrm{Hom}_A(N, N)$ such that θ is an isomorphism of A-algebras and ϕ is an isomorphism of A-modules such that $\phi(sn) = \theta(s)\phi(n)$ for every $s \in S$, $n \in N$.

Lemma 4.5.2 *Let R be a commutative unitary algebra over the field $\mathbb{C}$, and let M be an R-module. Then for every $A \in$* **Art**$_{\mathbb{C}}$ *the group $\exp(D^1_{\mathbb{C}}(R, M) \otimes \mathfrak{m}_A)$ is naturally isomorphic to the group $\mathrm{Aut}_A(R \otimes A, M \otimes A)$ of A-linear automorphisms of the pair $(R \otimes A, M \otimes A)$ lifting the identity on (R, M).*

Proof Let $R \oplus M$ be the trivial extension of R by M; it is a unitary commutative $\mathbb{C}$-algebra with product $(a, n)(b, m) = (ab, am + nb)$, for $a, b \in R$ and $n, m \in M$. There exists a natural inclusion

$$\mathrm{Hom}_{\mathbb{C}}(R, R) \times \mathrm{Hom}_{\mathbb{C}}(M, M) \subset \mathrm{Hom}_{\mathbb{C}}(R \oplus M, R \oplus M)$$

and we can immediately see that a couple $(h, u) \in \mathrm{Hom}_{\mathbb{C}}(R, R) \times \mathrm{Hom}_{\mathbb{C}}(M, M)$ is a $\mathbb{C}$-linear derivation (resp.: automorphism) of the pair (R, M) if and only if it is a $\mathbb{C}$-linear derivation (resp.: automorphism) of the trivial extension $R \oplus M$. Now the conclusion follows immediately from Proposition 3.4.3. □

The above algebraic construction extends naturally to sheaves of algebras and modules. For a complex manifold X and a sheaf of $\mathcal{O}_X$-modules $\mathcal{F}$, the derivations of the pair $(\mathcal{O}_X, \mathcal{F})$ is defined as the sheaf of Lie algebras

$$\mathcal{D}^1_X(\mathcal{F}) := \mathcal{D}^1_{\mathbb{C}}(\mathcal{O}_X, \mathcal{F}) \subset \mathcal{D}er_{\mathbb{C}}(\mathcal{O}_X, \mathcal{O}_X) \times \mathcal{H}om_{\mathbb{C}}(\mathcal{F}, \mathcal{F}).$$

By Lemma 2.3.6, we have an exact sequence of sheaves of Lie algebras

$$0 \to \mathcal{H}om_{\mathcal{O}_X}(\mathcal{F}, \mathcal{F}) \to \mathcal{D}^1_X(\mathcal{F}) \xrightarrow{p} \Theta_X = \mathcal{D}er_{\mathbb{C}}(\mathcal{O}_X, \mathcal{O}_X),$$

and if $\mathcal{F}$ is locally free, then p is surjective and $\mathcal{D}^1_X(\mathcal{F})$ is naturally isomorphic to the sheaf of first order differential operators with principal symbol. Similarly, for every $A \in \mathbf{Art}_{\mathbb{C}}$ we have a group

$$\mathrm{Aut}_A(\mathcal{O}_X \otimes A, \mathcal{F} \otimes A) \subset \mathcal{H}om_A(\mathcal{O}_X \otimes A, \mathcal{O}_X \otimes A) \times \mathcal{H}om_A(\mathcal{F} \otimes A, \mathcal{F} \otimes A)$$

and the exponential group $\exp(\Gamma(X, \mathcal{D}^1_X(\mathcal{F})) \otimes \mathfrak{m}_A)$ is naturally isomorphic to the subgroup

$$\{(\theta, \phi) \in \mathrm{Aut}_A(\mathcal{O}_X \otimes A, \mathcal{F} \otimes A) \mid (\theta, \phi) \equiv (\mathrm{Id}_{\mathcal{O}_X}, \mathrm{Id}_{\mathcal{F}}) \mod \mathfrak{m}_A\}.$$

Definition 4.5.3 Let X be a complex manifold and let $\mathcal{F}$ be a sheaf of $\mathcal{O}_X$-modules. A deformation of the pair $(X, \mathcal{F})$ over $A \in \mathbf{Art}_{\mathbb{C}}$ is the data of a deformation $X \xrightarrow{i} X_A \to \operatorname{Spec} A$ of X, of a sheaf $\mathcal{F}_A$ of $\mathcal{O}_{X_A}$-modules that is flat over A, and a morphism of sheaves of $\mathcal{O}_{X_A}$-modules $\mathcal{F}_A \to i_*\mathcal{F}$ inducing an isomorphism $\mathcal{F}_A \otimes_{\mathcal{O}_{X_A}} \mathcal{O}_X \to \mathcal{F}$.

Notice that the sheaf $i_*\mathcal{F}$ is nothing else than the sheaf $\mathcal{F}$ equipped with the structure of $\mathcal{O}_{X_A}$-module induced by the morphism $\mathcal{O}_{X_A} \to \mathcal{O}_X$. An isomorphism of two deformations $(X_A, \mathcal{F}_A)$ and $(X'_A, \mathcal{F}'_A)$ is a pair of commutative diagrams

$$\begin{array}{ccc} \mathcal{O}_{X_A} & \xrightarrow{\theta} & \mathcal{O}_{X'_A} \\ & \searrow \quad \swarrow & \\ & \mathcal{O}_X & \end{array} \qquad \begin{array}{ccc} \mathcal{F}_A & \xrightarrow{\phi} & \mathcal{F}'_A \\ & \searrow \quad \swarrow & \\ & i_*\mathcal{F} & \end{array}$$

where θ is an isomorphism of sheaves of A-algebras, and ϕ is an isomorphism of sheaves such that $\phi(af) = \theta(a)\phi(f)$ for $a \in \mathcal{O}_{X_A}$ and $f \in \mathcal{F}_A$.

We shall denote by $\mathrm{Def}_{(X,\mathcal{F})} \colon \mathbf{Art}_{\mathbb{C}} \to \mathbf{Set}$ the corresponding functor of Artin rings.

The same glueing argument used in previous sections gives the following result.

Theorem 4.5.4 *Let $\mathcal{F}$ be a sheaf of $\mathcal{O}_X$-modules on a complex manifold X. Given an open cover $\mathcal{U} = \{U_i\}$ of X, consider the semicosimplicial Lie algebra of Čech cochains*

$$\mathcal{D}^1_X(\mathcal{F}, \mathcal{U})\colon \quad \textstyle\prod_i \Gamma(U_i, \mathcal{D}^1_X(\mathcal{F})) \rightrightarrows \prod_{i,j} \Gamma(U_{ij}, \mathcal{D}^1_X(\mathcal{F})) \;\substack{\longrightarrow\\[-2pt]\longrightarrow\\[-2pt]\longrightarrow}\; \cdots .$$

Then there exists an injective natural transformation of functors of Artin rings

$$H^1_{\mathcal{D}^1_X(\mathcal{F},\mathcal{U})} \xrightarrow{\kappa} \mathrm{Def}_{(X,\mathcal{F})}.$$

For every $A \in \mathbf{Art}_{\mathbb{C}}$, the image of $\kappa\colon H^1_{\mathcal{D}^1_X(\mathcal{F},\mathcal{U})}(A) \to \mathrm{Def}_{(X,\mathcal{F})}(A)$ is the set of the isomorphism classes of deformations inducing the trivial deformation of the pair $(U_i, \mathcal{F}_{|U_i})$ for every i. If $\mathcal{F}$ is locally free of finite rank and $H^1(U_i, \mathcal{D}^1_X(\mathcal{F})) = 0$ for every i, then κ is an isomorphism of functors.

Proof In order to apply the same argument used in the proofs of Theorems 4.2.3, 4.3.8 and 4.4.3, we only need to prove that if $\mathcal{F}$ is locally free of finite rank, then every deformation of the pair $(X, \mathcal{F})$ is locally trivial; this follows from Lemma 4.2.4 since every deformation of X is locally trivial. □

Corollary 4.5.5 *Let $\mathcal{E}$ be a locally free sheaf of finite rank on a complex manifold X. Then*

$$T^1\mathrm{Def}_{(X,\mathcal{E})} = H^1(X, \mathcal{D}^1_X(\mathcal{E})),$$

and there exists a complete obstruction theory for the functor $\mathrm{Def}_{(X,\mathcal{E})}$ *with values in $H^2(X, \mathcal{D}^1_X(\mathcal{E}))$.*

Proof It is sufficient to repeat, mutatis mutandis, the proof of Corollary 4.3.9. □

Let $\mathcal{U}$ be a Stein covering of a complex manifold X and let $\mathcal{E}$ be a locally free sheaf. The surjective morphism of sheaves of Lie algebras $\mathcal{D}^1_X(\mathcal{E}) \to \Theta_X$ induces a morphism of semicosimplicial Lie algebras $\mathcal{D}^1_X(\mathcal{E}, \mathcal{U}) \to \Theta_X(\mathcal{U})$ and then a natural transformation $H^1_{\mathcal{D}^1_X(\mathcal{E},\mathcal{U})} \to H^1_{\Theta_X(\mathcal{U})}$ commuting with both κ and the forgetful morphism $\mathrm{Def}_{(X,\mathcal{E})} \to \mathrm{Def}_X$. Thus, if

$$H^1(X, \mathcal{D}^1_X(\mathcal{E})) \xrightarrow{\alpha} H^1(X, \Theta_X) \to \mathrm{Ext}^2_X(\mathcal{E}, \mathcal{E}) \to H^2(X, \mathcal{D}^1_X(\mathcal{E})) \xrightarrow{\beta} H^2(X, \Theta_X)$$

is the cohomology exact sequence of $0 \to \mathcal{H}om_{\mathcal{O}_X}(\mathcal{E}, \mathcal{E}) \to \mathcal{D}^1_X(\mathcal{E}) \to \Theta_X \to 0$, then α is the first order map $T^1\mathrm{Def}_{(X,\mathcal{E})} \to T^1\mathrm{Def}_X$ and β is a morphism of obstruction theories.

Corollary 4.5.6 *In the above situation, if* $\mathrm{Ext}^2_X(\mathcal{E}, \mathcal{E}) = 0$, *then the forgetful morphism* $\mathrm{Def}_{(X,\mathcal{E})} \to \mathrm{Def}_X$ *is smooth.*

Proof Apply the standard smoothness criterion. □

4.6 Exercises

4.6.1 (Flatness & integrability) Let

$$P_0 \xrightarrow{d_1} P_1 \xrightarrow{d_2} \cdots \xrightarrow{d_n} P_n \to 0, \tag{4.2}$$

be a finite complex of flat modules over a local Artin ring A with residue field $\mathbb{K}$. Prove that if

$$P_0 \otimes_A \mathbb{K} \xrightarrow{\overline{d_1}} P_1 \otimes_A \mathbb{K} \to \cdots \to P_n \otimes_A \mathbb{K} \to 0$$

is an exact sequence, then (4.2) is an exact sequence, the A-module $\ker(d_1)$ is flat and the natural map $\ker(d_1) \otimes_A \mathbb{K} \to \ker(\overline{d_1})$ is an isomorphism.

4.6.2 (Flatness & relations) Let

$$P \xrightarrow{f} Q \xrightarrow{g} R \xrightarrow{h} M \to 0 \tag{4.3}$$

be a complex of modules over a local Artin ring A with residue field $\mathbb{K}$. Assume that:

1. P, Q and R are flat;
2. $Q \xrightarrow{g} R \xrightarrow{h} M \to 0$ is an exact sequence;
3. $P \otimes_A \mathbb{K} \xrightarrow{\overline{f}} Q \otimes_A \mathbb{K} \xrightarrow{\overline{g}} R \otimes_A \mathbb{K} \xrightarrow{\overline{h}} M \otimes_A \mathbb{K} \to 0$ is exact.

Prove that M is flat and the complex (4.3) is an exact sequence.

4.6.3 Let $x_0, \ldots, x_n$, $n \geq 3$, be a system of homogeneous coordinates in the projective space $\mathbb{P}^n$. Using the Euler exact sequences of sheaves

$$0 \to \Omega^1_{\mathbb{P}^n} \to \bigoplus_{i=0}^{n} \mathcal{O}_{\mathbb{P}^n}(-1) \xrightarrow{(x_0,\ldots,x_n)} \mathcal{O}_{\mathbb{P}^n} \to 0,$$

compute the dimension of $H^i(\mathbb{P}^n, \Omega^1_{\mathbb{P}^n}(d))$ for every $0 \leq i < n$ and every integer d.

4.6.4 Let $L \subset \mathbb{P}^n$ be a projective subspace of codimension d. Prove that the normal sheaf of L inside $\mathbb{P}^n$ is $\mathcal{O}_L(1)^d$.

4.6.5 Let L be a projective line contained in a smooth surface $S \subset \mathbb{P}^n$ of degree d. Prove that $\mathcal{N}_{L|S} \simeq \mathcal{O}_L(2-d)$ and then $H^1(L, \mathcal{N}_{L|S}) = 0$ if and only if $d \leq 3$. Compare this fact with Corollary 4.4.8 and with the well known fact that the generic surface of degree $d \geq 4$ does not contain lines.

4.6.6 Let $Z = \{\sum_{i=0}^{n} x_i^d = 0\} \subset \mathbb{P}^n$ be the Fermat hypersurface of degree $d \geq 3$. Prove that $H^0(\Theta_{\mathbb{P}^n}(-\log Z)) = 0$ and deduce that the functor $\mathrm{Def}_{(Z,\mathbb{P}^n)}$ is homogeneous. Generalize to every smooth hypersurface $Z = \{f = 0\}$ of degree $d \geq 3$. (Hint: the partial derivatives of f are a regular sequence in the graded ring $\mathbb{C}[x_0, \ldots, x_n]$.)

4.6.7 We have already pointed out that the definition of the sheaf $\Theta_X(-\log Z)$ makes sense also for Z a singular subvariety of X and the same arguments used in Sect. 4.4 imply that the cosimplicial Lie algebra of its Čech cochains with respect to a Stein covering controls the deformations of the pair that are locally trivial.

Consider the pair (X, Z) where $X = \mathbb{P}^2$ and Z is the union of two distinct lines; prove that $H^1(X, \Theta_X(-\log Z)) = 0$, while the deformation of Z to a smooth conic gives a nontrivial deformation of the pair.

References

Some proofs of Sect. 4.1 can be simplified by using the properties of Tor functors, and Corollary 4.1.3 can be improved by saying that a module M over a local Artin ring A is free if and only if $\mathrm{Tor}_1^A(M, A/\mathfrak{m}_A) = 0$, cf. [7].

If X is a complex manifold, then for every $\mathcal{O}_X$-module $\mathcal{F}$ the functor $\mathrm{Def}_{\mathcal{F}}$ is a deformation functor, as proved for instance in [237, Sect. 3]. If $\mathcal{F}$ is a coherent sheaf on a smooth projective manifold, then it is still true that $T^1\mathrm{Def}_{\mathcal{F}} = \mathrm{Ext}^1_X(\mathcal{F}, \mathcal{F})$ and that there exists a complete obstruction theory with values in the kernel of the trace map $\mathrm{Ext}^2_X(\mathcal{F}, \mathcal{F}) \to H^2(X, \mathcal{O}_X)$, see e.g. [4, 73] and references therein.

The original reference for the stability theorem 4.4.8 is [144]; other proofs can be found in [203] and [242, Proposition 3.4.23].

For an invertible sheaf $\mathcal{L}$ on a complex manifold X, the exact sequence

$$E_{\mathcal{L}}: \qquad 0 \to \mathcal{H}om_{\mathcal{O}_X}(\mathcal{L}, \mathcal{L}) = \mathcal{O}_X \to \mathcal{D}^1_X(\mathcal{L}) \to \Theta_X \to 0$$

is called the *Atiyah extension* of $\mathcal{L}$ (cf. [120, 183], [242, p. 145]). The properties of this extension are studied by Atiyah in [9], where it is proved in particular that, if X is compact Kähler, then the image of its extension class

$$[E_{\mathcal{L}}] \in \mathrm{Ext}^1_X(\Theta_X, \mathcal{O}_X) = H^1(X, \Omega^1_X)$$

into the de Rham cohomology group $H^2(X, \mathbb{C})$ is equal to $2\pi i\, c_1(\mathcal{L})$.

The result of Exercise 4.6.1 is essentially the infinitesimal version of the Newlander–Niremberg integration theorem [206] applied to deformations of complex manifolds, see [124] for details. The result of Exercise 4.6.2 is well known and widely used in deformation theory of algebraic schemes, cf. [7].

Chapter 5
Differential Graded Lie Algebras

In this chapter we introduce the basic algebraic theory of differential graded vector spaces and differential graded Lie algebras over an arbitrary field. The application of these structures in deformation theory, for which the base field should be of characteristic 0, are postponed to Chap. 6. As in the previous chapters we assume that the reader has a basic knowledge of homological and commutative algebra, although some general and well known results are recalled in order to fix notation.

Throughout this chapter any vector space and any linear map is considered over a fixed field $\mathbb{K}$. Unless otherwise specified, by the symbol $\otimes$ we mean the tensor product $\otimes_{\mathbb{K}}$ over the field $\mathbb{K}$.

5.1 DG-Vector Spaces

By a graded vector space we mean a $\mathbb{K}$-vector space V together with a $\mathbb{Z}$-graded direct sum decomposition $V = \oplus_{n\in\mathbb{Z}} V^n$. Any V^n is a $\mathbb{K}$-vector space and any element $v \in V$ may be written uniquely as $v = \sum_n v_n$, where $v_n \in V^n$ for every $n \in \mathbb{Z}$, and all but a finite number of v_n are equal to 0; the nonzero components v_n are called the homogeneous components of v.

We shall say that a nonzero vector $v \in V$ is **homogeneous of degree** n, and we shall write $\deg(v; V) = n$, if $v \in V^n - \{0\}$. If there is no possibility of confusion about V we simply write either $\deg(v) = \deg(v; V)$ or $\overline{v} = \deg(v; V)$.

Many formulas in this book involve the degrees $\overline{v_1}, \overline{v_2}, \ldots$ of a finite sequence of vectors in some graded vector space V; when this occurs *we always assume*, even if not explicitly said, that $v_1, v_2, \ldots$ are nonzero homogeneous vectors.

A morphism of graded vector spaces $f\colon \oplus_{n\in\mathbb{Z}} V^n \to \oplus_{n\in\mathbb{Z}} W^n$ is a linear map such that $f(V^n) \subset W^n$ for every n.

M. Manetti, *Lie Methods in Deformation Theory*, Springer Monographs in Mathematics,
https://doi.org/10.1007/978-981-19-1185-9_5

Definition 5.1.1 A DG-vector space is a pair (V, d), where $V = \oplus_{n\in\mathbb{Z}} V^n$ is a graded vector space and $d\colon V \to V$ is a linear map, called a **differential**, with the properties that $d(V^n) \subset V^{n+1}$ for every n and $d^2 = d \circ d = 0$.

A morphism $f\colon (V, d_V) \to (W, d_W)$ of DG-vector spaces is a morphism of graded vector spaces commuting with the differentials: $d_W f = f d_V$.

The category of DG-vector spaces will be denoted by **DG**; a DG-vector space is also called a *cochain complex*.

Thus, giving a morphism $f\colon (V, d_V) \to (W, d_W)$ of DG-vector spaces is the same as giving a sequence of linear maps $f_n\colon V^n \to W^n$ such that $d_W f_n = f_{n+1} d_V$ for every n.

Remark 5.1.2 In the same way one can define dg-vector spaces, in which differentials have degree -1. In other words, we are adopting the convention that both prefixes dg and DG mean differential graded; the first is for chain complexes, and the second for cochain complexes. Sometimes, when the degree of the differential is clear from the context, we simply say differential graded in place of either DG or dg, and we say complex in place of either cochain or chain complex.

In the category **DG**, kernels, cokernels and exact sequences are defined in the usual way. For instance, for any DG-vector space V we have the short exact sequences

$$0 \to V^{\geq p} \to V \to V^{<p} \to 0, \qquad p \in \mathbb{Z},$$

where $V^{\geq p} = \oplus_{n\geq p} V^n$ and by $V^{<p} = \oplus_{n<p} V^n$ are the brutal truncations.

According to the standard notation of homological algebra, for any DG-vector space (V, d) we denote by $Z^*(V) = \ker d$ the graded subspace of cocycles, by $B^*(V) = d(V)$ the graded subspace of coboundaries and by $H^*(V) = Z^*(V)/B^*(V)$ the cohomology of V. For a morphism $f\colon V \to W$ of DG-vector spaces we use the same symbol to denote the induced morphism in cohomology $f\colon H^*(V) \to H^*(W)$.

The category **DG** is both complete and cocomplete; it is closed under small limits and small colimits, which are taken degreewise as the usual limit and colimit in the category of vector spaces. For instance, if $\{(V_i, d_i)\}$ is a family of DG-vector spaces, then its product is:

$$\prod_i V_i = \bigoplus_{n\in\mathbb{Z}} (\prod_i V_i)^n, \quad (\prod_i V_i)^n = \prod_i V_i^n, \quad d(\{v_i\}) = \{d_i(v_i)\}, \quad v_i \in V_i.$$

We can immediately see that $H^*(\prod_i V_i) = \prod_i H^*(V_i)$, while in general the cohomology functor does not commute with limits; given a pull-back diagram of DG-vector spaces

$$\begin{array}{ccc} M \times_N L & \xrightarrow{\bar{f}} & M \\ \downarrow{\scriptstyle \bar{g}} & & \downarrow{\scriptstyle g} \\ L & \xrightarrow{f} & N \end{array} \tag{5.1}$$

in general the map $H^*(M \times_N L) \to H^*(M) \times_{H^*(N)} H^*(L)$, induced by the commutative diagram

$$\begin{array}{ccc} H^*(M \times_N L) & \xrightarrow{\overline{f}} & H^*(M) \\ \downarrow{\scriptstyle \overline{g}} & & \downarrow{\scriptstyle g} \\ H^*(L) & \xrightarrow{f} & H^*(N), \end{array}$$

is not an isomorphism.

Lemma 5.1.3 *In the notation of* (5.1), *assume that both maps*

$$M \times L \to N, \qquad H^*(M) \times H^*(L) \to H^*(N), \qquad (m, l) \mapsto g(m) - f(l)$$

are surjective. Then $H^*(M \times_N L) = H^*(M) \times_{H^*(N)} H^*(L)$.

In particular, if $N = f(L) + g(M)$ *and* $f : H^*(L) \to H^*(N)$ *is bijective, then also* $\overline{f} : H^*(M \times_N L) \to H^*(M)$ *is bijective.*

Proof By assumption, the cohomology exact sequence of the short exact sequence of complexes

$$0 \to M \times_N L \to M \times L \xrightarrow{s} N \to 0, \qquad s(m, l) = g(m) - f(l),$$

splits into countably many short exact sequences

$$0 \to H^n(M \times_N L) \to H^n(M) \times H^n(L) \xrightarrow{s} H^n(N) \to 0, \qquad n \in \mathbb{Z},$$

and this is equivalent to the equality

$$H^*(M \times_N L) = H^*(M) \times_{H^*(N)} H^*(L).$$

□

A morphism of DG-vector spaces is called a **quasi-isomorphism**, or a **weak equivalence**, if it induces an isomorphism in cohomology. It follows immediately from Example 3.5.13 that the class of quasi-isomorphisms provides a homotopy structure in the category **DG**.

A DG-vector space (V, d) is called **acyclic** if $H^*(V) = 0$; equivalently a DG-vector space is acyclic if it is weakly equivalent to 0. A DG-vector space (V, d) is called **minimal** if $d = 0$. Every graded vector space may be interpreted as a minimal DG-vector space.

Lemma 5.1.4 (Homotopy classification) *Any DG-vector space is the direct sum of a minimal DG-vector space and of an acyclic DG-vector space.*

Proof Let (V, d) be a DG-vector space and choose for every n two direct sum decompositions of vector spaces

$$Z^n(V) = B^n(V) \oplus H^n, \qquad V^n = Z^n(V) \oplus C^n.$$

Since $d\colon C^n \to B^{n+1}(V)$ is an isomorphism for every n, the DG-vector space

$$(W, d), \qquad W^n = B^n(V) \oplus C^n,$$

is acyclic, $H = \oplus H^n$ is minimal and $(V, d) = (W, d) \oplus (H, 0)$. □

Given two DG-vector spaces V, W we may define their (total) **tensor product**

$$V \otimes W = \bigoplus_{n\in\mathbb{Z}} (V \otimes W)^n \in \mathbf{DG},$$

where

$$(V \otimes W)^n = \bigoplus_{i+j=n} V^i \otimes W^j, \qquad d(v \otimes w) = dv \otimes w + (-1)^{\bar{v}} v \otimes dw.$$

We can also define their (total) **Hom complex**

$$\mathrm{Hom}^*_{\mathbb{K}}(V, W) = \bigoplus_{n\in\mathbb{Z}} \mathrm{Hom}^n_{\mathbb{K}}(V, W) \in \mathbf{DG},$$

where

$$\begin{aligned}\mathrm{Hom}^n_{\mathbb{K}}(V, W) &= \{f\colon V \to W \text{ linear} \mid f(V^i) \subset W^{i+n} \ \forall i\} \\ &\cong \prod_{i\in\mathbb{Z}} \mathrm{Hom}_{\mathbb{K}}(V^i, W^{i+n}),\end{aligned}$$

$$d\colon \mathrm{Hom}^n_{\mathbb{K}}(V, W) \to \mathrm{Hom}^{n+1}_{\mathbb{K}}(V, W), \quad df(v) = d_W(f(v)) - (-1)^n f(d_V(v)).$$

Lemma 5.1.5 *A DG-vector space* (V, d) *is acyclic if and only if the identity is a coboundary in* $\mathrm{Hom}^*_{\mathbb{K}}(V, V)$.

Proof We want to prove that (V, d) is acyclic if and only if there exists a contracting homotopy $h \in \mathrm{Hom}^{-1}_{\mathbb{K}}(V, V)$ such that $hdv + dhv = v$ for every $v \in V$. If there exists h as above, for any cocycle $v \in Z^n(V)$, since $dv = 0$ we have

$$v = dhv + hdv = dhv = dw \in B^n(V), \qquad w = hv,$$

and then v is trivial in cohomology. Conversely, as in the proof of Lemma 5.1.4 we may choose for every n a direct sum decomposition of vector spaces

$$V^n = Z^n(V) \oplus C^n.$$

If (V, d) is acyclic, then $B^n(V) = Z^n(V)$ for every n and we may define the homotopy $h \in \mathrm{Hom}^{-1}_{\mathbb{K}}(V, V)$ by setting $h(C^n) = 0$ and $h\colon Z^n(V) \to C^{n-1}$ as the inverse of the bijective map $d\colon C^{n-1}\to Z^n(V)$. We can immediately verify that $dh+hd=\mathrm{Id}_V$. □

Theorem 5.1.6 (Künneth's formulas) *Given a DG-vector space V, consider its cohomology $H^*(V)=\bigoplus_n H^n(V)$ as a minimal DG-vector space. For any pair of DG-vector spaces V, W there exist two natural isomorphisms:*

$$H^*(V)\otimes H^*(W)\xrightarrow{\simeq} H^*(V\otimes W),$$
$$H^*(\mathrm{Hom}^*_{\mathbb{K}}(V,W))\xrightarrow{\simeq} \mathrm{Hom}^*_{\mathbb{K}}(H^*(V),H^*(W)).$$

Proof This is a well known result of basic homological algebra, see e.g. [269, 3.6]. For completeness sake we sketch a proof of the second formula; the proof of the first is essentially the same. Any element of $f\in Z^n(\mathrm{Hom}^*_{\mathbb{K}}(V,W))$ factors to a linear map $f\colon H^*(V)\to H^*(W)$ of degree n and there exists a natural morphism of graded vector spaces

$$H^*(\mathrm{Hom}^*_{\mathbb{K}}(V,W))\to \mathrm{Hom}^*_{\mathbb{K}}(H^*(V),H^*(W)). \tag{5.2}$$

If either $H^*(V)=0$ or $H^*(W)=0$ then also $H^*(\mathrm{Hom}^*_{\mathbb{K}}(V,W))=0$. In fact, if $H^*(W)=0$ then, by Lemma 5.1.5, the identity on W is a coboundary, i.e., there exists $h\in \mathrm{Hom}^{-1}_{\mathbb{K}}(W,W)$ such that $dh+hd=\mathrm{Id}_W$. It is easy to see that the linear map

$$k\colon \mathrm{Hom}^*_{\mathbb{K}}(V,W)\to \mathrm{Hom}^{*-1}_{\mathbb{K}}(V,W),\qquad k(f)=hf,$$

gives $kd+dk=\mathrm{Id}$ and then also the identity in $\mathrm{Hom}^*_{\mathbb{K}}(V,W)$ is a coboundary. The same argument applies if $H^*(V)=0$. By homotopy classification we have two direct sum decomposition of DG-vector spaces

$$V=H^*(V)\oplus C,\qquad W=H^*(W)\oplus D,$$

where C and D are acyclic. Therefore

$$\mathrm{Hom}^*_{\mathbb{K}}(V,W)=\mathrm{Hom}^*_{\mathbb{K}}(H^*(V),H^*(W))\oplus \mathrm{Hom}^*_{\mathbb{K}}(C,H^*(W))\oplus \mathrm{Hom}^*_{\mathbb{K}}(V,D),$$

and by the above remark, both $\mathrm{Hom}^*_{\mathbb{K}}(V,D)$ and $\mathrm{Hom}^*_{\mathbb{K}}(C,H^*(W))$ are acyclic. □

Corollary 5.1.7 *For a short exact sequence of DG-vector spaces*

$$0\to U\xrightarrow{f} V\xrightarrow{g} W\to 0 \tag{5.3}$$

the following conditions are equivalent:

1. *the sequence* (5.3) *is split exact: there exists a morphism of DG-vector spaces $s\colon W\to V$ such that $gs=\mathrm{Id}_W$;*
2. *$g\colon H^n(V)\to H^n(W)$ is surjective for every n;*
3. *$f\colon H^n(U)\to H^n(V)$ is injective for every n;*

4. *the morphism of DG-vector spaces* $\mathrm{Hom}^*_{\mathbb{K}}(W, U) \to \mathrm{Hom}^*_{\mathbb{K}}(W, V)$, $h \mapsto fh$, *is injective in cohomology.*

Proof The implication (1) ⇒ (2) is clear, the implication (2) ⇒ (3) follows from the long cohomology exact sequence and (3) ⇒ (4) follows from Künneth's formula.

Assume that the injective morphism of DG-vector spaces

$$\mathrm{Hom}^*_{\mathbb{K}}(W, U) \to \mathrm{Hom}^*_{\mathbb{K}}(W, V), \qquad h \mapsto fh,$$

is injective in cohomology. Let's choose a morphism $\alpha\colon W \to V$ of graded vector spaces such that $g\alpha = \mathrm{Id}_W$, and define $\beta = d\alpha - \alpha d$. Since $g\beta = 0$ we have $\beta = f\gamma$ for some $\gamma \in \mathrm{Hom}^1_{\mathbb{K}}(W, U)$. The morphism β is a coboundary in $\mathrm{Hom}^*_{\mathbb{K}}(W, V)$ and therefore also γ is coboundary in $\mathrm{Hom}^*_{\mathbb{K}}(W, U)$; choosing a morphism $\eta \in \mathrm{Hom}^0_{\mathbb{K}}(W, U)$ such that $d\eta - \eta d = \gamma$ we can immediately check that $s = \alpha - f\eta$ is a morphism of DG-vector spaces such that $gs = \mathrm{Id}_W$. □

5.2 Suspension and Mapping Cones

For any pair of DG-vector spaces V, W we choose as a *natural isomorphism* between $V \otimes W$ and $W \otimes V$ the twisting involution $\mathtt{tw}$ described in the following definition.

Definition 5.2.1 Given $V, W \in \mathbf{DG}$, the **twisting involution**

$$\mathtt{tw}\colon V \otimes W \to W \otimes V$$

is the unique linear map such that

$$\mathtt{tw}(v \otimes w) = (-1)^{\bar{v}\,\bar{w}} w \otimes v$$

for every pair of nonzero homogeneous vectors v, w.

The declaration of $\mathtt{tw}$ as a natural isomorphism leads naturally to the **Koszul signs convention**, in which graded commutation rules are generated by the twisting involution. More informally, to "get the signs right", whenever an "object of degree d passes on the other side of an object of degree h, a sign $(-1)^{dh}$ must be inserted".

For instance, if $f \in \mathrm{Hom}^*_{\mathbb{K}}(V, W)$ and $g \in \mathrm{Hom}^*_{\mathbb{K}}(H, K)$, then the Koszul rule of signs implies that $f \otimes g \in \mathrm{Hom}^*_{\mathbb{K}}(V \otimes H, W \otimes K)$ is defined by the formula:

$$(f \otimes g)(v \otimes h) = (-1)^{\bar{g}\,\bar{v}} f(v) \otimes g(h), \qquad v \in V,\ h \in H;$$

notice that $\mathtt{tw} \circ (f \otimes g) \circ \mathtt{tw} = (-1)^{\bar{f}\,\bar{g}} g \otimes f$. Similarly, the transpose of a linear map $f\colon V \to W$ is the map

$$f^T\colon \mathrm{Hom}^*_{\mathbb{K}}(W, \mathbb{K}) \to \mathrm{Hom}^*_{\mathbb{K}}(V, \mathbb{K}), \qquad f^T(\alpha) = (-1)^{\bar{f}\,\bar{\alpha}} \alpha \circ f.$$

It is important to notice that, by the Koszul signs convention, the differential on the tensor product $V \otimes W$ can be written as $d_V \otimes \mathrm{Id}_W + \mathrm{Id}_V \otimes d_W$.

Given a DG-vector space we can obtain infinitely many others by shifting the degrees by a fixed integer; in the *Grothendieck–Verdier formalism*, given a DG-vector space (V, d_V) and an integer p we define the DG-vector space $(V[p], d_{V[p]})$ by setting

$$V[p]^n = V^{n+p}, \qquad d_{V[p]} = (-1)^p d_V.$$

For instance, $\mathbb{K}[-n]$ is the complex that has $\mathbb{K}$ in degree n and 0 in degree $\neq n$.

The tautological map $s\colon V \to V[-1]$ of degree 1, defined in each degree n as the identity map $V^n \to V[-1]^{n+1} = V^n$ is called a **suspension**; more generally, for any integer p there exists a tautological morphism $s^{-p}\colon V \to V[p]$ of degree $-p$, defined in each degree n as the identity map $V^n \to V[p]^{n-p} = V^n$; the definition of $d_{V[p]}$ implies that s^{-p} is a cocycle in $\mathrm{Hom}^*_{\mathbb{K}}(V, V[p])$.

In the *Eilenberg–Moore formalism* the DG-vector space $V[p]$ is seen as the image of the bijective map $s^{-p}\colon V \to V[p]$:

$$V[p] = s^{-p}V = \{s^{-p}v \mid v \in V\},$$

and $s^p V = V[-p]$ is called the p-**fold suspension** of V, whereas $sV = V[-1]$ is simply called the suspension of V, and $s^{-1}V = V[1]$ is also called either the transgression or desuspension of V. There exist canonical isomorphisms:

$$\alpha\colon s^pV \otimes s^qW \to s^{p+q}(V \otimes W), \quad \alpha(s^pv \otimes s^qw) = (-1)^{q\overline{v}} s^{p+q}(v \otimes w),$$
$$\beta\colon s^p\mathrm{Hom}^*_{\mathbb{K}}(V, W) \to \mathrm{Hom}^*_{\mathbb{K}}(s^qV, s^{p+q}W), \quad \beta(s^pf)(s^qv) = (-1)^{q\overline{f}} s^{p+q} f(v), \tag{5.4}$$

obtained by the Koszul rule of sign, since s^p is an "object of degree p".

Definition 5.2.2 The **mapping cone** of a morphism $f\colon V \to W$ of DG-vector spaces is the graded vector space

$$\mathrm{cone}(f) = s^{-1}V \oplus W = V[1] \oplus W,$$

equipped with the differential

$$d(s^{-1}v + w) = -s^{-1}dv + f(v) + dw, \qquad v \in V, \quad w \in W.$$

More informally, by the Koszul sign convention, the differential in the mapping cone extends the differentials of V, W and it is determined by imposing that f is the differential of the transgression s^{-1}.

The inclusion $j\colon W \to \mathrm{cone}(f)$, and the projection $p\colon \mathrm{cone}(f) \to s^{-1}V$ give a short exact sequence of DG-vector spaces

$$0 \to W \xrightarrow{j} \mathrm{cone}(f) \xrightarrow{p} s^{-1}V \to 0$$

such that the connecting morphism $H^n(s^{-1}V) \simeq H^{n+1}(V) \to H^{n+1}(W)$ in the cohomology long exact sequence is naturally identified with the morphism induced by f in cohomology. In particular, $\mathrm{cone}(f)$ is acyclic if and only if f is a quasi-isomorphism. The projection $\mathrm{cone}(f) \to W$ is not a morphism of DG-vector spaces, while its composition with the quotient map $W \to \mathrm{coker}(f)$ gives a morphism of DG-vector spaces $\mathrm{cone}(f) \to \mathrm{coker}(f)$.

Definition 5.2.3 For a morphism of DG-vector spaces $f : V \to W$, its **mapping cocone** is the graded vector space

$$C(f) = V \times sW = V \times W[-1],$$

equipped with the differential

$$d(v, sw) = (dv, s(f(v) - dw)), \qquad v \in V,\ w \in W.$$

Notice that, via the natural isomorphism $(V \times W[-1])[1] \cong s^{-1}V \oplus W$, the mapping cocone of f is the suspension of the mapping cone of $-f$.

For any morphism of DG-vector spaces $f : V \to W$, the projection $p : C(f) \to V$ and the inclusion $i : W[-1] \to C(f)$ are morphisms of DG-vector spaces and we have a long exact cohomology sequence:

$$\cdots \to H^i(V) \xrightarrow{f} H^i(W) \xrightarrow{i} H^{i+1}(C(f)) \xrightarrow{p} H^{i+1}(V) \xrightarrow{f} H^{i+1}(W) \to \cdots \tag{5.5}$$

In particular, given a commutative square

$$\begin{array}{ccc} V & \xrightarrow{\alpha} & E \\ \downarrow{\scriptstyle f} & & \downarrow{\scriptstyle g} \\ W & \xrightarrow{\beta} & F \end{array}$$

if both α and β are quasi-isomorphisms, then also the induced map $C(f) \to C(g)$ is a quasi-isomorphism.

Lemma 5.2.4 *Let $f : V \to W$ be a morphism of DG-vector spaces:*

1. *if f is injective, then the natural projection* $\mathrm{cone}(f) \to \mathrm{coker}(f)$ *is a quasi-isomorphism;*
2. *if f is surjective, then the natural inclusion* $\ker(f) \to C(f)$ *is a quasi-isomorphism.*

Proof It is sufficient to observe that $\mathrm{coker}(f)$ is isomorphic to the cokernel of the natural map $\mathrm{cone}(\mathrm{Id}_V) \to \mathrm{cone}(f)$ and $\ker(f)$ is isomorphic to the kernel of the natural map $C(f) \to C(\mathrm{Id}_W)$. □

5.3 Filtered DG-Vector Spaces

Recall that a **decreasing filtration** of an abelian group A is a sequence of subgroups

$$F^{\bullet}A: \qquad \cdots \supset F^pA \supset F^{p+1}A \supset F^{p+2}A \supset \cdots \qquad p \in \mathbb{Z}.$$

Such a filtration is called **exhaustive** if $\bigcup_{p\in\mathbb{Z}} F^pA = A$; it is called **complete** if the natural map

$$A \to \lim_p \frac{A}{F^pA}$$

is an isomorphism. It should be noted that if $F^{\bullet}A$ is a complete filtration, then for every $n \in \mathbb{Z}$ and every sequence $a_i \in F^iA$, $i \geq n$, there exists a unique element, denoted by $\sum_{i=n}^{\infty} a_i \in F^nA$, such that

$$\sum_{i=n}^{\infty} a_i - \sum_{i=n}^{p} a_i \in F^{p+1}A, \qquad \text{for every} \quad p \geq n.$$

Example 5.3.1 Let A_n, $n \in \mathbb{Z}$, be any sequence of abelian groups and consider

$$A = \prod_{n\in\mathbb{Z}} A_n, \qquad F^pA = \ker(\pi\colon A \to \prod_{n<p} A_n) \cong \prod_{n\geq p} A_n,$$

where π denotes the natural projection. Then F^pA is a complete filtration and it is exhaustive if and only if $A_n = 0$ for $n << 0$. Similarly, if

$$B = \bigoplus_{n\in\mathbb{Z}} A_n, \qquad F^pB = \bigoplus_{n\geq p} A_n,$$

then F^pB is an exhaustive filtration which is complete if and only if $A_n = 0$ for $n >> 0$.

Let $f\colon A \to B$ be a morphism of abelian groups and assume that there exist two exhaustive and complete decreasing filtrations F^pA, F^pB, of A and B respectively, such that $f(F^pA) \subset F^pB$ for every p. Then for every sequence $a_p \in F^pA$, $p \geq n$, we have

$$f\left(\sum_{p=n}^{\infty} a_p\right) = \sum_{p=n}^{\infty} f(a_p)\,.$$

In fact, writing for every $m \geq n$

$$\sum_{p=n}^{\infty} a_p - \sum_{p=n}^{m} a_p = r_m \in F^{m+1}A,$$

we have

$$f(\sum_{p=n}^{\infty} a_p) - \sum_{p=n}^{m} f(a_p) = f(\sum_{p=n}^{m} a_p + r_m) - \sum_{p=n}^{m} f(a_p) = f(r_m) \in F^{m+1}B,$$

and therefore $f(\sum_{p=n}^{\infty} a_p) = \sum_{p=n}^{\infty} f(a_p)$.

Definition 5.3.2 A **filtered** DG-vector space is the data of a DG-vector space V and a decreasing filtration

$$F^\bullet V: \qquad \cdots \supset F^p V \supset F^{p+1} V \supset F^{p+2} V \supset \cdots, \qquad p \in \mathbb{Z}.$$

of DG-vector subspaces. Such a filtration is called exhaustive (resp.: complete) if for every degree n the filtration $F^p V^n$ is an exhaustive (resp.: complete) filtration of V^n. A morphism $f: (V, F^\bullet V) \to (W, F^\bullet W)$ of filtered DG-vector spaces is a morphism of DG-vector spaces $f: V \to W$ such that $f(F^p V) \subset F^p W$ for every p. A morphism of filtered DG-vector spaces $f: (V, F^\bullet V) \to (W, F^\bullet W)$ is called a strict morphism if $f: F^p V \to f(V) \cap F^p W$ is surjective for every p.

The category of filtered DG-vector spaces is additive, a kernel and cokernel exist for every morphism, but it is not an abelian category since not every monic (resp.: epi) is the kernel (resp.: cokernel) of its cokernel (resp.: kernel). Keep in mind the fact that the composition of strict morphisms is not strict in general.

Lemma 5.3.3 *Let $(V, F^\bullet V)$ be a filtered DG-vector space such that the filtration is complete and exhaustive. If $\dfrac{F^p V}{F^{p+1} V}$ is acyclic for every p, then also V is acyclic.*

Proof Let d be the differential of V; since $F^p V$ is a DG-vector subspace we have $d(F^p V^m) \subset F^p V^{m+1}$ and therefore, since for every m the filtration $F^p V^m$ is complete and exhaustive,

$$d(\sum_{p=n}^{\infty} v_p) = \sum_{p=n}^{\infty} d(v_p), \quad \text{for every } n \in \mathbb{Z},\ v_p \in F^p V^m.$$

Let $v \in V^m$ be such that $dv = 0$ and choose an integer n such that $v \in F^n V^m$. Since the quotient $\dfrac{F^n V}{F^{n+1} V}$ is acyclic there exists $a_n \in F^n V^{m-1}$ such that $v - da_n \in F^{n+1} V^m$. Now $d(v - da_n) = 0$ and, since $\dfrac{F^{n+1} V}{F^{n+2} V}$ is acyclic there exists $a_{n+1} \in F^{n+1} V^{m-1}$ such that $v - da_n - da_{n+1} \in F^{n+2} V^m$. Iterating the procedure we obtain an infinite sequence a_p, $p \geq n$ such that

$$v=\sum_{p=n}^{\infty} da_p = d(\sum_{p=n}^{\infty} a_p).$$

□

Theorem 5.3.4 (Comparison theorem) *Let $f\colon (V, F^\bullet V) \to (W, F^\bullet W)$ be a morphism of filtered DG-vector spaces and assume that both the filtrations $F^\bullet V$, $F^\bullet W$ are complete and exhaustive. Suppose that*

$$f\colon \frac{F^pV}{F^{p+1}V} \to \frac{F^pW}{F^{p+1}W}$$

is a quasi-isomorphism for every p. Then also $f\colon V \to W$ is a quasi-isomorphism.

Proof Let's denote for simplicity by $H = \operatorname{cone}(f)$ the mapping cone of f:

$$H^n = V^{n+1} \oplus W^n, \qquad d(v, w) = (-dv, f(v) + dw).$$

There exists a natural filtration on the mapping cone of f defined as

$$F^pH^n = F^pV^{n+1} \oplus F^pW^n.$$

We can immediately see that $F^\bullet H$ is complete and exhaustive and $dF^pH \subset F^pH$. Moreover, $\frac{F^pH}{F^{p+1}H}$ is the mapping cone of $f\colon \frac{F^pV}{F^{p+1}V} \to \frac{F^pW}{F^{p+1}W}$ and then $\frac{F^pH}{F^{p+1}H}$ is acyclic for every p. The conclusion now follows from Lemma 5.3.3. □

The above comparison theorem is usually applied to morphisms of double complexes; recall that a double complex $A^{*,*}$ of vector spaces is the data, for every $i, j \in \mathbb{Z}$, of a vector space $A^{i,j}$ and two linear maps

$$\partial\colon A^{i,j} \to A^{i+1,j}, \qquad \overline{\partial}\colon A^{i,j} \to A^{i,j+1},$$

such that $\partial^2 = \overline{\partial}^2 = \partial\overline{\partial} + \overline{\partial}\partial = 0$. Defining, for every $n \in \mathbb{Z}$,

$$\operatorname{Tot}^{\oplus}(A^{*,*})^n = \bigoplus_{i\in\mathbb{Z}} A^{i,n-i}, \qquad \operatorname{Tot}^{\prod}(A^{*,*})^n = \prod_{i\in\mathbb{Z}} A^{i,n-i},$$

by the universal properties of direct sums and products, the maps $\partial, \overline{\partial}$ induce the linear operators

$$\partial, \overline{\partial}\colon \operatorname{Tot}^{\oplus}(A^{*,*})^n \to \operatorname{Tot}^{\oplus}(A^{*,*})^{n+1}, \qquad \partial, \overline{\partial}\colon \operatorname{Tot}^{\prod}(A^{*,*})^n \to \operatorname{Tot}^{\prod}(A^{*,*})^{n+1},$$

which satisfy the condition $(\partial + \overline{\partial})^2 = 0$. Therefore we have two DG-vector spaces

$$(\operatorname{Tot}^{\oplus}(A^{*,*}), d = \partial + \overline{\partial}), \qquad (\operatorname{Tot}^{\prod}(A^{*,*}), d = \partial + \overline{\partial}),$$

carrying the filtrations

$$F^p\mathrm{Tot}^{\oplus}(A^{*,*})^n = \bigoplus_{i\ge p} A^{i,n-i}, \qquad F^p\mathrm{Tot}^{\prod}(A^{*,*})^n = \prod_{i\ge p} A^{i,n-i}, \qquad p \in \mathbb{Z},$$

such that

$$\frac{F^p\mathrm{Tot}^{\oplus}(A^{*,*})}{F^{p+1}\mathrm{Tot}^{\oplus}(A^{*,*})} = \frac{F^p\mathrm{Tot}^{\prod}(A^{*,*})}{F^{p+1}\mathrm{Tot}^{\prod}(A^{*,*})} = (A^{p,*}, \overline{\partial}).$$

According to Example 5.3.1 the filtration $F^p\mathrm{Tot}^{\oplus}$ is complete and exhaustive if and only if for every n we have $A^{i,n-i} = 0$ for $i >> 0$, while the filtration $F^p\mathrm{Tot}^{\prod}$ is complete and exhaustive if and only if for every n we have $A^{i,n-i} = 0$ for $i << 0$.

For instance, if $A^{i,j} = 0$ for every $i < 0$, then $F^p\mathrm{Tot}^{\prod}$ is complete and exhaustive, while if $A^{i,j} = 0$ for every $i > 0$, then $F^p\mathrm{Tot}^{\oplus}$ is complete and exhaustive.

5.4 Basic Homological Perturbation Theory

In this section we continue the homotopy classification of DG-vector spaces in a more detailed way with respect to Lemma 5.1.4 and we introduce the basic ideas of homological perturbation theory.

Definition 5.4.1 A **contraction** (of DG-vector spaces) is a diagram

$$M \underset{\pi}{\overset{\imath}{\rightleftarrows}} N \circlearrowleft h$$

where M, N are DG-vector spaces over a field $\mathbb{K}$, the maps $\imath, \pi$ are morphisms of DG-vector spaces and $h \in \mathrm{Hom}^{-1}_{\mathbb{K}}(N, N)$ is a map such that:

1. (deformation retraction) $\pi\imath = \mathrm{Id}_M$, $\imath\pi - \mathrm{Id}_N = d_N h + h d_N$,
2. (annihilation properties) $\pi h = h\imath = h^2 = 0$.

We shall call the morphism $\imath$ the *injection* of the contraction, π the *projection* of the contraction and h the *homotopy* of the contraction.

A motivation for introducing contractions is given by the following lemma.

Lemma 5.4.2 *Let $i\colon M \to N$ be an injective morphism of DG-vector spaces. Then i is a quasi-isomorphism if and only if it is the injection of a contraction.*

Proof If i is the injection of a contraction there exists a morphism of complexes $\pi\colon N \to M$ and a homotopy $h \in \mathrm{Hom}^{-1}_{\mathbb{K}}(N, N)$ such that:

$$\pi i = \mathrm{Id}_M, \quad hd + dh = i\pi - \mathrm{Id}_N, \quad hi = \pi h = h^2 = 0.$$

The first two conditions $\pi i = \mathrm{Id}_M$ and $hd + dh = i\pi - \mathrm{Id}_N$ imply that i, π are homotopy equivalences and then quasi-isomorphisms.

Conversely, assume that i is an injective quasi-isomorphism; it is not restrictive to assume that M is a DG-vector subspace of N and i is the inclusion. Since the quotient DG-vector space $V = N/M$ is acyclic, by Künneth's formula also $\mathrm{Hom}^*_{\mathbb{K}}(V, V)$ is acyclic and then there exists $\gamma \in \mathrm{Hom}^{-1}_{\mathbb{K}}(V, V)$ such that $\gamma d + d\gamma = \mathrm{Id}_V$. Let's write $\psi = \gamma d \gamma$, then we have

$$\psi d + d\psi = \mathrm{Id}_V, \qquad \psi^2 = 0.$$

In fact $d\gamma d = d(\mathrm{Id}_V - d\gamma) = d$ and then

$$\psi d + d\psi = \gamma d\gamma d + d\gamma d\gamma = \gamma d + d\gamma = \mathrm{Id}_V,$$

$$\psi^2 = \gamma d\gamma(\gamma d)\gamma = \gamma d\gamma(\mathrm{Id}_V - d\gamma)\gamma = \gamma d\gamma^2 - \gamma(d\gamma d)\gamma^2 = 0.$$

Let $\alpha \colon N \to V$ be the projection and let $\beta \in \mathrm{Hom}^0_{\mathbb{K}}(V, N)$ be any morphism of graded vector spaces such that $\alpha\beta = \mathrm{Id}_V$. The map $\rho = d\beta\psi + \beta\psi d \colon V \to N$ is a morphism of complexes, since $d\rho = \rho d = d\beta\psi d$, and

$$\alpha\rho = \alpha d\beta\psi + \alpha\beta\psi d = d\alpha\beta\psi + \alpha\beta\psi d = d\psi + \psi d = \mathrm{Id}_V.$$

Therefore $(\rho\alpha)^2 = \rho\alpha$ and $\pi = \mathrm{Id}_N - \rho\alpha \colon N \to M$ is a projection with kernel the image of ρ. Finally, setting $h = -\rho\psi\alpha$ we have

$$hi = -\rho\psi\alpha i = 0, \qquad \pi h = -\pi\rho\psi\alpha = 0, \quad h^2 = \rho\psi\alpha\rho\psi\alpha = \rho\psi^2\alpha = 0,$$

$$hd + dh = -\rho\psi\alpha d - d\rho\psi\alpha = -\rho(\psi d + d\psi)\alpha = -\rho\alpha = i\pi - \mathrm{Id}_N.$$

□

With a similar proof it is possible to prove that a surjective morphism $\pi \colon N \to M$ of DG-vector spaces is the projection of a contraction if and only if it is a quasi-isomorphism.

Definition 5.4.3 A morphism of contractions

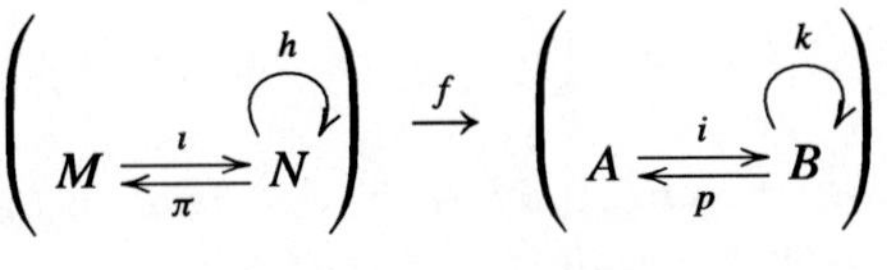

is a morphism of DG-vector spaces $f \colon N \to B$ such that $fh = kf$.

Given a morphism of contractions as in Definition 5.4.3 we denote by $\widehat{f} \colon M \to A$ the morphism $\widehat{f} = pf\imath$. Then the diagram

$$\begin{array}{ccccc} M & \xrightarrow{\imath} & N & \xrightarrow{\pi} & M \\ \downarrow{\scriptstyle \widehat{f}} & & \downarrow{\scriptstyle f} & & \downarrow{\scriptstyle \widehat{f}} \\ A & \xrightarrow{i} & B & \xrightarrow{p} & A \end{array}$$

is commutative. In fact:

$$i\widehat{f} = ipf\imath = f\imath + (d_B kf + kd_B f)\imath = f\imath + f(d_N h + hd_N)\imath = f\imath + f(\imath\pi - \mathrm{Id}_N)\imath = f\imath,$$

$$\begin{aligned} \widehat{f}\pi &= pf\imath\pi = pf(\mathrm{Id}_N + d_N h + hd_N) = pf + p(d_B k + kd_B)f \\ &= pf + p(ip - \mathrm{Id}_B) = pf. \end{aligned}$$

Given a contraction $M \underset{\pi}{\overset{\imath}{\rightleftarrows}} N \circlearrowright h$ and a morphism $\partial \in \mathrm{Hom}^1_{\mathbb{K}}(N, N)$, the ordinary homological perturbation theory consists in a series of statements about the linear maps:

$$\imath_\partial = \sum_{n\geq 0} (h\partial)^n \imath \ \in \mathrm{Hom}^0_{\mathbb{K}}(M, N), \tag{5.6}$$

$$\pi_\partial = \sum_{n\geq 0} \pi(\partial h)^n \ \in \mathrm{Hom}^0_{\mathbb{K}}(N, M), \tag{5.7}$$

$$D_\partial = \pi\partial\imath_\partial = \pi_\partial\partial\imath \ \in \mathrm{Hom}^1_{\mathbb{K}}(M, M), \tag{5.8}$$

$$h_\partial = \sum_{n\geq 0} (h\partial)^n h = \sum_{n\geq 0} h(\partial h)^n \ \in \mathrm{Hom}^{-1}_{\mathbb{K}}(N, N). \tag{5.9}$$

Clearly, in order to have the above maps properly defined we need to impose some extra assumption on ∂; usually this is done by considering filtered contractions of complete filtered DG-vector spaces or by imposing a sort of local nilpotency for the operators $h\partial$.

Here we consider this second situation, more precisely for a contraction

$$M \underset{\pi}{\overset{\imath}{\rightleftarrows}} N \circlearrowright h$$

we write

$$\mathcal{N}(N, h) = \{\partial \in \mathrm{Hom}^1_{\mathbb{K}}(N, N) \mid \cup_n \ker((h\partial)^n) = N\}.$$

Equivalently, a morphism ∂ belongs to $\mathcal{N}(N, h)$ if and only if for every $x \in N$ there exists a positive integer n such that $(h\partial)^n x = (\partial h)^n x = 0$.

It is plain that the maps $\imath_\partial$, π_∂, D_∂ and h_∂ are well defined for every $\partial \in \mathcal{N}(N, h)$. Moreover, they are functorial in the following sense: given a morphism of contractions

$$M \underset{\pi}{\overset{\imath}{\rightleftarrows}} N \circlearrowright h \xrightarrow{\ f\ } A \underset{p}{\overset{i}{\rightleftarrows}} B \circlearrowright k$$

and two elements $\partial \in \mathcal{N}(N, h)$, $\delta \in \mathcal{N}(B, k)$ such that $f\partial = \delta f$ we have

$$f\imath_\partial = \sum_{n\geq 0} f(h\partial)^n \imath = \sum_{n\geq 0}(k\delta)^n f\imath = \sum_{n\geq 0}(k\delta)^n i\hat{f} = i_\delta \hat{f}.$$

Similarly, we have $\hat{f}\pi_\partial = p_\delta f$, $\hat{f}D_\partial = D_\delta \hat{f}$ and $fh_\partial = k_\delta f$.

Lemma 5.4.4 *Let* $M \underset{\pi}{\overset{\imath}{\rightleftarrows}} N \circlearrowright h$ *be a contraction and* $\partial \in \mathcal{N}(N, h)$. *Then* $\imath_\partial$ *is injective and*

$$\pi_\partial \imath_\partial = \mathrm{Id}_M, \qquad \pi_\partial h_\partial = h_\partial \imath_\partial = h_\partial^2 = 0.$$

Proof Immediate consequence of annihilation properties. □

Definition 5.4.5 Let N be a DG-vector space. A **perturbation** of the differential d_N is a linear map $\partial \in \mathrm{Hom}^1_{\mathbb{K}}(N, N)$ such that $(d_N + \partial)^2 = 0$.

Theorem 5.4.6 (Ordinary perturbation lemma) *Let* $M \underset{\pi}{\overset{\imath}{\rightleftarrows}} N \circlearrowright h$ *be a contraction and let* $\partial \in \mathcal{N}(N, h)$ *be a perturbation of the differential* d_N. *Then* D_∂ *is a perturbation of* d_M *and*

$$(M, d_M + D_\partial) \underset{\pi_\partial}{\overset{\imath_\partial}{\rightleftarrows}} (N, d_N + \partial) \circlearrowright h_\partial$$

is a contraction.

Proof By Lemma 5.4.4 the proof of the ordinary perturbation lemma reduces to the completely straightforward verification of the two equalities

$$\pi_\partial(d_N + \partial) = (d_M + D_\partial)\pi_\partial, \quad \pi_\partial \imath_\partial - \mathrm{Id}_N = h_\partial(d_N + \partial) + (d_N + \partial)h_\partial.$$

□

After Theorem 5.4.6 we can give a more geometric interpretation of the maps $\imath_\partial$, π_∂ and D_∂, namely:

1. $\imath_\partial$ is the unique morphism of graded vector spaces whose image is a subcomplex of $(N, d_N + \partial)$ and satisfies the "gauge fixing" conditions:

$$h\imath_\partial = 0, \qquad \pi\imath_\partial = \mathrm{Id}_M.$$

2. π_∂ is the unique morphism of graded vector spaces whose kernel is a subcomplex of $(N, d_N + \partial)$ and such that

$$\pi_\partial h = 0, \qquad \pi_\partial \iota = \mathrm{Id}_M .$$

3. D_∂ is the unique map in $\mathrm{Hom}^1_{\mathbb{K}}(M, M)$ such that $(d_M + D_\partial)^2 = 0$ and

$$\iota_\partial \colon (M, d_M + D_\partial) \to (N, d_N + \partial), \quad \pi_\partial \colon (N, d_N + \partial) \to (M, d_M + D_\partial),$$

are morphisms of DG-vector spaces.

If $\iota_\partial = \sum_{n\geq 0}(h\partial)^n \iota$, then by Theorem 5.4.6 the image of ι_∂ is a subcomplex and the conditions $h^2 = h\iota = \pi h = 0$ give $h\iota_\partial = 0$, $\pi\iota_\partial = \mathrm{Id}_M$. Conversely, let $\iota_\partial \colon M \to N$ be such that $h\iota_\partial = 0, \pi\iota_\partial = \mathrm{Id}_M$ and the image of ι_∂ is $d_N + \partial$ invariant. Then we have $h(d_N + \partial)\iota_\partial = 0$, the morphism ι_∂ satisfies the equation

$$\iota_\partial = \iota_\partial + h d_N \iota_\partial + h\partial\iota_\partial = (\iota\pi - d_N h)\iota_\partial + h\partial\iota_\partial = \iota + (h\partial)\iota_\partial,$$

and therefore

$$\iota_\partial = (\mathrm{Id}_N - h\partial)^{-1}\iota = \sum_{n\geq 0}(h\partial)^n \iota .$$

The proof of items 2 and 3 is the same.

Remark 5.4.7 A deformation retraction data is a diagram

$$M \underset{\pi}{\overset{\iota}{\rightleftarrows}} N \circlearrowleft k$$

where M, N are DG-vector spaces, $k \in \mathrm{Hom}^{-1}_{\mathbb{K}}(N, N)$ and ι, π are morphisms of DG-vector spaces such that: $\pi\iota = \mathrm{Id}_M$, $\iota\pi - \mathrm{Id}_N = dk + kd$.

There is a canonical way to associate a contraction to all deformation retraction data; in the above setup, a straightforward and maximally boring computation shows that, denoting by $p = dk + kd = \iota\pi - \mathrm{Id}_N$ and by $h = -pkpdpkp$, the diagram

$$M \underset{\pi}{\overset{\iota}{\rightleftarrows}} N \circlearrowleft h$$

is a contraction.

5.5 Differential Graded Commutative Algebras

A **DG-algebra** (short for differential graded commutative algebra) is the data of a DG-vector space (A, d) equipped with a bilinear map

$$A \times A \to A, \qquad (a, b) \mapsto ab,$$

called a *product*, that satisfies the following conditions:

- (homogeneity of degree 0) $\overline{ab} = \overline{a} + \overline{b}$,
- (graded commutativity) $ab = (-1)^{\overline{a}\,\overline{b}} ba$,
- (graded Leibniz) $d(ab) = d(a)b + (-1)^{\overline{a}} a d(b)$,
- (associativity) $(ab)c = a(bc)$,
- $a^2 = 0$ for every homogeneous component a of odd degree.

Notice that the first three conditions above are equivalent to the fact that the linear map $A \otimes A \to A$, $a \otimes b \mapsto ab$, is a morphism of DG-vector spaces commuting with the twisting involution. Clearly, in characteristic $\neq 2$ the last condition is a consequence of the graded commutativity.

A morphism of DG-algebras is simply a morphism of DG-vector spaces commuting with products. The category of DG-algebras will be denoted by **DGA**. A DG-algebra A is called **unitary** if there exists an element $1 \in A^0$, called a unit, such that $1a = a1 = a$ for every $a \in A$. A DG-algebra A is called **nilpotent** if there exists an integer $n > 0$ such that every product of n elements in a is equal to 0.

Example 5.5.1 Every commutative $\mathbb{K}$-algebra can be considered as a DG-algebra concentrated in degree 0.

Example 5.5.2 The de Rham complex of a smooth manifold, equipped with the wedge product, is a DG-algebra.

Example 5.5.3 (*Koszul algebras*) Let V be a vector space, B a commutative $\mathbb{K}$-algebra and $f : V \to B$ a linear map. Consider the DG-vector space

$$A = \bigoplus_{n \leq 0} A^n, \qquad A^{-n} = B \otimes \left(\bigwedge^n V\right),$$

with the differential

$$d : B \otimes \bigwedge^i V \to B \otimes \bigwedge^{i-1} V,$$

defined by the formula

$$d(b \otimes (v_1 \wedge \cdots \wedge v_h)) = \sum_{j=1}^{h} (-1)^{j-1} b f(v_j) \otimes (v_1 \wedge \cdots \wedge \widehat{v_j} \wedge \cdots \wedge v_h).$$

We can immediately see that (A, d) is a DG-algebra.

Example 5.5.4 (*Polynomial DG-algebras*) Given a set of indeterminates $\{x_i\}$, $i \in I$, each one equipped with an integral degree $\overline{x_i} \in \mathbb{Z}$, we can consider the graded algebra $\mathbb{K}[\{x_i\}]$. As a graded vector space $\mathbb{K}[\{x_i\}]$ is generated by monomials in the indeterminates x_i subjected to the commuting relations $x_i x_j = (-1)^{\overline{x_i}\,\overline{x_j}} x_j x_i$ together with $x_i^2 = 0$ whenever $\overline{x_i}$ is odd. For instance, if $\overline{x} = 0$ and $\overline{y} = -1$ we have $\mathbb{K}[x, y] = \mathbb{K}[x]y \oplus \mathbb{K}[x]$, where $\mathbb{K}[x]$ is the usual polynomial ring. By the Leibniz rule every differential $d\colon \mathbb{K}[\{x_i\}] \to \mathbb{K}[\{x_i\}]$ is uniquely determined by the values dx_i. When every indeterminate has degree 0 or -1 we get a Koszul algebra; more precisely, denoting by x_i the indeterminates of degree 0 and by y_j the indeterminates of degree -1, we recover the algebra defined in Example 5.5.3, with $B = \mathbb{K}[\{x_i\}]$ and V the vector space generated by the set $\{y_j\}$.

Example 5.5.5 A particular type of polynomial DG-algebra is the de Rham complex of algebraic differential forms on the affine line, denoted by $\mathbb{K}[t, dt]$. We may write

$$\mathbb{K}[t, dt] = \mathbb{K}[t] \oplus \mathbb{K}[t]dt,$$

where t, dt are indeterminates of degree $\overline{t} = 0$, $\overline{dt} = 1$. The differential d is determined by the "obvious" equality $d(t) = dt$ and therefore $d(p(t) + q(t)dt) = p(t)'dt$. The inclusions

$$\mathbb{K} \to \mathbb{K}[dt] = \mathbb{K} \oplus \mathbb{K}dt \to \mathbb{K}[t, dt]$$

and the *evaluation maps*:

$$e_s\colon \mathbb{K}[t, dt] \to \mathbb{K}, \qquad p(t) + q(t)dt \mapsto p(s), \qquad s \in \mathbb{K},$$

are morphisms of DG-algebras.

By the Leibniz rule, for any DG-algebra A and every $x, y \in A$ we have:

1. if $dx = dy = 0$, then $d(xy) = 0$;
2. if $dy = 0$, then $(dx)y = d(xy)$.

Therefore the product on A factors through a product in cohomology, inducing on $H^*(A)$ a structure of a DG-algebra with trivial differential. If $f\colon A \to B$ is a morphism of DG-algebras, then $f\colon H^*(A) \to H^*(B)$ is a morphism of graded commutative algebras.

Definition 5.5.6 A **quasi-isomorphism** of DG-algebras is a morphism of DG-algebras that is a quasi-isomorphism of DG-vector spaces.

For instance, it is known from algebraic topology that if A is the de Rham algebra of an open convex subset of $\mathbb{R}^n$, then the inclusion of constant 0-forms $\mathbb{R} \hookrightarrow A$ is a quasi-isomorphism.

Lemma 5.5.7 *In characteristic 0, the inclusion* $i\colon \mathbb{K} \to \mathbb{K}[t, dt]$ *and the evaluation maps* $e_s\colon \mathbb{K}[t, dt] \to \mathbb{K}$, $s \in \mathbb{K}$, *are quasi-isomorphisms of differential graded algebras.*

Proof We have $e_s \circ i = \mathrm{Id}$ and then it is sufficient to prove that i is a quasi-isomorphism. This is obvious since every cocycle of $\mathbb{K}[t, dt]$ is of type $a + q(t)dt$ with $a \in \mathbb{K}$, and $q(t)dt$ is the differential of $\int_0^t q(s)ds$. □

The tensor product of two DG-algebras is still a DG-algebra. Clearly, we need to pay attention to the Koszul signs convention; if A, B are DG-algebras, then the product on $A \otimes B$ is defined as the linear extension of

$$(a_1 \otimes b_1)(a_2 \otimes b_2) = (-1)^{\overline{b_1}\,\overline{a_2}} a_1a_2 \otimes b_1b_2.$$

5.6 Differential Graded Lie Algebras

We are now ready to introduce one of the most important notions of this book.

Definition 5.6.1 A **differential graded Lie algebra** (also called DG-Lie algebra) is the data of a DG-vector space (L, d) with a bilinear bracket $[-, -]\colon L \times L \to L$ satisfying the following conditions:

1. $[-, -]$ is homogeneous graded skew-symmetric of degree 0. This means that:
   a. $[L^i, L^j] \subset L^{i+j}$,
   b. $[a, b] + (-1)^{\bar{a}\bar{b}}[b, a] = 0$ for every a, b homogeneous,
   c. $[a, a] = 0$ for every homogeneous vector a of even degree;
2. (Leibniz identity) $d[a, b] = [da, b] + (-1)^{\bar{a}}[a, db]$;
3. (Jacobi identity) every triple of homogeneous elements a, b, c satisfies the equality
$$[a, [b, c]] = [[a, b], c] + (-1)^{\bar{a}\,\bar{b}}[b, [a, c]];$$
4. (Bianchi identity) $[b, [b, b]] = 0$ for every homogeneous vector b of odd degree.

A DG-Lie algebra with trivial differential $d = 0$ is simply referred to as a **graded Lie algebra**, while a DG-Lie algebra is called **abelian** if its bracket is trivial. A morphism of differential graded Lie algebras is a morphism of DG-vector spaces commuting with brackets. The category of differential graded Lie algebras will be denoted by **DGLA**.

In characteristic $\neq 3$ the Bianchi identity is a consequence of the Jacobi identity; in fact if b is homogeneous of odd degree, then by the Jacobi identity we have

$$[b, [b, b]] = [[b, b], b] - [b, [b, b]] = -2[b, [b, b]].$$

On the other hand, it is easy to give examples in characteristic 3 where Jacobi does not imply Bianchi.

Example 5.6.2 If $L = \oplus L^i$ is a differential graded Lie algebra, then L^0 is a Lie algebra in the usual sense. Conversely, every Lie algebra can be considered as a DG-Lie algebra concentrated in degree 0.

The category **DGLA** of DG-Lie algebras is complete and the forgetful functor **DGLA** $\to$ **DG** commutes with limits. In particular, given two morphisms of DG-Lie algebras $f: L \to H$ and $g: M \to H$, their fibre product $L \times_H M$ in the category of DG-vector spaces is naturally a differential graded Lie algebra with bracket

$$[(a, x), (b, y)] = ([a, b], [x, y]).$$

Example 5.6.3 (*scalar extension*) Let A be a DG-algebra and L a differential graded Lie algebra. Then the DG-vector space $L \otimes A$ has a natural structure of a differential graded Lie algebra with bracket

$$[x \otimes a, y \otimes b] = (-1)^{\bar{a}\bar{y}}[x, y] \otimes ab.$$

Example 5.6.4 Let V be a DG-vector space. Then the Hom complex $\operatorname{Hom}^*_{\mathbb{K}}(V, V)$ has a natural structure of differential graded Lie algebra, with the bracket equal to the graded commutator:

$$[f, g] = fg - (-1)^{\bar{f}\bar{g}} gf.$$

Notice that the differential on $\operatorname{Hom}^*_{\mathbb{K}}(V, V)$ is equal to the adjoint operator $[d, -]$, where d is the differential of V.

Example 5.6.5 (*Derivations*) Let A be a DG-algebra over the field $\mathbb{K}$. The differential graded Lie algebra of derivations of A is the DG-Lie subalgebra of $\operatorname{Hom}^*_{\mathbb{K}}(A, A)$ defined by:

$$\operatorname{Der}^*_{\mathbb{K}}(A, A) = \bigoplus_n \operatorname{Der}^n_{\mathbb{K}}(A, A) \subset \operatorname{Hom}^*_{\mathbb{K}}(A, A),$$

$$\operatorname{Der}^n_{\mathbb{K}}(A, A) = \{\phi \in \operatorname{Hom}^n_{\mathbb{K}}(A, A) \mid \phi(ab) = \phi(a)b + (-1)^{n\bar{a}} a\phi(b)\}.$$

Notice that the differential of A is a derivation of degree +1. Similarly, if L is a differential graded Lie algebra, then $\operatorname{Der}^*_{\mathbb{K}}(L, L) = \bigoplus_n \operatorname{Der}^n_{\mathbb{K}}(L, L)$, where

$$\operatorname{Der}^n_{\mathbb{K}}(L, L) = \{\phi \in \operatorname{Hom}^n_{\mathbb{K}}(L, L) \mid \phi[a, b] = [\phi(a), b] + (-1)^{n\bar{a}}[a, \phi(b)]\}$$

is a DG-Lie subalgebra of $\operatorname{Hom}^*_{\mathbb{K}}(L, L)$.

Example 5.6.6 Let E be a holomorphic vector bundle on a complex manifold X. Then the Dolbeault complex

$$A^{0,*}_X(\mathcal{E}nd(E)) \colon \quad 0 \to A^{0,0}_X(\mathcal{E}nd(E)) \xrightarrow{\bar{\partial}} A^{0,1}_X(\mathcal{E}nd(E)) \xrightarrow{\bar{\partial}} \cdots$$

carries a natural structure of a DG-Lie algebra; if e, g are local holomorphic sections of $\mathcal{E}nd(E)$ and ϕ, ψ differential forms we have $\overline{\partial}(\phi e) = (\overline{\partial}\phi)e$ and $[\phi e, \psi g] = \phi \wedge \psi[e, g]$.

For any DG-Lie algebra $(L, d, [-, -])$ and every $x, y \in L$ we have:

1. if $dx = dy = 0$, then $d[x, y] = 0$;
2. if $dy = 0$, then $[dx, y] = d[x, y]$.

Therefore the bracket of L factors to a bracket in $H^*(L)$, inducing a graded Lie algebra structure on the cohomology of L. If $f : L \to M$ is a morphism of differential graded Lie algebras, then $f : H^*(L) \to H^*(M)$ is a morphism of graded Lie algebras.

Definition 5.6.7 A **quasi-isomorphism** of differential graded Lie algebras is a morphism of DG-Lie algebras which is a quasi-isomorphism of DG-vector spaces. Two differential graded Lie algebras are said to be **quasi-isomorphic** if they are equivalent under the equivalence relation generated by quasi-isomorphisms.

Thus, two differential graded Lie algebras L and M are quasi-isomorphic if and only if there exists a zigzag of finite length

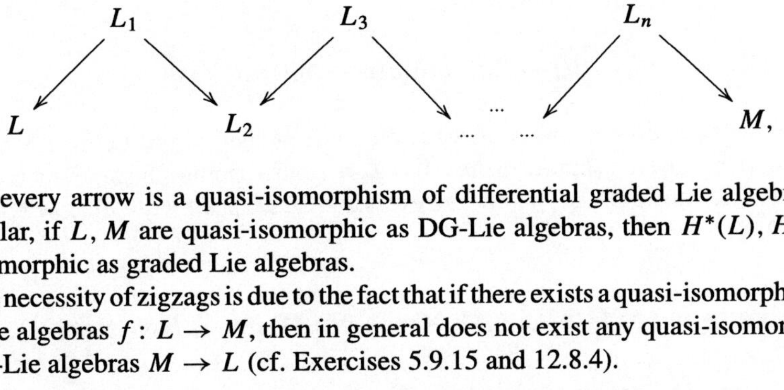

where every arrow is a quasi-isomorphism of differential graded Lie algebras. In particular, if L, M are quasi-isomorphic as DG-Lie algebras, then $H^*(L), H^*(M)$ are isomorphic as graded Lie algebras.

The necessity of zigzags is due to the fact that if there exists a quasi-isomorphism of DG-Lie algebras $f : L \to M$, then in general does not exist any quasi-isomorphism of DG-Lie algebras $M \to L$ (cf. Exercises 5.9.15 and 12.8.4).

Example 5.6.8 Let $f : L \to M$ be a surjective morphism of differential graded Lie algebras, so that $I = \ker f$ is a differential graded Lie ideal of L. The mapping cone $s^{-1}I \oplus L$ (Definition 5.2.2) of the inclusion $I \hookrightarrow L$, whose differential is

$$d(x + s^{-1}y) = dx + y - s^{-1}dy,$$

carries a structure of a DG-Lie algebra, with the bracket defined by the formula:

$$[x_1 + s^{-1}y_1, x_2 + s^{-1}y_2] = [x_1, x_2] + s^{-1}([y_1, x_2] + (-1)^{\overline{x_1}}[x_1, y_2]).$$

The inclusion $j : L \to s^{-1}I \oplus L$ is a morphism of differential graded Lie algebras, and the projection

$$f' : s^{-1}I \oplus L \to M, \qquad f'(x + s^{-1}y) = f(x),$$

is a surjective quasi-isomorphism of differential graded Lie algebras with kernel $s^{-1}I \oplus I = \operatorname{cone}(\mathrm{Id}_I)$. In particular, when $M = 0$, we get a DG-Lie algebra structure on $\operatorname{cone}(\mathrm{Id}_L)$.

Example 5.6.9 Denote by $\mathbb{K}[t, dt]$ the differential graded algebra of polynomial differential forms over the affine line (Example 5.5.5) and, for every differential graded Lie algebra L define

$$L[t, dt] = \mathbb{K}[t, dt] \otimes L.$$

As a graded vector space $L[t, dt]$ is generated by elements of the form $q(t)a + p(t)dt\, b$, for $p(t), q(t) \in \mathbb{K}[t]$ and $a, b \in L$. The differential and the bracket on $L[t, dt]$ are:

$$d(q(t)a + p(t)dt\, b) = q(t)(da) + q(t)'dt\, a - p(t)dt(db),$$

$$[q(t)a, h(t)c] = q(t)h(t)[a, c], \quad [q(t)dt\, a, h(t)c] = q(t)h(t)dt[a, c].$$

For every $s \in \mathbb{K}$, the evaluation morphism

$$e_s : L[t, dt] \to L, \quad e_s(aq(t) + bp(t)dt) = q(s)a,$$

is a morphism of differential graded Lie algebras. According to Lemma 5.5.7 and Künneth's formulas, in characteristic 0 it is also a quasi-isomorphism. For simplicity of notation, for any $a(t) \in L[t, dt]$ and every $s \in \mathbb{K}$ we shall write $a(s) = e_s(a(t))$.

5.7 Examples of DG-Lie Algebras via Derived Brackets

The theory of derived brackets is a rich subject of study belonging naturally to the theoretical framework of L_∞-algebra structures, which will be considered in Sect. 10.7. Here we anticipate some particular cases where the derived bracket construction gives some interesting examples of DG-Lie algebras.

Definition 5.7.1 Let V, W be two DG-vector spaces and let $\pi \in Z^1(\mathrm{Hom}^*_{\mathbb{K}}(W, V))$ be a cocycle of degree 1. The **derived bracket** of π on the shifted Hom complex $\mathrm{Hom}^*_{\mathbb{K}}(V, W)[-1]$ is defined as:

$$\mathrm{Hom}^{i-1}_{\mathbb{K}}(V, W) \times \mathrm{Hom}^{j-1}_{\mathbb{K}}(V, W) \xrightarrow{[-,-]_\pi} \mathrm{Hom}^{i+j-1}_{\mathbb{K}}(V, W),$$

$$[f, g]_\pi = f\pi g - (-1)^{ij} g\pi f.$$

Denoting by δ the canonical differential on $\mathrm{Hom}^*_{\mathbb{K}}(V, W)[-1] = \mathrm{Hom}^*_{\mathbb{K}}(V, W[-1])$:

$$\delta(f) = -d_W f - (-1)^i f d_V, \qquad f \in \mathrm{Hom}^{i-1}_{\mathbb{K}}(V, W) = (\mathrm{Hom}^*_{\mathbb{K}}(V, W)[-1])^i,$$

it is straightforward to check that $(\mathrm{Hom}^*_{\mathbb{K}}(V, W)[-1], \delta, [-, -]_\pi)$ is a differential graded Lie algebra; notice that $[-, -]_\pi$ is the graded commutator of the associative product $f * g = f\pi g$. Whenever $\pi = s \in \mathrm{Hom}^1_{\mathbb{K}}(s^{-1}V, V)$ is the suspension map we recover the usual structure of a DG-Lie algebra on the Hom complex $\mathrm{Hom}^*_{\mathbb{K}}(V, V) = \mathrm{Hom}^*_{\mathbb{K}}(V, s^{-1}V)[-1]$.

The derived bracket is functorial in the following natural sense: if $f: U \to W$ and $g: V \to Z$ are morphisms of DG-vector spaces, then

$$f_*: (\mathrm{Hom}^*_{\mathbb{K}}(V, U)[-1], \delta, [-, -]_{\pi f}) \to (\mathrm{Hom}^*_{\mathbb{K}}(V, W)[-1], \delta, [-, -]_\pi),$$

$$g^*: (\mathrm{Hom}^*_{\mathbb{K}}(Z, W)[-1], \delta, [-, -]_{g\pi}) \to (\mathrm{Hom}^*_{\mathbb{K}}(V, W)[-1], \delta, [-, -]_\pi),$$

are morphisms of differential graded Lie algebras.

Theorem 5.7.2 *In the setup of Definition 5.7.1, the quasi-isomorphism class of the differential graded Lie algebra* $(\mathrm{Hom}^*_{\mathbb{K}}(V, W)[-1], \delta, [-, -]_\pi)$ *is uniquely determined by the cohomology class of* π *in* $H^1(\mathrm{Hom}^*_{\mathbb{K}}(W, V)) = \mathrm{Hom}^1_{\mathbb{K}}(H^*(W), H^*(V))$.

In particular, if $\pi(H^*(W)) = 0 \subset H^*(V)$, *then* $(\mathrm{Hom}^*_{\mathbb{K}}(V, W)[-1], \delta, [-, -]_\pi)$ *is quasi-isomorphic to an abelian DG-Lie algebra.*

Proof There exist two quasi-isomorphisms of DG-vector spaces $f: H \to W$ and $g: V \to K$, where H, K have trivial differential; by functoriality and Künneth's formula the differential graded Lie algebra $(\mathrm{Hom}^*_{\mathbb{K}}(V, W)[-1], \delta, [-, -]_\pi)$ is quasi-isomorphic to $(\mathrm{Hom}^*_{\mathbb{K}}(K, H)[-1], \delta, [-, -]_{g\pi f})$ and the map $g\pi f: H \to K$ is uniquely determined by the cohomology class of π. Finally, if $\pi = 0$ then the bracket $[-, -]_\pi$ is trivial. □

Example 5.7.3 Let (V, d) be a DG-vector space, $W \subset V$ a DG-vector subspace and $A \subset V$ a graded vector subspace such that $V = W \oplus A$. Denoting by $p: V \to A$ and $p^\perp: V \to W$ the two projections, since $d(W) \subset W$ we have

$$p + p^\perp = \mathrm{Id}_V, \qquad pdp^\perp = 0, \qquad pdp = pd, \qquad (pd)^2 = pdpd = pd^2 = 0.$$

Therefore $pd: A \to A$ is a differential and $p: (V, d) \to (A, pd)$ is a morphism of DG-vector spaces. The linear map $p^\perp d: A \to W$ is a cocycle in $\mathrm{Hom}^*_{\mathbb{K}}(A, W)$ since

$$p^\perp = \mathrm{Id} - p, \qquad dp^\perp d + p^\perp dpd = d^2 - dpd + dpd - pdpd = 0,$$

and then the shifted Hom complex $\mathrm{Hom}^*_{\mathbb{K}}(W, A)[-1]$ becomes a differential graded Lie algebra when equipped with the derived bracket $[-, -]_{p^\perp d}$.

Example 5.7.4 Let V be a graded vector space equipped with two linear operators $\partial, \overline{\partial} \in \mathrm{Hom}^1_{\mathbb{K}}(V, V)$ satisfying the equalities:

$$\partial^2 = \overline{\partial}^2 = 0, \qquad [\partial, \overline{\partial}] = \partial\overline{\partial} + \overline{\partial}\partial = 0.$$

For instance, V may be either the $\mathrm{Tot}^{\oplus}$ or the $\mathrm{Tot}^{\prod}$ of a double complex $(A^{ij}, \partial, \overline{\partial})$.

Thus $(V, \partial + \overline{\partial})$ is a DG-vector space, and ∂V, $\ker \partial$ are DG-vector subspaces of V. Moreover, ∂ is a cocycle in $\mathrm{Hom}^*_{\mathbb{K}}(V, V)$ and then we can consider the differential graded Lie algebras

$$\mathrm{Hom}^*_{\mathbb{K}}(V, \ker \partial)[-1], \qquad \mathrm{Hom}^*_{\mathbb{K}}(V, V)[-1], \qquad \mathrm{Hom}^*_{\mathbb{K}}(\ker(\partial), \mathrm{coker}(\partial))[-1],$$

all of them equipped with the derived bracket $[-,-]_{\partial}$. Notice that $\mathrm{Hom}^*_{\mathbb{K}}(V, \ker \partial)[-1]$ is an abelian differential graded Lie algebra.

Lemma 5.7.5 *In the above situation the following conditions are equivalent:*

1. *∂V is an acyclic subcomplex of $(V, \partial + \overline{\partial})$;*
2. $\mathrm{Im}\, \overline{\partial}\partial = \ker \overline{\partial} \cap \mathrm{Im}\, \partial$.

If the above conditions hold, then the differential graded Lie algebras

$$\mathrm{Hom}^*_{\mathbb{K}}(V, V)[-1], \qquad \mathrm{Hom}^*_{\mathbb{K}}(\ker(\partial), \mathrm{coker}(\partial))[-1],$$

are quasi-isomorphic to the abelian DG-Lie algebra $\mathrm{Hom}^*_{\mathbb{K}}(V, \ker \partial)[-1]$.

Proof The equivalence of the two items is clear. The acyclicity of the complex ∂V implies immediately that the natural maps $\ker \partial \to V \to \mathrm{coker}\partial$ are quasi-isomorphisms, and then the natural morphisms

$$\mathrm{Hom}^*_{\mathbb{K}}(V, \ker \partial)[-1] \to \mathrm{Hom}^*_{\mathbb{K}}(V, V)[-1] \to \mathrm{Hom}^*_{\mathbb{K}}(\ker \partial, \mathrm{coker}\partial)[-1],$$

are quasi-isomorphisms of differential graded Lie algebras. □

In characteristic $\neq 2, 3$ the above examples can be seen as particular cases of the following general construction. Assume we have a differential graded Lie algebra $(M, d, [-,-])$, a DG-Lie subalgebra $L \subset M$ and a graded vector space $A \subset M$ such that:

1. $M = L \oplus A$ as graded vector spaces;
2. $[a, b] = 0$ for every $a, b \in A$;
3. $[da, b] \in A$ for every $a, b \in A$.

Denoting by $P \colon M \to A$ the projection with kernel L, consider the operators:

$$\delta \colon A^i \to A^{i+1}, \qquad \delta a = -Pda,$$

$$\{-,-\} \colon A^{i-1} \times A^{j-1} \to A^{i+j-1}, \qquad \{a, b\} = -(-1)^i [da, b].$$

Proposition 5.7.6 *In the above notation, if the characteristic of the base field if different from* 2, 3, *then* $(A[-1], \delta, \{-,-\})$ *is a differential graded Lie algebra.*

Proof We first notice that, since $P([L, L]) = 0$ and $[P(M), P(M)] = 0$, for every $x, y \in M$ we have

$$P[x, y] = P[(x - Px) + Px, (y - Py) + Py] = P[Px, y] + P[x, Py]. \quad (5.10)$$

Since $d(L) \subset L$ we have $Pd = PdP$ and then $\delta^2 = PdPd = Pd^2 = 0$. Next, for $a \in A^{i-1}$ and $b \in A^{j-1}$ we have

$$\begin{aligned} 0 = d[a, b] &= [da, b] + (-1)^{i-1}[a, db] = [da, b] - (-1)^{i-1+(i-1)j}[db, a] \\ &= -(-1)^i\{a, b\} + (-1)^{i-1+(i-1)j+j}\{b, a\} \end{aligned}$$

and multiplying for $(-1)^{i-1}$ we get $0 = \{a, b\} + (-1)^{ij}\{b, a\}$. Moreover, since

$$[Pda, b] = 0, \qquad 0 = d[Pda, b] = [dPda, b] + (-1)^i[Pda, db],$$

we have

$$\begin{aligned} \{\delta a, b\} &= P\{\delta a, b\} = (-1)^{i+1}P[dPda, b] = P[Pda, db], \\ \{a, \delta b\} &= -(-1)^{i(j+1)}\{\delta b, a\} = (-1)^i P[da, Pdb], \\ \delta\{a, b\} &= (-1)^i Pd[da, b] = P[da, db], \end{aligned}$$

and therefore by (5.10)

$$\delta\{a, b\} = P[da, db] = P[da, Pdb] + P[Pda, db] = \{\delta a, b\} + (-1)^i\{a, \delta b\}.$$

The proof of the Jacobi identity is the same and it is omitted. □

In order to see that the derived bracket $[-,-]_\pi$ of Definition 5.7.1 is a particular case of Proposition 5.7.6, consider the direct sum $M = L \oplus A$ where:

$$M = \mathrm{Hom}^*_{\mathbb{K}}(V \oplus W, V \oplus W), \quad A = \mathrm{Hom}^*_{\mathbb{K}}(V, W), \quad L = \{f \in M \mid f(V) \subset V\}.$$

Writing

$$\rho = \begin{pmatrix} d_V & -\pi \\ 0 & d_W \end{pmatrix} : V \oplus W \to V \oplus W,$$

we have $\rho^2 = 0$ and then the adjoint operator $d = [\rho, -]$ is a differential on the graded Lie algebra M such that $d(L) \subset L$. The verification that $\{-,-\} = [-,-]_\pi$ is completely straightforward.

5.8 Examples of DG-Lie Algebras via Graded Pre-Lie Algebras

The notion of a pre-Lie algebra (Definition 2.2.12) extends immediately to the graded case. Given a graded vector space V and a homogeneous bilinear map $*\colon V \times V \to V$ of degree 0, the associator of $*$ is the trilinear map:

$$A\colon V \times V \times V \to V, \qquad A(x, y, z) = (x * y) * z - x * (y * z).$$

The pair $(V, *)$ is called a graded right pre-Lie algebra if the associator is graded symmetric in the last two variables: $A(x, y, z) = (-1)^{\bar{y}\,\bar{z}} A(x, z, y)$.

The notion of left graded pre-Lie algebra is defined similarly by imposing the associator graded symmetric in the first two variables. It should be noted that if $(V, \triangleright)$ is a left graded pre-Lie algebra and

$$\triangleleft\colon V \times V \to V, \qquad x \triangleleft y = -(-1)^{\bar{x}\,\bar{y}} y \triangleright x,$$

then $(V, \triangleleft)$ is a right graded pre-Lie algebras with the same graded commutator.

Lemma 5.8.1 *Assume that $(V, *)$ is a graded pre-Lie algebra. Then the graded commutator*

$$[x, y] = x * y - (-1)^{\bar{x}\,\bar{y}} y * x$$

satisfies the graded Jacobi and the Bianchi identities.

Proof The only nontrivial part of the proof is the verification of the Bianchi identity. If x is a homogeneous component of odd degree we have $[x, x] = 2x * x$ and then

$$[[x, x], x] = 2A(x, x, x) = A(x, x, x) - (-1)^{\bar{x}\,\bar{x}} A(x, x, x) = 0.$$

□

Example 5.8.2 (*The Gerstenhaber bracket*) Let A be a vector space and, for every integer $n \geq 0$ let $V^n = \operatorname{Hom}_{\mathbb{K}}(\bigotimes^{n+1} A, A)$ be the space of multilinear maps

$$f\colon \underbrace{A \times \cdots \times A}_{n+1} \to A.$$

The **Gerstenhaber product** $\circ\colon V \times V \to V$, $V = \oplus_n V^n$, is defined for $f \in V^n$ and $g \in V^m$ by the formula:

$$\begin{aligned}(f \circ g)&(a_0, \ldots, a_{n+m}) \\ &= \sum_{i=0}^{n} (-1)^{im} f(a_0, \ldots, a_{i-1}, g(a_i, \ldots, a_{i+m}), a_{i+m+1}, \ldots, a_{n+m}).\end{aligned}$$

In particular, for $n = 0$ we recover the usual composition product. Although the Gerstenhaber product is not associative, it is easy to verify that $(V, \circ)$ is a graded right pre-Lie algebra, and then the graded commutator

$$[f, g] = f \circ g - (-1)^{\overline{f}\,\overline{g}} g \circ f,$$

called the **Gerstenhaber bracket**, satisfies the graded Jacobi identity. Notice that for an element $\mu \in V^1$ we have

$$\mu \circ \mu(a, b, c) = \mu(\mu(a, b), c) - \mu(a, \mu(b, c))$$

and therefore $\mu \circ \mu = 0$ if and only if $\mu\colon A \times A \to A$ is an associative product.

Sometimes it is useful to extend the Gerstenhaber product to $V^{-1} = A$ by setting $a \circ b = a \circ f = 0$ for every $a, b \in A$, $f \in V^n$, while for $b \in A$ and $f \in V^n$ we define $f \circ b \in V^{n-1}$ by the formula

$$f \circ b(a_0, \dots, a_{n-1}) = \sum_{i=0}^{n} (-1)^i f(a_0, \dots, a_{i-1}, b, a_i, \dots, a_{n-1}) \,.$$

Example 5.8.3 (*The Nijenhuis–Richardson bracket*) Let L be a vector space over a field $\mathbb{K}$ of characteristic $\neq 2$ and, for every integer $n \geq 0$ let $N^n = \operatorname{Hom}_{\mathbb{K}}(\bigwedge^{n+1} L, L)$. The **Nijenhuis–Richardson bracket** $[-, -]_{NR}\colon N^n \times N^m \to N^{n+m}$ is defined as

$$[f, g]_{NR} = f \barwedge g - (-1)^{\overline{f}\,\overline{g}} g \barwedge f \,,$$

where, for $f \in N^n$ and $g \in N^m$ we define

$$(f \barwedge g)(a_0, \dots, a_{n+m}) \\ = \sum_{\sigma \in S(m+1,n)} (-1)^{\sigma} f(g(a_{\sigma(0)}, \dots, a_{\sigma(m)}), a_{\sigma(m+1)}, \dots, a_{\sigma(n+m)}),$$

and the sum is taken over the set $S(m + 1, n)$ of permutations σ of $\{0, \dots, n + m\}$ such that $\sigma(0) < \sigma(1) < \cdots < \sigma(m)$ and $\sigma(m + 1) < \sigma(m + 2) < \cdots < \sigma(m + n)$. As in the previous example the associator of $\barwedge$ is graded symmetric in the last two variables and then $[-, -]_{NR}$ is a graded Lie bracket. Notice that for an element $m \in N^1$ we have $[m, m]_{NR} = 2m \barwedge m$ and

$$m \barwedge m(a, b, c) = m(m(a, b), c) + m(m(b, c), a) + m(m(c, a), b).$$

Therefore $[m, m]_{NR} = 0$ if and only if $m\colon L \times L \to L$ is a Lie bracket.

Example 5.8.4 (*The Hochschild DG-Lie algebra*) Let B be an associative algebra over a field $\mathbb{K}$ of characteristic $\neq 2$ and denote by $\mu\colon B \times B \to B$, $\mu(a, b) = ab$, the multiplication map. The graded vector space of Hochschild cochains of B with coefficients in B is

$$C^*(B,B)=\bigoplus_{n\ge 0} C^n(B,B),\qquad C^n(B,B)=\mathrm{Hom}_{\mathbb{K}}(B^{\otimes n},B).$$

According to Example 5.8.2, the space $\mathfrak{g}=C^*(B,B)[1]$, equipped with the Gerstenhaber bracket, is a graded Lie algebra. The Hochschild differential is defined in terms of the multiplication map μ as the inner derivation

$$d\colon \mathfrak{g}^n\to\mathfrak{g}^{n+1},\qquad d(f)=-[f,\mu].$$

In a more explicit form, for $f\in\mathfrak{g}^n=\mathrm{Hom}_{\mathbb{K}}(B^{\otimes n+1},B)$ we have:

$$\begin{aligned}df(a_0,\ldots,a_{n+1})=&\,a_0f(a_1,\ldots,a_{n+1})+(-1)^nf(a_0,\ldots,a_n)a_{n+1}\\&-\sum_{i=0}^n(-1)^if(a_0,\ldots,a_{i-1},a_ia_{i+1},a_{i+2},\ldots,a_{n+1}).\end{aligned}$$

Setting $\delta(f)=[\mu,f]=(-1)^nd(f)$, the Jacobi identity gives

$$\delta[f,g]=[\delta f,g]+(-1)^{\overline{f}}[f,\delta g].$$

The associativity of μ implies $\mu\circ\mu=\frac{1}{2}[\mu,\mu]=0$, and then

$$\delta^2(f)=[\mu,[\mu,f]]=\frac{1}{2}[[\mu,\mu],f]=0.$$

We have therefore proved that $(\mathfrak{g},\delta,[-,-])$ is a differential graded Lie algebra.

5.9 Exercises

5.9.1 Following the physics terminology, a *propagator* in a DG-vector space (V,d) is a "quasi-inverse" of the differential. More precisely, a propagator is a linear map $h\in\mathrm{Hom}^{-1}_{\mathbb{K}}(V,V)$ such that $h^2=0$ and $dhd=d$. Prove that:

1. propagators exist in every DG-vector space;
2. let h be a propagator in (V,d) and consider the graded vector subspaces

$$H=\{x\in V\mid dhx+hdx=0\},\qquad K=\{x\in V\mid hx=0,\ hdx=x\}.$$

Then $V=\ker d\oplus K$ and $\ker d=d(V)\oplus H$. In particular, $d(H)=0$ and the inclusion map $H\to V$ is a quasi-isomorphism.

5.9.2 (The coglueing theorem) Consider a commutative diagram of DG-vector spaces

$$\begin{CD}
A @>{p}>> B @<<< C\\
@V{f}VV @V{g}VV @V{h}VV\\
L @>{q}>> M @<<< N
\end{CD}$$

with p, q surjective and f, g, h quasi-isomorphisms. Prove that the induced map

$$A \times_B C \to L \times_M N$$

is a quasi-isomorphism.

5.9.3 Prove that for every pair of DG-vector spaces V, W there exists a natural isomorphism of vector spaces:

$$\operatorname{Hom}_{\mathbf{DG}}(V, W) = Z^0(\operatorname{Hom}^*_{\mathbb{K}}(V, W)),$$

and that a morphism of DG-vector spaces $\phi\colon V \to W$ belongs to $B^0(\operatorname{Hom}^*_{\mathbb{K}}(V, W))$ if and only if ϕ is trivial in cohomology.

5.9.4 Prove that there exists a canonical isomorphism

$$\operatorname{Hom}_{\mathbf{DG}}(V \otimes W, Z) = \operatorname{Hom}_{\mathbf{DG}}(V, \operatorname{Hom}^*_{\mathbb{K}}(W, Z)), \qquad V, W, Z \in \mathbf{DG}.$$

5.9.5 Let

$$\begin{CD}
V @>{\alpha}>> E\\
@V{f}VV @VV{g}V\\
W @>{\beta}>> F
\end{CD}$$

be a commutative diagram of DG-vector spaces, with f injective and g surjective. Prove that if either f or g is a quasi-isomorphism, then there exists a morphism of DG-vector spaces $\gamma\colon W \to E$ such that $g\gamma = \beta$ and $\gamma f = \alpha$.

5.9.6 Starting from the commutative diagram

$$\begin{CD}
0 @>>> 0 @>>> \mathbb{K} @>{\mathrm{Id}}>> \mathbb{K} @>>> 0\\
@VVV @VVV @VV{\mathrm{Id}}V @VVV @VVV\\
0 @>>> \mathbb{K} @>{\mathrm{Id}}>> \mathbb{K} @>>> 0 @>>> 0,
\end{CD}$$

give an example of a morphism $f\colon V \to W$ of acyclic DG-vector spaces such that its kernel, its cokernel and the subcomplex

$$D = \{(\alpha, \beta) \in \operatorname{Hom}^*_{\mathbb{K}}(V, V) \times \operatorname{Hom}^*_{\mathbb{K}}(W, W) \mid f\alpha = \beta f\}$$

are not acyclic.

5.9.7 Prove that the isomorphism α of Eq. (5.4) is associative, in the sense that the natural diagram

$$\begin{array}{ccc} s^pV\otimes s^qW\otimes s^rU & \xrightarrow{\alpha\otimes \mathrm{Id}} & s^{p+q}(V\otimes W)\otimes s^rU \\ \downarrow{\scriptstyle \mathrm{Id}\otimes\alpha} & & \downarrow{\scriptstyle \alpha} \\ s^pV\otimes s^{q+r}(W\otimes U) & \xrightarrow{\alpha} & s^{p+q+r}(V\otimes W\otimes U) \end{array}$$

is commutative.

5.9.8 Prove that for every pair of DG-vector spaces V, W and every integer n, the projection map $p\colon \operatorname{cone}(\mathrm{Id}_W)\to W$ induces a natural isomorphism of vector spaces:

$$\mathrm{Hom}_{\mathrm{DG}}(V[-n],\operatorname{cone}(\mathrm{Id}_W))=Z^n(\mathrm{Hom}^*_{\mathbb{K}}(V,\operatorname{cone}(\mathrm{Id}_W)))\cong \mathrm{Hom}^n_{\mathbb{K}}(V,W).$$

5.9.9 The **mapping cylinder** of a morphism $f\colon V\to W$ of DG-vector spaces is the DG-vector space $\mathrm{Cyl}(f)=s^{-1}V\oplus V\oplus W$, equipped with the differential

$$d(s^{-1}v+u+w)=-s^{-1}dv+(-v+du)+(f(v)+dw).$$

Prove that the map $p\colon \mathrm{Cyl}(f)\to W$, $p(s^{-1}v+u+w)=f(u)+w$, is a morphism of DG-vector spaces and there exist two short exact sequences

$$0\to V\xrightarrow{i}\mathrm{Cyl}(f)\to\operatorname{cone}(f)\to 0,\quad 0\to\operatorname{cone}(-\mathrm{Id}_V)\to\mathrm{Cyl}(f)\xrightarrow{p}W\to 0,$$

such that $f=pi$. Deduce that every morphism of DG-vector space is the composition of an injective morphism and a surjective quasi-isomorphism.

5.9.10 Let **C** be the category whose objects are the pair (V,h), where V is a DG-vector space and $h\in\mathrm{Hom}^{-1}_{\mathbb{K}}(V,V)$ satisfies the conditions $h^2=h+hdh=0$. Morphisms $(V,h)\to(W,k)$ are the morphisms $f\colon V\to W$ of DG-vector spaces such that $fh=kf$. Prove that **C** is equivalent to the category of contractions.

5.9.11 The composition of contractions is defined, when possible, as

$$M\underset{\pi}{\overset{\imath}{\rightleftarrows}}N\circlearrowright h\ \circ\ N\underset{p}{\overset{i}{\rightleftarrows}}P\circlearrowright k\ =\ M\underset{\pi p}{\overset{i\imath}{\rightleftarrows}}P\circlearrowright k+ihp\ .$$

Prove that the operators defined in formula (5.6) commute with compositions. More precisely, prove that for a composition of contractions as above and for any perturbation $\partial\in\mathrm{Hom}^1_{\mathbb{K}}(P,P)$ we have $(\imath i)_\partial=\imath_\partial i_{D_\partial}$, provided that all terms are properly defined and make sense.

5.9.12 Let A be a unitary DG-algebra with unit 1. Prove that $A = 0$ if and only if $1 = 0$ and deduce that $H^*(A) = 0$ if and only if 1 is a coboundary in A.

5.9.13 Give an example of DG-vector space over a field of characteristic 3, together with a skew-symmetric bracket satisfying the Jacobi identity but not the Bianchi identity.

5.9.14 Prove that a differential graded Lie algebra $(L, d, [-, -])$ is isomorphic to $(L, \alpha d, \beta[-, -])$ for every pair of invertible constants $\alpha, \beta \in \mathbb{K}$.

5.9.15 Denote by $\mathfrak{m}_n$ the maximal ideal of $\mathbb{K}[t]/(t^n)$ and define the Lie algebras $L = \mathfrak{sl}_2(\mathbb{K}) \otimes \mathfrak{m}_3$, $M = \mathfrak{sl}_2(\mathbb{K}) \otimes \mathfrak{m}_2$ and $f : L \to M$ the natural projection. Denote by $I = \ker(f)$, then by Example 5.6.8 there exists a surjective quasi-isomorphism of DG-Lie algebras $s^{-1}I \oplus L \to M$. Prove that there does not exist any quasi-isomorphism of DG-Lie algebras $M \to s^{-1}I \oplus L$.

5.9.16 Let $(L, d, [-, -])$ be a DG-Lie algebra over a field of characteristic $\neq 2$. Prove that the quotient vector space L^{-1}/dL^{-2}, equipped with the derived bracket

$$\{x, y\} = [x, dy] \qquad (\text{mod } dL^{-2})$$

is a Lie algebra.

5.9.17 (True or false?) Let X be a complex manifold with holomorphic tangent sheaf Θ_X. Let $Z \subset X$ a closed submanifold and denote by $\Theta_X(-\log Z) \subset \Theta_X$ the subsheaf of vector fields that are tangent to Z. Denote by $[-, -]$ the standard bracket in Θ_X.

In the complex $\mathcal{L}$: $\Theta_X(-\log Z) \xrightarrow{d} \Theta_X$, where $\Theta_X(-\log Z)$ is in degree 0, Θ_X is in degree $+1$ and d is the inclusion map, define a bracket $\{-, -\}$ by setting:

1. $\{a, b\} = [a, b] \in \Theta_X(-\log Z)$ if $a, b \in \Theta_X(-\log Z)$,
2. $\{a, b\} = \dfrac{1}{2}[a, b] \in \Theta_X$ if $a \in \Theta_X(-\log Z)$ and $b \in \Theta_X$,

(the $1/2$ factor is needed to make d a Lie derivation). Is it true that $(\mathcal{L}, d, \{-, -\})$ is a sheaf of DG-Lie algebras? Is the Leibniz condition satisfied? Is the Jacobi condition satisfied?

References

The Koszul rule of sign has been recognized as a standard convention since the very first works in homological algebra, cf. [37, Sect. IV.5]; for the Eilenberg–Moore formalism of p-fold suspensions we have followed [197].

The notion of contraction was introduced in [64, p. 81]. In the original definition, Eilenberg and Mac Lane do not require $h^2 = 0$; however, the computation in the proof of Lemma 5.4.2 shows that if h satisfies the remaining four conditions listed in Definition 5.4.1, then $h' = -hd_N h$ satisfies also the fifth, cf. Remark 5.4.7. We refer to the paper [121] for some historical facts and references for the ordinary perturbation lemma Theorem 5.4.6.

Standard and good references for graded commutative DG-algebras and their relation with rational homotopy theory (Sullivan model) are the books [102] by Griffiths and Morgan, and [72] by Félix, Halperin and Thomas.

Differential graded commutative algebras and their resolvents, i.e., a particular kind of cofibrant resolutions, also play an important role in deformation theory of (possibly singular) algebraic schemes and complex spaces [81, 187, 211, 212].

The deep interplay between the abstract notion of DG-Lie algebras and deformation theory was first recognized by Nijenhuis and Richardson in [208]; their basic observation was that a wide class of algebraic structures on a vector space can be defined by essentially the same equation in an appropriate graded Lie algebra. They consider only graded Lie algebras over fields of characteristic $\neq 2$ and they do not require the validity of Bianchi identity. The name Bianchi identity has a clear origin in the case where L is the space of linear endomorphisms of the graded vector space of differential forms with values in a vector bundle. If $D \in L^1$ is a connection with curvature $R = D^2 = \frac{1}{2}[D, D]$, then the usual Bianchi identity (see e.g. [143]) can be written as $[D, R] = 0$.

In the definition of a mapping cone and mapping cylinder we adopt the sign convention of [86], opposite to the one adopted in [269]; this choice leads to a simpler notion of Cartan homotopy in Chap. 8.

As far as we know, the derived bracket construction of Example 5.7.4 was first studied in [181] and used for the study of semiregularity maps in Kähler geometry, see Sect. 8.7. We refer to [157] for a good survey of the theory of derived brackets in the framework of differential graded associative and Lie algebras (references about derived brackets on L_∞-algebras will be given in Chap. 10).

The Gerstenhaber product was introduced in [87, p. 276] in order to define a graded Lie structure in the cohomology $H^*(A, A)$ of an associative algebra A with coefficients in A. This Lie structure, combined with the usual cup product, yields a structure nowadays called a *Gerstenhaber algebra*, see Definition 9.1.3. The Nijenhuis-Richardson bracket was introduced in [208, p. 408] and used in [210] in the study of deformations of Lie algebra structures.

Chapter 6
Maurer–Cartan Equation and Deligne Groupoids

In this chapter we introduce the natural transformation Def from the category of DG-Lie algebras, over a field of characteristic 0, to the category of deformation functors. Then we prove the homotopy invariance of Def, namely that for every quasi-isomorphism $L \to M$, the induced natural transformation $\operatorname{Def}_L \to \operatorname{Def}_M$ is an isomorphism. The explicit functorial construction $L \mapsto \operatorname{Def}_L$ is precisely the one involved in the general philosophy that, in characteristic 0, every deformation problem is controlled by a differential graded Lie algebra.

From now on, and throughout the rest of this book, every vector space is considered over a base field of characteristic 0; unless otherwise specified, by the symbol $\otimes$ we mean the tensor product over the base field.

6.1 Factorization and Homotopy Fibres of DG-Lie Morphisms

The aim of this section is to prove a list of useful algebraic results regarding DG-lie algebras over fields of characteristic 0. In particular, we prove that two DG-Lie algebras L, M are quasi-isomorphic if and only if there exists a span of two surjective quasi-isomorphisms $L \leftarrow H \to M$. Then we introduce the Thom–Whitney homotopy fibre of a morphism, which can be interpreted as the Lie analogue of the mapping cocone of a morphism of differential graded vector spaces.

For every category $\mathbf{C}$ we denote by $\operatorname{Map}\mathbf{C}$ the category whose objects are morphisms of $\mathbf{C}$ and whose morphisms are commutative squares. A functorial factorization in $\mathbf{C}$ is an ordered pair of functors

$$\alpha, \beta\colon \operatorname{Map}\mathbf{C} \to \operatorname{Map}\mathbf{C}$$

M. Manetti, *Lie Methods in Deformation Theory*, Springer Monographs in Mathematics,
https://doi.org/10.1007/978-981-19-1185-9_6

such that $f = \beta(f) \circ \alpha(f)$ for every $f \in \operatorname{Map} \mathbf{C}$.

Theorem 6.1.1 (The factorization lemma) *Every morphism $f : L \to M$ of differential graded Lie algebras over a field of characteristic 0 admits a functorial factorization*

$$f : L \xrightarrow{i} P(f) \xrightarrow{g} M$$

in the category of differential graded Lie algebras, where g is surjective and i is a right inverse of a surjective quasi-isomorphism (in particular i is an injective quasi-isomorphism).

Proof We have seen in Example 5.6.9 that the evaluation map $e_1 : M[t, dt] \to M$ is a surjective quasi-isomorphism and then also its pull-back

$$P(f) = \{(l, m(t)) \in L \times M[t, dt] \mid f(l) = m(1)\} \xrightarrow{p} L, \quad p(l, m(t)) = l,$$

is a surjective quasi-isomorphism. Now observe that

$$i : L \to P(f), \qquad i(l) = (l, f(l)),$$

is a right inverse of p and $f = gi$, where $g(l, m(t)) = m(0) = e_0(m(t))$. The surjectivity of g is simply proved by noticing that for every $m \in M$ we have

$$(0, (1-t)m) \in P(f), \quad g(0, (1-t)m) = e_0((1-t)m) = m\,.$$

The functoriality of the above factorization is clear; in particular, every commutative square of DG-Lie algebras

$$\begin{array}{ccc} L & \xrightarrow{f} & E \\ \downarrow & & \downarrow \\ M & \xrightarrow{g} & F \end{array}$$

gives a new commutative square

$$\begin{array}{ccccc} L & \longrightarrow & P(f) & \longrightarrow & E \\ \downarrow & & \downarrow & & \downarrow \\ M & \longrightarrow & P(g) & \longrightarrow & F. \end{array}$$

□

In the above explicit construction of the factorization, the choice of the evaluation maps e_1, e_0 is purely conventional and the same proof holds for every pair e_a, e_b with $a, b \in \mathbb{K}$, $a \neq b$. More generally, a factorization of a morphism $L \to M$ can be obtained from every factorization of the diagonal

$$\Delta \colon M \xrightarrow{\alpha} M^I \xrightarrow{\beta} M \times M,$$

such that α is a quasi-isomorphism of DG-Lie algebras and β is a surjective morphism of DG-Lie algebras; sometimes a factorization of Δ as above is called a path-object of M.

Corollary 6.1.2 *Two differential graded Lie algebras L, M over a field of characteristic 0 are quasi-isomorphic if and only if there exist two surjective quasi-isomorphisms of DG-Lie algebras $K \to L$, $K \to M$.*

Proof Given two differential graded Lie algebras L, M we shall write $L \sim M$ if there exists a DG-Lie algebra K and two surjective quasi-isomorphisms $K \to L$, $K \to M$. The relation $\sim$ is clearly reflexive, symmetric and transitive, since every diagram

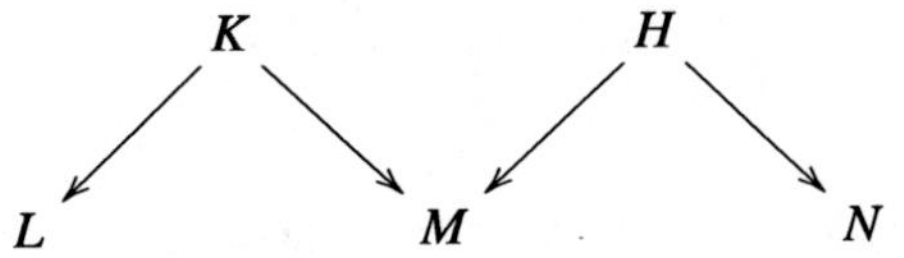

where every arrow is a surjective quasi-isomorphism can be completed to the diagram

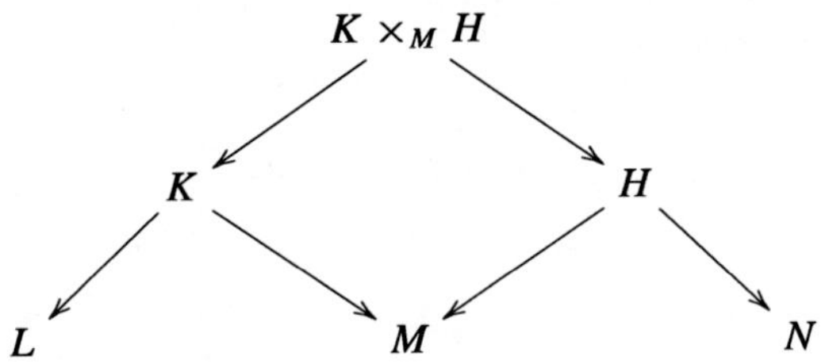

in which every arrow is again a surjective quasi-isomorphism of DG-Lie algebras. It is clear that if $L \sim M$, then L is quasi-isomorphic to M. Conversely, if $f : L \to M$ is a quasi-isomorphism, then the factorization lemma (Theorem 6.1.1) applied to f implies $L \sim M$. □

Corollary 6.1.3 *Let $f : L \to M$ be a morphism of differential graded Lie algebras over a field of characteristic 0:*

1. *If f is injective in cohomology and M is quasi-isomorphic to an abelian DG-Lie algebra, then also L is quasi-isomorphic to an abelian DG-Lie algebra.*
2. *If f is surjective in cohomology and L is quasi-isomorphic to an abelian DG-Lie algebra, then also M is quasi-isomorphic to an abelian DG-Lie algebra.*

Proof Notice first that every abelian DG-Lie algebra N is quasi-isomorphic to an abelian DG-Lie algebra with trivial differential. In fact every subcomplex of N is a DG-Lie subalgebra, and then it is sufficient to choose a subcomplex $H \subset Z^*(N)$ such that the inclusion $H \subset N$ is a quasi-isomorphism.

In case (1), since M is quasi-isomorphic to an abelian differential graded Lie algebra, by Corollary 6.1.2 there exists a diagram of DG-Lie algebras

$$\begin{array}{ccc} & & K & \xrightarrow{\gamma} & H \\ & & \downarrow{\scriptstyle g} & & \\ L & \xrightarrow{f} & M & & \end{array}$$

with g, γ surjective quasi-isomorphisms and H abelian with trivial differential. Considering the fibre product of f and g we get a new diagram

$$\begin{array}{ccccc} L \times_M K & \xrightarrow{\beta} & K & \xrightarrow{\gamma} & H \\ \downarrow{\scriptstyle \alpha} & & \downarrow{\scriptstyle g} & & \\ L & \xrightarrow{f} & M & & \end{array}$$

with α, g, γ quasi-isomorphisms and then β is injective in cohomology. It is now sufficient to consider a suitable projection $\delta \colon H \to E$ of graded vector spaces such that the composition $\delta\gamma\beta$ is a quasi-isomorphism.

In case (2), since L is quasi-isomorphic to an abelian differential graded Lie algebra there exists a diagram

$$\begin{array}{ccccc} K & \xrightarrow{\gamma} & L & \xrightarrow{f} & M \\ \downarrow{\scriptstyle g} & & & & \\ H & & & & \end{array}$$

with g, γ surjective quasi-isomorphisms and H abelian with trivial differential. Let's choose any morphism $i \colon V \to H$ of graded abelian Lie algebras such that the composition

$$V \xrightarrow{i} H \xrightarrow{g^{-1}} H^*(K) \xrightarrow{\gamma} H^*(L) \xrightarrow{f} H^*(M)$$

is an isomorphism. Taking the fibre product of g and i we get a diagram

$$\begin{array}{ccccccc} V \times_H K & \xrightarrow{\bar{i}} & K & \xrightarrow{\gamma} & L & \xrightarrow{f} & M \\ \downarrow{\scriptstyle \bar{g}} & & \downarrow{\scriptstyle g} & & & & \\ V & \xrightarrow{i} & H & & & & \end{array}$$

with $g, \bar{g}$ and $f\gamma\bar{i}$ quasi-isomorphisms of differential graded Lie algebras. □

Let $f \colon L \xrightarrow{i} P \xrightarrow{p} M$ be a factorization of a morphism of differential graded Lie algebras with p surjective and i a quasi-isomorphism. The kernel of p is called the (abstract) homotopy fibre of f and plays an important role in deformation theory. It is easy to prove that its quasi-isomorphism class is independent of the choice of the factorization; in fact, when $f: L \to M$ is surjective we have a commutative diagram of DG-Lie algebras with exact rows

$$\begin{array}{ccccccccc} 0 & \longrightarrow & \ker(f) & \longrightarrow & L & \xrightarrow{f} & M & \longrightarrow & 0 \\ & & \downarrow{\scriptstyle i} & & \downarrow{\scriptstyle i} & & \| & & \\ 0 & \longrightarrow & \ker(p) & \longrightarrow & P & \xrightarrow{p} & M & \longrightarrow & 0 \end{array}$$

and by the five-lemma also the restriction $i: \ker(f) \to \ker(p)$ is a quasi-isomorphism. In general, given two factorizations $L \xrightarrow{i} P \xrightarrow{p} M$, $L \xrightarrow{j} Q \xrightarrow{q} M$ of the same morphism $pi = qj$, we have a pull-back diagram

$$\begin{array}{ccc} P \times_M Q & \xrightarrow{q^*} & P \\ {\scriptstyle p^*}\downarrow & \searrow^{s} & \downarrow{\scriptstyle p} \\ Q & \xrightarrow{q} & M \end{array}$$

where every map is surjective. Taking any factorization

$$(i, j): L \xrightarrow{k} R \xrightarrow{r} P \times_M Q$$

with k quasi-isomorphism and r surjective, since the composite maps p^*r and q^*r are still surjective, we have two commutative diagrams

$$\begin{array}{ccc} L & \xrightarrow{i} & P \\ {\scriptstyle k}\downarrow & \nearrow^{q^*r} & \downarrow{\scriptstyle p} \\ R & \xrightarrow{sr} & M, \end{array} \qquad \begin{array}{ccc} L & \xrightarrow{j} & Q \\ {\scriptstyle k}\downarrow & \nearrow^{p^*r} & \downarrow{\scriptstyle q} \\ R & \xrightarrow{sr} & M, \end{array}$$

giving two quasi-isomorphisms $\ker(sr) \to \ker(p)$, $\ker(sr) \to \ker(q)$. In a similar way we can define homotopy fibre products and homotopy equalizers.

For the application we have in mind it is often more convenient to consider the following functorial constructions for the homotopy fibre and the homotopy equalizer.

Definition 6.1.4 The (Thom–Whitney) **homotopy fibre** of a morphism $f: L \to M$ of DG-Lie algebras is defined as

$$K(f) = \{(l, m(t)) \in L \times M[t, dt] \mid m(0) = 0,\ m(1) = f(l)\}.$$

In other words, $K(f) = \ker g$, where $f = gi$ is the explicit factorization given in the proof of Theorem 6.1.1. The construction of the Thom-Whitney homotopy fibre is functorial; every diagram of differential graded Lie algebras

$$\begin{array}{ccc} L & \xrightarrow{i} & E \\ \downarrow{\scriptstyle f} & & \downarrow{\scriptstyle g} \\ M & \xrightarrow{h} & F \end{array}$$

induces in the obvious way two morphisms of DG-Lie algebras $K(f) \to K(g)$ and $K(i) \to K(h)$.

Definition 6.1.5 Let $h, g: L \to M$ be two morphisms of differential graded Lie algebras. Their (Thom–Whitney) **homotopy equalizer** is the DG-Lie algebra:

$$K(h, g) = \{(l, m(t)) \in L \times M[t, dt] \mid m(0) = h(l),\ m(1) = g(l)\}.$$

In particular, $K(0, f) = K(f)$ is the homotopy fibre of a morphism f.

In the notation of the above definition, let $C(g - h) = L \times sM$ be the mapping cocone of the morphism of DG-vector spaces $g - h: L \to M$ (Definition 5.2.3). Then there exists a canonical contraction

$$C(g-h) \underset{\pi}{\overset{\imath}{\rightleftarrows}} K(h, g) \circlearrowleft k$$

where

$$\imath: C(g-h) \to K(h, g), \qquad \imath(x, sm) = (x, tg(x) + (1-t)h(x) + dt\, m),$$

$$\pi: K(h, g) \to C(g-h), \qquad \pi(l, p(t)m_1 + q(t)dtm_2) = \left(l, sm_2 \int_0^1 q(t)dt\right).$$

The homotopy k is defined by the formula

$$k: K(h, g) \to K(h, g), \qquad k(l, a(t, dt)) = \left(0, t\int_0^1 a(s, ds) - \int_0^t a(s, ds)\right),$$

where the operator $\int_0^t: M[t, dt] \to M[t, dt]$ is intended as

$$\int_0^t \otimes \mathrm{Id}_M : \mathbb{K}[t, dt] \otimes M \to \mathbb{K}[t, dt] \otimes M.$$

The map $\imath$ is a morphism of DG-vector spaces since

$$d\imath(x, sm) = (dx, tg(dx) + (1-t)h(dx) + dt(g(x) - h(x) - dm))$$
$$= \imath(dx, s(g(x) - h(x) - dm)) = \imath d(x, sm).$$

and it is straightforward to check that $\pi\imath = \mathrm{Id}$, $\imath\pi - \mathrm{Id} = dk + kd$ and $k\imath = \pi k = k^2 = 0$. In particular, $\imath$ is a quasi-isomorphism and therefore, by Eq. (5.5), the natural projection $K(h, g) \to L$, and the embedding

$$M[-1] \to K(h, g), \qquad m \mapsto (0, dt\, m),$$

give a cohomology long exact sequence

$$\cdots H^i(L) \xrightarrow{g-h} H^i(M) \to H^{i+1}(K(h, g)) \to H^{i+1}(L) \xrightarrow{g-h} H^{i+1}(M) \cdots \quad (6.1)$$

Theorem 6.1.6 *Let $h, g\colon L \to M$ be morphisms of differential graded Lie algebras such that $g - h\colon H^*(L) \to H^*(M)$ is injective. Then, the homotopy equalizer*

$$K(h, g) = \{(l, m(t)) \in L \times M[t, dt] \mid m(0) = h(l),\ m(1) = g(l)\}$$

is quasi-isomorphic to an abelian DG-Lie algebra.

Proof Considering the DG-vector space $M[-1]$ as an abelian DG-Lie algebras, by Corollary 6.1.3 it is sufficient to prove that the morphism of differential graded Lie algebras

$$\alpha\colon M[-1] \to K(h, g), \qquad \alpha(m) = (0, dt\, m),$$

is surjective in cohomology. By the exact sequence (6.1) the surjectivity of α in cohomology is equivalent to the injectivity of $g - h$ in cohomology. □

For reference purposes, it is useful to write a separate statement for the following easy consequences of the above computations.

Corollary 6.1.7 *Let $f\colon L \to M$ be a morphism of differential graded Lie algebras; denote by $K(f)$ its Thom–Whitney homotopy fibre and by $C(f) = L \times sM$ its mapping cocone. Then:*

1. *the maps*

$$K(f) \to C(f), \qquad (l, m(t)) \mapsto \left(l, s\int_0^1 m(t)\right),$$

$$C(f) \to K(f), \qquad (l, sm) \mapsto (l, tf(l) + dt\, m),$$

   are quasi-isomorphisms of complexes;
2. *if $f\colon L \to M$ is surjective, then the natural map*

$$\ker(f) \to K(f), \qquad a \mapsto (a, 0),$$

is a quasi-isomorphism of differential graded Lie algebras;

3. *if $f : L \to M$ is injective, then the natural map*

$$C(f) \to s\,\mathrm{coker}(f), \qquad (a, sb) \mapsto sb \pmod{sL},$$

is a quasi-isomorphism of DG-vector spaces;

4. *if $f : H^*(L) \to H^*(M)$ is injective, then $K(f)$ is quasi-isomorphic to an abelian DG-Lie algebra.*

Proof The first item has been already proved for homotopy equalizers, while the last item is an immediate consequence of Theorem 6.1.6. For the second and third items it is sufficient to apply Lemma 5.2.4 and observe that the natural inclusion $\ker(f) \to K(f)$ factors through $C(f)$. □

Every commutative diagram of differential graded Lie algebras

$$\begin{array}{ccc} L & \xrightarrow{i} & E \\ {\scriptstyle f}\downarrow & & \downarrow{\scriptstyle g} \\ M & \xrightarrow{h} & F \end{array} \tag{6.2}$$

induces in the obvious way a morphism of DG-Lie algebras $K(f) \to K(g)$ and also a morphism of DG-vector spaces $C(f) \to C(g)$. If $C(f) \to C(g)$ is a quasi-isomorphism, e.g. if both i and h are quasi-isomorphisms, then also $K(f) \to K(g)$ is a quasi-isomorphism.

For later use it is helpful to introduce also the (Thom–Whitney) homotopy fiber product.

Definition 6.1.8 The Thom–Whitney **homotopy fibre product** of two morphisms $f : L \to H, g : M \to H$ of differential graded Lie algebras is defined by the formula

$$L \times_H^h M := \{(l, m, h(t)) \in L \times M \times H[t, dt] \mid h(0) = f(l),\ h(1) = g(m)\}.$$

It is plain that $0 \times_H^h M = K(g)$ and that $L \times_H^h M$ is the homotopy equalizer of the two morphisms $fp_L, gp_M : L \times M \to H$, where $p_L : L \times M \to L$ and $p_M : L \times M \to M$ are the projections. In particular, $L \times_H^h M$ is quasi-isomorphic, as a complex, to the mapping cocone of the map $L \times M \to H$, $(l, m) \mapsto g(m) - f(l)$.

6.2 Formal and Homotopy Abelian DG-Lie Algebras

The general principle, proved in the following sections, that "quasi-isomorphic DG-Lie algebras give the same deformation theory" suggests that we assign a special role to those differential graded Lie algebras having a simple structure up to quasi-isomorphisms. In this section we consider DG-Lie algebras that are quasi-isomorphic to those having either a trivial bracket or a trivial differential.

Definition 6.2.1 A differential graded Lie algebra is called **homotopy abelian** if it is quasi-isomorphic to an abelian DG-Lie algebra.

The name homotopy abelian is derived from the fact that the class of quasi-isomorphisms contain isomorphisms and satisfies the 2 out of 6 property, thus giving a homotopical structure on the category **DGLA** (cf. Definition 3.5.11).

A necessary condition for a differential graded Lie algebra L to be homotopy abelian is that the cohomology graded Lie algebra $H^*(L)$ is abelian; keep in mind that, as we shall see in Example 6.2.5, this condition is not sufficient.

Definition 6.2.2 A differential graded Lie algebra L is called **formal** if it is quasi-isomorphic to its cohomology graded Lie algebra $H^*(L)$, intended as a DG-Lie algebra with trivial differential.

The above considerations imply that a DG-Lie algebra is homotopy abelian if and only if it is formal and $H^*(L)$ is abelian. In view of this simple fact, the Examples 6.2.5 and 6.2.6 below show that there exist non-formal differential graded Lie algebras. It should be noted that every DG-Lie algebra is quasi-isomorphic to its cohomology as a DG-vector space, and then it is important to stress that in Definition 6.2.2 a DG-Lie algebra L is formal if and only if it is quasi-isomorphic to $H^*(L)$ as DG-Lie algebra.

Example 6.2.3 Every DG-Lie algebra L with at most one nontrivial cohomology group is formal. In fact, assume that $H^i(L) = 0$ for every $i \neq n$; if $n > 0$ let's choose a direct sum decomposition $L^n = B^n(L) \oplus C^n$ and consider the DG-Lie subalgebra

$$C \subset L, \qquad C = C^n \oplus_{i>n} L^i,$$

while for $n \leq 0$ consider the DG-Lie subalgebra

$$C \subset L, \qquad C = Z^n(L) \oplus_{i<n} L^i.$$

In both cases, the inclusion $C \to L$ and the projection $C \to H^n(L)$ are quasi-isomorphisms of DG-Lie algebras.

Example 6.2.4 For every DG-vector space V, the differential graded Lie algebra $\operatorname{Hom}^*_{\mathbb{K}}(V, V)$ is formal. In fact, for every index i we may choose a vector subspace $H^i \subset Z^i(V)$ such that the projection $H^i \to H^i(V)$ is bijective. Then the inclusion of DG-vector space $H = \oplus H^i \to V$ is a quasi-isomorphism.

The subspace $K = \{f \in \operatorname{Hom}^*_{\mathbb{K}}(V, V) \mid f(H) \subset H\}$ is a differential graded Lie subalgebra and there exists a commutative diagram of complexes with exact rows

$$\begin{array}{ccccccccc}
0 & \longrightarrow & K & \xrightarrow{\alpha} & \mathrm{Hom}^*_{\mathbb{K}}(V,V) & \longrightarrow & \mathrm{Hom}^*_{\mathbb{K}}(H,V/H) & \longrightarrow & 0 \\
 & & \downarrow{\scriptstyle\beta} & & \downarrow{\scriptstyle\gamma} & & \| & & \\
0 & \longrightarrow & \mathrm{Hom}^*_{\mathbb{K}}(H,H) & \longrightarrow & \mathrm{Hom}^*_{\mathbb{K}}(H,V) & \longrightarrow & \mathrm{Hom}^*_{\mathbb{K}}(H,V/H) & \longrightarrow & 0.
\end{array}$$

The maps α and β are morphisms of differential graded Lie algebras. By Künneth's formula the complex $\mathrm{Hom}^*_{\mathbb{K}}(H, V/H)$ is acyclic and γ is a quasi-isomorphism, therefore also α and β are quasi-isomorphisms.

To decide whether or not a differential graded Lie algebra is formal is usually difficult. Perhaps the simplest necessary condition for formality is the vanishing of **triple Massey powers**. Given a differential graded Lie algebra L, an odd integer n and an element $\theta \in H^n(L)$ such that $[\theta, \theta] = 0$ in $H^{2n}(L)$, its triple Massey power is the subset

$$[\theta, \theta, \theta] \subset \frac{H^{3n-1}(L)}{[\theta, H^{2n-1}(L)]}$$

defined in the following way: for every cocycle $x \in Z^n(L)$ representing the cohomology class θ, choose an element $p \in L^{2n-1}$ such that $dp = [x, x]$. Then $d[p, x] = [dp, x] = [[x, x], x] = 0$ and it makes sense to consider the element

$$[x, x, x] = \text{class of } [p, x] \in \frac{H^{3n-1}(L)}{[\theta, H^{2n-1}(L)]}.$$

The element $[x, x, x]$ is independent of the choice of p; if $dq = dp = [x, x]$, then $p - q$ gives a cohomology class in $H^{2n-1}(L)$ and therefore $[p, x] - [q, x] \in [\theta, H^{2n-1}(L)]$. Then we define $[\theta, \theta, \theta]$ as the set of all elements $[x, x, x]$, where x varies on all the cocycles representing θ.

If L has a trivial differential, then $[\theta, \theta] = 0$ implies $[x, x] = 0$ and then, choosing $p = 0$ we get $[\theta, \theta, \theta] = \{0\}$.

If $f : L \to M$ is a morphism of DG-Lie algebras then, in the above setup, $[f(x), f(x), f(x)]$ may be defined as the class of $[f(p), f(x)]$ and therefore

$$[f(x), f(x), f(x)] = f([x, x, x]) \in \frac{H^{3n-1}(M)}{[f(\theta), H^{2n-1}(M)]}.$$

If f is a quasi-isomorphism we have $[\theta, \theta] = 0$ if and only if $[f(\theta), f(\theta)] = 0$. If f is a surjective quasi-isomorphism, since $Z(L) \to Z(M)$ is surjective, we have $[\theta, \theta, \theta] = \{0\}$ if and only if $[f(\theta), f(\theta), f(\theta)] = \{0\}$. Thus by Corollary 6.1.2 we have that in a formal differential graded Lie algebra every triple Massey power vanishes.

Here are two basic examples of differential graded Lie algebra with nontrivial triple Massey powers, and hence not formal.

Example 6.2.5 Consider the DG-Lie algebra $L = \oplus L^i$ where:

1. $L^i = \mathbb{K}$ for $i = 3, 5, 6, 8$ and $L^i = 0$ otherwise;
2. $d\colon L^5 \to L^6$ is the identity;
3. $[-,-]\colon L^i \times L^j \to L^{i+j}$ is the product in $\mathbb{K}$ for $(i, j) = (3, 3), (3, 5), (5, 3)$ and 0 otherwise.

Then $H^n(L) = \mathbb{K}$ for $n = 3, 8$ and $H^n(L) = 0$ otherwise; if $\theta \in L^3 = H^3(L) = \mathbb{K}$ then $[\theta, \theta] = d\theta^2$ and therefore $[\theta, \theta, \theta]=\{\theta^3\} \subset L^8=H^8(L)=H^8(L)/[\theta, H^5(L)]$.

Example 6.2.6 For every integer $n \geq 1$, define a DG-Lie algebra $A_n = \oplus_i A_n^i$ in the following way: $A_n^i = 0$ for every $i \neq 1, 2$, $A_n^1 \cong \mathbb{K}^n$ with basis $e_1, \dots, e_n$ and $A_n^2 \cong \mathbb{K}^n$ with basis $h_1, \dots, h_n$. The differential and the bracket are defined by the formulas:

1. $de_1 = 0$ and $[e_1, e_1] = -2h_1$;
2. $de_i = h_{i-1}$ and $[e_1, e_i] = [e_i, e_1] = -h_i$ for every $i > 1$;
3. $[e_i, e_j] = 0$ for every $i, j > 1$.

In particular, for $n = 2$ we have

$$de_1 = 0, \quad de_2 = -[e_1, e_1] = h_1, \quad [e_1, e_2] = -h_2, \quad [e_2, e_2] = 0,$$

$[e_1, H^1(A_2)] = 0$, $[e_1, e_1, e_1] = [-e_2, e_1] = h_2$ and then the triple Massey power of the cocycle e_1 is not trivial. We shall prove in Sect. 6.7 that A_n is not formal for every $n \geq 2$, although triple Massey powers vanish whenever $n > 2$.

Notice also that in both the above two examples the bracket is trivial in cohomology.

6.3 Maurer–Cartan Equation and Gauge Action

The classical Maurer–Cartan equation, discovered by Maurer (1879) and Cartan (1904) as the equation satisfied by the canonical left invariant differential form on a Lie group, can be immediately abstracted to every DG-Lie algebra.

Definition 6.3.1 The **Maurer–Cartan equation** of a differential graded Lie algebra L is

$$da + \frac{1}{2}[a, a] = 0, \qquad a \in L^1.$$

The set of its solutions will be denoted by $\mathrm{MC}(L) \subset L^1$. We shall also refer to $\mathrm{MC}(L)$ as the set of Maurer–Cartan elements of L.

The notion of a nilpotent Lie algebra extends naturally to the differential graded case; in particular for every differential graded Lie algebra L and every proper ideal I of an Artin local $\mathbb{K}$-algebra, the DG-Lie algebra $L \otimes I$ is nilpotent.

Assume now that L is a nilpotent DG-Lie algebra, then L^0 is a nilpotent Lie algebra and we can consider its exponential group $\exp(L^0)$. By the Jacobi identity, for every $a \in L^0$ the corresponding adjoint operator

$$\operatorname{ad} a\colon L \to L, \qquad \operatorname{ad} a = [a, -], \qquad (\operatorname{ad} a)b = [a, b],$$

is a nilpotent derivation of degree 0 and then its exponential

$$e^{\operatorname{ad} a}\colon L \to L, \qquad e^{\operatorname{ad} a}(b) = \sum_{n \geq 0} \frac{(\operatorname{ad} a)^n}{n!}(b),$$

is an isomorphism of graded Lie algebras; for every $b, c \in L$ we have

$$e^{\operatorname{ad} a}([b, c]) = [e^{\operatorname{ad} a}(b), e^{\operatorname{ad} a}(c)]\,.$$

In particular, the quadratic cone $\{b \in L^1 \mid [b, b] = 0\}$ is stable under the adjoint action of $\exp(L^0)$.

The **gauge action** can be considered as a perturbation of the adjoint action that depends on the differential d of L; if $d = 0$ then the gauge action is the same as the adjoint action. It is deeply useful and instructive to introduce the gauge action first in the particular case when the differential d is an adjoint operator.

Example 6.3.2 Let $(M, d, [-, -])$ be a DG-Lie algebra and assume that there exists $u \in M^1$ such that $dx = [u, x]$ for every $x \in M$. Given $a \in M^0$ such that the adjoint operator $\operatorname{ad} a\colon M \to M$ is nilpotent, for every $x \in M^1$ define

$$e^a * x = e^{\operatorname{ad} a}(u + x) - u. \tag{6.3}$$

The element $e^a * x$ is independent of the choice of u: in fact we may write

$$e^a * x = x + \sum_{n>0} \frac{[a, -]^n}{n!}(u + x) = x + \sum_{n \geq 0} \frac{(\operatorname{ad} a)^n}{(n+1)!}([a, u + x])$$

and since $[a, u] = -[u, a] = -da$ we have

$$e^a * x = x + \sum_{n \geq 0} \frac{(\operatorname{ad} a)^n}{(n+1)!}([a, x] - da). \tag{6.4}$$

Since the right side of (6.4) involves only the DG-Lie structure, it is natural to use such a formula to define the gauge action in general. The proof that this is indeed an action follows from the following general construction: given a differential graded Lie algebra $(L, [-, -], d)$ we can construct a new DG-Lie algebra $(L', [-, -]', d')$ by setting $(L')^i = L^i$ for every $i \neq 1$, $(L')^1 = L^1 \oplus \mathbb{K}d$ (here d is considered as a formal symbol of degree 1) with the bracket and the differential defined as:

$$[a + vd, b + wd]' = [a, b] + vd(b) - (-1)^{\bar{a}} wd(a), \quad d'(a + vd) = [d, a + vd]' = d(a).$$

It is easy to prove by induction on n that $d(L^{[n]}) \subset L^{[n]}$ and $(L')^{[2n]} \subset L^{[n]}$ for every $n \geq 1$; in particular, if L is nilpotent, then also L' is nilpotent.

The natural inclusion $L \subset L'$ is a morphism of DG-Lie algebras; denote by ϕ the affine embedding $\phi\colon L^1 \to (L')^1$, $\phi(x) = x + d$. Since $[L^0, L']' \subset L$, the image of ϕ is stable under the adjoint action and then the following definition makes sense.

Definition 6.3.3 Let L be a nilpotent differential graded Lie algebra. The **gauge action** $\exp(L^0) \times L^1 \xrightarrow{*} L^1$ is defined, in the above notation, as

$$e^a * x = \phi^{-1}(e^{\operatorname{ad} a}(\phi(x))) = e^{\operatorname{ad} a}(x + d) - d,$$

where the rightmost expression is obviously intended in L'. More explicitly:

$$\begin{aligned} e^a * x &= \sum_{n\geq 0} \frac{1}{n!}(\operatorname{ad} a)^n(x) + \sum_{n\geq 1} \frac{1}{n!}(\operatorname{ad} a)^n(d) \\ &= \sum_{n\geq 0} \frac{1}{n!}(\operatorname{ad} a)^n(x) - \sum_{n\geq 1} \frac{1}{n!}(\operatorname{ad} a)^{n-1}(da) \\ &= x + \sum_{n\geq 0} \frac{(\operatorname{ad} a)^n}{(n+1)!}([a, x] - da). \end{aligned}$$

Since $\operatorname{ad} a$ is nilpotent, the operator $\sum_{n\geq 0} \dfrac{(\operatorname{ad} a)^n}{(n+1)!} = \dfrac{e^{\operatorname{ad} a} - 1}{\operatorname{ad} a}$ is invertible and we can write

$$[a, x] - da = \frac{\operatorname{ad} a}{e^{\operatorname{ad} a} - 1}(e^a * x - x) = \sum_{n=0}^{\infty} \frac{B_n}{n!}(\operatorname{ad} a)^n(e^a * x - x). \tag{6.5}$$

The fact that $*$ is a right action, or equivalently that $e^a * (e^b * x) = (e^a e^b) * x$, follows from Remark 2.5.8 together with the fact that the image of the Lie morphism $\operatorname{ad}\colon L^0 \to \operatorname{Hom}^0_{\mathbb{K}}(L, L)$ is contained in the nilpotent ideal

$$\{f \in \operatorname{Hom}^0_{\mathbb{K}}(L, L) \mid f(L^{[n]}) \subset L^{[n+1]} \quad \forall\, n > 0\}.$$

Lemma 6.3.4 *Let L be a nilpotent differential graded Lie algebra. Then:*

1. *the set of Maurer–Cartan elements is stable under the gauge action;*
2. *$e^a * x = x$ if and only if $da + [x, a] = (d + \operatorname{ad} x)a = 0$;*
3. *for every $x \in \operatorname{MC}(L)$ and every $u \in L^{-1}$ we have $e^{du+[x,u]} * x = x$.*

Proof (1) For an element $a \in L^1$ we have

$$d(a) + \frac{1}{2}[a, a] = 0 \iff [\phi(a), \phi(a)]' = 0$$

and the quadratic cone $\{b + d \in (L^1)' \mid [b + d, b + d]' = 0\}$ is stable under the adjoint action of $\exp(L^0)$.

(2) We have $[x, a] = -[a, x]$, $da + [x, a] = 0$ if and only if $[a, x] - da = 0$ and the proof follows from (6.5).

(3) Setting $a = [x, u] + du$ we have

$$[x, a] + da = [x, [x, u]] + [x, du] + d[x, u] = \frac{1}{2}[[x, x], u] + [dx, u] = 0.$$

□

It is plain that the Maurer–Cartan equation and gauge action commute with morphisms of differential graded Lie algebras.

Example 6.3.5 Let R be a commutative unitary $\mathbb{K}$-algebra, $I \subset R$ a nilpotent ideal, M an R-module and $m \in M$. The complex

$$I \xrightarrow{d} M, \qquad d(x) = -mx,$$

with I in degree 0 and M in degree +1, carries a natural structure of nilpotent differential graded Lie algebras, where $[a, x] = ax = -[x, a]$ for every $a \in I$, $x \in M$, $[I, I] = 0$ and $[M, M] = 0$. Every element of M satisfies the Maurer–Cartan equation, while for $x \in M$ and $a \in I$, we have

$$e^a * x = x + \sum_{n \geq 0} \frac{[a, -]^n}{(n+1)!}([a, x] + ma) = e^a(m + x) - m.$$

Therefore, for $x, y \in M$ and $a \in I$, we have $e^a * x = y$ if and only if $e^a(m + x) = m + y$.

Remark 6.3.6 For every $a \in L^0$, $x \in L^1$, the polynomial $\gamma(t) = e^{ta} * x \in L^1 \otimes \mathbb{K}[t]$ is the solution of the Cauchy problem

$$\frac{d\gamma(t)}{dt} = [a, \gamma(t)] - da, \qquad \gamma(0) = x.$$

6.4 Deformation Functors Associated to a Differential Graded Lie Algebra

In order to introduce the basic ideas of the use of differential graded Lie algebras in deformation theory we begin with an example where the technical difficulties are reduced to a minimum. Consider a finite complex of vector spaces

$$(V, \overline{\partial}): \quad 0 \to V^0 \xrightarrow{\overline{\partial}} V^1 \xrightarrow{\overline{\partial}} \cdots \xrightarrow{\overline{\partial}} V^n \to 0.$$

Given an Artin local $\mathbb{K}$-algebra A with maximal ideal $\mathfrak{m}_A$ and residue field $\mathbb{K}$, we define a deformation of $(V, \overline{\partial})$ over A as a complex of A-modules of the form

$$0 \to V^0 \otimes A \xrightarrow{\overline{\partial}_A} V^1 \otimes A \xrightarrow{\overline{\partial}_A} \cdots \xrightarrow{\overline{\partial}_A} V^n \otimes A \to 0$$

such that its residue modulo $\mathfrak{m}_A$ gives the original complex $(V, \overline{\partial})$. Since $A = \mathbb{K} \oplus \mathfrak{m}_A$, the last condition is equivalent to

$$\overline{\partial}_A = \overline{\partial} + \xi, \qquad \text{with } \xi \in \operatorname{Hom}^1(V, V) \otimes \mathfrak{m}_A.$$

The "integrability" condition $\overline{\partial}_A^2 = 0$ becomes

$$0 = (\overline{\partial} + \xi)^2 = \overline{\partial}\xi + \xi\overline{\partial} + \xi^2 = d\xi + \frac{1}{2}[\xi, \xi],$$

where d and $[-, -]$ are the differential and the bracket on the differential graded Lie algebra $\operatorname{Hom}^*_{\mathbb{K}}(V, V) \otimes \mathfrak{m}_A$. Two deformations $\overline{\partial}_A, \overline{\partial}'_A$ are isomorphic if there exists a commutative diagram

$$\begin{array}{ccccccccc}
0 \longrightarrow & V^0 \otimes A & \xrightarrow{\overline{\partial}_A} & V^1 \otimes A & \xrightarrow{\overline{\partial}_A} \cdots \xrightarrow{\overline{\partial}_A} & V^n \otimes A & \longrightarrow 0 \\
& \downarrow \phi_0 & & \downarrow \phi_1 & & \downarrow \phi_n & \\
0 \longrightarrow & V^0 \otimes A & \xrightarrow{\overline{\partial}'_A} & V^1 \otimes A & \xrightarrow{\overline{\partial}'_A} \cdots \xrightarrow{\overline{\partial}'_A} & V^n \otimes A & \longrightarrow 0
\end{array}$$

such that every ϕ_i is an isomorphism of A-modules whose specialization to the residue field is the identity. Therefore we can write $\phi := \sum_i \phi_i = \operatorname{Id} + \eta$, where $\eta \in \operatorname{Hom}^0(V, V) \otimes \mathfrak{m}_A$ and, since $\mathbb{K}$ is assumed of characteristic 0, we can take the logarithm and write $\phi = e^a$ for some $a \in \operatorname{Hom}^0(V, V) \otimes \mathfrak{m}_A$. The commutativity of the diagram is therefore given by the equation $\overline{\partial}'_A = e^a \circ \overline{\partial}_A \circ e^{-a}$. Writing $\overline{\partial}_A = \overline{\partial} + \xi$, $\overline{\partial}'_A = \overline{\partial} + \xi'$ and using the relation $e^a \circ b \circ e^{-a} = e^{\operatorname{ad} a}(b)$ we get

$$\xi' = e^{\operatorname{ad} a}(\overline{\partial} + \xi) - \overline{\partial} = \xi + \frac{e^{\operatorname{ad} a} - 1}{\operatorname{ad} a}([a, \xi] + [a, \overline{\partial}]) = \xi + \sum_{n=0}^{\infty} \frac{(\operatorname{ad} a)^n}{(n+1)!}([a, \xi] - da).$$

In particular, both the integrability and the isomorphism conditions are entirely written in terms of the DG-Lie structure of $\operatorname{Hom}^*(V, V) \otimes \mathfrak{m}_A$, and more precisely in terms of the Maurer–Cartan equation and gauge action, respectively. This leads to the following general construction.

Given a differential graded Lie algebra $L = \oplus L^i$ over a field $\mathbb{K}$ of characteristic 0, we can define the following three functors:

1. The **exponential functor** $\exp_L\colon \mathbf{Art}_{\mathbb{K}} \to \mathbf{Grp}$,
$$\exp_L(A) = \exp(L^0 \otimes \mathfrak{m}_A).$$
We can immediately see that $\exp_L$ is smooth.
2. The **Maurer–Cartan functor** $\mathrm{MC}_L\colon \mathbf{Art}_{\mathbb{K}} \to \mathbf{Set}$ defined by
$$\mathrm{MC}_L(A) = \mathrm{MC}(L \otimes \mathfrak{m}_A) = \left\{ x \in L^1 \otimes \mathfrak{m}_A \;\middle|\; dx + \frac{1}{2}[x,x] = 0 \right\}.$$
3. The gauge action of the group $\exp(L^0 \otimes \mathfrak{m}_A)$ on the set $\mathrm{MC}(L \otimes \mathfrak{m}_A)$ is functorial in A and gives an action of the group functor $\exp_L$ on the Maurer–Cartan functor MC_L. The quotient functor $\mathrm{Def}_L = \mathrm{MC}_L / \exp_L$ is called the **deformation functor** associated to L; for every $A \in \mathbf{Art}_{\mathbb{K}}$ we have
$$\mathrm{Def}_L(A) = \frac{\mathrm{MC}(L \otimes \mathfrak{m}_A)}{\exp(L^0 \otimes \mathfrak{m}_A)} = \frac{\{x \in L^1 \otimes \mathfrak{m}_A \mid dx + \frac{1}{2}[x,x] = 0\}}{\text{gauge action}}.$$

Both MC_L and Def_L are deformation functors in the sense of Definition 3.2.5; in fact MC_L commutes with fibre products, while, according to Corollary 3.2.9, Def_L is a deformation functor and the projection $\mathrm{MC}_L \to \mathrm{Def}_L$ is smooth.

The reader should pay attention to the difference between the deformation functor Def_L associated to a DG-Lie algebra L and the functor of deformations of L; the latter is associated to a different differential graded Lie algebra.

Lemma 6.4.1 *In the above setup, if $L \otimes \mathfrak{m}_A$ is abelian then* $\mathrm{Def}_L(A) = H^1(L) \otimes \mathfrak{m}_A$. *In particular*
$$T^1\mathrm{Def}_L = \mathrm{Def}_L(\mathbb{K}[\epsilon]) = H^1(L) \otimes \mathbb{K}\epsilon, \qquad \epsilon^2 = 0.$$

Proof If $L \otimes \mathfrak{m}_A$ is abelian then the Maurer–Cartan equation reduces to $dx = 0$ and therefore $\mathrm{MC}_L(A) = Z^1(L) \otimes \mathfrak{m}_A$. If $a \in L^0 \otimes \mathfrak{m}_A$ and $x \in L^1 \otimes \mathfrak{m}_A$ we have
$$e^a * x = x + \sum_{n \geq 0} \frac{\mathrm{ad}(a)^n}{(n+1)!}([a,x] - da) = x - da$$
and then $\mathrm{Def}_L(A) = \dfrac{Z^1(L \otimes \mathfrak{m}_A)}{d(L^0 \otimes \mathfrak{m}_A)} = H^1(L \otimes \mathfrak{m}_A) = H^1(L) \otimes \mathfrak{m}_A.$ □

It is clear that every morphism $\alpha\colon L \to N$ of differential graded Lie algebras induces two natural transformations $\exp_L \to \exp_N$ and $\mathrm{MC}_L \to \mathrm{MC}_N$. These transformations are compatible with the gauge actions and therefore induce a morphism between the deformation functors $\alpha\colon \mathrm{Def}_L \to \mathrm{Def}_N$.

Obstructions for $\mathbf{MC}_L$ and $\mathbf{Def}_L$

Let L be a differential graded Lie algebra. We want to show that the deformation functor MC_L carries a natural complete obstruction theory with values in the vector space $H^2(L)$.

Let's consider a small extension in $\mathbf{Art}_{\mathbb{K}}$

$$e\colon \qquad 0 \to M \to A \to B \to 0, \qquad\qquad \mathfrak{m}_A M = 0,$$

and let $x \in \mathrm{MC}_L(B) = \{x \in L^1 \otimes \mathfrak{m}_B \mid dx + \frac{1}{2}[x,x] = 0\}$ be a Maurer–Cartan element. We define an obstruction $v_e(x) \in H^2(L \otimes M) = H^2(L) \otimes M$ in the following way: first take a lifting $\tilde{x} \in L^1 \otimes \mathfrak{m}_A$ of x and consider $h = d\tilde{x} + \frac{1}{2}[\tilde{x},\tilde{x}] \in L^2 \otimes M$; we have

$$dh = d^2\tilde{x} + [d\tilde{x},\tilde{x}] = [h,\tilde{x}] - \frac{1}{2}[[\tilde{x},\tilde{x}],\tilde{x}].$$

Since $[L^2 \otimes M, L^1 \otimes \mathfrak{m}_A]=0$ we have $[h,\tilde{x}] = 0$; by Bianchi's identity $[[\tilde{x},\tilde{x}],\tilde{x}] = 0$ and then $dh = 0$. Define $v_e(x)$ as the class of h in $H^2(L \otimes M) = H^2(L) \otimes M$; the first property to prove is that $v_e(x)$ is independent of the choice of the lifting $\tilde{x}$; every other lifting is of the form $\tilde{y} = \tilde{x} + z$, $z \in L^1 \otimes M$ and then

$$d\tilde{y} + \frac{1}{2}[\tilde{y},\tilde{y}] = d\tilde{x} + dz + \frac{1}{2}([\tilde{x},\tilde{x}] + [\tilde{x},z] + [z,\tilde{y}]) = h + dz.$$

It is evident from the above computation that $(H^2(L), v_e)$ is a complete obstruction theory for the functor MC_L.

Example 6.4.2 Let L be a DG-Lie algebra such that $[L^1, L^1] \cap Z^2(L) \subset B^2(L)$. Then the functor MC_L is unobstructed; this follows from the above explicit description of obstruction maps since the above condition implies

$$[L^1 \otimes \mathfrak{m}_A, L^1 \otimes \mathfrak{m}_A] \cap Z^2(L \otimes \mathfrak{m}_A) \subset B^2(L \otimes \mathfrak{m}_A)$$

for every $A \in \mathbf{Art}_{\mathbb{K}}$, cf. Example 6.4.6.

It is not difficult to see that the above condition is sufficient but not necessary for the smoothness of MC_L. On the other hand, we shall see that the weaker condition $[Z^1(L), Z^1(L)] \subset B^2(L)$ is necessary but not sufficient for the smoothness of MC_L.

Lemma 6.4.3 *The complete obstruction theory described above for the functor* MC_L *is invariant under the gauge action and then it is also a complete obstruction theory for* Def_L.

Proof Since the projection $\mathrm{MC}_L \to \mathrm{Def}_L$ is smooth, it is sufficient to apply the general properties of universal obstruction theories. However, it is useful to give also a direct and elementary proof of this fact. Let

$$e\colon \quad 0 \to M \to A \to B \to 0$$

be a small extension and let x, y be two gauge equivalent solutions of the Maurer–Cartan equation in $L \otimes \mathfrak{m}_B$; choosing a lifting $\tilde{x} \in L^1 \otimes \mathfrak{m}_A$ of x, by construction the obstruction $v_e(x)$ is the cohomology class of $h = d\tilde{x} + \frac{1}{2}[\tilde{x}, \tilde{x}]$. Thus, it is sufficient to prove that there exists a lifting $\tilde{y}$ of y such that

$$d\tilde{y} + \frac{1}{2}[\tilde{y}, \tilde{y}] = h.$$

Let $a \in L^0 \otimes \mathfrak{m}_B$ be such that $e^a * x = y$, choose a lifting $\tilde{a} \in L^0 \otimes \mathfrak{m}_A$ and define $e^{\tilde{a}} * \tilde{x} = \tilde{y}$. Then

$$\begin{aligned} d\tilde{y} + \frac{1}{2}[\tilde{y}, \tilde{y}] &= \frac{1}{2}[\tilde{y} + d, \tilde{y} + d]' = \frac{1}{2}[e^{\mathrm{ad}\,\tilde{a}}(\tilde{x} + d), e^{\mathrm{ad}\,\tilde{a}}(\tilde{x} + d)]' \\ &= \frac{1}{2}e^{\mathrm{ad}\,\tilde{a}}[\tilde{x} + d, \tilde{x} + d]' = e^{\mathrm{ad}\,\tilde{a}}(h) = h. \end{aligned}$$

□

Theorem 6.4.4 *Let* $f\colon L \to N$ *be a morphism of differential graded Lie algebras such that* $f\colon H^1(L) \to H^1(N)$ *is surjective and* $f\colon H^2(L) \to H^2(N)$ *is injective. Then the morphism* $f\colon \mathrm{Def}_L \to \mathrm{Def}_N$ *is smooth.*

Proof Since $H^1(L)$ is the tangent space of Def_L and $H^2(L)$ is a complete obstruction space, it is sufficient to apply the standard smoothness criterion (Theorem 3.6.5). □

Theorem 6.4.5 *Let* $f\colon L \to N$ *be a morphism of differential graded Lie algebras with homotopy fibre* $K(f)$*:*

1. *if* Def_L *is smooth and* $f\colon H^1(L) \to H^1(N)$ *is surjective, then* Def_N *is smooth;*
2. *if* Def_N *is smooth and* $f\colon H^2(L) \to H^2(N)$ *is injective, then* Def_L *is smooth;*
3. *if* $f\colon H^1(L) \to H^1(N)$ *is injective, then* $\mathrm{Def}_{K(f)}$ *is smooth.*

Proof The first two items follow immediately from the standard smoothness criterion (Theorem 3.6.5). The map $f\colon H^2(L) \to H^2(N)$ commutes with obstruction maps, thus every obstruction of Def_L is mapped into an obstruction of Def_N.

Therefore if Def_N is smooth every obstruction of Def_L belongs to the kernel of f; if $f\colon H^2(L) \to H^2(N)$ is injective and Def_N is smooth then every obstruction of Def_L is trivial.

The last item is a consequence of the first applied to the morphism of DG-Lie algebras

$$N[-1] \to K(f), \qquad n \mapsto (0, dt\, n),$$

introduced in Theorem 6.1.6. □

Example 6.4.6 We have seen in Example 6.4.2 that if L is a DG-Lie algebra such that $[L^1, L^1] \cap Z^2(L) \subset B^2(L)$, then Def_L is a smooth functor. We can give a different proof by introducing the DG-Lie subalgebra $N \subset L$ defined as:

1. $N^i = 0$ for every $i \leq 0$,
2. $N^1 = L^1$,
3. $N^2 = [L^1, L^1] + B^2(L)$,
4. $N^i = L^i$ for every $i > 2$.

By assumption $H^2(N) = 0$ and then Def_N is smooth. Since $H^1(N) \to H^1(L)$ is surjective, the morphism $\mathrm{Def}_N \to \mathrm{Def}_L$ is smooth.

Example 6.4.7 Let L be a differential graded Lie algebra. Choose a vector space decomposition $N^1 \oplus B^1(L) = L^1$ and consider the DG-Lie subalgebra $N = \oplus N^i \subset L$, where $N^i = 0$ for $i < 1$ and $N^i = L^i$ for $i > 1$. The natural inclusion $N \to L$ gives an isomorphism $H^i(N) \to H^i(L)$ for every $i \geq 1$. According to Theorem 6.4.4 the natural transformation $\mathrm{MC}_N = \mathrm{Def}_N \to \mathrm{Def}_L$ is smooth and induces an isomorphism on tangent spaces $T^1\mathrm{Def}_N = T^1\mathrm{Def}_L$. Notice that $\mathrm{Def}_N = \mathrm{MC}_N$ is a homogeneous deformation functor.

Corollary 6.4.8 *Let L be a differential graded Lie algebra over a field $\mathbb{K}$ of characteristic 0. For every $A \in \mathbf{Art}_{\mathbb{K}}$ and every $x(t) \in \mathrm{MC}(L[t, dt] \otimes \mathfrak{m}_A)$ there exists a unique element $p(t) \in L^0[t] \otimes \mathfrak{m}_A$ such that $p(0) = 0$ and $x(t) = e^{p(t)} * x(0)$.*

Proof The brutal truncation $L^{\geq 0} = \oplus_{i \geq 0} L^i$ is a DG-Lie subalgebra of L and $\mathrm{MC}_{L[t,dt]} = \mathrm{MC}_{L^{\geq 0}[t,dt]}$. Thus it is not restrictive to assume that $L = L^{\geq 0}$ and therefore that $L[t, dt]^0 = L^0[t]$.

The inclusion $L \to L[t, dt]$ is a quasi-isomorphism and then the natural transformation $\mathrm{Def}_L \to \mathrm{Def}_{L[t,dt]}$ is smooth, hence surjective and there exists $z \in \mathrm{MC}_L(A)$ that is gauge equivalent to $x(t)$ in the DG-Lie algebra $L[t, dt] \otimes \mathfrak{m}_A$. This means that there exists $a(t) \in L[t, dt]^0 \otimes \mathfrak{m}_A = L^0[t] \otimes \mathfrak{m}_A$ such that $e^{a(t)} * z = x(t)$. Now $x(0) = e^{a(0)} * z$ and then $x = e^{a(t) \bullet (-a(0))} * x(0)$. Notice finally that for the polynomial $p(t) = a(t) \bullet (-a(0))$ we have $p(0) = 0$.

Assume now that $e^{p(t)} * x(0) = e^{q(t)} * x(0)$ for some $p(t), q(t) \in L^0[t] \otimes \mathfrak{m}_A$, with $p(0) = q(0) = 0$. Setting $r(t) = (-p(t)) \bullet q(t)$ we have $r(0) = 0$, $e^{r(t)} * x(0) = x(0)$ and then $[r(t), x(0)] = d(r(t))$ by Lemma 6.3.4. Since $[r(t), x(0)]$ does not contain terms involving dt, the polynomial $r(t)$ is constant, i.e., $r(t) \in L^0 \otimes \mathfrak{m}_A$ and then $r(t) = 0$. □

6.5 Deligne Groupoids

For simplicity of notation, if **C** is a category we shall write $x \in \mathbf{C}$ if x is an object and $f \in \mathrm{Mor}_{\mathbf{C}}$ if f is a morphism. For every small category **A** and every category **C** we shall denote by $\mathbf{C}^{\mathbf{A}}$ the category of (covariant) functors $\mathbf{A} \to \mathbf{C}$.

We have observed in Sect. 3.8 that the category $\mathbf{Art}_{\mathbb{K}}$ is equivalent to a small full subcategory **A**. In particular, for every category **C** it is safe to consider the category of functors $F\colon \mathbf{Art}_{\mathbb{K}} \to \mathbf{C}$, which is equivalent to $\mathbf{C}^{\mathbf{A}}$ and whose morphisms are the natural transformations.

If the category **C** has a final object $*$, in order to simplify the terminology we shall call a formal pointed object in **C** every covariant functor $F\colon \mathbf{Art}_{\mathbb{K}} \to \mathbf{C}$ such that $F(\mathbb{K}) = *$. For instance, a formal pointed set is a functor of Artin rings according to Definition 3.2.1. Don't confuse the above notion of formal pointed DG-Lie algebras with the notion of formality of differential graded Lie algebras (Definition 6.2.2).

Definition 6.5.1 A **groupoid** is a small category such that every morphism is an isomorphism. Morphisms of groupoids are the functors and the category of groupoids is denoted by **Grpd**.

For every groupoid G it is standard to denote by $\pi_0(G)$ the set of isomorphism classes of objects and, and for every $g \in G$ by $\pi_1(G, g) := \mathrm{Mor}_G(g, g)$. Notice that $\pi_1(G, g)$ is a group with unit element equal to the identity.

Definition 6.5.2 An **equivalence of groupoids** is a morphism of groupoids that is also an equivalence of categories.

It is useful to observe that a morphism of groupoids $f\colon G \to H$ is an equivalence if and only if $f\colon \pi_0(G) \to \pi_0(H)$ is bijective and $f\colon \pi_1(G, g) \to \pi_1(H, f(g))$ is an isomorphism for every $g \in G$.

Example 6.5.3 Given a group G acting on a set X, the **action groupoid** X/G is defined in the following way: the set of objects is X, while the morphisms between two objects $x, y \in X$ are the elements $g \in G$ such that $g(x) = y$.

For every nilpotent differential graded Lie algebra L we shall denote by $C(L)$ the action groupoid of the gauge action of $\exp(L^0)$ on the set of Maurer–Cartan element. This means that the set of objects is $\mathrm{MC}(L)$, and

$$\mathrm{Mor}_{C(L)}(x, y) = \{e^a \in \exp(L^0) \mid e^a * x = y\}, \qquad x, y \in \mathrm{MC}(L).$$

Definition 6.5.4 Let L be a nilpotent differential graded Lie algebra. The **irrelevant stabilizer** of a Maurer–Cartan element $x \in \mathrm{MC}(L)$ is defined as the subgroup (see Lemma 6.3.4):

$$I(x) = \{e^{du+[x,u]} \mid u \in L^{-1}\} \subset \mathrm{Mor}_{C(L)}(x, x).$$

Lemma 6.5.5 *Let L be a nilpotent differential graded Lie algebra, $a \in L^0$ and $x \in \mathrm{MC}(L)$. Then*

$$e^a I(x) e^{-a} = I(y), \qquad \textit{where} \qquad y = e^a * x.$$

In particular, $I(x)$ is a normal subgroup of $\mathrm{Mor}_{C(L)}(x, x)$ and there exists a natural isomorphism

$$\frac{\mathrm{Mor}_{C(L)}(x, y)}{I(x)} = \frac{\mathrm{Mor}_{C(L)}(x, y)}{I(y)},$$

with $I(x)$ and $I(y)$ acting by composition in the obvious way.

Proof Recall that for every $a \in L^0$, the map $e^{\operatorname{ad} a} \colon L^0 \to L^0$ is an isomorphism of Lie algebras; therefore it induces an automorphism of the exponential group

$$e^{\operatorname{ad} a} \colon \exp(L^0) \to \exp(L^0), \qquad e^{\operatorname{ad} a}(e^b) = e^{e^{\operatorname{ad} a}(b)}.$$

Thus, for every $a, b \in L^0$ we have

$$e^a e^b e^{-a} = e^{\operatorname{ad} a}(e^b) = e^c, \qquad \text{with} \quad c = e^{\operatorname{ad} a}(b).$$

Therefore, for every $u \in L^{-1}$ we get $e^a e^{[x,u]+du} e^{-a} = e^c$, where, setting $v = e^{\operatorname{ad} a}(u)$ and $y = e^a * x$, we have

$$c = e^{\operatorname{ad} a}([x, u] + du) = e^{\operatorname{ad} a}([x + d, u]') = [e^{\operatorname{ad} a}(x + d), v]' = [y, v] + dv.$$

□

Definition 6.5.6 The **Deligne groupoid** of a nilpotent differential graded Lie algebra L is the groupoid $\mathrm{Del}(L)$ having as objects the Maurer–Cartan elements of L and as morphisms

$$\mathrm{Mor}_{\mathrm{Del}(L)}(x, y) := \frac{\mathrm{Mor}_{C(L)}(x, y)}{I(x)} = \frac{\mathrm{Mor}_{C(L)}(x, y)}{I(y)}, \qquad x, y \in \mathrm{MC}(L).$$

In order to verify that $\mathrm{Del}(L)$ is a groupoid we only need to verify that the (associative) multiplication map

$$\mathrm{Mor}_{C(L)}(y, z) \times \mathrm{Mor}_{C(L)}(x, y) \to \mathrm{Mor}_{C(L)}(x, z), \qquad (e^a, e^b) \mapsto e^a e^b,$$

factors to a morphism

$$\frac{\mathrm{Mor}_{C(L)}(y, z)}{I(y)} \times \frac{\mathrm{Mor}_{C(L)}(x, y)}{I(x)} \to \frac{\mathrm{Mor}_{C(L)}(x, z)}{I(x)},$$

and this follows immediately from Lemma 6.5.5.

Lemma 6.5.7 *Let M be a nilpotent differential graded Lie algebra. Assume that in $M[t, dt]$ we have*

$$e^{q(t)+h(t)dt} * x = x$$

for some $x \in \mathrm{MC}(M)$, $q(t) \in M^0[t]$ and $h(t) \in M^{-1}[t]$. Then $e^{q(1)}e^{-q(0)} \in I(x)$.

Proof Notice first that both $e^{q(1)}$ and $e^{q(0)}$ belong to the stabilizer of x. Setting

$$e^{p(t)+k(t)dt} = e^{q(t)+h(t)dt}e^{-q(0)}$$

we have $p(0) = 0$ and we want to prove that $e^{p(1)} \in I(x)$. According to Lemma 6.3.4 we have

$$[x, p(t) + k(t)dt] + d(p(t) + k(t)dt) = 0. \tag{6.6}$$

If $p(t) = \sum_{i\geq 0} p_i t^{i+1}$ and $k(t) = \sum_{i\geq 0} k_i t^i dt$, Eq. (6.6) gives

$$[x, k_i dt] = -((i+1)p_i + dk_i)dt, \qquad [x, k_i] + dk_i = -(i+1)p_i,$$

and therefore we may write

$$p(1) = \sum_{i\geq 0} p_i = [x, v] + dv, \quad \text{where} \quad v = -\sum_{i\geq 0} \frac{k_i}{i+1}.$$

□

Definition 6.5.8 Given a pair of morphisms $h, g\colon L \to M$ of DG-Lie algebras define two functors $\mathrm{MC}_{h,g}, \mathrm{Def}_{h,g}\colon \mathbf{Art}_{\mathbb{K}} \to \mathbf{Set}$ by setting:

$$\mathrm{MC}_{h,g}(A) = \left\{(x, e^a) \in \mathrm{MC}_L(A) \times \exp(M^0 \otimes \mathfrak{m}_A) \mid e^a * h(x) = g(x)\right\},$$
$$\mathrm{Def}_{h,g}(A) = \frac{\mathrm{MC}_{h,g}}{\sim},$$

where $(x, e^a) \sim (y, e^b)$ if there exists $\alpha \in L^0 \otimes \mathfrak{m}_A$ such that $e^\alpha * x = y$ and the diagram

$$\begin{array}{ccc} h(x) & \xrightarrow{e^a} & g(x) \\ \downarrow{\scriptstyle e^{h(\alpha)}} & & \downarrow{\scriptstyle e^{g(\alpha)}} \\ h(y) & \xrightarrow{e^b} & g(y) \end{array}$$

is commutative in the Deligne groupoid of $M \otimes \mathfrak{m}_A$.

Notice that for $M = 0$ we recover the usual functor Def_L.

Lemma 6.5.9 *In the setup of Definition 6.5.8, consider the homotopy equalizer*

$$K := \{(x, a(t)) \in L \times M[t, dt] \mid a(0) = h(x),\ a(1) = g(x)\}.$$

Then there exists a natural surjective morphism of functors

$$\Phi\colon \operatorname{Def}_K \to \operatorname{Def}_{h,g}.$$

If in addition the map $(h, g)\colon L^{-1} \to M^{-1} \times M^{-1}$, $u \mapsto (h(u), g(u))$, *is surjective then* Φ *is an isomorphism.*

Remark 6.5.10 In Corollary 6.6.4 we shall be able to prove that Φ is always an isomorphism, without any assumption on the map $(h, g)\colon L^{-1} \to M^{-1} \times M^{-1}$.

Proof Given an element $(x, a(t)) \in \mathrm{MC}_K(A)$, $A \in \mathbf{Art}_{\mathbb{K}}$, we have $a(t) \in \mathrm{MC}_{M[t,dt]}(A)$ and by Corollary 6.4.8 there exists a unique $p(t) \in M^0[t] \otimes \mathfrak{m}_A$ such that $p(0) = 0$ and

$$a(t) = e^{p(t)} * a(0) = e^{p(t)} * h(x).$$

In particular $g(x) = a(1) = e^{p(1)} * h(x)$ and we may define a natural transformation

$$\widetilde{\Phi}\colon \mathrm{MC}_K \to \mathrm{MC}_{h,g}, \qquad (x, a(t)) \mapsto (x, e^{p(1)}),$$

which is surjective since $(x, e^{ta} * h(x)) \in \mathrm{MC}_K(A)$ whenever $(x, e^a) \in \mathrm{MC}_{(h,g)}(A)$. If $p(t), q(t) \in M^0[t] \otimes \mathfrak{m}_A$ and $p(0) = q(0) = 0$, $p(1) = q(1)$, then $(x, e^{p(t)} * h(x))$ and $(x, e^{q(t)} * h(x))$ are gauge equivalent; in fact if $s(t) \in M^0[t] \otimes \mathfrak{m}_A$ is defined by the relation $e^{s(t)}e^{p(t)} = e^{q(t)}$ then $s(0) = s(1) = 0$ and therefore

$$(0, s(t)) \in K^0 \otimes \mathfrak{m}_A, \qquad e^{(0,s(t))} * (x, e^{p(t)} * h(x)) = (x, e^{q(t)} * h(x)).$$

In other words $\widetilde{\Phi}$ is injective up to gauge equivalence in K.

Assume now that $(x, e^{p(t)} * h(x))$ and $(y, e^{q(t)} * h(y))$ are gauge equivalent in $\mathrm{MC}_K(A)$, with $p(0) = q(0) = 0$; this means that there exist

$$u \in L^0 \otimes \mathfrak{m}_A, \quad r(t) + s(t)dt \in M[t, dt]^0 \otimes \mathfrak{m}_A$$

such that

$$r(0) = h(u), \quad r(1) = g(u), \quad e^u * x = y, \quad e^{r(t)+s(t)dt}e^{p(t)} * h(x) = e^{q(t)} * h(y)$$

and then

$$e^{-h(u)}e^{-q(t)}e^{r(t)+s(t)dt}e^{p(t)} * h(x) = e^{-h(u)} * h(y) = h(x).$$

Since $r(0) = h(u)$ we have by Lemma 6.5.7 that

$$e^{-h(u)}e^{-q(1)}e^{r(1)}e^{p(1)} = e^{-h(u)}e^{-q(1)}e^{g(u)}e^{p(1)} \in I(h(x))$$

and then $(x, e^{p(1)}) \sim (y, e^{q(1)})$. This proves that Φ factors to a surjective natural transformation

$$\Phi \colon \mathrm{Def}_K \to \mathrm{Def}_{h,g}.$$

Finally, we prove the injectivity of Φ under the additional technical assumption that $(h, g) \colon L^{-1} \to M^{-1} \times M^{-1}$ is surjective. Since $\widetilde{\Phi}$ is injective up to gauge, it is sufficient to prove that if $(x, e^{p(t)} * h(x)) \in \mathrm{MC}_K(A)$, $p(0) = 0$, and $(x, e^{p(1)}) \sim (y, e^b)$ then there exists a polynomial $q(t) \in M^0[t] \otimes \mathfrak{m}_A$ such that $q(0) = 0, q(1) = b$ and $(y, e^{q(t)} * h(y)) \in \mathrm{MC}_K(A)$ is gauge equivalent to $(x, e^{p(t)} * h(x))$.

The assumption $(x, e^{p(1)}) \sim (y, e^b)$ means that there exist $\alpha \in L^0 \otimes \mathfrak{m}_A$ and $v \in M^{-1} \otimes \mathfrak{m}_A$ such that

$$e^{\alpha} * x = y, \qquad e^{g(\alpha)} e^{[g(x),v]+dv} e^{p(1)} e^{-h(\alpha)} = e^b.$$

Let $u \in L^{-1} \otimes \mathfrak{m}_A$ such that $h(u) = 0$, $g(u) = v$, and define $\beta \in L^0 \otimes \mathfrak{m}_A$, $q(t) \in M^0[t] \otimes \mathfrak{m}_A$ by the formulas:

$$e^{\beta} = e^{\alpha} e^{[x,u]+du}, \qquad e^{q(t)} = e^{tg(\beta)+(1-t)h(\beta)} e^{p(t)} e^{-h(\beta)}.$$

Then we have $q(0) = 0$, $h(\beta) = h(\alpha)$,

$$e^{\beta} * x = y, \quad e^{q(1)} = e^{g(\beta)} e^{p(1)} e^{-h(\beta)} = e^b, \quad (\beta, tg(\beta) + (1-t)h(\beta)) \in K^0 \otimes \mathfrak{m}_A,$$

and, since $h(x) = e^{-h(\alpha)} * h(y) = e^{-h(\beta)} * h(y)$ we have the gauge equivalence

$$e^{(\beta, tg(\beta)+(1-t)h(\beta))} * (x, e^{p(t)} * h(x)) = (y, e^{q(t)} * h(y)).$$

□

Remark 6.5.11 It is easy to see that $\mathrm{Def}_{h,g}$ is a deformation functor with tangent and obstruction spaces equal to $H^1(C(g-h))$ and $H^2(C(g-h))$ respectively.

The fact that $\mathrm{Def}_{h,g}$ is a deformation functor follows from Corollary 3.2.9, while it is straightforward to prove the equality $T^1\mathrm{Def}_{h,g} = H^1(C(h-g))$. Let

$$0 \to I \to A \to B \to 0$$

be a small extension and $(\hat{x}, e^{\hat{a}}) \in \mathrm{MC}(L \otimes \mathfrak{m}_B) \times \exp(M^0 \otimes \mathfrak{m}_B)$ be such that $e^{\hat{a}} * h(\hat{x}) = g(\hat{x})$. Choose a lifting $(x, e^a) \in L^1 \otimes \mathfrak{m}_A \times \exp(M^0 \otimes \mathfrak{m}_A)$ and consider the elements $r = dx + \frac{1}{2}[x, x] \in L^2 \otimes I$, $s = e^a * h(x) - g(x) \in M^1 \otimes I$ and $t = (r, s) \in C(h-g)^2 \otimes I$.

We already know that $dr = 0$ and we want to prove that $dt = 0$. Since

$$\begin{aligned}
g(r) &= dg(x) + \frac{1}{2}[g(x), g(x)] = d(e^a * h(x)) - ds + \frac{1}{2}[e^a * h(x), e^a * h(x)] \\
&= \frac{1}{2}[e^a * h(x) + d, e^a * h(x) + d]' - ds \\
&= \frac{1}{2}[e^{\operatorname{ad} a}(h(x) + d), e^{\operatorname{ad} a}(h(x) + d)]' - ds \\
&= \frac{1}{2}e^{\operatorname{ad} a}[h(x) + d, h(x) + d]' - ds = e^{\operatorname{ad} a}h(r) - ds = h(r) - ds,
\end{aligned}$$

we have $(h - g)r - ds = 0$ and then t is a cocycle in $C(h - g) \otimes I$.

If x is replaced with $x + u$, $u \in L^1 \otimes I$ and a is replaced with $a + v$, $v \in M^0 \otimes I$, the element (r, s) will be replaced with $(r + du, s + (h - g)u - dv)$. This implies that the cohomology class of t in $H^2(C(h - g)) \otimes I$ is well defined and is a complete obstruction.

6.6 Homotopy Invariance of Deformation Functors and Deligne Groupoids

We shall say that a functor $F\colon \mathbf{DGLA} \to \mathbf{C}$ is homotopy invariant if for every quasi-isomorphism f of differential graded Lie algebras, the morphism $F(f)$ is an isomorphism in the category $\mathbf{C}$. The main result of this section will be the proof that the functor

$$\operatorname{Def}\colon \mathbf{DGLA} \to \{\text{Deformation functors}\}$$

is homotopy invariant.

Lemma 6.6.1 *Let $f\colon L \to M$ be a morphism of differential graded Lie algebras. Assume that the induced map $f\colon H^i(L) \to H^i(M)$ is:*

1. *surjective for $i = 0$,*
2. *injective for $i = 1$.*

Then $f\colon \operatorname{Def}_L \to \operatorname{Def}_M$ is injective.

Proof Denote by $p_0, p_1\colon L \times L \to L$ the projections and by K, E the homotopy equalizers of the two pairs of maps $p_0, p_1\colon L \times L \to L$ and $fp_0, fp_1\colon L \times L \to M$ respectively. By Lemma 6.5.9, the commutative diagram of differential graded Lie algebras

$$\begin{array}{ccccc}
L & \xleftarrow{p_0} & L \times L & \xrightarrow{p_1} & L \\
\downarrow{\scriptstyle f} & & \downarrow{\scriptstyle \mathrm{Id}} & & \downarrow{\scriptstyle f} \\
M & \xleftarrow[fp_0]{} & L \times L & \xrightarrow[fp_1]{} & M
\end{array}$$

induces a commutative diagram of natural transformations

$$\begin{array}{ccc}
\operatorname{Def}_K & \xrightarrow{\ \gamma\ } & \operatorname{Def}_E \\
\Big\downarrow{\scriptstyle \Phi} & & \Big\downarrow{\scriptstyle \Phi} \\
\operatorname{Def}_{p_0,p_1} & \xrightarrow{\ \eta\ } & \operatorname{Def}_{fp_0,fp_1}\,.
\end{array}$$

Now, the diagram with exact rows

$$\begin{array}{ccccccc}
H^0(L) & \longrightarrow & H^1(C(p_0-p_1)) & \longrightarrow & H^1(L\times L) & \longrightarrow & H^1(L) \\
\Big\downarrow{\scriptstyle f} & & \Big\downarrow & & \Big\| & & \Big\downarrow{\scriptstyle f} \\
H^0(M) & \longrightarrow & H^1(C(fp_0-fp_1)) & \longrightarrow & H^1(L\times L) & \longrightarrow & H^1(M)
\end{array}$$

implies that $f\colon H^1(C(p_1-p_0)) \to H^1(C(fp_1-fp_0))$ is surjective, while the diagram

$$\begin{array}{ccccccc}
H^1(L\times L) & \longrightarrow & H^1(L) & \longrightarrow & H^2(C(p_0-p_1)) & \longrightarrow & H^2(L\times L) \\
\Big\| & & \Big\downarrow{\scriptstyle f} & & \Big\downarrow & & \Big\| \\
H^1(L\times L) & \longrightarrow & H^1(M) & \longrightarrow & H^2(C(fp_0-fp_1)) & \longrightarrow & H^2(L\times L)
\end{array}$$

implies that $f\colon H^2(C(p_1-p_0)) \to H^2(C(fp_1-fp_0))$ is injective. Since K and E are quasi-isomorphic to $C(p_1-p_0)$ and $C(fp_1-fp_0)$ respectively, we have that γ is smooth and then, since Φ is surjective, also η is surjective.

Given $A \in \mathbf{Art}_{\mathbb{K}}$ we need to prove that if $x, y \in \mathrm{MC}_L(A)$ and there exists $b \in M^0 \otimes \mathfrak{m}_A$ such that $e^b * f(x) = f(y)$, then x is gauge equivalent to y. For later use in Lemma 6.6.6 we prove the stronger fact that there exists $c \in L^0 \otimes \mathfrak{m}_A$ such that $e^c * x = y$ and $e^{f(c)}$ is equal to e^b in the Deligne groupoid of $M \otimes \mathfrak{m}_A$. Since $(x, y, e^b) \in \operatorname{Def}_{fp_0,fp_1}(A)$ and η is surjective, there exists $(u, v, e^a) \in \operatorname{Def}_{p_0,p_1}(A)$ such that $\eta(u, v, e^a) = (x, y, e^b)$, i.e., $(u, v, e^{f(a)}) \sim (x, y, e^b)$. This means that $e^a * u = v$ and there exists $\alpha, \beta \in L^0 \otimes \mathfrak{m}_A$ such that $e^\alpha * u = x$, $e^\beta * v = y$, and the diagram

$$\begin{array}{ccc}
f(u) & \xrightarrow{\ e^{f(a)}\ } & f(v) \\
\Big\downarrow{\scriptstyle e^{f(\alpha)}} & & \Big\downarrow{\scriptstyle e^{f(\beta)}} \\
f(x) & \xrightarrow{\ e^{b}\ } & f(y)
\end{array}$$

is commutative in the Deligne groupoid of $M \otimes \mathfrak{m}_A$; it is now sufficient to define $e^c = e^\beta e^a e^{-\alpha}$. □

Theorem 6.6.2 *Let $f\colon L \to M$ be a morphism of differential graded Lie algebras. Assume that the morphism $f\colon H^i(L) \to H^i(M)$ is:*

1. *surjective for $i = 0$,*

2. *bijective for* $i = 1$,
3. *injective for* $i = 2$.

Then $f: \mathrm{Def}_L \to \mathrm{Def}_M$ *is an isomorphism of functors.*

Proof Immediate from Theorem 6.4.4 and Lemma 6.6.1. □

Corollary 6.6.3 *Let* $L \to N$ *be a quasi-isomorphism of differential graded Lie algebras. Then the induced morphism* $\mathrm{Def}_L \to \mathrm{Def}_N$ *is an isomorphism.*

One of the wrong interpretations of Corollary 6.6.3 asserts that if $L \to N$ is a quasi-isomorphism of nilpotent DG-Lie algebras then $\mathrm{MC}(L)/\exp(L^0) \to \mathrm{MC}(N)/\exp(N^0)$ is a bijection. This is false in general; consider for instance $L = 0$ and $N = \oplus N^i$ with $N^i = \mathbb{C}$ for $i = 1, 2$, $N^i = 0$ for $i \neq 1, 2$, $d: N^1 \to N^2$ the identity and $[a, b] = ab$ for $a, b \in N^1 = \mathbb{C}$.

Corollary 6.6.4 *Given two morphisms of DG-Lie algebras* $h, g: L \to M$, *denote by*

$$K = \{(x, m(t)) \in L \times M[t, dt] \mid m(0) = h(x),\ m(1) = g(x)\}$$

their homotopy equalizer. Then the canonical natural transformation

$$\Phi: \mathrm{Def}_K \to \mathrm{Def}_{h,g}$$

of Lemma 6.5.9 is an isomorphism of deformation functors.

Proof By Lemma 6.5.9 we already know that Φ is surjective and it is an isomorphism when the map $(h, g): L^{-1} \to M^{-1} \times M^{-1}$ is surjective. According to Corollary 6.6.3 it is sufficient to prove that there exists a quasi-isomorphism of differential graded Lie algebras $f: L \to N$ and two DG-Lie morphisms $r, s: N \to M$ such that $h = rf$, $g = sf$ and $(r, s): N^{-1} \to M^{-1} \times M^{-1}$ is surjective. In fact this situation gives a commutative diagram

$$\begin{array}{ccc} \mathrm{Def}_K & \xrightarrow{f} & \mathrm{Def}_E \\ \downarrow{\scriptstyle \Phi_K} & & \downarrow{\scriptstyle \Phi_E} \\ \mathrm{Def}_{h,g} & \xrightarrow{\eta} & \mathrm{Def}_{r,s} \end{array}$$

where E is the homotopic equalizer of r, s. By previous results f and Φ_E are isomorphisms and then Φ_K must be injective. A possible choice of N, r, s, f as above is:

$$N = \{(x, a(t), b(t)) \in L \times M[t, dt] \times M[t, dt] \mid a(0) = h(x),\ b(0) = g(x)\},$$
$$f(x) = (x, h(x), g(x)), \quad r(x, a(t), b(t)) = a(1), \quad s(x, a(t), b(t)) = b(1).$$

□

Definition 6.6.5 Let **Grpd** be the category of groupoids. The **(formal) Deligne groupoid** of a differential graded Lie algebra L is the formal pointed groupoid

$$\mathrm{Del}_L \colon \mathbf{Art}_{\mathbb{K}} \to \mathbf{Grpd}, \qquad \mathrm{Del}_L(A) = \mathrm{Del}(L \otimes \mathfrak{m}_A).$$

In other words, for every Artin local $\mathbb{K}$-algebra A with residue field $\mathbb{K}$ and maximal ideal $\mathfrak{m}_A$, $\mathrm{Del}_L(A)$ is the Deligne groupoid (Definition 6.5.6) of the nilpotent DG-Lie algebra $L \otimes \mathfrak{m}_A$.

It is immediate from definitions that the composition of Del_L with the functor

$$\pi_0 \colon \mathbf{Grpd} \to \mathbf{Set}, \qquad \pi_0(G) = \{\text{connected components of } G\},$$

is the deformation functor Def_L associated to the DG-Lie algebra L.

Lemma 6.6.6 *Let $f \colon L \to M$ be a morphism of differential graded Lie algebras such that the map $f \colon H^i(L) \to H^i(M)$ is:*

1. *surjective for $i = 0$,*
2. *injective for $i = 1$.*

Then for every $A \in \mathbf{Art}_{\mathbb{K}}$ the morphism

$$f \colon \mathrm{Del}_L(A) \to \mathrm{Del}_M(A)$$

is full.

Proof This is exactly the same as the proof of Lemma 6.6.1. □

Lemma 6.6.7 *Let x be a Maurer-Cartan element of a nilpotent differential graded Lie algebra L. Then the morphism $\delta = d + [x, -] \colon L \to L$ is a differential and there exists a natural isomorphism*

$$\exp(H^0(L, \delta)) = \mathrm{Mor}_{\mathrm{Del}(L)}(x, x).$$

Proof By the Jacobi identity the adjoint operator $[x, -]$ is a derivation of degree $+1$ and then also δ is a derivation. Moreover,

$$\delta^2(u) = \delta(du + [x, u]) = [x, du] + d[x, u] + [x, [x, u]] = [dx, u] + \frac{1}{2}[[x, x], u] = 0.$$

Since δ is a derivation, the Leibniz rule implies that $Z^0(L, \delta) = \{a \in L^0 \mid da + [x, a] = 0\}$ is a Lie subalgebra. Moreover, for every $a \in Z^0(L, \delta)$ and every $u \in L^{-1}$ we have

$$\delta([a, u]) = [\delta a, u] + [a, \delta u] = [a, \delta u],$$

thus $\delta(L^{-1})$ is a Lie ideal of $Z^0(L, \delta)$ and $I(x) = \exp(\delta L^{-1})$ is the irrelevant stabilizer. We have already seen in Lemma 6.3.4 that $\mathrm{Mor}_{C(L)}(x, x) = \exp(Z^0(L, \delta))$ and therefore

$$\mathrm{Mor}_{\mathrm{Del}(L)}(x, x) = \frac{\exp(Z^0(L, \delta))}{I(x)} = \exp(H^0(L, \delta)).$$

□

Lemma 6.6.8 *Let $f: L \to M$ be a morphism of differential graded Lie algebras such that the map $f: H^i(L) \to H^i(M)$ is:*

1. *surjective for $i = -1$,*
2. *injective for $i = 0$.*

Then for every $A \in \mathbf{Art}_{\mathbb{K}}$ and every $x \in \mathrm{MC}(L \otimes \mathfrak{m}_A)$, the map

$$f: \mathrm{Mor}_{\mathrm{Del}_L(A)}(x, x) \to \mathrm{Mor}_{\mathrm{Del}_M(A)}(f(x), f(x))$$

is injective.

Proof According to Theorem 4.1.1 the morphism of complexes

$$(L \otimes \mathfrak{m}_A, d + [x, -]) \to (M \otimes \mathfrak{m}_A, d + [f(x), -])$$

induces an injective map on their H^0 and then it is sufficient to apply Lemma 6.6.7. □

Theorem 6.6.9 *Let $f: L \to M$ be a morphism of differential graded Lie algebras such that the map $f: H^i(L) \to H^i(M)$ is:*

1. *surjective for $i = -1$,*
2. *bijective for $i = 0, 1$,*
3. *injective for $i = 2$.*

Then for every $A \in \mathbf{Art}_{\mathbb{K}}$ the morphism

$$f: \mathrm{Del}_L(A) \to \mathrm{Del}_M(A)$$

is an equivalence of groupoids.

Proof Immediate from Theorem 6.6.2, Lemmas 6.6.6 and 6.6.8. □

6.7 Further Examples of Nonformal DG-Lie Algebras

In Examples 6.2.5 and 6.2.6 we described examples of differential graded Lie algebras that are not formal. Here we illustrate some further examples which are more interesting from the point of view of deformation theory. The starting point is the obvious consideration that formality has relevant consequences on the associated Maurer-Cartan and deformation functors.

Proposition 6.7.1 *Let L be a formal DG-Lie algebra. Then for every $n \geq 3$ the two maps*

$$\mathrm{Def}_L(\mathbb{K}[t]/(t^3)) \to \mathrm{Def}_L(\mathbb{K}[t]/(t^2)), \qquad \mathrm{Def}_L(\mathbb{K}[t]/(t^n)) \to \mathrm{Def}_L(\mathbb{K}[t]/(t^2)),$$

have the same image. If the map $\mathrm{Def}_L(\mathbb{K}[t]/(t^3)) \to \mathrm{Def}_L(\mathbb{K}[t]/(t^2))$ *is surjective then* Def_L *is smooth.*

Proof According to Corollary 6.6.3 we may assume that L has a trivial differential and therefore that the Maurer–Cartan equation is $[x, x] = 0, x \in L^1$. Therefore $tx_1 \in \mathrm{Def}_L(\mathbb{K}[t]/(t^2))$ lifts to $\mathrm{Def}_L(\mathbb{K}[t]/(t^3))$ if and only if there exists $x_2 \in L^1$ such that

$$t^2[x_1, x_1] \equiv [tx_1 + t^2x_2, tx_1 + t^2x_2] \equiv 0 \pmod{t^3} \iff [x_1, x_1] = 0,$$

while $[x_1, x_1] = 0$ implies that $tx_1 \in \mathrm{Def}_L(\mathbb{K}[t]/(t^n))$ for every $n \geq 3$. The statement about smoothness follows from the fact that $\mathrm{Def}_L(\mathbb{K}[t]/(t^3)) \to \mathrm{Def}_L(\mathbb{K}[t]/(t^2))$ is surjective if and only if the bracket $L^1 \times L^1 \to L^2$ is trivial. □

The formality of L does not imply that $\mathrm{Def}_L(\mathbb{K}[t]/(t^n)) \to \mathrm{Def}_L(\mathbb{K}[t]/(t^3))$ is surjective for $n \geq 3$. The reader can easily verify that for a generic DG-vector space V the DG-Lie algebra $L = \mathrm{Hom}^*_{\mathbb{K}}(V, V)$ is formal and the map $\mathrm{Def}_L(\mathbb{K}[t]/(t^{n+1})) \to \mathrm{Def}_L(\mathbb{K}[t]/(t^n))$ is not surjective for every $n \geq 2$.

Example 6.7.2 We define the **Iwasawa DG-algebra** as the polynomial DG-algebra $R = \mathbb{K}[\omega_1, \omega_2, \omega_3]$, where

$$\overline{\omega_1} = \overline{\omega_2} = \overline{\omega_3} = 1, \quad d\omega_1 = d\omega_2 = 0, \quad d\omega_3 = -\omega_1\omega_2.$$

Then $d(\omega_i\omega_3) = 0$ for every i and there exists an obvious injective quasi-isomorphism of DG-vector spaces $j\colon H^*(R) \hookrightarrow R$ whose image is the graded vector subspace spanned by $1, \omega_1, \omega_2, \omega_1\omega_3, \omega_2\omega_3, \omega_1\omega_2\omega_3$. However, j is not a morphism of algebras, since $\omega_1\omega_2$ is trivial in cohomology.

Proposition 6.7.3 *Let $\mathfrak{n}_3(\mathbb{K})$ be the Lie algebra of strictly upper triangular 3×3 matrices and let R be the Iwasawa DG-algebra defined above. Then:*

1. *the differential graded Lie algebra $\mathfrak{n}_3(\mathbb{K}) \otimes R$ is formal and the functor $\mathrm{Def}_{\mathfrak{n}_3(\mathbb{K})\otimes R}$ is smooth;*
2. *the differential graded Lie algebra $\mathfrak{sl}_2(\mathbb{K}) \otimes R$ is not formal and the functor $\mathrm{Def}_{\mathfrak{sl}_2(\mathbb{K})\otimes R}$ is obstructed.*

Moreover, R is not formal as a DG-algebra and $L \times (\mathfrak{sl}_2(\mathbb{K}) \otimes R)$ is not formal for every differential graded Lie algebra L.

Proof Let's denote by $C \subset R$ the (acyclic) DG-vector subspace spanned by $\omega_3, \omega_1\omega_2$ and by $I \subset \mathfrak{n}_3(\mathbb{K})$ the Lie ideal of matrices of type

$$\begin{pmatrix} 0 & 0 & t \\ 0 & 0 & 0 \\ 0 & 0 & 0 \end{pmatrix}, \qquad t \in \mathbb{K}.$$

Since $I = [\mathfrak{n}_3(\mathbb{K}), \mathfrak{n}_3(\mathbb{K})]$ and $[I, \mathfrak{n}_3(\mathbb{K})] = 0$, the subcomplex $I \otimes C$ is an acyclic DG-Lie ideal of $\mathfrak{n}_3(\mathbb{K}) \otimes R$. The formality of $\mathfrak{n}_3(\mathbb{K}) \otimes R$ is now an immediate consequence of the easy facts that the projection

$$\pi \colon \mathfrak{n}_3(\mathbb{K}) \otimes R \to \frac{\mathfrak{n}_3(\mathbb{K}) \otimes R}{I \otimes C}$$

and

$$\pi \circ (\mathrm{Id} \otimes j) \colon \mathfrak{n}_3(\mathbb{K}) \otimes H^*(R) \to \frac{\mathfrak{n}_3(\mathbb{K}) \otimes R}{I \otimes C}$$

are quasi-isomorphisms of differential graded Lie algebras. The smoothness of $\mathrm{Def}_{\mathfrak{n}_3(\mathbb{K}) \otimes R}$ follows from the fact that the Maurer–Cartan equation in the DG-Lie algebra $\mathfrak{n}_3(\mathbb{K}) \otimes H^*(R)$ is trivial. Notice that the bracket in $H^*(\mathfrak{n}_3(\mathbb{K}) \otimes R)$ is not trivial and then $\mathfrak{n}_3(\mathbb{K}) \otimes R$ is not homotopy abelian.

Next, we shall use Proposition 6.7.1 in order to prove that $M = \mathfrak{sl}_2(\mathbb{K}) \otimes R$ is not formal, although the same fact can be proved by computing triple Massey powers. More precisely we shall prove that there exists an element in $\mathrm{MC}_M(\mathbb{K}[t]/(t^2))$ that lifts to $\mathrm{MC}_M(\mathbb{K}[t]/(t^3))$ but does not lift to $\mathrm{MC}_M(\mathbb{K}[t]/(t^4))$. Denote by u, v, h the standard basis of $\mathfrak{sl}_2(\mathbb{K})$:

$$[u, v] = h, \qquad [h, u] = 2u, \qquad [h, v] = -2v,$$

and consider the element

$$\xi = u\omega_1 t + v\omega_2 t - h\omega_3 t^2 \in \mathrm{MC}_M(\mathbb{K}[t]/(t^3)) \subset M^1 \otimes \mathbb{K}[t]/(t^3).$$

A generic element of $M^1 \otimes \mathbb{K}[t]/(t^4)$ lifting $u\omega_1 t + v\omega_2 t \in \mathrm{MC}_M(\mathbb{K}[t]/(t^2))$ may be written as

$$\eta = u\omega_1 t + v\omega_2 t + (a\omega_1 + b\omega_2 + c\omega_3)t^2 + (d\omega_1 + e\omega_2 + f\omega_3)t^3,$$

with $a, b, c, d, e, f \in \mathfrak{sl}_2(\mathbb{K})$. Assume that η satisfies the Maurer-Cartan equation, since

$$d\eta = -c\omega_1\omega_2 t^2 - f\omega_1\omega_2 t^3, \qquad \frac{1}{2}[\eta, \eta] = h\omega_1\omega_2 t^2 + (\cdots)t^3,$$

we must have $c = h$; therefore the $\mathfrak{sl}_2(\mathbb{K})$-coefficient of $\omega_1\omega_3 t^3$ in $\frac{1}{2}[\eta, \eta]$ is equal to $[u, c] = [u, h] = -2u \neq 0$ and this gives a contradiction. For every differential graded Lie algebra L we have $\mathrm{Def}_{L\times M} = \mathrm{Def}_L \times \mathrm{Def}_M$ and the same proof as above shows that also $L \times M$ is not formal. □

Example 6.7.4 (*The trivial bracket in cohomology does not imply the smoothness of deformation functors*) For every integer $n \geq 3$ there exists a differential graded Lie algebra L such that:

1. $\mathrm{Def}_L(\mathbb{K}[t]/(t^n)) \to \mathrm{Def}_L(\mathbb{K}[t]/(t^2))$ is surjective;
2. $\mathrm{Def}_L(\mathbb{K}[t]/(t^{n+1})) \to \mathrm{Def}_L(\mathbb{K}[t]/(t^2))$ is not surjective;
3. $H^i(L) = 0$ for every $i \neq 1, 2$;
4. $H^1(L) = H^2(L) = \mathbb{K}$ and the bracket is trivial in cohomology.

According to Proposition 6.7.1 such a DG-Lie algebra is not formal, and the induced bracket in $H^*(L)$ is trivial, because for every element $x \in H^1(L)$ the obstruction to the lifting to $\mathrm{Def}_L(\mathbb{K}[t]/(t^3))$ is exactly $[x, x]/2$.

We provide two classes of examples: in the first we take, in the notation of Example 6.2.6, the DG-Lie algebra $L = A_{n-1}$, and it is straightforward to see that for every Artin local ring A we have

$$\mathrm{Def}_L(A) = \mathrm{MC}_L(A) = \left\{ \sum_{i=1}^{n} e_i \otimes x^i \mid x \in \mathfrak{m}_A,\ x^n = 0 \right\}.$$

In the second class of examples, consider the polynomial DG-algebra $\mathbb{K}[x, y]$, where x has degree 0, the degree of y is -1, and $dy = x^n$. The DG-Lie algebra

$$M = \mathrm{Der}^*_{\mathbb{K}}(\mathbb{K}[x, y], \mathbb{K}[x, y]) = M^{-1} \oplus M^0 \oplus M^1$$

is concentrated in the degrees $-1, 0, +1$; the $\mathbb{K}[x]$-module M^0 is free with basis $\frac{\partial}{\partial x}$ and $y\frac{\partial}{\partial y}$, while M^{-1} and M^1 are the free $\mathbb{K}[x]$-modules of rank 1 generated by $y\frac{\partial}{\partial x}$ and $\frac{\partial}{\partial y}$ respectively. We can write $d = x^n\frac{\partial}{\partial y} \in M^1$ and $\delta = [d, -] \colon M \to M$ is the differential of the DG-Lie algebra M. The graded subspace

$$\overline{M} = \{\alpha \in M \mid \alpha(\mathbb{K}[x, y]) \subset (x, y)\}$$

is a differential graded Lie subalgebra, and it is straightforward to check that

$$\left[\frac{\partial}{\partial x}, -\right]^n (d) = n!\,\frac{\partial}{\partial y} \notin \overline{M}$$

and $[\alpha, -]^m(d) \in \overline{M}$ for every $m < n$ and every $\alpha \in M^0$. Let L be the homotopy fibre of the inclusion $\overline{M} \subset M$. According to Corollary 6.6.4 the functor Def_L is a smooth quotient of the functor

$$\mathbf{Art}_{\mathbb{K}} \ni A \mapsto F(A) = \{\alpha \in M^0 \otimes \mathfrak{m}_A \mid e^{\alpha} * 0 \in \overline{M} \otimes \mathfrak{m}_A\}$$

and from the explicit formula

$$e^{\alpha} * 0 = \sum_{m \geq 0} \frac{[\alpha, -]^m}{(m+1)!}([\alpha, d])$$

it follows that $F(\mathbb{K}[t]/(t^{i+1})) \to F(\mathbb{K}[t]/(t^2))$ is surjective if and only if $i < n$. Since $n \geq 3$ by assumption, this implies in particular that the bracket is trivial on $H^1(L)$. By Corollary 6.1.7 $H^i(L) = H^{i-1}(M/\overline{M})$; then $H^1(L) = H^2(L) = \mathbb{K}$ and $H^i(L) = 0$ for every $i \neq 1, 2$.

6.8 Exercises

6.8.1 Prove that the Maurer–Cartan functor of the DG-Lie algebra A_n (Example 6.2.6) is pro-represented by the Artin local ring $\mathbb{K}[t]/(t^{n+1})$.

6.8.2 Consider the DG-Lie algebra $L = L^1 \oplus L^2$ with basis $x, y \in L^1$, $u, v \in L^2$, with differential $dx = 0, dy = u$ and with bracket $[x, x] = u, [x, y] = 0, [y, y] = v$. Prove that the map $\operatorname{Def}_L(\mathbb{K}[t]/(t^4)) \to \operatorname{Def}_L(\mathbb{K}[t]/(t^2))$ is surjective and that the map $\operatorname{Def}_L(\mathbb{K}[t]/(t^5)) \to \operatorname{Def}_L(\mathbb{K}[t]/(t^2))$ is not surjective.

6.8.3 For every integer $n \geq 4$, describe a DG-Lie algebra D_n such that $D_n^i = 0$ for $i \neq 1, 2$, $D_n^1 \cong \mathbb{K}^{n-1}$, $D_n^2 \cong \mathbb{K}^{n-1}$ and its Maurer–Cartan functor is pro-represented by the complete local ring $\mathbb{K}[[x, y]]/(yx^2 + y^{n-1})$.

6.8.4 Let L be the DG-Lie algebra generated as a graded vector space by three generators a, b, c of degree $\overline{a} = \overline{b} = 1, \overline{c} = 2$, equipped with the differential $da = 0$, $db = c$ and with the bracket $[a, a] = c$, $[a, b] = [b, b] = 0$. Let M be the DG-Lie algebra generated by b and let $f : L \to M$ be the natural projection. Prove that the image of $f : \operatorname{Def}_L \to \operatorname{Def}_M$ is not a deformation functor.

6.8.5 Prove that for every DG-Lie algebra M, the kernel of the Lie morphism

$$M[t, dt] \to M \times M, \qquad p(t) \mapsto (p(0), p(1)),$$

is homotopy abelian. Deduce that for every morphism $f : L \to M$ of differential graded Lie algebras, the homotopy fibre of the natural projection $K(f) \to L$ is homotopy abelian.

6.8.6 Let $f : L \to H$, $g : M \to H$ be two morphisms of differential graded Lie algebras over a field of characteristic 0, and denote by

$$L \times_H^h M := \{(l, m, h(t)) \in L \times M \times H[t, dt] \mid h(0) = f(l),\ h(1) = g(m)\}$$

its Thom–Whitney homotopy fibre product. Prove that the natural projections

$$L \times^h_H M \to L, \qquad L \times^h_H M \to M,$$

are surjective morphisms of differential graded Lie algebras inducing a commutative diagram in cohomology:

$$\begin{array}{ccc} H^*(L \times^h_H M) & \longrightarrow & H^*(M) \\ \downarrow & & \downarrow g \\ H^*(L) & \xrightarrow{f} & H^*(H) \end{array}$$

Prove that there exists a long exact sequence

$$\cdots \to H^{i-1}(H) \to H^i(L \times^h_H M) \to H^i(L) \oplus H^i(M) \xrightarrow{f-g} H^i(H) \to \cdots$$

and deduce that if the morphism $f\colon L \to H$ is a quasi-isomorphism, then the projection $L \times^h_H M \to M$ is a quasi-isomorphism.

6.8.7 Let A be a unitary commutative $\mathbb{K}$-algebra and consider the complex

$$V: \qquad A \xrightarrow{d} A,$$

where the differential d is the multiplication by an element of A that is not a divisor of zero. Denote by $L \subset \mathrm{Hom}^*_{\mathbb{K}}(V, V)$ the DG-Lie subalgebra of A-linear endomorphisms. Prove that the inclusion $Z^0(L) \oplus L^1 \hookrightarrow L$ is surjective in cohomology and that L is homotopy abelian over $\mathbb{K}$.

6.8.8 (cf. [195, Sect. 3.1.2]) Let V be a graded vector space and $L = V \oplus sV \simeq V \oplus V[-1]$ the mapping cocone of the identity; the differential on L is thus given by the formula $d(x + sy) = sx$. Every homogeneous bilinear map $\circ\colon V \times V \to V$ of degree 0 can be extended naturally to L by the formulas:

$$x \circ sy = sx \circ sy = 0, \qquad sx \circ y = s(x \circ y), \qquad x, y \in V.$$

Denoting by $[-, -]$ the graded commutator of $\circ$, prove that:

1. $(L, d, [-, -])$ is a DG-Lie algebra if and only if $(V, \circ)$ is a graded right pre-Lie algebra;
2. there exists a natural bijection between the set of right pre-Lie algebra structures on V and the set of Lie brackets on (L, d) such that $[V, V] \subset V$, $[sV, V] \subset sV$ and $[sV, sV] = 0$;
3. if $(L, d, [-, -])$ is a DG-Lie algebra, an element $x + sy \in L^1$ satisfies the Maurer–Cartan equation if and only if $x \circ x = 0$, $x + y \circ x = 0$.

6.8.9 (Ken Brown's lemma) Let $F\colon \mathbf{DGLA} \to \mathbf{C}$ be a functor with the property that if $f\colon L \to M$ is a surjective quasi-isomorphism, then $F(f)\colon F(L) \to F(M)$ is an isomorphism. Prove that $F(g)$ is an isomorphism for every quasi-isomorphism $g\colon H \to K$.

6.8.10 (Wawrik criterion) In the situation of Example 6.4.7, assume that for every surjective morphism $A \to B$ in $\mathbf{Art}_{\mathbb{K}}$ and every $x \in \mathrm{MC}_N(A)$ the natural map

$$H^0(L \otimes \mathfrak{m}_A, d + [x, -]) \to H^0(L \otimes \mathfrak{m}_B, d + [x, -])$$

is surjective. Prove that $\mathrm{MC}_N \to \mathrm{Def}_L$ is an isomorphism. We refer to [237, 268] for the analogous criterion for the functor of deformations of a manifold.

6.8.11 Let $L = \oplus L^i$ be a differential graded Lie algebra, denote by

$$N^0 = \{x \in L^0 \mid dx = 0,\ [x, y] = 0 \text{ for every } y \in L^1\}$$

and assume that the natural map $N^0 \to H^0(L)$ is surjective. Prove, by using the same argument of Example 6.4.7, that Def_L is a homogeneous deformation functor.

6.8.12 Let L be a DG-Lie algebra and consider the functor of inner automorphisms:

$$\mathrm{Inn}_L\colon \mathbf{Art}_{\mathbb{K}} \to \mathbf{Set}, \qquad \mathrm{Inn}_L(A) = \exp(H^0(L) \otimes \mathfrak{m}_A).$$

Denoting, as usual, by $K(f)$ the homotopy fibre of a morphism of differential graded Lie algebras $f\colon L \to M$, prove that there exists a sequence of morphisms of functors of Artin rings

$$\mathrm{Inn}_L \xrightarrow{f} \mathrm{Inn}_M \to \mathrm{Def}_{K(f)} \to \mathrm{Def}_L \xrightarrow{f} \mathrm{Def}_M$$

inducing an exact sequence on tangent spaces. In particular, $\mathrm{Inn}_M \to \mathrm{Def}_{K(f)}$ is smooth if and only if $H^1(L) \xrightarrow{f} H^1(M)$ is injective (cf. Theorem 6.4.5).

References

The basic facts about differential graded Lie algebras proved in Sect. 6.1 imply that in characteristic 0, the category **DGLA** is a *pointed category of fibrant objects* as defined by Brown in [29], where surjective morphisms are fibrations, quasi-isomorphisms are weak equivalences and $L[t, dt] \xrightarrow{(e_0, e_1)} L \times L$ is the path object of L. The name factorization lemma used in Theorem 6.1.1 is therefore intended in the sense of categories of fibrant objects and not in the sense of model categories.

Some sets of formality criteria for DG-Lie algebras, based on a deep study of A_∞ and L_∞ minimal models (see Chap. 12), are given in the recent papers [14, 137, 173, 185].

The treatment of triple Massey powers is taken from [239]; we also refer to this paper for the definition of higher Massey powers $[\theta, \ldots, \theta]$. A graded Lie algebra B is called intrinsically formal if every DG-Lie algebra L with $H^*(L) = B$ is formal; the Examples 6.2.5 and 6.2.6 show essentially that the general abelian graded Lie algebra is not intrinsically formal. It may be interesting to notice that, on the other hand, for every graded vector space V, the graded Lie algebra $\operatorname{Hom}^*_{\mathbb{K}}(V, V)$ is intrinsically formal [185].

The Maurer–Cartan equation for a general differential graded Lie algebra L has been introduced in the papers [208, 209]. In the same papers one can find the construction of the algebra L', the description of the gauge equivalence, and the first relations with deformation theory. The Deligne groupoid, has been introduced in [54, 94], under the assumption that the differential graded Lie algebra has no elements of negative degree; in the same papers it is proved the homotopy invariance of deformation functors for positively graded DG-Lie algebras.

The name *Maurer-Cartan equation*, assigned to the equation $dx + \frac{1}{2}[x, x] = 0$ in a generic DG-Lie algebra, was used by Drinfeld in [60] and obtained a wide popularity after [151, 153]. In the previous contributions to the subject (e.g. [54, 94, 208–210, 238]), the same equation was called the *deformation equation*.

In a letter to Breen on February 28, 1994, Deligne defined a 2-groupoid associated to a generic DG-Lie algebra; taking 1-morphisms up to 2-morphism equivalence we obtain the Deligne groupoid as defined here, and coincide with the groupoid described in [151, lecture of September 13, 1994] and used by Kontsevich for proving the homotopy invariance of deformation functors; another proof based on the same idea is written in the expository paper [176]. For a more recent treatment of the Deligne groupoid see also [32, 275].

Later on this book we shall see an alternative proof of the homotopy invariance of deformation functors which is taken from [153] and it is based on the homotopy classification of L_∞-algebras. Another proof, based on the Schlessinger-type theory of extended deformation functors, is contained in [178, 180].

Proposition 6.7.1 is equivalent to the fact that in a formal DG-Lie algebra, if $\theta \in H^1(L)$ and $[\theta, \theta] = 0$, then the zero vector 0 belongs to every higher Massey power $[\theta, \ldots, \theta]$, provided that it is defined; for more details and references see e.g. [183, Remark 2.5].

In the notation of Proposition 6.7.3, the non-formality of the DG-Lie algebra $\mathfrak{sl}_2(\mathbb{K}) \otimes R$ was essentially proved by Douady [59, 184] by using triple Massey powers.

One of the goals of Example 6.7.4 is to warn the reader about a mistake that appeared several times in the literature, cf. also Remark 8.4.5.

The differential graded Lie algebra L described in Example 6.7.4 has a clear geometrical meaning: it is the DG-Lie algebra controlling the embedded deformations of the closed point inside the unreduced affine scheme $\operatorname{Spec}(\mathbb{K}[x]/(x^n))$. This interpretation is beyond the goal of this book and will be discussed elsewhere; the interested reader can find an almost complete proof in [132, 187].

Part of the material of this chapter is also treated, with a different approach, in [32].

Chapter 7
Totalization and Descent of Deligne Groupoids

This chapter is devoted to two of the main results of this book, namely Hinich's theorem on descent of Deligne groupoids and Whitney's integration theorem. In doing this a certain background on simplicial methods is needed and briefly reviewed in the first part of the chapter. Unless otherwise specified, we work on a fixed field $\mathbb{K}$ of characteristic 0.

7.1 Simplicial and Cosimplicial Objects

Let $\mathbf{\Delta}$ be the category of finite ordinals; the objects in this category are the ordered sets $[n] = \{0 < 1 < \cdots < n\}$, $n \geq 0$, and the morphisms are the nondecreasing maps; it is notationally convenient to denote $[n] = \emptyset$ for every $n < 0$.

Among the morphisms in the category $\mathbf{\Delta}$ we have the **face** maps

$$\delta_k \colon [n-1] \to [n], \qquad \delta_k(p) = \begin{cases} p & \text{if } p < k, \\ p+1 & \text{if } p \geq k, \end{cases} \qquad k = 0, \ldots, n,$$

and the **degeneracy** maps

$$\sigma_k \colon [n+1] \to [n], \qquad \sigma_k(p) = \begin{cases} p & \text{if } p \leq k, \\ p-1 & \text{if } p > k, \end{cases} \qquad k = 0, \ldots, n.$$

Equivalently, $\delta_k \colon [n-1] \to [n]$ is the unique injective monotone map whose image misses k and $\sigma_k \colon [n+1] \to [n]$ is the unique surjective monotone map hitting k twice. It is easy to see that every morphism $f \colon [n] \to [m]$ in $\mathbf{\Delta}$ has a unique representation of the form

$$f = \delta_{i_1} \cdots \delta_{i_k} \sigma_{j_1} \cdots \sigma_{j_h} \tag{7.1}$$

with

M. Manetti, *Lie Methods in Deformation Theory*, Springer Monographs in Mathematics,
https://doi.org/10.1007/978-981-19-1185-9_7

$$m \geq i_1 > i_2 > \cdots > i_k \geq 0, \qquad 0 \leq j_1 < \cdots < j_h < n.$$

Every composition of face and degeneracy maps can be transformed in the form (7.1) by using a finite number of times the **cosimplicial identities**:

$$\delta_i\delta_j = \delta_{j+1}\delta_i \quad \text{for} \quad i \leq j;$$
$$\sigma_i\sigma_j = \sigma_j\sigma_{i+1} \quad \text{for} \quad i \geq j;$$
$$\sigma_i\delta_j = \begin{cases} \delta_{j-1}\sigma_i & \text{for } j > i+1, \\ \mathrm{Id} & \text{for } j = i, i+1, \\ \delta_j\sigma_{i-1} & \text{for } j < i. \end{cases} \tag{7.2}$$

Definition 7.1.1 Let **C** be any category:

1. a *cosimplicial* object in **C** is a functor $A\colon \mathbf{\Delta} \to \mathbf{C}$;
2. a *simplicial* object in **C** is a functor $A_\bullet\colon \mathbf{\Delta}^{op} \to \mathbf{C}$.

In other words, a cosimplicial (resp.: simplicial) object in **C** is a covariant (resp.: contravariant) functor $\mathbf{\Delta} \to \mathbf{C}$.

For notational simplicity, given a cosimplicial object $A\colon \mathbf{\Delta} \to \mathbf{C}$ we write $A_n = A([n])$ for every $n \geq 0$, and for every monotone map $f\colon [n] \to [m]$ we also use the same symbol f to denote the operator $A(f)\colon A_n \to A_m$. Similarly, for a simplicial object $A_\bullet\colon \mathbf{\Delta}^{op} \to \mathbf{C}$ we write $A_n = A_\bullet([n])$ for every $n \geq 0$ and for every monotone map $f\colon [n] \to [m]$ we use the symbol f^* to denote the operator $A_\bullet(f)\colon A_m \to A_n$.

A morphism $\phi\colon A \to B$ of cosimplicial objects is a natural transformation of the corresponding functors $A, B\colon \mathbf{\Delta} \to \mathbf{C}$; equivalently it is a sequence of morphisms $\phi_n \in \mathrm{Mor}_{\mathbf{C}}(A_n, B_n)$ such that $f\phi_n = \phi_m f$ for every monotone map $f\colon [n] \to [m]$. Morphisms of simplicial objects are defined similarly.

Example 7.1.2 An abstract simplicial complex is a pair (I, K) where I is a set and K is a family of nonempty finite subsets of I such that:

1. $\{x\} \in K$ for every $x \in I$;
2. if $A \in K$ and $\emptyset \neq B \subset A$ then also $B \in K$.

Every abstract simplicial complex K gives a simplicial set $K_\bullet$ by taking

$$K_n = \{\alpha\colon [n] \to I \mid \alpha([n]) \in K\}.$$

A monotone map $f\colon [m] \to [n]$ induces by right composition an obvious map $f^*\colon K_m \to K_n$.

Example 7.1.3 The nerve (also called classifying space) of a small category **C** is the simplicial set $B\mathbf{C}_\bullet$ defined in the following way: $B\mathbf{C}_n$ is the set of diagrams

$$C_0 \xrightarrow{\alpha_1} C_1 \to \cdots \to C_{n-1} \xrightarrow{\alpha_n} C_n$$

of n composable morphisms in **C**. Equivalently, $B\mathbf{C}_n$ is the set of functors $[n] \to \mathbf{C}$, where the category structure in the ordered set $[n]$ is defined in the obvious way. In particular, $B\mathbf{C}_0$ is the set of objects of **C**.

Every monotone map $f: [n] \to [m]$ gives the operator

$$f^*: B\mathbf{C}_m \to B\mathbf{C}_n,$$
$$f^*(C_0 \xrightarrow{\alpha_1} \cdots \xrightarrow{\alpha_m} C_m) = C_{f(0)} \xrightarrow{\beta_1} C_{f(1)} \to \cdots \xrightarrow{\beta_n} C_{f(n)},$$

where the morphism β_k is the composition of the identity and the morphisms α_i, with $f(k-1) < i \leq f(k)$. For example, the face maps $\delta_0, \delta_1: [0] \to [1]$ give:

$$\delta_0^*(A \xrightarrow{f} B) = B, \qquad \delta_1^*(A \xrightarrow{f} B) = A.$$

Example 7.1.4 (*Combinatorial standard simplex*) For every integer $p \geq 0$, the simplicial set $\Delta[p]_\bullet$ is defined by setting

$$\Delta[p]_n = \operatorname{Mor}_{\mathbf{\Delta}}([n], [p]), \qquad n \geq 0,$$

and every map $f \in \operatorname{Mor}_{\mathbf{\Delta}}([n], [m])$ induces by right composition a morphism

$$f^*: \Delta[p]_m \to \Delta[p]_n.$$

Every monotone map $g: [p] \to [q]$ induces by left composition a morphism of simplicial sets $g: \Delta[p]_\bullet \to \Delta[q]_\bullet$. Thus the family $\{\Delta[p]_\bullet\}$, $p \geq 0$, is a cosimplicial object in the category of simplicial sets.

More generally, for every subset $S \subset [p]$ we may define a simplicial set $\Delta[p, S]_\bullet$ by setting

$$\Delta[p, S]_n = \{f \in \operatorname{Mor}_{\mathbf{\Delta}}([n], [p]) \mid S \not\subset f([n])\}.$$

It is worth noticing that for every simplicial set $X_\bullet$ and every $p \geq 0$ there exists a natural bijection between X_p and the collection of morphisms of simplicial sets $\Delta[p]_\bullet \to X_\bullet$; in fact, every morphism $\alpha: \Delta[p]_\bullet \to X_\bullet$ of simplicial sets is uniquely determined by the element $x = \alpha(\mathrm{Id}_{[p]}) \in X_p$, since for every monotone map $f: [n] \to [p]$, $f \in \Delta[p]_n$, we have $f = f^*\mathrm{Id}_{[p]}$ and then

$$\alpha(f) = \alpha(f^*\mathrm{Id}_{[p]}) = f^*\alpha(\mathrm{Id}_{[p]}) = f^*(x) \in X_n.$$

Example 7.1.5 (*Affine standard simplex*) Let $\mathbb{K}$ be a field. Define the affine standard n-simplex over $\mathbb{K}$ as the affine space

$$\Delta^n_{\mathbb{K}} = \{(t_0, \ldots, t_n) \in \mathbb{K}^{n+1} \mid t_0 + t_1 + \cdots + t_n = 1\}.$$

By definition, the vertices of $\Delta^n_{\mathbb{K}}$ are the points

$$e_0 = (1, 0, \dots, 0), \quad e_1 = (0, 1, \dots, 0), \quad \dots, \quad e_n = (0, 0, \dots, 1).$$

The family $\{\Delta^n_{\mathbb{K}}\}$, $n \geq 0$, carries a canonical structure of a cosimplicial affine space; for every monotone map $f : [n] \to [m]$, the operator $f : \Delta^n_{\mathbb{K}} \to \Delta^m_{\mathbb{K}}$ is defined as the unique affine map such that $f(e_i) = e_{f(i)}$. Equivalently,

$$f(t_0, \dots, t_n) = \left(\sum_{\{j \mid f(j)=0\}} t_j, \dots, \sum_{\{j \mid f(j)=m\}} t_j \right), \quad \text{(where } \sum_{\emptyset} t_j = 0).$$

For later use it is useful to point out that for the face map $\delta_k : [n] \to [n+1]$ we have $\delta_k(t_0, \dots, t_n) = (t_0, \dots, t_{k-1}, 0, t_k, \dots, t_n)$.

Example 7.1.6 (*Topological standard simplex*) For every $n \geq 0$ define the topological standard n-simplex as the topological space

$$\Delta^n_{\text{top}} = \{(t_0, \dots, t_n) \in \mathbb{R}^{n+1} \mid 0 \leq t_i \leq 1, \ t_0 + t_1 + \cdots + t_n = 1\}.$$

The argument of Example 7.1.5 shows that the family $\{\Delta^n_{\text{top}}\}$, $n \geq 0$, is a cosimplicial topological space.

7.2 Kan Complexes

Let $X_\bullet$ be a simplicial set with face operators $\delta_k^* : X_n \to X_{n-1}$, $n > 0$. Dualizing the cosimplicial identities (7.2) we obtain in particular the relation

$$\delta_i^* \delta_j^* = \delta_{j-1}^* \delta_i^*, \qquad \text{for every } i < j, \tag{7.3}$$

and then for every $n \geq 2$ and every $x \in X_n$, the sequence

$$x_0 = \delta_0^* x, \qquad x_1 = \delta_1^* x, \qquad \dots, \qquad x_n = \delta_n^* x \ \in X_{n-1},$$

satisfies the equalities $\delta_i^* x_j = \delta_{j-1}^* x_i$ for every $i < j$.

Definition 7.2.1 A simplicial set $X_\bullet$ is called a **Kan complex** if for every $n \geq 2$, every $k = 0, \dots, n$ and every sequence $x_0, \dots, x_{k-1}, x_{k+1}, \dots, x_n \in X_{n-1}$ such that

$$\delta_i^* x_j = \delta_{j-1}^* x_i \qquad \text{for every } \ 0 \leq i < j \leq n, \quad i, j \neq k,$$

there exists $x \in X_n$ such that $\delta_i^* x = x_i$ for every $i \neq k$.

Notice that every face operator δ_k^* on a simplicial set is a surjective map since it has either σ_k^* or σ_{k-1}^* as a right inverse. In particular, $\delta_0^*, \delta_1^*: X_1 \to X_0$ are always surjective and this justifies the condition $n \geq 2$ in Definition 7.2.1.

Example 7.2.2 For every $p > 0$, the combinatorial standard simplex $\Delta[p]_\bullet$ is not a Kan complex; consider for instance $k = n = 2$ and the two monotone maps

$$x_0, x_1 : [1] \to [p], \qquad x_0(0) = 0, \quad x_0(1) = x_1(0) = x_1(1) = 1 .$$

They satisfy the relation $\delta_0^* x_1(0) = x_1(1) = 1 = x_0(1) = \delta_0^* x_0(0)$ and, if there exists a map $x : [2] \to [p]$ such that $\delta_0^* x = x_0$ and $\delta_1^* x = x_1$, we have

$$x(0) = \delta_1^* x(0) = x_1(0) = 1, \qquad x(1) = \delta_0^* x(0) = x_0(0) = 0,$$

and therefore x cannot be monotone.

Example 7.2.3 Let $\mathbf{C}$ be a small category. Then its nerve $B\mathbf{C}_\bullet$ is a Kan complex if and only if $\mathbf{C}$ is a groupoid.

One implication is easy: if $B\mathbf{C}_\bullet$ is a Kan complex, for every morphism $A \xrightarrow{f} B$ consider the sequence $x_0 = f$, $x_1 = \mathrm{Id}_B \in B\mathbf{C}_1$. Since $\delta_0^* x_1 = B = \delta_0^* x_0$ there exists $x \in B\mathbf{C}_2$ such that $\delta_0^* x = x_0$ and $\delta_1^* x = x_1$ and this implies that x must be equal to $B \xrightarrow{g} A \xrightarrow{f} B$, with $fg = \mathrm{Id}_B$. Taking $x_1 = \mathrm{Id}_A$ and $x_2 = f$, the same argument implies that there exists $h : B \to A$ such that $hf = \mathrm{Id}_A$ and therefore f is an isomorphism. For the, more difficult, proof of the other implication we refer to [92, Lemma I.3.5].

Sometimes a Kan complex is also called either a **fibrant simplicial set** or an **∞-groupoid**.

Theorem 7.2.4 (Moore) *Every simplicial group is a Kan complex.*

Proof Let $G_\bullet$ be a simplicial object in the category of groups and consider a sequence $x_0, \ldots, x_{k-1}, x_{k+1}, \ldots, x_n \in G_{n-1}$, $n \geq 2$, $n \geq k \geq 0$, such that $\delta_i^* x_j = \delta_{j-1}^* x_i$ for every $i, j \neq k$, $i < j$.

The first step of the proof is to find an element $u \in G_n$ such that $\delta_i^* u = x_i$ for every $i < k$; obviously it is not restrictive here to assume $k > 0$. To this end we construct, recursively, a sequence $u_r \in G_n$, $r = 0, \ldots, k-1$ such that $\delta_i^* u_r = x_i$ for every $i \leq r$. For $r = 0$ we can take $u_0 = \sigma_0^* x_0$, since $\sigma_0 \delta_0 = \mathrm{Id}$ and then $\delta_0^* \sigma_0^* x_0 = x_0$. If $0 < r < k$ and u_{r-1} is defined, we set

$$u_r = u_{r-1} \cdot (\sigma_r^* \delta_r^* u_{r-1})^{-1} \cdot (\sigma_r^* x_r),$$

where $\cdot$ and $(-)^{-1}$ denote the multiplication and the inverse on the group G_n. Then for $i < r$ we have

$$\delta_i^* u_r = (\delta_i^* u_{r-1}) \cdot (\delta_i^* \sigma_r^* \delta_r^* u_{r-1})^{-1} \cdot (\delta_i^* \sigma_r^* x_r).$$

Since $\sigma_r\delta_i = \delta_i\sigma_{r-1}$ we have $\delta_i^*\sigma_r^* = \sigma_{r-1}^*\delta_i^*$ and

$$\begin{aligned}\delta_i^* u_r &= (\delta_i^* u_{r-1}) \cdot (\sigma_{r-1}^* \delta_i^* \delta_r^* u_{r-1})^{-1} \cdot (\sigma_{r-1}^* \delta_i^* x_r)\\ &= (\delta_i^* u_{r-1}) \cdot (\sigma_{r-1}^* \delta_{r-1}^* \delta_i^* u_{r-1})^{-1} \cdot (\sigma_{r-1}^* \delta_{r-1}^* x_i)\\ &= (\delta_i^* u_{r-1}) \cdot (\sigma_{r-1}^* \delta_{r-1}^* x_i)^{-1} \cdot (\sigma_{r-1}^* \delta_{r-1}^* x_i)\\ &= \delta_i^* u_{r-1} = x_i .\end{aligned}$$

Finally,

$$\delta_r^* u_r = (\delta_r^* u_{r-1}) \cdot (\delta_r^* \sigma_r^* \delta_r^* u_{r-1})^{-1} \cdot (\delta_r^* \sigma_r^* x_r) = (\delta_r^* u_{r-1}) \cdot (\delta_r^* u_{r-1})^{-1} \cdot x_r = x_r .$$

In the second and last step of the proof we construct recursively a sequence $v_0, v_1, \ldots, v_{n-k} \in G_n$ such that $\delta_i^* v_q = x_i$ if either $i < k$ or $i > n - q$. To this end it is sufficient to define recursively $v_0 = u$ and, for $q > 0$,

$$v_q = v_{q-1} \cdot (\sigma_{n-q}^* \delta_{n-q+1}^* v_{q-1})^{-1} \cdot (\sigma_{n-q}^* x_{n-q+1}).$$

A computation as above concludes the proof. □

Definition 7.2.5 Given a simplicial group $G_\bullet$, its **Moore complex**

$$N(G_\bullet): \qquad 0 \leftarrow NG_0 \xleftarrow{d} NG_1 \xleftarrow{d} NG_2 \xleftarrow{d} \cdots$$

is defined as $NG_0 = G_0$,

$$NG_n = \bigcap_{i>0} \ker(\delta_i^* \colon G_n \to G_{n-1}), \qquad n > 0,$$

and the morphism d is the restriction of δ_0^*.

The proof that d is well defined and d^2 is the trivial morphism follows immediately from the equalities $\delta_i^*\delta_0^* = \delta_0^*\delta_{i+1}^*$. For simplicial abelian groups, and more generally for simplicial objects in any abelian category, the Moore complex is also called the **normalized chain complex**.

Definition 7.2.6 A simplicial set $X_\bullet$ is called an **acyclic Kan complex** if:

1. the map $X_1 \to X_0 \times X_0$, $x \mapsto (\delta_0^* x, \delta_1^* x)$, is surjective;
2. for every $n \geq 2$ and every sequence $x_0, \ldots, x_n \in X_{n-1}$ such that

$$\delta_i^* x_j = \delta_{j-1}^* x_i \qquad \text{for every} \quad i < j,$$

there exists $x \in X_n$ such that $\delta_i^* x = x_i$ for every i.

Every acyclic Kan complex is also a Kan complex; in fact, given a sequence

$$x_0, \ldots, x_{k-1}, x_{k+1}, \ldots, x_n \in X_{n-1}$$

such that $\delta_i^* x_j = \delta_{j-1}^* x_i$ for $i, j \neq k$, $i < j$, we may consider the sequence $y_0, \ldots, y_{n-1} \in X_{n-2}$:

$$y_i = \begin{cases} \delta_{k-1}^* x_i & 0 \le i < k, \\ \delta_k^* x_{i+1} & k \le i \le n-1. \end{cases}$$

A straightforward computation shows that $\delta_i^* y_j = \delta_{j-1}^* y_i$ for every $0 \le i < j \le n-1$ and, since $X_\bullet$ is an acyclic Kan complex, there exists $x_k \in X_{n-1}$ such that $\delta_i^* x_k = y_i$ for every i. Thus $\delta_i^* x_j = \delta_{j-1}^* x_i$ for every $0 \le i < j \le n$. Using again the fact that $X_\bullet$ is an acyclic Kan complex, there exists $x \in X_n$ such that $\delta_i^* x = x_i$ for every i.

Corollary 7.2.7 *A simplicial group is an acyclic Kan complex if and only if its Moore complex is an exact sequence.*

Proof Let $G_\bullet$ be a simplicial group such that $N(G_\bullet)$ is exact and consider a sequence $x_0, \ldots, x_n \in G_{n-1}$, $n \ge 1$, such that

$$\delta_i^* x_j = \delta_{j-1}^* x_i \qquad \text{for every} \quad 0 \le i < j \le n.$$

By Theorem 7.2.4 the simplicial group $G_\bullet$ is a Kan complex and then there exists $u \in G_n$ such that $\delta_i^* u = x_i$ for every $i > 0$. Denoting by e the unit elements of the groups G_n, since

$$\delta_i^*(x_0 \cdot (\delta_0^* u)^{-1}) = \delta_i^* x_0 \cdot (\delta_i^* \delta_0^* u)^{-1} = \delta_0^* x_{i+1} \cdot (\delta_0^* \delta_{i+1}^* u)^{-1} = e, \qquad i \ge 0,$$

the element $v = x_0 \cdot (\delta_0^* u)^{-1}$ belongs to NG_{n-1} and $dv = e$. Thus there exists $w \in G_n$ such that $\delta_0^* w = v$ and $\delta_i^* w = e$ for $i > 0$; therefore $\delta_i^*(w \cdot u) = x_i$ for every i, and this proves that $G_\bullet$ is an acyclic Kan complex. The proof of the opposite implication follows immediately from the definition of acyclic Kan complexes. □

7.3 Differential Forms on Standard Simplices

As usual, we work over a fixed field $\mathbb{K}$ of characteristic 0. For every integer $n \ge 0$ we shall denote by

$$\Omega_n = \bigoplus_{p=0}^{n} \Omega_n^p = \frac{\mathbb{K}[t_0, \ldots, t_n, dt_0, \ldots, dt_n]}{(1 - \sum t_i, \sum dt_i)}$$

the differential graded algebra of polynomial differential forms on the affine standard n-simplex $\Delta_{\mathbb{K}}^n$. In other words, Ω_n is the quotient of the polynomial differential graded commutative algebra $\mathbb{K}[t_0, \ldots, t_n, dt_0, \ldots, dt_n]$ by the differential ideal generated by the relation $\sum t_i = 1$. Notice that the affine isomorphism

$$\Delta^n_{\mathbb{K}} \to \mathbb{K}^n, \qquad (t_0, t_1, \ldots, t_n) \mapsto (t_1, \ldots, t_n)$$

induces an isomorphism of unitary differential graded algebras

$$\mathbb{K}[t, dt]^{\otimes n} \simeq \mathbb{K}[t_1, \ldots, t_n, dt_1, \ldots, dt_n] \to \Omega_n. \tag{7.4}$$

In particular, by Lemma 5.5.7 and Künneth's formula we have $H^0(\Omega_n) = \mathbb{K}$ and $H^i(\Omega_n) = 0$ for $i \neq 0$.

Since every affine map $f\colon \Delta^n_{\mathbb{K}} \to \Delta^m_{\mathbb{K}}$ induces by pull-back a morphism of differential graded algebras $f^*\colon \Omega_m \to \Omega_n$, we have that the sequence $\Omega_\bullet = \{\Omega_n\}$ is a simplicial DG-algebra. In particular, the face operators δ^*_k are given by pull-back of forms under the inclusion of standard simplices

$$(t_0, \ldots, t_{n-1}) \mapsto (t_0, \ldots, t_{k-1}, 0, t_k, \ldots, t_{n-1}),$$

and then $\delta^*_k\colon \Omega_n \to \Omega_{n-1}$ is the unique morphism of differential graded algebras such that

$$\delta^*_k(t_i) = t_i, \quad i < k; \qquad \delta^*_k(t_k) = 0; \qquad \delta^*_k(t_i) = t_{i-1}, \quad i > k.$$

Lemma 7.3.1 *The simplicial DG-algebra $\Omega_\bullet$ is an acyclic Kan complex.*

Proof Since $\Omega_\bullet$ is a simplicial abelian group, by Corollary 7.2.7 it is sufficient to show that the Moore complex $N(\Omega_\bullet)$ is acyclic. For every $n > 0$ let B_n be the DG-algebra of algebraic differential forms on the open subset $\{(t_0, \ldots, t_n) \in \Delta^n_{\mathbb{K}} \mid t_0 \neq 1\}$; more precisely

$$B_n = \left\{ \frac{\phi}{(1-t_0)^s} \mid \phi \in \Omega_n,\ s \geq 0 \right\}$$

is the ring of fractions with respect to the powers of $1 - t_0 = \sum_{i=1}^n t_i$; notice that

$$d\left(\frac{1}{1-t_0}\right) = \frac{dt_0}{(1-t_0)^2}.$$

Next, consider the following morphisms of DG-algebras:

$$i_n\colon \Omega_n \to B_n, \qquad i_n(t_i) = t_i,$$

$$h_n\colon \Omega_{n-1} \to B_n, \qquad h_n(t_i) = \frac{t_{i+1}}{(1-t_0)},$$

and we leave to the reader the easy verification that i_n is injective (Exercise 7.9.5). Since, for every $\phi \in \Omega_n$ we have

$$\delta^*_0((1-t_0)\phi) = \delta^*_0\phi, \qquad \delta^*_i((1-t_0)\phi) = (1-t_0)\delta^*_i\phi, \quad i > 0,$$

we can define the morphisms of DG-algebras $\delta_i^*\colon B_n \to B_{n-1}$ by setting

$$\delta_0^* \frac{\phi}{(1-t_0)^s} = \delta_0^*\phi, \qquad \delta_i^* \frac{\phi}{(1-t_0)^s} = \frac{\delta_i^*\phi}{(1-t_0)^s}, \quad i > 0.$$

We can immediately verify the validity of the following identities:

1. $i_{n-1}\delta_i^* = \delta_i^* i_n \colon \Omega_n \to B_{n-1}$;
2. $\delta_0^* h_n = i_{n-1} \colon \Omega_{n-1} \to B_{n-1}$;
3. $\delta_i^* h_n = h_{n-1}\delta_{i-1}^* \colon \Omega_{n-1} \to B_{n-1},\ i > 0$.

We are now ready to prove that the Moore complex $N(\Omega_\bullet)$ is acyclic; for simplicity of notation we identify every Ω_n with its image under the injective morphism i_n. Let $\phi \in \Omega_{n-1}$ be such that $\delta_i^*\phi = 0$ for every i. Choosing a sufficiently large integer m we have

$$\psi = (1-t_0)^m h_n(\phi) \in \Omega_n$$

and then

$$\delta_0^*\psi = \delta_0^* h_n(\phi) = \phi,$$

while, for $i > 0$,

$$\delta_i^*\psi = (1-t_0)^m \delta_i^* h_n(\phi) = (1-t_0)^m h_{n-1}\delta_{i-1}^*(\phi) = 0.$$

□

Consider the sequence of linear maps

$$\int_{\Delta^n_{\mathbb{K}}} \colon \Omega_n \to \mathbb{K}, \qquad n \geq 0,$$

that are (uniquely) determined by the following three conditions:

1. $\displaystyle\int_{\Delta^0_{\mathbb{K}}} \colon \Omega_n \xrightarrow{\simeq} \mathbb{K}$ is the evaluation at $t_0 = 1$, $dt_0 = 0$;
2. $\displaystyle\int_{\Delta^n_{\mathbb{K}}} \eta = 0$ if $\eta \in \Omega_n^p$ and $p \neq n$;
3. $\displaystyle\int_{\Delta^n_{\mathbb{K}}} t_1^{k_1} \cdots t_n^{k_n} dt_1 \wedge \cdots \wedge dt_n = \frac{k_1! \cdots k_n!}{(k_1 + \cdots + k_n + n)!}$.

The isomorphisms (7.4) imply that the sequence of maps $\int_{\Delta^n_{\mathbb{K}}}$ is properly defined. Notice that for $\mathbb{K} = \mathbb{R}$, these maps correspond to the usual integration of forms on the standard topological simplices, see Exercise 7.9.3.

Proposition 7.3.2 *Let $\mathbb{K}$ be a field of characteristic 0. Then:*

1. $$\int_{\Delta^n_{\mathbb{K}}} t_0^{k_0} t_1^{k_1} \cdots t_n^{k_n} dt_1 \wedge \cdots \wedge dt_n = \frac{k_0!\, k_1! \cdots k_n!}{(k_0 + k_1 + \cdots + k_n + n)!};$$

2. $$\int_{\Delta^n_{\mathbb{K}}} t_0^{k_0} t_1^{k_1} \cdots t_n^{k_n} dt_0 \wedge \cdots \widehat{dt_i} \cdots \wedge dt_n = (-1)^i \frac{k_0!\, k_1! \cdots k_n!}{(k_0 + k_1 + \cdots + k_n + n)!};$$

3. *(Stokes formula) for every $n > 0$ and $\omega \in \Omega_n^{n-1}$, we have*

$$\int_{\Delta^n_{\mathbb{K}}} d\omega = \sum_{k=0}^{n} (-1)^k \int_{\Delta^{n-1}_{\mathbb{K}}} \delta_k^* \omega.$$

Proof We first prove by induction on k_0 the formula

$$\int_{\Delta^n_{\mathbb{K}}} t_0^{k_0} t_1^{k_1} \cdots t_n^{k_n} dt_1 \wedge \cdots \wedge dt_n = \frac{k_0!\, k_1! \cdots k_n!}{(k_0 + k_1 + \cdots + k_n + n)!}.$$

Assume $k_0 > 0$ and write $a = (k_0 - 1)! k_1! \cdots k_n!$, $b = k_0 + k_1 + \cdots + k_n + n$. Since

$$t_0^{k_0} t_1^{k_1} \cdots t_n^{k_n} = t_0^{k_0 - 1} t_1^{k_1} \cdots t_n^{k_n} (1 - \sum_{i=1}^{n} t_i),$$

the inductive assumption gives

$$\begin{aligned} \int_{\Delta^n_{\mathbb{K}}} t_0^{k_0} t_1^{k_1} \cdots t_n^{k_n} dt_1 \wedge \cdots \wedge dt_n &= \frac{a}{(b-1)!} - \sum_{i=1}^{n} \frac{a}{b!}(k_i + 1) \\ &= \frac{a}{(b-1)!} - \frac{a}{b!}(b - k_0) = \frac{k_0 a}{b!}. \end{aligned}$$

The second formula follows immediately from the first in view of the relation $\sum dt_i = 0$. Finally, by linearity it is sufficient to prove the Stokes formula for differential forms of type

$$\omega = t_1^{k_1} \cdots t_n^{k_n} dt_1 \wedge \cdots \wedge \widehat{dt_i} \wedge \cdots \wedge dt_n.$$

Up to permutation of indices, we may assume $i = n$. Assume first $k_n = 0$, i.e.,

$$\omega = t_1^{k_1} \cdots t_{n-1}^{k_{n-1}} dt_1 \wedge \cdots \wedge dt_{n-1}.$$

In this case, $d\omega = 0$, $\delta_k^* \omega = 0$ for every $k \neq 0, n$, and

$$\delta_0^* \omega = t_0^{k_1} \cdots t_{n-2}^{k_{n-1}} dt_0 \wedge \cdots \wedge dt_{n-2} = (-1)^{n-1} t_0^{k_1} \cdots t_{n-2}^{k_{n-1}} dt_1 \wedge \cdots \wedge dt_{n-1},$$

$$\delta_n^*\omega = t_1^{k_1} \cdots t_{n-1}^{k_{n-1}} dt_1 \wedge \cdots \wedge dt_{n-1};$$

therefore

$$\int_{\Delta_{\mathbb{K}}^{n-1}} \delta_0^*\omega + (-1)^n \int_{\Delta_{\mathbb{K}}^{n-1}} \delta_n^*\omega = 0.$$

Next, assume $k_n > 0$, then $\delta_k^*\omega = 0$ for every $k \neq 0$, and

$$\int_{\Delta_{\mathbb{K}}^n} d\omega = \int_{\Delta^n} (-1)^{n-1} k_n t_1^{k_1} \cdots t_n^{k_n-1} dt_1 \wedge \cdots \wedge dt_n = \frac{(-1)^{n-1} k_1! \cdots k_n!}{(k_1 + \cdots + k_n + n - 1)!},$$

$$\begin{aligned}\int_{\Delta_{\mathbb{K}}^{n-1}} \delta_0^*\omega &= \int_{\Delta_{\mathbb{K}}^{n-1}} t_0^{k_1} \cdots t_{n-1}^{k_n} dt_0 \wedge \cdots \wedge dt_{n-2} \\ &= (-1)^{n-1} \int_{\Delta_{\mathbb{K}}^{n-1}} t_0^{k_1} \cdots t_{n-1}^{k_n} dt_1 \wedge \cdots \wedge dt_{n-1} \\ &= \frac{(-1)^{n-1} k_1! \cdots k_n!}{(k_1 + \cdots + k_n + n - 1)!}.\end{aligned}$$

□

7.4 Cochains and Totalization of Semicosimplicial DG-Vector Spaces

As in Sect. 2.6, we shall denote by $\vec{\mathbf{\Delta}}$ the subcategory of $\mathbf{\Delta}$ with the same objects, and with arrows the strictly increasing maps.

Definition 7.4.1 Let $\mathbf{C}$ be any category:

1. a *semicosimplicial* object in $\mathbf{C}$ is a functor $A\colon \vec{\mathbf{\Delta}} \to \mathbf{C}$;
2. a *semisimplicial* object in $\mathbf{C}$ is a functor $A_\bullet\colon \vec{\mathbf{\Delta}}^{op} \to \mathbf{C}$;

A morphism of semi(co)simplicial objects is a natural transformation of functors. Notice that every (co)simplicial object is semi(co)simplicial in the obvious way.

Since every morphism in $\vec{\mathbf{\Delta}}$ may be written as a composition of the identity and face maps, a semicosimplicial object A is the same as a diagram

$$A\colon \qquad A_0 \rightrightarrows A_1 \substack{\rightarrow\\\rightarrow\\\rightarrow} A_2 \substack{\rightarrow\\\rightarrow\\\rightarrow\\\rightarrow} \cdots,$$

where, for every $n > 0$, there are $n + 1$ morphisms (the face operators)

$$\delta_k\colon A_{n-1} \to A_n, \qquad k = 0, \ldots, n,$$

such that

$$\delta_i \delta_j = \delta_{j+1}\delta_i \quad \text{for every} \quad i \le j \iff j+1 > i.$$

Similarly, a morphism $\eta\colon A \to B$ of semicosimplicial objects is the data of a sequence of morphisms $\eta_n\colon A_n \to B_n$ such that

$$\delta_k \eta_n = \eta_{n+1}\delta_k\colon A_n \to B_{n+1}, \qquad n \ge 0, \;\; 0 \le k \le n+1.$$

Let

$$V\colon \qquad V_0 \rightrightarrows V_1 \;\overset{\rightarrow}{\rightrightarrows}\; V_2 \;\overset{\rightrightarrows}{\rightrightarrows}\; \cdots,$$

be a semicosimplicial DG-vector space. As in Definition 2.6.2 we get a complex in the abelian category of DG-vector spaces

$$0 \to V_0 \xrightarrow{\delta} V_1 \xrightarrow{\delta} \cdots, \qquad \delta = \sum_{i=0}^{n}(-1)^i\delta_i\colon V_{n-1} \to V_n,$$

where the vanishing of δ^2 follows immediately from the cosimplicial identities (7.2).

The above differential δ defines in the obvious way a differential of degree $+1$ on the graded vector space $\prod_{n=0}^{\infty} V_n[-n]$. The same graded vector space carries the product differential $d = \prod d_{V_n[-n]}$, and the fact that δ is a morphism of DG-vector spaces implies that $d\delta + \delta d = 0$.

Definition 7.4.2 In the above notation, the **cochain complex** of a semicosimplicial DG-vector space V is the DG-vector space

$$C(V) = \left(\prod_{n\ge 0} V_n[-n],\, d+\delta\right).$$

More explicitly, we have:

$$C(V) = \bigoplus_{p\in\mathbb{Z}} C(V)^p, \qquad C(V)^p = \prod_{n\ge 0} V_n[-n]^p = \prod_{n\ge 0} V_n^{p-n},$$

and for every $v = (v_0, v_1, v_2, \ldots) \in C(V)^p$ we have

$$dv = (d_{V_0}v_0, -d_{V_1}v_1, d_{V_2}v_2, \ldots), \qquad \delta v = (0, \delta v_0, \delta v_1, \delta v_2, \ldots).$$

The construction of the cochain complex is functorial; every morphism of semicosimplicial DG-vector spaces $f\colon V \to W$ induces, in the obvious way, a morphism of DG-vector spaces $f\colon C(V) \to C(W)$.

Lemma 7.4.3 *Let $f: V \to W$ be a morphism of semicosimplicial DG-vector spaces. If $f: V_n \to W_n$ is a quasi-isomorphism for every $n \geq 0$, then also the map*

$$f: C(V) \to C(W)$$

is a quasi-isomorphism.

Proof The decreasing filtrations

$$F^p C(V) = C(V_{\geq p}) = \prod_{n \geq p} V_n[-n], \qquad F^p C(W) = C(W_{\geq p}) = \prod_{n \geq p} W_n[-n],$$

are complete and exhaustive. Since

$$\frac{F^p C(V)}{F^{p+1} C(V)} = V_p[-p], \qquad \frac{F^p C(W)}{F^{p+1} C(W)} = W_p[-p],$$

the result follows from Theorem 5.3.4. □

Definition 7.4.4 The (Thom–Whitney) **totalization** of a semicosimplicial DG-vector space

$$V: \qquad V_0 \rightrightarrows V_1 \substack{\rightarrow\\\rightarrow\\\rightarrow} V_2 \substack{\rightarrow\\\rightarrow\\\rightarrow\\\rightarrow} \cdots,$$

is the DG-vector space

$$\mathrm{Tot}(V) = \left\{ (x_n) \in \prod_{n \geq 0} \Omega_n \otimes V_n \mid (\delta_k^* \otimes \mathrm{Id}) x_n = (\mathrm{Id} \otimes \delta_k) x_{n-1} \ \forall 0 \leq k \leq n \right\}$$

where the differential in $\mathrm{Tot}(V)$ is induced by the differential on $\prod_{n \geq 0} \Omega_n \otimes V_n$.

In the literature the symbol Tot denotes a large number of different objects, thus the above notation may be potentially misleading; several authors use the symbols Tot_{TW} (or more generally the subscript TW) to denote Thom–Whitney totalization. Our main interest is to have notation as light as possible and apply the totalization to semicosimplicial DG-algebras and to semicosimplicial DG-Lie algebras; in these cases the Thom-Whitney totalization is the same as the totalization procedure for complete simplicial categories. For DG-vector spaces this is not true, since in this case the standard semicosimplicial totalization is isomorphic to the cochain complex.

Theorem 7.4.5 (Whitney) *For every semicosimplicial DG-vector space V, the linear map*

$$I: \mathrm{Tot}(V) \to C(V)$$

defined degreewise as

$$\mathrm{Tot}(V)^p \xrightarrow{\text{inclusion}} \prod_n \left(\bigoplus_i \Omega_n^i \otimes V_n^{p-i}\right) \xrightarrow{\prod_n \int_{\Delta^n_{\mathbb{K}}} \otimes \mathrm{Id}_{V_n}} \prod_n V_n^{p-n} = C(V)^p,$$

is a quasi-isomorphism of differential graded vector spaces.

We shall call the quasi-isomorphism I the **Whitney integration map.**

Proof For the clarity of exposition we only prove here that I is a morphism of DG-vector spaces. We shall prove that I is a quasi-isomorphism later on, as a consequence of Theorem 7.7.5.

Denoting by $I_n \in \mathrm{Hom}^{-n}_{\mathbb{K}}(\mathrm{Tot}(V), V_n)$ the components of I, for every $n \geq 0$ we have

$$I_n\colon \mathrm{Tot}(V) \xrightarrow{\text{projection}} \Omega_n \otimes V_n \xrightarrow{\int_{\Delta_n} \otimes \mathrm{Id}_{V_n}} V_n,$$

and the fact that I is a morphism of differential graded vector spaces is equivalent to the sequence of equalities

$$I_n(dx) = \delta(I_{n-1}(x)) + (-1)^n d_{V_n}(I_n(x)), \qquad n \geq 0, \quad x \in \mathrm{Tot}(V). \tag{7.5}$$

The proof of (7.5) is tedious but completely straightforward. For every $x \in \mathrm{Tot}(V)$ we may write

$$x = (x_0, x_1, \ldots), \quad x_n = \sum_j \eta_{n,j} \otimes v_{n,j}, \qquad \eta_{n,j} \in \Omega_n, \quad v_{n,j} \in V_n,$$

$$dx = (dx_0, dx_1, \ldots), \quad dx_n = \sum_j d\eta_{n,j} \otimes v_{n,j} + \sum_j (-1)^{\overline{\eta_{n,j}}} \eta_{n,j} \otimes d_{V_n}(v_{n,j}).$$

Since $\int_{\Delta^n_{\mathbb{K}}} \eta \neq 0$ only if $\overline{\eta} = n$, by the Stokes formula for every $n \geq 0$ we have

$$\begin{aligned}
I_n(dx) &= \sum_j \int_{\Delta^n_{\mathbb{K}}} d\eta_{n,j} \otimes v_{n,j} + (-1)^n \sum_j \int_{\Delta^n_{\mathbb{K}}} \eta_{n,j} \otimes d_{V_n} v_{n,j} \\
&= \sum_{k=0}^{n} (-1)^k \sum_j \int_{\Delta^{n-1}_{\mathbb{K}}} \delta_k^* \eta_{n,j} \otimes v_{n,j} + (-1)^n \sum_j \int_{\Delta^n_{\mathbb{K}}} \eta_{n,j} \otimes d_{V_n} v_{n,j} \\
&= \sum_j \int_{\Delta^{n-1}_{\mathbb{K}}} \eta_{n,j} \otimes \sum_{k=0}^{n} (-1)^k \delta_k v_{n,j} + (-1)^n \sum_j \int_{\Delta^n_{\mathbb{K}}} \eta_{n,j} \otimes d_{V_n} v_{n,j} \\
&= \delta(I_{n-1}(x)) + (-1)^n d_{V_n}(I_n(x)).
\end{aligned}$$

□

When A is a semicosimplicial algebra (either associative or Lie), then $\mathrm{Tot}(A)$ inherits a natural structure of an algebra and, via the quasi-isomorphism I, this structure induces a multiplicative structure in the cohomology of $C(A)$.

Corollary 7.4.6 *Let $f\colon V \to W$ be a morphism of semicosimplicial DG-vector spaces. If $f\colon V_n \to W_n$ is a quasi-isomorphism for every $n \geq 0$, then also the map $f\colon \mathrm{Tot}(V) \to \mathrm{Tot}(W)$ is a quasi-isomorphism.*

Proof Immediate from Theorem 7.4.5 and Lemma 7.4.3. □

Theorem 7.4.7 *Let $K \xrightarrow{g} V \xrightarrow{f} W$ be two morphisms of semicosimplicial differential graded vector spaces such that for every $n \geq 0$ the sequence*

$$0 \to K_n \xrightarrow{g} V_n \xrightarrow{f} W_n \to 0$$

is exact. Then we have a short exact sequence of DG-vector spaces

$$0 \to \mathrm{Tot}(K) \xrightarrow{g} \mathrm{Tot}(V) \xrightarrow{f} \mathrm{Tot}(W) \to 0.$$

Proof The only nontrivial assertion is the surjectivity of $\mathrm{Tot}(V) \xrightarrow{f} \mathrm{Tot}(W)$. Given an element $(w_0, w_1, \ldots) \in \mathrm{Tot}(W)$ we prove by induction on $n \geq 1$ there exists an element $(v_0, \ldots, v_{n-1}) \in \prod_{i<n} \Omega_i \otimes V_i$ such that

$$f(v_i) = w_i, \qquad \delta_k v_i = \delta_k^* v_{i+1}, \tag{7.6}$$

for every i and k such that the Eq. (7.6) make sense.

The case $n = 1$ is nothing else than the surjectivity of f. Let $(v_0, \ldots, v_{n-1})$ be as above and we choose $z \in \Omega_n \otimes V_n$ such that $f(z) = w_n$. Consider the elements

$$k_i = \delta_i^* z - \delta_i v_{n-1} \in \Omega_{n-1} \otimes K_n, \qquad i = 0, \ldots, n.$$

For every $0 \leq i < j \leq n$ we have:

$$\delta_i^* k_j = \delta_i^* \delta_j^* z - \delta_i^* \delta_j v_{n-1} = \delta_i^* \delta_j^* z - \delta_j \delta_i^* v_{n-1} = \delta_i^* \delta_j^* z - \delta_j \delta_i v_{n-2}.$$

Similarly, we have $\delta_{j-1}^* k_i = \delta_{j-1}^* \delta_i^* z - \delta_i \delta_{j-1} v_{n-2}$ and then $\delta_i^* k_j = \delta_{j-1}^* k_i$ for every $i < j$.

According to Lemma 7.3.1 the simplicial vector space $\Omega_\bullet \otimes K_n$ is an acyclic Kan complex; in particular, there exists $k \in \Omega_n \otimes K_n$ such that $\delta_i^* k = k_i$. Setting $v_n = z - k$ we have

$$f(v_n) = w_n, \qquad \delta_i^*(v_n) = \delta_i v_{n-1},$$

and we can conclude the proof by induction. □

The functor Tot commutes with direct products; if V, W are semicosimplicial DG-vector spaces, then $\mathrm{Tot}(V \times W)$ is naturally isomorphic to $\mathrm{Tot}(V) \times \mathrm{Tot}(W)$. Similarly, the functor Tot commutes with tensor products in the sense described by the following lemma.

Lemma 7.4.8 *Let V be a semicosimplicial DG-vector space and let A be a DG-vector space. Define the semicosimplicial DG-vector spaces $V \otimes A$ and $A \otimes V$ in the natural way, i.e.,*

$$(A \otimes V)_n = A \otimes V_n, \qquad (V \otimes A)_n = V_n \otimes A.$$

Then there exist natural isomorphisms

$$\mathrm{Tot}(A \otimes V) \cong A \otimes \mathrm{Tot}(V), \qquad \mathrm{Tot}(V \otimes A) \cong \mathrm{Tot}(V) \otimes A.$$

Proof It is sufficient to consider the isomorphisms induced by the associativity of tensor products $(\Omega_n \otimes V_n) \otimes A = \Omega_n \otimes (V_n \otimes A)$ and by the Koszul rule of signs

$$A \otimes (\Omega_n \otimes V_n) \cong \Omega_n \otimes (A \otimes V_n), \qquad a \otimes \omega \otimes v \mapsto (-1)^{\bar{a}\,\bar{\omega}} \omega \otimes a \otimes v.$$

□

Remark 7.4.9 Let

$$V: \qquad V_0 \rightrightarrows V_1 \ \substack{\longrightarrow\\[-1ex]\longrightarrow\\[-1ex]\longrightarrow}\ V_2 \ \substack{\longrightarrow\\[-1ex]\longrightarrow\\[-1ex]\longrightarrow\\[-1ex]\longrightarrow}\ \cdots$$

be a semicosimplicial DG-vector space, and let $\alpha\colon W \to V_0$ be a morphism of DG-vector spaces such that $\delta_0\alpha = \delta_1\alpha$. A simple application of cosimplicial identities shows that the composition $f\alpha\colon W \to V_n$ is independent of $f\colon [0] \to [n]$, and then $f\alpha = \delta_0^n\alpha\colon W \to V_n$. This immediately implies that the map

$$e\colon W \to \mathrm{Tot}(V), \quad e(w) = (1 \otimes \alpha(w), 1 \otimes \delta_0\alpha(w), 1 \otimes \delta_0^2\alpha(w), 1 \otimes \delta_0^3\alpha(w), \ldots),$$

is properly defined; in fact the image of the map e is contained in $\mathrm{Tot}(V)$ whenever $\delta_k\delta_0^n\alpha = \delta_0^{n+1}\alpha$ for every $n \geq 0$, $0 \leq k \leq n+1$. For later use we point out that the composition of e with the Whitney integration map $I\colon \mathrm{Tot}(V) \to C(V)$ is the composition of α with the natural inclusion $V_0 \to C(V)$.

Remark 7.4.10 The totalization of semicosimplicial DG-Lie algebras commutes with Thom–Whitney homotopy fibres. In fact, if $f_n\colon \mathfrak{g}_n \to \mathfrak{h}_n$ are the components of a morphism $f\colon \mathfrak{g} \to \mathfrak{h}$ of semicosimplicial differential graded Lie algebras, then the morphism f induces both a totalization morphism $f\colon \mathrm{Tot}(\mathfrak{g}) \to \mathrm{Tot}(\mathfrak{h})$ and a semicosimplicial homotopy fibre

$$K: \qquad K(f_0) \rightrightarrows K(f_1) \ \substack{\longrightarrow\\[-1ex]\longrightarrow\\[-1ex]\longrightarrow}\ K(f_2) \ \substack{\longrightarrow\\[-1ex]\longrightarrow\\[-1ex]\longrightarrow\\[-1ex]\longrightarrow}\ \cdots,$$

where $K(f_i)$ is the Thom–Whitney homotopy fibre of f_i. It is completely straightforward to check that, by Lemma 7.4.8 we have $\mathrm{Tot}(\mathfrak{h})[t, dt] \cong \mathrm{Tot}(\mathfrak{h}[t, dt])$ and then an isomorphism

$$\mathrm{Tot}(\mathfrak{g}) \times \mathrm{Tot}(\mathfrak{h})[t, dt] \cong \mathrm{Tot}(\mathfrak{g} \times \mathfrak{h}[t, dt])$$

inducing by restriction an isomorphism $K(f) \cong \mathrm{Tot}(K)$.

The construction of Example 2.6.3 extends naturally to any sheaf $\mathcal{L}^*$ of DG-vector spaces on a topological space X. Let $\mathcal{U} = \{U_i\}$ be an open covering of X; then it is defined the semicosimplicial DG-vector space of Čech (hyper)cochains:

$$\mathcal{L}^*(\mathcal{U}): \qquad \prod_i \mathcal{L}^*(U_i) \rightrightarrows \prod_{i,j} \mathcal{L}^*(U_{ij}) \overrightarrow{\rightrightarrows} \prod_{i,j,k} \mathcal{L}^*(U_{ijk}) \overrightarrow{\overrightarrow{\rightrightarrows}} \cdots .$$

In this situation, we shall also write $\mathrm{Tot}(\mathcal{U}, \mathcal{L}^*)$ and $C(\mathcal{U}, \mathcal{L}^*)$ to denote $\mathrm{Tot}(\mathcal{L}^*(\mathcal{U}))$ and $C(\mathcal{L}^*(\mathcal{U}))$ respectively. Notice that, when $\mathcal{L}^*$ is bounded below, the complex $C(\mathcal{U}, \mathcal{L}^*)$ computes, by definition, the Čech hypercohomology groups $\mathbb{H}^*(\mathcal{U}, \mathcal{L}^*)$. For later use we point out the following application of Lemma 7.4.3 and Theorem 7.4.5: if $\mathcal{L}^*(U_{i_0 \cdots i_k})$ is an acyclic DG-vector space for every $i_0, \ldots, i_k$, then also $\mathrm{Tot}(\mathcal{U}, \mathcal{L}^*)$ is acyclic.

7.5 Semicosimplicial Groupoids

By definition, a semicosimplicial groupoid is a semicosimplicial object in the category of groupoids **Grpd**. Although every groupoid is a small category, equality of morphisms in **Grpd** is intended in the strict sense.

Let

$$G: \qquad G_0 \rightrightarrows G_1 \overrightarrow{\rightrightarrows} G_2 \overrightarrow{\overrightarrow{\rightrightarrows}} \cdots$$

be a semicosimplicial groupoid with face operators $\delta_i : G_n \to G_{n+1}$. Then the cosimplicial identities $\delta_l \delta_k = \delta_{k+1} \delta_l$ for $l \leq k$ gives in particular:

$$\delta_0\delta_0 = \delta_1\delta_0, \quad \delta_0\delta_1 = \delta_2\delta_0, \quad \delta_1\delta_1 = \delta_2\delta_1.$$

Mimicking the construction of non-abelian 1-cocycles (Sect. 3.7) we define $Z^1(G)$ as the set of pairs (l, m) with $l \in G_0$ and $m : \delta_0 l \to \delta_1 l$ a morphism in G_1 such that the *cocycle diagram*

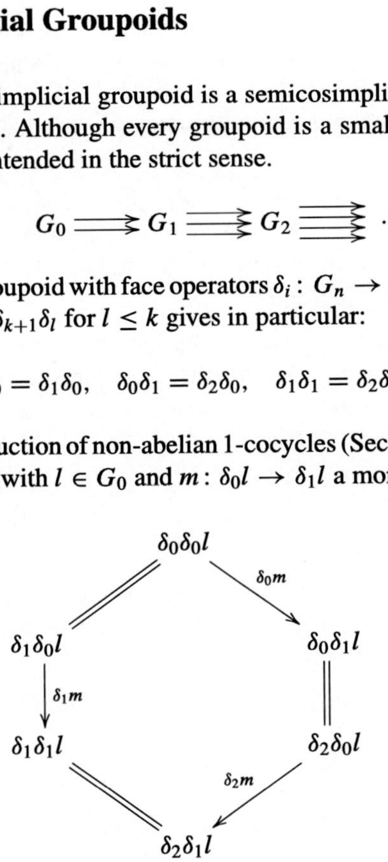

(7.7)

is commutative in G_2, or equivalently such that $(\delta_0 m)(\delta_1 m)^{-1}(\delta_2 m)$ is the identity on $\delta_0\delta_1 l$.

Definition 7.5.1 The **total groupoid** $\mathrm{Tot}(G)$ is defined in the following way:

1. the objects of $\mathrm{Tot}(G)$ are the pairs (l, m) that belong to the above defined set $Z^1(G)$;
2. the morphisms between (l_0, m_0) and (l_1, m_1) are the morphisms a in G_0 between l_0 and l_1 such that the diagram

$$\begin{array}{ccc} \delta_0 l_0 & \xrightarrow{m_0} & \delta_1 l_0 \\ {\scriptstyle \delta_0 a}\downarrow & & \downarrow{\scriptstyle \delta_1 a} \\ \delta_0 l_1 & \xrightarrow{m_1} & \delta_1 l_1 \end{array}$$

is commutative in G_1.

The above construction is functorial; every morphism $f: G \to H$ of semicosimplicial groupoids induces a morphism of groupoids $f: \mathrm{Tot}(G) \to \mathrm{Tot}(H)$.

It is immediate from the definition that the total groupoid $\mathrm{Tot}(G)$ depends only on G_0, G_1, G_2. As for the deformation functor associated to a semicosimplicial Lie algebra $\mathfrak{g}$ that depends only on $\mathfrak{g}_0, \mathfrak{g}_1, \mathfrak{g}_2$ and the deformation functor associated to a DG-Lie algebra L that depends only on L^0, L^1, L^2, the total groupoid can be considered as the truncation of a more general construction (not considered here).

Example 7.5.2 In the above setup, assume $G_i = G$ and δ_i the identity for every i. Then $(l, m) \in Z^1(G)$ if and only if $m: l \to l$ and $m^2 = m$. Therefore $(l, m) \in Z^1(G)$ if and only if the isomorphism m is the identity 1_l. Every morphism $a: l_0 \to l_1$ in G gives a morphism in the total groupoid $a: (l_0, 1_{l_0}) \to (l_1, 1_{l_1})$ and then there exists a natural isomorphism $G \cong \mathrm{Tot}(G)$.

Example 7.5.3 Every group G can be considered as a groupoid with one object $*$ having an endomorphism group equal to G. Thus every semicosimplicial group can be considered as a semicosimplicial groupoid. If G is a semicosimplicial group, then up to some canonical isomorphism:

1. the set of objects of $\mathrm{Tot}(G)$ is $Z^1(G) = \{g \in G_1 \mid (\delta_0 g)(\delta_1 g)^{-1}(\delta_2 g) = 1\}$;
2. the morphisms in $\mathrm{Tot}(G)$ between $g_0, g_1 \in Z^1(G)$ are the elements $a \in G_0$ such that $g_1(\delta_0 a) = (\delta_1 a) g_0$.

Every set can be considered as a groupoid having only the identities as morphisms. For reference purposes, it is useful to state as a separate lemma the following easy fact about semicosimplicial sets.

Lemma 7.5.4 *The total groupoid of a semicosimplicial set*

$$X: \quad X_0 \rightrightarrows X_1 \;\substack{\longrightarrow\\[-1ex]\longrightarrow\\[-1ex]\longrightarrow}\; X_2 \;\substack{\longrightarrow\\[-1ex]\longrightarrow\\[-1ex]\longrightarrow\\[-1ex]\longrightarrow}\; \cdots$$

is a set canonically isomorphic to the equalizer of $\delta_0, \delta_1: X_0 \to X_1$.

Proof Given a semicosimplicial set X as above, it is immediate from the definition that its total groupoid has only the identities as morphisms; then it is still a set whose elements are

$$\mathrm{Tot}(X) = \{x \in X_0 \mid \delta_0 x = \delta_1 x\}.$$

□

Proposition 7.5.5 *Let* $f: G \to H$ *be a morphism of semicosimplicial groupoids. If for every* $n = 0, 1, 2$ *the component* $f_n: G_n \to H_n$ *is an equivalence of groupoids, then also* $f: \mathrm{Tot}(G) \to \mathrm{Tot}(H)$ *is an equivalence of groupoids.*

Proof The only nontrivial assertion to prove is that $f: \mathrm{Tot}(G) \to \mathrm{Tot}(H)$ is essentially surjective on objects. We first note that, since $f_2: G_2 \to H_2$ is fully faithful, given $a \in G_0$ and $n \in \mathrm{Mor}_{G_1}(\delta_0 a, \delta_1 a)$ we have

$$(a, n) \in \mathrm{Tot}(G) \iff (f_0(a), f_1(n)) \in \mathrm{Tot}(H).$$

Let (l, m) be an object of $\mathrm{Tot}(H)$; since $f_0: G_0 \to H_0$ is an equivalence there exists $a \in G_0$ such that $\mathrm{Mor}_{H_0}(f_0(a), l) \neq \emptyset$. Choose $s \in \mathrm{Mor}_{H_0}(f_0(a), l)$ and denote by $\widetilde{m} \in \mathrm{Mor}_{H_1}(f_1(\delta_0 a), f_1(\delta_1 a))$ the unique arrow making the diagram

$$\begin{array}{ccc} f(\delta_0 a) & \xrightarrow{\widetilde{m}} & f(\delta_1 a) \\ {\scriptstyle \delta_0 s}\downarrow & & \downarrow{\scriptstyle \delta_1 s} \\ \delta_0 l & \xrightarrow{m} & \delta_1 l \end{array}$$

commutative. Then $(f_0(a), \widetilde{m}) \in \mathrm{Tot}(H)$ and if $\widetilde{m} = f_1(n)$ with $n \in \mathrm{Mor}_{G_1}(\delta_0 a, \delta_1 a)$, then $f(a, n) = (f_0(a), \widetilde{m})$ is isomorphic to (l, m) and this proves that the morphism $f: \mathrm{Tot}(G) \to \mathrm{Tot}(H)$ is essentially surjective. □

Remark 7.5.6 If G is a cosimplicial groupoid, with degeneracies $\sigma_i: G_{n+1} \to G_n$, then for every $(l, m) \in Z^1(G)$ we have $\sigma_0(m) = 1$, as follows immediately by applying σ_0^2 to the diagram (7.7).

7.6 Descent of Deligne Groupoids

For every semicosimplicial differential graded Lie algebra

$$\mathfrak{g}: \qquad \mathfrak{g}_0 \rightrightarrows \mathfrak{g}_1 \substack{\longrightarrow\\[-1ex]\longrightarrow\\[-1ex]\longrightarrow} \mathfrak{g}_2 \substack{\longrightarrow\\[-1ex]\longrightarrow\\[-1ex]\longrightarrow\\[-1ex]\longrightarrow} \cdots,$$

its Thom–Whitney–Sullivan totalization

$$\mathrm{Tot}(\mathfrak{g}) = \left\{ (x_n) \in \prod_{n\geq 0} \Omega_n \otimes \mathfrak{g}_n \mid (\delta_k^* \otimes \mathrm{Id})x_n = (\mathrm{Id} \otimes \delta_k)x_{n-1}\ \forall 0 \leq k \leq n \right\}$$

is a DG-Lie subalgebra of the product $\prod_{n\geq 0} \Omega_n \otimes \mathfrak{g}_n$ and therefore carries a natural structure of a differential graded Lie algebra.

Given $\mathfrak{g}$ as above and an Artin local $\mathbb{K}$-algebra $A \in \mathbf{Art}_{\mathbb{K}}$, we can consider the semicosimplicial nilpotent differential graded Lie algebra

$$\mathfrak{g} \otimes \mathfrak{m}_A\colon \quad \mathfrak{g}_0 \otimes \mathfrak{m}_A \rightrightarrows \mathfrak{g}_1 \otimes \mathfrak{m}_A \substack{\longrightarrow\\[-1ex]\longrightarrow\\[-1ex]\longrightarrow} \mathfrak{g}_2 \otimes \mathfrak{m}_A \substack{\longrightarrow\\[-1ex]\longrightarrow\\[-1ex]\longrightarrow\\[-1ex]\longrightarrow} \cdots, \tag{7.8}$$

and therefore also the semicosimplicial groupoid

$$\mathrm{Del}(\mathfrak{g} \otimes \mathfrak{m}_A)\colon\ \mathrm{Del}(\mathfrak{g}_0 \otimes \mathfrak{m}_A) \rightrightarrows \mathrm{Del}(\mathfrak{g}_1 \otimes \mathfrak{m}_A) \substack{\longrightarrow\\[-1ex]\longrightarrow\\[-1ex]\longrightarrow} \mathrm{Del}(\mathfrak{g}_2 \otimes \mathfrak{m}_A) \substack{\longrightarrow\\[-1ex]\longrightarrow\\[-1ex]\longrightarrow\\[-1ex]\longrightarrow}\ . \tag{7.9}$$

This we may define a functor

$$\mathrm{Del}_{\mathfrak{g}}\colon \mathbf{Art}_{\mathbb{K}} \to \mathbf{Grpd}^{\vec{\Delta}}, \qquad \mathrm{Del}_{\mathfrak{g}}(A) = \mathrm{Del}(\mathfrak{g} \otimes \mathfrak{m}_A),$$

from the category of Artin local $\mathbb{K}$-algebras to the category of semicosimplicial groupoids. Equivalently, $\mathrm{Del}_{\mathfrak{g}}$ is the semicosimplicial object

$$\mathrm{Del}_{\mathfrak{g}}\colon \quad \mathrm{Del}_{\mathfrak{g}_0} \rightrightarrows \mathrm{Del}_{\mathfrak{g}_1} \substack{\longrightarrow\\[-1ex]\longrightarrow\\[-1ex]\longrightarrow} \mathrm{Del}_{\mathfrak{g}_2} \substack{\longrightarrow\\[-1ex]\longrightarrow\\[-1ex]\longrightarrow\\[-1ex]\longrightarrow} \cdots,$$

in the category $\mathbf{Grpd}^{\mathbf{Art}_{\mathbb{K}}}$ of (groupoid valued) functors of Artin rings.

Our next goal is to show that, if every DG-Lie algebra $\mathfrak{g}_n$ is concentrated in nonnegative degrees, then for every $A \in \mathbf{Art}_{\mathbb{K}}$ there exists a natural equivalence of groupoids

$$\Phi\colon \mathrm{Del}(\mathrm{Tot}(\mathfrak{g}) \otimes \mathfrak{m}_A) \xrightarrow{\sim} \mathrm{Tot}(\mathrm{Del}(\mathfrak{g} \otimes \mathfrak{m}_A)),$$

called a **descent map**.

The definition of the descent map requires a number of preliminary results. First of all, for simplicity of notation, for every DG-Lie algebra L and every $n \geq 0$ we always identify L with the subset $\mathbb{K} \otimes L$ of $\Omega_n \otimes L$. For every $x \in \Omega_n \otimes L$ and every $s_0, \ldots, s_n \in \mathbb{K}$ such that $\sum s_i = 1$ we shall denote by $x(s_0, \ldots, s_n) \in L$ the image of x under the evaluation map $e \otimes \mathrm{Id}_L\colon \Omega_n \otimes L \to \mathbb{K} \otimes L = L$, where $e\colon \Omega_n \to \mathbb{K}$ is the unique morphism of algebras such that $e(t_i) = s_i$. Our first result is an analogue of Corollary 6.4.8.

Lemma 7.6.1 *Let $H = \oplus_{i\geq 0} H^i$ be a DG-Lie algebra concentrated in nonnegative degrees, let $n \in \mathbb{N}$, $A \in \mathbf{Art}_{\mathbb{K}}$ and consider a Maurer–Cartan element $x \in \mathrm{MC}_{\Omega_n \otimes H}(A)$. Then there exists a unique element*

$$p \in \Omega_n^0 \otimes H^0 \otimes \mathfrak{m}_A$$

such that $(\delta_0^*)^n p = p(0, \dots, 0, 1) = 0$ *and*

$$x = e^p * x(0, \dots, 0, 1) = e^p * (\sigma_0^*)^n (\delta_0^*)^n x.$$

Proof The proof is essentially the same as that of Corollary 6.4.8, where the same result is proved for $n = 1$ without any restriction on the DG-Lie algebra H.

Since the natural inclusion $H \to \Omega_n \otimes H$ is a quasi-isomorphism of DG-Lie algebras, there exists a Maurer–Cartan element $z \in \mathrm{MC}_H(A)$ and another element $p \in \Omega_n^0 \otimes H^0 \otimes \mathfrak{m}_A$ such that $x = e^p * z$. Possibly replacing z with $e^{p(0,\dots,0,1)} * z$ it is not restrictive to assume $p(0, \dots, 0, 1) = 0$. The unicity is proved in the same way as in Corollary 6.4.8. □

Setup 7.6.2 *We consider a fixed semicosimplicial DG-Lie algebra* $\mathfrak{g}$ *such that every* $\mathfrak{g}_n$ *is trivial in negative degrees, i.e.,* $\mathfrak{g}_n = \oplus_{i\geq 0}\mathfrak{g}_n^i$ *for every n. Let* $A \in \mathbf{Art}_{\mathbb{K}}$ *be a fixed Artin local* $\mathbb{K}$*-algebra and write for simplicity* $L_n = \mathfrak{g}_n \otimes \mathfrak{m}_A$. □

In the above setup, for every Maurer–Cartan element $x = (x_0, x_1, x_2, \dots) \in \mathrm{MC}(\mathrm{Tot}(L))$, since $\delta_0^n x_0 = (\delta_0^*)^n x_n \in \Omega_0 \otimes L_n$ for every n, according to Lemma 7.6.1 there exists a unique sequence $p_n \in \Omega_n^0 \otimes L_n^0$, $n \geq 0$, such that

$$x_n = e^{p_n} * \delta_0^n x_0, \qquad (\delta_0^*)^n p_n = p_n(0, \dots, 1) = 0;$$

and therefore every element of MC(Tot(L)) can be written in a unique way in the following canonical form;

$$(x_0,\ e^{p_1} * \delta_0 x_0,\ e^{p_2} * \delta_0^2 x_0,\ \dots), \qquad p_n \in \Omega_n^0 \otimes L_n, \quad p_n(0, \dots, 0, 1) = 0. \tag{7.10}$$

In the convention of the above setup we can rewrite the semicosimplicial nilpotent differential graded Lie algebra (7.8) as

$$L:\quad L_0 \rightrightarrows L_1 \substack{\rightarrow\\ \rightarrow\\ \rightarrow} L_2 \substack{\rightarrow\\ \rightarrow\\ \rightarrow\\ \rightarrow} \cdots,$$

and the semicosimplicial groupoid (7.9) as

$$\mathrm{Del}(L):\quad \mathrm{Del}(L_0) \rightrightarrows \mathrm{Del}(L_1) \substack{\rightarrow\\ \rightarrow\\ \rightarrow} \mathrm{Del}(L_2) \substack{\rightarrow\\ \rightarrow\\ \rightarrow\\ \rightarrow} \cdots.$$

Lemma 7.6.3 *In Setup 7.6.2, let*

$$x = (x_0,\ e^{p_1} * \delta_0 x_0,\ e^{p_2} * \delta_0^2 x_0,\ \dots) \in \mathrm{MC}(\mathrm{Tot}(L))$$

be a solution of the Maurer–Cartan equation written in canonical form (7.10). *Then:*

1. $x_0 \in \mathrm{MC}(L_0)$;
2. *if* $a = \delta_1^* p_1 = p_1(1, 0) \in L_1^0$, *then*

$$e^{\delta_1 a} = e^{\delta_2 a} e^{\delta_0 a} \in \exp(L_2^0).$$

Proof The first item is obvious since the projection $\mathrm{Tot}(L) \to L_0$ is a morphism of DG-Lie algebras. For every $i = 0, 1, 2$ we have

$$e^{\delta_i^* p_2} * \delta_0^2 x_0 = \delta_i^* x_2 = \delta_i x_1 = \delta_i (e^{p_1} * \delta_0 x_0) = e^{\delta_i p_1} * \delta_i \delta_0 x_0 \in \Omega_1 \otimes L_2.$$

If $i = 0, 1$ then $\delta_i^* p_2(0, 1) = p_2(0, 0, 1) = 0$, $\delta_i p_1(0, 1) = 0$ and $\delta_i \delta_0 x_0 = \delta_0^2 x_0$. By Lemma 7.6.1 we have

$$\delta_i^* p_2 = \delta_i p_1, \qquad i = 0, 1,$$

and then

$$\begin{aligned} p_2(1, 0, 0) &= \delta_1^* p_2(1, 0) = \delta_1 p_1(1, 0) = \delta_1 a, \\ p_2(0, 1, 0) &= \delta_0^* p_2(1, 0) = \delta_0 p_1(1, 0) = \delta_0 a. \end{aligned}$$

For $i = 2$ we have $\delta_2 p_1(0, 1) = 0$ and $\delta_2^* p_2(0, 1) = p_2(0, 1, 0) = \delta_0 a$. On the other HAND $\delta_1 x_0 = \delta_1^* x_1 = e^a * \delta_0 x_0$ and then

$$\begin{aligned} e^{\delta_2 p_1} * \delta_2 \delta_0 x_0 = e^{\delta_2^* p_2} * \delta_0^2 x_0 &= e^{\delta_2^* p_2} * \delta_0 (e^{-a} * \delta_1 x_0) = (e^{\delta_2^* p_2} e^{-\delta_0 a}) * \delta_0 \delta_1 x_0 \\ &= (e^{\delta_2^* p_2} e^{-\delta_0 a}) * \delta_2 \delta_0 x_0. \end{aligned}$$

Again by Lemma 7.6.1 we have $e^{\delta_2^* p_2} e^{-\delta_0 a} = e^{\delta_2 p_1}$ and computing at the point $(1, 0)$ we get $\delta_2^* p_2(1, 0) = p_2(1, 0, 0) = \delta_1 a$, $e^{\delta_1 a} e^{-\delta_0 a} = e^{\delta_2 a}$. □

The first consequence of the canonical form of Maurer–Cartan elements and Lemma 7.6.3 is that the following descent map is properly defined:

$$\Phi \colon \mathrm{MC}(\mathrm{Tot}(L)) \to Z^1(\mathrm{Del}(L)),$$

$$\Phi(x_0, e^{p_1} * \delta_0 x_0, \ldots) = (x_0, e^{\delta_1^* p_1}) = (x_0, e^{p_1(1,0)}).$$

Given two canonical forms of Maurer–Cartan elements $x = (x_0, e^{p_1} * \delta_0 x_0, \ldots)$, $y = (y_0, e^{q_1} * \delta_0 y_0, \ldots)$ of $\mathrm{Tot}(L)$ and an element $a = (a_0, a_1, \ldots) \in \mathrm{Tot}(L)^0$ such that $e^a * x = y$, the diagram

$$\begin{array}{ccc} \delta_0 x_0 & \xrightarrow{\delta_1^* p_1} & \delta_1 x_0 \\ {\scriptstyle \delta_0 a_0}\downarrow & & \downarrow{\scriptstyle \delta_1 a_0} \\ \delta_0 y_0 & \xrightarrow{\delta_1^* q_1} & \delta_1 y_0 \end{array}$$

is commutative in $\mathrm{Del}(L_1)$. In fact $e^{a_0} * x_0 = y_0$,

$$y_1 = e^{a_1} * x_1 = e^{a_1} e^{p_1} * \delta_0 x_0, \qquad y_1 = e^{q_1} * \delta_0 y_0 = e^{q_1} e^{\delta_0 a_0} * \delta_0 x_0,$$

$$x_1 = e^{-a_1} * y_1 = e^{-a_1} e^{q_1} * \delta_0 y_0 = e^{-a_1} e^{q_1} e^{\delta_0 a_0} * \delta_0 x_0,$$

and by Lemma 7.6.1 $e^{p_1} = e^{-a_1}e^{q_1}e^{\delta_0 a_0}$. Applying the operator δ_1^* we obtain

$$e^{\delta_1 a_0}e^{\delta_1^* p_1} = e^{\delta_1^* a_1}e^{\delta_1^* p_1} = e^{\delta_1^* q_1}e^{\delta_0 a_0}.$$

Finally, it is clear that if e^a belongs to the irrelevant stabilizer $I(x)$ then a_0 belongs to $I(x_0)$. Thus we have proved that the descent map Φ factors to a morphism of groupoids

$$\Phi\colon \mathrm{Del}(\mathrm{Tot}(L)) \to \mathrm{Tot}(\mathrm{Del}(L)).$$

We are now ready to prove the main result of this section.

Theorem 7.6.4 (Hinich's theorem on descent of Deligne groupoids) *In Setup 7.6.2, the descent map*

$$\Phi\colon \mathrm{Del}(\mathrm{Tot}(L)) \to \mathrm{Tot}(\mathrm{Del}(L))$$

is an equivalence of groupoids.

Proof It is not restrictive to assume $\mathfrak{g}_n = 0$ for every $n > 2$. In fact, if $\mathfrak{g}_{[0,2]}$ is the semicosimplicial DG-Lie algebra obtained from $\mathfrak{g}$ by annihilating $\mathfrak{g}_n$ for every $n > 2$ we have $\mathrm{Tot}(\mathrm{Del}(L)) = \mathrm{Tot}(\mathrm{Del}(L_{[0,2]}))$. On the other hand, Theorem 7.4.5 implies that the morphism

$$H^i(\mathrm{Tot}(\mathfrak{g})) \to H^i(\mathrm{Tot}(\mathfrak{g}_{[0,2]}))$$

is bijective for $i \le 1$ and injective for $i = 2$, giving therefore an equivalence of groupoids $\mathrm{Del}(\mathrm{Tot}(L)) \to \mathrm{Del}(\mathrm{Tot}(L_{[0,2]}))$ by Theorem 6.6.9.

Now, assuming $L_n = 0$ for every $n > 2$ we shall prove that the descent map $\Phi\colon \mathrm{MC}(\mathrm{Tot}(L)) \to Z^1(\mathrm{Del}(L))$ is surjective. Let $(x_0, e^a) \in Z^1(\mathrm{Del}(L))$ and define $x_1 = e^{t_0 a} * \delta_0 x_0$. We have $\delta_i^* x_1 = \delta_i x_0$; we want to prove that there exists $q \in \Omega_2^0 \otimes L_2^0$ such that $q(0,0,1) = 0$ and $(x_0, x_1, e^q * \delta_0^2 x_0) \in \mathrm{Tot}(L)$.

The three conditions $\delta_i^* x_2 = \delta_i x_1$, $i = 0, 1, 2$, give:

1. for $i = 0$, $e^{\delta_0^* q} * \delta_0^2 x_0 = e^{\delta_0(t_0 a)} * \delta_0\delta_0 x_0 = \delta_0\delta_0 x_0$ and then $\delta_0^* q = 0$;
2. for $i = 1$, $e^{\delta_1^* q} * \delta_0^2 x_0 = e^{\delta_1(t_0 a)} * \delta_1\delta_0 x_0 = e^{t_0\delta_1 a} * \delta_0\delta_0 x_0$ and then $\delta_1^* q = t_0\delta_1 a$;
3. for $i = 2$, $e^{\delta_2^* q} * \delta_0^2 x_0 = e^{t_0\delta_2 a} * \delta_2\delta_0 x_0 = e^{t_0\delta_2 a} * \delta_0\delta_1 x_0 = e^{t_0\delta_2 a}e^{\delta_0 a} * \delta_0\delta_0 x_0$ and then $\delta_2^* q = (t_0\delta_2 a) \bullet (\delta_0 a)$.

The acyclicity of the Kan complex $\Omega_\bullet \otimes L_2^0$ implies that a polynomial q with the above properties exists; notice that $\delta_0^* q = 0$ implies in particular that $q(0,0,1) = 0$.

For proving that Φ is fully faithful it is not restrictive to assume $L_n = 0$ for every $n > 1$, since both morphisms

$$\mathrm{Tot}(\mathrm{Del}(L)) \to \mathrm{Tot}(\mathrm{Del}(L_{[0,1]})), \quad \mathrm{Del}(\mathrm{Tot}(L)) \to \mathrm{Del}(\mathrm{Tot}(L_{[0,1]})),$$

are fully faithful; again we have used here that the differential graded Lie algebras $\mathfrak{g}_n$ are not negatively graded. Let $(x, e^p * \delta_0 x), (y, e^q * \delta_0 y) \in \mathrm{MC}(\mathrm{Tot}(L))$ with

$\delta_0^* p = \delta_0^* q = 0$ an let $a \in L_0^0$ such that $e^a \colon (x, e^{\delta_1^* p}) \to (y, e^{\delta_1^* q})$ is a morphism in $\operatorname{Tot}(\operatorname{Del}(L))$; this means that

$$e^a * x = y, \qquad e^{\delta_1^* q} e^{\delta_0 a} = e^{\delta_1 a} e^{\delta_1^* p}.$$

Defining $h(t) \in \Omega_1^0 \otimes L_1^0$ by the equation

$$e^h = e^q e^{\delta_0 a} e^{-p}$$

we have $\delta_0^* h = \delta_0 a$, $\delta_1^* h = \delta_1 a$, therefore $(a, h) \in \operatorname{Tot}(L)^0$ and

$$e^{(a,h)} * (x, e^p * \delta_0 x) = (y, e^q * \delta_0 y).$$

Since $\Phi(e^{(a,h)}) = e^a$ we have proved that Φ is full. Now let $(x, e^p * \delta_0 x) \in \operatorname{MC}(\operatorname{Tot}(L))$ and $(0, h) \in \operatorname{Tot}(L)^0$ such that $e^{(0,h)} * (x, e^p * \delta_0 x) = (x, e^p * \delta_0 x)$. Since $\delta_0^* h = 0$, the unicity of p gives $e^h e^p = e^p$ and then also $h = 0$. This proves that Φ is fully faithful. □

Corollary 7.6.5 *Let $\mathfrak{g}$ be a semicosimplicial differential graded Lie algebra such that every $\mathfrak{g}_n$ is nonnegatively graded. Then the descent map*

$$\Phi \colon \operatorname{Def}_{\operatorname{Tot}(\mathfrak{g})} \to \pi_0 \operatorname{Tot}(\operatorname{Del}_{\mathfrak{g}})$$

is an isomorphism of functors of Artin rings.

Proof Immediate from the Theorem 7.6.4 and the equalities

$$\pi_0 \operatorname{Tot}(\operatorname{Del}_{\mathfrak{g}})(A) = \pi_0 \operatorname{Tot}(\operatorname{Del}(\mathfrak{g} \otimes \mathfrak{m}_A)), \quad \operatorname{Tot}(\mathfrak{g} \otimes \mathfrak{m}_A) = \operatorname{Tot}(\mathfrak{g}) \otimes \mathfrak{m}_A,$$

for every $A \in \mathbf{Art}_{\mathbb{K}}$. □

Given a semicosimplicial Lie algebra

$$\mathfrak{g} \colon \quad \mathfrak{g}_0 \rightrightarrows \mathfrak{g}_1 \substack{\longrightarrow\\[-1em] \longrightarrow \\[-1em] \longrightarrow} \mathfrak{g}_2 \substack{\longrightarrow\\[-1em] \longrightarrow \\[-1em] \longrightarrow \\[-1em] \longrightarrow} \cdots,$$

in Sect. 3.7 we have defined the functors

$$Z_{\mathfrak{g}}^1, H_{\mathfrak{g}}^1 \colon \mathbf{Art}_{\mathbb{K}} \to \mathbf{Set},$$

and it is an immediate consequence of Example 7.5.3 that for every $A \in \mathbf{Art}_{\mathbb{K}}$ the set $Z_{\mathfrak{g}}^1(A)$ is naturally isomorphic to the set of objects of the total groupoid $\operatorname{Tot}(\operatorname{Del}(\mathfrak{g} \otimes \mathfrak{m}_A))$, while

$$H_{\mathfrak{g}}^1(A) = \pi_0 \operatorname{Tot}(\operatorname{Del}(\mathfrak{g} \otimes \mathfrak{m}_A)).$$

Corollary 7.6.6 *For every semicosimplicial Lie algebra $\mathfrak{g}$ there exists a natural isomorphism of functors*

$$H^1_{\mathfrak{g}} = \operatorname{Def}_{\operatorname{Tot}(\mathfrak{g})}\,.$$

Thus, every deformation functor of the form $H^1_{\mathfrak{g}}$ is governed by a differential graded Lie algebra via solutions of the Maurer–Cartan equation and gauge equivalence.

Proof Immediate consequence of Corollary 7.6.5 and equality $\pi_0\operatorname{Tot}(\operatorname{Del}_{\mathfrak{g}}) = H^1_{\mathfrak{g}}$. □

Remark 7.6.7 The descent map Φ commutes with augmentations. More precisely, in the situation of Corollary 7.6.5, let $\epsilon\colon \mathfrak{h}_0 \to \mathfrak{g}_0$ be a morphism of differential graded Lie algebras such that $\delta_0\epsilon = \delta_1\epsilon$, then by Proposition 7.7.6 we have a morphism of differential graded Lie algebras:

$$\beta\colon \mathfrak{h}_0 \to \operatorname{Tot}(\mathfrak{g}),\quad x \mapsto (\epsilon x, \delta_0\epsilon x, \delta_0^2\epsilon x, \ldots).$$

For every $A \in \mathbf{Art}_{\mathbb{K}}$ there exists a natural morphism of groupoids (Example 7.5.2)

$$\alpha\colon \operatorname{Del}(\mathfrak{h}_0 \otimes \mathfrak{m}_A) \to \operatorname{Tot}(\operatorname{Del}(\mathfrak{g} \otimes \mathfrak{m}_A)),$$
$$\operatorname{MC}(\mathfrak{h}_0 \otimes \mathfrak{m}_A) \ni l \mapsto (l, 1) \in Z^1(\operatorname{Del}(\mathfrak{g} \otimes \mathfrak{m}_A)).$$

We can immediately check that α is the composition of

$$\operatorname{Del}(\mathfrak{h}_0 \otimes \mathfrak{m}_A) \xrightarrow{\beta} \operatorname{Del}(\operatorname{Tot}(\mathfrak{g} \otimes \mathfrak{m}_A)) \xrightarrow{\Phi} \operatorname{Tot}(\operatorname{Del}(\mathfrak{g} \otimes \mathfrak{m}_A)).$$

Corollary 7.6.6 applies in particular to the deformation problems considered in Chap. 4 and controlled by a semicosimplicial Lie algebra. For reference purposes we consider here only deformations of locally free sheaves, while deformations of complex manifolds will be considered in Chap. 8.

Theorem 7.6.8 *Let X be a complex manifold, $\mathcal{U} = \{U_i\}$ a Stein covering of X and $\mathcal{F}$ a locally free sheaf of $\mathcal{O}_X$-modules. Consider the sheaf of Lie algebras $\mathcal{E} = \mathcal{H}om_{\mathcal{O}_X}(\mathcal{F}, \mathcal{F})$ and the semicosimplicial Lie algebra of Čech cochains (see Example 2.6.3)*

$$\mathcal{E}(\mathcal{U}):\quad \prod_i \mathcal{E}(U_i) \rightrightarrows \prod_{i,j} \mathcal{E}(U_{ij}) \substack{\longrightarrow\\\longrightarrow\\\longrightarrow} \prod_{i,j,k} \mathcal{E}(U_{ijk}) \substack{\longrightarrow\\\longrightarrow\\\longrightarrow\\\longrightarrow} \cdots\,.$$

Then there exists a natural isomorphism of functors of Artin rings

$$\operatorname{Def}_{\operatorname{Tot}(\mathcal{E}(\mathcal{U}))} \xrightarrow{\simeq} \operatorname{Def}_{\mathcal{F}}.$$

Proof By Lemma 4.2.4 the sheaf $\mathcal{F}$ has only trivial deformations over every open subset U_i and then, by Theorem 4.2.3 we have an isomorphism $H^1_{\mathcal{E}(\mathcal{U})} \xrightarrow{\simeq} \operatorname{Def}_{\mathcal{F}}$. It is now sufficient to apply Corollary 7.6.6. □

7.7 Homotopy Operators and Elementary Forms

We have defined Ω_n as the DG-algebras of polynomial forms on standard simplexes and seen that for every $n \geq 0$ the natural inclusion $\mathbb{K} \to \Omega_n$ is a quasi-isomorphism of DG-algebras, and therefore also a homotopy equivalence of DG-vector spaces. On the other hand it is easy to see that there does not exist any family of retractions $r_n : \Omega_n \to \mathbb{K}$ commuting with face operators δ_i^*, and this may be a serious obstacle in certain mathematical constructions, for instance in those involved in the study of the totalization functor.

As a partial solution of the above problem, in this section we define a "sufficiently small" simplicial DG-vector subspace $\mathcal{E}_\bullet \subset \Omega_\bullet$ together with a simplicial contraction

$$\mathcal{E}_\bullet \underset{\pi}{\overset{\imath}{\rightleftarrows}} \Omega_\bullet \circlearrowleft h\,,$$

which is a simplicial object in the category of contractions (Definition 5.4.1). In order to avoid superfluous computations we only prove here the deformation retraction equalities $\pi\imath = \mathrm{Id}$ and $dh + hd = \imath\pi - \mathrm{Id}$, which are completely sufficient for our purposes, and we refer to the existing literature for the proof of the annihilation properties $\pi h = h\imath = h^2 = 0$.

Lemma 7.7.1 *There exists a family of linear operators*

$$h_f \in \mathrm{Hom}_{\mathbb{K}}^{-n-1}(\Omega_m, \Omega_m), \qquad \textit{for all } n, m \geq 0,\; f \in \mathrm{Mor}_{\mathbf{\Delta}}([n], [m]),$$

with the following properties:

1. *if either $n \geq m$ or f is not injective, then $h_f = 0$;*
2. *for every $f \in \mathrm{Mor}_{\mathbf{\Delta}}([n], [m])$ and $g \in \mathrm{Mor}_{\mathbf{\Delta}}([m], [p])$ we have*

$$g^* \circ h_{gf} = h_f \circ g^* : \Omega_p \to \Omega_m;$$

3. *for every $f \in \mathrm{Mor}_{\mathbf{\Delta}}([n], [m])$ and $\eta \in \Omega_m$ we have*

$$[h_f, d](\eta) = h_f(d\eta) + (-1)^n dh_f(\eta) = \int_{\Delta^n_{\mathbb{K}}} f^*\eta - \sum_{k=0}^{n} (-1)^k h_{f\delta_k}(\eta),$$

where for $n = 0$ we have (by convention) $h_{f\delta_0} = \mathrm{Id}$.

Proof When we write $f : [n] \to [m]$ we shall always mean that f is a morphism in the category $\mathbf{\Delta}$. It is useful to add the object $[-1] = \emptyset$ to the category $\mathbf{\Delta}$, so that for every $m \geq -1$ the set $\mathrm{Mor}_{\mathbf{\Delta}}([-1], [m])$ contains exactly one element.

Since the field $\mathbb{K}$ is fixed, for simplicity we denote by $\Delta^n = \Delta^n_{\mathbb{K}}$ the standard affine simplex of dimension n:

$$\Delta^n \subset \bigoplus_{i=0}^{n} \mathbb{K}e_i, \qquad \Delta^n = \left\{ \sum_{i=0}^{n} t_i e_i \mid \sum t_i = 1 \right\}.$$

Its affine cone is the affine space $\text{cone}(\Delta^n)$ defined as

$$\text{cone}(\Delta^n) \subset \bigoplus_{i=-1}^{n} \mathbb{K}e_i, \qquad \text{cone}(\Delta^n) = \{se_{-1} + \sum_{i=0}^{n} t_i e_i \mid s + \sum t_i = 1\},$$

or equivalently as

$$\text{cone}(\Delta^n) = \{(s, t_0, t_1, \dots, t_n) \in \mathbb{K}^{n+2} \mid s + \sum t_i = 1\}.$$

Notice that Δ^n and $\text{cone}(\Delta^n)$ make sense also for $n = -1$, with $\Delta^{-1} = \emptyset$ and $\text{cone}(\Delta^{-1}) = \{e_{-1}\}$. According to Example 7.1.5, every $f : [n] \to [m]$ gives an affine morphism $f : \Delta^n \to \Delta^m$ uniquely determined by the condition $f(e_i) = e_{f(i)}$, $i = 0. \dots, n$. Consequently, also the whole sequence $\text{cone}(\Delta^n)$, $n \geq 0$, has a natural structure of a cosimplicial affine space, where every $f : [n] \to [m]$ gives the affine morphism

$$f : \text{cone}(\Delta^n) \to \text{cone}(\Delta^m), \qquad f(e_{-1}) = e_{-1}, \quad f(e_i) = e_{f(i)}, \quad i = 0, \dots, n.$$

In order to define the homotopy operators h_f, for every $f : [n] \to [m]$ we shall consider the affine map

$$\widehat{f} : \text{cone}(\Delta^n) \times \Delta^m \to \Delta^m, \qquad \widehat{f}((s, t_0, \dots, t_n), v) = sv + \sum_{i=0}^{n} t_i e_{f(i)}. \tag{7.11}$$

Denoting by

$$B_n = \frac{\mathbb{K}[s, t_0, \dots, t_n, ds, dt_0, \dots, dt_n]}{(s + \sum t_i - 1, ds + \sum dt_i)}$$

the de Rham algebra of $\text{cone}(\Delta^n)$, the affine isomorphism

$$c : \Delta^{n+1} \to \text{cone}(\Delta^n), \qquad c(e_i) = e_{i-1},$$

induces an isomorphism of DG-algebras $c^* : B_n \to \Omega_{n+1}$ and then we can consider the transfer of the integration operator

$$\Psi_n : B_n \to \mathbb{K}, \qquad \Psi_n(\eta) = \int_{\Delta^{n+1}} c^*\eta,$$

which is a linear map of degree $-n-1$. The pull-back of differential forms under the affine map (7.11) gives the morphisms of DG-algebras

$$\widehat{f}^*\colon \Omega_m \to B_n \otimes \Omega_m,$$

and we define the operator $h_f \in \mathrm{Hom}_{\mathbb{K}}^{-n-1}(\Omega_m, \Omega_m)$ as the composition

$$h_f\colon \Omega_m \xrightarrow{\widehat{f}^*} B_n \otimes \Omega_m \xrightarrow{\Psi_n \otimes \mathrm{Id}} \Omega_m\,. \tag{7.12}$$

In order to prove item (1), if $n \geq m$ then h_f has degree $-n-1$ and therefore $h_f(\Omega_m) = 0$ for reasons of degree. If f is not injective, then $n > 0$ and there exists an index $0 \leq i < n$ such that $f(i) = f(i+1)$; this implies that $f = g\sigma_i$, where

$$g\colon [n-1] \to [m], \qquad g(j) = f(j), \text{ for } j \leq i; \qquad g(j) = f(j+1) \text{ for } j \geq i.$$

Then we have a commutative diagram of affine maps

$$\begin{array}{ccc} \mathrm{cone}(\Delta^n) \times \Delta^m & \xrightarrow{\widehat{f}} & \Delta^m. \\ \Big\downarrow{\scriptstyle \sigma_i \times \mathrm{Id}} & \nearrow{\scriptstyle \widehat{g}} & \\ \mathrm{cone}(\Delta^{n-1}) \times \Delta^m & & \end{array}$$

Since Ψ_n vanishes on the image of $\sigma_i^*\colon B_{n-1} \to B_n$, we have proved that $h_f = 0$.

As regards item (2), for every $f\colon [n] \to [m]$ and every $g\colon [m] \to [p]$, we have a commutative diagram of affine spaces

$$\begin{array}{ccc} \mathrm{cone}(\Delta^n) \times \Delta^m & \xrightarrow{\widehat{f}} & \Delta^m \\ \Big\downarrow{\scriptstyle \mathrm{Id} \times g} & & \Big\downarrow{\scriptstyle g} \\ \mathrm{cone}(\Delta^n) \times \Delta^p & \xrightarrow{\widehat{gf}} & \Delta^p \end{array}$$

that induces a commutative diagram

$$\begin{array}{ccccc} \Omega_p & \xrightarrow{\widehat{gf}^*} & B_n \otimes \Omega_p & \xrightarrow{\Psi_n \otimes \mathrm{Id}} & \Omega_p \\ \Big\downarrow{\scriptstyle g^*} & & \Big\downarrow{\scriptstyle \mathrm{Id} \otimes g^*} & & \Big\downarrow{\scriptstyle g^*} \\ \Omega_m & \xrightarrow{\widehat{f}^*} & B_n \otimes \Omega_m & \xrightarrow{\Psi_n \otimes \mathrm{Id}} & \Omega_m \end{array} \qquad \Rightarrow \quad g^* h_{gf} = h_f g^*.$$

For the proof of Item (3) it is useful to introduce the face operator

$$\delta_{-1}\colon \Delta^n \to \mathrm{cone}(\Delta^n), \qquad \delta_{-1}(t_0, \ldots, t_n) = (0, t_0, \ldots, t_n).$$

The Stokes formula in $\mathrm{cone}(\Delta^n)$ becomes

$$\Psi_n(d\beta) = \int_{\Delta^n} \delta_{-1}^* \beta - \sum_{k=0}^{n} (-1)^k \Psi_{n-1}(\delta_k^* \beta), \qquad \beta \in B_n.$$

Given $f\colon [n] \to [m]$ and $\eta \in \Omega_m$, if

$$\widehat{f}^* \eta = \sum_i \beta_i \otimes \alpha_i, \qquad \beta_i \in B_n,\ \alpha_i \in \Omega_m,$$

then:

$$dh_f(\eta) = d \sum_i \Psi_n(\beta_i) \alpha_i = \sum_i \Psi_n(\beta_i)\, d\alpha_i,$$

$$\widehat{f}^*(d\eta) = d\widehat{f}^*(\eta) = \sum_i d\beta_i \otimes \alpha_i + \sum_i (-1)^{\overline{\beta_i}} \beta_i \otimes d\alpha_i,$$

$$h_f(d\eta) = \sum_i \Psi_n(d\beta_i) \otimes \alpha_i + (-1)^{n+1} \sum_i \Psi_n(\beta_i) \otimes d\alpha_i.$$

Therefore by the Stokes formula

$$h_f(d\eta) + (-1)^n dh_f(\eta) = \sum_i \Psi_n(d\beta_i) \otimes \alpha_i$$

$$= \sum_i \left(\int_{\Delta^n} \delta_{-1}^* \beta_i \right) \otimes \alpha_i - \sum_{k=0}^{n} (-1)^k \sum_i \Psi_{n-1}(\delta_k^* \beta_i) \otimes \alpha_i.$$

The commutative diagram

$$\begin{array}{ccc}
\Delta^n \times \Delta^m & \xrightarrow{\ p\ } & \Delta^n \\
\Big\downarrow{\scriptstyle \delta_{-1} \times \mathrm{Id}} & & \Big\downarrow{\scriptstyle f} \\
\mathrm{cone}(\Delta^n) \times \Delta^m & \xrightarrow{\ \widehat{f}\ } & \Delta^m
\end{array}$$

where p is the projection on the first factor, gives

$$(\delta_{-1}^* \times \mathrm{Id}) \widehat{f}^*(\eta) = p^* f^*(\eta) = f^*(\eta) \otimes 1,$$

and then

$$\sum_i \left(\int_{\Delta^n} \delta_{-1}^* \beta_i \right) \otimes \alpha_i = \int_{\Delta^n} f^* \eta.$$

The commutative diagram

$$\begin{array}{ccc} \mathrm{cone}(\Delta^n) \times \Delta^m & \xrightarrow{\widehat{f}} & \Delta^m \\ \uparrow{\scriptstyle \delta_k \times \mathrm{Id}} & \nearrow_{\widehat{f\delta_k}} & \\ \mathrm{cone}(\Delta^{n-1}) \times \Delta^m & & \end{array}$$

gives

$$\begin{aligned} \sum_i \Psi_{n-1}(\delta_k^* \beta_i) \otimes \alpha_i &= (\Psi_{n-1} \otimes \mathrm{Id})(\delta_k^* \otimes \mathrm{Id})\widehat{f}^*(\eta) \\ &= (\Psi_{n-1} \otimes \mathrm{Id})\widehat{f\delta_k}^*(\eta) = h_{f\delta_k}(\eta), \end{aligned}$$

and therefore

$$h_f(d\eta) + (-1)^n dh_f(\eta) = \int_{\Delta^n} f^*\eta - \sum_{k=0}^{n} (-1)^k h_{f\delta_k}(\eta),$$

concluding the proof of the lemma. □

The proof of Lemma 7.7.1 is undoubtedly tedious and hard to remember; fortunately in the sequel of this book, as well in most applications, the explicit description of the operators h_f is not relevant and it is sufficient to be aware of the properties 1, 2 and 3 described in the statement.

Definition 7.7.2 (Whitney) For every morphism $f : [n] \to [m]$ in the category $\mathbf{\Delta}$, define the *elementary form* $\omega_f \in \Omega_m^n$ by the formula:

$$\omega_f = n! \sum_{i=0}^{n} (-1)^i t_{f(i)}\, dt_{f(0)} \wedge \cdots \wedge \widehat{dt_{f(i)}} \wedge \cdots \wedge dt_{f(n)}.$$

For instance if $f : [0] \to [m]$, then $\omega_f = t_{f(0)} \in \Omega_m^0$. It is clear from the definition that if f is not injective, then $\omega_f = 0$. The graded vector subspaces spanned by elementary forms will be denoted by

$$\mathcal{E}_m = \mathrm{Span}\{\omega_f \mid f \in \Delta[m]_\bullet\} \subset \Omega_m, \qquad m \geq 0.$$

Lemma 7.7.3 *The elementary forms ω_f, with f injective, are linearly independent and their span $\mathcal{E}_\bullet$ is a simplicial DG-vector subspace of $\Omega_\bullet$. Moreover they satisfy the following properties:*

1. *if $f : [n] \to [m]$ is injective, then*

$$\int_{\Delta_{\mathbb{K}}^n} f^*\omega_f = 1;$$

2. *for every injective map* $f\colon [n] \to [m]$ *and every* $g\colon [p] \to [m]$ *we have*

$$g^*\omega_f = \sum_{\{h:[n]\to[p]\,|\, f=gh\}} \omega_h;$$

3. *for every* $f\colon [n] \to [m]$ *we have*

$$d\omega_f = \sum_k (-1)^k \sum_{\{g:[n+1]\to[m]\,|\, g\delta_k=f\}} \omega_g.$$

Proof We first prove item (1). Since

$$f^*\omega_f = n!\sum_{k=0}^{n}(-1)^k t_k dt_0 \wedge \cdots \wedge \widehat{dt_k} \wedge \cdots \wedge dt_n,$$

using the equalities $dt_0 = -\sum_{i>0} dt_i$, $\sum_i t_i = 1$ we obtain

$$\begin{aligned} f^*\omega_f &= n!\left(t_0 dt_1 \wedge \cdots \wedge dt_n - \sum_{k=1}^{n}(-1)^k t_k dt_k \wedge \cdots \wedge \widehat{dt_k} \wedge \cdots \wedge dt_n\right) \\ &= n!\,(t_0 + \cdots + t_n) dt_1 \wedge \cdots \wedge dt_n = n!\, dt_1 \wedge \cdots \wedge dt_n\,, \end{aligned}$$

and then

$$\int_{\Delta^n_{\mathbb{K}}} f^*\omega_f = n! \int_{\Delta^n_{\mathbb{K}}} dt_1 \wedge \cdots \wedge dt_n = 1.$$

In particular, $f^*\omega_f \neq 0$ for every injective map f.

As regards item (2), for given maps $f\colon [n] \to [m]$, $g\colon [p] \to [m]$ and for every $i = 0, \ldots, n$ we write by $P_i = \{j \in [p] \mid g(j) = f(i)\}$. Since f is assumed to be injective we have $P_i \cap P_j = \emptyset$ for every $i \neq j$ and there is a natural bijection between the cartesian product $P_0 \times \cdots \times P_n$ and the set of maps $h\colon [n] \to [p]$ such that $gh = f$. On the other hand

$$g^*(t_{f(i)}) = \sum_{j\in P_i} t_j, \qquad g^*(dt_{f(i)}) = \sum_{j\in P_i} dt_j,$$

$$\begin{aligned} g^*\omega_f &= n!\sum_{k=0}^{n}(-1)^k \left(\sum_{i\in P_k} t_i\right)\left(\sum_{i\in P_0} dt_i\right) \wedge \cdots \widehat{\left(\sum_{i\in P_k} dt_i\right)} \cdots \wedge \left(\sum_{i\in P_n} dt_i\right) \\ &= n! \sum_{(i_0,\ldots,i_n)\in P_0\times\cdots\times P_n} \sum_{k=0}^{n} (-1)^k t_{i_k} dt_{i_0} \wedge \cdots \widehat{dt_{i_k}} \cdots \wedge dt_{i_n}. \end{aligned}$$

Now the formula $g^*\omega_f = \sum_{\{h:[n]\to[p]|f=gh\}} \omega_h$ is completely straightforward.

In order to prove (3), for every finite sequence of indices $0 \le i_0, i_1, \dots, i_n \le m$ we write

$$\omega_{i_0,\dots,i_n} = n! \sum_{k=0}^{n} (-1)^k t_{i_k} dt_{i_0} \wedge \cdots \wedge \widehat{dt_{i_k}} \wedge \cdots \wedge dt_{i_n}.$$

Then $\omega_{i_0,\dots,i_n}$ is alternating on the indices and we have

$$d\omega_{i_0,\dots,i_n} = (n+1)!\, dt_{i_0} \wedge \cdots \wedge dt_{i_n} = \sum_{i=0}^{m} \omega_{i,i_0,\dots,i_n}.$$

In fact

$$d\omega_{i_0,\dots,i_n} = n! \sum_{k=0}^{n} dt_{i_0} \wedge \cdots \wedge dt_{i_k} \wedge \cdots \wedge dt_{i_n} = (n+1)!\, dt_{i_0} \wedge \cdots \wedge dt_{i_n},$$

while

$$\begin{aligned}\sum_{i=0}^{m} \omega_{i,i_0,\dots,i_n} &= (n+1)! \sum_{i=0}^{m} t_i\, dt_{i_0} \wedge \cdots \wedge dt_{i_n} - (n+1) \sum_{i=0}^{m} dt_i \wedge \omega_{i_0,\dots,i_n} \\ &= (n+1)!\, dt_{i_0} \wedge \cdots \wedge dt_{i_n}.\end{aligned}$$

Since $\omega_{i_0,\dots,i_n} = 0$ whenever $i_h = i_k$ for some $h \ne k$, it is now sufficient to observe that for every $f : [n] \to [m]$ we have

$$\begin{aligned}\sum_{i=0}^{m} \omega_{i,f(0),\dots,f(n)} &= \sum_{k=0}^{n} (-1)^k \sum_{f(k-1)<i<f(k)} \omega_{f(0),\dots,f(k-1),i,f(k),\dots,f(n)} \\ &= \sum_{k} (-1)^k \sum_{\{g|g\delta_k=f\}} \omega_g.\end{aligned}$$

The fact that the linear span $\mathcal{E}_\bullet$ of elementary forms is a simplicial DG-vector subspace of $\Omega_\bullet$ follows immediately from items (2) and (3). If $f, g : [n] \to [m]$ are two injective maps, according to items (1) and (2) we have $f^*\omega_g \ne 0$ if and only if $f = g$. Assume that $\sum a_g \omega_g = 0$, where $a_g \in \mathbb{K}$ and the sum is taken over the set of all strictly monotone maps $g : [n] \to [m]$; then for every strictly monotone map $f : [n] \to [m]$ we have $0 = f^*(\sum a_g \omega_g) = a_f f^*\omega_f$ and therefore $a_f = 0$. We have proved the linear independence of the elementary forms ω_f for f injective. □

In the following pages, for simplicity of notation, we shall write

$$\overrightarrow{\Delta}(n,m)=\operatorname{Mor}_{\overrightarrow{\Delta}}([n],[m])=\{f\colon [n]\to[m]\mid f(i-1)<f(i),\quad i=1,\dots,n\}.$$

Lemma 7.7.3 implies in particular that the set $\{\omega_f \mid f\in\overrightarrow{\Delta}(n,m)\}$ is an unordered basis of $\mathcal{E}^n_m\subset\Omega^n_m$, and therefore $\dim_{\mathbb{K}}\mathcal{E}^n_m=\binom{m+1}{n+1}$.

Theorem 7.7.4 *For every integer $m\ge 0$ consider the linear operators:*

$$\pi_m\in\operatorname{Hom}^0_{\mathbb{K}}(\Omega_m,\Omega_m),\qquad \pi_m(\eta)=\sum_{n=0}^{m}\sum_{f\in\overrightarrow{\Delta}(n,m)}\left(\int_{\Delta^n_{\mathbb{K}}} f^*\eta\right)\omega_f,$$

$$h_m\in\operatorname{Hom}^{-1}_{\mathbb{K}}(\Omega_m,\Omega_m),\qquad h_m(\eta)=\sum_{n=0}^{m}\sum_{f\in\overrightarrow{\Delta}(n,m)}\omega_f\wedge h_f(\eta).$$

Then:

1. *for every $\eta\in\Omega_m$ we have $\int_{\Delta^m}\pi_m(\eta)=\int_{\Delta^m}\eta$;*
2. *π_m is a projector onto the subspace generated by elementary forms; equivalently, $\pi_m^2=\pi_m$ and $\pi_m(\Omega_m)=\mathcal{E}_m$;*
3. *$h_md+dh_m=\pi_m-\mathrm{Id}$;*
4. *for every $g\in\operatorname{Mor}_{\Delta}([p],[m])$ we have*

$$\pi_pg^*=g^*\pi_m,\qquad h_pg^*=g^*h_m,$$

and then the two sequences $\{\pi_m\}$ and $\{h_m\}$ provide simplicial endomorphisms of the simplicial DG-vector space $\Omega_\bullet$.

Proof The first item is an immediate consequence of the formulas

$$\int_{\Delta^m_{\mathbb{K}}}\omega_{\mathrm{Id}_{[m]}}=1,\qquad \int_{\Delta^m_{\mathbb{K}}}\omega_f=0,\qquad f\in\overrightarrow{\Delta}(n,m),\quad f\ne \mathrm{Id}_{[m]},$$

and the second item is an immediate consequence of the formulas

$$\int_{\Delta^n_{\mathbb{K}}} f^*\omega_f=1,\qquad \int_{\Delta^n_{\mathbb{K}}} f^*\omega_g=0,\qquad f\in\overrightarrow{\Delta}(n,m),\quad f\ne g.$$

For the third we have

$$
\begin{aligned}
&h_m(d\eta) + dh_m(\eta)\\
&= \sum_{n=0}^{m} \sum_{f\in\vec{\Delta}(n,m)} d\omega_f \wedge h_f(\eta) + \sum_{n=0}^{m} \sum_{f\in\vec{\Delta}(n,m)} \omega_f \wedge ((-1)^n dh_f(\eta) + h_f(d\eta))\\
&= \sum_{n=0}^{m} \sum_{f\in\vec{\Delta}(n,m)} d\omega_f \wedge h_f(\eta) + \sum_{n=0}^{m} \sum_{f\in\vec{\Delta}(n,m)} \omega_f \wedge \left(\int_{\Delta^n_{\mathbb{K}}} f^*\eta - \sum_{k=0}^{n}(-1)^k h_{f\delta_k}(\eta)\right)\\
&= \pi_m(\eta) + \sum_{n=0}^{m} \sum_{f\in\vec{\Delta}(n,m)} d\omega_f \wedge h_f(\eta) + \sum_{n=0}^{m} \sum_{f\in\vec{\Delta}(n,m)} \omega_f \wedge \left(-\sum_{k=0}^{n}(-1)^k h_{f\delta_k}(\eta)\right).
\end{aligned}
$$

Since $h_\emptyset = \mathrm{Id}$ and $\sum_{f\in\vec{\Delta}(0,m)} \omega_f = \sum_{i=0}^m t_i = 1$ we have

$$
\sum_{f\in\vec{\Delta}(0,m)} \omega_f \wedge \left(-\sum_{k=0}^{0}(-1)^k h_{f\delta_k}(\eta)\right) = -\eta,
$$

and then

$$
\begin{aligned}
h_m(d\eta) + dh_m(\eta) - \pi_m(\eta) + \eta = &\sum_{n=0}^{m} \sum_{f\in\vec{\Delta}(n,m)} d\omega_f \wedge h_f(\eta)\\
&- \sum_{n=1}^{m} \sum_{f\in\vec{\Delta}(n,m)} \omega_f \wedge \sum_{k=0}^{n}(-1)^k h_{f\delta_k}(\eta).
\end{aligned}
$$

The vanishing of the right side in the above formula follows from Lemma 7.7.3, since

$$
\begin{aligned}
\sum_{n=0}^{m} \sum_{f\in\vec{\Delta}(n,m)} d\omega_f \wedge h_f(\eta) &= \sum_{n=0}^{m-1} \sum_{f\in\vec{\Delta}(n,m)} d\omega_f \wedge h_f(\eta)\\
&= \sum_{n=0}^{m-1} \sum_{f\in\vec{\Delta}(n,m)} \sum_{k=0}^{n}(-1)^k \sum_{\{g\mid f=g\delta_k\}} \omega_g \wedge h_{g\delta_k}(\eta)\\
&= \sum_{n=1}^{m} \sum_{g\in\vec{\Delta}(n,m)} \sum_{k=0}^{n}(-1)^k \omega_g \wedge h_{g\delta_k}(\eta).
\end{aligned}
$$

The formula $\pi_p g^* = g^*\pi_m$ follows immediately from the formulas $h_p g^* = g^* h_m$, $dh_m + h_m d = \pi_m - \mathrm{Id}$ and $dh_p + h_p d = \pi_p - \mathrm{Id}$. The formula $h_p g^* = g^* h_m$ is a consequence of Lemma 7.7.1:

$$\begin{aligned} g^*h_m(\eta) &= \sum_{n=0}^{m} \sum_{f\in\vec{\Delta}(n,m)} g^*(\omega_f)\wedge g^*h_f(\eta) \\ &= \sum_{n=0}^{m} \sum_{f\in\vec{\Delta}(n,m)} \sum_{\{h\in\vec{\Delta}(n,p)\mid f=gh\}} \omega_h\wedge g^*h_f(\eta) \\ &= \sum_{n=0}^{m} \sum_{h\in\vec{\Delta}(n,p)} \omega_h\wedge g^*h_{gh}(\eta) = \sum_{n=0}^{m} \sum_{h\in\vec{\Delta}(n,p)} \omega_h\wedge h_h(g^*\eta) = h_p g^*(\eta). \end{aligned}$$

□

In Sect. 7.4 we introduced, for every semicosimplicial DG-vector space V, the Whitney integration map

$$I\colon \operatorname{Tot}(V)\to C(V),$$

defined degreewise as

$$\operatorname{Tot}(V)^p \xrightarrow{\text{inclusion}} \prod_{n=0}^{\infty}(\Omega_n\otimes V_n)^p \xrightarrow{\prod_n \int_{\Delta^n_{\mathbb{K}}}\otimes \mathrm{Id}_{V_n}} \prod_n V_n^{p-n} = C(V)^p.$$

The fact, stated in Theorem 7.4.5, that I is a quasi-isomorphism of DG-vector spaces, is now proved as an immediate consequence of the following more general result.

Theorem 7.7.5 (Dupont) *Let $\{h_n\}$, $n\ge 0$, be the simplicial homotopy of $\Omega_\bullet$ defined in Theorem 7.7.4 and let V be a semicosimplicial DG-vector space. Then the linear operator of degree -1,*

$$h=\prod_n (h_n\otimes \mathrm{Id}_{V_n})\colon \operatorname{Tot}(V)\to\operatorname{Tot}(V),$$

is properly defined and there exists an injective quasi-isomorphism of DG-vector spaces $E\colon C(V)\to\operatorname{Tot}(V)$ such that

$$IE=\mathrm{Id}_{C(V)},\qquad EI-\mathrm{Id}_{\operatorname{Tot}(V)}=hd+dh, \tag{7.13}$$

where d is the differential of $\operatorname{Tot}(V)$. *In particular, I is a quasi-isomorphisms of DG-vector spaces.*

Proof Mimicking the construction of $\operatorname{Tot}(V)$ with the simplicial DG-vector space of elementary forms $\mathcal{E}_\bullet$ instead of $\Omega_\bullet$, we get a DG-subspace $\mathcal{E}(V)\subset\operatorname{Tot}(V)$:

$$\mathcal{E}(V)=\left\{(x_n)\in\prod_{n\ge 0}\mathcal{E}_n\otimes V_n \mid (\delta_k^*\otimes\mathrm{Id})x_n=(\mathrm{Id}\otimes\delta_k)x_{n-1}\ \forall 0\le k\le n\right\}.$$

Let h_n, π_n be the operators introduced in Theorem 7.7.4; since $h_p g^* = g^* h_m$ and $\pi_p g^* = g^* \pi_m$ for every $g \in \mathrm{Mor}_{\mathbf{\Delta}}([p],[m])$, the operator

$$\pi = \prod_n (\pi_n \otimes \mathrm{Id}_{V_n})\colon \mathrm{Tot}(V) \to \mathrm{Tot}(V),$$

is well defined and it is a projector onto $\mathcal{E}(V)$. Similarly, the operator

$$h = \prod_n (h_n \otimes \mathrm{Id}_{V_n})\colon \mathrm{Tot}(V) \to \mathrm{Tot}(V),$$

is well defined and satisfies the equality $dh + hd = \pi - \mathrm{Id}_{\mathrm{Tot}(V)}$. In particular, $\pi\colon \mathrm{Tot}(V) \to \mathcal{E}(V)$ is a surjective quasi-isomorphism of DG-vector spaces. Consider now the morphism

$$\phi\colon \mathcal{E}(V) \to C(V)$$

defined degreewise as

$$\mathcal{E}(V)^p \xrightarrow{\text{inclusion}} \prod_{n\ge 0} (\mathcal{E}_n \otimes V_n)^p \xrightarrow{\prod_n \int_{\Delta^n} \otimes \mathrm{Id}_{V_n}} \prod_n V_n^{p-n} = C(V)^p,$$

then $I = \phi \circ \pi$; this follows directly from the fact that for every $\eta \in \Omega_n$ we have $\int_{\Delta^m} \pi_m(\eta) = \int_{\Delta^m} \eta$. We already know that I, π are morphisms of DG-vector spaces and, since $\pi\colon \mathrm{Tot}(V) \to \mathcal{E}(V)$ is surjective, also ϕ is a morphism of differential graded vector spaces. In order to conclude the proof we shall prove that ϕ is an isomorphism of DG-vector spaces; to this end we consider the morphism of graded vector spaces $E\colon C(V) \to \prod_m \mathcal{E}_m \otimes V_m$, whose mth component $E_m\colon C(V) \to \mathcal{E}_m \otimes V_m$ is defined by the formula:

$$E_m(v) = \sum_{n=0}^{m} \sum_{f \in \overrightarrow{\Delta}(n,m)} \omega_f \otimes f(v_n), \qquad v = (v_0, v_1, \ldots). \tag{7.14}$$

For every $g \in \overrightarrow{\Delta}(p,m)$ we have

$$\begin{aligned}(g^* \otimes \mathrm{Id})E_m(v) &= \sum_{n\ge 0} \sum_{f \in I(n,m)} g^*\omega_f \otimes f(v_n)\\ &= \sum_{n\ge 0} \sum_{f \in I(n,m)} \sum_{\{h \mid f = gh\}} \omega_h \otimes gh(v_n)\\ &= \sum_{n\ge 0} \sum_{h \in \overrightarrow{\Delta}(n,p)} \omega_h \otimes gh(v_n) = (\mathrm{Id} \otimes g)E_p(v),\end{aligned}$$

and then $E\colon C(V) \to \mathcal{E}(V)$. We now prove that E is the inverse of ϕ, and immediately implies the deformation retraction identities (7.13).

Since for every $f \in \overrightarrow{\Delta}(n, m)$ we have

$$\int_{\Delta^m_{\mathbb{K}}} \omega_f = \begin{cases} 0 & \text{if } f \neq \mathrm{Id}, \\ 1 & \text{if } f = \mathrm{Id}, \end{cases}$$

it follows that $\phi \circ E = \mathrm{Id}$ and then that ϕ is surjective.

Conversely, for every $x = (x_0, x_1, \ldots) \in \mathcal{E}(V)$ we may write, for every m,

$$x_m = \sum_{p\geq 0} \sum_{f\in\overrightarrow{\Delta}(p,m)} \omega_f \otimes v_f, \qquad v_f \in V_m,$$

and the same computation as above shows that $\phi(x) = (v_{\mathrm{Id}_0}, v_{\mathrm{Id}_1}, \ldots)$, where Id_n is the identity on $[n]$. Now, for every $g \in \overrightarrow{\Delta}(n, m)$ we have $(g^* \otimes \mathrm{Id})x_m = (\mathrm{Id} \otimes g)x_n$ and, since

$$(g^* \otimes \mathrm{Id})x_m = \sum_{p} \sum_{f\in\overrightarrow{\Delta}(p,m)} g^*\omega_f \otimes v_f = \sum_{p} \sum_{h\in\overrightarrow{\Delta}(p,n)} \omega_h \otimes v_{gh},$$

$$(\mathrm{Id} \otimes g)x_n = \sum_{p\leq n} \sum_{h\in\overrightarrow{\Delta}(p,n)} \omega_h \otimes g(v_h),$$

we get $v_{gh} = g(v_h)$, and in particular $v_g = g(v_{\mathrm{Id}_n})$. Thus x is uniquely determined by the elements v_{Id_n} and then ϕ is bijective with inverse E. □

The explicit definition of the morphism E given in the proof of Theorem 7.7.5 it is sometimes useful in the applications, for instance via the following proposition.

Proposition 7.7.6 *Let*

$$V\colon \qquad V_0 \underset{\delta_0}{\overset{\delta_1}{\rightrightarrows}} V_1 \;\substack{\rightarrow\\\rightarrow\\\rightarrow}\; V_2 \;\substack{\rightarrow\\\rightarrow\\\rightarrow\\\rightarrow}\; \cdots$$

be a semicosimplicial DG-vector space, and let $\alpha\colon W_0 \to V_0$ be a morphism of DG-vector spaces such that $\delta_0\alpha = \delta_1\alpha$. Then the composite map

$$\beta\colon W_0 \xrightarrow{\alpha} V_0 \xrightarrow{\text{inclusion}} C(V) \xrightarrow{E} \mathrm{Tot}(V),$$

is equal to

$$\beta(w) = (1 \otimes \alpha(w), 1 \otimes \delta_0\alpha(w), 1 \otimes \delta_0^2\alpha(w), 1 \otimes \delta_0^3\alpha(w), \ldots).$$

Proof By the definition of E, the component of $\beta(w)$ in $\Omega_n \otimes V_n$ is equal to $\sum_{f\in\overrightarrow{\Delta}(0,n)} \omega_f \otimes f(\alpha(w))$; according to Remark 7.4.9 we have

$$\sum_{f\in\vec{\Delta}(0,n)} \omega_f \otimes f(\alpha(w)) = \sum_{f\in\vec{\Delta}(0,n)} \omega_f \otimes \delta_0^n(\alpha(w))$$
$$= \sum_{i=0}^{n} t_i \otimes \delta_0^n(\alpha(w)) = 1 \otimes \delta_0^n(\alpha(w)).$$

□

7.8 Cosimplicial Totalization and Cup Product

If L is a semicosimplicial DG-algebra (resp.: DG-Lie algebra), then $\mathrm{Tot}(A)$ has a natural structure of a DG-algebra (resp.: DG-Lie algebra) with the same cohomology of the cochain complex $C(L)$. In particular, $H^*(C(L))$ carries a natural structure of graded commutative (resp.: Lie) algebra, and it is possible to prove that this structure is equivalent to the classical cup product defined by simplicial methods [61, Thm. 2.33].

Consider for instance the case where L is a semicosimplicial Lie algebra with cochain complex

$$C(L):\qquad L_0 \xrightarrow{\delta} L_1 \xrightarrow{\delta} \cdots,\qquad \delta = \sum_i (-1)^i \delta_i.$$

In this case, according to Definition 2.6.5, the cup product

$$\cup\colon H^*(C(L)) \times H^*(C(L)) \to H^*(C(L))$$

is induced by the bilinear map

$$\cup\colon L_a \times L_b \to L_{a+b},\qquad x \cup y = [i(x), s(y)],$$

where i and s are the operators induced by the monotone maps:

$$i\colon [a] \to [a+b],\qquad i(h) = h,\quad h = 0,\ldots,a;$$

$$s\colon [b] \to [a+b],\qquad s(k) = a+k,\quad k = 0,\ldots,b.$$

Theorem 7.8.1 *Let L be the semicosimplicial Lie algebra over a field of characteristic 0. Then for every pair of cocycles $x \in L_a$, $y \in L_b$, we have*

$$x \cup y \equiv I([E(x), E(y)]) \pmod{\delta L_{a+b-1}},$$

where $I\colon \mathrm{Tot}(L) \to C(L)$ is the Whitney integration map and $E\colon C(L) \to \mathrm{Tot}(L)$ is the morphisms of DG-vector spaces defined in Theorem 7.7.5.

Proof For simplicity of exposition we do not prove the above theorem in full generality; since our interest in the cup product is related to the primary obstruction map, we only give a proof in the case $a = b = 1$. Since we have already proved in (2.10) that the bilinear map

$$\cup\colon H^1(C(L)) \times H^1(C(L)) \to H^2(C(L))$$

is symmetric, it is enough to prove that for every $x \in L_1$ such that $\delta(x) = 0$ we have $x \cup x \equiv I([E(x), E(x)]) = I([E_2(x), E_2(x)]), \pmod{\delta L_1}$. We have

$$E_2(x) = \sum_{f \in \vec{\Delta}(1,2)} \omega_f \otimes f(x) = \sum_{k=0}^{2} \omega_{\delta_k} \otimes \delta_k(x),$$

and therefore

$$\begin{aligned} I([E_2(x), E_2(x)]) &= \sum_{h,k=0}^{2} \left(\int_{\Delta^2_{\mathbb{K}}} \omega_{\delta_h} \wedge \omega_{\delta_k} \right) [\delta_h x, \delta_k x] \\ &= 2 \sum_{h>k} \left(\int_{\Delta^2_{\mathbb{K}}} \omega_{\delta_h} \wedge \omega_{\delta_k} \right) [\delta_h x, \delta_k x]. \end{aligned}$$

Since

$$\omega_{\delta_0} = t_1 dt_2 - t_2 dt_1, \qquad \omega_{\delta_1} = t_0 dt_2 - t_2 dt_0, \qquad \omega_{\delta_2} = t_0 dt_1 - t_1 dt_0,$$

a straightforward computation gives:

$$\int_{\Delta^2_{\mathbb{K}}} \omega_{\delta_2} \wedge \omega_{\delta_0} = \int_{\Delta^2_{\mathbb{K}}} \omega_{\delta_2} \wedge \omega_{\delta_1} = \int_{\Delta^2_{\mathbb{K}}} \omega_{\delta_1} \wedge \omega_{\delta_0} = \frac{1}{6},$$

and then

$$I([E_2(x), E_2(x)]) = \frac{1}{3}([\delta_2 x, \delta_0 x] + [\delta_2 x, \delta_1 x] + [\delta_1 x, \delta_0 x]).$$

Using the equality $\delta_1 x = \delta_0 x + \delta_2 x$ we finally get $I([E_2(x), E_2(x)]) = [\delta_2 x, \delta_0 x]$. □

We conclude this chapter by explaining, without proofs, how cochains and totalization work in the cosimplicial case. If V is a cosimplicial DG-vector space, it makes sense to define its **normalized cochain complex** $N(V)$ as the graded vector subspace $N(V) \subset C(V)$ defined as the kernel of all degeneration operators σ_i. Keep in mind that

$$K_n = \{v \in V_n \mid \sigma_i(v) = 0 \; \forall \, i\}$$

is not a semicosimplicial object, although $\delta(K_n) \subset K_{n+1}$. For the proof of the following nontrivial theorem we refer to [66, 92, 269].

Theorem 7.8.2 *In the notation above, the normalized cochain complex $N(V)$ is a DG-vector subspace of $C(V)$ and the inclusion $N(V) \to C(V)$ is a quasi-isomorphism.*

The Thom–Whitney–Sullivan (cosimplicial) totalization of a cosimplicial DG-vector space V is defined as

$$\widehat{\mathrm{Tot}}(V) = \left\{ (x_n) \in \prod_{n\geq 0} \Omega_n \otimes V_n \mid (f^* \otimes \mathrm{Id})x_n = (\mathrm{Id} \otimes f)x_m \ \forall f\colon [m] \to [n] \right\}.$$

Clearly, $\widehat{\mathrm{Tot}}(V) \subset \mathrm{Tot}(V)$, and the restriction to $\widehat{\mathrm{Tot}}(V)$ of the Whitney integration map I is a surjective quasi-isomorphism onto the normalized cochain complex $N(V)$; here the proof is the same as the semicosimplicial case. Therefore, for every cosimplicial DG-vector space V the inclusion $\widehat{\mathrm{Tot}}(V) \subset \mathrm{Tot}(V)$ is a quasi-isomorphism.

Conversely, for a semicosimplicial DG-vector space V, its **Kan extension** is the cosimplicial DG-vector space KV is defined as:

$$KV_n = \prod \{V_{p,f} \mid 0 \leq p \leq n,\ f\colon [n] \to [p] \text{ surjective}\},$$

where every $V_{p,f}$ is a copy of V_p labelled f. Every monotone map $\alpha\colon [m] \to [n]$ induces a morphism $\alpha\colon KV_n \to KV_m$, uniquely determined by the property that for every surjective map $f\colon [n] \to [p]$, we have a commutative diagram

$$\begin{array}{ccc} KV_m & \xrightarrow{\alpha} & KV_n\ , \\ \downarrow & & \downarrow \\ V_{q,\eta} & \xrightarrow{\epsilon} & V_{p,f} \end{array}$$

where the vertical arrows are the projections and

$$f\alpha\colon [m] \xrightarrow{\eta} [q] \xrightarrow{\epsilon} [p], \qquad \eta \text{ surjective}, \quad \epsilon \text{ injective monotone}.$$

is the epi-monic factorization of $f\alpha$. It is easy to see that:

1. the natural projection $KV \to V$ is a semicosimplicial morphism;
2. the cochain complex $C(V)$ is isomorphic to the normalized cochain complex $N(KV)$;
3. there exist natural quasi-isomorphisms

$$N(KV) \to C(KV) \to C(V), \qquad \widehat{\mathrm{Tot}}(KV) \to \mathrm{Tot}(KV) \to \mathrm{Tot}(V).$$

7.9 Exercises

7.9.1 In the situation of Theorem 7.6.8, assume that every $f \in \mathrm{Hom}_X(\mathcal{F}, \mathcal{F})$ is a multiple of the identity. Use the criterion of Exercise 6.8.11 in order to prove that $\mathrm{Def}_{\mathcal{F}}$ is a homogeneous deformation functor.

7.9.2 Prove that every morphism $f\colon [n] \to [m]$ in $\mathbf{\Delta}$ has a unique representation

$$f = \delta_{i_1} \cdots \delta_{i_k} \sigma_{j_1} \cdots \sigma_{j_h}, \tag{7.15}$$

where

$$m \geq i_1 > i_2 > \cdots > i_k \geq 0, \qquad 0 \leq j_1 < \cdots < j_h < n.$$

Deduce that f admits a unique *epi-monic factorization*, $f = \epsilon\eta$, with $\eta\colon [n] \to [p]$ surjective and $\epsilon\colon [p] \to [m]$ injective monotone, cf. [175, p. 173], [194, p. 4], [269, Lemma 8.1.2].

7.9.3 Prove that for $\mathbb{K} = \mathbb{R}$ the operator $\int_{\Delta^n}$ is equal to the usual integration on the topological standard simplex Δ^n_{top}. More generally, for every positive real number s prove that

$$\int_{\Delta^n_{\mathrm{top}}(s)} t_1^{k_1} \cdots t_n^{k_n} dt_1 \wedge \cdots \wedge dt_n = \frac{k_1! \cdots k_n!}{(k_1 + \cdots + k_n + n)!} s^{k_1+\cdots+k_n+n},$$

where

$$\Delta^n_{\mathrm{top}}(s) = \{(t_0, \ldots, t_n) \mid t_i \geq 0, \sum t_i = s\}.$$

(Hint: prove the formula $\int_0^s t^a (s-t)^b dt = a!\, b!\, s^{a+b+1}/(a+b+1)!$.)

7.9.4 Let

$$A_0 \rightrightarrows A_1 \;\substack{\longrightarrow\\[-2pt]\longrightarrow\\[-2pt]\longrightarrow}\; A_2 \;\substack{\longrightarrow\\[-2pt]\longrightarrow\\[-2pt]\longrightarrow\\[-2pt]\longrightarrow}\; \cdots,$$

be a semicosimplicial object in any category, and let $\alpha\colon B_0 \to A_0$ be a morphism such that $\delta_0\alpha = \delta_1\alpha$. Prove that

$$f\alpha = g\alpha\colon B_0 \to A_n \quad \text{for every} \quad f, g\colon [0] \to [n].$$

7.9.5 Prove that the natural morphism of DG-algebras

$$\mathbb{K}[t_1, \ldots, t_n, dt_1, \ldots, dt_n] \to \frac{\mathbb{K}[t_1, \ldots, t_n, T, dt_1, \ldots, dt_n]}{(1 - T(\sum t_i))},$$

where $dT = -T^2(\sum dt_i)$, is injective. (Hint: consider the change of variables $x_i = t_i$ for $i < n$, $x_n = \sum t_i$.)

7.9.6 Describe the Moore complex $N(\mathcal{E}_\bullet)$ of the simplicial DG-vector space of Whitney elementary forms.

7.9.7 Given a simplicial abelian group $A_\bullet$, its chain complex $C(A_\bullet)$ is defined as

$$0 \leftarrow A_0 \xleftarrow{d} A_1 \xleftarrow{d} A_2 \xleftarrow{d} \cdots,$$

where $d = \sum_{i=0}^{n}(-1)^i \delta_i^*\colon A_n \to A_{n-1}$. The subcomplex of degenerate chains $D(A_\bullet) \subset C(A_\bullet)$ is defined by the formula

$$D(A_\bullet)_n = \sum_{k=0}^{n-1} \sigma_k^* A_{n-1} \subset A_n.$$

Prove that $d^2 = 0$, that the Moore complex $N(A_\bullet)$ and $D(A_\bullet)$ are subcomplexes of $C(A_\bullet)$, and that $C(A_\bullet) = N(A_\bullet) \oplus D(A_\bullet)$. Thus $N(A_\bullet)$ is isomorphic to the complex of nondegenerate chains $C(A_\bullet)/D(A_\bullet)$, cf. [269, Lemma 8.3.7].

7.9.8 (Cosimplicial Kan extension) Let $\mathbf{C}$ be a category closed under finite products and denote by $\mathbf{C}^{\Delta}$ and $\mathbf{C}^{\vec{\Delta}}$ the categories of cosimplicial and semicosimplicial objects in $\mathbf{C}$ respectively; denote by $F\colon \mathbf{C}^{\Delta} \to \mathbf{C}^{\vec{\Delta}}$ the natural restriction functor. Given a semicosimplicial object $C \in \mathbf{C}^{\vec{\Delta}}$, for every $n \geq 0$ define

$$KC_n = \prod_{\substack{0\leq p\leq n \\ f\colon [n]\to[p] \text{ surjective}}} C_{p,f},$$

where every $C_{p,f}$ is a copy of C_p. Every monotone map $\alpha\colon [n] \to [m]$ induces a morphism $\alpha\colon KC_n \to KC_m$ such that, for every surjective morphism $f\colon [m] \to [p]$, the f-component of α is:

$$\alpha_f\colon KC_n \to C_{p,f}, \qquad \alpha_f = \epsilon \circ \pi_\eta,$$

where $f\alpha\colon [n] \xrightarrow{\eta} [q] \xrightarrow{\epsilon} [p]$ is the epi-monic factorization of the morphism $f\alpha$ and $\pi_\eta\colon KC_n \to C_{q,\eta}$ is the projection. Prove that:

1. KC is a cosimplicial object in $\mathbf{C}$ and the "Kan extension" $K\colon \mathbf{C}^{\vec{\Delta}} \to \mathbf{C}^{\Delta}$ is a functor;
2. for every $S \in \mathbf{C}^{\Delta}$ there exists a natural morphism of cosimplicial objects $\alpha_S\colon S \to KFS$;
3. for every $S \in \mathbf{C}^{\Delta}$ and every $C \in \mathbf{C}^{\vec{\Delta}}$ the map

$$\operatorname{Mor}(FS, C) \to \operatorname{Mor}(S, KC), \qquad \gamma \mapsto K(\gamma) \circ \alpha_S,$$

is bijective. Determine moreover the morphism $FKC \to C$ corresponding to the identity in KC.

In other words, the Kan extension K is the right adjoint of the restriction functor F.

References

This chapter contains only a small part of simplicial theory; many other interesting results are omitted and we refer to [51, 85, 86, 92, 194, 269] for a deeper study of simplicial methods.

The notion of a simplicial set was introduced in [66] as a purely algebraic model capturing those topological spaces that are weakly homotopy equivalent to simplicial complexes. Simplicial and cosimplicial sets arise naturally in nonabelian cohomology (see e.g. [2, 135, 218]) and this explains the reason why there is a growing contingent of researchers interested in simplicial sets for applications in algebraic geometry and deformation theory.

The original reference for Theorem 7.2.4 and Definition 7.2.5 is [196]. The proof of Theorem 7.2.4 presented here is taken almost literally from [194]; the proof of Lemma 7.3.1 is based on the proof of [27, Prop.1.1], cf. also [72, Lemma 10.7].

We refer to [28] for the definition and first properties of the totalization functor in a complete simplicial category. According to Sullivan [251, p. 300], the totalization should be interpreted, at least heuristically, as the smallest natural differential graded multiplicative structure giving the correct cohomology of a semicosimplicial algebra.

Finally, the cosimplicial Kan extension (Exercise 7.9.8) is a particular case of a more general theorem about the existence of certain adjoint functors [175, Chapter X].

In the treatment of the total groupoid of a semicosimplicial groupoid we have followed [78, 112, 132]). Theorem 7.6.4 on descent of Deligne groupoids was first proved by Hinich in [112] for cosimplicial DG-Lie algebras; however, the semicosimplicial version follows immediately from the cosimplicial case in view of the Kan extension (Exercise 7.9.8) and Theorem 7.8.2, cf. [78]. In [73] a version of Theorem 7.6.4 has been proved under the weaker assumption that $H^i(\mathfrak{g}_n) = 0$ for every n and every $i < 0$; this assumption is sufficient for the majority of applications in classical algebraic geometry, e.g., deformation of (possibly singular) algebraic schemes, coherent sheaves etc.

The homotopy operators h_f were introduced by Dupont in [61] in a rather different way, cf. [32, 90, 205]; the equivalence of the construction given in the proof of Lemma 7.7.1 with the original one requires some hard computation and has been proved by Lunardon [172]. We also refer to [32, 90] for the proof of the annihilation properties $h\imath = \pi h = h^2 = 0$. The elementary forms of Definition 7.7.2 were introduced by Whitney [272, p. 139]. Theorems 7.4.5 and 7.7.5 were proved, respectively, in [272] and [61] in a topological setting; the algebraic version presented here is based on the papers [90, 205].

Chapter 8
Deformations of Complex Manifolds and Holomorphic Maps

In this chapter we work over the field of complex numbers $\mathbb{C}$ and we study deformations of complex manifolds and holomorphic maps from the point of view of DG-Lie algebras.

The first results come immediately from Hinich's theorem on descent of Deligne groupoids; putting together Theorem 4.3.8 and Corollary 7.6.6 we get a DG-Lie algebra controlling deformations of a complex manifold X.

Theorem 8.0.1 *Let Θ_X be the holomorphic tangent sheaf of a complex manifold X and let $\mathcal{U} = \{U_i\}$ be an open covering of X such that $H^1(U_i, \Theta_X) = 0$ for every i. Consider the semicosimplicial Lie algebra of Čech cochains*

$$\Theta_X(\mathcal{U}): \quad \prod_i \Theta_X(U_i) \rightrightarrows \prod_{i,j} \Theta_X(U_{ij}) \rightrightarrows \prod_{i,j,k} \Theta_X(U_{ijk}) \rightrightarrows \cdots$$

and its totalization $\mathrm{Tot}(\mathcal{U}, \Theta_X)$. *Then there exists a natural isomorphism of functors of Artin rings*

$$\mathrm{Def}_X \cong \mathrm{Def}_{\mathrm{Tot}(\mathcal{U},\Theta_X)}.$$

Similarly, by Theorem 4.4.3 and Corollary 7.6.6 we have:

Theorem 8.0.2 *Let $\Theta_X(-\log Z)$ be the sheaf of holomorphic vector fields on a complex manifold X that are tangent to a closed submanifold Z, and let $\mathcal{U} = \{U_i\}$ be an open covering of X such that $H^1(U_i, \Theta_X(-\log Z)) = 0$ for every i. Consider the semicosimplicial Lie algebra of Čech cochains*

$$\Theta_X(-\log Z, \mathcal{U}): \quad \prod_i \Theta_X(-\log Z)(U_i) \rightrightarrows \prod_{i,j} \Theta_X(-\log Z)(U_{ij}) \rightrightarrows$$

and its totalization $\mathrm{Tot}(\mathcal{U}, \Theta_X(-\log Z))$. *Then there exists a natural isomorphism of functors of Artin rings*

M. Manetti, *Lie Methods in Deformation Theory*, Springer Monographs in Mathematics,
https://doi.org/10.1007/978-981-19-1185-9_8

$$\mathrm{Def}_{(X,Z)} \cong \mathrm{Def}_{\mathrm{Tot}(\mathcal{U},\Theta_X(-\log Z))}.$$

Throughout all this chapter and unless otherwise stated, for every complex manifold X and every closed submanifold $Z \subset X$ we shall use the letter χ to denote the inclusion of sheaves of Lie algebras $\chi\colon \Theta_X(-\log Z) \to \Theta_X$ as well as all the induced maps at the levels of global sections, Čech cochains, Thom–Whitney totalizations etc.

For an open covering $\mathcal{U}$ satisfying the assumptions of Theorems 8.0.1 and 8.0.2, the natural forgetful morphism $\mathrm{Def}_{(X,Z)} \to \mathrm{Def}_X$ is induced by the inclusion of semicosimplicial Lie algebras $\chi\colon \Theta_X(-\log Z, \mathcal{U}) \to \Theta_X(\mathcal{U})$, while the forgetful morphism $\mathrm{Def}_{(X,Z)} \to \mathrm{Def}_Z$ is induced by the morphism of semicosimplicial Lie algebras $\Theta_X(-\log Z, \mathcal{U}) \xrightarrow{r} \Theta_Z(\mathcal{V})$, where $\mathcal{V} = \{Z \cap U \mid U \in \mathcal{U}\}$ and r denotes the restriction to Z.

It is useful to notice that the assumptions of Theorems 8.0.1 and 8.0.2 are always satisfied whenever $\mathcal{U}$ is a Stein covering of X. Moreover, if $\mathcal{U}$ and $\mathcal{V}$ are Stein coverings of X, then the two DG-Le algebras $\mathrm{Tot}(\mathcal{U}, \Theta_X)$ and $\mathrm{Tot}(\mathcal{V}, \Theta_X)$ are quasi-isomorphic; according to Theorem 7.4.5 and Leray's theorem on acyclic covers, a possible quasi-isomorphism is given by the span

$$\mathrm{Tot}(\mathcal{U}, \Theta_X) \xleftarrow{\text{projection}} \mathrm{Tot}(\mathcal{U} \cup \mathcal{V}, \Theta_X) \xrightarrow{\text{projection}} \mathrm{Tot}(\mathcal{V}, \Theta_X).$$

8.1 Embedded Deformations of Submanifolds

Let X be a complex manifold and let $Z \subset X$ be a closed submanifold defined by a sheaf of ideals $\mathcal{I} \subset \mathcal{O}_X$.

Definition 8.1.1 The **Hilbert functor** $\mathrm{Hilb}^Z_X\colon \mathbf{Art}_\mathbb{C} \to \mathbf{Set}$ of Z in X is defined, for every $A \in \mathbf{Art}_\mathbb{C}$, as

$$\mathrm{Hilb}^Z_X(A) = \{A\text{-flat ideal sheaves } \mathcal{I}_A \subset \mathcal{O}_X \otimes_\mathbb{C} A \text{ such that } \mathcal{I}_A \otimes_A \mathbb{C} = \mathcal{I}\}.$$

Equivalently, $\mathrm{Hilb}^Z_X(A)$ is the set of closed subschemes $Z_A \subset X \times \mathrm{Spec}\, A$, flat over $\mathrm{Spec}\, A$ and such that $Z_A \cap X = Z$. The elements of $\mathrm{Hilb}^Z_X(A)$ are also called **embedded deformations** of Z in X. Every embedded deformation of Z in X can be viewed as a deformation of the pair (X, Z) where the deformation of X is trivial. Keep in mind the fact that two distinct embedded deformations may give isomorphic deformations of pairs and the natural transformation of functor $\mathrm{Hilb}^Z_X \to \mathrm{Def}_{(X,Z)}$ is generally neither injective nor surjective.

If $\mathcal{U} = \{U_i\}$ is an open covering of X, then every embedded deformation of Z over A is the data of a collection of ideal sheaves $\mathcal{I}_{A,i} \in \mathrm{Hilb}^{Z\cap U_i}_{U_i}(A)$ having the same restrictions on double intersections $U_{ij} = U_i \cap U_j$. Thus, according to Lemma 7.5.4 we may describe $\mathrm{Hilb}^Z_X(A)$ as the totalization of a semicosimplicial groupoid:

$$\mathrm{Hilb}_X^Z(A) = \mathrm{Tot}\left(\prod_i \mathrm{Hilb}_{U_i}^{Z\cap U_i}(A) \rightrightarrows \prod_{i,j} \mathrm{Hilb}_{U_{ij}}^{Z\cap U_{ij}}(A) \substack{\rightarrow\\[-1em]\rightarrow\\[-1em]\rightarrow} \cdots \right).$$

Recall that we have a short exact sequence

$$0 \to \Theta_X(-\log Z) \xrightarrow{\chi} \Theta_X \to \mathcal{N}_{Z|X} \to 0,$$

where $\mathcal{N}_{Z|X} = \mathcal{H}om_{\mathcal{O}_X}\left(\mathcal{I}, \dfrac{\mathcal{O}_X}{\mathcal{I}}\right)$ denotes the normal sheaf of Z in X and with the inclusion χ is a morphism of sheaves of Lie algebras.

Theorem 8.1.2 *Let X be a complex manifold and let $Z \subset X$ be a smooth closed submanifold. For any open subset $U \subset X$ denote by $H(U)$ the (Thom–Whitney) homotopy fibre of the inclusion of Lie algebras*

$$\chi\colon \Gamma(U, \Theta_X(-\log Z)) \hookrightarrow \Gamma(U, \Theta_X).$$

Let $\mathcal{U} = \{U_i\}$ be a Stein covering of X and consider the semicosimplicial DG-Lie algebra

$$H(\mathcal{U})\colon \quad \prod_i H(U_i) \rightrightarrows \prod_{i,j} H(U_{ij}) \substack{\rightarrow\\[-1em]\rightarrow\\[-1em]\rightarrow} \prod_{i,j,k} H(U_{ijk}) \substack{\rightarrow\\[-1em]\rightarrow\\[-1em]\rightarrow\\[-1em]\rightarrow} \cdots .$$

Then there exists a natural isomorphism of functors of Artin rings

$$\mathrm{Hilb}_X^Z \cong \mathrm{Def}_{\mathrm{Tot}(H(\mathcal{U}))}.$$

In particular, the tangent space of Hilb_X^Z *is isomorphic to $H^0(Z, \mathcal{N}_{Z|X})$ and there exists a complete obstruction theory with values in $H^1(Z, \mathcal{N}_{Z|X})$.*

Proof Since every homotopy fibre $H(U)$ is a differential graded Lie algebra concentrated in degrees 0 and 1, according to Proposition 7.5.5 and Theorem 7.6.4 it is sufficient to prove that, if X is a Stein manifold, then there exists a natural equivalence of formal pointed groupoids

$$\mathrm{Del}_{H(X)} \simeq \mathrm{Hilb}_X^Z .$$

Assume therefore that X is a Stein manifold; considering $H(X)$ as the totalization of the semicosimplicial Lie algebra

$$\chi\colon \quad \Gamma(X, \Theta_X(-\log Z)) \underset{0}{\overset{\chi}{\rightrightarrows}} \Gamma(X, \Theta_X) \substack{\rightarrow\\[-1em]\rightarrow\\[-1em]\rightarrow} 0 \cdots ,$$

by Theorem 7.6.4 we have an equivalence of formal pointed groupoids

$$\mathrm{Del}_{H(X)} \xrightarrow{\simeq} \mathrm{Tot}(\mathrm{Del}_\chi)$$

and $\operatorname{Tot}(\mathrm{Del}_X)$ is described in the following way: for every $A \in \mathbf{Art}_{\mathbb{C}}$ the objects of $\operatorname{Tot}(\mathrm{Del}_X)(A)$ are the elements $e^\eta \in \exp(\Gamma(X, \Theta_X) \otimes \mathfrak{m}_A)$, while the morphisms between two objects e^η, e^ρ are the elements $e^\alpha \in \exp(\Gamma(X, \Theta_X(-\log Z)) \otimes \mathfrak{m}_A)$ such that $e^\rho = e^\eta e^\alpha$. Thus there is a well defined natural morphism

$$h\colon \operatorname{Tot}(\mathrm{Del}_X) \to \mathrm{Hilb}_X^Z, \quad \operatorname{Tot}(\mathrm{Del}_X)(A) \ni e^\eta \mapsto h(e^\eta) = e^\eta(\mathcal{I} \otimes A).$$

Since every object of $\operatorname{Tot}(\mathrm{Del}_X)(A)$ has only the identity as an endomorphism, the morphism h is fully faithful. Since X is Stein, $H^1(X, \Theta_X(-\log Z)) = 0$ and every deformation of the pair (X, Z) over A is isomorphic to the trivial deformation; therefore the group $\exp(\Gamma(X, \Theta_X) \otimes \mathfrak{m}_A)$ acts transitively on $\mathrm{Hilb}_{Z|X}(A)$ and then h is surjective.

For the computation of tangent and obstructions, it is sufficient to note that if $U \subset X$ is a Stein open subset, then $H(U)$ is quasi-isomorphic to $\Gamma(U \cap Z, \mathcal{N}_{Z|X})[-1]$ and then $H^i(\operatorname{Tot}(H(\mathcal{U}))) = H^{i-1}(Z, \mathcal{N}_{Z|X})$ for every i. □

Remark 8.1.3 In the notation of Theorem 8.1.2, the homotopy fibre $H(U_{i_0 i_1 \cdots i_n})$ is acyclic whenever $U_{i_k} \cap Z = \emptyset$ for some $k = 0, \ldots, n$.

Therefore, denoting by $\mathcal{V} = \{U \in \mathcal{U} \mid U \cap Z \neq \emptyset\}$, the natural projection of semicosimplicial DG-Lie algebras $H(\mathcal{U}) \to H(\mathcal{V})$ induces a surjective quasi-isomorphism $\operatorname{Tot}(H(\mathcal{U})) \to \operatorname{Tot}(H(\mathcal{V}))$ and then an isomorphism of functors

$$\mathrm{Hilb}_X^Z \cong \mathrm{Def}_{\operatorname{Tot}(H(\mathcal{U}))} \cong \mathrm{Def}_{\operatorname{Tot}(H(\mathcal{V}))}.$$

In particular, for the validity of Theorem 8.1.2 it is not necessary that $\mathcal{U}$ is a cover of X, and it is sufficient to require that $\mathcal{U}$ is a family of open Stein subsets of X covering Z.

Corollary 8.1.4 *Let $\mathcal{U} = \{U_i\}$ be a Stein covering of a smooth manifold X and let $Z \subset X$ be a smooth closed submanifold. Consider the two semicosimplicial Lie algebras of Čech cochains*

$$\Theta(\mathcal{U})\colon \quad \prod_i \Theta(U_i) \rightrightarrows \prod_{i,j} \Theta(U_{ij}) \substack{\longrightarrow\\[-1em] \longrightarrow\\[-1em] \longrightarrow} \prod_{i,j,k} \Theta(U_{ijk}) \substack{\longrightarrow\\[-1em] \longrightarrow\\[-1em] \longrightarrow\\[-1em] \longrightarrow} \cdots,$$

$$\Theta_X(-\log Z, \mathcal{U})\colon \quad \prod_i \Theta_X(-\log Z)(U_i) \rightrightarrows \prod_{i,j} \Theta_X(-\log Z)(U_{ij}) \substack{\longrightarrow\\[-1em] \longrightarrow\\[-1em] \longrightarrow},$$

and let K be the homotopy fibre of the morphism between their totalizations

$$\operatorname{Tot}(\mathcal{U}, \Theta_X(-\log Z)) \to \operatorname{Tot}(\mathcal{U}, \Theta_X).$$

Then there exists a natural isomorphism of functors of Artin rings

$$\mathrm{Hilb}_X^Z \cong \mathrm{Def}_K.$$

Proof As pointed out in Remark 7.4.10 the Thom–Whitney-Sullivan totalization commutes with homotopy fibres. □

In some concrete cases, it may be useful to replace the differential graded Lie algebra described in Theorem 8.1.2 with a quasi-isomorphic DG-Lie algebra. Consider for instance the case where Z is a hypersurface in X, defined as the divisor of a section $z \in \mathcal{O}_X(Z)$. The complex of invertible sheaves

$$\mathcal{L}\colon\quad \mathcal{O}_X \xrightarrow{-z} \mathcal{O}_X(Z), \tag{8.1}$$

where $\mathcal{O}_X$ is in degree 0 and $\mathcal{O}_X(Z)$ is in degree $+1$, carries a natural structure of sheaf of differential graded Lie algebras, where $[\mathcal{O}_X, \mathcal{O}_X] = 0$, $[\mathcal{O}_X(Z), \mathcal{O}_X(Z)] = 0$ and $[a, b] = ab = -[b, a]$ for every $a \in \mathcal{O}_X$, $b \in \mathcal{O}_X(Z)$.

Using the computation of Example 6.3.5 and the same ideas used in Theorem 8.1.2, it is easy to prove that, for every Stein covering $\mathcal{U} = \{U_i\}$, the differential graded Lie algebra $\operatorname{Tot}(\mathcal{U}, \mathcal{L})$ controls the embedded deformations of Z; details of the proof are left as an exercise.

Notice that the inclusion $\mathcal{O}_X(Z)[-1] \to \mathcal{L}$ is a morphism of sheaves of DG-Lie algebras. The differential graded Lie algebra $\operatorname{Tot}(\mathcal{U}, \mathcal{O}_X(Z)[-1])$ is abelian, $H^1(\operatorname{Tot}(\mathcal{U}, \mathcal{O}_X(Z)[-1])) = H^0(X, \mathcal{O}_X(Z))$ and then there exists a natural transformation

$$H^0(X, \mathcal{O}_X(Z)) \otimes \mathfrak{m}_A \to \operatorname{Hilb}^Z_X(A), \qquad A \in \mathbf{Art}_{\mathbb{C}}\ ;$$

its image corresponds geometrically to the deformations of Z inside its complete linear system.

Theorem 8.1.5 (Severi, Kodaira) *Let Z be a smooth hypersurface of a complex manifold X. Then the obstructions to embedded deformations of Z in X are contained in the kernel of the morphism $H^1(Z, \mathcal{N}_{Z|X}) \to H^2(X, \mathcal{O}_X)$ induced by the short exact sequence of sheaves*

$$0 \to \mathcal{O}_X \to \mathcal{O}_X(Z) \to \mathcal{O}_Z(Z) = \mathcal{N}_{Z|X} \to 0.$$

Proof We have proved that the Hilbert functor Hilb^Z_X is controlled by the sheaf of DG-Lie algebras $\mathcal{L} = \{\mathcal{O}_X \to \mathcal{O}_X(Z)\}$. It is now sufficient to observe that $\mathcal{O}_X$ is a sheaf of abelian Lie algebras and hence the DG-Lie algebra $\operatorname{Tot}(\mathcal{U}, \mathcal{O}_X)$ is abelian, the projection $\mathcal{L} \to \mathcal{O}_X$ is a Lie morphism and then for every Stein covering $\mathcal{U}$ the obstructions of $\operatorname{Def}_{\operatorname{Tot}(\mathcal{U},\mathcal{L})}$ are contained in the kernel of

$$H^1(Z, \mathcal{O}_Z(Z)) = H^2(\operatorname{Tot}(\mathcal{U}, \mathcal{L})) \to H^2(\operatorname{Tot}(\mathcal{U}, \mathcal{O}_X)) = H^2(X, \mathcal{O}_X).$$

□

Remark 8.1.6 Let $\mathcal{L}$ be the sheaf of DG-Lie algebra defined in (8.1). It is not surprising that if Z is smooth, then $\operatorname{Tot}(\mathcal{U}, \mathcal{L})$ is quasi-isomorphic to the homotopy fibre of $\mathrm{Tot}(\mathcal{U}, \Theta_X(-\log Z)) \to \mathrm{Tot}(\mathcal{U}, \Theta_X)$; for the brevity of exposition we only give here a sketch of the proof.

Consider the sheaf of DG-Lie algebras in degrees 0, 1:

$$\mathcal{M}\colon \quad \mathcal{D}^1_{\mathbb{C}}(\mathcal{O}_X, \mathcal{O}_X(Z)) \xrightarrow{d} \mathcal{O}_X(Z), \qquad d(h,u) = -u(z), \tag{8.2}$$

where $\mathcal{D}^1_{\mathbb{C}}(\mathcal{O}_X, \mathcal{O}_X(Z))$ is the sheaf of derivation of pairs (Definition 2.3.5) with bracket extended by setting $[\mathcal{O}_X(Z), \mathcal{O}_X(Z)] = 0$ and $[(h,u), b] = u(b)$ for every $(h,u) \in \mathcal{D}^1_{\mathbb{C}}(\mathcal{O}_X, \mathcal{O}_X(Z))$, $b \in \mathcal{O}_X(Z)$.

The short exact sequence $0 \to \mathcal{O}_X \to \mathcal{D}^1_{\mathbb{C}}(\mathcal{O}_X, \mathcal{O}_X(Z)) \to \Theta_X \to 0$ induces a short exact sequence of sheaves of DG-Lie algebras $0 \to \mathcal{L} \to \mathcal{M} \to \Theta_X \to 0$ and then $\mathcal{L}$ is the homotopy fibre of $\mathcal{M} \to \Theta_X$. If e is a local frame for $\mathcal{O}_X(Z)$ and $z = ef$ for a local equation $f = 0$ of Z, then $d(h,u) = -u(ef) = 0$ if and only if f divides $h(f)$, or equivalently if and only if $h \in \Theta_X(-\log Z)$, and $u(e) = -h(f)/f$. Hence the sheaf $\Theta_X(-\log Z)$ is canonically isomorphic to the kernel of the morphism $\mathcal{D}^1_{\mathbb{C}}(\mathcal{O}_X, \mathcal{O}_X(Z)) \xrightarrow{d} \mathcal{O}_X(Z)$; on the other hand it is easy to see that the smoothness of Z implies that d is surjective and hence that $\Theta_X(-\log Z)$ is quasi-isomorphic to $\mathcal{M}$. Therefore, when Z is smooth, the homotopy fibre of $\mathcal{M} \to \Theta_X$ is quasi-isomorphic to the homotopy fibre of $\Theta_X(-\log Z) \to \Theta_X$.

In the proof of Theorem 8.1.5 we have implicitly used the following abstract algebraic argument. Let $F\colon \mathbf{Art}_{\mathbb{K}} \to \mathbf{Set}$ be the functor of infinitesimal deformations of some geometric object. If such object is "reasonable", then F is a deformation functor and it is provided of a natural and geometrically defined complete obstruction theory (V, v_e) and with (universal) obstruction space O_F canonically isomorphic to a subspace of V. As an example, the functor Def_X of deformations of a complex manifold X has a natural obstruction theory with values in $H^2(X, \Theta_X)$ but, in general, the obstruction space O_{Def_X} is properly contained in $H^2(X, \Theta_X)$; for instance, if X is a complex torus of dimension n we shall see in Example 8.3.3 that Def_X is smooth, while $H^2(X, \Theta_X)$ has dimension $n^2(n-1)/2$. A similar situation happens for smooth surfaces in $\mathbb{P}^3$ of degree at least 6 (Example 4.4.6).

A natural way to obtain information about the obstruction space O_F is by annihilation maps; we shall say that a linear map $\omega\colon V \to W$ annihilates the obstructions of F if $\omega(O_F) = 0$ or equivalently, if $\omega v_e = 0$ for every principal small extension e. The possibility to describe a deformation functor in the form Def_L for some differential graded Lie algebra L gives a simple and powerful way to construct maps $\omega\colon H^2(L) \to W$ that annihilate obstructions.

The idea, already used in the proof of Corollary 4.2.5, is easy: assume there is given a morphism $f\colon L \to M$ of differential graded Lie algebras, then the morphism in cohomology $f\colon H^2(L) \to H^2(M)$ is compatible with the natural transformation $f\colon \mathrm{Def}_L \to \mathrm{Def}_M$ and with obstruction maps. Therefore if $\omega\colon H^2(M) \to W$ annihilates the obstructions of Def_M, the composition ωf annihilates the obstructions of Def_L. The best situation is when Def_M is unobstructed (e.g. if M is homotopy abelian) and therefore f itself annihilates the obstructions of Def_L. Notice that this procedure is purely formal and it is not necessary for the functor Def_M to have any geometrical meaning.

8.2 Deformations of Holomorphic Maps

Every holomorphic map of complex manifolds $f: X \to Y$ leads naturally to the following four deformation functors:

1. the functor Def_f of deformations of f between trivial families;
2. the functor $\mathrm{Def}_{f,Y}$ of deformations of f with fixed domain;
3. the functor $\mathrm{Def}_{X,f}$ of deformations of f with fixed codomain;
4. the functor $\mathrm{Def}_{X,f,Y}$ of all deformations of f.

These deformation problems where studied intensively in the seventies by the Japanese mathematician Horikawa. The goal of this section is to introduce the above functors and to describe their controlling differential graded Lie algebras. We only give a precise definition of Def_f and $\mathrm{Def}_{X,f,Y}$, leaving to the reader the (obvious) formulation of the intermediate cases $\mathrm{Def}_{f,Y}$ and $\mathrm{Def}_{X,f}$.

Remark 8.2.1 It will be clear very soon that a good framework for studying the above functors would be the category of sheaves of DG-Lie algebras on the site of pair (U, V), with U, V open subsets of X, Y respectively such that $f(U) \subset V$ (cf. Exercise 8.9.1). However, we do not follow this approach in order to avoid any kind of topos theoretic machinery and we adopt a more basic treatment, very close to the original approaches by Kodaira and Horikawa.

Lemma 8.2.2 *Let $f: X \to Y$ be a holomorphic map of complex manifolds, denote by $Z \subset X \times Y$ the graph of f and by $i: X \to X \times Y$, $i(x) = (x, f(x))$, the natural isomorphism between X and Z. Then there exists an exact sequence of sheaves*

$$0 \to \Theta_{X\times Y}(-\log Z) \xrightarrow{\chi} \Theta_{X\times Y} \xrightarrow{r} i_*f^*\Theta_Y \to 0, \tag{8.3}$$

and therefore $H^j(Z, N_{Z|X\times Y}) = H^j(X, f^*\Theta_Y)$ for every j.

Proof The proof is completely straightforward. For later use and reference purposes it is useful to write explicitly the map r in terms of two systems of local holomorphic coordinates x_i on X and y_j on Y, with the morphism f described locally by the equations $y_j = f_j(x)$. Then a local frame for Θ_Z is given by the vector fields

$$i_*\frac{\partial}{\partial x_i} = \frac{\partial}{\partial x_i} + \sum_j \frac{\partial f_j(x)}{\partial x_i}\frac{\partial}{\partial y_j}$$

and it is sufficient to define the map r by the formulas

$$r\left(\frac{\partial}{\partial y_i}\right)=\frac{\partial}{\partial y_i},\quad r\left(\frac{\partial}{\partial x_i}\right)=-\sum_j \frac{\partial f_j(x)}{\partial x_i}\frac{\partial}{\partial y_j}\,. \tag{8.4}$$

□

8.2.1 Deformations of Holomorphic Maps Between Trivial Families

Let's begin with the precise definition of the functor Def_f of deformations of f between trivial families.

Definition 8.2.3 Let $f\colon X\to Y$ be a holomorphic map of complex manifolds and $A\in\mathbf{Art}_{\mathbb{C}}$. A **deformation of** f over $\operatorname{Spec} A$ between trivial families is a commutative diagram of complex spaces

$$\begin{array}{ccc} X\times\operatorname{Spec}A & \xrightarrow{\quad f_A\quad} & Y\times\operatorname{Spec}A \\ & \searrow \quad \swarrow & \\ & \operatorname{Spec}A & \end{array}$$

where the diagonal arrows are the projection maps and such that the restriction of f_A to the fibres over the closed point coincides with f. The associated deformation functor of Artin rings will be denoted by

$$\mathrm{Def}_f\colon \mathbf{Art}_{\mathbb{C}}\to\mathbf{Set},\qquad \mathrm{Def}_f(A)=\left\{\begin{array}{c}\text{Deformations of } f \text{ over } \operatorname{Spec}A\\ \text{between trivial families}\end{array}\right\}.$$

Our starting point for the study of the functor Def_f is the following theorem, asserting that, via the natural isomorphism

$$(X\times\operatorname{Spec}A)\times_{\operatorname{Spec}A}(Y\times\operatorname{Spec}A)=X\times Y\times\operatorname{Spec}A,$$

the graph of a deformation of f is an embedded deformation of the graph of f, and conversely.

Theorem 8.2.4 *Let $Z\subset X\times Y$ be the graph of a holomorphic map $f\colon X\to Y$ of complex manifolds. Then there exists a natural isomorphism*

$$\mathrm{Def}_f=\mathrm{Hilb}^Z_{X\times Y}$$

between the functor of deformations of f between trivial families and the functor of embedded deformations of Z in $X\times Y$.

Proof Let's first give a detailed proof that the graph of a deformation of f is an embedded deformation of Z. The question is purely local, choose a point $p \in X$, a system of holomorphic local coordinates $x_1, \dots, x_n$ on X centred at p and a system of holomorphic local coordinates $y_1, \dots, y_m$ on Y centred at $f(p)$. In these coordinates the function f is locally defined by a system of equations

$$y_i = f_i(x_1, \dots, x_n), \qquad f_i \in \mathbb{C}\{x_1, \dots, x_n\}, \quad i = 1, \dots, m,$$

and the ideal sheaf $\mathcal{I}$ of the graph Z is locally generated by the m functions

$$y_i - f_i(x_1, \dots, x_n), \qquad i = 1, \dots, m.$$

Every deformation of f over $A \in \mathbf{Art}_{\mathbb{C}}$ is locally uniquely determined by m equations

$$y_i = f_i + g_i, \quad \text{with } g_i \in \mathbb{C}\{x_1, \dots, x_n\} \otimes \mathfrak{m}_A, \qquad i = 1, \dots, m.$$

Then locally the ideal $\mathcal{I}_A \subset \mathcal{O}_{X\times Y} \otimes A$ generated by $y_i - f_i - g_i$ is equal to

$$e^{-\sum g_i \frac{\partial}{\partial y_i}}(\mathcal{I} \otimes A),$$

and in particular $\mathcal{I}_A$ is a sheaf of free A-modules. This proves that the graph construction gives a natural transformation of functors

$$G \colon \mathrm{Def}_f \to \mathrm{Hilb}^Z_{X\times Y}.$$

By the implicit function theorem the morphism of analytic algebras

$$\frac{A\{x_1, \dots, x_n, y_1, \dots, y_n\}}{(y_i - f_i - g_i)} \xrightarrow{x_i \mapsto x_i,\ y_i \mapsto f_i + g_i} A\{x_1, \dots, x_n\}$$

is an isomorphism and then

$$A\{x_1, \dots, x_n\} \cap (y_1 - f_1 - g_1, \dots, y_m - f_m - g_m) = 0.$$

This implies the injectivity of G since, if

$$(y_1 - f_1 - g_1, \dots, y_m - f_m - g_m) = (y_1 - f_1 - h_1, \dots, y_m - f_m - h_m)$$

for $g_i, h_i \in \mathbb{C}\{x_1, \dots, x_n\} \otimes \mathfrak{m}_A$, then

$$g_i - h_i \in (y_1 - f_1 - g_1, \dots, y_m - f_m - g_m) \cap A\{x_1, \dots, x_n\} = 0.$$

The surjectivity of G follows from the fact that the ideal of any embedded deformation of the graph can be locally written in the form

$$e^{-\sum g_i \frac{\partial}{\partial y_i}}(I \otimes A), \qquad g_1, \ldots, g_n \in \mathcal{O}_X \otimes \mathfrak{m}_A.$$

In fact, the embedded deformations are controlled by the homotopy fibre of the morphism $\Theta_{X\times Y}(-\log Z) \to \Theta_{X\times Y}$ and, in an open Stein neighbourhood of the point $(p, f(p)) \in Z$, the commutative diagram of Lie algebras

$$\begin{array}{ccc} 0 & \longrightarrow & \Theta_{X\times Y}(-\log Z) \\ \downarrow & & \downarrow \\ \oplus_i \mathcal{O}_X \dfrac{\partial}{\partial y_i} & \longrightarrow & \Theta_{X\times Y} \end{array}$$

induces, by Lemma 8.2.2, a quasi-isomorphism between the homotopy fibres of the vertical arrows.

□

Corollary 8.2.5 *Let $f: X \to Y$ be a homolorphic map of complex manifolds. Then Def_f is a deformation functor with tangent space $H^0(X, f^*\Theta_Y)$ and admits a complete obstruction theory with values in $H^1(X, f^*\Theta_Y)$.*

Proof Denoting by $Z \subset X \times Y$ the graph of f, there exists a factorization

$$f: X \xrightarrow{i} Z \xrightarrow{p} Y, \qquad i(x) = (x, f(x)), \quad p(x, y) = y,$$

and $p^*\Theta_Y = \mathcal{N}_{Z|X\times Y}$. Since i is an isomorphism we have

$$H^k(Z, \mathcal{N}_{Z|X\times Y}) = H^k(X, i^*\mathcal{N}_{Z|X\times Y}) = H^k(X, f^*\Theta_Y),$$

and the conclusion follows from Theorems 8.2.4 and 8.1.2. □

We may also use Theorem 8.1.2 and Remark 8.1.3 for giving an explicit description of the differential graded Lie algebra governing the functor Def_f. If $U \subset X$ and $V \subset Y$ are Stein open subsets such that $f(U) \subset V$, we have a short exact sequence of global sections

$$0 \to \Gamma(U \times V, \Theta_{X\times Y}(-\log Z)) \xrightarrow{\chi} \Gamma(U \times V, \Theta_{X\times Y}) \xrightarrow{r} \Gamma(U, f^*\Theta_Y) \to 0,$$

and we denote by $K(\chi, U, V)$ the homotopy fibre of the above Lie morphism χ. We have already seen that, if both X and Y are Stein manifolds, then the differential graded Lie algebra $K(\chi, X, Y)$ governs the functor Def_f.

Let $\mathcal{U} = \{(U_i, V_i)\}$ be a family of pairs such that: U_i is a Stein open subset of X, and V_i is a Stein open subset of Y, $f(U_i) \subset V_i$, and $\bigcup_i U_i = X$. Then the graph Z is contained in the union $\cup_i U_i \times V_i$ and the above construction gives a semicosimplicial differential graded Lie algebra:

$$K(\chi, \mathcal{U}): \quad \prod_i K(\chi, U_i, V_i) \rightrightarrows \prod_{i,j} K(\chi, U_{ij}, V_{ij}) \rightrightarrows\!\!\!\rightarrow \prod_{i,j,k} K(\chi, U_{ijk}, V_{ijk})$$

It is now an immediate consequence of Remark 8.1.3 and Theorem 8.1.2 that the totalization of $K(\chi, \mathcal{U})$ is a DG-Lie algebra controlling the functor Def_f.

Example 8.2.6 Let C be a smooth compact Riemann surface and let $f: C \to \mathbb{P}^n$ be a morphism. If the pull-back $f^*\mathcal{O}_{\mathbb{P}^n}(1)$ of the hyperplane line bundle is not special, for instance if $\deg(f) > 2g(C) - 1$, then Def_f is unobstructed. In fact, by assumption $H^1(C, f^*\mathcal{O}_{\mathbb{P}^n}(1)) = 0$, the pull-back of the Euler exact sequence gives

$$0 \to \mathcal{O}_C \to \bigoplus_{i=0}^{n} f^*\mathcal{O}_{\mathbb{P}^n}(1) \to f^*\Theta_{\mathbb{P}^n} \to 0,$$

and then $H^1(C, f^*\Theta_{\mathbb{P}^n}) = 0$.

8.2.2 *Unrestricted Deformations of Holomorphic Maps*

We are now ready to give the precise definition of the functor $\mathrm{Def}_{X,f,Y}$ of all deformations of a holomorphic map $f: X \to Y$.

Definition 8.2.7 Let $f: X \to Y$ be a holomorphic map of complex manifolds and $A \in \mathbf{Art}_{\mathbb{C}}$. A deformation of f over Spec A is a commutative diagram of complex spaces

$$\begin{array}{ccccc} X & \xrightarrow{f} & Y & \longrightarrow & \operatorname{Spec}\mathbb{C} \\ \downarrow{\scriptstyle i} & & \downarrow{\scriptstyle j} & & \downarrow \\ X_A & \xrightarrow{f_A} & Y_A & \xrightarrow{\mu} & \operatorname{Spec} A \end{array}$$

where $X \xrightarrow{i} X_A \xrightarrow{\mu f_A} \operatorname{Spec} A$ and $Y \xrightarrow{j} Y_A \xrightarrow{\mu} \operatorname{Spec} A$ are deformations of X and Y, respectively. Two deformations $X_A \xrightarrow{f_A} Y_A \xrightarrow{\mu} \operatorname{Spec} A$ and $X'_A \xrightarrow{f'_A} Y'_A \xrightarrow{\mu'} \operatorname{Spec} A$ are isomorphic if there exists a commutative diagram

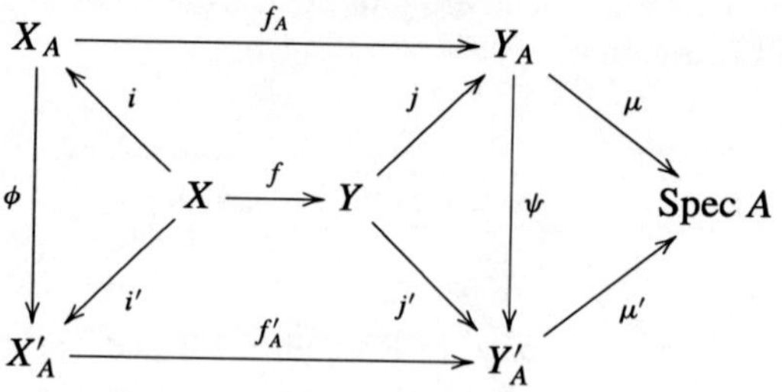

with ϕ and ψ analytic isomorphisms.

The functor $\mathrm{Def}_{X,f,Y}\colon \mathbf{Art}_{\mathbb{C}} \to \mathbf{Set}$ is then defined as

$$\mathrm{Def}_{X,f,Y}(A) = \left\{ \begin{array}{c} \text{isomorphism classes of deformations} \\ \text{of } f \text{ over Spec } A \end{array} \right\}.$$

For instance, if X is empty then $\mathrm{Def}_{X,f,Y} = \mathrm{Def}_Y$, while if Y is a point, then $\mathrm{Def}_{X,f,Y} = \mathrm{Def}_X$.

Given any complex manifold X, denote by $\mathrm{Aut}_X\colon \mathbf{Art}_{\mathbb{C}} \to \mathbf{Grp}$ the homogeneous group functor

$$\mathrm{Aut}_X(A) = \{A\text{-linear isomorphisms of } \mathcal{O}_U \otimes A \text{ lifting the identity on } \mathcal{O}_U\}.$$

We have already proved in Lemma 3.4.5 that the exponential of vector fields gives an isomorphism $\mathrm{Aut}_X(A) \simeq \exp(\Gamma(X, \Theta_X) \otimes \mathfrak{m}_A)$.

Given any holomorphic map $f\colon X \to Y$, there exists a natural action of the group functor $\mathrm{Aut}_Y \times \mathrm{Aut}_X$ on the functor Def_f, and two deformations belong to the same orbit if and only if they have the same image under the natural transformation $\mathrm{Def}_f \to \mathrm{Def}_{X,f,Y}$.

We leave to the reader to give the precise definition of the functors $\mathrm{Def}_{f,Y}$ (when X is fixed and Y varies) and $\mathrm{Def}_{X,f}$ (when Y is fixed and X varies). There exists a clear natural transformation $\mathrm{Def}_f \to \mathrm{Def}_{X,f}$ whose image is the set of (isomorphism class of) deformations inducing trivial deformations of X. Keep in mind that this map is not injective in general, since two different elements of $\mathrm{Def}_f(A)$ may differ by an A-linear automorphism of $X \times \mathrm{Spec}\, A$ and then give isomorphic deformations in $\mathrm{Def}_{X,f}(A)$. Similar considerations hold for the other three natural transformations $\mathrm{Def}_f \to \mathrm{Def}_{f,Y}$, $\mathrm{Def}_{X,f} \to \mathrm{Def}_{X,f,Y}$ and $\mathrm{Def}_{f,Y} \to \mathrm{Def}_{X,f,Y}$.

Some Remarks About f-related Vector Fields

As usual, let $f\colon X \to Y$ be a holomorphic map of complex manifolds and let $U \subset X$ and $V \subset Y$ be two open subsets such that $f(U) \subset V$. Recall from Sect. 1.3 that two vector fields $\eta \in \Gamma(U, \Theta_X)$, $\mu \in \Gamma(V, \Theta_Y)$ are called f-related if $df(\eta(p)) =$

$\mu(f(p))$ for every point $p \in U$, or equivalently if

$$f_*(\eta) = f^*(\mu) \in \Gamma(U, f^*\Theta_Y).$$

Hence the space of f-related vector fields on (U, V) is the kernel of the linear map

$$\Gamma(U, \Theta_X) \times \Gamma(V, \Theta_Y) \xrightarrow{f^*-f_*} \Gamma(U, f^*\Theta_Y),$$

and it is also a Lie subalgebra of $\Gamma(U, \Theta_X) \times \Gamma(V, \Theta_Y)$.

It should be noted that if X is a closed submanifold of Y and f is the inclusion, then two vector fields $\eta \in \Gamma(X \cap V, \Theta_X)$ and $\mu \in \Gamma(V, \Theta_Y)$ are f-related if and only if $\mu \in \Gamma(V, \Theta_Y(-\log X))$ and $\eta = \mu_{|X}$. This fact suggests that the Lie algebra of f-related vector fields plays a central role in deformation theory of holomorphic maps; this happens to be true only when $f^* - f_*$ is surjective and in general it should be replaced with the whole complex of vector spaces

$$\Theta_f(U, V)\colon \quad \Gamma(U, \Theta_X) \times \Gamma(V, \Theta_Y) \xrightarrow{f^*-f_*} \Gamma(U, f^*\Theta_Y),$$

where $\Gamma(U, \Theta_X) \times \Gamma(V, \Theta_Y)$ is in degree 0 and $\Gamma(U, f^*\Theta_Y)$ is in degree 1. Unfortunately, the complex $\Theta_f(U, V)$ does not carry any natural structure of DG-Lie algebras and the next step is to define a new DG-Lie algebra $L(U, f, V)$ that is canonically quasi-isomorphic to $\Theta_f(U, V)$ when U, V are Stein open subsets. From a heuristic point of view this new DG-Lie algebra is the local model for derived f-related vector fields and we shall prove that the DG-Lie algebra controlling $\mathrm{Def}_{X,f,Y}$ may be constructed as a suitable totalization of a semicosimplicial DG-Lie algebra of type $L(U, f, V)$.

In order to define $L(U, f, V)$ in a precise way we need the following characterization of f-related vector fields. Let $p_X\colon X \times Y \to X$ and $p_Y\colon X \times Y \to Y$ be the projections, then there exists an isomorphism $(p_{X*}, p_{Y*})\colon \Theta_{X\times Y} \to p_X^*\Theta_X \oplus p_Y^*\Theta_Y$ that gives a morphism of Lie algebras

$$\Gamma(U, \Theta_X) \times \Gamma(V, \Theta_Y) \xrightarrow{h} \Gamma(U \times V, \Theta_{X\times Y}), \qquad h(\eta, \mu) = p_X^*\eta + p_Y^*\mu.$$

Lemma 8.2.8 *Let $Z \subset X \times Y$ be the graph of a holomorphic map $f\colon X \to Y$ and let $U \subset X$, $V \subset Y$ be open subsets.*

Then a pair of vector fields $(\eta, \mu) \in \Gamma(U, \Theta_X) \times \Gamma(V, \Theta_Y)$ is f-related if and only if $h(\eta, \mu) \in \Gamma(U \times V, \Theta_{X\times Y}(-\log Z))$.

Proof The computation is purely local; it is plain that, in the notation of (8.4), we have

$$rh = f^* - f_*\colon \Gamma(U, \Theta_X) \times \Gamma(V, \Theta_Y) \to \Gamma(U, f^*\Theta_Y).$$

Now the conclusion follows immediately from Lemma 8.2.2. □

Definition 8.2.9 In the above situation, define the DG-Lie algebra $L(U, f, V)$ as the Thom-Whitney homotopy fibre product of the diagram of Lie algebras:

$$\begin{array}{ccc} & & \Gamma(U \times V, \Theta_{X\times Y}(-\log Z)) \\ & & \downarrow \chi \\ \Gamma(U, \Theta_X) \times \Gamma(V, \Theta_Y) & \xrightarrow{h} & \Gamma(U \times V, \Theta_{X\times Y}). \end{array}$$

More concretely, see Definition 6.1.8, the elements of the DG-Lie algebra $L(U, f, V)$ are the quadruples $(\xi, \eta, \mu, \rho(t))$ with:

$$\xi \in \Gamma(U \times Y, \Theta_{X\times Y}(-\log Z)), \quad \eta \in \Gamma(U, \Theta_X),$$

$$\mu \in \Gamma(V, \Theta_Y), \quad \rho(t) \in \Gamma(U \times V, \Theta_{X\times Y})[t, dt]$$

such that $\rho(0) = \chi(\xi)$ and $\rho(1) = h(\eta, \mu) = p_X^*\eta + p_Y^*\mu$.

Notice that if (η, μ) is a pair of f-related vector fields and $\xi = h(\eta, \mu)$, then $(\xi, \eta, \mu, \xi) \in L(U, f, V)$. By Whitney's Theorem 7.4.5 the DG-Lie algebra $L(U, f, V)$ is quasi-isomorphic, as a complex, to the complex in degrees 0, 1

$$\Gamma(U \times V, \Theta_{X\times Y}(-\log Z)) \times \Gamma(U, \Theta_X) \times \Gamma(V, \Theta_Y) \xrightarrow{h-\chi} \Gamma(U \times V, \Theta_{X\times Y}).$$

Since χ is always injective, it follows that $L(U, f, V)$ is also quasi-isomorphic, as a complex, to the complex in degrees 0, 1

$$\Gamma(U, \Theta_X) \times \Gamma(V, \Theta_Y) \xrightarrow{h} \operatorname{coker}\chi.$$

8.2.3 *The Stein Case*

Consider first the case when X and Y are Stein manifolds and denote by

$$G(X, f, Y) \colon \mathbf{Art}_{\mathbb{C}} \to \mathbf{Grpd}$$

the (formal pointed) action groupoid having Def_f as objects and $\mathrm{Aut}_X \times \mathrm{Aut}_Y$ as morphisms. We consider also the case where $X = \emptyset$, in this case Aut_X is the trivial group and $G(\emptyset, f, Y)$ is equivalent to the groupoid of deformations of Y.

According to Corollary 4.3.9 every infinitesimal deformation of a Stein manifold is trivial; this implies that the morphism $\mathrm{Def}_f \to \mathrm{Def}_{X,f,Y}$ is surjective and then

$$\pi_0(G(X, f, Y)) \simeq \mathrm{Def}_{X,f,Y}.$$

Similarly, if $G(X, f)$ (resp.: $G(f, Y)$, $G(f)$) is the formal pointed groupoid having Def_f as objects and Aut_X (resp.: Aut_Y, Id) as morphisms, we have:

$$\pi_0(G(X, f)) \simeq \mathrm{Def}_{X,f}, \qquad \pi_0(G(f, Y)) \simeq \mathrm{Def}_{f,Y}, \qquad \pi_0(G(f)) \simeq \mathrm{Def}_f.$$

We have seen at the end of Sect. 8.2.1 that the discrete groupoid $G(f)$ is equivalent to the totalization of the Deligne groupoid of the semicosimplicial Lie algebra

$$\Gamma(X \times Y, \Theta_{X\times Y}(-\log Z)) \underset{0}{\overset{\chi}{\rightrightarrows}} \Gamma(X \times Y, \Theta_{X\times Y}) \substack{\longrightarrow\\ \longrightarrow\\ \longrightarrow} 0\cdots,$$

where Z is the graph of f and χ is the inclusion map. The tangent map of the natural morphism $\mathrm{Aut}_X \times \mathrm{Aut}_Y \to \mathrm{Aut}_{X\times Y}$ is the morphism of Lie algebras

$$\Gamma(X, \Theta_X) \times \Gamma(Y, \Theta_Y) \xrightarrow{h} \Gamma(X \times Y, \Theta_{X\times Y}), \qquad h(\eta, \mu) = p_X^*\eta + p_Y^*\mu,$$

where p_X and p_Y are the projections. Thus, the groupoid $G(X, f, Y)$ is the totalization of the Deligne groupoid of the semicosimplicial Lie algebra

$$\Gamma(X \times Y, \Theta_{X\times Y}(-\log Z)) \times \Gamma(X, \Theta_X) \times \Gamma(Y, \Theta_Y) \underset{h}{\overset{\chi}{\rightrightarrows}} \Gamma(X \times Y, \Theta_{X\times Y}). \tag{8.5}$$

Since the DG-Lie algebra $L(X, f, Y)$ coincides by definition with the Thom–Whitney–Sullivan totalization of (8.5), by Theorem 7.6.4 the DG-Lie algebra $L(X, f, Y)$ governs the functor $\mathrm{Def}_{X,f,Y}$. Since X and Y are Stein manifolds, the cokernel of χ is isomorphic to $\Gamma(X, f^*\Theta_Y)$ and the cohomology of $L(X, f, Y)$ is the same as the cohomology of the complex of vector spaces

$$\Theta_f(X, Y)\colon \quad \Gamma(X, \Theta_X) \times \Gamma(Y, \Theta_Y) \xrightarrow{f^* - f_*} \Gamma(X, f^*\Theta_Y), \tag{8.6}$$

where $\Gamma(X, \Theta_X) \times \Gamma(Y, \Theta_Y)$ is in degree 0 and $\Gamma(X, f^*\Theta_Y)$ is in degree 1.

8.2.4 The General Case

Let $f\colon X \to Y$ be a holomorphic map of complex manifolds and let $\mathcal{U} = \{(U_i, V_i)\}$ be a family of pairs of open subsets such that U_i is a (possibly empty) Stein open subset of X, V_i is a Stein open subset of Y and $f(U_i) \subset V_i$ for every index i; we also require that $\bigcup_i U_i = X$ and $\bigcup_i V_i = Y$. The usual glueing procedure shows that the π_0 of the totalization of the formal pointed semicosimplicial groupoid

$$G(f,\mathcal{U})\colon\quad \prod_i G(U_i,f,V_i) \rightrightarrows \prod_{i,j} G(U_{ij},f,V_{ij}) \substack{\rightarrow\\ \rightarrow\\ \rightarrow} \prod_{i,j,k} G(U_{ijk},f,V_{ijk})$$

is isomorphic to the functor of deformations of f:

$$\pi_0(\mathrm{Tot}(G(f,\mathcal{U}))) \simeq \mathrm{Def}_{X,f,Y}.$$

Again by Hinich's Theorem 7.6.4, the totalization of the semicosimplicial differential graded Lie algebra

$$L(f,\mathcal{U})\colon\quad \prod_i L(U_i,f,V_i) \rightrightarrows \prod_{i,j} L(U_{ij},f,V_{ij}) \substack{\rightarrow\\ \rightarrow\\ \rightarrow} \prod_{i,j,k} L(U_{ijk},f,V_{ijk})$$

governs the deformations of f. Equivalently, there exists a natural isomorphism of functors $\mathrm{Def}_{\mathrm{Tot}(L(f,\mathcal{U}))} \simeq \mathrm{Def}_{X,f,Y}$.

Since for every index i there exists a canonical surjective quasi-isomorphism of complexes $L(U_i,f,V_i) \to \Theta_f(U_i,V_i)$, the cohomology of $\mathrm{Tot}(L(f,\mathcal{U}))$ is the same as the cohomology of the cochain complex of the semicosimplicial DG-vector space

$$\Theta_f(\mathcal{U})\colon\quad \prod_i \Theta_f(U_i,V_i) \rightrightarrows \prod_{i,j} \Theta_f(U_{ij},V_{ij}) \substack{\rightarrow\\ \rightarrow\\ \rightarrow} \prod_{i,j,k} \Theta_f(U_{ijk},V_{ijk})$$

or, equivalently, to the total cohomology of the double complex

$$\begin{array}{ccccc}
\prod_i \Gamma(U_i,f^*\Theta_Y) & \xrightarrow{\ \delta\ } & \prod_{i,j} \Gamma(U_{ij},f^*\Theta_Y) & \xrightarrow{\ \delta\ } & \cdots \\
\Big\uparrow{\scriptstyle f^*-f_*} & & \Big\uparrow{\scriptstyle f^*-f_*} & & \\
\prod_i \Gamma(U_i,\Theta_X)\times\Gamma(V_i,\Theta_Y) & \xrightarrow{\ \delta\ } & \prod_{i,j} \Gamma(U_{ij},\Theta_X)\times\Gamma(V_{ij},\Theta_Y) & \xrightarrow{\ \delta\ } & \cdots
\end{array} \tag{8.7}$$

where δ is the Čech differential. Thus we have proved the following result:

Theorem 8.2.10 *Let $f\colon X\to Y$ be a holomorphic map and denote by T_f^* the cohomology of the double complex (8.7). Then the deformation functor $\mathrm{Def}_{X,f,Y}$ has tangent space equal to T_f^1 and has a complete obstruction theory with values in T_f^2. Moreover, there exists a long exact sequence*

$$\cdots \to T_f^i \to H^i(X,\Theta_X)\oplus H^i(Y,\Theta_Y) \xrightarrow{f^*-f_*} H^i(X,f^*\Theta_Y) \to T_f^{i+1} \to \cdots. \tag{8.8}$$

Example 8.2.11 (Deformations of closed embeddings) If $i\colon X \to Y$ is a closed embedding of complex manifolds we have two geometrically clear natural transformations

$$\mathrm{Hilb}_Y^X \to \mathrm{Def}_{(Y,X)} \to \mathrm{Def}_{X,i,Y}.$$

We claim that the natural transformation $\mathrm{Def}_{(Y,X)} \to \mathrm{Def}_{X,i,Y}$ is induced by a quasi-isomorphism of DG-Lie algebras, and hence $\mathrm{Def}_{(Y,X)} \simeq \mathrm{Def}_{X,i,Y}$.

In the above notation we can choose $\mathcal{U} = \{(U_i, V_i)\}$, where $\{V_i\}$ is a Stein covering of Y and $U_i = X \cap V_i$ for every i. According to Theorem 8.0.2 it is sufficient to prove that for every Stein open subset $V \subset Y$, writing $U = X \cap V$, there exists a canonical quasi-isomorphism of DG-Lie algebras

$$\Phi\colon \Gamma(V, \Theta_Y(-\log X)) \to L(U, i, V).$$

Denote as usual by $Z \subset U \times V$ the graph of the closed immersion $i\colon U \to V$ and by $p_U\colon U \times V \to U$, $p_V\colon U \times V \to V$ the projections. Then the restriction of vector fields to U gives a natural Lie morphism

$$\Gamma(V, \Theta_Y(-\log X)) \to \Gamma(U, \Theta_X), \quad \eta \mapsto \eta_{|X}.$$

An easy computation in local coordinates shows that the map

$$h\colon \Gamma(V, \Theta_Y(-\log X)) \to \Gamma(U \times V, \Theta_{X\times Y}(-\log Z)), \quad h(\eta) = p_U^*(\eta_{|X}) + p_V^*(\eta),$$

is a properly defined morphism of Lie algebras. Then, in the notation of Definition 8.2.9, the DG-Lie morphism Φ is defined by setting

$$\Phi(\eta) = (h(\eta), \eta_{|U}, \eta, h(\eta)).$$

Since U is a closed submanifold of the Stein manifold V we have an exact sequence of vector spaces

$$\begin{array}{c} 0 \to \Gamma(V, \Theta_Y(-\log X)) \to \Gamma(U, \Theta_X) \times \Gamma(V, \Theta_Y) \xrightarrow{i^*-i_*} \Gamma(U, i^*\Theta_Y) \to 0 \\ \eta \mapsto (\eta_{|U}, \eta) \end{array}$$

and this, together with (8.6), immediately implies that Φ is a quasi-isomorphism.

8.2.5 Horikawa's Theorems

Every holomorphic map $f\colon X \to Y$ of complex manifolds induces a commutative diagram of functors

$$\begin{array}{ccccc} & & \mathrm{Def}_f & & \\ & \swarrow & & \searrow & \\ & \mathrm{Def}_{X,f} & & \mathrm{Def}_{f,Y} & \\ \swarrow & & \searrow \quad \swarrow & & \searrow \\ \mathrm{Def}_X & \longleftarrow & \mathrm{Def}_{X,f,Y} & \longrightarrow & \mathrm{Def}_Y \end{array}$$

where every natural transformation is defined in the obvious way. The modern interpretation of the classical Horikawa's theorems is a list of statements about the smoothness of these natural transformations.

Definition 8.2.12 (Horikawa) A holomorphic map $f : X \to Y$ is called:

- **stable** under deformations, if $\mathrm{Def}_{X,f,Y} \to \mathrm{Def}_Y$ is a smooth morphism;
- **costable** under deformations, if $\mathrm{Def}_{X,f,Y} \to \mathrm{Def}_X$ is a smooth morphism.

The standard smoothness criterion and Theorem 8.2.10 immediately give the following results.

Theorem 8.2.13 *In the setup of Theorem 8.2.10:*

1. *if $T^1_f \to H^1(Y, \Theta_Y)$ is surjective and $T^2_f \to H^2(Y, \Theta_Y)$ is injective, then f is stable under deformations;*
2. *if $T^1_f \to H^1(X, \Theta_X)$ is surjective and $T^2_f \to H^2(X, \Theta_X)$ is injective, then f is costable under deformations.*

Corollary 8.2.14 (Horikawa) *Let $f : X \to Y$ be a holomorphic map of complex manifolds:*

1. *if $H^1(X, \Theta_X) \xrightarrow{f_*} H^1(X, f^*\Theta_Y)$ is surjective and $H^2(X, \Theta_X) \xrightarrow{f_*} H^2(X, f^*\Theta_Y)$ is injective, then f is stable under deformations;*
2. *if $H^1(Y, \Theta_Y) \xrightarrow{f^*} H^1(X, f^*\Theta_Y)$ is surjective and $H^2(Y, \Theta_Y) \xrightarrow{f^*} H^2(X, f^*\Theta_Y)$ is injective, then f is costable under deformations.*

Proof Immediate from Theorem 8.2.13 and from an easy diagram chasing on the long exact sequence (8.8):

$$T^1_f \to H^1(X, \Theta_X) \oplus H^1(Y, \Theta_Y) \xrightarrow{f^*-f_*} H^1(X, f^*\Theta_Y) \to T^2_f$$

$$\to H^2(X, \Theta_X) \oplus H^2(Y, \Theta_Y) \xrightarrow{f^*-f_*} H^2(X, f^*\Theta_Y) \to \cdots .$$

□

Example 8.2.15 (Deformation stability and costability of submanifolds) Assume that $f : X \hookrightarrow Y$ is a closed regular embedding, denote by $\mathcal{I}_X \subset \mathcal{O}_Y$ the ideal sheaf of X and by $\mathcal{N}_{X|Y} = \mathcal{H}om_{\mathcal{O}_Y}(\mathcal{I}_X, \mathcal{O}_X)$ the normal sheaf. In this situation, the above corollary gives:

1. [144, Theorem 4] if $H^1(X, \mathcal{N}_{X|Y}) = 0$, then $f: X \to Y$ is stable under deformations;
2. [119, Theorem 8.3] if $H^2(Y, \mathcal{I}_X\Theta_Y) = 0$, then $f: X \to Y$ is costable under deformations.

The first item is the Kodaira's stability theorem already proved in Corollary 4.4.8; both items are an immediate consequence of Corollary 8.2.14 and of the exact sequences of sheaves

$$0 \to \Theta_X \to f^*\Theta_Y \to \mathcal{N}_{X|Y} \to 0, \qquad 0 \to \mathcal{I}_X\Theta_Y \to \Theta_Y \to f^*\Theta_Y \to 0.$$

The computation of Example 4.4.6 shows therefore that the inclusion $S_d \subset \mathbb{P}^3$ of a smooth surface of degree $d \geq 5$ in the projective space is costable. It is possible to prove, with a different proof, that the inclusion $S_d \subset \mathbb{P}^3$ is costable also for $d \leq 3$. On the other hand, it is known that every smooth quartic S_4 deforms to non-projective K3 surfaces and then the inclusion $S_4 \subset \mathbb{P}^3$ is not costable.

Remark 8.2.16 For every closed embedding $f: X \hookrightarrow Y$, the cohomology group $H^1(X, \mathcal{N}_{X|Y})$ contains both the obstructions to the smoothness of the functor Hilb_Y^X and the obstructions to the costability of f. While it is almost immediate from the definition that if f is costable then Hilb_Y^X is smooth, the converse is generally false, showing therefore that $H^1(X, \mathcal{N}_{X|Y})$ is generally redundant as obstruction space for Hilb_Y^X.

As an example consider the embedding of a smooth rational curve X of self-intersection -2 in a smooth algebraic surface Y. Then $H^0(X, \mathcal{N}_{X|Y}) = 0$ and the functor Hilb_Y^X is trivial, while $H^1(X, \mathcal{N}_{X|Y}) \neq 0$ and in general the inclusion map is not costable, cf. Example 1.4.7.

Example 8.2.17 (Costability of fibrations) Let $f: X \to Y$ be a holomorphic map such that $f_*\mathcal{O}_X = \mathcal{O}_Y$. Then the natural transformation $\mathrm{Def}_{X,f,Y} \to \mathrm{Def}_X$ is injective. If in addition $H^0(Y, R^1f_*\mathcal{O}_X \otimes \Theta_Y) = 0$, then $\mathrm{Def}_{X,f,Y} \to \mathrm{Def}_X$ is an isomorphism.

In fact, by the projection formula we have $R^if_*(f^*\Theta_Y) = R^if_*\mathcal{O}_X \otimes_{\mathcal{O}_Y} \Theta_Y$ for every i; in particular, $f_*(f^*\Theta_Y) = \Theta_Y$ and then $f^*: H^0(Y, \Theta_Y) \to H^0(X, f^*\Theta_Y)$ is an isomorphism; by the Leray spectral sequence we have a long exact sequence

$$\begin{aligned} 0 \to H^1(Y, \Theta_Y) \to H^1(X, f^*\Theta_Y) &\to H^0(Y, R^1f_*(f^*\Theta_Y)) \\ \to H^2(Y, \Theta_Y) \to H^2(X, f^*\Theta_Y) &\to \cdots \end{aligned}$$

and then $f^*: H^1(Y, \Theta_Y) \to H^1(X, f^*\Theta_Y)$ is injective. Now the same argument as used in the proof of Corollary 8.2.14 shows that $T_f^0 \to H^0(X, \Theta_X)$ is surjective and $T_f^1 \to H^1(X, \Theta_X)$ is injective. The injectivity of $\mathrm{Def}_{X,f,Y} \to \mathrm{Def}_X$ now follows from Lemma 6.6.1. If moreover $H^0(Y, R^1f_*(f^*\Theta_Y)) = 0$, then the smoothness of the natural transformation $\mathrm{Def}_{X,f,Y} \to \mathrm{Def}_X$ is an immediate consequence of Horikawa's costability theorem.

Example 8.2.18 (Costability of monoidal transformations) As a particular case of Example 8.2.17 we have that if $f: X \to Y$ is a map such that $f_*\mathcal{O}_X = \mathcal{O}_Y$ and $R^1 f_*\mathcal{O}_X = 0$, then the natural transformation $\mathrm{Def}_{X,f,Y} \to \mathrm{Def}_X$ is an isomorphism. These assumptions are satisfied for instance when X is the blowing-up of Y along a smooth submanifold.

Example 8.2.19 (Costability of Albanese fibrations) Let $f: X \to Y$ be a holomorphic map of compact complex manifolds such that $f_*\mathcal{O}_X = \mathcal{O}_Y$, Y is a smooth curve of positive genus g and $\dim H^1(X, \mathcal{O}_X) = \dim H^1(Y, \mathcal{O}_Y) = g > 0$. Then $\mathrm{Def}_{X,f,Y} \to \mathrm{Def}_X$ is an isomorphism. In fact, the first terms of Leray's spectral sequence

$$0 \to H^1(Y, \mathcal{O}_Y) \to H^1(X, \mathcal{O}_X) \to H^0(Y, R^1 f_*(\mathcal{O}_X)) \to H^2(Y, \mathcal{O}_Y) \to \cdots$$

give $H^0(Y, R^1 f_*\mathcal{O}_X) = 0$. Since Ω^1_Y is generated by global sections, there exists a short exact sequence

$$0 \to \mathcal{F} \to H^0(Y, \Omega^1_Y) \otimes \mathcal{O}_Y \to \Omega^1_Y \to 0$$

of locally free sheaves on Y and therefore also an injective morphism of sheaves

$$R^1 f_*\mathcal{O}_X \otimes \Theta_Y \to R^1 f_*\mathcal{O}_X \otimes H^0(Y, \Omega^1_Y)^\vee.$$

Hence $H^0(Y, R^1 f_*\mathcal{O}_X \otimes \Theta_Y) \subset H^0(Y, R^1 f_*\mathcal{O}_X) \otimes H^0(Y, \Omega^1_Y)^\vee = 0$ and the costability under deformations of f follows from Example 8.2.17.

8.3 The Kodaira–Spencer Algebra

On any complex manifold X, the Lie bracket on the holomorphic tangent sheaf Θ_X extends naturally to the sheaf $\mathcal{A}^{0,0}_X(\Theta_X)$ of differentiable vector fields of type $(1, 0)$. In a system of local holomorphic coordinates $z_1, \ldots, z_n$ we have

$$\left[f\frac{\partial}{\partial z_i}, g\frac{\partial}{\partial z_j}\right] = f\frac{\partial g}{\partial z_i}\frac{\partial}{\partial z_j} - g\frac{\partial f}{\partial z_j}\frac{\partial}{\partial z_i}, \qquad f, g \in \mathcal{A}^{0,0}_X.$$

This bracket is bilinear over the sheaf $\overline{\Omega^0} = \ker(\partial: \mathcal{A}^{0,0}_X \to \mathcal{A}^{1,0}_X)$ of antiholomorphic functions and admits a unique extension

$$[-, -]: \mathcal{A}^{0,i}_X(\Theta_X) \times \mathcal{A}^{0,j}_X(\Theta_X) \to \mathcal{A}^{0,i+j}_X(\Theta_X) \tag{8.9}$$

which is bilinear over the sheaf $\overline{\Omega^*_X} = \ker(\partial: \mathcal{A}^{0,*}_X \to \mathcal{A}^{1,*}_X)$ of antiholomorphic differential forms. More explicitly, in local holomorphic coordinates $z_1, \ldots, z_n$, if $f, g \in \mathcal{A}^{0,0}_X$, $\alpha, \beta \in \mathcal{A}^{0,*}_X$ and $\partial\alpha = \partial\beta = 0$, we have

$$\left[f\alpha\frac{\partial}{\partial z_i}, g\beta\frac{\partial}{\partial z_j}\right] = \alpha\wedge\beta\left(f\frac{\partial g}{\partial z_i}\frac{\partial}{\partial z_j} - g\frac{\partial f}{\partial z_j}\frac{\partial}{\partial z_i}\right). \tag{8.10}$$

Definition 8.3.1 The **Kodaira–Spencer** DG-Lie algebra $(KS_X, -\overline{\partial}, [-,-])$ of a complex manifold X is the graded vector space

$$KS_X = \oplus_i KS_X^i, \qquad KS_X^i = A_X^{0,i}(\Theta_X) = \Gamma\left(X, \mathcal{A}_X^{0,i}(\Theta_X)\right),$$

equipped with the opposite of the Dolbeault differential and with the standard bracket (8.9)–(8.10).

The multiplication of any homogeneous element x for $(-1)^{\overline{x}}$ gives a DG-Lie isomorphism $(KS_X, -\overline{\partial}, [-,-]) \simeq (KS_X, \overline{\partial}, [-,-])$ and then the choice of $-\overline{\partial}$ as differential is purely conventional and it is motivated by the Gerstenhaber structure on polyvector fields, that we analyse in Chap. 9, and by holomorphic Cartan homotopy formulas, cf. Corollary 8.5.5.

Theorem 8.3.2 *Let X be a complex manifold. Then the infinitesimal deformations of X are controlled by the Kodaira–Spencer differential graded Lie algebra KS_X. In other words, there exists an isomorphism of functors of Artin rings*

$$\mathrm{Def}_{KS_X} \cong \mathrm{Def}_X.$$

Proof Let $\mathcal{U} = \{U_i\}$ be a Stein covering of X. By Theorem 8.0.1 the totalization $\mathrm{Tot}(\mathcal{U}, \Theta_X)$ of the semicosimplicial Lie algebra

$$\Theta_X(\mathcal{U}): \quad \prod_i \Theta_X(U_i) \rightrightarrows \prod_{i,j} \Theta_X(U_{ij}) \rightrightarrows\!\!\!\!\!\rightarrow \prod_{i,j,k} \Theta_X(U_{ijk}) \rightrightarrows\!\!\!\!\rightrightarrows \cdots$$

controls the infinitesimal deformations of X, in the sense that there exists a natural isomorphism of functors of Artin rings

$$\mathrm{Def}_X \cong \mathrm{Def}_{\mathrm{Tot}(\mathcal{U},\Theta_X)}.$$

Therefore for proving the theorem it is sufficient to show that the DG-Lie algebra KS_X is quasi-isomorphic to $\mathrm{Tot}(\mathcal{U}, \Theta_X)$. Consider the semicosimplicial differential graded Lie algebra

$$K(\mathcal{U}): \quad \prod_i KS_{U_i} \rightrightarrows \prod_{i,j} KS_{U_{ij}} \rightrightarrows\!\!\!\!\!\rightarrow \prod_{i,j,k} KS_{U_{ijk}} \rightrightarrows\!\!\!\!\rightrightarrows \cdots,$$

then the natural morphism

$$KS_X \to C(K(\mathcal{U})) \tag{8.11}$$

is a quasi-isomorphism of complexes. In fact, the above map is compatible with the filtrations $KS_X^{\ge p} = \oplus_{i\ge p} A_X^{0,i}(\Theta_X)$ and $C(K(\mathcal{U})^{\ge p})$, where

$$K(\mathcal{U})^{\geq p}:\quad \prod_i KS_{U_i}^{\geq p} \rightrightarrows \prod_{i,j} KS_{U_{ij}}^{\geq p} \rightrightarrows \prod_{i,j,k} KS_{U_{ijk}}^{\geq p} \rightrightarrows \cdots,$$

and the quotient complex $C(K(\mathcal{U})^{\geq p})/C(K(\mathcal{U})^{\geq p+1})$ is the Čech complex of the fine sheaf $\mathcal{A}_X^{0,i}(\Theta_X)$ with respect to the open cover $\mathcal{U}$. Thus for every p the map

$$\frac{KS_X^{\geq p}}{KS_X^{\geq p+1}} \to \frac{C(K(\mathcal{U})^{\geq p})}{C(K(\mathcal{U})^{\geq p+1})}$$

is a quasi-isomorphism and the conclusion follows by Theorem 5.3.4. Every finite intersection $U_{i_1 i_2 \cdots i_k}$ is a Stein manifold and then the natural inclusions

$$\Theta_X(U_{i_1 i_2 \cdots i_k}) \to KS_{U_{i_1 i_2 \cdots i_k}}$$

are quasi-isomorphisms of DG-Lie algebras. According to Theorem 7.4.5 the induced map

$$\mathrm{Tot}(\mathcal{U}, \Theta_X) \to \mathrm{Tot}(K(\mathcal{U}))$$

is a quasi-isomorphism of differential graded Lie algebras. Finally, according to Proposition 7.7.6 the maps

$$KS_X \xrightarrow{\text{restriction}} KS_{U_{i_1 i_2 \cdots i_k}} \xrightarrow{1\otimes -} \Omega_k \otimes KS_{U_{i_1 i_2 \cdots i_k}}$$

induce a quasi-isomorphism of differential graded Lie algebras

$$KS_X \to \mathrm{Tot}(K(\mathcal{U})).$$

□

There exist many concrete cases where it is more convenient to use the Kodaira–Spencer algebra instead of the totalization algebra $\mathrm{Tot}(\mathcal{U}, \Theta_X)$ for the analysis of deformations.

Example 8.3.3 (Deformations of complex tori) Let $X = \mathbb{C}^q/\Gamma$ be a complex torus of dimension q. Denote by $B^i \subset A_X^{0,i}$ the subspace of invariant forms of type $(0, i)$: if $z_1, \ldots, z_q$ are linear coordinates on $\mathbb{C}^q$, then B^1 is the vector space generated by $d\overline{z_1}, \ldots, d\overline{z_q}$, $B^i = \bigwedge^i B^1$ and the inclusion $B^i \subset A_X^{0,i}$ induces an isomorphism $B^i \simeq H^i(X, \mathcal{O}_X)$; for a proof of this fact see for instance [101, p. 301]. The natural inclusion

$$B = \bigoplus_{i\geq 0} B^i = \bigwedge^* B^1 \xrightarrow{\imath} A_X^{0,*}$$

is a quasi-isomorphism of differential graded algebras and therefore, since the tangent bundle of a complex torus is trivial, the inclusion

$$B \otimes H^0(X, \Theta_X) \xrightarrow{\imath\otimes \mathrm{Id}} A_X^{0,*}(\Theta_X) = KS_X$$

is a quasi-isomorphism of differential graded Lie algebras. Since the invariant vector fields $\dfrac{\partial}{\partial z_i}$ are a basis of $H^0(X, \Theta_X)$, the bracket on $B \otimes H^0(X, \Theta_X)$ is trivial and then the Kodaira–Spencer algebra of X is quasi-isomorphic to an abelian DG-Lie algebra of finite dimension. In particular, the functor Def_X is smooth and pro-representable.

8.4 Deformations of Products and Examples of Obstructed Manifolds

Let X, Y be two complex manifolds, then there is a naturally defined morphism of deformation functors

$$\alpha\colon \operatorname{Def}_X \times \operatorname{Def}_Y \to \operatorname{Def}_{X\times Y}.$$

In fact, given two deformations

$$X \to X_A \to \operatorname{Spec} A, \qquad Y \to Y_A \to \operatorname{Spec} A,$$

over the same basis, their fibre product

$$X \times Y \to X_A \times_{\operatorname{Spec} A} Y_A \to \operatorname{Spec} A$$

is a deformation of the product.

In general, the morphism α is not surjective; as an example, let $X = E$ be an elliptic curve and $Y = \mathbb{P}^1$; then there exists at least one surjective homomorphism $\pi_1(E) \xrightarrow{g} \mathbb{Z}$ and a holomorphic map $\mathbb{C} \times Y \to Y$ giving a nontrivial one-parameter subgroup $\{\theta_t\}$, $t \in \mathbb{C}$, of holomorphic automorphisms of Y. Therefore we get a holomorphic family of representations

$$\rho_t\colon \pi_1(E) \to \operatorname{Aut}(\mathbb{P}^1), \qquad \rho_t(\gamma) = \theta_t^{g(\gamma)} = \theta_{tg(\gamma)}, \qquad t \in \mathbb{C},$$

inducing a family of locally trivial analytic $\mathbb{P}^1$-bundles over E, parametrized by t, and isomorphic to $E \times \mathbb{P}^1$ for $t = 0$.

It is easy to describe α in terms of morphisms of differential graded Lie algebras. Assume for simplicity X, Y compact connected and denote by $p_X\colon X \times Y \to X$, $p_Y\colon X \times Y \to Y$ the projections; if $\mathcal{E}$ and $\mathcal{F}$ are locally free sheaves on X and Y respectively, we write $\mathcal{E} \boxtimes \mathcal{F} = p_X^*\mathcal{E} \otimes p_Y^*\mathcal{F}$ and by Künneth's formula [105, Theorem 6.7.8] we have

$$H^i(X \times Y, \mathcal{E} \boxtimes \mathcal{F}) = \bigoplus_j H^j(X, \mathcal{E}) \otimes H^{i-j}(Y, \mathcal{F}).$$

Since $\Theta_{X\times Y}$ is the direct sum of $p_X^*\Theta_X = \Theta_X \boxtimes \mathcal{O}_Y$ and $p_Y^*\Theta_Y = \mathcal{O}_X \boxtimes \Theta_Y$ we have:

$$H^i(X\times Y, \Theta_{X\times Y}) = H^i(X\times Y, p_X^*\Theta_X) \oplus H^i(X\times Y, p_Y^*\Theta_Y),$$
$$H^i(X\times Y, p_X^*\Theta_X) = \bigoplus_j H^j(X, \Theta_X)\otimes H^{i-j}(Y, \mathcal{O}_Y),$$
$$H^i(X\times Y, p_Y^*\Theta_Y) = \bigoplus_j H^j(X, \mathcal{O}_X)\otimes H^{i-j}(Y, \Theta_Y).$$

The isomorphism $\Theta_{X\times Y} = p_X^*\Theta_X \oplus p_Y^*\Theta_Y$ allows us to define, in the obvious way, two injective morphisms of differential graded Lie algebras

$$p_X^* \colon KS_X \to KS_{X\times Y}, \qquad p_Y^* \colon KS_Y \to KS_{X\times Y}.$$

Notice that p_X^* is injective in cohomology and the image of $H^i(X, \Theta_X)$ is the subspace $H^i(X, \Theta_X)\otimes H^0(Y, \mathcal{O}_Y) \subset H^i(X\times Y, p^*\Theta_X)$; similarly for the morphism p_Y^*. Finally, since $[p_X^*\eta, p_Y^*\mu] = 0$ for every $\eta \in KS_X$, $\mu \in KS_Y$, we have a morphism of differential graded Lie algebras

$$p_X^* \times p_Y^* \colon KS_X \times KS_Y \to KS_{X\times Y} \tag{8.12}$$

inducing α at the level of associated deformation functors.

Lemma 8.4.1 *In the above situation, assume X, Y compact and connected. Then: the morphism α is injective; the morphism α is an isomorphism if and only if*

$$H^0(X, \Theta_X)\otimes H^1(Y, \mathcal{O}_Y) = H^1(X, \mathcal{O}_X)\otimes H^0(Y, \Theta_Y) = 0.$$

Proof Since $H^0(X, \mathcal{O}_X) = H^0(Y, \mathcal{O}_Y) = \mathbb{C}$, by Künneth's formula we have:

1. $H^0(KS_{X\times Y}) = H^0(KS_X) \oplus H^0(KS_Y)$;
2. $H^1(KS_{X\times Y})$ is the direct sum of $H^1(KS_X) \oplus H^1(KS_Y)$ and of

$$H^0(X, \Theta_X)\otimes H^1(Y, \mathcal{O}_Y) \oplus H^1(X, \mathcal{O}_X)\otimes H^0(Y, \Theta_Y);$$

3. the map $H^2(KS_X) \oplus H^2(KS_Y) \to H^2(KS_{X\times Y})$ is injective.

Thus α is injective by Lemma 6.6.1. If α is an isomorphism then, looking at first order deformations, we have $H^0(X, \Theta_X)\otimes H^1(Y, \mathcal{O}_Y) = H^1(X, \mathcal{O}_X)\otimes H^0(Y, \Theta_Y) = 0$, and conversely, the bijectivity of α follows by Theorem 6.6.2. □

The condition $H^0(X, \Theta_X)\otimes H^1(Y, \mathcal{O}_Y) = H^1(X, \mathcal{O}_X)\otimes H^0(Y, \Theta_Y) = 0$ of Lemma 8.4.1 is satisfied in most cases; for instance, by a theorem of Matsumura [192] we have that $H^0(X, \Theta_X) = 0$ for every compact manifold of general type X.

Whenever the conditions of Lemma 8.4.1 are not satisfied, for the study of $\mathrm{Def}_{X\times Y}$ it is useful to introduce the DG-algebra

$$B_X^* = \{\phi \in A_X^{0,*} \mid \partial\phi = 0\}$$

of antiholomorphic differential forms on a complex manifold X. In the above setup we can define two morphisms

$$h_1 \colon KS_X \otimes B_Y^* \to KS_{X\times Y}, \qquad h_1(\phi \otimes \eta) = p^*(\phi) \wedge q^*(\eta),$$

$$h_2 \colon B_X^* \otimes KS_Y \to KS_{X\times Y}, \qquad h_2(\phi \otimes \eta) = p^*(\phi) \wedge q^*(\eta).$$

It is straightforward to check that h_1, h_2 are morphisms of differential graded Lie algebras and that the image of h_1 commutes with the image of h_2. This implies that the morphism (8.12) extends naturally to a morphism of differential graded Lie algebras

$$h \colon (KS_X \otimes B_Y^*) \times (B_X^* \otimes KS_Y) \to KS_{X\times Y} \tag{8.13}$$

The following result implies that the Lie morphism h has good properties whenever X and Y are compact Kähler manifolds (Appendix C).

Theorem 8.4.2 *For every pair of compact Kähler manifolds X, Y the morphism*

$$(8.13) \qquad h \colon (KS_X \otimes B_Y^*) \times (B_X^* \otimes KS_Y) \to KS_{X\times Y}$$

is an injective quasi-isomorphism of differential graded Lie algebras. Considering $H^*(X, \mathcal{O}_X)$ and $H^*(Y, \mathcal{O}_Y)$ as graded commutative algebras (equipped with cup product), there exists an isomorphism of functors

$$\mathrm{Def}_{X\times Y} \cong \mathrm{Def}_{H^*(Y,\mathcal{O}_Y)\otimes \mathrm{KS}_X} \times \mathrm{Def}_{H^*(X,\mathcal{O}_X)\otimes \mathrm{KS}_Y}.$$

Proof According to Corollary C.5.4, if X is compact Kähler, then $B_X^i \subset A_X^{0,i}$ is a set of representatives for the Dolbeault cohomology group $H^i(X, \mathcal{O}_X)$, and therefore B_X^* is isomorphic to $H^*(X, \mathcal{O}_X)$ as a DG-algebra. Now, Künneth's formula implies immediately that the morphism (8.13) is a quasi-isomorphism and therefore

$$\mathrm{Def}_{X\times Y} \cong \mathrm{Def}_{\mathrm{KS}_X\otimes B_Y^*} \times \mathrm{Def}_{B_X^*\otimes \mathrm{KS}_Y} \cong \mathrm{Def}_{H^*(Y,\mathcal{O}_Y)\otimes \mathrm{KS}_X} \times \mathrm{Def}_{H^*(X,\mathcal{O}_X)\otimes \mathrm{KS}_Y}.$$

□

Corollary 8.4.3 *Let X, Y be compact Kähler manifolds such that both KS_X and KS_Y are formal DG-Lie algebras. Then also $KS_{X\times Y}$ is a formal DG-Lie algebra.*

Proof If L is a formal differential graded Lie algebra then also $B \otimes L$ is formal for every graded commutative algebra B (with trivial differential). The direct product of formal DG-Lie algebras is formal. □

8.4.1 Two Examples of Obstructed Manifolds

The above computations suggest that the functor $\mathrm{Def}_{X\times Y}$ may be obstructed even in the case that Def_X and Def_Y are both smooth. This was already observed by Kodaira and Spencer in [150, Sect. B.3], where they proved that projective spaces $\mathbb{P}^n$ and complex tori $(\mathbb{C}^q/\Gamma)$ have unobstructed deformations, while the product $(\mathbb{C}^q/\Gamma)\times\mathbb{P}^n$ has obstructed deformations for every $q\geq 2$ and every $n\geq 1$.

To prove the above fact as a consequence of Theorem 8.4.2 it is sufficient to observe that if $X=\mathbb{C}^q/\Gamma$ and $Y=\mathbb{P}^n$ then $H^*(Y,\mathcal{O}_Y)\otimes KS_X=KS_X$ and we have already proved that KS_X is homotopy abelian.

Moreover, the inclusion $H^0(Y,\Theta_Y)=\mathfrak{sl}_{n+1}(\mathbb{C})\to KS_Y$ is a quasi-isomorphism of DG-Lie algebras, by Künneth's formula the same holds for the inclusion

$$H^*(X,\mathcal{O}_X)\otimes\mathfrak{sl}_{n+1}(\mathbb{C})\to H^*(X,\mathcal{O}_X)\otimes KS_Y$$

and then

$$\mathrm{Def}_{H^*(X,\mathcal{O}_X)\otimes KS_Y}=\mathrm{Def}_{H^*(X,\mathcal{O}_X)\otimes\mathfrak{sl}_{n+1}(\mathbb{C})}\,.$$

Since $H^*(X,\mathcal{O}_X)=\bigwedge^* H^1(X,\mathcal{O}_X)$ and $\dim H^1(X,\mathcal{O}_X)=q$ we have that the Maurer–Cartan locus of the DG-Lie algebra $H^*(X,\mathcal{O}_X)\otimes\mathfrak{sl}_{n+1}(\mathbb{C})$ is naturally isomorphic to the commuting variety

$$C(q,\mathfrak{sl}_{n+1}(\mathbb{C}))=\{(A_1,\ldots,A_q)\in\mathfrak{sl}_{n+1}(\mathbb{C})^{\oplus q}\mid [A_i,A_j]=0\quad\forall\, i,j\},$$

which is clearly singular whenever $q\geq 2$ and $n\geq 1$.

It is known, see [108] and references therein, that $C(2,\mathfrak{sl}_n(\mathbb{C}))$ is irreducible for every n, while if $q\geq 3$, then the variety $C(q,\mathfrak{sl}_n(\mathbb{C}))$ is reducible for n sufficiently large. Here we show, by a simple argument of basic algebraic geometry, that if $n\geq 4$ and $q\geq 3+8/(n-3)$, then the commuting variety $C(q,\mathfrak{sl}_n(\mathbb{C}))$ is reducible. In fact, assume $n\geq 4$, $C(q,\mathfrak{sl}_n(\mathbb{C}))$ irreducible and consider the projection on the first factor

$$\pi\colon C(q,\mathfrak{sl}_n(\mathbb{C}))\to C(1,\mathfrak{sl}(n))=\mathfrak{sl}_n(\mathbb{C}).$$

Let $D\in\mathfrak{sl}_n(\mathbb{C})$ be a diagonal matrix with distinct eigenvalues, then every matrix commuting with D must be diagonal. Therefore the fibre $\pi^{-1}(D)$ is irreducible of dimension $(n-1)(q-1)$ and the dimension of $C(q,\mathfrak{sl}_n(\mathbb{C}))$ is less than or equal to $n^2-1+(n-1)(q-1)$. On the other hand, let r be the integral part of $n/2$ and let $N\subset\mathfrak{sl}_n(\mathbb{C})$ be the closed subset of matrices A such that $A^2=0$; equivalently $A\in N$ if and only if there exists a vector subspace $V\subset\mathbb{C}^n$ of dimension r such that $\operatorname{Im}A\subset V\subset\ker A$. It is easy to see that N is irreducible and of dimension $2r(n-r)$ and therefore, for a generic point $A\in N$, we have

$$\dim\pi^{-1}(A)<n^2-1-2r(n-r)+(n-1)(q-1).$$

After a possible change of basis, every $A \in N$ belongs to the space

$$H = \{(h_{ij}) \in \mathfrak{sl}_n(\mathbb{C}) \mid h_{ij} \neq 0 \text{ only if } i > r,\ j \leq r\}.$$

Since H is an abelian subalgebra of $\mathfrak{sl}_n(\mathbb{C})$ we have $\{A\} \times H^{\oplus q-1} \subset \pi^{-1}(A)$. In particular,

$$r(n-r)(q-1) = \dim H^{\oplus q-1} < n^2 - 1 + (n-1)(q-1) - 2r(n-r),$$

$$r(n-r)(q+1) < n^2 - 1 + (n-1)(q-1)\,.$$

The assumption $n \geq 4$ implies $r(n-r) \geq (n^2-1)/4$, hence $r(n-r)/(n-1) \geq (n+1)/4$,

$$(q+1)\frac{n+1}{4} \leq \frac{r(n-r)}{n-1}(q+1) < n+1+(q-1),$$

and therefore $q < (3n-1)/(n-3) = 3 + 8/(n-3)$.

Remark 8.4.4 Denote by $\mathfrak{R}(\mathbb{Z}^q, \mathrm{PGL}(n,\mathbb{C}))$ the complex algebraic variety of group representations $\rho\colon \mathbb{Z}^q \to \mathrm{PGL}(n,\mathbb{C})$. Then the exponential gives an isomorphism between the germ at 0 of the commuting variety $C(q, \mathfrak{sl}_n(\mathbb{C}))$ and a neighbourhood of the trivial representation in $\mathfrak{R}(\mathbb{Z}^q, \mathrm{PGL}(n,\mathbb{C}))$.

The above example of obstructed manifolds is discussed, with a different approach, also in [59] together with another example showing that Corollary 8.4.3 fails without the Kähler assumption. The non-Kähler manifold involved is the Iwasawa manifold X [101, p. 444], defined as the quotient of the group of matrices of type

$$\begin{pmatrix} 1 & z_1 & z_3 \\ 0 & 1 & z_2 \\ 0 & 0 & 1 \end{pmatrix}, \qquad z_1, z_2, z_3 \in \mathbb{C},$$

by the right action of the cocompact subgroup of matrices with coefficients in the Gauss integers. Denoting by R the Iwasawa DG-algebra introduced in Example 6.7.2, by a nontrivial result by Nakamura [202, p. 96], cf. [95, Lemma 6.5], the morphism of DG-algebras

$$j\colon R \to A_X^{0,*}, \qquad j(\omega_1) = d\overline{z_1}, \quad j(\omega_2) = d\overline{z_2}, \quad j(\omega_3) = d\overline{z_3} - \overline{z_1}d\overline{z_2},$$

is a quasi-isomorphism. Since X is a parallelisable manifold, the morphism of DG-Lie algebras $H^0(X, \Theta_X) \otimes R \to A_X^{0,*}(\Theta_X)$ is a quasi-isomorphism. In view of the isomorphism of Lie algebras

$$\mathfrak{n}_3(\mathbb{C}) \simeq H^0(X, \Theta_X), \qquad \begin{pmatrix} 0 & a & c \\ 0 & 0 & b \\ 0 & 0 & 0 \end{pmatrix} \mapsto a\frac{\partial}{\partial z_1} + b\left(\frac{\partial}{\partial z_2} + z_1\frac{\partial}{\partial z_3}\right) + c\frac{\partial}{\partial z_3},$$

we get that the Kodaira–Spencer algebra of the Iwasawa manifold X is quasi-isomorphic to the formal DG-Lie algebra $\mathfrak{n}_3(\mathbb{C}) \otimes R$.

Consider now $Y = \mathbb{P}^1$, then $H^*(Y, \Theta_Y) = H^0(Y, \Theta_Y) \simeq \mathfrak{sl}_2(\mathbb{C})$ and therefore the Kodaira–Spencer algebra KS_Y is quasi-isomorphic to the Lie algebra $\mathfrak{sl}_2(\mathbb{C})$. Since every differential form in the image of j is antiholomorphic, as in (8.13) we can define a morphism of differential graded Lie algebras

$$(KS_X \otimes B_Y^*) \times (R \otimes KS_Y) \to KS_{X\times Y} \tag{8.14}$$

that, by Künneth's formula is a quasi-isomorphism. Thus the Kodaira–Spencer algebra of $X \times Y$ is quasi-isomorphic to

$$(\mathfrak{n}_3(\mathbb{C}) \otimes R) \times (\mathfrak{sl}_2(\mathbb{C}) \otimes R).$$

According to Proposition 6.7.3 the differential graded Lie algebra $KS_{X\times Y}$ is not formal and the functor $\operatorname{Def}_{X\times Y}$ is obstructed.

Remark 8.4.5 The Kähler assumption is not a sufficient condition for the formality of the Kodaira–Spencer algebra of a compact manifold. For instance, Vakil proved in [259] that for every analytic singularity $(U, 0)$ defined over $\mathbb{Z}$ there exists a complex surface S with very ample canonical bundle such that its local moduli space is analytically isomorphic to the germ at 0 of $U \times \mathbb{C}^n$ for some integer $n \geq 0$. Choosing $U = \{(x, y) \in \mathbb{C}^2 \mid xy(x - y) = 0\}$ and taking S as above, the Kodaira–Spencer algebra of S cannot be formal; in fact, every first order deformation lifts to a deformation over $\mathbb{C}[t]/(t^3)$ and then the quadratic map

$$H^1(S, \Theta_S) \to H^2(S, \Theta_S), \qquad \theta \mapsto [\theta, \theta],$$

is trivial.

Moreover, since such surface is of general type, we have $H^0(S, \Theta_S) = 0$ and therefore this example also implies that, for a smooth projective manifold X, the triviality of the bracket on the graded Lie algebra $H^*(X, \Theta_X)$ is not sufficient to ensure the smoothness of Def_X.

In the case of embedded deformations, as far as I know, the first example of smooth submanifold with obstructed Hilbert functor is due to Severi and Zappa [276]; it is a smooth elliptic curve having obstructed embedded deformations inside a smooth ruled surface. A detailed analysis of this example is beyond the goal of this book and we give here only a short description, referring to [39, 201] for more information. Let E be an elliptic curve and let $0 \to \mathcal{O}_E \to V \xrightarrow{p} \mathcal{O}_E \to 0$ be an exact sequence of sheaves such that $H^0(V) = H^1(V) = \mathbb{C}$. Then $X = \mathbb{P}(V) \to E$ is a smooth ruled surface and $Z = \{x \in X \mid p(x) = 0\}$ is a smooth elliptic curve with trivial normal bundle. Therefore $H^0(Z, N_{Z|X}) = H^1(Z, N_{Z|X}) = \mathbb{C}$ and it is not difficult to prove that the obstruction map $\theta\colon H^0(Z, N_{Z|X}) \to H^1(Z, N_{Z|X})$ of the small extension

$$0 \to \mathbb{C} \xrightarrow{t^2} \mathbb{C}[t]/(t^3) \to \mathbb{C}[t]/(t^2) \to 0$$

is a nondegenerate quadratic map and then $\theta(x) = 0$ if and only if $x = 0$.

Notice that, if $H^1(X, \mathcal{O}_X) = 0$, then for every hypersurface $Z \subset X$ the natural map $H^0(X, \mathcal{O}_X(Z)) \to H^0(Z, \mathcal{N}_{Z|X})$ is surjective and therefore Hilb_X^Z is unobstructed (every first order deformation lifts to a deformation inside the complete linear system). In particular, every projective hypersurface has unobstructed embedded deformations.

Mumford [200] gave the first example of smooth submanifold of the projective space with obstructed embedded deformations; it is the curve $C \subset \mathbb{P}^3$ described in Example 1.7.8.

8.5 Holomorphic Cartan Homotopy Formulas

This section is almost entirely devoted to proving a technical result known as "holomorphic Cartan homotopy formulas", which will be used several times in this book. For the reader's convenience, and for fixing notation, we begin with a short review of inner products in multilinear algebra.

8.5.1 *Inner Products*

Let E be a vector space over a field $\mathbb{K}$ (of any characteristic), let $E^\vee$ be its algebraic dual and denote by

$$E^\vee \times E \xrightarrow{\lrcorner} \mathbb{K}, \qquad u \lrcorner v = u(v),$$

the natural pairing. For every integer $n \geq 0$ there exists a natural extension of $\lrcorner$ to a bilinear map

$$E^\vee \times \bigwedge^{n+1} E \xrightarrow{\lrcorner} \bigwedge^n E,$$

defined on decomposable multivectors by the formula

$$u \lrcorner (v_0 \wedge \cdots \wedge v_n) = \sum_{i=0}^{n} (-1)^i (u \lrcorner v_i)\, v_0 \wedge \cdots \wedge \widehat{v_i} \wedge \cdots \wedge v_n.$$

It is notationally convenient to introduce the linear map

$$i\colon E^\vee \to \mathrm{Hom}_{\mathbb{K}}^{-1}\left(\bigwedge^* E, \bigwedge^* E\right), \qquad i_u(w) = u \lrcorner w,$$

where $\bigwedge^* E = \oplus_{n\geq 0}\bigwedge^n E$. The graded vector space $\bigwedge^* E$ is endowed with a natural structure of graded $\mathbb{K}$-algebra; we can immediately observe that, for every $u \in E^\vee$, the operator $\boldsymbol{i}_u$ is a derivation of degree -1:

$$\boldsymbol{i}_u(\alpha \wedge \beta) = \boldsymbol{i}_u(\alpha) \wedge \beta + (-1)^{\overline{\alpha}}\, \alpha \wedge \boldsymbol{i}_u(\beta), \qquad \alpha, \beta \in \bigwedge^* E. \tag{8.15}$$

Lemma 8.5.1 *There exists a unique extension of the operator* $\boldsymbol{i}$ *to a sequence of linear maps*

$$\boldsymbol{i} \colon \bigwedge^k E^\vee \to \operatorname{Hom}_{\mathbb{K}}^{-k}\left(\bigwedge^* E, \bigwedge^* E\right), \qquad k \geq 0,$$

such that $\boldsymbol{i}_{u\wedge v} = \boldsymbol{i}_u \circ \boldsymbol{i}_v$ *for every* $u, v \in \bigwedge^* E^\vee$.

Proof The unicity is clear; notice that if $u \in \mathbb{K} = \bigwedge^0 E^\vee$ then $\boldsymbol{i}_u$ is the scalar multiplication by u. For the existence it is sufficient to prove that for every $u \in E^\vee$, every $n \geq 0$ and every $w \in \bigwedge^n E$ we have $\boldsymbol{i}_u(\boldsymbol{i}_u(w)) = 0$; this is obvious if $n \leq 1$, while for $n > 1$ it is not restrictive to assume $w = w_1 \wedge w_2$ with $w_1 \in E$ and $w_2 \in \bigwedge^{n-1} E$. By induction $\boldsymbol{i}_u(\boldsymbol{i}_u(w_1)) = \boldsymbol{i}_u(\boldsymbol{i}_u(w_2)) = 0$ and then

$$\begin{aligned}
\boldsymbol{i}_u(\boldsymbol{i}_u(w)) &= \boldsymbol{i}_u(\boldsymbol{i}_u(w_1) \wedge w_2 - w_1 \wedge \boldsymbol{i}_u(w_2)) \\
&= \boldsymbol{i}_u(w_1) \wedge \boldsymbol{i}_u(w_2) - \boldsymbol{i}_u(w_1) \wedge \boldsymbol{i}_u(w_2) = 0.
\end{aligned}$$

□

For later use we point out that $[\boldsymbol{i}_u, \boldsymbol{i}_v] = 0$ for every $u, v \in \bigwedge^* E^\vee$; in fact

$$[\boldsymbol{i}_u, \boldsymbol{i}_v] = \boldsymbol{i}_u \circ \boldsymbol{i}_v - (-1)^{\overline{u}\,\overline{v}}\boldsymbol{i}_v \circ \boldsymbol{i}_u = \boldsymbol{i}_{u\wedge v} - (-1)^{\overline{u}\,\overline{v}}\boldsymbol{i}_{v\wedge u} = \boldsymbol{i}_{u\wedge v - (-1)^{\overline{u}\,\overline{v}} v\wedge u} = 0.$$

Definition 8.5.2 In the notation of Lemma 8.5.1 the map $\boldsymbol{i}$ is called the **inner product** or the **convolution** operator. We shall also write $u \lrcorner w = \boldsymbol{i}_u(w)$ for every pair of vectors $u \in \bigwedge^* E^\vee$ and $v \in \bigwedge^* E$.

We leave as an exercise the proof that for every nonnegative integer k the composition

$$\boldsymbol{i} \colon \bigwedge^k E^\vee \to \operatorname{Hom}_{\mathbb{K}}^{-k}\left(\bigwedge^* E, \bigwedge^* E\right) \to \operatorname{Hom}_{\mathbb{K}}\left(\bigwedge^k E, \bigwedge^0 E\right)$$

is injective; when E is finite-dimensional the above map gives a canonical isomorphism $\bigwedge^k E^\vee \simeq (\bigwedge^k E)^\vee$.

Lemma 8.5.3 *Let* E *be a vector space of finite dimension* n *over a field* $\mathbb{K}$ *and let* $\tau \in \bigwedge^n E^\vee$, $\omega \in \bigwedge^n E$ *be such that* $\tau \lrcorner \omega = 1$. *Then for every* $k = 0, \ldots, n$ *the linear maps*

$$f \colon \bigwedge^k E^\vee \to \bigwedge^{n-k} E, \qquad f(u) = u \lrcorner \omega,$$

$$g\colon \bigwedge^{n-k} E \to \bigwedge^{k} E^\vee, \qquad g(v) = v \lrcorner \tau,$$

are the inverse of each other. Equivalently, we have

$$(u \lrcorner \omega) \lrcorner \tau = u, \qquad (v \lrcorner \tau) \lrcorner \omega = v, \quad \textit{for every } u \in \bigwedge^{*} E^\vee,\ v \in \bigwedge^{*} E.$$

In particular, for $k = 0$ *we have* $1 = gf(1) = g(\omega) = \omega \lrcorner \tau$.

Proof We prove that $gf(v) = v$ for every $v \in \bigwedge^k E^\vee$; the proof that fg is the identity is the same and it is left to the reader. By linearity it is not restrictive to assume $v = x_k \wedge x_{k-1} \wedge \cdots \wedge x_1$ for $x_1, \ldots, x_k \in E^\vee$ linearly independent. Choosing a completion of these vectors to a basis $x_1, \ldots, x_n$ of $E^\vee$, there exists an invertible scalar $a \in \mathbb{K}$ such that

$$\tau = a(x_n \wedge \cdots \wedge x_1), \qquad \omega = \frac{1}{a}(e_1 \wedge \cdots \wedge e_n),$$

where $e_1, \ldots, e_n$ is the dual basis. We have $f(v) = v \lrcorner \omega = \dfrac{1}{a}(e_{k+1} \wedge \cdots \wedge e_n)$ and then

$$gf(v) = f(v) \lrcorner \tau = x_k \wedge x_{k-1} \wedge \cdots \wedge x_1 = v.$$

□

8.5.2 Holomorphic Cartan Homotopy Formulas

The inner product makes sense also for vector bundles and locally free sheaves. In particular, for every complex manifold X we can define the inner product on the exterior powers of the holomorphic tangent sheaf:

$$\boldsymbol{i}\colon \bigwedge^{k} \Theta_X \to \mathcal{H}om^*_{\mathcal{O}_X}(\Omega^*_X, \Omega^*_X), \qquad k \geq 0,$$

and their $\mathcal{A}^{0,*}_X$-linear extensions

$$\boldsymbol{i}\colon \mathcal{A}^{0,*}_X\Big(\bigwedge^{k} \Theta_X\Big) \to \mathcal{H}om^*_{\mathcal{A}^{0,0}_X}(\mathcal{A}^{*,*}_X, \mathcal{A}^{*,*}_X), \qquad k \geq 0.$$

For $k = 1$, according to (8.15) we have:

$$\boldsymbol{i}\colon \mathcal{A}^{0,*}_X(\Theta_X) \to \mathcal{D}er^*_{\mathcal{A}^{0,0}_X}(\mathcal{A}^{*,*}_X, \mathcal{A}^{*,*}_X),$$

and, in local coordinates, if $\eta = \sum_i \phi_i \dfrac{\partial}{\partial z_i}$, with $\phi_i \in \mathcal{A}^{0,p}_X$, then

$$\boldsymbol{i}_\eta(\omega) = \sum_i \phi_i \wedge \left(\frac{\partial}{\partial z_i} \lrcorner \omega\right).$$

Therefore $\boldsymbol{i}_\eta$ is the unique derivation of bidegree $(-1, p)$ of the sheaf $\mathcal{A}_X^{*,*}$ such that

$$\boldsymbol{i}_\eta(dz_i) = \phi_i = \eta \lrcorner dz_i.$$

Proposition 8.5.4 (Holomorphic Cartan homotopy formulas) *Let X be a complex manifold. Then for every pair of germs $\eta, \mu \in \mathcal{A}_X^{0,*}(\Theta_X)$ we have:*

$$[\boldsymbol{i}_\eta, \boldsymbol{i}_\mu] = 0, \qquad \boldsymbol{i}_{\overline{\partial}\eta} = [\overline{\partial}, \boldsymbol{i}_\eta], \qquad \boldsymbol{i}_{[\eta,\mu]} = [[\boldsymbol{i}_\eta, \partial], \boldsymbol{i}_\mu] = [\boldsymbol{i}_\eta, [\partial, \boldsymbol{i}_\mu]].$$

Proof Let $z_1, \ldots, z_n$ be a system of holomorphic local coordinates; since all the expressions involve derivations of $\mathcal{A}_X^{*,*}$ vanishing on $\mathcal{A}_X^{0,*}$ it is sufficient to check their validity when applied to the holomorphic differentials $dz_1, \ldots, dz_n$. By linearity it is not restrictive to assume

$$\eta = \alpha f \frac{\partial}{\partial z_i}, \qquad \mu = \beta g \frac{\partial}{\partial z_j},$$

where $\alpha \in \overline{\Omega_X^p}$, $\beta \in \overline{\Omega_X^q}$ are antiholomorphic differential forms and $f, g \in \mathcal{A}_X^{0,0}$.

Since $[\boldsymbol{i}_\eta, \boldsymbol{i}_\mu]$ is a derivation of bidegree $(-2, p+q)$, it is clear that it annihilates every dz_i. Next, we have $\overline{\partial}\eta = \overline{\partial}(f\alpha)\frac{\partial}{\partial z_i}$ and then

$$\boldsymbol{i}_{\overline{\partial}\eta}(dz_i) = \overline{\partial}(f\alpha), \qquad [\overline{\partial}, \boldsymbol{i}_\eta](dz_i) = \overline{\partial}(\boldsymbol{i}_\eta(dz_i)) = \overline{\partial}(f\alpha),$$

while for every $h \neq i$ we have

$$\boldsymbol{i}_{\overline{\partial}\eta}(dz_h) = 0, \qquad [\overline{\partial}, \boldsymbol{i}_\eta](dz_h) = 0,$$

and therefore $\boldsymbol{i}_{\overline{\partial}\eta} = [\overline{\partial}, \boldsymbol{i}_\eta]$. Finally,

$$[\eta, \mu] = \alpha \wedge \beta \left(f \frac{\partial g}{\partial z_i}\frac{\partial}{\partial z_j} - g\frac{\partial f}{\partial z_j}\frac{\partial}{\partial z_i}\right),$$

and the verification of the n equalities $\boldsymbol{i}_{[\eta,\mu]}(dz_h) = [[\boldsymbol{i}_\eta, \partial], \boldsymbol{i}_\mu](dz_h)$, $h = 1, \ldots, n$, is completely straightforward. Notice that the equality $[[\boldsymbol{i}_\eta, \partial], \boldsymbol{i}_\mu] = [\boldsymbol{i}_\eta, [\partial, \boldsymbol{i}_\mu]]$ is a formal consequence of the Jacobi formula applied to the equality $[\partial, [\boldsymbol{i}_\eta, \boldsymbol{i}_\mu]] = 0$.

□

Corollary 8.5.5 *The map*

$$\boldsymbol{l}\colon KS_X \to \mathrm{Hom}^*_{\mathcal{O}_X}(\mathcal{A}_X^{0,*}, \mathcal{A}_X^{0,*}), \qquad \boldsymbol{l}_\eta = [\partial, \boldsymbol{i}_\eta] = (-1)^{\overline{\eta}}\boldsymbol{i}_\eta \circ \partial,$$

is an injective morphism of DG-Lie algebras. In particular, $\eta \in A_X^{0,1}(\Theta_X)$ satisfies the Maurer–Cartan equation $\overline{\partial}\eta = \frac{1}{2}[\eta, \eta]$ (recall that the differential in KS_X is $-\overline{\partial}$) if and only if $(\overline{\partial} + \boldsymbol{l}_\eta)^2 = 0$.

Proof The fact that $\boldsymbol{l}$ is a Lie morphism is equivalent to the equalities

$$\boldsymbol{l}_{-\overline{\partial}\eta} = [\overline{\partial}, \boldsymbol{l}_\eta], \qquad \boldsymbol{l}_{[\eta,\mu]} = [\boldsymbol{l}_\eta, \boldsymbol{l}_\mu], \qquad \eta, \mu \in KS_X,$$

that are a straightforward consequence of the holomorphic Cartan homotopy formulas (Proposition 8.5.4). For the injectivity, if in local coordinates $\eta = \sum_i \phi_i \dfrac{\partial}{\partial z_i}$, then $\boldsymbol{l}_\eta(z_i) = (-1)^{\overline{\eta}}\phi_i$. For the last part it is sufficient to observe that

$$(\overline{\partial} + \boldsymbol{l}_\eta)^2 = \overline{\partial}\boldsymbol{l}_\eta + \boldsymbol{l}_\eta\overline{\partial} + \boldsymbol{l}_\eta^2 = [\overline{\partial}, \boldsymbol{l}_\eta] + \frac{1}{2}[\boldsymbol{l}_\eta, \boldsymbol{l}_\eta] = \boldsymbol{l}_{-\overline{\partial}\eta + [\eta,\eta]/2}.$$

□

Don't confuse the operator $\boldsymbol{l}_\eta$ with the **holomorphic Lie derivative** $\mathcal{L}_\eta$; the latter satisfies the equation $\mathcal{L}_\eta = [\boldsymbol{i}_\eta, \partial] = (-1)^{\overline{\eta}}\boldsymbol{l}_\eta$.

Remark 8.5.6 In the notation of Corollary 8.5.5, for an Artin local $\mathbb{C}$-algebra A and an element $\eta \in \mathrm{MC}_{KS_X}(A) \subset \mathrm{KS}_X^1 \otimes \mathfrak{m}_A$ we write

$$\mathcal{O}_\eta = \ker\left(\mathcal{A}_X^{0,0} \otimes A \xrightarrow{\overline{\partial}+\boldsymbol{l}_\eta} \mathcal{A}_X^{0,1} \otimes A\right) = \{f \in \mathcal{A}_X^{0,0} \mid \overline{\partial} f = \eta \lrcorner \partial f\}.$$

Then the sheaf $\mathcal{O}_\eta$ is a deformation of $\mathcal{O}_X$ and the map $\eta \mapsto \mathcal{O}_\eta$ explicates the isomorphism $\mathrm{Def}_{KS_X} \cong \mathrm{Def}_X$ of Theorem 8.3.2.

To see this, fix a Stein covering $\mathcal{U} = \{U_i\}$ of X. Then, since the functor $\mathrm{Def}_{KS_{U_i}}$ is trivial for every i, on every open subset U_i the section η is gauge equivalent to 0. We may choose vector fields $a_i \in A_{U_i}^{0,0}(\Theta_X) \otimes \mathfrak{m}_A$ such that $e^{a_i} * \eta = 0$ and then, by Corollary 8.5.5:

$$e^{\boldsymbol{l}_{a_i}} * \boldsymbol{l}_\eta = 0 \iff \overline{\partial} = e^{\boldsymbol{l}_{a_i}}(\overline{\partial} + \boldsymbol{l}_\eta)e^{-\boldsymbol{l}_{a_i}}.$$

The sheaf of A-algebras $\mathcal{O}_\eta$ is locally isomorphic to $\mathcal{O}_X \otimes A$ since

$$\mathcal{O}_{\eta|U_i} = \ker(\overline{\partial} + \boldsymbol{l}_\eta) = \ker e^{\boldsymbol{l}_{a_i}}(\overline{\partial} + \boldsymbol{l}_\eta) = \ker \overline{\partial} e^{\boldsymbol{l}_{a_i}} = e^{-\boldsymbol{l}_{a_i}}(\ker \overline{\partial}) = e^{-\boldsymbol{l}_{a_i}}(\mathcal{O}_{U_i} \otimes A),$$

and then $\mathcal{O}_\eta$ is a deformation of $\mathcal{O}_X$.

According to Lemma 6.6.7, over the open set U_{ij} we have

$$\eta_{|U_{ij}} = e^{-a_i} * 0 = e^{-a_j} * 0 \Rightarrow (e^{a_i}e^{-a_j}) * 0 = 0 \Rightarrow e^{a_i}e^{-a_j} = e^{\phi_{ij}},$$

where $\phi_{ij} \in \Theta_X(U_{ij}) \otimes \mathfrak{m}_A$. Therefore O_η is the deformation obtained as the image of the cocycle $\{e^{\phi_{ij}}\}$ under the isomorphism $H^1_{\Theta_X(\mathcal{U})} \xrightarrow{\kappa} \mathrm{Def}_X$ of Theorem 4.3.8. Finally, we leave as an exercise the easy verification that the images of η and $\{e^{\phi_{ij}}\}$ under the natural maps

$$\mathrm{Del}(KS_X) \to \mathrm{Tot}(\mathrm{Del}(K(\mathcal{U}))), \qquad \mathrm{Tot}(\mathrm{Del}(\Theta_X(\mathcal{U}))) \to \mathrm{Tot}(\mathrm{Del}(K(\mathcal{U}))),$$

are isomorphic.

8.6 Ambient Cohomology Annihilates Obstructions

Let X be a complex manifold and denote by $V = (A_X^{*,*}, \overline{\partial})$ the DG-vector space of differential forms on X, equipped with the $\overline{\partial}$ differential; by Dolbeault's theorem we have $H^i(V) = \oplus_j H^j(X, \Omega_X^{i-j})$. The Cartan homotopy formula $\boldsymbol{i}_{-\overline{\partial}u} = -[\overline{\partial}, \boldsymbol{i}_u]$ implies in particular that the convolution operator $\boldsymbol{i}\colon KS_X \to \mathrm{Hom}^*_{\mathbb{K}}(V, V)[-1]$ is a morphism of DG-vector spaces and then it induces a morphism in cohomology

$$\boldsymbol{i}\colon H^i(X, \Theta_X) \to \prod_{p,q} \mathrm{Hom}_{\mathbb{K}}(H^p(X, \Omega_X^q), H^{p+i}(X, \Omega_X^{q-1})). \tag{8.16}$$

Equivalently, the morphisms (8.16) is induced in the obvious way via cup product and contraction

$$H^i(X, \Theta_X) \otimes H^p(X, \Omega_X^q) \xrightarrow{\cup} H^{p+i}(X, \Theta_X \otimes \Omega_X^q) \xrightarrow{\lrcorner} H^{p+i}(X, \Omega_X^{q-1}).$$

Theorem 8.6.1 (Kodaira principle) *Let X be a complex manifold such that the subcomplex $\partial A_X^{*,*}$ of ∂-exact forms is acyclic. Then the inner product in Dolbeault cohomology*

$$\boldsymbol{i}\colon H^2(X, \Theta_X) \to \prod_{p,q} \mathrm{Hom}_{\mathbb{K}}(H^p(X, \Omega_X^q), H^{p+2}(X, \Omega_X^{q-1})), \qquad \boldsymbol{i}_\eta(\omega) = \eta \lrcorner \omega,$$

annihilates every obstruction to deformations of X.

Proof As above we write $V = (A_X^{*,*}, \overline{\partial})$. Since the $\overline{\partial}$ and the de Rham differential $d = \partial + \overline{\partial}$ induce the same differentials on $\ker \partial$, $\partial A_X^{*,*}$ and $\operatorname{coker}\partial$, the natural maps $\ker \partial \to V$ and $V \to \operatorname{coker}\partial$ are morphisms of DG-vector spaces. The differential $\partial\colon A_X^{*,*} \to A_X^{*+1,*}$ factors to a linear map

$$\partial \in \mathrm{Hom}^1_{\mathbb{K}}(\operatorname{coker}\partial, \ker \partial)$$

such that $\overline{\partial}\partial + \partial\overline{\partial} = 0$ and then we may consider the derived bracket $[-,-]_\partial$ of Example 5.7.4 in order to give a differential graded Lie algebra structure on $\mathrm{Hom}^*_{\mathbb{K}}(\ker\partial, \mathrm{coker}\partial)[-1]$. Recall that the differential and the derived bracket are defined respectively as:

$$\delta f = -(\overline{\partial} f - (-1)^{i-1} f\overline{\partial}), \quad f \in \mathrm{Hom}^{i-1}_{\mathbb{K}}(\ker\partial, \mathrm{coker}\partial);$$

$$[f, g]_\partial = f\partial g - (-1)^{ij} g\partial f, \qquad \begin{array}{l} f \in \mathrm{Hom}^{i-1}_{\mathbb{K}}(\ker\partial, \mathrm{coker}\partial), \\ g \in \mathrm{Hom}^{j-1}_{\mathbb{K}}(\ker\partial, \mathrm{coker}\partial). \end{array}$$

The short exact sequences of DG-vector spaces

$$0 \to \ker\partial \to V \xrightarrow{\partial} \partial A^{*,*}_X[1] \to 0, \qquad 0 \to \partial A^{*,*}_X \to V \to \mathrm{coker}\partial \to 0,$$

show that $\ker\partial \to V$ and $V \to \mathrm{coker}\partial$ are quasi-isomorphisms and then

$$\begin{aligned} H^i(\mathrm{Hom}^*_{\mathbb{K}}(\ker\partial, \mathrm{coker}\partial)[-1]) &= H^{i-1}(\mathrm{Hom}^*_{\mathbb{K}}(V, V)) \\ &= \oplus_{p,q}\mathrm{Hom}_{\mathbb{K}}(H^p(X, \Omega^q_X), H^{p+i}(X, \Omega^{q-1}_X)). \end{aligned}$$

Moreover, also $\ker\partial \to \mathrm{coker}\partial$ is a quasi-isomorphism and therefore the induced morphism of differential graded Lie algebras

$$\mathrm{Hom}^*_{\mathbb{K}}(\ker\partial, \ker\partial)[-1] \to \mathrm{Hom}^*_{\mathbb{K}}(\ker\partial, \mathrm{coker}\partial)[-1]$$

is a quasi-isomorphism. Since the derived bracket $[-,-]_\partial$ is trivial in the subcomplex $\mathrm{Hom}^*_{\mathbb{K}}(\ker\partial, \ker\partial)[-1]$ we have that the DG-Lie algebra

$$L := \mathrm{Hom}^*_{\mathbb{K}}(\ker\partial, \mathrm{coker}\partial)[-1]$$

is homotopy abelian and then every obstruction of Def_L vanishes.

The composite morphism of DG-vector spaces

$$\boldsymbol{i}\colon KS_X \to \mathrm{Hom}^*_{\mathbb{K}}(V, V)[-1] \to L = \mathrm{Hom}^*_{\mathbb{K}}(\ker\partial, \mathrm{coker}\partial)[-1]$$

is a morphism of differential graded Lie algebras. In fact for every $u \in A^{0,i}_X(\Theta_X)$, $v \in A^{0,j}_X(\Theta_X)$ we have

$$\begin{aligned} \boldsymbol{i}_{[u,v]} &= [[\boldsymbol{i}_u, \partial], \boldsymbol{i}_v] = [\boldsymbol{i}_u, \partial]\boldsymbol{i}_v - (-1)^{i(j-1)}\boldsymbol{i}_v[\boldsymbol{i}_u, \partial] \\ &= \boldsymbol{i}_u\partial\boldsymbol{i}_v + (-1)^{i(j-1)+i-1}\boldsymbol{i}_v\partial\boldsymbol{i}_u \pm \partial\boldsymbol{i}_u\boldsymbol{i}_v \pm \boldsymbol{i}_v\boldsymbol{i}_u\partial \\ &= \boldsymbol{i}_u\partial\boldsymbol{i}_v - (-1)^{ij}\boldsymbol{i}_v\partial\boldsymbol{i}_u = [\boldsymbol{i}_u, \boldsymbol{i}_v]_\partial. \end{aligned}$$

We conclude the proof by using the general argument already used in the proofs of Corollary 4.2.5 and of Theorem 8.1.5: since Def_L is smooth, the natural transformation $\mathrm{Def}_{KS_X} \to \mathrm{Def}_L$ annihilates every obstruction and then every obstruction of Def_{KS_X} is contained in the kernel of $H^2(KS_X) \to H^2(L)$. □

Corollary 8.6.2 *Let X be a compact Kähler manifold. Then the morphism*

$$\boldsymbol{i}\colon H^2(X, \Theta_X) \to \prod_{p,q} \mathrm{Hom}_{\mathbb{K}}(H^p(X, \Omega_X^q), H^{p+2}(X, \Omega_X^{q-1})), \quad \boldsymbol{i}_\eta(\omega) = \eta \lrcorner \omega,$$

annihilates every obstruction to deformations of X.

Proof Every compact Kähler manifold satisfies the $\partial\overline{\partial}$-lemma (Corollary C.5.7), and in particular the equality $\mathrm{Im}\,\overline{\partial}\partial = \ker \overline{\partial} \cap \mathrm{Im}\,\partial$. Therefore the subcomplex $\mathrm{Im}\,\partial = \partial A_X^{*,*}$ is acyclic and the proof follows from Theorem 8.6.1. □

Since on every compact Kähler manifold X the cohomology $H^*(X, \mathbb{C})$ is isomorphic to the direct sum of the groups $H^p(X, \Omega_X^q)$; a short way to rephrase the above corollary is by saying that in compact Kähler manifolds the ambient cohomology annihilates obstructions to deformations.

Corollary 8.6.3 (Classical Bogomolov–Tian–Todorov (BTT) theorem) *Let X be a compact Kähler manifold with trivial canonical bundle. Then* Def_X *is smooth.*

Proof Let n be the dimension of X and let $\omega \in H^0(X, \Omega_X^n)$ be a nowhere vanishing holomorphic form. Then the map

$$\Theta_X \to \Omega_X^{n-1}, \qquad \eta \mapsto \eta \lrcorner \omega,$$

is an isomorphism of sheaves and then the map

$$\boldsymbol{i}\colon H^2(X, \Theta_X) \to \mathrm{Hom}_{\mathbb{K}}(H^0(X, \Omega_X^n), H^2(X, \Omega_X^{n-1})), \qquad \boldsymbol{i}_\eta(\omega) = \eta \lrcorner \omega,$$

is injective. The conclusion follows from Corollary 8.6.2. □

Remark 8.6.4 The proof of Corollary 8.6.3 shows that, if X has a trivial canonical bundle, then the Lie morphism

$$\boldsymbol{i}\colon KS_X \to \mathrm{Hom}^*_{\mathbb{K}}(\ker \partial, \mathrm{coker}\partial)\,[-1]$$

in injective in cohomology. According to Corollary 6.1.3 the Kodaira–Spencer algebra KS_X is quasi-isomorphic to an abelian differential graded Lie algebra.

In Sect. 8.8, after a deeper algebraic analysis of holomorphic Cartan homotopy formulas, we shall be able to improve Theorem 8.6.1 by replacing the acyclicity of the complex of ∂-exact forms with the weaker assumption that the Hodge to de Rham spectral sequence degenerates at the first page. In Sect. 13.5 we shall be able to interpret geometrically the Lie morphism $\mathrm{Def}_{KS_X} \to \mathrm{Def}_L$ of Theorem 8.6.1 in terms of infinitesimal variations of Hodge structures.

8.7 Semi-Regularity Maps

Let X be a compact complex manifold of dimension n and let $Z \subset X$ be a closed smooth submanifold of codimension $p > 0$. Following Severi [246], Kodaira and Spencer [149] and Bloch [24], in this section we define a sequence of linear maps

$$\pi_i \colon H^i(Z, \mathcal{N}_{Z|X}) \to H^{p+i}(X, \Omega_X^{p-1}), \qquad i \geq 0,$$

whose first two elements are called:

- π_0, the infinitesimal Abel–Jacobi map of Z in X;
- π_1, the semi-regularity map of Z in X.

The submanifold Z is called semi-regular if π_1 is injective.

For $p = 1$ the definition of the maps π_i is very easy; in fact Z is a smooth divisor, $\mathcal{N}_{Z|X} = \mathcal{O}_Z(Z)$, $\Omega_X^{p-1} = \mathcal{O}_X$ and the maps π_i are the connecting morphisms in the cohomology long exact sequence associated to the short exact sequence of sheaves

$$0 \to \mathcal{O}_X \to \mathcal{O}_X(Z) \to \mathcal{O}_Z(Z) \to 0.$$

For $p > 1$ the general definition of the above maps requires local cohomology theory and it is not considered here. However, *when X is compact*, we also have the following equivalent and more elementary definition: denoting by $i \colon Z \to X$ the embedding we have two short exact sequences of sheaves of DG-vector spaces,

$$\begin{gathered} 0 \to \mathcal{I}_Z^* \to \Omega_X^* \xrightarrow{i^*} \Omega_Z^* \to 0, \\ 0 \to \Theta_X(-\log Z) \to \Theta_X \to \mathcal{N}_{Z|X} \to 0. \end{gathered} \tag{8.17}$$

A simple computation shows that the contraction map $\Theta_X \otimes \Omega_X^* \xrightarrow{\lrcorner} \Omega_X^{*-1}$ restricts to $\Theta_X(-\log Z) \otimes \mathcal{I}_Z^* \xrightarrow{\lrcorner} \mathcal{I}_Z^{*-1}$ and then, since $\Omega_Z^{n-p+1} = 0$ we have $\Omega_X^{n-p+1} = \mathcal{I}_Z^{n-p+1}$ and therefore the contraction map $\Theta_X \otimes \Omega_X^{n-p+1} \xrightarrow{\lrcorner} \Omega_X^{n-p}$ factors through a pairing

$$\mathcal{N}_{Z|X} \otimes \Omega_X^{n-p+1} \xrightarrow{\lrcorner} \Omega_Z^{n-p}.$$

Passing to cohomology we get a sequence of bilinear maps

$$H^i(Z, \mathcal{N}_{Z|X}) \otimes H^{n-p-i}(X, \Omega_X^{n-p+1}) \xrightarrow{\lrcorner} H^{n-p}(Z, \Omega_Z^{n-p}) \xrightarrow{\int_Z} \mathbb{C},$$

and by Serre duality we get:

$$\pi_i \colon H^i(Z, \mathcal{N}_{Z|X}) \to H^{n-p-i}(X, \Omega_X^{n-p+1})^\vee = H^{p+i}(X, \Omega_X^{p-1}).$$

Recall that, according to Theorem 8.1.2, for a smooth submanifold $Z \subset X$, the obstructions to embedded deformations of Z in X are contained in $H^1(Z, \mathcal{N}_{Z|X})$.

We are now ready to extend Theorem 8.1.5 to submanifolds of arbitrary codimension of compact Kähler manifolds. Before doing this we introduce the Dolbeault analogs of (8.17), i.e., the two exact sequences of DG-vector spaces

$$0 \to I_Z^* \to A_X^{*,*} \xrightarrow{i^*} A_Z^{*,*} \to 0, \quad 0 \to L \xrightarrow{\eta_X} A_X^{0,*}(\Theta_X) \to A_Z^{0,*}(\mathcal{N}_{Z|X}) \to 0.$$

Notice that the sheaf $\Theta_X(-\log Z)$ is not locally free for $p > 1$ and the DG-vector space L is not its Dolbeault resolution (even for $p = 1$). However, there exists a natural isomorphism $H^i(L) = H^i(X, \Theta_X(-\log Z))$ for every i.

Lemma 8.7.1 *In the notation above, L is a differential graded Lie subalgebra of $KS_X = A_X^{0,*}(\Theta_X)$ and the functor Hilb_X^Z is controlled by the homotopy fibre of η_X.*

Proof The proof that L is a DG-Lie subalgebra is an easy and straightforward computation in local coordinates. The proof that Hilb_X^Z is controlled by the homotopy fibre of η_X follows closely the proof of Theorem 8.3.2. Given an open Stein subset $U \subset X$ we have a commutative diagram with exact rows:

$$\begin{array}{ccccccccc}
0 & \longrightarrow & \Gamma(U, \Theta_X(-\log Z)) & \xrightarrow{\chi_U} & \Gamma(U, \Theta_X) & \longrightarrow & \Gamma(U \cap Z, \mathcal{N}_{Z|X}) & \longrightarrow & 0 \\
 & & \downarrow & & \downarrow & & \downarrow & & \\
0 & \longrightarrow & L_U & \xrightarrow{\eta_U} & A_U^{0,*}(\Theta_X) & \longrightarrow & A_{U\cap Z}^{0,*}(\mathcal{N}_{Z|X}) & \longrightarrow & 0.
\end{array}$$

Denoting, as in Theorem 8.1.2, by $H(U)$ the homotopy fibre of χ_U and by $K(U)$ the homotopy fibre of η_U, the above diagram gives a natural quasi-isomorphism $H(U) \to K(U)$.

We have already seen that, if $\mathcal{U}$ is a Stein open cover of X, then the differential graded Lie algebra $\mathrm{Tot}(H(\mathcal{U}))$ controls the functor Hilb_X^Z and the natural map $\mathrm{Tot}(H(\mathcal{U})) \to \mathrm{Tot}(K(\mathcal{U}))$ is a quasi-isomorphism. On the other hand, by Proposition 7.7.6 the maps

$$K(X) \xrightarrow{\text{restriction}} K(U_{i_1 i_2 \cdots i_k}) \xrightarrow{1\otimes -} \Omega_k \otimes K(U_{i_1 i_2 \cdots i_k})$$

induce a quasi-isomorphism of differential graded Lie algebras

$$K(X) \to \mathrm{Tot}(K(\mathcal{U})).$$

□

Theorem 8.7.2 *Let Z be a smooth closed submanifold of codimension p of a compact Kähler manifold X. Then the obstructions to embedded deformations of Z in X are annihilated by the semi-regularity map*

$$\pi_1 \colon H^1(Z, \mathcal{N}_{Z|X}) \to H^{p+1}(X, \Omega_X^{p-1}).$$

Proof We have already seen in the proof of Theorem 8.6.1 that the operator of inner product

$$\boldsymbol{i}\colon KS_X \to \operatorname{Hom}^*_{\mathbb{K}}(\ker\partial, \operatorname{coker}\partial)[-1]$$

is a morphism of DG-Lie algebras, where the codomain is equipped with the derived bracket $[-,-]_\partial$. Since $\boldsymbol{i}_\alpha(I^*_Z) \subset I^*_Z$ for every $\alpha \in L$, we have a commutative diagram of differential graded Lie algebras

$$\begin{array}{ccc} L & \xrightarrow{\quad i \quad} & K \\ \Big\downarrow{\scriptstyle \eta_X} & & \Big\downarrow{\scriptstyle j} \\ A^{0,*}_X(\Theta_X) & \xrightarrow{\ i\ } & \operatorname{Hom}^*_{\mathbb{K}}(\ker\partial, \operatorname{coker}\partial)[-1] \end{array}$$

where

$$K = \left\{ f \in \operatorname{Hom}^*_{\mathbb{K}}(\ker\partial, \operatorname{coker}\partial)[-1] \;\middle|\; f(I^*_Z \cap \ker\partial) \subset \frac{I^*_Z}{I^*_Z \cap \partial A^{*,*}_X} \right\}.$$

Notice that the cokernel of j is $\operatorname{Hom}^*_{\mathbb{K}}\left(I^*_Z \cap \ker\partial, \dfrac{A^{*,*}_Z}{\partial A^{*,*}_Z}\right)[-1]$. Moreover, we have another commutative diagram

$$\begin{array}{ccc} \{f \in K \mid f(\partial A^{*,*}_X) = 0\} & \xrightarrow{\quad \alpha \quad} & K \\ \Big\downarrow{\scriptstyle h} & & \Big\downarrow{\scriptstyle j} \\ \operatorname{Hom}^*_{\mathbb{K}}\left(\dfrac{\ker\partial}{\partial A^{*,*}_X}, \operatorname{coker}\partial\right)[-1] & \xrightarrow{\ \beta\ } & \operatorname{Hom}^*_{\mathbb{K}}(\ker\partial, \operatorname{coker}\partial)[-1] \end{array}$$

where the DG-Lie algebras in the first column are both abelian and the horizontal maps are the natural injection; in particular the Thom–Whitney homotopy fibre of h is abelian. In order to prove that α and β are quasi-isomorphisms we show that their cokernels are acyclic, provided that the manifold X is compact Kähler. By the $\partial\overline{\partial}$-lemma the complex $\partial A^{*,*}_X$ is acyclic and then also

$$\operatorname{coker}(\beta) = \operatorname{Hom}^*_{\mathbb{K}}(\partial A^{*,*}_X, \operatorname{coker}\partial)[-1]$$

is acyclic. Moreover, also Z is compact Kähler, by the $\partial\overline{\partial}$-lemma the complex $\partial A^{*,*}_Z$ is acyclic and by the short exact sequence

$$0 \to I^*_Z \cap \partial A^{*,*}_X \to \partial A^{*,*}_X \to \partial A^{*,*}_Z \to 0$$

also the complex $I_Z \cap \partial A^{*,*}_X$ is acyclic. Next, we have

$$\operatorname{coker}(\alpha) = \left\{ f \in \operatorname{Hom}^*_{\mathbb{K}}(\partial A^{*,*}_X, \operatorname{coker}\partial)[-1] \mid f(I^*_Z \cap \partial A^{*,*}_X) \subset \frac{I^*_Z}{I^*_Z \cap \partial A^{*,*}_X} \right\}$$

and then the cokernel of α is the kernel of the surjective morphism of acyclic complexes

$$\operatorname{Hom}^*_{\mathbb{K}}(\partial A^{*,*}_X, \operatorname{coker}\partial)[-1] \to \operatorname{Hom}^*_{\mathbb{K}}\left(I^*_Z \cap \partial A^{*,*}_X, \frac{A^{*,*}_Z}{\partial A^{*,*}_Z}\right)[-1].$$

Thus the homotopy fibre of h is a homotopy abelian differential graded Lie algebra which is quasi-isomorphic to the homotopy fibre of j and then the morphism $\boldsymbol{i}$ from the homotopy fibre of η_X to the homotopy fibre of j annihilates every obstruction of the functor Hilb^Z_X. In particular, since $A^{n-p+1,*}_X \subset I^*_Z$ we have that the map

$$\boldsymbol{i}\colon H^1(Z, \mathcal{N}_{Z|X}) \to H^2\left(\operatorname{Hom}^*_{\mathbb{K}}\left(A^{n-p+1,*}_X \cap \ker\partial, \frac{A^{n-p,*}_Z}{\partial A^{n-p-1,*}_Z}\right)[-1]\right)$$

annihilates the obstructions to embedded deformations. The semi-regularity map

$$\pi_1\colon H^1(Z, \mathcal{N}_{Z|X}) \to H^{n-p-1}(X, \Omega^{n-p+1}_X)^\vee$$

is just the composition of the above map $\boldsymbol{i}$ with the map obtained by choosing a harmonic embedding

$$H^{n-p-1}(X, \Omega^{n-p+1}_X) \hookrightarrow A^{n-p+1,n-p-1}_X \cap \ker\partial$$

and performing the integration of forms

$$\frac{A^{n-p,n-p}_Z}{\partial A^{n-p-1,n-p}_Z} \xrightarrow{\int_Z} \mathbb{C}.$$

□

8.8 Cartan Homotopies and the Abstract BTT Theorem

A long list of mathematical results, not only in deformation theory, follow from either classical or holomorphic Cartan homotopy formulas. It is therefore natural to seek for an abstract algebraic concept encoding these concepts; the notion of Cartan homotopy goes into this direction and allows us to prove an useful abstract version of the BTT theorem (Theorem 8.8.3).

Definition 8.8.1 Let L and M be two differential graded Lie algebras. A linear map

$$\boldsymbol{i} \in \mathrm{Hom}^{-1}_{\mathbb{K}}(L, M), \qquad a \mapsto \boldsymbol{i}_a,$$

is called a **Cartan homotopy** if, for every $a, b \in L$, we have

$$[\boldsymbol{i}_a, \boldsymbol{i}_b] = 0, \qquad \boldsymbol{i}_{[a,b]} = [\boldsymbol{i}_a, d_M \boldsymbol{i}_b]. \tag{8.18}$$

The **boundary** of a Cartan homotopy $\boldsymbol{i}$ is the morphism of DG-vector spaces

$$\boldsymbol{l} = d_M \boldsymbol{i} + \boldsymbol{i} d_L \colon L \to M.$$

Notice that the boundary of a Cartan homotopy $\boldsymbol{i}$ is, by definition, the differential of $\boldsymbol{i}$ in the DG-vector space $\mathrm{Hom}^{-1}_{\mathbb{K}}(L, M)$; in particular, as a morphism of cochain complexes, the boundary of a Cartan homotopy is homotopic to the trivial map. It is plain that the equations (8.18) are equivalent to the set of equations

$$[\boldsymbol{i}_a, \boldsymbol{i}_b] = 0, \quad \boldsymbol{l}_a = d_M \boldsymbol{i}_a + \boldsymbol{i}_{d_L a}, \quad \boldsymbol{i}_{[a,b]} = [\boldsymbol{i}_a, \boldsymbol{l}_b], \qquad \forall\, a, b \in L. \tag{8.19}$$

In fact, if $[\boldsymbol{i}_a, \boldsymbol{i}_b] = 0$ for every a, b, then $[\boldsymbol{i}_a, \boldsymbol{i}_{d_L b}] = 0$ and therefore

$$\boldsymbol{i}_{[a,b]} = [\boldsymbol{i}_a, d_M \boldsymbol{i}_b] = [\boldsymbol{i}_a, d_M \boldsymbol{i}_b + \boldsymbol{i}_{d_L b}] = [\boldsymbol{i}_a, \boldsymbol{l}_b].$$

Example 8.8.2 Let $A^{*,*}_X$ and KS_X be respectively the de Rham complex and the Kodaira–Spencer algebra of a complex manifold X. Then the convolution operator (Sect. 8.5)

$$\boldsymbol{i} \colon KS_X \to \mathrm{Hom}^*_{\mathbb{K}}(A^{*,*}_X, A^{*,*}_X)$$

is a Cartan homotopy. In fact, by holomorphic Cartan homotopy formulas (Proposition 8.5.4), for every $\eta, \mu \in KS_X$ we have:

$$[\boldsymbol{i}_\eta, \boldsymbol{i}_\mu] = 0, \qquad \boldsymbol{i}_{\overline{\partial}\eta} = [\overline{\partial}, \boldsymbol{i}_\eta], \qquad \boldsymbol{i}_{[\eta,\mu]} = [\boldsymbol{i}_\eta, [\partial, \boldsymbol{i}_\mu]]$$

and then

$$[\boldsymbol{i}_\eta, [d, \boldsymbol{i}_\mu]] = [\boldsymbol{i}_\eta, [\partial, \boldsymbol{i}_\mu] + [\overline{\partial}, \boldsymbol{i}_\mu]] = [\boldsymbol{i}_\eta, [\partial, \boldsymbol{i}_\mu] + \boldsymbol{i}_{\overline{\partial}\mu}] = [\boldsymbol{i}_\eta, [\partial, \boldsymbol{i}_\mu]] = \boldsymbol{i}_{[\eta,\mu]}.$$

Notice that the differential of KS_X is $-\overline{\partial}$ and then the boundary of $\boldsymbol{i}$ is

$$\boldsymbol{l}_\eta = [d, \boldsymbol{i}_\eta] + \boldsymbol{i}_{-\overline{\partial}\eta} = [\partial + \overline{\partial}, \boldsymbol{i}_\eta] - [\overline{\partial}, \boldsymbol{i}_\eta] = [\partial, \boldsymbol{i}_\eta].$$

Let $\boldsymbol{i} \in \mathrm{Hom}^{-1}_{\mathbb{K}}(L, M)$ be a Cartan homotopy with boundary $\boldsymbol{l}$, then the linear map $\boldsymbol{i} \colon L \to M[-1]$ is a morphism of DG-vector spaces if and only if $\boldsymbol{l} = 0$. If $H \subset M$

is a DG-Lie subalgebra such that $\boldsymbol{l}(L) \subset H$, then the map $\boldsymbol{i}\colon L \to M/H[-1]$ is morphism of DG-vector spaces; this follows immediately from the definition of $\boldsymbol{l}$, since the differential on $M/H[-1]$ is induced by $-d_M$. In particular, $\boldsymbol{i}$ induces a sequence of morphisms of cohomology groups

$$\boldsymbol{i}\colon H^i(L) \to H^i(M/H[-1]) = H^{i-1}(M/H).$$

We are now ready to write the main result of this section.

Theorem 8.8.3 (Abstract BTT theorem) *Let $\boldsymbol{i} \in \operatorname{Hom}^{-1}_{\mathbb{K}}(L, M)$ be a Cartan homotopy with boundary $\boldsymbol{l}$ and let $H \subset M$ be a differential graded Lie subalgebra of M such that $\boldsymbol{l}(L) \subset H$.*

1. *If the natural map $H^1(H) \to H^1(M)$ is injective, then the obstructions of* Def_L *are contained in the kernel of $\boldsymbol{i}\colon H^2(L) \to H^1(M/H)$.*
2. *If the maps $H^1(H) \to H^1(M)$ and $\boldsymbol{i}\colon H^2(L) \to H^1(M/H)$ are injective, then the functor* Def_L *is smooth.*
3. *If the maps $H^i(H) \to H^i(M)$ and $\boldsymbol{i}\colon H^i(L) \to H^{i-1}(M/H)$ are injective for every i, then the DG-Lie algebra L is homotopy abelian.*

We postpone the proof of the abstract BTT theorem to the end of this section, after some applications to deformation theory of complex manifolds and after the proof of some general properties of Cartan homotopies.

In our first application of the abstract BTT theorem we improve Corollary 8.6.2 by replacing the Kähler assumption with the degeneration of the Hodge to de Rham spectral sequence. For a complex manifold X the Hodge filtration of the de Rham complex is the decreasing finite filtration:

$$F_X^p = \bigoplus_{i \geq p} A_X^{i,*}.$$

Notice that if the subcomplex $\partial A_X^{*,*}$ of ∂-exact forms is acyclic, then the inclusion $F_X^{p+1} \to F_X^p$ is injective in cohomology for every p. In fact if $\omega \in F_X^p$ and $d\omega \in F_X^{p+1}$, we may write $\omega = u + \omega'$ with $\omega' \in F_X^{p+1}$, $u \in A_X^{p,*}$ and we have $\overline{\partial} u = 0$, $du = \partial u$. If the inclusion $\ker \partial \to A_X^{*,*}$ is a quasi-isomorphism, then we can write $u = h + \overline{\partial} v$ with $h, v \in A_X^{p,*}$, $\partial h = \overline{\partial} h = 0$ and then

$$du = \partial u = \partial\overline{\partial} v = d(-\partial v), \qquad d\omega = d(\omega' - \partial v) \in dF_X^{p+1}.$$

Corollary 8.8.4 (Kodaira principle) *Let X be a complex manifold and let p be a positive integer such that the three inclusions*

$$F_X^{p+1} \to F_X^p \to F_X^{p-1} \to F_X^0 = A_X^{*,*}$$

are injective in cohomology. Then the morphism

$$\boldsymbol{i}\colon H^2(X,\Theta_X)\to\prod_q \operatorname{Hom}_{\mathbb{K}}(H^q(X,\Omega_X^p),H^{q+2}(X,\Omega_X^{p-1})),\qquad \boldsymbol{i}_\eta(\omega)=\eta\lrcorner\omega,$$

annihilates every obstruction to deformations of X.

Proof It is useful to write the above morphism as

$$\boldsymbol{i}\colon H^2(X,\Theta_X)\to \operatorname{Hom}^1_{\mathbb{K}}(H^*(X,\Omega_X^p[-p]),H^*(X,\Omega_X^{p-1}[-p+1])).$$

Since $F_X^{p+1}\to F_X^p$ is injective in cohomology the natural morphism of graded vector spaces $H^*(F_X^p)\to H^*(X,\Omega_X^p[-p])$ is surjective. Since $F_X^{p-1}\to A_X^{*,*}$ is injective in cohomology, also the natural morphism of graded vector spaces

$$H^*(X,\Omega_X^{p-1}[-p+1]))\to H^*(A_X^{*,*}/F_X^p)$$

is injective. It is therefore sufficient to prove that obstructions are annihilated by the map

$$\boldsymbol{i}\colon H^2(X,\Theta_X)\to H^1(\operatorname{Hom}^*_{\mathbb{K}}(F_X^p,A_X^{*,*}/F_X^p)),\qquad \boldsymbol{i}_\eta(\omega)=\eta\lrcorner\omega.$$

By Example 8.8.2 the convolution operator

$$\boldsymbol{i}\colon KS_X\to \operatorname{Hom}^*_{\mathbb{K}}(A_X^{*,*},A_X^{*,*})$$

is a Cartan homotopy with boundary $\eta\mapsto \boldsymbol{l}_\eta=[\partial,\boldsymbol{i}_\eta]$ and then $\boldsymbol{l}_\eta(F_X^p)\subset F_X^p$ for every $\eta\in KS_X$. In other words, the image of $\boldsymbol{l}$ is contained in the DG-Lie subalgebra

$$H=\{\phi\in \operatorname{Hom}^*_{\mathbb{K}}(A_X^{*,*},A_X^{*,*})\mid \phi(F_X^p)\subset F_X^p\}.$$

Since $F_X^p\to A_X^{*,*}$ is injective in cohomology, also the inclusion $H\hookrightarrow \operatorname{Hom}^*_{\mathbb{K}}(A_X^{*,*},A_X^{*,*})$ is injective in cohomology and $\operatorname{Hom}^*_{\mathbb{K}}(A_X^{*,*},A_X^{*,*})/H=\operatorname{Hom}^*_{\mathbb{K}}(F_X^p,A_X^{*,*}/F_X^p)$. Now the conclusion follows immediately from Theorem 8.8.3. □

Corollary 8.8.5 (Classical BTT theorem) *Let X be a complex manifold of dimension n with trivial canonical bundle such that the inclusions $F_X^n\hookrightarrow F_X^{n-1}\hookrightarrow A_X^{*,*}$ are injective in cohomology. Then KS_X is homotopy abelian and* Def_X *is smooth.*

Proof Since $F_X^{n+1}=0\to F_X^n\to F_X^{n-1}\to A_X^{*,*}$ are injective in cohomology, by Corollary 8.8.4 it is sufficient to prove that the map

$$H^*(X,\Theta_X)\xrightarrow{\boldsymbol{i}}\operatorname{Hom}^*_{\mathbb{C}}\left(H^*(X,\Omega_X^n),H^*(X,\Omega_X^{n-1})\right)$$

is injective. Let $\omega\in H^0(X,\Omega_X^n)$ be a holomorphic volume form, then

$$\Theta_X \to \Omega_X^{n-1}, \qquad a \mapsto \boldsymbol{i}_a(\omega) = a \lrcorner \omega,$$

is an isomorphism of sheaves, $H^*(X, \Theta_X) \cong H^*(X, \Omega_X^{n-1})$ and the composition of $\boldsymbol{i}$ with the evaluation at ω

$$\operatorname{Hom}^*_{\mathbb{C}}\left(H^*(X, \Omega_X^n), H^*(X, \Omega_X^{n-1})\right) \to H^*(X, \Omega_X^{n-1}),$$

is an isomorphism. In particular, $\boldsymbol{i}$ is injective. □

Lemma 8.8.6 *Let $\boldsymbol{i}\colon L \to M$ be a Cartan homotopy with boundary $\boldsymbol{l}$:*

$$\boldsymbol{l}\colon L \to M, \qquad \boldsymbol{l}_a = d_M \boldsymbol{i}_a + \boldsymbol{i}_{d_L a}\,.$$

Then:

1. $\boldsymbol{i}_{[a,b]} = [\boldsymbol{i}_a, \boldsymbol{l}_b] = (-1)^{\bar{a}}[\boldsymbol{l}_a, \boldsymbol{i}_b]$ *for every $a, b \in L$;*
2. $\boldsymbol{l}$ *is a morphism of differential graded Lie algebras;*
3. $[\boldsymbol{i}_a, [\boldsymbol{i}_b, \boldsymbol{l}_c]] = 0$ *for every $a, b, c \in L$.*

Proof (1) We already seen in Eq. (8.19) that $\boldsymbol{i}_{[a,b]} = [\boldsymbol{i}_a, \boldsymbol{l}_b]$. Moreover, since $[\boldsymbol{i}_a, \boldsymbol{i}_b] = [\boldsymbol{i}_{d_L a}, \boldsymbol{i}_b] = 0$ we have

$$\boldsymbol{i}_{[a,b]} = [\boldsymbol{i}_a, d_M \boldsymbol{i}_b] = (-1)^{\bar{a}}([d_M \boldsymbol{i}_a, \boldsymbol{i}_b] - d_M[\boldsymbol{i}_a, \boldsymbol{i}_b]) = (-1)^{\bar{a}}[\boldsymbol{l}_a, \boldsymbol{i}_b].$$

(2) It is clear from the definition that $\boldsymbol{l}$ is a morphism of complexes. We have

$$\begin{aligned}\boldsymbol{l}_{[a,b]} &= d_M \boldsymbol{i}_{[a,b]} + \boldsymbol{i}_{d_L[a,b]} = d_M[\boldsymbol{i}_a, d_M \boldsymbol{i}_b] + \boldsymbol{i}_{[d_L a, b]} + (-1)^{\bar{a}} \boldsymbol{i}_{[a, d_L b]}\\ &= [d_M \boldsymbol{i}_a, d_M \boldsymbol{i}_b] + [\boldsymbol{i}_{d_L a}, d_M \boldsymbol{i}_b] + [\boldsymbol{l}_a, \boldsymbol{i}_{d_L b}] = [\boldsymbol{l}_a, d_M \boldsymbol{i}_b] + [\boldsymbol{l}_a, \boldsymbol{i}_{d_L b}]\\ &= [\boldsymbol{l}_a, \boldsymbol{l}_b].\end{aligned}$$

For (3) every a, b and c, we have $[\boldsymbol{i}_a, [\boldsymbol{i}_b, \boldsymbol{l}_c]] = [\boldsymbol{i}_a, \boldsymbol{i}_{[b,c]}] = 0$. □

Remark 8.8.7 It is plain from the definition that Cartan homotopies are stable under composition with morphisms of differential graded Lie algebras. More precisely, if $f\colon L' \to L$ and $g\colon M \to M'$ are morphisms of differential graded Lie algebras and $\boldsymbol{i}\colon L \to M$ is a Cartan homotopy, then also $g\boldsymbol{i}f\colon L' \to M'$ is a Cartan homotopy.

Example 8.8.8 Let L be a differential graded Lie algebra. We have seen in Example 5.6.8 that the mapping cone of the identity $\operatorname{cone}(\mathrm{Id}_L) = s^{-1}L \oplus L$ carries a structure of an acyclic differential graded Lie algebra such that the natural inclusion $j\colon L \to \operatorname{cone}(\mathrm{Id}_L)$ is a morphism of DG-Lie algebras and where, for every $x, y \in L$,

$$d(s^{-1}x) = x - s^{-1}dx, \qquad [s^{-1}x, y] = s^{-1}[x, y], \qquad [s^{-1}x, s^{-1}y] = 0\,.$$

We can immediately check that the linear map of degree -1:

$$\boldsymbol{u}_L : L \to \text{cone}(\text{Id}_L), \qquad \boldsymbol{u}_L(x) = s^{-1}x,$$

is a Cartan homotopy with boundary j. We shall call $\boldsymbol{u}_L$ the **universal Cartan homotopy** of L.

Lemma 8.8.9 *Let L, M be DG-Lie algebras, and denote by $\boldsymbol{u}_L : L \to \text{cone}(\text{Id}_L)$ the universal Cartan homotopy of L (Example 8.8.8). Then for every Cartan homotopy $\boldsymbol{i} : L \to M$ there exists a unique morphism of DG-Lie algebras $f : \text{cone}(\text{Id}_L) \to M$ such that $\boldsymbol{i} = f\boldsymbol{u}_L$.*

Proof We have already pointed out that the composition of a Cartan homotopy with a morphism of DG-Lie algebras is still a Cartan homotopy. Conversely, let $\boldsymbol{i} : L \to M$ be a Cartan homotopy with boundary $\boldsymbol{l}$ and consider the morphism of graded vector spaces

$$f : \text{cone}(\text{Id}_L) \to M, \qquad f(x + s^{-1}y) = \boldsymbol{l}_x + \boldsymbol{i}_y.$$

The formulas

$$\boldsymbol{l}_x = \boldsymbol{i}_{dx} + d\boldsymbol{i}_x, \qquad \boldsymbol{l}_{[x,y]} = [\boldsymbol{l}_x, \boldsymbol{l}_y], \qquad \boldsymbol{i}_{[x,y]} = [\boldsymbol{i}_x, \boldsymbol{l}_y], \qquad 0 = [\boldsymbol{i}_x, \boldsymbol{i}_y],$$

imply that f is a morphism of DG-Lie algebras such that $\boldsymbol{i} = f\boldsymbol{u}_L$. This morphism is unique since the condition $f\boldsymbol{u}_L = \boldsymbol{i}$ implies that $f(s^{-1}x) = \boldsymbol{i}_x$ and the equality $df = fd$ gives

$$f(j(x)) = f(d(s^{-1}x) + s^{-1}(dx)) = df(s^{-1}x) + s^{-1}(dx)) = d\boldsymbol{i}_x + \boldsymbol{i}_{dx} = \boldsymbol{l}_x.$$

□

Consider now the situation of Theorem 8.8.3, where the given data are a Cartan homotopy $\boldsymbol{i} \in \text{Hom}_{\mathbb{K}}^{-1}(L, M)$ with boundary $\boldsymbol{l}$ and a DG-Lie subalgebra $H \subset M$ such that $\boldsymbol{l}(L) \subset H$.

We have already observed that the complex $M/H[-1]$ does not have, in general, a natural DG-Lie structure; hovever, if

$$K(H \hookrightarrow M) = \{m(t) \in M[t, dt] \mid m(0) = 0, \ m(1) \in H\}$$

is the Thom–Whitney homotopy fibre of the inclusion $H \hookrightarrow M$, by Corollary 6.1.7 there exists a canonical quasi-isomorphism of DG-vector spaces

$$\int_0^1 : K(H \hookrightarrow M) \to M/H[-1], \qquad m(t) \mapsto \int_0^1 m(t) \pmod{H[-1]}.$$

Lemma 8.8.10 *In the above situation, the morphism $\boldsymbol{i} : L \to M/H[-1]$ lifts canonically to a commutative diagram of DG-vector spaces*

$$
\begin{array}{ccccc}
 & & L' & \xrightarrow{\;g\;} & K(H \hookrightarrow M) \\
 & \swarrow^{p} & \uparrow{\scriptstyle\alpha} & & \downarrow{\scriptstyle\int_0^1} \\
L & \xleftarrow[\mathrm{Id}_L]{} & L & \xrightarrow{\;i\;} & \dfrac{M}{H}[-1]
\end{array}
$$

such that L' is a DG-Lie algebra, g is a morphism of DG-Lie algebras and p is a surjective quasi-isomorphism of DG-Lie algebras. In particular, g and $\boldsymbol{i}$ induce isomorphic maps in cohomology.

Proof Denote by L' the Thom–Whitney homotopy fibre of the inclusion of DG-Lie algebras $L \subseteq \operatorname{cone}(\mathrm{Id}_L) = L \oplus s^{-1}L$, namely:

$$L' = \{m(t) \in \operatorname{cone}(\mathrm{Id}_L)[t, dt] \mid m(0) = 0,\ m(1) \in L\}.$$

Define then

$$p\colon L' \to L, \qquad p(m(t)) = m(1),$$
$$\alpha\colon L \to L', \qquad \alpha(x) = tx + dt\, s^{-1}x,$$

and it is straightforward to check that α is a morphism of complexes such that $p\alpha = \mathrm{Id}_L$. Notice that p is a quasi-isomorphism since it is the morphism between homotopy fibres induced by the commutative diagram

$$
\begin{array}{ccc}
L & \xrightarrow{\;\mathrm{Id}_L\;} & L \\
\downarrow{\scriptstyle j} & & \downarrow \\
\operatorname{cone}(\mathrm{Id}_L) & \longrightarrow & 0,
\end{array}
$$

where the horizontal maps are quasi-isomorphisms of DG-Lie algebras. Then g is defined as the morphism between homotopy fibres induced by the commutative diagram

$$
\begin{array}{ccc}
L & \xrightarrow{\;l\;} & H \\
\downarrow{\scriptstyle j} & & \downarrow \\
\operatorname{cone}(\mathrm{Id}_L) & \xrightarrow{\;f\;} & M,
\end{array}
\qquad f(s^{-1}x) = \boldsymbol{i}_x,
$$

provided by Lemma 8.8.9. Finally, we have the equality $g\alpha(x) = t\boldsymbol{l}_x + dt\boldsymbol{i}_x$ and therefore $\int_0^1 g\alpha = \boldsymbol{i}$. □

We are now ready to prove Theorem 8.8.3. Denoting by $K = K(H \hookrightarrow M)$ the homotopy fibre of the inclusion $H \subset M$, if $H^1(H) \to H^1(M)$ is injective then Def_K is smooth by Theorem 6.4.5, while if $H^*(H) \to H^*(M)$ is injective then K is homotopy abelian by Corollary 6.1.7.

By Lemma 8.8.10 there exists a morphism of DG-Lie algebras $g\colon L' \to K$ and a quasi-isomorphism of DG-Lie algebras $p\colon L' \to L$ inducing an isomorphism of functors $p\colon \operatorname{Def}_{L'} \xrightarrow{\simeq} \operatorname{Def}_L$. Moreover, since p is the inverse of α in cohomology, it also induces an isomorphism

$$p\colon \ker\big(g\colon H^*(L') \to H^*(K)\big) \xrightarrow{\simeq} \ker(i\colon H^*(L) \to H^*(M/H[-1])). \quad (8.20)$$

Therefore, if $H^1(H) \to H^1(M)$ is injective then Def_K is smooth and the obstructions of $\operatorname{Def}_{L'}$ are contained in the kernel of $g\colon H^2(L') \to H^2(K)$.

If the map $H^*(H) \to H^*(M)$ is injective then K is homotopy abelian and then, if $g\colon H^*(L') \to H^*(K)$ is injective, then also L' is homotopy abelian by Corollary 6.1.3. The proof of Theorem 8.8.3 is now concluded in view of the isomorphism (8.20).

8.9 Exercises

8.9.1 Let $f\colon X \to Y$ be a continuous map of topological spaces and consider the category **C** having as objects the pairs of open subsets (U, V) with $U \subset X$, $V \subset Y$ and such that $f(U) \subset V$; morphisms in **C** are the inclusions of pairs. Given an object (U, V) we shall say that a family of inclusions $\{(U_i, V_i) \to (U, V)\}$ is a covering if $U = \cup U_i$ and $V = \cup V_i$. Prove that this setup defines a Grothendieck topology. (For the definition of Grothendieck topology see e.g. [68].)

8.9.2 Let $f\colon X \to Y$ be a holomorphic map of complex manifolds and assume that the map

$$H^0(X, \Theta_X) \oplus H^0(Y, \Theta_Y) \xrightarrow{f^*-f_*} H^0(X, f^*\Theta_Y)$$

is surjective. Prove that the natural transformation of functors

$$\operatorname{Aut}_Y \times \operatorname{Aut}_X \to \operatorname{Def}_f, \qquad (\alpha, \beta) \mapsto \alpha f \beta^{-1},$$

is surjective on tangent spaces and deduce that the functor Def_f is smooth.

8.9.3 Let Y be a smooth complex projective manifold of dimension 3 such that $H^2(Y, \mathcal{O}_Y) = 0$ and let $X \subset Y$ be a sufficiently ample smooth hypersurface. Prove, using Example 8.2.15, that the inclusion $X \subset Y$ is both stable and costable, cf. [119, Lemma 10.1]. Deduce that X is obstructed if and only if Y is obstructed.

8.9.4 Let $f\colon X \to Y$ be a holomorphic map of complex manifolds. Assume that Def_f is smooth and $H^1(Y, \Theta_Y) \to H^1(X, f^*\Theta_Y)$ is injective. Prove that every deformation of f inducing a trivial deformation of X also induces a trivial deformation of Y.

8.9.5 Show that the result of Example 8.2.19 is generally false if $Y = \mathbb{P}^1$.

8.9.6 Let $f: \mathbb{P}(E) \to Y$ be the projectivization of a holomorphic vector bundle $E \to Y$. Prove that f is costable under deformations.

8.9.7 Let $0 \to E \to F \to G \to 0$ be an exact sequence of vector spaces with G of finite dimension n. Use inner products in order to define, for every integer $0 \le a \le \dim E$, a natural surjective linear map $\bigwedge^{a+n} F \to \bigwedge^a E \otimes \bigwedge^n G$.

8.9.8 (Self-duality formula) Let E be a vector space of finite dimension n and let $x \in \bigwedge^a E^\vee$, $y \in \bigwedge^b E$, $\tau \in \bigwedge^n E^\vee$, with $a \le b \le n$. Prove that

$$x \wedge (y \lrcorner \tau) = (x \lrcorner y) \lrcorner \tau \in \bigwedge^{n+a-b} E^\vee.$$

8.9.9 Prove that for every vector space E of finite dimension n and every integer $a = 0, \ldots, n$, the contraction operator defines a natural isomorphism

$$\bigwedge^a E \xrightarrow{i} \bigwedge^n E \otimes \bigwedge^{n-a} E^\vee, \qquad i(v) = \omega \otimes (v \lrcorner \tau),$$

where $(\omega, \tau) \in \bigwedge^n E \times \bigwedge^n E^\vee$ is any pair satisfying $\omega \lrcorner \tau = 1$. Compare this fact with the canonical isomorphism $\bigwedge^{n-a} E^\vee \simeq (\bigwedge^{n-a} E)^\vee$ and with the nondegeneracy of the wedge product $\bigwedge^{n-a} E \times \bigwedge^a E \xrightarrow{\wedge} \bigwedge^n E$.

References

As far as I know, the validity of Theorem 8.0.1 was first suggested by Drinfeld in [60]; the proof given here is due to Hinich and Schechtman [235]. The analogue result for embedded deformations (Theorem 8.1.2) is taken from [126]. The underlying philosophy is that deformations of a complex manifold are controlled by DG-Lie representatives of complexes computing the cohomology of the tangent sheaf; the same idea is used, with different approaches, also in [215, 222].

Most of the results of Sect. 8.2 are contained in the original papers by Horikawa [117–119], while the long exact sequence (8.8) is taken from [203, Proposition 3.6.9] and [219]. A trascendental approach to Horikawa's theorems via DG-Lie algebras is contained in [124], while the algebraic approach presented here seems to be new. We refer to [219] for a discussion and an extension of Horikawa's theorems to holomorphic maps between singular compact complex spaces. A nice review of Horikawa's theory and some applications to manifolds of Albanese general type is given in [41]. A good reference for (projective and non-projective) K3 surfaces is [16].

The primary reason for the use of $-\overline{\partial}$ as differential in the Kodaira–Spencer DG-Lie algebra is the transformation of the usual Newlander–Nirenberg integrability equation $\overline{\partial}\varphi = \frac{1}{2}[\varphi, \varphi]$ (see e.g. [145, 5.84], [39, 2.5]) into the Maurer–Cartan equation. Other good reasons are Corollary 8.5.5 and the results of Chap. 13. The argument of Example 8.3.3 has been extended to complex nilmanifolds in [229–231].

The results of Sect. 8.4 are largely taken from [183, 184]. The construction of Remark 8.5.6 has been studied extensively by Iacono [124, 125] for proving the equality $\mathrm{Def}_{KS_X} = \mathrm{Def}_X$ as a consequence of Horikawa's theorems.

The fact that the trivial torus bundle $(\mathbb{C}^q/\Gamma) \times \mathbb{P}^n$ has obstructed deformations is a classical result by Kodaira. A study of deformations of general holomorphic principal torus bundles has been given in [42, 43].

There are several different ways in the literature to denote the contraction operator $\lrcorner$ and the inner product $\boldsymbol{i}$; here we have followed the notation of [58, 75, 159, 233].

The proof of Theorem 8.6.1 is taken from [182]. The name "Kodaira principle" is a short version of "ambient cohomology of a Kähler manifold annihilates obstructions", which has been known (at least informally and for obstructions of special kind) and exploited since the pioneering work of Kodaira, cf. [48]. It is based on the idea that certain natural pairings between ambient Hodge classes and obstructions measure nothing more than the obstructions to deforming Hodge structures inside the period domain and therefore must vanish; a rigorous proof based on this idea has been given in [75] and will be reproduced here in Chap. 13.

Theorem 8.7.2 was almost proved in [24]; more precisely, Bloch proved that the semi-regularity map annihilates every obstruction arising from a small extension $0 \to J \to A \to B \to 0$ in $\mathbf{Art}_{\mathbb{C}}$ such that the differential map $d\colon J \to \Omega_{A/\mathbb{C}} \otimes_A B$ is injective. As we shall see in Appendix B, this result is not sufficient for proving the annihilation of all obstructions, although it is sufficient for proving that, if Z is semi-regular, then Hilb^Z_X is smooth. Notice that, according to Theorem 8.1.5, the Theorem 8.7.2 holds for $p = 1$ without the Kähler assumption; this discrepancy has been clarified in [132], where it is proved, in particular, that for every manifold X and every smooth submanifold Z of codimension p the obstructions to embedded deformations are annihilated by the composition of the semi-regularity map and the hypercohomology truncation map $H^{p+1}(X, \Omega^{p-1}_X) \to \mathbb{H}^{2p}(X, \Omega^{<p}_X)$.

The statement of Theorem 8.7.2 is also present in the paper [221] but the proposed proof is affected by some mistakes: the first (ibidem, end of page 810) is that the bracket introduced in the complex $T' \to T$ does not satisfy the Jacobi identity and thus it does not give a DG-Lie structure. It is also easy to see (cf. Exercise 14.6.4) that in the same situation there does not exist any natural Lie bracket and this has been one of the main motivations for the paper [74], where a natural L_∞ structure is introduced (see Sect. 14.5). There is also a second, more subtle, controversial point concerning Lemma 1 on page 819: given two DG-Lie algebras L, M and a morphism of DG-vector spaces $f\colon L \to M$, if f is a morphism in the derived category of DG-Lie algebras then f commutes with brackets in cohomology, but the converse is generally false.

Lemma 8.7.1 and the proof of Theorem 8.7.2 proposed here are taken from [181]. It is easy to define the semi-regularity map for an arbitrary holomorphic map of complex manifolds; the analogue of Theorem 8.7.2 for holomorphic maps of compact Kähler manifolds has been proved in [127].

The name "Cartan homotopy formulas" clearly derives from Cartan's paper [34]. Besides Cartan homotopies, introduced in [75], there exists in the literature several abstract algebraic notions encoding Cartan homotopy formulas; for instance, the notions of an operation of a Lie algebra [98], of a complex over a Lie algebra [58] and of a noncommutative calculus [255] go in this direction.

The abstract Bogomolov–Tian–Todorov theorem was essentially proved in [130] as a preliminary lemma for an algebraic proof of the standard Bogomolov–Tian–Todorov theorem. For a survey of several applications of Theorem 8.8.3 we refer to [129].

Chapter 9
Poisson, Gerstenhaber and Batalin–Vilkovisky Algebras

In several interesting deformation problems the controlling differential graded Lie algebra admits an enhanced algebraic structure, and this is usually useful in the study of its homotopy type and of its associated deformation functor.

In this chapter we consider the case where, in addition to the Lie bracket, there exists a graded commutative product of integral degree that satisfies certain relations called *Poisson identities*; we obtain in this way the classes of (differential) graded Poisson and Gerstenhaber algebras. In this chapter we also introduce Batalin–Vilkovisky algebras, a particular type of graded Poisson algebras where the Lie bracket may be defined in terms of the commutative product and of a differential operator of second order.

We also describe some geometric situations that are related to the above algebraic notions and interesting from the point of view of deformation theory; in particular we shall see that the de Rham complex of either a complex manifold with trivial canonical bundle or a holomorphic Poisson manifold carries a structure of differential Batalin–Vilkovisky algebra that is easily seen to be homotopy abelian under some mild additional assumptions.

9.1 Graded Poisson and Gerstenhaber Algebras

The classical notion of Poisson algebras (Definition 2.2.15) extends naturally to the graded case; we are also interested in a shifted version of Poisson algebra that extends the classical one and includes, as a particular case, the usual notion of Gerstenhaber algebra. Unless otherwise specified, every algebra is considered over a fixed field $\mathbb{K}$ of characteristic 0.

M. Manetti, *Lie Methods in Deformation Theory*, Springer Monographs in Mathematics,
https://doi.org/10.1007/978-981-19-1185-9_9

Definition 9.1.1 A **graded Poisson algebra of degree** n is the data of a graded commutative algebra A equipped with a Lie bracket $[-,-]$ of degree $-n$ that is a biderivation of the product, namely:

1. $[-,-]\colon A^i \times A^j \to A^{i+j-n}$;
2. the pair $(A[n], [-,-])$ is a graded Lie algebra;
3. for any $a \in A$ the adjoint operators $[a,-]$ is a derivation of A; equivalently

$$[a, bc] = [a, b]c + (-1)^{(\overline{a}-n)\overline{b}} b[a, c]$$

for every $a, b, c \in A$, where $\overline{a}$ denotes the degree of a in A and hence $\overline{a} - n$ is the degree of a in $A[n]$).

The Leibniz identities for the biderivation $[-,-]$:

$$[a, bc] = [a, b]c + (-1)^{(\overline{a}-n)\overline{b}} b[a, c], \quad [ab, c] = a[b, c] + (-1)^{(\overline{c}-n)\overline{b}} [a, c]b$$

are called *Poisson identities* for n even and *odd Poisson identities* for n odd.

A morphism of graded Poisson algebras (of the same degree) is a linear map of degree 0 commuting with products and brackets. A derivation of a graded Poisson algebra is a linear map that is a derivation for both the product and the bracket. More explicitly, if A is a Poisson algebra of degree n, a map $f \in \mathrm{Hom}^k_{\mathbb{K}}(A, A)$ is a derivation if and only if for every $a, b \in A$

$$f(ab) = f(a)b + (-1)^{k\overline{a}} af(b), \quad f[a, b] = [f(a), b] + (-1)^{k(\overline{a}-n)}[a, f(b)].$$

Remark 9.1.2 Let $A = \oplus A^i$ be a graded Poisson algebra of degree n. Then $\oplus A^{-i}$ is a graded Poisson algebra of degree $-n$.

In concrete cases, the most relevant graded Poisson algebras are the ones of degree $-1, 0, +1$. Graded Poisson algebras of degree zero are also called **Poisson algebras,** while we shall call the graded Poisson algebras of degree -1 **odd Poisson algebras.**

Definition 9.1.3 A **Gerstenhaber algebra** is a graded Poisson algebra of degree $+1$.

In the literature, some authors define Gerstenhaber algebras as graded Poisson algebras of degree -1, i.e., in our terminology, the odd Poisson algebras. However, according to Remark 9.1.2 there exists a natural bijection between Gerstenhaber and odd Poisson algebras.

The following lemma is particularly useful for checking the validity of the Leibniz and Jacobi identities in graded Poisson algebras.

Lemma 9.1.4 *Let n be an integer and $A = \oplus A^i$ a graded commutative algebra equipped with a bracket $[-,-]$ such that:*

- $[-,-]\colon A^i \times A^j \to A^{i+j-n}$;
- $[b,a] = -(-1)^{(\overline{a}-n)(\overline{b}-n)}[a,b]$;
- $[a,bc] = [a,b]c + (-1)^{(\overline{a}-n)\overline{b}}b[a,c]$.

Let $\mathcal{A} \subset A$ be a set of homogeneous generators of the algebra A. Then:

1. *a derivation $d \in \mathrm{Der}^k_{\mathbb{K}}(A,A)$ is also a derivation with respect to the bracket $[-,-]$ if and only if*

$$d[a,b] = [da,b] + (-1)^{k(\overline{a}-n)}[a,db]$$

for every $a, b \in \mathcal{A}$;

2. *the bracket $[-,-]$ satisfies the Jacobi identity in $A[n]$ if and only if*

$$[[a,b],c] = [a,[b,c]] - (-1)^{(\overline{a}-n)(\overline{b}-n)}[b,[a,c]]$$

for every $a, b, c \in \mathcal{A}$.

Proof For every $a \in A$ write $L_a = [a,-] \in \mathrm{Der}^{\overline{a}-n}_{\mathbb{K}}(A,A)$. Given $a \in \mathcal{A}$, the condition $d[a,b] = [da,b] + (-1)^{k(\overline{a}-n)}[a,db]$ for $b \in \mathcal{A}$ implies that the two derivations $[d, L_a]$ and L_{da} are equal on $\mathcal{A}$ and then $[d, L_a] = L_{da}$.

Thus for every $a \in \mathcal{A}$ and every $b \in A$ we have

$$d[a,b] = [da,b] + (-1)^{k(\overline{a}-n)}[a,db],$$

and

$$\begin{aligned} a &= d[b,a] - (-1)^{k(\overline{b}-n)}[b,da] \\ &= -(-1)^{(\overline{a}-n)(\overline{b}-n)}d[a,b] + (-1)^{k(\overline{b}-n)+(\overline{a}+k-n)(\overline{b}-n)}[da,b] \\ &= -(-1)^{(\overline{a}-n)(\overline{b}-n)}(d[a,b] - [da,b]) \\ &= -(-1)^{(\overline{a}-n)(\overline{b}+k-n)}[a,db] = [db,a] \\ &= L_{db}(a). \end{aligned}$$

Therefore the two derivations $[d, L_b]$ and L_{db} are equal on $\mathcal{A}$ and then $[d, L_b] = L_{db}$ for every $b \in A$, i.e., d is a derivation with respect to the bracket.

The proof of (2) is made by applying (1) twice, first with $d = L_a$, $a \in \mathcal{A}$, and then with $d = L_b$, $b \in A$. □

Example 9.1.5 (Schouten algebras) A particular class of Gerstenhaber algebras are the so called **Schouten algebras**; here the bracket is usually called a **Schouten bracket**. Let A^0 be a commutative $\mathbb{K}$-algebra and let $T \subset \mathrm{Der}_{\mathbb{K}}(A^0, A^0)$ be an A^0-submodule such that $[f,g] = fg - gf \in T$ for every $f, g \in T$. Then consider the exterior algebra $(A, \wedge)$ generated by T:

$$A = \bigoplus_{i \geq 0} A^i, \qquad A^i = \bigwedge\nolimits^i_{A^0} T.$$

Then there exists a unique structure of a Gerstenhaber algebra $(A, \wedge, [-,-])$ such that for every $h \in A^0$, $\xi, \eta \in A^1 = T$ we have:

$$[\xi, h] = -[h, \xi] = \xi(h), \qquad [\xi, \eta] = \xi\eta - \eta\xi.$$

The unicity is clear since $A[1]^{-2} = 0$ implies $[a, b] = 0$ for every $a, b \in A^0$ and A is generated by $A^0 \cup T$. The odd Poisson identity forces the bracket to be equal to the bilinear extension of:

$$[\xi_0 \wedge \cdots \wedge \xi_n, h] = \sum_{i=0}^{n} (-1)^{n-i} \xi_i(h) \xi_0 \wedge \cdots \wedge \widehat{\xi_i} \wedge \cdots \wedge \xi_n,$$

$$[\xi_0 \wedge \cdots \wedge \xi_n, \zeta_0 \wedge \cdots \wedge \zeta_m]$$
$$= \sum_{i=0}^{n}\sum_{j=0}^{m} (-1)^{i+j} [\xi_i, \zeta_j] \wedge \xi_0 \wedge \cdots \wedge \widehat{\xi_i} \wedge \cdots \wedge \xi_n \wedge \zeta_0 \wedge \cdots \wedge \widehat{\zeta_j} \wedge \cdots \wedge \zeta_m,$$

for $h \in A^0$ and $\xi_0, \ldots, \xi_n, \zeta_0, \ldots, \zeta_m \in T$. As regards the existence we need to show that $[-,-]$ satisfies the Jacobi identity

$$[[a, b], c] = [a, [b, c]] - (-1)^{(\overline{a}-1)(\overline{b}-1)}[b, [a, c]].$$

Since the Jacobi identity is invariant under permutation it is not restrictive to assume $\overline{a} \geq \overline{b} \geq \overline{c}$. If $1 \geq \overline{a} \geq \overline{b} \geq \overline{c} \geq 0$ there are 4 possible cases, where the Jacobi identity is satisfied for the reasons described in the table:

$\overline{a}$	$\overline{b}$	$\overline{c}$	Jacobi identity holds because
0	0	0	all terms are $= 0$
1	0	0	all terms are $= 0$
1	1	0	definition of $[-,-]$ on T
1	1	1	Jacobi identity on T

The general case is a consequence of the second item of Lemma 9.1.4.

Definition 9.1.6 Let X be a differentiable manifold with tangent bundle T_X and denote by $\mathcal{T}_X^i$ the sheaf of sections of $\bigwedge^i T_X$. The sheaf of graded algebras

$$\mathcal{T}_X^* = \bigoplus_{i \geq 0} \mathcal{T}_X^i$$

is called the sheaf of **polyvector fields** on X.

It is plain that the construction of Example 9.1.5 applies to polyvector fields; then $\mathcal{T}_X^*$ is a sheaf of Gerstenhaber algebras. The corresponding bracket $[-,-]_{SN}$ is called the **Schouten–Nijenhuis bracket**.

If $z_1, \ldots, z_n$ are local coordinates, then a local set of generators of $\mathcal{T}_X^b$ is given by the polyvector fields $\dfrac{\partial}{\partial z_I} = \dfrac{\partial}{\partial z_{i_1}} \wedge \cdots \wedge \dfrac{\partial}{\partial z_{i_b}}$, where $I = (i_1, \ldots, i_b)$ is a multi-index of cardinality $|I| = b$. Since

$$\left[\frac{\partial}{\partial z_i}, \frac{\partial}{\partial z_j}\right]_{SN} = 0, \qquad \left[\frac{\partial}{\partial z_i}, g\right]_{SN} = \frac{\partial g}{\partial z_i} = dg \lrcorner \frac{\partial}{\partial z_i},$$

the odd Poisson identity implies that for every pair of multi-indices I, $H \subset \{1, 2, \ldots, n\}$ and every pair of functions f, g we have:

$$\begin{aligned}\left[\frac{\partial}{\partial z_I}, g\right]_{SN} &= (-1)^{|I|-1}\left(dg \lrcorner \frac{\partial}{\partial z_I}\right),\\ \left[f\frac{\partial}{\partial z_I}, g\frac{\partial}{\partial z_H}\right]_{SN} &= (-1)^{|I|-1} f\left(dg \lrcorner \frac{\partial}{\partial z_I}\right) \wedge \frac{\partial}{\partial z_H} - g\frac{\partial}{\partial z_I} \wedge \left(df \lrcorner \frac{\partial}{\partial z_H}\right).\end{aligned} \tag{9.1}$$

There exists a version of Cartan homotopy formulas also for polyvector fields; for a differentiable manifold X we shall denote by $\mathcal{A}_X^p$ the sheaf of differential p-forms on X and by $d \colon \mathcal{A}_X^p \to \mathcal{A}_X^{p+1}$ the de Rham differential; recall that every global polyvector field $\eta \in \Gamma(X, \bigwedge^p T_X)$ gives a morphism of sheaves

$$\boldsymbol{i}_\eta \colon \mathcal{A}_X^q \to \mathcal{A}_X^{q-p}, \qquad \boldsymbol{i}_\eta(\alpha) = \eta \lrcorner \alpha.$$

Lemma 9.1.7 (Cartan homotopy formulas) *For every $\eta, \mu \in \mathcal{T}_X^*$ we have*

$$[\boldsymbol{i}_\eta, \boldsymbol{i}_\mu] = 0, \qquad \boldsymbol{i}_{[\eta,\mu]_{SN}} = [[\boldsymbol{i}_\eta, d], \boldsymbol{i}_\mu] = [\boldsymbol{i}_\eta, [d, \boldsymbol{i}_\mu]],$$

where $[-,-]$ denotes the graded commutator bracket in $\mathcal{H}om_{\mathbb{K}}^*(\mathcal{A}_X^*, \mathcal{A}_X^*)$.

Proof Since $\boldsymbol{i}_{\eta\wedge\mu} = \boldsymbol{i}_\eta \boldsymbol{i}_\mu$, the morphism

$$\boldsymbol{i} \colon \mathcal{T}_X^{-*} \to \mathcal{H}om_{\mathbb{K}}^*(\mathcal{A}_X^*, \mathcal{A}_X^*)$$

is a morphism of sheaves of associative algebras an then commutes with graded commutators: $[\boldsymbol{i}_\eta, \boldsymbol{i}_\mu] = \boldsymbol{i}_{\eta\wedge\mu - (-1)^{\overline{\eta}\,\overline{\mu}}\mu\wedge\eta} = 0$. By the Jacobi identity we have

$$[[\boldsymbol{i}_\eta, d], \boldsymbol{i}_\mu] = [\boldsymbol{i}_\eta, [d, \boldsymbol{i}_\mu]] \pm [d, [\boldsymbol{i}_\eta, \boldsymbol{i}_\mu]] = [\boldsymbol{i}_\eta, [d, \boldsymbol{i}_\mu]]$$

and

$$[\boldsymbol{i}_\eta, [d, \boldsymbol{i}_\mu]] = -(-1)^{\overline{\eta}(\overline{\mu}+1)}[[d, \boldsymbol{i}_\mu], \boldsymbol{i}_\eta] = -(-1)^{(\overline{\eta}+1)(\overline{\mu}+1)}[[\boldsymbol{i}_\mu, d], \boldsymbol{i}_\eta].$$

Thus in order to prove the lemma it is sufficient to prove that the bilinear map

$$\Phi(\eta, \mu) = \boldsymbol{i}_{[\eta,\mu]_{SN}} - [\boldsymbol{i}_\eta, [d, \boldsymbol{i}_\mu]]$$

is trivial. Using again the equality $[\boldsymbol{i}_\eta, \boldsymbol{i}_\mu] = 0$ it is easy to check that Φ satisfies the odd Poisson identities:

$$\Phi(\xi, \eta_1 \wedge \eta_2) = \Phi(\xi, \eta_1)\boldsymbol{i}_{\eta_2} + (-1)^{(\overline{\xi}+1)\overline{\eta_1}}\boldsymbol{i}_{\eta_1}\Phi(\xi, \eta_2),$$

$$\Phi(\xi_1 \wedge \xi_2, \eta) = \boldsymbol{i}_{\xi_1}\Phi(\xi_2, \eta) + (-1)^{\overline{\xi_2}(\overline{\eta}+1)}\Phi(\xi_1, \eta)\boldsymbol{i}_{\eta_2}.$$

Therefore it is sufficient to prove that $\Phi(\eta, \mu) = 0$ whenever $\eta, \mu \in \mathcal{T}_X^{\leq 1}$; since $\boldsymbol{i}(\mathcal{T}_X^0) = 0$ the only nontrivial case to check is when $\eta, \mu \in \mathcal{T}_X^1$.

Assume then that η and μ are vector fields; in this case $[\eta, \mu]_{SN} = [\eta, \mu]$ is the usual bracket and $\boldsymbol{i}_\eta, \boldsymbol{i}_\mu, \boldsymbol{i}_{[\eta,\mu]}$ and $[\boldsymbol{i}_\eta, [d, \boldsymbol{i}_\mu]]$ are derivations of $\mathcal{A}_X^*$ of degree -1 vanishing on $\mathcal{A}_X^0$. By the Leibniz rule it is therefore sufficient to check that the operator $\Phi(\eta, \mu)$ annihilates every exact 1-form df. Since $\eta(f) = \boldsymbol{i}_\eta(df)$ and $[\eta, \mu](f) = \eta(\mu(f)) - \mu(\eta(f))$ we have

$$[\boldsymbol{i}_\eta, [d, \boldsymbol{i}_\mu]](df) = \boldsymbol{i}_\eta d\boldsymbol{i}_\mu(df) - \boldsymbol{i}_\mu d\boldsymbol{i}_\eta(df) = \eta(\mu(f)) - \mu(\eta(f)) = \boldsymbol{i}_{[\eta,\mu]}(df).$$

□

Sometimes it is useful to introduce the **Lie derivative** $\mathcal{L}_\eta$ along a polyvector field η by the formula $\mathcal{L}_\eta := [\boldsymbol{i}_\eta, d]$; it should be noted that the Lie derivative differs from the boundary $\boldsymbol{l}_\eta = [d, \boldsymbol{i}_\eta]$ by the sign $(-1)^{\overline{\eta}}$. Then the Cartan homotopy formulas may be written as:

$$[\boldsymbol{i}_\eta, \boldsymbol{i}_\mu] = 0, \qquad \mathcal{L}_\eta = [\boldsymbol{i}_\eta, d], \qquad \boldsymbol{i}_{[\eta,\mu]_{SN}} = [\mathcal{L}_\eta, \boldsymbol{i}_\mu].$$

By the Jacobi identity we have

$$[\mathcal{L}_\eta, d] = [[\boldsymbol{i}_\eta, d], d] = 0, \qquad \mathcal{L}_{[\eta,\mu]_{SN}} = [[\mathcal{L}_\eta, \boldsymbol{i}_\mu], d] = [\mathcal{L}_\eta, \mathcal{L}_\mu],$$

and then $\mathcal{L}\colon \mathcal{T}_X^{\overline{-}*} \to \mathcal{H}om_{\mathbb{K}}^*(\mathcal{A}_X^*, \mathcal{A}_X^*)$ is a morphism of sheaves of DG-Lie algebras.

Remark 9.1.8 For every function $f \in \mathcal{A}_X^0$, every covector field $\omega \in \mathcal{A}_X^1$ and every pair of vector fields $\eta, \mu \in \mathcal{T}_X^1$, the Cartan homotopy formulas give

$$\boldsymbol{i}_\eta(df) = \eta(f), \qquad \boldsymbol{i}_\eta\boldsymbol{i}_\mu(d\omega) = (\boldsymbol{i}_\eta d\boldsymbol{i}_\mu - \boldsymbol{i}_\mu d\boldsymbol{i}_\eta - \boldsymbol{i}_{[\eta,\mu]})\omega. \tag{9.2}$$

Notice that the formulas (9.2) may be used in order to define the de Rham differential $d\colon \mathcal{A}_X^1 \to \mathcal{A}_X^2$ without using local coordinates.

Example 9.1.9 (Gerstenhaber) For any associative $\mathbb{K}$-algebra A with multiplication map $\mu\colon A \times A \to A$ we introduced in Example 5.8.4 the Hochshild cochain complex

$$C^*(A,A)=\bigoplus_{n\geq 0} C^n(A,A), \qquad C^n(A,A)=\operatorname{Hom}_{\mathbb{K}}(A^{\otimes n},A),$$

where the differential is $d=-[-,\mu]$ and

$$[-,-]\colon C^n(A,A)\times C^m(A,A)\to C^{n+m-1}(A,A)$$

is the Gerstenhaber bracket (Example 5.8.2). The cohomology $H^*(A,A)$ of the Hochshild cochain complex is called the **Hochshild cohomology of** A.

The cup product $f\cup g\in C^{n+m}(A,A)$ of $f\in C^n(A,A)$ and $g\in C^m(A,A)$ is defined by the formula:

$$f\cup g(a_1,\dots,a_{n+m})=(-1)^{nm}f(a_1,\dots,a_n)g(a_{n+1},\dots,a_{n+m}).$$

After a nontrivial computation, it is proved in [87] that the cup product is graded commutative in cohomology and the Gerstenhaber bracket induces on the Hochshild cohomology a structure of a Gerstenhaber algebra.

Definition 9.1.10 A **differential graded Poisson algebra** is the data of a graded Poisson algebra and a differential of the underlying graded vector space that is a derivation of degree $+1$ for the commutative product and for the Lie bracket.

The notion of differential Gerstenhaber algebras is implicitly contained in Definition 9.1.10. Notice that we have two natural forgetful functors:

$$\begin{array}{ccc} \text{Differential Gerstenhaber algebras} \to \mathbf{DGA} & \qquad & (A,d,[-,-]) \mapsto (A,d) \\ \downarrow & & \downarrow \\ \mathbf{DGLA}, & & (A[1],-d,[-,-]). \end{array}$$

Example 9.1.11 Given any Gerstenhaber algebra $(A,[-,-])$ and an element $\pi\in A^2$ such that $[\pi,\pi]=0\in A^3$ we have that $d=[\pi,-]$ is a differential giving a structure of differential Gerstenhaber algebra.

9.2 The Koszul–Tian–Todorov Lemma

The Gerstenhaber algebra of polyvector fields makes sense also in the holomorphic setting. Given a complex manifold X with holomorphic tangent sheaf Θ_X, the Schouten–Nijenhuis bracket induces on $\bigwedge^*\Theta_X=\bigoplus_{i\geq 0}\bigwedge^i\Theta_X$ a structure of a sheaf of Gerstenhaber algebras. By (9.1), in local holomorphic coordinates $z_1,\dots,z_n$, for every pair of multiindices I, H we have:

$$\left[f\frac{\partial}{\partial z_I}, g\frac{\partial}{\partial z_H}\right]_{SN} = (-1)^{|I|-1} f\left(\partial g \lrcorner \frac{\partial}{\partial z_I}\right) \wedge \frac{\partial}{\partial z_H} - g\frac{\partial}{\partial z_I} \wedge \left(\partial f \lrcorner \frac{\partial}{\partial z_H}\right). \tag{9.3}$$

We can immediately see that the Schouten–Nijenhuis bracket is $\overline{\mathcal{O}_X}$-bilinear and then, as in the case of the Kodaira–Spencer algebra, there exists a natural extension to the Dolbeault resolution, namely:

$$[-,-]_{SN}\colon \mathcal{A}_X^{0,p}(\bigwedge^i \Theta_X) \times \mathcal{A}_X^{0,q}(\bigwedge^j \Theta_X) \to \mathcal{A}_X^{0,p+q}(\bigwedge^{i+j-1} \Theta_X),$$

$$\left[f d\overline{z}_J\frac{\partial}{\partial z_I}, g d\overline{z}_K\frac{\partial}{\partial z_H}\right]_{SN} = (-1)^{|K|(|I|-1)} d\overline{z}_J \wedge d\overline{z}_K \left[f\frac{\partial}{\partial z_I}, g\frac{\partial}{\partial z_H}\right]_{SN}. \tag{9.4}$$

Since the Dolbeault differential is a derivation for the Schouten–Nijenhuis bracket we have that $(\mathcal{A}_X^{0,*}(\bigwedge^* \Theta_X), \overline{\partial}, [-,-]_{SN})$ is a sheaf of differential Gerstenhaber algebras.

Let n be the dimension of X, then for every $a = 0,\ldots,n$ the inner product (Sect. 8.5) gives an isomorphism of locally free sheaves

$$i\colon \bigwedge^a \Theta_X \to \mathcal{H}om_{\mathcal{O}_X}(\Omega_X^n, \Omega_X^{n-a}) = \Omega_X^{n-a}(K_X^\vee),$$

where $K_X^\vee$ is the dual of the canonical bundle $K_X = \Omega_X^n$. The above isomorphism of locally free sheaves extends to their Dolbeault's resolutions

$$i\colon \left(\mathcal{A}_X^{0,*}(\bigwedge^a \Theta_X), \overline{\partial}\right) \to \left(\mathcal{A}_X^{n-a,*}(K_X^\vee), \overline{\partial}\right).$$

In local holomorphic coordinates, if ω is a local frame of K_X and τ a local frame of $K_X^\vee$ such that $\tau \lrcorner \omega = 1$, then

$$i\left(d\overline{z}_J\frac{\partial}{\partial z_I}\right) = d\overline{z}_J \wedge i\left(\frac{\partial}{\partial z_I}\right) = d\overline{z}_J \wedge \left(\frac{\partial}{\partial z_I} \lrcorner \omega\right) \otimes \tau.$$

Given a Hermitian metric h on the line bundle $K_X^\vee$ we denote by $D = D' + \overline{\partial}$ the unique Hermitian connection on $K_X^\vee$ compatible with the complex structure. It is well known that the (1, 0) component $D'\colon \mathcal{A}_X^{a,b}(K_X^\vee) \to \mathcal{A}_X^{a+1,b}(K_X^\vee)$ of the connection is defined locally as

$$D'(\phi \otimes \tau) = (\theta \wedge \phi + \partial\phi) \otimes \tau, \qquad \phi \in \mathcal{A}_X^{a,b},$$

where $\theta = \partial \log(|\tau|^2) = \partial \log(h(\tau,\tau))$ is the connection form of the frame τ; observe that for every invertible complex number c, the frames τ and $c\tau$ have the same connection form.

Consider now the $\mathbb{C}$-linear endomorphism Δ of $\mathcal{A}_X^{0,*}(\bigwedge^* \Theta_X)$ defined as:

$$\Delta\colon \mathcal{A}_X^{0,b}\left(\bigwedge^a \Theta_X\right) \to \mathcal{A}_X^{0,b}\left(\bigwedge^{a-1} \Theta_X\right), \qquad \Delta = -\,\boldsymbol{i}^{-1} \circ D' \circ \boldsymbol{i}.$$

Equivalently, Δ is uniquely determined by the equation:

$$\boldsymbol{i}_{\Delta(\phi)}(\eta) = -D'(\boldsymbol{i}_\phi(\eta)), \qquad \eta \in \mathcal{A}_X^{n,0}, \quad \phi \in \mathcal{A}_X^{0,*}(\bigwedge^* \Theta_X). \tag{9.5}$$

Since $(D')^2 = 0$ and $[D', \overline{\partial}] = D'\overline{\partial} + \overline{\partial} D'$ is the curvature of the metric h, it is plain that $\Delta^2 = 0$ and $[\overline{\partial}, \Delta] = 0$ if and only if the metric h is flat.

Lemma 9.2.1 *In the above situation and local holomorphic coordinates $z_1, \ldots, z_n$, we have*

$$\Delta\left(f d\overline{z}_J \frac{\partial}{\partial z_I}\right) = -(-1)^{|J|} d\overline{z}_J \left((\theta f + \partial f) \lrcorner \frac{\partial}{\partial z_I}\right), \qquad f \in \mathcal{A}^{0,0},$$

where θ is the connection form of the frame $\tau = \dfrac{\partial}{\partial z_1} \wedge \cdots \wedge \dfrac{\partial}{\partial z_n}$ and the right hand side is considered equal to 0 when $I = \emptyset$.

Proof Since $D'(d\overline{z}_J \phi \otimes \tau) = (-1)^{|J|} d\overline{z}_J D'(\phi \otimes \tau)$ it is not restrictive to assume $J = \emptyset$ and $I \neq \emptyset$. Considering the holomorphic n-form $\omega = dz_n \wedge \cdots \wedge dz_1$, we have $\tau \lrcorner \omega = \omega \lrcorner \tau = 1$ and then Equation (9.5) for the polyvector field $f \dfrac{\partial}{\partial z_I}$ and the form $\eta = \omega$ gives

$$-\Delta\left(f \frac{\partial}{\partial z_I}\right) \lrcorner \omega = \theta \wedge \left(f \frac{\partial}{\partial z_I} \lrcorner \omega\right) + \partial\left(f \frac{\partial}{\partial z_I} \lrcorner \omega\right) = (\theta f + \partial f) \wedge \left(\frac{\partial}{\partial z_I} \lrcorner \omega\right).$$

According to Lemma 8.5.3 we have

$$\begin{aligned} -\Delta\left(f \frac{\partial}{\partial z_I}\right) &= -\left(\Delta\left(f \frac{\partial}{\partial z_I}\right) \lrcorner \omega\right) \lrcorner \tau = \left((\theta f + \partial f) \wedge \left(\frac{\partial}{\partial z_I} \lrcorner \omega\right)\right) \lrcorner \tau \\ &= (\theta f + \partial f) \lrcorner \left(\left(\frac{\partial}{\partial z_I} \lrcorner \omega\right) \lrcorner \tau\right) = (\theta f + \partial f) \lrcorner \frac{\partial}{\partial z_I}. \end{aligned}$$

□

Theorem 9.2.2 (Koszul–Tian–Todorov lemma) *In the notation above, for every choice of the Hermitian connection on the anticanonical bundle, and for every $\alpha \in \mathcal{A}_X^{0,a}(\bigwedge^p \Theta_X)$, $\beta \in \mathcal{A}_X^{0,b}(\bigwedge^q \Theta_X)$ we have:*

$$[\alpha, \beta]_{SN} = (-1)^{a+p} \left(\Delta(\alpha \wedge \beta) - \Delta(\alpha) \wedge \beta\right) - \alpha \wedge \Delta(\beta).$$

Proof The proof is a straightforward computation in local coordinates and we give only a sketch. We first notice that it is not restrictive to assume $a = b = 0$ and

$\alpha = f\dfrac{\partial}{\partial z_I}$, $\beta = g\dfrac{\partial}{\partial z_J}$ with $|I| = p$, $|J| = q$. Thus, by (9.3) we have

$$\begin{aligned}[\alpha, \beta]_{SN} &= \left[f\frac{\partial}{\partial z_I}, g\frac{\partial}{\partial z_H}\right]_{SN} \\ &= (-1)^{p-1} f\left(\partial g \lrcorner \frac{\partial}{\partial z_I}\right) \wedge \frac{\partial}{\partial z_J} - g\frac{\partial}{\partial z_I} \wedge \left(\partial f \lrcorner \frac{\partial}{\partial z_J}\right)\end{aligned}$$

and by Lemma 9.2.1

$$\Delta(\alpha) = \Delta\left(f\frac{\partial}{\partial z_I}\right) = -(\theta f + \partial f)\lrcorner \frac{\partial}{\partial z_I},$$

$$\Delta(\beta) = \Delta\left(g\frac{\partial}{\partial z_J}\right) = -(\theta g + \partial g)\lrcorner \frac{\partial}{\partial z_J},$$

$$\Delta(\alpha \wedge \beta) = \Delta\left(fg\frac{\partial}{\partial z_I} \wedge \frac{\partial}{\partial z_J}\right) = -(\theta fg + (\partial f)g + f(\partial g))\lrcorner \left(\frac{\partial}{\partial z_I} \wedge \frac{\partial}{\partial z_J}\right).$$

Now, in order to conclude the proof it is sufficient to observe that the operator $(\theta fg + (\partial f)g + f(\partial g))\lrcorner$ is a derivation of degree -1. □

We are now ready to give another proof of the BTT theorem (Corollary 8.6.3).

Corollary 9.2.3 (Bogomolov–Tian–Todorov) *Let X be a compact Kähler manifold with trivial canonical bundle. Then KS_X is quasi-isomorphic to an abelian DG-Lie algebra and therefore* Def_X *is smooth.*

Proof Since X has a nowhere vanishing holomorphic n-form ω ($n = \dim X$) we can set on $K_M^\vee$ the trivial Hermitian metric induced by the isomorphism $\omega\colon K_M^\vee \to \mathcal{O}_M$. In this case $D' = \partial$ and then

$$\boldsymbol{i}\colon (A_X^{0,*}(\wedge^*\Theta_X), -\Delta, \overline{\partial}) \to (A_X^{*,*}, \partial, \overline{\partial})$$

is an isomorphism of double complexes. The $\partial\overline{\partial}$-lemma implies that the image of Δ is an acyclic subcomplex of $(A_X^{0,*}(\wedge^*\Theta_X), \overline{\partial})$ and then the inclusion

$$L = (\ker \Delta \cap A_X^{0,*}(\Theta_X)) \to A_X^{0,*}(\Theta_X) = KS_X$$

is a quasi-isomorphism of complexes. On the other hand, the Koszul-Tian-Todorov lemma implies that L is a differential graded Lie subalgebra of KS_X. To conclude the proof we observe that the projection

$$L \to M = \frac{\ker \Delta \cap A_X^{0,*}(\Theta_X))}{\Delta(A_X^{0,*}(\Theta_X^{\wedge 2}))}$$

is a quasi-isomorphism of complexes and, again by the Koszul-Tian-Todorov lemma, it is a morphism of DG-Lie algebras whenever M is equipped with the trivial bracket. □

9.3 Koszul Braces and Differential Operators

The notion of differential operators on a commutative algebra (Sect. 2.3) extends naturally to the graded case; given a graded commutative algebra A over a field $\mathbb{K}$ we may consider A as an abelian graded Lie subalgebra of $\operatorname{Hom}^*_{\mathbb{K}}(A, A)$, where every $a \in A$ is identified with the operator of left multiplication:

$$a\colon A \to A, \qquad a(b) = ab\,.$$

For every integer k we denote by

$$\operatorname{Diff}^k_{A/\mathbb{K}}(A, A) \subset \operatorname{Hom}^*_{\mathbb{K}}(A, A)$$

the graded subspace of differential operators of order $\le k$; it is defined recursively by setting $\operatorname{Diff}^k_{A/\mathbb{K}}(A, A) = 0$ for $k < 0$ and

$$\operatorname{Diff}^k_{A/\mathbb{K}}(A, A) = \{f \in \operatorname{Hom}^*_{\mathbb{K}}(A, A) \mid [f, a] \in \operatorname{Diff}^{k-1}_{A/\mathbb{K}}(A, A), \quad \forall\, a \in A\}$$

for $k \ge 0$. The same computation of Sect. 2.3 shows that for $f \in \operatorname{Diff}^n_{A/\mathbb{K}}(A, A)$ and $g \in \operatorname{Diff}^m_{A/\mathbb{K}}(A, A)$ we have:

$$fg \in \operatorname{Diff}^{n+m}_{A/\mathbb{K}}(A, A) \quad \text{and} \quad [f, g] \in \operatorname{Diff}^{n+m-1}_{A/\mathbb{K}}(A, A). \tag{9.6}$$

In particular, $\operatorname{Diff}^1_{A/\mathbb{K}}(A, A)$ and $\operatorname{Diff}_{A/\mathbb{K}}(A, A) = \bigcup_k \operatorname{Diff}^k_{A/\mathbb{K}}(A, A)$ are graded Lie subalgebras of $\operatorname{Hom}^*_{\mathbb{K}}(A, A)$.

For any linear operator $f \in \operatorname{Hom}^*_{\mathbb{K}}(A, A)$ we define an infinite sequence of multilinear maps

$$f_n\colon \underbrace{A \times \cdots \times A}_{n} \to \operatorname{Hom}^*_{\mathbb{K}}(A, A), \qquad n \ge 0,$$

by the recursive formula

$$f_0 = f, \qquad f_n(a_1, \ldots, a_n) = [f_{n-1}(a_1, \ldots, a_{n-1}), a_n], \qquad n > 0.$$

In particular, $f_1(a) = [f, a]$ and we may give the following equivalent definition of differential operators: for a linear operator $f \in \operatorname{Hom}^*_{\mathbb{K}}(A, A)$ we have $f \in \operatorname{Diff}^k_{A/\mathbb{K}}(A, A)$ if and only if $f_1(a) = [f, a] \in \operatorname{Diff}^{k-1}_{A/\mathbb{K}}(A, A)$ for every a. By induction on k we have therefore that $f \in \operatorname{Diff}^k_{A/\mathbb{K}}(A, A)$ if and only if $f_{k+1} = 0$.

It is easy to see that f_n is graded symmetric for every $n \geq 2$; by induction on n it is sufficient to prove that

$$f_{n+2}(a_1, \ldots, a_n, b, c) = (-1)^{\overline{b}\,\overline{c}} f_{n+2}(a_1, \ldots, a_n, c, b).$$

This follows from the graded commutativity of A, since

$$\begin{aligned} f_{n+2}(a_1, \ldots, a_n, b, c) &= [[f_n(a_1, \ldots, a_n), b], c] \\ &= [f_n(a_1, \ldots, a_n), [b, c]] + (-1)^{\overline{b}\,\overline{c}}[[f_n(a_1, \ldots, a_n), c], b] \\ &= (-1)^{\overline{b}\,\overline{c}} f_{n+2}(a_1, \ldots, a_n, c, b). \end{aligned}$$

Definition 9.3.1 (Koszul braces) In the notation above, assume that the algebra A has a unit $1 \in A^0$. For every $f \in \operatorname{Hom}^*_{\mathbb{K}}(A, A)$ and every $n \geq 0$ the nth **Koszul brace** of f is the graded symmetric multilinear map

$$\Phi^n_f \colon A \times \cdots \times A \to A, \qquad \Phi^n_f(a_1, \ldots, a_n) = f_n(a_1, \ldots, a_n)(1).$$

For instance, $\Phi^0_f = f(1)$,

$$\Phi^1_f(a) = f(a) - f(1)a, \quad \Phi^2_f(a, b) = f(ab) - f(a)b - (-1)^{\overline{a}\,\overline{f}} af(b) + f(1)ab.$$

Moreover, for every $n > 0$ we have

$$\begin{aligned} \Phi^{n+1}_f(a_1, \ldots, a_{n-1}, b, c) &= \Phi^n_f(a_1, \ldots, a_{n-1}, bc) - \Phi^n_f(a_1, \ldots, a_{n-1}, b)c \\ &\quad - (-1)^{\overline{f(a_1,\ldots,a_{n-1})}\,\overline{b}}\, b\, \Phi^n_f(a_1, \ldots, a_{n-1}, c) \end{aligned}$$

and then $\Phi^{n+1}_f = 0$ if and only if Φ^n_f is a multiderivation of A, that is a derivation in each variable.

Lemma 9.3.2 *Let A be a unitary graded commutative algebra. For a linear operator $f \in \operatorname{Hom}^*_{\mathbb{K}}(A, A)$ and an integer $k \geq 0$, the following conditions are equivalent:*

1. $f \in \operatorname{Diff}^k_{A/\mathbb{K}}(A, A)$,
2. $\Phi^n_f = 0$ *for every* $n > k$,
3. $\Phi^{k+1}_f = 0$.

If $k > 0$ then the above conditions are also equivalent to:

4. Φ^k_f *is a multiderivation of A.*

Proof The only nontrivial implication is 9.3.2 ⇒ 9.3.2. This is trivially true for $k = 0$, and the general case is proved by induction on k by using the formula

$$\Phi^{k+1}_f(a, a_1, \ldots, a_k) = \Phi^k_{f_1(a)}(a_1, \ldots, a_k).$$

□

Example 9.3.3 In the notation of Sect. 9.2, the operator Δ is a differential operator of second order of degree -1; in fact by the Koszul–Tian–Todorov lemma (Theorem 9.2.2) we have

$$\Phi^2_\Delta(\alpha, \beta) = (-1)^{\overline{\alpha}}[\alpha, \beta]_{SN}$$

and then Φ^2_Δ is a biderivation in view of the odd Poisson identities.

Corollary 9.3.4 *Let A be a unitary graded commutative algebra. For a fixed integer $k \geq 2$ and an endomorphism $f \in \mathrm{Hom}^*_{\mathbb{K}}(A, A)$ the following conditions are equivalent:*

1. (the $(2^k - 1)$- terms relation)

$$\Phi^k_f(a_1, \dots, a_k) - (-1)^k f(1)a_1 \cdots a_k = 0, \qquad \textit{for every} \quad a_1, \dots, a_k \in A;$$

2. $f \in \mathrm{Diff}^{k-1}_{A/\mathbb{K}}(A, A)$ *and* $f(1) = 0$.

Proof The implication *2.* $\Rightarrow$ *1.* follows from Lemma 9.3.2.

Conversely, assume that $\Phi^k_f(a_1, \dots, a_k) = (-1)^k f(1)a_1 \cdots a_k$ for every $a_1, \dots, a_k$. Then $f(1) = (-1)^k \Phi^k_f(1, 1, \dots, 1) = 0$ and therefore $\Phi^k_f = 0$. By Lemma 9.3.2 we have $f \in \mathrm{Diff}^{k-1}_{A/\mathbb{K}}(A, A)$. □

Since $\Phi^2_f(a, b) - f(1)ab = f(ab) - f(a)b - (-1)^{\overline{a}\,\overline{f}} af(b)$, for $k = 2$ the above corollary says that f is a derivation if and only if $f \in \mathrm{Diff}^1_{A/\mathbb{K}}(A, A)$ and $f(1) = 0$. Since

$$\begin{aligned}\Phi^3_f(a, b, c) + f(1)abc = f(abc) &+ f(a)bc + (-1)^{\overline{a}\,\overline{b}} f(b)ac + (-1)^{\overline{c}(\overline{a}+\overline{b})} f(c)ab \\ &- f(ab)c - (-1)^{\overline{a}(\overline{b}+\overline{c})} f(bc)a - (-1)^{\overline{b}\overline{c}} f(ac)b,\end{aligned}$$

for $k = 3$ the above corollary takes the more explicit form:

Corollary 9.3.5 *Let A be a unitary graded commutative algebra. For an endomorphism $f \in \mathrm{Hom}^*_{\mathbb{K}}(A, A)$ the following conditions are equivalent:*

1. (the seven-terms relation) *for every a, b, c we have*

$$\begin{aligned}f(abc) = f(ab)c &+ (-1)^{\overline{a}(\overline{b}+\overline{c})} f(bc)a + (-1)^{\overline{b}\overline{c}} f(ac)b \\ &- f(a)bc - (-1)^{\overline{a}\,\overline{b}} f(b)ac - (-1)^{\overline{c}(\overline{a}+\overline{b})} f(c)ab;\end{aligned}$$

2. $f \in \mathrm{Diff}^2_{A/\mathbb{K}}(A, A)$ *and* $f(1) = 0$.

Remark 9.3.6 One of the reasons for the interest concerning Koszul braces in mathematics and physics is given by the following simple observation: in the above setup, assume either that $a \in A^0$ is a nilpotent element or ignore any convergence issue. Then

$$f(e^a) = 0 \iff \sum_{n\geq 0} \frac{1}{n!}\Phi^n_f(a, \dots, a) = 0.$$

In fact, identifying every $a \in A$ with the corresponding operator of left multiplication $A \to A$, for every $n \geq 0$ we have $f_n(a, \dots, a) = (-\operatorname{ad} a)^n f$ and therefore

$$e^{-a} f e^a = e^{-\operatorname{ad} a}(f) = \sum_{n\geq 0} \frac{1}{n!} f_n(a, \dots, a) \in \operatorname{Hom}^*_{\mathbb{K}}(A, A),$$

$$e^{-a} f(e^a) = e^{-a} f e^a(1) = \sum_{n\geq 0} \frac{1}{n!}\Phi^n_f(a, \dots, a) \in A.$$

The clear advantage is that, whenever f is a differential operator of order $\leq k$, the equation $f(e^a) = 0$ is equivalent to a polynomial equation in a of degree $\leq k$.

9.4 Batalin–Vilkovisky Algebras

The following definition is motivated, among the other things, by the Koszul–Tian–Todorov lemma.

Definition 9.4.1 Let A be a graded commutative algebra over a field $\mathbb{K}$, let n be an odd integer and let $[-, -]$ be a skew-symmetric bracket on the shifted graded vector space $A[n]$. A linear operator $\Delta \in \operatorname{Hom}^{-n}_{\mathbb{K}}(A, A)$ is called a generator for the bracket $[-, -]$ if for every pair of homogeneous elements $a \in A^{\overline{a}}$, $b \in A^{\overline{b}}$, we have:

$$[a, b] = (-1)^{\overline{a}}(\Delta(ab) - \Delta(a)b) - a\Delta(b).$$

The generator Δ is said to be exact if $\Delta^2 = 0$.

In other words, the bracket measures the distance for the generator to be a derivation. If a generator exists, then every other generator is obtained by adding a derivation of degree $-n$.

As an example, by the Koszul–Tian–Todorov lemma, for a complex manifold X the Gerstenhaber algebra $A^{0,*}_X(\bigwedge^* \Theta_X)$ admits an exact generator.

We are interested in understanding when a linear operator generates a bracket inducing a graded Poisson structure. The following lemma gives an answer in the unitary case.

Lemma 9.4.2 *Let A be a unitary graded commutative algebra, let n be an odd integer and $\Delta \in \operatorname{Hom}^{-n}_{\mathbb{K}}(A, A)$. The following conditions are equivalent:*

1. *A is a graded Poisson algebra of degree n when equipped with the bracket*

$$[a, b] = (-1)^{\overline{a}}(\Delta(ab) - \Delta(a)b) - a\Delta(b);$$

2. $\Delta(1) = 0$ *and* $\Delta, \Delta^2 \in \mathrm{Diff}^2_{A/\mathbb{K}}(A, A)$.

Proof We first show that $[-, -]$ is a biderivation, or equivalently that $[-, -]$ satisfies the odd Poisson identity, if and only if $\Delta(1) = 0$ and $\Delta \in \mathrm{Diff}^2_{A/\mathbb{K}}(A, A)$. By definition we have $\Delta(1) = 0$ if and only if

$$[a, b] = (-1)^{\overline{a}}\Phi^2_\Delta(a, b),$$

where Φ^2_Δ is the bilinear Koszul brace of Δ. If $[-, -]$ is a biderivation, then $0 = [1, 1] = -\Delta(1)$ and therefore also Φ^2_Δ is a biderivation. Conversely, if $\Delta(1) = 0$ and Φ^2_Δ is a biderivation then also $[-, -]$ is a biderivation. The conclusion now follows from Lemma 9.3.2.

Assuming $\Delta(1) = 0$ and $\Delta \in \mathrm{Diff}^2_{A/\mathbb{K}}(A, A)$, a tedious but completely straightforward computation (cf. Eq. (10.20)) gives

$$[[a, b], c] + (-1)^{(\overline{a}-n)(\overline{b}+\overline{c})}[[b, c], a] + (-1)^{(\overline{c}-n)(\overline{a}+\overline{b})}[[c, a], b] = (-1)^{\overline{b}-n}\Phi^3_{\Delta^2}(a, b, c)$$

and then $[-, -]$ satisfies the Jacobi identity in $A[n]$ if and only if $\Phi^3_{\Delta^2} = 0$, or equivalently if and only if $\Delta^2 \in \mathrm{Diff}^2_{A/\mathbb{K}}(A, A)$. □

It is instructive to give also a (simpler) proof of the Lemma 9.4.2 above under the additional assumption $\Delta^2 = 0$.

Lemma 9.4.3 *Let A be a unitary graded commutative algebra, let n be an odd integer and let* $\Delta \in \mathrm{Hom}^{-n}_{\mathbb{K}}(A, A)$ *be a differential operator of order* ≤ 2 *such that* $\Delta(1) = 0$ *and* $\Delta^2 = 0$. *Then the pair* $(A, [-, -])$ *is a graded Poisson algebra of degree n, where*

$$[a, b] = (-1)^{\overline{a}}(\Delta(ab) - \Delta(a)b) - a\Delta(b).$$

In addition, Δ *is a derivation of the graded Lie algebra* $(A[n], [-, -])$.

Proof We first prove that Δ is a derivation of the (magmatic) algebra $(A[n], [-, -])$. Since $\Delta^2 = 0$ we have

$$\Delta[a, b] = -(-1)^{\overline{a}}\Delta(\Delta(a)b) - \Delta(a\Delta(b)),$$
$$[\Delta(a), b] = (-1)^{\overline{\Delta(a)}}\Delta(\Delta(a)b) - \Delta(a)\Delta(b),$$
$$[a, \Delta(b)] = (-1)^{\overline{a}}(\Delta(a\Delta(b)) - \Delta(a)\Delta(b)).$$

Since $\overline{\Delta(a)} = \overline{a} - n$ and n is odd, the above expressions give the Leibniz condition:

$$\Delta[a, b] = [\Delta(a), b] + (-1)^{\overline{a}-n}[a, \Delta(b)].$$

Then we prove the Jacobi identity as a consequence of the Leibniz identity and the vanishing of Δ^2. In fact, since Δ is a differential operator of order ≤ 2, the bracket $[-,-]$ is a biderivation of A, it satisfies the odd Poisson identity and then

$$\begin{aligned}[a,\Delta(bc)] &= (-1)^{\overline{a}}(\Delta[a,bc]-[\Delta(a),bc])\\ &= (-1)^{\overline{a}}(\Delta([a,b]c)+(-1)^{(\overline{a}-n)\overline{b}}\Delta(b[a,c])-[\Delta(a),b]c-(-1)^{\overline{a}\,\overline{b}}b[\Delta(a),c]).\end{aligned}$$

Since

$$\begin{aligned}[a,[b,c]] &= -[a,b\Delta(c)]+(-1)^{\overline{b}}([a,\Delta(bc))]-[a,\Delta(b)c]),\\ [[a,b],c] &= -[a,b]\Delta(c)-(-1)^{\overline{a}+\overline{b}}(\Delta([a,b]c)-\Delta([a,b])c),\\ [b,[a,c]] &= -b\Delta([a,c])+(-1)^{\overline{b}}(\Delta(b[a,c])-\Delta(b)[a,c]),\end{aligned}$$

an easy computation gives

$$[a,[b,c]] = [[a,b],c]+(-1)^{(\overline{a}-n)(\overline{b}-n)}[b,[a,c]].$$

□

Definition 9.4.4 Let n be an odd integer; a graded Batalin–Vilkovisky algebra of degree n is a graded Poisson algebra of degree n equipped with an exact generator. A **Batalin–Vilkovisky algebra** is a graded Batalin–Vilkovisky algebra of degree $+1$ (with an exact generator of degree -1).

Thus, in the unitary case, a graded Batalin–Vilkovisky algebra is just a graded commutative unitary algebra equipped with a second order differential operator Δ of degree -1 such that $\Delta(1)=0$ and $\Delta^2=0$.

Example 9.4.5 In the polynomial algebra $\mathbb{K}[z_1,\ldots,z_n]$, where each z_i has integral degree, we denote by $\dfrac{\partial}{\partial z_i}$ the unique derivation of A such that $\dfrac{\partial z_i}{\partial z_i}=1$ and $\dfrac{\partial z_j}{\partial z_i}=0$ for $i\neq j$; clearly the degree of $\dfrac{\partial}{\partial z_i}$ is the opposite of the degree of z_i.

Consider now the algebra $A=\mathbb{K}[x_1,\ldots,x_n,y_1,\ldots,y_n]$, where $\overline{x_i}=0$ and $\overline{y_i}=1$. The second order differential operator

$$\Delta = -\sum_{i=1}^{n}\frac{\partial}{\partial x_i}\frac{\partial}{\partial y_i}$$

has degree -1 and satisfies the conditions $\Delta(1)=0$ and $\Delta^2=0$. In fact

$$\Delta^2=\sum_{i,j=1}^{n}\frac{\partial}{\partial x_i}\frac{\partial}{\partial y_i}\frac{\partial}{\partial x_j}\frac{\partial}{\partial y_j}=\sum_{i,j=1}^{n}\frac{\partial}{\partial x_i}\frac{\partial}{\partial x_j}\frac{\partial}{\partial y_i}\frac{\partial}{\partial y_j}$$

and it is sufficient to observe that

$$\frac{\partial}{\partial y_i}\frac{\partial}{\partial y_i}=0, \qquad \frac{\partial}{\partial y_i}\frac{\partial}{\partial y_j}=-\frac{\partial}{\partial y_j}\frac{\partial}{\partial y_i}, \qquad \frac{\partial}{\partial x_i}\frac{\partial}{\partial x_j}=\frac{\partial}{\partial x_j}\frac{\partial}{\partial x_i}.$$

Thus the pair (A, Δ) is a Batalin–Vilkovisky algebra. By the odd Poisson identities, the bracket generated by Δ is uniquely determined by the relations

$$[x_i, x_j] = 0, \qquad [y_i, x_j] = \delta_{ij}, \qquad [y_i, y_j] = 0$$

and then coincides with the Schouten bracket induced by the identification of y_i with the derivation $\dfrac{\partial}{\partial x_i}\colon A^0 \to A^0$. If we identify A with the polynomial de Rham algebra of $\mathbb{K}^n$, where y_i correspond to dx_i, then $-\Delta(df)$ is the classical Laplacian of $f \in \mathbb{K}[x_1, \ldots, x_n]$.

Definition 9.4.6 Let n be an odd integer; a differential graded Batalin–Vilkovisky algebra of degree n is triple (A, d, Δ), where (A, d) is a DG-algebra, (A, Δ) is a graded Poisson algebra of degree n with an exact generator Δ and $[d, \Delta] = 0$. A **differential Batalin–Vilkovisky algebra** is a differential graded Batalin–Vilkovisky algebra of degree $+1$ (with an exact generator of degree -1).

9.5 Poisson and Symplectic Manifolds

In this section we introduce Poisson manifolds, which are the geometric counterpart of Poisson algebras. Among the possible equivalent definitions, the most convenient for our goals is given in terms of Poisson bivector fields.

Definition 9.5.1 (Poisson bivector fields) A bivector field on a differentiable manifold X is a global section of $\bigwedge^2 T_X$. A **Poisson bivector field** is a bivector field π that satisfies the integrability condition $[\pi, \pi]_{SN} = 0$.

For instance, over $\mathbb{R}^n$, every bivector field with constant coefficients

$$\sum a_{ij}\frac{\partial}{\partial z_i}\wedge\frac{\partial}{\partial z_j}, \qquad a_{ij} \in \mathbb{R},$$

is a Poisson bivector field. On a manifold of dimension 2 every bivector field is Poisson.

Lemma 9.5.2 *A bivector field π on a differentiable manifold X is Poisson if and only if the bilinear map*

$$\{-,-\}\colon \mathcal{A}^0_X \times \mathcal{A}^0_X \to \mathcal{A}^0_X, \qquad \{f, g\} = \pi \lrcorner (df \wedge dg) = \boldsymbol{i}_\pi(df \wedge dg),$$

satisfies the Jacobi identity.

Proof The question can be solved by carrying a computation in a system of local coordinates $z_1, \ldots, z_n$. Wriing $\pi^{ij} = \{z_i, z_j\}$, by the formula $\{f, g\} = \pi \lrcorner (df \wedge dg)$ we get

$$\pi = -\frac{1}{2}\sum_{i,j} \pi^{ij} \frac{\partial}{\partial z_i} \wedge \frac{\partial}{\partial z_j}, \qquad \pi^{ij} + \pi^{ji} = 0.$$

We prove that both the conditions described in the lemma are equivalent to

$$\sum_i \pi^{ij} \frac{\partial \pi^{hk}}{\partial z_i} + \pi^{ih} \frac{\partial \pi^{kj}}{\partial z_i} + \pi^{ik} \frac{\partial \pi^{jh}}{\partial z_i} = 0 \quad \text{for every } j, h, k. \tag{9.7}$$

We have:

$$-4[\pi, \pi]_{SN} = -\sum_{i,j,h,k} \left[\pi^{ij} \frac{\partial}{\partial z_i} \wedge \frac{\partial}{\partial z_j}, \pi^{hk} \frac{\partial}{\partial z_h} \wedge \frac{\partial}{\partial z_k}\right]_{SN}$$

$$= \sum_{i,j,h,k} \pi^{ij} \left(\frac{\partial \pi^{hk}}{\partial z_i} \frac{\partial}{\partial z_j} - \frac{\partial \pi^{hk}}{\partial z_j} \frac{\partial}{\partial z_i}\right) \wedge \frac{\partial}{\partial z_h} \wedge \frac{\partial}{\partial z_k}$$

$$+ \pi^{hk} \frac{\partial}{\partial z_i} \wedge \frac{\partial}{\partial z_j} \wedge \left(\frac{\partial \pi^{ij}}{\partial z_h} \frac{\partial}{\partial z_k} - \frac{\partial \pi^{hk}}{\partial z_k} \frac{\partial}{\partial z_h}\right)$$

$$= 2\sum_{i,j,h,k} \pi^{ij} \frac{\partial \pi^{hk}}{\partial z_i} \frac{\partial}{\partial z_j} \wedge \frac{\partial}{\partial z_h} \wedge \frac{\partial}{\partial z_k} + 2\sum_{i,j,h,k} \pi^{hk} \frac{\partial \pi^{ij}}{\partial z_h} \frac{\partial}{\partial z_i} \wedge \frac{\partial}{\partial z_j} \wedge \frac{\partial}{\partial z_k}.$$

Exchanging the ordered pair (i, j) with (h, k) in the rightmost summand we obtain

$$-[\pi, \pi]_{SN} = \sum_{i,j,h,k} \left(\pi^{ij} \frac{\partial \pi^{hk}}{\partial z_i}\right) \frac{\partial}{\partial z_j} \wedge \frac{\partial}{\partial z_h} \wedge \frac{\partial}{\partial z_k},$$

and therefore we have $[\pi, \pi]_{SN} = 0$ if and only if

$$\sum_i \pi^{ij} \frac{\partial \pi^{hk}}{\partial z_i} + \pi^{ih} \frac{\partial \pi^{kj}}{\partial z_i} + \pi^{ik} \frac{\partial \pi^{jh}}{\partial z_i} = 0$$

for every j, h, k. For every pair of differentiable functions f, g we have

$$\{f, g\} = -\frac{1}{2}\sum_{i,j} \pi^{ij} \left(\frac{\partial f}{\partial z_j} \frac{\partial g}{\partial z_i} - \frac{\partial f}{\partial z_i} \frac{\partial g}{\partial z_j}\right) = \sum_{i,j} \pi^{ij} \frac{\partial f}{\partial z_i} \frac{\partial g}{\partial z_j}$$

and then the Leibniz rule implies that $\{-, -\}$ satisfies the Poisson identity

$$\{r, st\} = s\{r, t\} + \{r, s\}t.$$

Defining the Jacobiator of $\{-,-\}$ by the formula

$$J(r,s,t)=\{r,\{s,t\}\}+\{s,\{t,r\}\}+\{t,\{r,s\}\},$$

we can immediately observe that Equation (9.7) is equivalent to

$$-J(z_j,z_h,z_k)=\{\{z_h,z_k\},z_j\}+\{\{z_k,z_j\},z_h\}+\{\{z_j,z_h\},z_k\}=0,\quad \forall\ j,h,k.$$

It is a formal consequence of the Poisson identity that J is multiderivation and satisfies the Leibniz rule in each variable. Thus, by the Taylor expansion, the value of $J(r,s,t)$ at a point $p=(z_1^0,\dots,z_n^0)$ is equal to

$$J(r,s,t)(p)=\sum_{j,h,k}\frac{\partial r}{\partial z_j}\frac{\partial s}{\partial z_h}\frac{\partial t}{\partial z_k}J(z_j-z_j^0,z_h-z_h^0,z_k-z_k^0)(p)$$

and the conclusion follows from the equality $J(z_j-z_j^0,z_h-z_h^0,z_k-z_k^0)=J(z_j,z_h,z_k)$. □

In other words, Lemma 9.5.2 says that $[\pi,\pi]_{SN}=0$ if and only if $(\mathcal{A}_X^0,\{-,-\})$ is a sheaf of Poisson algebras.

Definition 9.5.3 A (differentiable) **Poisson manifold** is a pair (X,π), where X is a differentiable manifold and $\pi\in\Gamma(X,\bigwedge^2 T_X)$ is a Poisson bivector field.

Historically, Poisson manifolds where introduced as a natural generalization of symplectic manifolds; every differential 2-form $\omega\in A_X^2$ on a differentiable manifold X induces a morphism of sheaves

$$\omega^\flat\colon \mathcal{T}_X^1\to\mathcal{A}_X^1,\qquad \omega^\flat(\eta)=\eta\lrcorner\omega=\boldsymbol{i}_\eta(\omega).$$

The form ω is said to be nondegenerate if $\omega^\flat$ is an isomorphism; in this case the dimension of X must be even and there exists a unique bivector field $\pi\in H^0(X,\bigwedge^2 T_X)$ such that

$$(\omega^\flat)^{-1}(\alpha)\lrcorner\beta=\pi\lrcorner(\alpha\wedge\beta),\qquad \text{for every}\quad \alpha,\beta\in\mathcal{A}_X^1.$$

In local coordinates $z_1,\dots,z_n$, writing

$$\omega=\frac{1}{2}\sum_{i,j}\omega_{ij}\,dz_i\wedge dz_j,\qquad \omega_{ij}=-\omega_{ji},$$

we get

$$\omega^\flat\left(\frac{\partial}{\partial z_i}\right)=\sum_j\omega_{ij}\,dz_j$$

and the matrix (ω_{ij}) is invertible. Denoting by π^{ij} the coefficients of the inverse matrix we have

$$\pi^{ij} = -\pi^{ji}, \qquad (\omega^{\flat})^{-1}(dz_i) = \sum_j \pi^{ij}\frac{\partial}{\partial z_j},$$

and the formulas $\pi^{ij} = (\omega^{\flat})^{-1}(dz_i) \lrcorner dz_j = \pi \lrcorner (dz_i \wedge dz_j)$ imply that

$$\pi = -\frac{1}{2}\sum_{i,j} \pi^{ij}\frac{\partial}{\partial z_i} \wedge \frac{\partial}{\partial z_j}.$$

Lemma 9.5.4 *In the notation above* $[\pi, \pi]_{SN} = 0$ *if and only if* $d\omega = 0$.

Proof The question is purely local, working in local coordinates we may write

$$\omega = \frac{1}{2}\sum_{i,j} \omega_{ij}\, dz_i \wedge dz_j, \qquad \pi = -\frac{1}{2}\sum_{i,j} \pi^{ij}\frac{\partial}{\partial z_i} \wedge \frac{\partial}{\partial z_j},$$

where

$$\pi^{ij} = -\pi^{ji}, \qquad \omega_{ij} = -\omega_{ji}, \qquad \sum_h \omega_{ih}\pi^{hi} = \delta_{ij}, \qquad \forall\, i, j.$$

We have already proved that $[\pi, \pi]_{SN} = 0$ is equivalent to

$$\sum_i \pi^{ij}\frac{\partial \pi^{hk}}{\partial z_i} + \pi^{ih}\frac{\partial \pi^{kj}}{\partial z_i} + \pi^{ik}\frac{\partial \pi^{jh}}{\partial z_i} = 0, \quad \forall\, j, h, k. \tag{9.8}$$

Since ω_{ij} is an invertible matrix the above equation is equivalent to

$$\sum_{i,j,h,k} \omega_{aj}\omega_{bh}\omega_{ck}\left(\pi^{ij}\frac{\partial \pi^{hk}}{\partial z_i} + \pi^{ih}\frac{\partial \pi^{kj}}{\partial z_i} + \pi^{ik}\frac{\partial \pi^{jh}}{\partial z_i}\right) = 0, \qquad \forall\, a, b, c. \tag{9.9}$$

Using the relation

$$\sum_{i,j} \omega_{aj}\pi^{ij}\frac{\partial \pi^{hk}}{\partial z_i} = -\sum_i\sum_j \omega_{aj}\pi^{ji}\frac{\partial \pi^{hk}}{\partial z_i} = -\frac{\partial \pi^{hk}}{\partial z_a}$$

we get

$$\sum_{i,j,h,k} \omega_{aj}\omega_{bh}\omega_{ck}\left(\pi^{ij}\frac{\partial \pi^{hk}}{\partial z_i} + \pi^{ih}\frac{\partial \pi^{kj}}{\partial z_i} + \pi^{ik}\frac{\partial \pi^{jh}}{\partial z_i}\right)$$

$$= \sum_{h,k} \omega_{bh}\frac{\partial \pi^{hk}}{\partial z_a}\omega_{kc} + \sum_{j,k} \omega_{ck}\frac{\partial \pi^{kj}}{\partial z_b}\omega_{ja} + \sum_{j,h} \omega_{aj}\frac{\partial \pi^{jh}}{\partial z_c}\omega_{hb}$$

$$= \sum_{h,k} \omega_{bh}\frac{\partial \pi^{hk}}{\partial z_a}\omega_{kc} + \omega_{ch}\frac{\partial \pi^{hk}}{\partial z_b}\omega_{ka} + \omega_{ah}\frac{\partial \pi^{hk}}{\partial z_c}\omega_{kb}.$$

On the other hand

$$0 = \frac{\partial}{\partial z_a}\sum_k \pi^{hk}\omega_{kc} = \sum_k \frac{\partial \pi^{hk}}{\partial z_a}\omega_{kc} + \pi^{hk}\frac{\partial \omega_{kc}}{\partial z_a},$$

$$\sum_{h,k} \omega_{bh}\frac{\partial \pi^{hk}}{\partial z_a}\omega_{kc} = -\sum_{h,k} \omega_{bh}\pi^{hk}\frac{\partial \omega_{kc}}{\partial z_a} = -\frac{\partial \omega_{bc}}{\partial z_a},$$

and therefore (9.8) is equivalent to

$$\frac{\partial \omega_{bc}}{\partial z_a} + \frac{\partial \omega_{ca}}{\partial z_b} + \frac{\partial \omega_{ab}}{\partial z_c} = 0, \quad \forall\, a, b, c \iff d\omega = 0.$$

□

Definition 9.5.5 A pair (X, ω), where X is a differentiable manifold and ω is a closed nondegenerate 2-form is called a **symplectic manifold**.

Lemma 9.5.6 *Let X be a differentiable manifold. Then for every polyvector field $\eta \in \Gamma(X, \bigwedge^p T_X)$ the linear maps*

$$\boldsymbol{i}_\eta : A_X^* \to A_X^*, \qquad \mathcal{L}_\eta = [\boldsymbol{i}_\eta, d] : A_X^* \to A_X^*,$$

are differential operators of order $\le p$.

Proof By linearity it is not restrictive to assume $\eta = \eta_1 \wedge \cdots \wedge \eta_p$, where every η_i is a vector field. Now d and every $\boldsymbol{i}_{\eta_i}$ are derivations and hence differential operators of order ≤ 1; the conclusion follows from the formulas (9.6) and $\boldsymbol{i}_\eta = \boldsymbol{i}_{\eta_1}\boldsymbol{i}_{\eta_2}\cdots \boldsymbol{i}_{\eta_p}$. □

Thus, if (X, π) is a Poisson manifold, the Lie derivative $\mathcal{L}_\pi$ is a differential operator of order ≤ 2, of degree -1 and such that $\mathcal{L}_\pi^2 = \frac{1}{2}[\mathcal{L}_\pi, \mathcal{L}_\pi] = \frac{1}{2}\mathcal{L}_{[\pi,\pi]_{SN}} = 0$, $[d, \mathcal{L}_\pi] = 0$. This implies that the de Rham complex of a Poisson manifold carries a natural structure of a differential graded Batalin–Vilkovisky algebra. In fact, according to Lemma 9.4.3 the operator $\mathcal{L}_\pi$ induces a structure of a differential Batalin–Vilkovisky algebra on the de Rham complex of X. The associated bracket on $A_X^*[1]$, called **Koszul bracket** of the Poisson manifold, is given by the formula:

$$[\alpha, \beta]_\pi = (-1)^p(\mathcal{L}_\pi(\alpha \wedge \beta) - \mathcal{L}_\pi(\alpha) \wedge \beta) - \alpha \wedge \mathcal{L}_\pi(\beta), \quad \alpha \in A_X^p,\ \beta \in A_X^*.$$

Notice that if $d\alpha = d\beta = 0$ then the differential form $[\alpha, \beta]_\pi$ is exact, as follows immediately from the expression of $[-, -]_\pi$ in terms of $\boldsymbol{i}_\pi$ and d:

$$\begin{aligned}[\alpha, \beta]_\pi = (-1)^{\overline{\alpha}}(\boldsymbol{i}_\pi d(\alpha \wedge \beta) - d\boldsymbol{i}_\pi(\alpha \wedge \beta) + d(\boldsymbol{i}_\pi(\alpha)) \wedge \beta - \boldsymbol{i}_\pi(d\alpha) \wedge \beta)\\ - \alpha \wedge \boldsymbol{i}_\pi(d\beta) - \alpha \wedge d(\boldsymbol{i}_\pi(\beta)).\end{aligned}$$

When α and β are 1-forms we have $\boldsymbol{i}_\pi(\alpha) = \boldsymbol{i}_\pi(\beta) = 0$ and the above formula reduces to:

$$[\alpha, \beta]_\pi = \boldsymbol{i}_\pi(d\alpha) \wedge \beta - \alpha \wedge \boldsymbol{i}_\pi(d\beta) + d\boldsymbol{i}_\pi(\alpha \wedge \beta) - \boldsymbol{i}_\pi d(\alpha \wedge \beta), \quad \alpha, \beta \in A_X^1. \tag{9.10}$$

9.6 Holomorphic Poisson and Symplectic Manifolds

The notions of Poisson and symplectic manifolds have a natural analogue in the holomorphic setting.

Definition 9.6.1 A **holomorphic Poisson manifold** is a pair (X, π) where X is a complex manifold and $\pi \in H^0(X, \bigwedge^2 \Theta_X)$ satisfies the integrability equation $[\pi, \pi]_{SN} = 0$ in $H^0(X, \bigwedge^3 \Theta_X)$.

The proof of Lemma 9.5.2 works also in the holomorphic case and shows that the integrability condition $[\pi, \pi]_{SN} = 0$ is equivalent to the Jacobi identity for the Poisson bracket $\{f, g\} = \boldsymbol{i}_\pi(df \wedge dg)$ on the sheaf of holomorphic functions.

Example 9.6.2 Let X be a complex manifold.

1. If X has dimension 2, then $H^0(X, \bigwedge^3 \Theta_X) = 0$ and therefore every global section of $\bigwedge^2 \Theta_X = K_X^\vee$ gives a holomorphic Poisson structure.
2. If $A \subseteq H^0(X, \Theta_X)$ is an abelian Lie subalgebra, then every element in the image of $\bigwedge^2 A \to H^0(X, \bigwedge^2 \Theta_X)$ is a holomorphic Poisson bivector fields. In particular, every toric manifold of dimension $n \geq 2$ admits nontrivial holomorphic Poisson structures.

Example 9.6.3 Let X be a complex manifold of dimension 3. We have already observed that the inner product gives an isomorphism of locally free sheaves

$$i\colon \bigwedge^2 \Theta_X \to \mathcal{H}om_{\mathcal{O}_X}(\Omega_X^3, \Omega_X^1) = \Omega_X^1(K_X^\vee).$$

If α is a section of $\Omega_X^1(K_X^\vee)$ then it is not possible, in general, to define its de Rham differential $d\alpha$. However, the element $\alpha \wedge d\alpha$ is well defined as a section of $\Omega_X^3(K_X^\vee \otimes K_X^\vee) \simeq K_X^\vee$; more generally, for every complex manifold X, every odd positive integer p, every holomorphic line bundle $\mathcal{M}$ on X and every section $\alpha \in H^0(X, \Omega_X^p \otimes \mathcal{M})$ there is a properly defined global section $\alpha \wedge d\alpha \in H^0(X, \Omega_X^{2p+1} \otimes \mathcal{M}^{\otimes 2})$. Locally, if s is a local frame of $\mathcal{M}$ and $\alpha = \phi \otimes s \in \Omega_X^p \otimes \mathcal{M}$ we have $\phi \wedge \phi = 0$ and

$$\alpha \wedge d\alpha = (\phi \wedge d\phi) \otimes s^2 \in \Omega_X^{2p+1} \otimes \mathcal{M}^{\otimes 2}$$

is properly defined.

We prove that, via the isomorphism $\boldsymbol{i}\colon H^0(X, \bigwedge^2 \Theta_X) \to H^0(X, \Omega_X^1(K_X^\vee))$, a section $\alpha \in H^0(X, \Omega_X^1(K_X^\vee))$ corresponds to a Poisson bivector field if and only if $\alpha \wedge d\alpha = 0$: since the question is purely local we can assume that there exists a holomorphic volume form $\omega \in \Omega_X^3$ and a polyvector field $\tau \in \bigwedge^3 \Theta_X$ such that $\tau \lrcorner \omega = 1$. Let $\pi \in \bigwedge^2 \Theta_X$ and write $\alpha = \boldsymbol{i}_\pi(\omega) \otimes \tau$. By Cartan formulas we have

$$\boldsymbol{i}_{[\pi,\pi]_{SN}}(\omega) = [[\boldsymbol{i}_\pi, d], \boldsymbol{i}_\pi](\omega) = 2\boldsymbol{i}_\pi d \boldsymbol{i}_\pi(\omega)$$

and then $[\pi, \pi]_{SN} = 0$ if and only if the function $\boldsymbol{i}_\pi d \boldsymbol{i}_\pi(\omega) = \pi \lrcorner d(\pi \lrcorner \omega)$ vanishes. On the other hand, by duality $\pi \lrcorner d(\pi \lrcorner \omega) = d(\pi \lrcorner \omega) \lrcorner \pi$ and by Lemma 8.5.3

$$(d(\pi \lrcorner \omega) \wedge (\pi \lrcorner \omega)) \lrcorner \tau = d(\pi \lrcorner \omega) \lrcorner ((\pi \lrcorner \omega) \lrcorner \tau) = d(\pi \lrcorner \omega) \lrcorner \pi.$$

Thus the condition $\alpha \wedge d\alpha = 0$ is the same as $d(\pi \lrcorner \omega) \wedge (\pi \lrcorner \omega) = 0$.

It is interesting to point out that, if $K_X^\vee = \mathcal{L}^{\otimes 2}$ for some line bundle $\mathcal{L}$ on X, e.g. if $X = \mathbb{P}^3$ and $\mathcal{L} = \mathcal{O}(2)$, then for every $\beta, \gamma \in H^0(X, \mathcal{L})$ the section $\alpha = \beta d\gamma - \gamma d\beta \in H^0(X, \Omega_X^1(K_X^\vee))$ is well defined and satisfies $\alpha \wedge d\alpha = 0$.

Example 9.6.4 (Holomorphic symplectic manifolds) A pair (X, ω), where X is a complex manifold and ω is a closed nondegenerate holomorphic 2-form is called a **holomorphic symplectic manifold**. Lemma 9.5.4 is valid also in the holomorphic case and implies that every holomorphic symplectic manifold is a holomorphic Poisson manifold.

Definition 9.6.5 The **anchor map** of a holomorphic Poisson manifold (X, π) is the morphism of sheaves of $\mathcal{O}_X$-modules:

$$\pi^\#\colon \Omega_X^1 \to \Theta_X, \qquad \pi^\#(\alpha) \lrcorner \beta = \pi \lrcorner (\alpha \wedge \beta) = \boldsymbol{i}_\pi(\alpha \wedge \beta), \qquad \alpha, \beta \in \Omega_X^1.$$

It is plain that a holomorphic Poisson manifold is holomorphic symplectic if and only if the anchor map is an isomorphism of sheaves. In a system of local coordinates $z_1, \ldots, z_n$, if $\pi^{ij} = \{z_i, z_j\} = \boldsymbol{i}_\pi(dz_i \wedge dz_j)$, then

$$\pi = -\frac{1}{2}\sum_{i,j} \pi^{ij} \frac{\partial}{\partial z_i} \wedge \frac{\partial}{\partial z_j} \quad \text{and} \quad \pi^\#(dz_i) = \sum_j \pi^{ij} \frac{\partial}{\partial z_j}. \tag{9.11}$$

Given $\alpha \in \Omega^1_X$ and denoting by $\alpha\wedge$ the left multiplication operator, we have the equality

$$\boldsymbol{i}_{\pi^\#(\alpha)} = [\boldsymbol{i}_\pi, \alpha\wedge] \in \mathcal{H}om^*_{\mathcal{O}_X}(\Omega^*_X, \Omega^*_X). \tag{9.12}$$

In fact both sides are differential operators of order ≤ 1 of the graded commutative algebra Ω^*_X, taking the same values in $\Omega^{\le 1}_X$.

As in the differentiable case, every holomorphic Poisson bivector π induces a Koszul bracket $[-,-]_\pi$ of degree -1 on the sheaf Ω^*_X of holomorphic differential forms:

$$[\alpha,\beta]_\pi = (-1)^p\left(\mathcal{L}_\pi(\alpha\wedge\beta) - \mathcal{L}_\pi(\alpha)\wedge\beta\right) - \alpha\wedge\mathcal{L}_\pi(\beta), \quad \alpha\in\Omega^p_X, \beta\in\Omega_X.$$

Lemma 9.6.6 (Magri formulas) *In the above setup, for every $\alpha, \beta \in \Omega^1_X$ and every $f \in \mathcal{O}_X$ we have:*

1. $[\alpha,\beta]_\pi = \mathcal{L}_{\pi^\#(\alpha)}(\beta) - \mathcal{L}_{\pi^\#(\beta)}(\alpha) - d\boldsymbol{i}_\pi(\alpha\wedge\beta)$,
2. $[\alpha,\beta]_\pi = \boldsymbol{i}_{\pi^\#(\alpha)}(d\beta) - \boldsymbol{i}_{\pi^\#(\beta)}(d\alpha) + d\boldsymbol{i}_\pi(\alpha\wedge\beta)$,
3. $[\alpha, f\beta]_\pi = f[\alpha,\beta]_\pi + (\pi^\#(\alpha)\lrcorner df)\beta$.

Proof The equality $\boldsymbol{i}_{\pi^\#(\alpha)} = [\boldsymbol{i}_\pi, \alpha\wedge]$ gives in particular

$$\boldsymbol{i}_{\pi^\#(\alpha)}(d\beta) = \boldsymbol{i}_\pi(\alpha\wedge d\beta) - \alpha\wedge\boldsymbol{i}_\pi(d\beta).$$

Using this, the proof of items (1) and (2) follows immediately from Equation (9.10). Next, since $\boldsymbol{i}_\pi(\alpha) = 0$ for every $\alpha\in\Omega^1_X$, the seven-terms relation (Corollary 9.3.5) for the second order differential operator $\boldsymbol{i}_\pi$ gives

$$\boldsymbol{i}_\pi(\alpha\wedge\beta\wedge\gamma) = \boldsymbol{i}_\pi(\alpha\wedge\beta)\gamma + \boldsymbol{i}_\pi(\beta\wedge\gamma)\alpha - \boldsymbol{i}_\pi(\alpha\wedge\gamma)\beta, \quad \alpha,\beta,\gamma\in\Omega^1_X.$$

In particular

$$\begin{aligned}
[\alpha, f\beta]_\pi &= \boldsymbol{i}_{\pi^\#(\alpha)}(d(f\beta)) - \boldsymbol{i}_{\pi^\#(f\beta)}(d\alpha) + d\boldsymbol{i}_\pi(f\alpha\wedge\beta)\\
&= f[\alpha,\beta]_\pi + \boldsymbol{i}_{\pi^\#(\alpha)}(df\wedge\beta) + df\wedge\boldsymbol{i}_\pi(\alpha\wedge\beta)\\
&= f[\alpha,\beta]_\pi + \boldsymbol{i}_\pi(\alpha\wedge df\wedge\beta) - \boldsymbol{i}_\pi(df\wedge\beta)\alpha + \boldsymbol{i}_\pi(\alpha\wedge\beta)df\\
&= f[\alpha,\beta]_\pi + \boldsymbol{i}_\pi(\alpha\wedge df)\beta = f[\alpha,\beta]_\pi + \boldsymbol{i}_{\pi^\#(\alpha)}(df)\beta.
\end{aligned}$$

□

Theorem 9.6.7 *In the above setup, for every $\alpha, \beta \in \Omega^1_X$ we have*

$$\pi^\#([\alpha,\beta]_\pi) = [\pi^\#(\alpha), \pi^\#(\beta)]$$

and then the anchor map $\pi^\#\colon \Omega^1_X \to \Theta_X$ is a morphism of sheaves of Lie algebras.

Proof It is sufficient to prove the formula

$$\boldsymbol{i}_{\pi^\#([\alpha,\beta]_\pi)} = \boldsymbol{i}_{[\pi^\#(\alpha),\pi^\#(\beta)]} = [[\boldsymbol{i}_{\pi^\#(\alpha)}, d], \boldsymbol{i}_{\pi^\#(\beta)}]$$

that, by (9.12), is equivalent to

$$[\boldsymbol{i}_\pi, [\alpha, \beta]_\pi\wedge] = [[[\boldsymbol{i}_\pi, \alpha\wedge], d], [\boldsymbol{i}_\pi, \beta\wedge]]. \tag{9.13}$$

Since both sides of (9.13) are differential operators of order ≤ 1 and degree -1 on the sheaf of graded algebras Ω^*_X, they vanish on Ω^0_X and then they are $\mathcal{O}_X$-linear derivations of degree -1. Thus to prove (9.13) it is sufficient to check that both sides of the equality take the same value on every exact form $\gamma = dh$, $h \in \mathcal{O}_X$. Assume first that $\alpha = df$ and $\beta = dg$ are exact forms, with $f, g \in \mathcal{O}_X$. Then $[\alpha, \beta]_\pi = d\boldsymbol{i}_\pi(df \wedge dg) = d\{f, g\}$,

$$[\boldsymbol{i}_\pi, [\alpha, \beta]_\pi\wedge]\gamma = \boldsymbol{i}_\pi(d\{f, g\} \wedge dh) = \{\{f, g\}, h\},$$

$$\begin{aligned}
[[[\boldsymbol{i}_\pi, \alpha\wedge], d], [\boldsymbol{i}_\pi, \beta\wedge]]\gamma &= [[\boldsymbol{i}_\pi, \alpha\wedge], d]([\boldsymbol{i}_\pi, \beta\wedge]\gamma) - [\boldsymbol{i}_\pi, \beta\wedge]([[\boldsymbol{i}_\pi, \alpha\wedge], d]\gamma)\\
&= [[\boldsymbol{i}_\pi, \alpha\wedge], d](\{g, h\}) - [\boldsymbol{i}_\pi, \beta\wedge](d\{f, h\})\\
&= \{f, \{g, h\}\} - \{g, \{f, h\}\},
\end{aligned}$$

and the conclusion follows from Lemma 9.5.2.

Thus we have proved the theorem for exact forms; by linearity and skew-symmetry, in order to conclude the proof it is sufficient to show that if $\pi^\#([\alpha, \beta]_\pi) = [\pi^\#(\alpha), \pi^\#(\beta)]$, then $\pi^\#([\alpha, f\beta]_\pi) = [\pi^\#(\alpha), \pi^\#(f\beta)]$ for every holomorphic function $f \in \mathcal{O}_X$. According to Lemma 9.6.6 we have:

$$\begin{aligned}
\pi^\#([\alpha, f\beta]_\pi) &= \pi^\#(f[\alpha, \beta]_\pi + (\pi^\#(\alpha)\lrcorner df)\beta)\\
&= f\pi^\#([\alpha, \beta]_\pi) + (\pi^\#(\alpha)\lrcorner df)\pi^\#(\beta),\\
[\pi^\#(\alpha), \pi^\#(f\beta)] &= [\pi^\#(\alpha), f\pi^\#(\beta)] = f\pi^\#([\alpha, \beta]_\pi) + (\pi^\#(\alpha)\lrcorner df)\pi^\#(\beta),
\end{aligned}$$

where the last equality follows from the standard properties of the bracket of vector fields. □

We are now ready to apply the previous computation to deformation theory.

Theorem 9.6.8 (Bogomolov–Hitchin) *Let (X, π) be a compact Kähler holomorphic Poisson manifold. Then for every cohomology class $\omega \in H^1(X, \Omega^1_X)$, the first order deformation $\pi^\#(\omega) \in H^1(X, \Theta_X)$ can be extended to a deformation over $\mathbb{C}[t]/(t^n)$, for every $n > 0$.*

In particular, if the anchor map $\pi^\#\colon H^1(X, \Omega^1_X) \to H^1(X, \Theta_X)$ is surjective, then the functor Def_X *is smooth.*

Proof Since π is holomorphic, the Koszul bracket on Ω^1_X can be extended to the Dolbeault resolution $\mathcal{A}^{0,*}_X(\Omega^1_X) = \mathcal{A}^{1,*}_X$ by the formula

$$[\alpha, \beta]_\pi = \boldsymbol{i}_\pi(\partial\alpha) \wedge \beta - \alpha \wedge \boldsymbol{i}_\pi(\partial\beta) + \partial \boldsymbol{i}_\pi(\alpha \wedge \beta) - \boldsymbol{i}_\pi \partial(\alpha \wedge \beta), \qquad (9.14)$$

giving a structure of a sheaf of differential graded Lie algebras on $\mathcal{A}_X^{0,*}(\Omega_X^1)$; similarly, the anchor map gives a morphism of differential graded Lie algebras

$$L := A_X^{0,*}(\Omega_X^1) \xrightarrow{\pi^\#} A_X^{0,*}(\Theta_X) = KS_X,$$

and then a morphism of deformation functors

$$\pi^\# \colon \mathrm{Def}_L \to \mathrm{Def}_{KS_X} = \mathrm{Def}_X.$$

Thus, in order to prove the theorem it is sufficient to prove that Def_L is smooth; in fact we prove the stronger result that L is homotopy abelian. To this end consider the morphisms of differential graded Lie algebras

$$L \xleftarrow{\text{inclusion}} A_X^{1,*} \cap \ker \partial \xrightarrow{\text{projection}} \frac{A_X^{1,*} \cap \ker \partial}{\partial A_X^{0,*}}.$$

By the $\partial\bar{\partial}$-lemma the subcomplexes $\partial A_X^{0,*}$ and $\partial A_X^{1,*}$ are acyclic and therefore the above maps are quasi-isomorphisms. Finally, the bracket on the rightmost differential graded Lie algebra is trivial by (9.14). □

9.7 Exercises

9.7.1 In the situation of Example 9.1.9, for every $f, g \colon A \to A$ compute the three associators

$$\mu \circ (f \circ g) - (\mu \circ f) \circ g, \quad f \circ (\mu \circ g) - (f \circ \mu) \circ g, \quad f \circ (g \circ \mu) - (f \circ g) \circ \mu,$$

and deduce that the (cohomological) cup product

$$\cup \colon H^1(A, A) \times H^1(A, A) \to H^2(A, A)$$

is skew-symmetric.

9.7.2 Let A be a unitary graded commutative algebra over the field $\mathbb{K}$ and consider the graded vector space

$$B = \bigoplus_{n \geq 0} \frac{\mathrm{Diff}^n_{A/\mathbb{K}}(A, A)}{\mathrm{Diff}^{n-1}_{A/\mathbb{K}}(A, A)}.$$

Show that B carries a natural structure of a graded Poisson algebra of degree 0.

9.7.3 Show that not every graded Poisson algebra of odd degree admits a generator. For instance, consider Lie algebra $L \neq 0$ and the trivial extension $A = \mathbb{K} \oplus L[1]$.

9.7.4 Let (A, d, Δ) be a differential graded Batalin–Vilkovisky algebra of degree -1. Prove that $(A[-1], d + \hbar\Delta, [-, -])$ is a DG-Lie algebra for every $\hbar \in \mathbb{K}$.

9.7.5 In the notation of Example 9.4.5, compute the squares Q_i^2 of the (inhomogeneous) operators

$$Q_i : A \to A, \qquad Q_i = y_i \frac{\partial}{\partial x_i} + \frac{\partial}{\partial y_i}, \qquad i = 1, \ldots, n.$$

References

The terminology adopted in Sect. 9.1 is borrowed from [45]. The Schouten bracket was introduced by Schouten in [240] while the Jacobi identity was proved 15 years later by Nijenhuis [207]. The nowcalled Gerstenhaber algebras were first studied in [87] as a structure on the cohomology of an associative ring.

The theorem of Bogomolov–Tian–Todorov was first proved by Bogomolov [25] for holomorphic symplectic manifolds and then, independently, by Tian [257] and Todorov [258] for Kähler manifolds with trivial canonical bundle; their proofs, albeit different, use the same general ideas as used here in the proof of Corollary 9.2.3. Both papers [257, 258] contain a proof of Theorem 9.2.2, commonly known today as the Tian–Todorov lemma, although previously proved by Koszul in [160][Prop. 2.3]. The proof of Corollary 9.2.3 given here follows essentially [95].

The Bogomolov–Tian–Todorov theorem also holds for Kähler manifolds with torsion canonical bundle; this has been proved by Ran [220] and Kawamata [141] by using the T^1-lifting property, while a proof via DG-Lie algebras is proposed in [179].

Koszul braces (the name is taken from [190]), also called either higher antibrackets or higher Koszul brackets, were introduced in [160] and recently generalized to associative unitary algebras in [11, 21, 186, 188].

Concrete examples of Batalin–Vilkovisky (BV) algebras arising from string theory were studied in 1981 by Batalin and Vilkovisky, while the abstract definition of BV algebra given in this book was proposed in [89, 168] (see also [228, 250, 274]). The equation involving $\Phi^3_{\Delta^2}$ in the proof of Lemma 9.4.2 is also present, without proof, in [160][Lemme 1.5]; a nice and elegant proof of such an equation, inspired by [76][Theorem 4.3], is proposed here in Exercise 10.8.6.

The first published definition of Poisson manifolds was given in [169], although the same notion is implicit is several previous papers; a good reference for Magri formulas is [158][p. 54]. A good introductory textbook on Poisson manifolds is [260]. We also mention the notes [67] for a treatment of Poisson manifolds together with deformation theory of associative algebras via DG-Lie and L_∞-algebras.

For symplectic manifolds, both the Poisson bracket $[-, -]_\pi$ and Magri formulas were well known much before in the framework of Hamiltonian dynamical systems [1]. It has been proved by Sharygin and Talalaev [247] that for every Poisson manifold

(X, π), the DG-Lie algebra $(A_X^*[1], -d, [-, -]_\pi)$ is quasi-isomorphic to an abelian DG-Lie algebra; we prove this fact in Theorem 13.6.8 and we also refer to [76, 129] for different proofs.

Theorem 9.6.8 was proved by Bogomolov [25] for holomorphic symplectic manifolds and then extended by Hitchin [116]. The proof given here is based on the ideas of [76] and can be improved by showing that every first order deformation induced by the anchor map may be extended to a deformation of the pair (X, π) to any order [12, 116].

Chapter 10
L_∞-Algebras

The aim of this chapter is to present L_∞-algebras as a natural generalization of differential graded Lie algebras, and to extend Maurer–Cartan and deformation functors to them. It is easy to give the definition of L_∞-algebras; it is sufficient to modify the notion of a differential graded Lie algebra by imposing that the Jacobi identity holds only up to a hierarchy of higher homotopies. However, the real utility of L_∞-algebras is given by the notion of L_∞-morphisms, which is quite subtle and that we shall give in Chap. 12 after a long and necessary list of algebraic preliminaries.

It happens that almost all the formulas involved in the definition and properties of L_∞-morphisms become much more readable and easy to manage after a degree shifting of the underlying graded vector space; this leads naturally to the notion of $L_\infty[1]$-algebras. It will be immediately clear from the definition that the categories of L_∞ and $L_\infty[1]$-algebras are canonically equivalent.

10.1 Symmetric Powers and Koszul Sign

Given a graded vector space V, the twist map

$$\mathtt{tw}\colon V\otimes V\to V\otimes V,\qquad \mathtt{tw}(v\otimes w)=(-1)^{\overline{v}\,\overline{w}}w\otimes v,$$

extends naturally, for every $n\geq 0$, to a right action of the symmetric group Σ_n of permutations of the set $\{1,\ldots,n\}$ on the nth tensor power of V:

$$\mathtt{tw}\colon V^{\otimes n}\times\Sigma_n\to V^{\otimes n}.$$

More explicitly, for $v_1,\ldots,v_n$ homogeneous components and $\sigma\in\Sigma_n$ we have:

$$\mathtt{tw}(v_1\otimes\cdots\otimes v_n,\sigma)=\pm(v_{\sigma(1)}\otimes\cdots\otimes v_{\sigma(n)}),$$

M. Manetti, *Lie Methods in Deformation Theory*, Springer Monographs in Mathematics,
https://doi.org/10.1007/978-981-19-1185-9_10

where the above sign $\pm$ is equal to the signature of the restriction of σ to the subset of indices i such that v_i has odd degree.

Definition 10.1.1 Given a permutation $\sigma \in \Sigma_n$, a graded vector space V and a sequence of n nontrivial homogeneous vectors $v_1, \ldots, v_n$, the **Koszul sign**

$$\varepsilon(\sigma, V; v_1, \ldots, v_n) = \pm 1$$

is defined by the relation

$$\mathtt{tw}(v_1 \otimes \cdots \otimes v_n, \sigma) = \varepsilon(\sigma, V; v_1, \ldots, v_n)\, v_{\sigma(1)} \otimes \cdots \otimes v_{\sigma(n)}.$$

The **antisymmetric Koszul sign** $\chi(\sigma, V; v_1, \ldots, v_n) = \pm 1$ is the product of the Koszul sign and the signature of the permutation:

$$\chi(\sigma, V; v_1, \ldots, v_n) = (-1)^{\sigma} \varepsilon(\sigma, V; v_1, \ldots, v_n).$$

By convention, if $v_i = 0$ for some index i we set

$$\varepsilon(\sigma, V; v_1, \ldots, v_n) = \chi(\sigma, V; v_1, \ldots, v_n) = 0.$$

For instance, if $v_1, \ldots, v_n$ are nontrivial homogeneous, $\sigma(n) = 1$ and $\sigma(i) = i + 1$ for $1 \leq i < n$ we have

$$\varepsilon(\sigma, V; v_1, \ldots, v_n) = (-1)^{\overline{v_1}(\overline{v_2}+\cdots+\overline{v_n})},$$
$$\chi(\sigma, V; v_1, \ldots, v_n) = (-1)^{n-1+\overline{v_1}(\overline{v_2}+\cdots+\overline{v_n})}.$$

For the Koszul sign we adopt the same convention used for the degree; when the Koszul sign $\varepsilon(\sigma, V; v_1, \ldots, v_n)$ appears in some formula, *we always assume*, even if not explicitly said, that $v_1, \ldots, v_n$ are homogeneous. The same convention is assumed also for the antisymmetric Koszul sign.

For notational simplicity we shall write $\varepsilon(\sigma; v_1, \ldots, v_n)$ or $\varepsilon(\sigma)$ for the Koszul sign when there is no possible confusion about V and $v_1, \ldots, v_n$; similarly, we write $\chi(\sigma; v_1, \ldots, v_n)$ and $\chi(\sigma)$ for the antisymmetric Koszul sign.

Notice that for every $\sigma, \tau \in \Sigma_n$ we have

$$\begin{aligned}\mathtt{tw}(v_1 \otimes \cdots \otimes v_n, \sigma\tau) &= \mathtt{tw}(\mathtt{tw}(v_1 \otimes \cdots \otimes v_n, \sigma), \tau) \\ &= \varepsilon(\sigma; v_1, \ldots, v_n)\mathtt{tw}(v_{\sigma(1)} \otimes \cdots \otimes v_{\sigma(n)}, \tau)\end{aligned}$$

and therefore

$$\varepsilon(\sigma\tau; v_1, \ldots, v_n) = \varepsilon(\sigma; v_1, \ldots, v_n)\varepsilon(\tau; v_{\sigma(1)}, \ldots, v_{\sigma(n)}). \tag{10.1}$$

$$\chi(\sigma\tau; v_1, \ldots, v_n) = \chi(\sigma; v_1, \ldots, v_n)\chi(\tau; v_{\sigma(1)}, \ldots, v_{\sigma(n)}). \tag{10.2}$$

It is plain that if $f: V \to W$ is an *injective* linear map of *even degree* between graded vector spaces, then

$$\varepsilon(\sigma, W; f(v_1), \ldots, f(v_n)) = \varepsilon(\sigma, V; v_1, \ldots, v_n),$$

while for an injective linear map of odd degree the behaviour of the Koszul sign is described by the following lemma.

Lemma 10.1.2 *Let $h: V \to W$ be an injective linear map of odd degree between graded vector spaces. Given homogeneous vectors $v_1, \ldots, v_n \in V$ and $\sigma \in \Sigma_n$ we have:*

$$\begin{aligned}\chi(\sigma, W; h(v_1), \ldots, h(v_n)) &= (-1)^{\sum_{i=1}^{n}(n-i)(\overline{v_{\sigma(i)}}-\overline{v_i})}\varepsilon(\sigma, V; v_1, \ldots, v_n)\\ &= (-1)^{\sum_{i=1}^{n} i\,(\overline{v_{\sigma(i)}}-\overline{v_i})}\,\varepsilon(\sigma, V; v_1, \ldots, v_n).\end{aligned}$$

Proof According to Eqs. (10.1) and (10.2) it is sufficient to check the formula when σ is a transposition exchanging two consecutive elements; this is easy and left to the reader. □

Definition 10.1.3 Let V, W be graded vector spaces. A multilinear map

$$f: V \times \cdots \times V \to W$$

is called (graded) **symmetric** if for every sequence of homogeneous vectors $v_1, \ldots, v_n$ and for every $\sigma \in \Sigma_n$ we have

$$f(v_{\sigma(1)}, \ldots, v_{\sigma(n)}) = \varepsilon(\sigma) f(v_1, \ldots, v_n).$$

The **symmetric powers** of a graded vector space V are defined as

$$V^{\odot n} = \frac{V^{\otimes n}}{I_n},$$

where I_n is the graded vector subspace generated by the homogeneous vectors

$$v_1 \otimes \cdots \otimes v_n - \varepsilon(\sigma) v_{\sigma(1)} \otimes \cdots \otimes v_{\sigma(n)}, \qquad v_i \in V, \quad \sigma \in \Sigma_n. \tag{10.3}$$

If $\pi: V^{\otimes n} \to V^{\odot n}$ is the natural projection we shall write

$$v_1 \odot \cdots \odot v_n = \pi(v_1 \otimes \cdots \otimes v_n).$$

It is clear by definition that

$$v_{\sigma(1)} \odot \cdots \odot v_{\sigma(n)} = \varepsilon(\sigma)\, v_1 \odot \cdots \odot v_n \quad \text{for every} \quad \sigma \in \Sigma_n.$$

By definition, there exists a canonical isomorphism between the vector space of linear maps $f\colon V^{\odot n} \to W$ and the vector space of graded symmetric multilinear maps $V \times \cdots \times V \to W$. By a slight abuse of notation we shall frequently write $f(v_1, \ldots, v_n)$ with the same meaning as $f(v_1 \odot \cdots \odot v_n)$.

Another equally useful definition of symmetric powers is given in term of symmetric algebra. As in Sect. 2.4, for a graded vector space V over a field $\mathbb{K}$, the **tensor algebra** generated by V is the graded vector space

$$T(V) = \bigoplus_{n\geq 0} V^{\otimes n},$$

equipped with the concatenation product, defined as the bilinear extension of the obvious map

$$(v_1 \otimes \cdots \otimes v_a)(w_1 \otimes \cdots \otimes w_b) = v_1 \otimes \cdots \otimes v_a \otimes w_1 \otimes \cdots \otimes w_b.$$

The algebra $T(V)$ is graded associative and unitary, with unit $1 \in \mathbb{K} = V^{\otimes 0}$. The **symmetric algebra** generated by V is the quotient algebra

$$S(V) = \frac{T(V)}{I},$$

where I is the ideal generated by the elements $u \otimes v - (-1)^{\overline{u}\,\overline{v}} v \otimes u \in V^{\otimes 2}$, for all pairs of homogeneous vectors $u, v \in V$. Since I is a proper graded ideal of $T(V)$ we have

$$S(V) = \bigoplus_{n\geq 0} \frac{V^{\otimes n}}{I \cap V^{\otimes n}},$$

and therefore the symmetric algebra is graded associative and unitary.

Lemma 10.1.4 *In the above notation $I \cap V^{\otimes n} = I_n$ and then $S(V) = \oplus_{n\geq 0} V^{\odot n}$. In particular, the product in $S(V)$ is the $\odot$-concatenation product and the symmetric algebra is graded commutative.*

Proof Denoting by $P \subset \Sigma_n$ the subset of the $n-1$ transpositions of two consecutive numbers, it is plain that $I \cap V^{\otimes n}$ is the vector subspace of $V^{\otimes n}$ generated by the vectors (10.3) when $\sigma \in P$, and then $I \cap V^{\otimes n} \subset I_n$. On the other hand the subset P generates the symmetric group and therefore $I_n \subset I \cap V^{\otimes n}$.

The equality $I \cap V^{\otimes n} = I_n$ gives $V^{\odot n} = V^{\otimes n}/(I \cap V^{\otimes n})$ and implies that the projection map

$$\pi\colon \oplus_{n\geq 0} V^{\otimes n} \to \oplus_{n\geq 0} V^{\odot n}, \qquad \pi(v_1 \otimes \cdots \otimes v_n) = v_1 \odot \cdots \odot v_n,$$

is a morphism of graded algebras. Since the product in $T(V)$ is the bilinear extension of the $\otimes$-concatenation, we have that the product in $S(V)$ is the bilinear extension of the $\odot$-concatenation:

$$(u_1 \odot \cdots \odot u_n)(v_1 \odot \cdots \odot v_m) = u_1 \odot \cdots \odot u_n \odot v_1 \odot \cdots \odot v_m .$$

The graded commutativity of the product in $S(V)$ is clear. □

Notice that if $\{x_i\}$ is a homogeneous basis of a graded vector space V, then there exists a clear isomorphism of graded algebras $\mathbb{K}[\{x_i\}] \cong S(V)$, see Example 5.5.4.

Definition 10.1.5 Let V, W be graded vector spaces. A multilinear map

$$f : V \times \cdots \times V \to W$$

is called (graded) **skew-symmetric** if for every sequence of homogeneous vectors $v_1, \ldots, v_n$ and for every $\sigma \in \Sigma_n$ we have

$$f(v_{\sigma(1)}, \ldots, v_{\sigma(n)}) = \chi(\sigma) f(v_1, \ldots, v_n).$$

The **exterior powers** of a graded vector space V are defined as

$$V^{\wedge n} = \frac{V^{\otimes n}}{J_n},$$

where J_n is the graded vector subspace generated by the vectors

$$v_1 \otimes \cdots \otimes v_n - \chi(\sigma) v_{\sigma(1)} \otimes \cdots \otimes v_{\sigma(n)}, \qquad v_i \in V, \quad \sigma \in \Sigma_n .$$

If $\pi : V^{\otimes n} \to V^{\wedge n}$ is the natural projection we shall write

$$v_1 \wedge \cdots \wedge v_n = \pi(v_1 \otimes \cdots \otimes v_n).$$

Notice that $I_n = J_n = 0$ for $i = 0, 1$ and then $V^{\odot 0} = V^{\wedge 0} = V^{\otimes 0} = \mathbb{K}$, $V^{\odot 1} = V^{\wedge 1} = V^{\otimes 1} = V$. Other names for the exterior powers are **alternating powers** and **Grassmann powers**.

Remark 10.1.6 The twist action on $\mathrm{Hom}^*_{\mathbb{K}}(V, W)^{\otimes n}$ is compatible with the conjugate of the twist action on $\mathrm{Hom}^*_{\mathbb{K}}(V^{\otimes n}, W^{\otimes n})$. This means that

$$\sigma_{\mathtt{tw}}(f_1 \otimes \cdots \otimes f_n) = \sigma_{\mathtt{tw}} \circ (f_1 \otimes \cdots \otimes f_n) \circ \sigma_{\mathtt{tw}}^{-1},$$

where $\circ$ is the composition product and $\sigma_{\mathtt{tw}} = \mathtt{tw}(-, \sigma^{-1})$.

Every differential d on a graded vector space V induces naturally various differentials on tensor, symmetric and exterior powers of V. For instance we have

$$d \colon V^{\otimes 2} \to V^{\otimes 2}, \qquad d(u \otimes v) = d(u) \otimes v + (-1)^{\overline{u}} u \otimes d(v),$$

$$d \colon V^{\odot 2} \to V^{\odot 2}, \qquad d(u \odot v) = d(u) \odot v + (-1)^{\overline{u}} u \odot d(v),$$

$$d\colon V^{\wedge 2} \to V^{\wedge 2}, \qquad d(u \wedge v) = d(u) \wedge v + (-1)^{\overline{u}} u \wedge d(v).$$

Since the maps

$$V^{\odot 2} \to V^{\otimes 2}, \qquad u \odot v \mapsto \frac{1}{2}\left(u \otimes v + (-1)^{\overline{u}\,\overline{v}} v \otimes u\right),$$

$$V^{\wedge 2} \to V^{\otimes 2}, \qquad u \odot v \mapsto \frac{1}{2}\left(u \otimes v - (-1)^{\overline{u}\,\overline{v}} v \otimes u\right),$$

are morphisms of DG-vector spaces, it follows that $V^{\odot 2}$ and $V^{\wedge 2}$ are direct summand DG-vector subspaces of $V^{\otimes 2}$; in particular, $H^*(V^{\odot 2})$ and $H^*(V^{\wedge 2})$ are direct summands of $H^*(V^{\otimes 2})$.

More generally, keeping in mind the Leibniz fomula and Koszul rule of signs, for every $n > 1$ we have:

$$d\colon V^{\otimes n} \to V^{\otimes n}, \quad d(v_1 \otimes \cdots \otimes v_n) = \sum_{i=1}^{n} (-1)^{\overline{v_1}+\cdots+\overline{v_{i-1}}} v_1 \otimes \cdots \otimes d(v_i) \otimes \cdots \otimes v_n,$$

$$d\colon V^{\odot n} \to V^{\odot n}, \quad d(v_1 \odot \cdots \odot v_n) = \sum_{i=1}^{n} (-1)^{\overline{v_1}+\cdots+\overline{v_{i-1}}} v_1 \odot \cdots \odot d(v_i) \odot \cdots \odot v_n,$$

$$d\colon V^{\wedge n} \to V^{\wedge n}, \quad d(v_1 \wedge \cdots \wedge v_n) = \sum_{i=1}^{n} (-1)^{\overline{v_1}+\cdots+\overline{v_{i-1}}} v_1 \wedge \cdots \wedge d(v_i) \wedge \cdots \wedge v_n.$$

It is immediate from the above formulas that if $f\colon V \to W$ is a morphism of DG-vector spaces, then also

$$f^{\otimes n}\colon V^{\otimes n} \to W^{\otimes n}, \qquad f^{\otimes n}(v_1 \otimes \cdots \otimes v_n) = f(v_1) \otimes \cdots \otimes f(v_n),$$

$$f^{\odot n}\colon V^{\odot n} \to W^{\odot n}, \qquad f^{\odot n}(v_1 \otimes \cdots \otimes v_n) = f(v_1) \odot \cdots \odot f(v_n),$$

$$f^{\wedge n}\colon V^{\wedge n} \to W^{\wedge n}, \qquad f^{\wedge n}(v_1 \otimes \cdots \otimes v_n) = f(v_1) \wedge \cdots \wedge f(v_n),$$

are morphisms of DG-vector spaces.

Finally, a tedious but straightforward computation shows that for every n, the symmetrization map

$$N\colon V^{\odot n} \to V^{\otimes n}, \qquad N(v_1 \odot \cdots \odot v_n) = \sum_{\sigma \in \Sigma_n} \epsilon(\sigma) v_{\sigma(1)} \otimes \cdots \otimes v_{\sigma(n)}$$

is a morphism of DG-vector spaces such that $\pi N = n!\,\mathrm{Id}$, cf. Lemma 11.4.4, and then $V^{\odot n}$ is a (differential, graded) direct summand of $V^{\otimes n}$.

Proposition 10.1.7 *Let $f: V \to W$ be a morphism of DG-vector spaces over a field of characteristic 0. If $f: H^*(V) \to H^*(W)$ is injective (resp.: surjective) then*

$$f^{\otimes n}: H^*(V^{\otimes n}) \to H^*(W^{\otimes n}), \qquad f^{\odot n}: H^*(V^{\odot n}) \to H^*(W^{\odot n}),$$

are injective (surjective) for every $n > 0$. In particular, for every n there exists an isomorphism $H^*(V)^{\odot n} \simeq H^*(V^{\odot n})$.

Proof The part about tensor powers follows from Künneth's formula $H^*(V^{\otimes n}) \simeq H^*(V)^{\otimes n}$. The part about symmetric powers follows from the fact that $f^{\otimes n}$ and $f^{\odot n}$ commute with the projection $\pi: V^{\otimes n} \to V^{\odot n}$ and the symmetrization map $N: V^{\odot n} \to V^{\otimes n}$. For the last part it is sufficient to recall that by Lemma 5.1.4 there exists a quasi-isomorphism of DG-vector spaces $H^*(V) \to V$. □

Example 10.1.8 Every contraction (Definition 5.4.1)

$$M \underset{\pi}{\overset{\imath}{\rightleftarrows}} N \circlearrowleft h$$

extends canonically to two different contractions between tensor algebras:

$$T(M) \underset{T(\pi)}{\overset{T(\imath)}{\rightleftarrows}} T(N) \circlearrowleft k,l \; .$$

It is not restrictive to assume that $\imath$ is the inclusion morphism of a differential graded vector subspace $M \subset N$. Then the homotopy l is defined by the formula $l = \sum_{i=1}^{n} (\imath\pi)^{\otimes i-1} \otimes h \otimes \mathrm{Id}_N^{\otimes n-i}$; for instance

$$l(x_1 \otimes x_2) = h(x_1) \otimes x_2 + (-1)^{\overline{x_1}} \imath\pi(x_1) \otimes h(x_2),$$

and we leave as an easy exercise the proof that

$$T(M) \underset{T(\pi)}{\overset{T(\imath)}{\rightleftarrows}} T(N) \circlearrowleft l$$

is a contraction.

For the definition of k, let V be the kernel of π. Then $N = M \oplus V$, $h(M) = 0$ and $-h: V \to V$ is a contracting homotopy, i.e., $(dh + hd)(v) = -v$ for every $v \in V$. Every element of $T(N)$ is a finite sum of elementary tensors of type

$$x_1 \otimes \cdots \otimes x_n, \quad \text{with either } x_i \in M \text{ or } x_i \in V. \tag{10.4}$$

For an element as in (10.4) define $k(x_1 \otimes \cdots \otimes x_n) = 0$ if $x_i \in M$ for every i, and

$$k(x_1 \otimes \cdots \otimes x_n) = \frac{1}{p}\sum_{i=1}^{n}(-1)^{\overline{x_1}+\cdots+\overline{x_{i-1}}} x_1 \otimes \cdots \otimes h(x_i) \otimes \cdots \otimes x_n,$$

where p is the number of vectors x_i belonging to V. The proof that

$$k^2 = kT(\imath) = T(\pi)k = 0, \qquad dk + kd = T(\imath)T(\pi) - \mathrm{Id}_{T(N)},$$

is completely straightforward.

10.2 Formal Neighbourhoods of Graded Vector Spaces

Denote by **NCGA** the category of commutative graded algebras that are nilpotent and finite-dimensional as a vector space over $\mathbb{K}$. The same argument as used in Sect. 3.1 shows that **NCGA** is equivalent to a small category (cf. the proof of Theorem 10.2.2 below) and we may consider the category of functors **NCGA** $\to$ **Set**.

For every graded vector space V we define its **formal neighbourhood** as the functor

$$\widehat{V}\colon \mathbf{NCGA} \to \mathbf{Set}, \qquad \widehat{V}(A) = (V \otimes A)^0.$$

A morphism of formal neighbourhood is a natural transformation of functors. Notice that for every integer n, the formal neighbourhood of $V[n] = \mathbb{K}[n] \otimes V$ is naturally isomorphic to the functor $A \mapsto (V \otimes A)^n$.

Remark 10.2.1 Every proper ideal of a local Artin $\mathbb{K}$-algebra $B \in \mathbf{Art}_{\mathbb{K}}$ belongs to **NCGA**. Conversely, every algebra $\mathfrak{m}$ in **NCGA** concentrated in degree 0 is the maximal ideal of $\mathbb{K} \oplus \mathfrak{m} \in \mathbf{Art}_{\mathbb{K}}$. For any (non-graded) vector space V and any $A \in \mathbf{NCGA}$ we have $\widehat{V}(A) = V \otimes A^0$; therefore the formal neighbourhood of V is uniquely determined by the functor of Artin rings

$$\mathbf{Art}_{\mathbb{K}} \ni B \mapsto V \otimes \mathfrak{m}_B$$

already considered in Example 3.2.2.

Every morphism of graded vector spaces $f\colon V \to W$ induces naturally a morphism of formal neighbourhood $\widehat{f}\colon \widehat{V} \to \widehat{W}$; for every $A \in \mathbf{NCGA}$ the map of sets $\widehat{f}_A\colon (V \otimes A)^0 \to (W \otimes A)^0$ is defined as

$$\widehat{f}_A\left(\sum v_i \otimes a_i\right) = \sum f(v_i) \otimes a_i, \qquad v_i \in V,\ \ a_i \in A,\ \ \overline{v_i} + \overline{a_i} = 0.$$

More generally, for every positive integer n, any morphism of graded vector spaces $f\colon V^{\odot n} \to W$ induces a morphism of formal neighbourhood $\widehat{f}\colon \widehat{V} \to \widehat{W}$ defined by

$$\widehat{f}_A\left(\sum v_i \otimes a_i\right) = f_A\left(\frac{1}{n!}\left(\sum v_i \otimes a_i\right)^{\odot n}\right), \quad v_i \in V,\ a_i \in A,\ \overline{v_i} + \overline{a_i} = 0,$$

where $f_A\colon (V \otimes A)^{\odot n} \to W \otimes A$ is the linear extension of the map

$$f_A(v_1 \otimes a_1, \dots, v_n \otimes a_n) = (-1)^{\sum_{i<j} \overline{a_i}\,\overline{v_j}} f(v_1, \dots, v_n) \otimes a_1 \cdots a_n.$$

Since every $A \in \mathbf{NCGA}$ is nilpotent, we may extend by linearity the above construction to every morphism of graded vector space $f\colon \bigoplus_{n=1}^{\infty} V^{\odot n} \to W$; more precisely we have

$$\widehat{f}_A\left(\sum v_i \otimes a_i\right) = \sum_{n=1}^{\infty} f_A\left(\frac{(\sum v_i \otimes a_i)^{\odot n}}{n!}\right) \quad v_i \in V,\ a_i \in A,\ \overline{v_i} + \overline{a_i} = 0. \tag{10.5}$$

The vector space structure on W induces a vector space structure of the set of natural transformation $\mathrm{Mor}(\widehat{V}, \widehat{W})$ between formal neighbourhoods and the formula (10.5) defines a linear map,

$$\mathrm{Hom}^0_{\mathbb{K}}\left(\bigoplus_{n=1}^{\infty} V^{\odot n}, W\right) \to \mathrm{Mor}\left(\widehat{V}, \widehat{W}\right), \qquad f \mapsto \widehat{f}. \tag{10.6}$$

Theorem 10.2.2 *For every pair of graded vector spaces V, W the linear map* (10.6) *is bijective.*

Proof For a finite sequence of indeterminates $x_1, \dots, x_n$ of integral degree let's denote by $\Lambda(x_1, \dots, x_n) \in \mathbf{NCGA}$ the ideal generated by $x_1, \dots, x_n$ in the graded algebra

$$\frac{\mathbb{K}[x_1, \dots, x_n]}{(x_1^2, \dots, x_n^2)} = \bigotimes_{i=1}^{n} \frac{\mathbb{K}[x_i]}{(x_i^2)}.$$

The major step in the proof of the theorem is given by the following lemma. □

Lemma 10.2.3 *Let n be a positive integer and let $\gamma\colon \widehat{V} \to \widehat{W}$ be a natural transformation such that $\gamma_A = 0$ whenever $A = \Lambda(x_1, \dots, x_s)$ with $s < n$. Then there exists a unique morphism of graded vector space $f\colon V^{\odot n} \to W$ such that $\gamma_A = \widehat{f}_A$ whenever $A = \Lambda(x_1, \dots, x_s)$ with $s \le n$.* □

Proof Let's first prove the unicity of f. Let $v_1, \dots, v_n \in V$ be homogeneous elements and consider n indeterminates $x_1, \dots, x_n$ of degree $\overline{x_i} = -\overline{v_i}$. Taking $A = \Lambda(x_1, \dots, x_n)$ we have

$$\begin{aligned}\gamma_A(v_1\otimes x_1+\cdots+v_n\otimes x_n)&=\widehat{f_A}(v_1\otimes x_1+\cdots+v_n\otimes x_n)\\ &=\frac{1}{n!}f_A((v_1\otimes x_1+\cdots+v_n\otimes x_n)^{\odot n})\\ &=(-1)^{\sum_{i<j}\overline{x_i}\,\overline{v_j}}f(v_1,\ldots,v_n)\otimes x_1\cdots x_n.\end{aligned} \tag{10.7}$$

As regards the existence, consider the projections

$$A\to A_i:=A/(x_i)=\Lambda(x_1,\ldots,\widehat{x_i},\ldots,x_n).$$

Then the assumption that $\gamma_{A_i}=0$ implies that

$$\gamma_A(v_1\otimes x_1+\cdots+v_n\otimes x_n)=w\otimes x_1\cdots x_n$$

for some $w\in W$ and then (10.7) can be also used to define properly a graded symmetric map $f\colon V\times\cdots\times V\to W$. To conclude we need to prove that f is multilinear, and by symmetry it is enough to see that for every $a,b\in\mathbb{K}$ and $u,v\in V$ of the same degree we have

$$f(au+bv,v_2,\ldots,v_n)=af(u,v_2,\ldots,v_n)+bf(v,v_2,\ldots,v_n). \tag{10.8}$$

Let y,z be formal variables of degree $\overline{y}=\overline{z}=\overline{x_1}$ and consider the morphism of graded algebras

$$B:=\Lambda(y,z,x_2,\ldots,x_n)\xrightarrow{p}A=\Lambda(x_1,x_2,\ldots,x_n),$$

$$p(y)=ax_1,\qquad p(z)=bx_1,\qquad p(x_i)=x_i,\quad i>1.$$

Notice that $p(yz)=abx_1^2=0$. As above, the functoriality of γ with respect to the quotients of B gives

$$\begin{aligned}&(-1)^{\sum_{i<j}\overline{x_i}\,\overline{v_j}}\gamma_B(u\otimes y+v\otimes z+v_2\otimes x_2+\cdots+v_n\otimes x_n)\\ &\quad=f(u,v_2,\ldots,v_n)\otimes yx_2\cdots x_n+f(v,v_2,\ldots,v_n)\otimes zx_2\cdots x_n\\ &\quad\quad+w\otimes yzx_2\cdots x_n\end{aligned}$$

for some $w\in W$, and functoriality with respect to the morphism $p\colon B\to A$ gives (10.8). □

Proof We are now ready to prove Theorem 10.2.2. Since the field $\mathbb{K}$ is assumed of characteristic 0, if $s>0$ and $t,x_1,\ldots,x_s$ are indeterminates of the same even degree, then the morphism of graded algebras

$$\frac{\mathbb{K}[t]}{(t^{s+1})}\to\frac{\mathbb{K}[x_1,\ldots,x_s]}{(x_1^2,\ldots,x_s^2)},\qquad t\mapsto x_1+\cdots+x_s,$$

is well defined and injective. Therefore every $A \in \mathbf{NCGA}$ fits into a diagram

$$A \xleftarrow{p} B \xrightarrow{i} \Lambda(x_1, \dots, x_n), \qquad \overline{x_i} \in \mathbb{Z},$$

where p is surjective, B is the ideal generated by $t_1, \dots, t_r$ in a graded algebra of type

$$\bigotimes_{i=1}^{r} \frac{\mathbb{K}[t_i]}{(t_i^{s_i+1})}, \qquad \overline{t_i} \in \mathbb{Z},$$

and i is injective. Clearly, $p(t_1), \dots, p(t_s)$ is a set of homogeneous generators of A, $s_i > 0$ and $s_i = 1$ whenever the degree of t_i is odd.

Thus, if Λ is the full subcategory of **NCGA** of all algebras of type $\Lambda(x_1, \dots, x_n)$, every natural transformation of formal neighbourhoods is uniquely determined by its restriction to Λ.

Let $f \in \operatorname{Hom}^0_{\mathbb{K}}\left(\bigoplus_{n=1}^{\infty} V^{\odot n}, W\right)$ be such that $\widehat{f}_A = 0$ for every $A \in \Lambda$; by induction on n and Lemma 10.2.3 we obtain $f(V^{\odot n}) = 0$ for every n and therefore $f = 0$.

Conversely, given $\gamma \in \operatorname{Mor}\left(\widehat{V}, \widehat{W}\right)$, again by induction on n and Lemma 10.2.3 we can find a sequence of maps $f_n \in \operatorname{Hom}^0_{\mathbb{K}}\left(V^{\odot n}, W\right)$ such that the morphism $\gamma - \sum_{i=1}^{n} \widehat{f_i}$ vanishes on every algebra of type $\Lambda(x_1, \dots, x_s)$ with $s \leq n$. Therefore $\gamma = \widehat{f}$, where $f = \sum_{n=1}^{\infty} f_n$. □

A quite surprising consequence of Theorem 10.2.2 is that the usual composition map

$$\operatorname{Hom}^0_{\mathbb{K}}(V, W) \times \operatorname{Hom}^0_{\mathbb{K}}(W, U) \to \operatorname{Hom}^0_{\mathbb{K}}(V, U),$$

has a nontrivial extension to an associative product

$$\operatorname{Hom}^0_{\mathbb{K}}\left(\bigoplus_{n=1}^{\infty} V^{\odot n}, W\right) \times \operatorname{Hom}^0_{\mathbb{K}}\left(\bigoplus_{n=1}^{\infty} W^{\odot n}, U\right) \to \operatorname{Hom}^0_{\mathbb{K}}\left(\bigoplus_{n=1}^{\infty} V^{\odot n}, U\right),$$

which corresponds to the composition of morphisms in the category of formal neighbourhoods. We shall give the explicit description of this product later on, by showing that the category of formal neighbourhoods is equivalent to the category of reduced symmetric graded coalgebras (Theorem 11.6.3).

10.3 A Simple Model of Infinity Structure

The aim if this is to introduce certain typical properties of L_∞-algebras in an easier framework where technical difficulties are reduced to the minimum. As usual every graded vector space is considered over a fixed field $\mathbb{K}$ of characteristic 0.

Definition 10.3.1 A **DG$_\infty$-vector space** over $\mathbb{K}$, denoted by $(V, d_1, d_2, \ldots)$, is the data of a graded vector space V and a sequence $d_i \in \operatorname{Hom}^1_{\mathbb{K}}(V, V)$, $i \geq 1$, of linear maps such that

$$\frac{1}{2}\sum_{i=1}^{k}[d_{k-i+1}, d_i] = \sum_{i=1}^{k} d_{k-i+1}d_i = 0 \qquad \text{for every } \; k > 0. \tag{10.9}$$

For low values of k, the Equation (10.9) becomes:

$$\begin{aligned} d_1^2 &= 0 && k=1\\ d_1d_2 + d_2d_1 &= 0 && k=2\\ d_1d_3 + d_2^2 + d_3d_1 &= 0 && k=3\\ d_1d_4 + d_2d_3 + d_3d_2 + d_4d_1 &= 0 && k=4\\ \cdots\ &\text{et cetera.} \end{aligned}$$

The first of the above equations tells us that (V, d_1) is a DG-vector space, the second implies that the linear map d_2 factors to linear map $\overline{d_2}\colon H^*(V, d_1) \to H^*(V, d_1)$ of degree 1, while the third equation tells us that $\overline{d_2}^2 = 0$ and therefore that $(H^*(V, d_1), \overline{d_2})$ is a DG-vector space.

Example 10.3.2 If $d_i = 0$ for every $i > 1$, then (10.9) reduces to the single equation $d_1^2 = 0$. Thus every DG-vector spaces (V, d) can be interpreted as the DG$_\infty$-vector space $(V, d, 0, 0, 0, \ldots)$ and then there exists a natural bijection between DG-vector spaces and DG$_\infty$-vector space $(V, d_1, \ldots)$ such that $d_i = 0$ for every $i > 1$.

Example 10.3.3 If $d_i = 0$ for every $i > 2$, then (10.9) reduces to:

$$d_1^2 = 0, \qquad d_1d_2 + d_2d_1 = 0, \qquad d_2^2 = 0.$$

Therefore there exists a natural bijection between double complexes and DG$_\infty$-vector spaces $(V, d_1, d_2, 0, 0, 0, \ldots)$ such that $d_i = 0$ for every $i > 1$.

Example 10.3.4 Let $V = \oplus_{n=0}^{\infty} V_n$ be the direct sum of a sequence of graded vector spaces $V_1.V_2, \ldots$ and let $d\colon V \to V$ be a differential preserving the decreasing filtration $F^pV = \oplus_{n\geq p} V_n$. Writing

$$d = \sum_{i\geq 1} d_i, \quad \text{with} \quad d_i(V_n) \subseteq V_{n+i-1} \quad \text{for every } i, n,$$

we have that $(V, d_1, d_2, \ldots)$ is a DG$_\infty$-vector space.

We denote by $\mathbf{DG}_\infty$ the category of DG$_\infty$-vector spaces, where the morphisms and composition rule are described in the following definition.

Definition 10.3.5 A DG_∞-morphism

$$f = (f_1, f_2, \dots): (V, d_1, d_2, \dots) \to (W, \delta_1, \delta_2, \dots)$$

of DG_∞-vector spaces is given by a sequence of morphisms of graded vector spaces $f_i: V \to W$ such that for every positive integer k we have

$$\sum_{i+j=k+1} f_i d_j = \sum_{i+j=k+1} \delta_j f_i. \tag{10.10}$$

Composition of DG_∞-morphisms is defined by the "Cauchy product" formula

$$fg = (f_1, f_2, \dots)(g_1, g_2, \dots) = \left(f_1 g_1, \; f_1 g_2 + f_2 g_1, \; \dots, \; \sum_{i+j=k+1} f_i g_j, \dots \right).$$

It is straightforward to see that the composition is associative and properly defined. The identity morphism on a DG_∞-vector space V is given by the sequence $(\mathrm{Id}_V, 0, 0, \dots)$. It is a simple consequence of (10.10) that for every DG_∞-morphism

$$f = (f_1, f_2, \dots): (V, d_1, d_2, \dots) \to (W, \delta_1, \delta_2, \dots),$$

the map $f_1: (V, d_1) \to (W, \delta_1)$ is a morphism of DG-vector spaces, as well as the map $f_1: (H^*(V, d_1), \overline{d_2}) \to (H^*(W, \delta_1), \overline{\delta_2})$. Conversely, if $f_1: (V, d_1) \to (W, \delta_1)$ is a morphism of cochain complexes, then

$$(f_1, 0, 0, \dots): (V, d_1, 0, \dots) \to (W, \delta_1, 0, \dots)$$

is a DG_∞-morphism. In other words we have two natural functors:

$$\begin{aligned} i: \mathbf{DG} \to \mathbf{DG}_\infty, &\qquad (V, d) \mapsto (V, d, 0, \dots), \\ \mathbf{DG}_\infty \to \mathbf{DG}, &\qquad (V, d_1, d_2, \dots) \mapsto (V, d_1). \end{aligned}$$

It is also useful and instructive to describe $\mathbf{DG}_\infty$ as a subcategory of $\mathbf{DG}$ via the faithful functor

$$B: \mathbf{DG}_\infty \to \mathbf{DG}, \qquad B(V, d_1, d_2, \dots) = (V[t], D),$$

where $V[t] = V \otimes \mathbb{K}[t]$ is the graded vector space of polynomials with coefficients in V and the differential $D \in \mathrm{Hom}^1_{\mathbb{K}}(V[t], V[t])$ is defined by the formula

$$D(vt^n) = \sum_{i=0}^{n} d_{i+1}(v) t^{n-i}. \tag{10.11}$$

Notice that the condition $D^2 = 0$ is equivalent to (10.9) for every k.

Given two DG$_\infty$-vector spaces $(V, d_1, \ldots)$ and $(W, \delta_1, \ldots)$, let $D\colon V[t] \to V[t]$ and $\Delta\colon W[t] \to W[t]$ be the differentials defined as above. Then a sequence of linear maps $f_1, f_2, \ldots \in \operatorname{Hom}^0_{\mathbb{K}}(V, W)$ gives a DG$_\infty$-morphism

$$(f_1, f_2, \ldots)\colon (V, d_1, d_2, \ldots) \to (W, \delta_1, \delta_2, \ldots)$$

if and only if the map

$$F\colon (V[t], D) \to (W[t], \Delta), \qquad F(vt^n) = \sum_{i=0}^{n} f_{i+1}(v)t^{n-i}, \tag{10.12}$$

is a morphism of DG-vector spaces, cf. Example 11.2.3.

Defining the functor B on morphisms according to (10.12), it is very easy to prove that B is a faithful functor.

10.4 L_∞-Algebras

We are now ready to define L_∞-algebras. For the reader's convenience we recall the definition of shuffle permutations.

Definition 10.4.1 (Shuffles) Given two integers $p, q \geq 0$, a (p, q)**-shuffle** is a permutation σ of the set $\{1, \ldots, p+q\}$ such that

$$\sigma(1) < \sigma(2) < \cdots < \sigma(p) \quad \text{and} \quad \sigma(p+1) < \cdots < \sigma(p+q).$$

The subset of (p, q)-shuffles is denoted by $S(p, q) \subset \Sigma_{p+q}$.

For example, for $p + q \leq 2$ we have $S(0, 0) = \Sigma_0$, $S(1, 0) = S(0, 1) = \Sigma_1$, $S(1, 1) = \Sigma_2$. The three (2, 1)-shuffles are:

$$(1, 2, 3), \qquad (1, 3, 2), \qquad (2, 3, 1).$$

Notice also that $S(0, n) = S(n, 0) = \{\mathrm{Id}\}$ for every $n \geq 0$. The cardinality of $S(p, q)$ is $\binom{p+q}{p}$ and the (p, q)-shuffles are a system of representatives for the left cosets of the canonical embedding of $\Sigma_p \times \Sigma_q$ inside Σ_{p+q}. More precisely, for every $\eta \in \Sigma_{p+q}$ there exists a unique decomposition $\eta = \sigma\tau$ with $\sigma \in S(p, q)$ and $\tau \in \Sigma_p \times \Sigma_q$. Following Stasheff, we call the inverse of a shuffle an unshuffle.

Before giving the definition of L_∞-algebra, we note that the axioms of a differential graded Lie algebra $(L, d, [-, -])$ over a field of characteristic 0:

1. $d(d(x_1)) = 0$;
2. $d[x_1, x_2] - \left([dx_1, x_2] - (-1)^{\overline{x_1}\,\overline{x_2}}[dx_2, x_1]\right) = 0$;

3. $[[x_1, x_2], x_3] - (-1)^{\overline{x_2}\,\overline{x_3}}[[x_1, x_3], x_2] + (-1)^{\overline{x_1}(\overline{x_2}+\overline{x_3})}[[x_2, x_3], x_1] = 0$;

may be written, by using the formalism of shuffles and the antisymmetric Koszul sign, respectively as:

1. $\sum_{\sigma \in S(1,0)} d(d(x_{\sigma(1)})) = 0$;

2. $\sum_{\sigma \in S(2,0)} \chi(\sigma) d[x_{\sigma(1)}, x_{\sigma(2)}] - \sum_{\sigma \in S(1,1)} \chi(\sigma)[dx_{\sigma(1)}, x_{\sigma(2)}] = 0$;

3. $\sum_{\sigma \in S(2,1)} \chi(\sigma)[[x_{\sigma(1)}, x_{\sigma(2)}], x_{\sigma(3)}] = 0.$

Definition 10.4.2 (L_∞-algebra) An L_∞ **structure** (short for L_∞-algebra structure) on a graded vector space L is a sequence of graded skew-symmetric maps

$$l_n \colon L^{\wedge n} \to L, \qquad \deg(l_n) = 2 - n, \quad n > 0,$$

such that for every $n > 0$ and every sequence $x_1, \ldots, x_n$ of homogeneous vectors of L we have:

$$\sum_{k=1}^{n} (-1)^{n-k} \sum_{\sigma \in S(k,n-k)} \chi(\sigma)\, l_{n-k+1}(l_k(x_{\sigma(1)}, \ldots, x_{\sigma(k)}), x_{\sigma(k+1)}, \ldots, x_{\sigma(n)}) = 0. \tag{10.13}$$

An L_∞-**algebra** $(L, l_1, l_2, \ldots)$ is a graded vector space L equipped with an L_∞ structure $l_1, l_2, \ldots$.

Notice that, since every l_k is skew-symmetric, Equation (10.13) may be also written as

$$\sum_{k=1}^{n} (-1)^{n-k} \sum_{\sigma \in \Sigma_n} \frac{\chi(\sigma)}{k!(n-k)!} l_{n-k+1}(l_k(x_{\sigma(1)}, \ldots, x_{\sigma(k)}), x_{\sigma(k+1)}, \ldots, x_{\sigma(n)}) = 0.$$

In the literature, L_∞-algebras are also called either **strong homotopy Lie algebras** or **Sugawara algebras**.

Remark 10.4.3 Notice that:

1. Every differential graded Lie algebra $(L, d, [-,-])$ is also an L_∞-algebra with $l_1 = d$, $l_2 = [-,-]$ and $l_n = 0$ for every $n > 2$.
2. For an L_∞-algebra $(L, l_1, l_2, \ldots)$ we have $\deg(l_1) = 1$ and $l_1^2 = 0$. Therefore (L, l_1) is a DG-vector space.
3. For an L_∞-algebra $(L, l_1, l_2, \ldots)$ the morphism l_2 induces a structure of graded Lie algebra in the cohomology of the complex (L, l_1). In fact, if $l_1(x) = l_1(y) = 0$ then $l_1(l_2(x, y)) = \pm l_2(l_1(x), y) \pm l_2(l_1(y), x) = 0$. The equation

$$\sum_{k=1}^{3}(-1)^{3-k}\sum_{\sigma\in S(k,3-k)}\chi(\sigma)\,l_{4-k}(l_k(x_{\sigma(1)},\ldots,x_{\sigma(k)}),x_{\sigma(k+1)},\ldots,x_{\sigma(3)})=0$$

implies that if $l_1(x_i)=0$ then

$$\sum_{\sigma\in S(2,1)}\chi(\sigma)\,l_2(l_2(x_{\sigma(1)},x_{\sigma(2)}),x_{\sigma(3)})$$

belongs to the image of l_1 and then the bracket

$$[-,-]\colon H^*(L,l_1)\times H^*(L,l_1)\to H^*(L,l_1),\qquad [x,y]=l_2(x,y),$$

satisfies the graded Jacobi identity.

Example 10.4.4 Let L be a graded vector space and let $f\colon L^0\times L^0\times L^0\to L^{-1}$ be a skew-symmetric map. Then the sequence of maps:

$$l_n(x_1,\ldots,x_n)=\begin{cases} f(x_1,x_2,x_3) & \text{if } n=3 \text{ and } x_i\in L^0,\\ 0 & \text{otherwise,}\end{cases}$$

gives an L_∞ structure on L.

Definition 10.4.5 A **linear morphism** $f\colon (L,l_1,l_2,l_3,\ldots)\to(H,h_1,h_2,h_3,\ldots)$ of L_∞-algebras is a linear map $f\colon L\to H$ of degree 0 such that

$$fl_n(x_1,\ldots,x_n)=h_n(f(x_1),\ldots,f(x_n))$$

for every $n>0$ and every $x_1,\ldots,x_n\in L$.

As an example, every morphism of differential graded Lie algebras is a linear morphism of L_∞-algebras.

Do not confuse the notion of linear morphism of L_∞-algebras with the notion (that we shall give later) of L_∞-morphisms. Every linear morphism is also an L_∞-morphism but the converse is not true. In the literature, linear morphisms of L_∞-algebras are sometimes called *strict morphisms*; the reader must pay attention to the fact that the strict L_∞-morphisms defined in the following chapters of this book are not linear in general, although they are equivalent to linear morphisms up to L_∞-isomorphisms, see next Lemma 12.1.4.

Example 10.4.6 (Finite products of L_∞-algebras) Given two L_∞-algebras $(L,l_1,l_2,\ldots)$ and $(H,h_1,h_2,\ldots)$, their product is defined in the obvious way:

$$(L,l_1,l_2,\ldots)\times(H,h_1,h_2,\ldots):=(L\times H,l_1\times h_1,l_2\times h_2,\ldots),$$

where

$$l_n \times h_n((v_1, w_1), \dots, (v_n, w_n)) = (l_n(v_1, \dots, v_n), h_n(w_1, \dots, w_n)).$$

We can immediately see that it is a product in the category of L_∞-algebras and linear morphisms.

Remark 10.4.7 The scalar extension for DG-Lie algebras (Example 5.6.3) generalizes immediately to L_∞-algebras. Given an L_∞-algebra $(L, l_1, l_2, \dots)$ and a graded *commutative* DG-algebra (A, d), we may define an L_∞-algebra $(L \otimes A, m_1, m_2, \dots)$ by setting:

$$m_1(v \otimes a) = l_1(v) \otimes a + (-1)^{\overline{v}} v \otimes d(a),$$

$$m_n(v_1 \otimes a_1, \dots, v_n \otimes a_n) = (-1)^{\sum_{i<j} \overline{a_i}\,\overline{v_j}} l_n(v_1, \dots, v_n) \otimes a_1 \cdots a_n, \qquad n > 1. \tag{10.14}$$

Notice that $(-1)^{\sum_{i<j} \overline{a_i}\,\overline{v_j}}$ is the Koszul sign relating the sequences $v_1, a_1, v_2, a_2, \dots, v_n, a_n$ and $v_1, \dots, v_n, a_1, \dots, a_n$. The proof that the maps (10.14) satisfy the Eq. (10.13) is an easy consequence of the graded commutativity of A and the Leibniz formula for the differential d.

This construction is functorial in the category of L_∞-algebras and linear morphisms. Moreover, every morphism of DG-algebras $A \to B$ induces a linear morphism of L_∞-algebras $L \otimes A \to L \otimes B$. In a similar way, for L and A as above we may define the L_∞-algebra $A \otimes L$; for simplicity of notation we often write $L[t, dt]$ for the L_∞-algebra $\mathbb{K}[t, dt] \otimes L$.

10.5 Maurer–Cartan and Deformation Functors

Most of the notions about DG-Lie algebras extend to the framework of L_∞-algebras. For instance, the descending central series $L^{[n]}$ of an L_∞-algebra $(L, l_1, l_2, \dots)$ is defined recursively as $L^{[1]} = L$ and

$$L^{[n]} = \operatorname{Span}\{l_k(x_1, \dots, x_k) \mid k \geq 2,\ x_i \in L^{[n_i]},\ 0 < n_i < n,\ n_1 + \cdots + n_k \geq n\}.$$

An L_∞-algebra $(L, l_1, l_2, \dots)$ is called **nilpotent** if $L^{[n]} = 0$ for $n >> 0$; notice that $L^{[n]} = 0$ implies in particular that $l_k = 0$ for every $k \geq n$.

For instance, if L is any L_∞-algebra and $A \in \mathbf{Art}_{\mathbb{K}}$ we have $(L \otimes \mathfrak{m}_A)^{[n]} \subset L \otimes \mathfrak{m}_A^n$ and then $L \otimes \mathfrak{m}_A$ is nilpotent.

Definition 10.5.1 A Maurer–Cartan element in a nilpotent L_∞-algebra $(L, l_1, l_2, \dots)$ is a vector $x \in L^1$ that satisfies the **Maurer–Cartan equation**:

$$\sum_{n>0} \frac{1}{n!} l_n(x, x, \dots, x) = 0\,. \tag{10.15}$$

The subset of Maurer-Cartan elements will be denoted by MC(L).

Thus for every L_∞-algebra L it makes sense to consider the Maurer-Cartan functor

$$\mathrm{MC}_L\colon \mathbf{Art}_{\mathbb{K}} \to \mathbf{Set}, \qquad \mathrm{MC}_L(A) = \mathrm{MC}(L \otimes \mathfrak{m}_A),$$

where the L_∞ structure on $L \otimes \mathfrak{m}_A$ is given by scalar extension. It is clear that MC_L commutes with fibre products in $\mathbf{Art}_{\mathbb{K}}$ and therefore it is a deformation functor.

Lemma 10.5.2 *Let $L = (L, l_1, l_2, \dots)$ be an L_∞-algebra. Then the tangent space of* MC_L *is $Z^1(L, l_1)$ and there exists a natural complete obstruction theory for* MC_L, *with values in $H^2(L, l_1)$.*

Proof The first part is clear, since the Maurer–Cartan equation reduces to $l_1(x) = 0$ whenever $\mathfrak{m}_A^2 = 0$. Assume that $0 \to I \to A \to B \to 0$ is a small extension in $\mathbf{Art}_{\mathbb{K}}$; given a Maurer–Cartan element $y \in \mathrm{MC}(L \otimes \mathfrak{m}_B)$, choose a lifting $x \in L^1 \otimes \mathfrak{m}_A$ of it and consider its value under the Maurer–Cartan functional

$$h(x) = \sum_{n>0} \frac{1}{n!} l_n(x, x, \dots, x) \in L^2 \otimes I.$$

We want to prove that $l_1(h(x)) = 0$ and that its cohomology class in $H^2(L, l_1) \otimes I$ is the required obstruction. For every $n > 0$ we have

$$0 = \sum_{k=1}^{n} (-1)^{n-k} \sum_{\sigma \in S(k,n-k)} \chi(\sigma)\, l_{n-k+1}(l_k(x, \dots, x), x, \dots, x)$$

and then, since x has odd degree, we have $\chi(\sigma) = 1$ for every permutation σ and the above equation simplifies to

$$0 = \sum_{k=1}^{n} (-1)^{n-k} \binom{n}{k} l_{n-k+1}(l_k(x, \dots, x), x, \dots, x).$$

Dividing for $n!$ we get

$$0 = \sum_{k=1}^{n} (-1)^{n-k} \frac{l_{n-k+1}}{(n-k)!} \left(\frac{l_k}{k!}(x, \dots, x), x, \dots, x \right) = 0$$

and summing over all $n > 0$ we obtain (setting $a = n - k + 1$)

$$\begin{aligned}
0 &= \sum_{a,k>0} (-1)^{a-1} \frac{l_a}{(a-1)!} \left(\frac{l_k}{k!}(x, \ldots, x), x, \ldots, x \right) \\
&= \sum_{a>0} (-1)^{a-1} \frac{l_a}{(a-1)!} (h(x), x, \ldots, x) \\
&= l_1(h(x)),
\end{aligned}$$

where the last equality follows from the fact that $h(x) \in L \otimes I$.

For every $s \in L^1 \otimes I$ we have $h(x+s) = h(x) + l_1(s)$ and then the cohomology class of $h(x)$ in $H^2(L, l_1) \otimes I$ depends only on y; it is plain that y admits a lifting to $\mathrm{MC}_L(A)$ if and only if $h(x)$ is a coboundary in the complex $(L \otimes I, l_1)$. □

Every linear morphism $f : L \to M$ of L_∞-algebras induces a natural transformation of Maurer–Cartan functors that commutes with the obstruction maps introduced in the proof of Lemma 10.5.2.

Proposition 10.5.3 *Let $f : (L, l_1, l_2, \ldots) \to (M, m_1, m_2, \ldots)$ be a linear morphism of L_∞-algebras. Suppose that:*

1. *$f : L^0 \to M^0$ is surjective,*
2. *$f : H^1(L, l_1) \to H^1(M, m_1)$ is surjective,*
3. *$f : H^2(L, l_1) \to H^2(M, m_1)$ is injective.*

Then $f : \mathrm{MC}_L \to \mathrm{MC}_M$ is a smooth morphism of deformation functors.

Proof The first two conditions imply that $f : Z^1(L, l_1) \to Z^1(M, m_1)$ is surjective and then the proposition follows from the standard smoothness criterion (Theorem 3.6.5). □

Definition 10.5.4 Let $(L, l_1, \ldots)$ be an L_∞-algebra, $A \in \mathbf{Art}_{\mathbb{K}}$ and $x, y \in \mathrm{MC}_L(A)$. We shall say that x and y are **homotopy equivalent** if there exists a Maurer–Cartan element $\xi \in \mathrm{MC}_{L[t,dt]}(A)$ such that $\xi(0) = x$ and $\xi(1) = y$. We denote by

$$\mathrm{Def}_L : \mathbf{Art}_{\mathbb{K}} \to \mathbf{Set}$$

the quotient of MC_L under the equivalence relation generated[1] by homotopy equivalence.

More precisely, by functoriality of scalar extension, the maps $e_0, e_1 : \mathbb{K}[t, dt] \to \mathbb{K}$ extend to linear morphisms of L_∞-algebras $e_0, e_1 : L[t, dt] = \mathbb{K}[t, dt] \otimes L \to L$ and two elements $x, y \in \mathrm{MC}_L(A)$ are homotopy equivalent if there exists $\xi \in \mathrm{MC}_{L[t,dt]}(A)$ such that $e_0(\xi) = x$ and $e_1(\xi) = y$.

Since $e_0, e_1 : (L[t, dt], l_1) \to (L, l_1)$ are both surjective quasi-isomorphisms, according to Proposition 10.5.3 the natural transformations $e_0, e_1 : \mathrm{MC}_{L[t,dt]} \to \mathrm{MC}_L$ are smooth and then Def_L is a deformation functor by Theorem 3.2.8. It is

[1] We shall prove later (Lemma 13.1.3) that homotopy equivalence is already an equivalence relation.

easy to see that the tangent space of the functor Def_L is $H^1(L, l_1)$. This follows from Lemma 10.5.2, from the fact the morphisms of DG-vector spaces

$$e_1, e_0 \colon L[t, dt] \to L$$

are homotopic via the homotopy $\int_0^1 \otimes \mathrm{Id}_L$ and observing that for $z \in Z^1(L)$ and $u \in L^0$ we have $z + tl_1(u) + dt\, u \in Z^1(L[t, dt])$.

A priori the above notation Def_L for the functor of Maurer–Cartan elements up to homotopy may be conflictual with the use of the same symbol for the functor of Maurer–Cartan elements up to gauge equivalence. However, this is not the case in view of the following proposition.

Proposition 10.5.5 *Let L be a differential graded Lie algebra. Then for every local Artin $\mathbb{K}$-algebra $A \in \mathbf{Art}_{\mathbb{K}}$ the homotopy equivalence in* $\mathrm{MC}_L(A)$ *is the same as the gauge equivalence and then the Definition 10.5.4 of the functor* Def_L *coincides with the one given in Sect. 6.4.*

Proof It is sufficient to prove that $x, y \in \mathrm{MC}_L(A)$ are gauge equivalent if and only if they are homotopy equivalent. So first, assume $e^a * x = y$ for some $a \in L^0 \otimes \mathfrak{m}_A$; then we can consider $z(t) = e^{ta} * x \in \mathrm{MC}_{L[t,dt]}(A)$ and therefore $z(0) = x$ and $z(1) = y$. Conversely, assume that $z(0) = x$ and $z(1) = y$ for some $z(t) \in \mathrm{MC}_{L[t,dt]}(A)$; by Corollary 6.4.8 there exists $p(t) \in L^0[t]$ such that $p(0) = 0$ and $z(t) = e^{p(t)} * x$. Then $y = z(1) = e^{p(1)} * x$ and this implies that y is gauge equivalent to x. □

Remark 10.5.6 As a consequence of Proposition 10.5.5 we have that the bifunctor

$$\operatorname{Def} \colon \mathbf{DGLA} \times \mathbf{Art}_{\mathbb{K}} \to \mathbf{Set}$$

is completely determined by the Maurer–Cartan bifunctor

$$\mathrm{MC} \colon \mathbf{DGLA} \times \mathbf{Art}_{\mathbb{K}} \to \mathbf{Set}.$$

10.6 Décalage Isomorphisms and $L_\infty[1]$-Algebras

Following the notation introduced in Sect. 5.1, for a graded vector space V we denote by $sV \cong V[-1]$ its suspension, where s is considered both as a formal symbol of degree 1 and as the tautological linear isomorphism $s \colon V \to V[-1]$ of degree 1.

Definition 10.6.1 The **décalage isomorphisms** of a graded vector spaces V are the linear isomorphisms

$$\text{déc} \colon \operatorname{Hom}^n_{\mathbb{K}}(V^{\otimes k}, V) \xrightarrow{\simeq} \operatorname{Hom}^{n-k+1}_{\mathbb{K}}((sV)^{\otimes k}, sV), \qquad n, k \in \mathbb{Z},\ k \geq 0,$$

defined by imposing the commutativity of the diagrams

$$\begin{array}{ccc} V^{\otimes k} & \xrightarrow{\quad f \quad} & V \\ \big\downarrow{\scriptstyle s^{\otimes k}} & & \big\downarrow{\scriptstyle s} \\ (sV)^{\otimes k} & \xrightarrow{\quad \text{déc}(f) \quad} & sV \end{array}$$

or, equivalently, by the formula

$$\text{déc}(f)(sv_1, \ldots, sv_k) = (-1)^{\sum_i (k-i)\overline{v_i}} sf(v_1, \ldots, v_k), \qquad \overline{v_i} = \deg(v_i; V).$$

The importance of décalage isomorphisms relies essentially on the following proposition.

Proposition 10.6.2 *The décalage isomorphisms exchange symmetric maps into skew-symmetric maps and conversely. For every $n, k \in \mathbb{Z}$, $k \geq 0$, we have two linear isomorphisms:*

$$\begin{aligned} \text{déc}\colon \operatorname{Hom}^n_{\mathbb{K}}(V^{\odot k}, V) &\xrightarrow{\simeq} \operatorname{Hom}^{n-k+1}_{\mathbb{K}}((sV)^{\wedge k}, sV), \\ \text{déc}\colon \operatorname{Hom}^n_{\mathbb{K}}(V^{\wedge k}, V) &\xrightarrow{\simeq} \operatorname{Hom}^{n-k+1}_{\mathbb{K}}((sV)^{\odot k}, sV). \end{aligned} \tag{10.16}$$

Proof Immediate consequence of Lemma 10.1.2. □

The equations (10.13) take a new and simpler form after a décalage isomorphism. It is convenient to encode these new equations into a new algebraic structure, clearly equivalent to the old one.

Definition 10.6.3 An $L_\infty[1]$ **structure** (short for $L_\infty[1]$-algebra structure) on a graded vector space V is a sequence of graded symmetric linear maps

$$q_n\colon V^{\odot n} \to V, \qquad \deg(q_n) = 1, \quad n > 0,$$

such that for every $n > 0$ and every sequence $v_1, \ldots, v_n$ of homogeneous vectors of V we have:

$$\sum_{k=1}^{n} \sum_{\sigma \in S(k,n-k)} \varepsilon(\sigma)\, q_{n-k+1}(q_k(v_{\sigma(1)}, \ldots, v_{\sigma(k)}), v_{\sigma(k+1)}, \ldots, v_{\sigma(n)}) = 0. \tag{10.17}$$

An $L_\infty[1]$ **-algebra** $(V, q_1, q_2, \ldots)$ is a graded vector space equipped with an $L_\infty[1]$ structure.

A linear morphism $f\colon (V, q_1, q_2, \ldots) \to (W, r_1, r_2, \ldots)$ of $L_\infty[1]$-algebras is a linear map $f\colon V \to W$ of degree 0 such that

$$fq_n(x_1, \ldots, x_n) = r_n(f(x_1), \ldots, f(x_n))$$

for every $n > 0$ and every $x_1, \ldots, x_n \in V$. For notational convenience we shall frequently denote an $L_\infty[1]$-algebra $(V, q_1, q_2, \ldots)$ by the pair (V, q), where

$$q \in \operatorname{Hom}^1_{\mathbb{K}}(\oplus_{n\geq 1} V^{\odot n}, V), \qquad q = \sum_{n=1}^{\infty} q_n.$$

If $(V, q_1, q_2, \ldots)$ is an $L_\infty[1]$-algebra, then equation (10.17) gives $q_1^2 = 0$ and therefore (V, q_1) is a DG-vector space. By an $L_\infty[1]$ structure on a DG-vector space (V, d) we intend a sequence of maps $q_n \in \operatorname{Hom}^1_{\mathbb{K}}(V^{\odot n}, V)$, $n \geq 2$, such that $(V, d, q_2, \ldots)$ is an $L_\infty[1]$-algebra.

Lemma 10.6.4 *For every graded vector space V, the opposite of the décalage isomorphisms give a canonical bijection from the set of $L_\infty[1]$ structures on V and the set of L_∞ structures on $sV = V[-1]$:*

$$-\text{déc}\colon \{L_\infty[1] \textit{ structures on } V\} \xrightarrow{\simeq} \{L_\infty \textit{ structures on } sV\}.$$

More explicitly, in the notation above, the bijection is given by:

$$l_k = -\text{déc}(q_k), \qquad l_k(sv_1, \ldots, sv_k) = -(-1)^{\sum_i (k-i)\overline{v_i}} s q_k(v_1, \ldots, v_k).$$

Proof Immediate consequence of Lemma 10.1.2. Notice that for every $k > 0$ we have a commutative diagram

$$\begin{array}{ccc} V^{\otimes k} & \xrightarrow{q_k} & V \\ \Big\downarrow{\scriptstyle s^{\otimes k}} & & \Big\downarrow{\scriptstyle s} \\ (sV)^{\otimes k} & \xrightarrow{-l_k} & sV. \end{array}$$

□

Since $V \cong (sV)[1]$ the above lemma says that there is a bijection between L_∞ structures on L and $L_\infty[1]$ structures on $L[1]$. Very often, in the literature an L_∞ structure on a graded vector space L is defined directly as an $L_\infty[1]$ structure on $L[1]$.

Definition 10.6.5 Given two maps $f \in \operatorname{Hom}^{\overline{f}}_{\mathbb{K}}(V^{\odot n+1}, V)$, $g \in \operatorname{Hom}^{\overline{g}}_{\mathbb{K}}(V^{\odot m+1}, V)$, with $n, m \geq -1$, their **Nijenhuis–Richardson bracket** is defined as

$$[f, g]_{NR} = f \barwedge g - (-1)^{\overline{f}\,\overline{g}} g \barwedge f \in \operatorname{Hom}^*_{\mathbb{K}}(V^{\odot n+m+1}, V),$$

where $f \barwedge g = 0$ for $n = -1$, $f \barwedge g = f \circ g$ for $n = 0$, and

$$f \barwedge g(v_0, \ldots, v_{n+m}) = \sum_{\sigma \in S(m+1,n)} \epsilon(\sigma) f(g(v_{\sigma(0)}, \ldots, v_{\sigma(m)}), v_{\sigma(m+1)}, \ldots, v_{\sigma(m+n)})$$

for $n > 0$ and $v_0, \dots, v_{n+m} \in V$.

The evaluation at $1 \in \mathbb{K} = V^{\odot 0}$ gives a natural isomorphism

$$\mathrm{Hom}^*_{\mathbb{K}}(V^{\odot 0}, V) \to V, \qquad g \mapsto g(1)$$

and, whenever $m = -1$ the above formula for $f \barwedge g$ should be read as

$$f \barwedge g(v_1, \dots, v_n) = f(g(1), v_1, \dots, v_n).$$

We can immediately see that if $V = L[1]$, with L a vector space, then via décalage isomorphisms, Definition 10.6.5 is equivalent to the one given in Example 5.8.3. An equivalent formulation, intrinsically equipped with the proof of the Jacobi identity, will be given in Lemma 11.5.11.

As in the case of the Gerstenhaber bracket (Example 5.8.2), the product $\barwedge$ is not associative and defines a structure of graded right pre-Lie algebra; it is easy to see that the Nijenhuis–Richardson bracket is the graded symmetrization of the Gerstenhaber bracket.

It is important to observe that a sequence of maps $q_k \in \mathrm{Hom}^1_{\mathbb{K}}(V^{\odot k}, V)$, $k \geq 1$, gives an $L_\infty[1]$ structure on V if and only if for every $n > 0$ we have

$$\sum_{a+b=n+1} [q_a, q_b]_{NR} = 0\,.$$

In fact

$$\begin{aligned}\frac{1}{2}\sum_{k=1}^{n}[q_{n-k+1}, q_k]_{NR} &= \frac{1}{2}\sum_{k=1}^{n} q_{n-k+1} \barwedge q_k + \frac{1}{2}\sum_{k=1}^{n} q_k \barwedge q_{n-k+1}\\ &= \sum_{k=1}^{n}(q_{n-k+1} \barwedge q_k)\end{aligned}$$

and we recover exactly the left side of (10.17).

Given an $L_\infty[1]$-algebra $(V, q_1, q_2, \dots)$ and a graded commutative DG-algebra (A, d), as in the case of L_∞-algebras we can consider the scalar extension and obtain an $L_\infty[1]$-algebra $(V \otimes A, r_1, r_2, \dots)$ by setting:

$$r_1(v \otimes a) = q_1(v) \otimes a + (-1)^{\overline{v}} v \otimes d(a),$$

$$r_n(v_1 \otimes a_1, \dots, v_n \otimes a_n) = (-1)^{\sum_{i<j}\overline{a_i}\,\overline{v_j}} q_n(v_1, \dots, v_n) \otimes a_1 \cdots a_n, \quad n > 1. \tag{10.18}$$

We leave to the reader the straightforward verification that the maps $r_1, r_2, \dots$ satisfy equations (10.17), and that the scalar extensions of L_∞ and $L_\infty[1]$-algebras commute with the bijective correspondence of Lemma 10.6.4.

We shall say that an $L_\infty[1]$-algebra $(V, q_1, q_2, \ldots)$ is nilpotent if the associated L_∞-algebra $(sV, l_1, l_2, \ldots)$ is nilpotent: in this case we have in particular $q_n = 0$ for $n >> 0$.

Since for every vector $v \in V^0 = (sV)^1$ and every positive integer n we have the equality $sq_n(v, v, \ldots, v) = -l_n(sv, \ldots, sv)$, the Maurer–Cartan equation on a nilpotent $L_\infty[1]$-algebra $(V, q) = (V, q_1, q_2, \ldots)$ becomes:

$$\sum_{n>0} \frac{1}{n!} q_n(v^{\odot n}) = \sum_{n>0} \frac{1}{n!} q_n(v, v, \ldots, v) = 0, \qquad v \in V^0. \tag{10.19}$$

10.7 Derived Brackets

It is not surprising to observe that there exist lots of examples of L_∞-algebras beyond differential graded Lie algebras. In concrete cases L_∞-algebras appear mainly in two standard constructions: via derived brackets and via homotopy transfer. In this section we explain the first situation, while homotopy transfer will be studied in Chap. 12. More precisely, in this section we illustrate Voronov's extension of the classical Koszul brackets (Definition 9.3.1) and we recover the construction of DG-Lie algebras given in Sect. 5.7 as a very particular case.

As explained at the beginning of the chapter, for computational purposes it is convenient to work with $L_\infty[1]$ structures. As usual we work over a fixed field $\mathbb{K}$ of characteristic 0 and the starting data is described in the following setup.

Setup *There is given a triple* (M, A, P), *where* M *is a graded Lie algebra (whose Lie bracket is denoted by* $[-, -]$*),* $A \subset M$ *is an abelian graded Lie subalgebra and* $P \in \mathrm{Hom}^0_{\mathbb{K}}(M, A)$ *is a projection such that its kernel* $\ker P$ *is a graded Lie subalgebra of* M.

Setup 10.7 provides a direct sum decomposition of the graded vector space M, namely:

$$M = \ker P \oplus A, \quad A = \{x \in M \mid P(x) = x\},$$

with $\ker P$ and A graded Lie subalgebras of M and $[x, y] = 0$ for every $x, y \in A$. Obviously the triple (M, A, P) is redundant data since A is determined by P; however, it is useful to keep A as part of the data in view of the following definition.

Definition 10.7.1 (*Derived brackets*) Let M, $A \subset M$ and $P\colon M \to A$ be as in Setup 10.7. For every homogeneous derivation $f \in \mathrm{Der}^*_{\mathbb{K}}(M, M)$ define recursively two sequences of graded symmetric multilinear maps:

$$\begin{aligned} &f_n\colon A^{\odot n} \to M, \quad f_1 = f_{|A}, \quad f_n(a_1, \ldots, a_n) = [f_{n-1}(a_1, \ldots, a_{n-1}), a_n], \\ &\{\cdots\}^f_n\colon A^{\odot n} \to A, \quad \{a_1, \ldots, a_n\}^f_n = P f_n(a_1, \ldots, a_n), \qquad n > 0. \end{aligned}$$

The multilinear map $\{\cdots\}^f_n$ is called the nth **derived bracket** of $f \in \mathrm{Der}^*_{\mathbb{K}}(M, M)$.

The graded symmetry of f_n, and hence of $\{\cdots\}_n^f$, is a simple consequence of the Leibniz and the Jacobi identities; by induction on $k \geq 0$ it is sufficient to prove that

$$f_{k+2}(a_1, \ldots, a_k, b, c) = (-1)^{\bar{b}\bar{c}} f_{k+2}(a_1, \ldots, a_k, c, b).$$

Since $b, c \in A$ we have $[b, c] = 0$. For $k = 0$ we have

$$0 = f([b, c]) = [f(b), c] - (-1)^{\bar{b}\bar{c}}[f(c), b] = f_2(b, c) - (-1)^{\bar{b}\bar{c}} f_2(c, b),$$

while for $k > 0$ by Jacobi identity we have:

$$\begin{aligned} 0 = [f_k(a_1, \ldots, a_k), [b, c]] &= [[f_k(a_1, \ldots, a_k), b], c] - (-1)^{\bar{b}\bar{c}}[[f_k(a_1, \ldots, a_k), c], b] \\ &= f_{k+2}(a_1, \ldots, a_k, b, c) - (-1)^{\bar{b}\bar{c}} f_{k+2}(a_1, \ldots, a_k, c, b). \end{aligned}$$

It should be noted that, since A is abelian, if $f(A) \subset A$ then $f_n = 0$ for every $n \geq 2$, while if P is a morphism of graded Lie algebras then $\{\cdots\}_n^f = 0$ for every $n \geq 2$.

Definition 10.7.2 (*Derived brackets of inner derivations*) Let M, A and $P \colon M \to A$ as in Setup 10.7. For every homogeneous element $m \in \ker P$ and every integer $n > 0$ we denote by $\{\cdots\}_n^m$ the nth derived bracket of the inner derivation $[m, -] \in \operatorname{Der}_{\mathbb{K}}^{\overline{m}}(M, M)$, i.e., $\{\cdots\}_n^m := \{\cdots\}_n^{[m,-]}$.

Example 10.7.3 Let A be a unitary graded commutative algebra and consider the graded Lie algebra $M = A[[t]]$, equipped with the Wronskian bracket

$$[p(t), q(t)] = p(t)\frac{q(t)}{dt} - \frac{p(t)}{dt}q(t).$$

(In practice we identify M with the graded Lie algebra of derivations of $A[[t]]$ of type $p(t)\frac{d}{dt}$.) The subspace $A \subset A[[t]]$ of all constant power series is an abelian graded Lie subalgebra and the triple (M, A, P), where $P(q(t)) = q(0)$ satisfies the condition of Setup 10.7.

Then for every $a_1, \ldots, a_n \in A$ we have $\{a_1, \ldots, a_n\}_n^{e^{-t}} = a_1 \cdots a_n$. In fact $[e^{-t}a, b] = e^{-t}ab$ for every $a, b \in A$. Therefore, taking the inner derivation

$$f = [e^{-t}, -] \colon M \to M, \qquad f(p(t)) = [e^{-t}, p(t)],$$

we have $f_1(a) = [e^{-t}, a] = e^{-t}a$, $f_2(a, b) = [e^{-t}a, b] = e^{-t}ab$ and by induction $f_n(a_1, \ldots, a_n) = e^{-t}a_1 \cdots a_n$.

Example 10.7.4 Let A be a unitary graded commutative algebra and write $M = \operatorname{Hom}_{\mathbb{K}}^*(A, A)$. We may consider A as an abelian graded subalgebra of M, where every $a \in A$ is identified with the operator

$$a \colon A \to A, \qquad a(b) = ab.$$

If $P: M \to A$ is defined as $P(m) = m(1)$, then the triple (M, A, P) satisfies the condition of Setup 10.7. Given $m \in M$ such that $m(1) = 0$ and Φ^n_m are the Koszul braces (Definition 9.3.1), it is immediate from the definition that for every $n \geq 0$ we have $\Phi^n_m(a_1, \ldots, a_n) = \{a_1, \ldots, a_n\}^m_n$.

Example 10.7.5 Let A be a graded vector space. Then any sequence of linear maps $q_n: A^{\odot n} \to A$, $n > 0$, can be obtained as the sequence of derived brackets of a suitable inner derivation.

In fact, one can consider the graded Lie algebra $M = \mathrm{Hom}^*_{\mathbb{K}}(\oplus_{n\geq 0} A^{\odot n}, A)$ equipped with the Nijenhuis–Richardson bracket, the abelian subalgebra

$$\{f \in M \mid f(A^{\odot n}) = 0, \ \forall\, n > 0\} \cong A$$

and the projection $P: M \to A$, $P(f) = f(1)$. Given $m: A^{\odot n} \to A$, for every $a \in A$ we have $[m, a] = 0$ if $n = 0$, while for $n > 0$ we have

$$[m, a]: A^{\odot n-1} \to A, \qquad [m, a](b_2, \ldots, b_n) = m(a, b_2, \ldots, b_n).$$

Therefore $\{\cdots\}^m_i = 0$ for $i \neq n$, $\{\cdots\}^m_n = m$ and this proves that the map

$$M \to M, \qquad m \mapsto \sum_{n=0}^{\infty} \{\cdots\}^m_n,$$

is the identity. In particular, denoting by $q = \sum q_i \in \ker P$, we get $\{\cdots\}^q_n = q_n$ for every n.

Lemma 10.7.6 *Let (M, A, P) be as in Setup 10.7. Then:*

1. $P^2 = P$;
2. $[Px, Py] = 0$ *for every* $x, y \in M$;
3. $P[x, y] = P[Px, y] + P[x, Py] = P[x, Py] - (-1)^{\bar{x}\,\bar{y}} P[x, Py]$ *for every* $x, y \in M$;
4. *for a linear endomorphism* $f \in \mathrm{Hom}^*_{\mathbb{K}}(M, M)$, *we have* $f(\ker P) \subset \ker P$ *if and only if* $Pf = PfP$.

Proof The abelianity of A immediately gives the second item and the equivalence of the third item with the equality $P[x - Px, y - Py] = 0$ that is satisfied since $M = \ker P$ is a graded Lie subalgebra of M.

Given $f \in \mathrm{Hom}^*_{\mathbb{K}}(M, M)$, if $Pf = PfP$, for every $x \in \ker P$ we have $Pf(x) = PfP(x) = 0$ and then $f(x) \in \ker P$. Conversely, if $f(\ker P) \subset \ker P$, then for every $x \in M$ we have $x - Px \in \ker P$ and therefore $Pf(x - Px) = Pf(x) - PfP(x) = 0$. □

The relation between derived brackets and $L_\infty[1]$ structures is based on the following theorem.

Theorem 10.7.7 *In the situation of Setup 10.7 and in the notation of Definition 10.7.1, denote by* $L \subset \operatorname{Der}^*(M, M)$ *the graded Lie subalgebra of derivations* f *such that* $f(\ker P) \subset \ker P$. *Then, for every* $f, g \in L$ *and every* $n > 0$ *we have*

$$\{\cdots\}_n^{[f,g]} = \sum_{i=1}^{n} \left[\{\cdots\}_i^f , \{\cdots\}_{n+1-i}^g \right]_{NR} .$$

In other words, the liner map

$$L \to \operatorname{Hom}_{\mathbb{K}}^*(\oplus_{n=0}^{\infty} A^{\odot n}, A), \qquad f \mapsto \sum_{n=1}^{\infty} \{\cdots\}_n^f ,$$

is a morphism of graded Lie algebras.

Proof The proof is completely elementary but requires some hard explicit computations; for notational simplicity, in the following formulas we denote by $\pm_K$ the appropriate Koszul sign. The first step of the proof is to prove that, for every $f, g \in \operatorname{Der}_{\mathbb{K}}^*(M, M)$, every $n > 0$ and every $a_1, \ldots, a_n \in A$ we have

$$\begin{aligned} &[f, g]_n(a_1, \ldots, a_n) \\ &= \sum_{k=0}^{n} \sum_{\sigma \in S(k,n-k)} \pm_K [f_k(a_{\sigma(1)}, \ldots, a_{\sigma(k)}), g_{n-k}(a_{\sigma(k+1)}, \ldots, a_{\sigma(n)})], \end{aligned}$$

where we intend that

$$[f_0(\emptyset), a] = f(a), \qquad [b, g_0(\emptyset)] = -(-1)^{\bar{b}\,\bar{g}} g(b).$$

For $n = 1$ we have:

$$\begin{aligned} [f, g]_1(a) &= f(g(a)) - (-1)^{\bar{f}\,\bar{g}} g(f(a)) = [f_0, g_1(a)] - (-1)^{\bar{f}\,\bar{g}} [g_0, f_1(a)] \\ &= [f_0, g_1(a)] + (-1)^{\bar{a}\,\bar{g}} [f_1(a), g_0]. \end{aligned}$$

For $n > 1$ we have $[f, g]_n(a_1, \ldots, a_n) = [[f, g]_{n-1}(a_1, \ldots, a_{n-1}), a_n]$ and by induction:

$$\begin{aligned}
&[f, g]_n(a_1, \ldots, a_n)\\
&=\sum_{k=0}^{n-1} \sum_{\sigma\in S(k,n-1-k)} \pm_K [[f_k(a_{\sigma(1)}, \ldots, a_{\sigma(k)}), g_{n-1-k}(a_{\sigma(k+1)}, \ldots, a_{\sigma(n-1)})], a_n]\\
&=\sum_{k=0}^{n-1} \sum_{\sigma\in S(k,n-1-k)} \pm_K [f_{k+1}(a_{\sigma(1)}, \ldots, a_{\sigma(k)}, a_n), g_{n-1-k}(a_{\sigma(k+1)}, \ldots, a_{\sigma(n-1)})]\\
&\quad+\sum_{k=0}^{n-1} \sum_{\sigma\in S(k,n-1-k)} \pm_K [f_k(a_{\sigma(1)}, \ldots, a_{\sigma(k)}), g_{n-k}(a_{\sigma(k+1)}, \ldots, a_{\sigma(n-1)}, a_n)]\\
&=\sum_{k=1}^{n} \sum_{\sigma\in S(k,n-k),\sigma(n)<n} \pm_K [f_k(a_{\sigma(1)}, \ldots, a_{\sigma(k)}), g_{n-k}(a_{\sigma(k+1)}, \ldots, a_{\sigma(n)})]\\
&\quad+\sum_{k=0}^{n-1} \sum_{\sigma\in S(k,n-k),\sigma(n)=n} \pm_K [f_k(a_{\sigma(1)}, \ldots, a_{\sigma(k)}), g_{n-k}(a_{\sigma(k+1)}, \ldots, a_{\sigma(n)})].
\end{aligned}$$

The second step is to prove that, if $f, g \in L$ then the equality

$$\begin{aligned}
P[f_n(a_1, \ldots, a_n), g_m(b_1, \ldots, b_m)] &= P[\{a_1, \ldots, a_n\}^f_n, g_m(b_1, \ldots, b_m)]\\
&\quad+ P[f_n(a_1, \ldots, a_n), \{b_1, \ldots, b_m\}^g_m]
\end{aligned}$$

holds for every $n, m \geq 0$. For $n, m > 0$ this follows from Lemma 10.7.6, while for $n = 0$ we have:

$$\begin{aligned}
P[f_0, g_m(b_1, \ldots, b_m)] &= Pf(g_m(b_1, \ldots, b_m))\\
&= Pf\,P(g_m(b_1, \ldots, b_m)) = P[f_0, \{b_1, \ldots, b_m\}^g_m].
\end{aligned}$$

Since

$$P[f_n(a_1, \ldots, a_n), \{b_1, \ldots, b_m\}^g_m] = Pf_{n+1}(a_1, \ldots, a_n, \{b_1, \ldots, b_m\}^g_m)$$

we have

$$\begin{aligned}
P[f_n(a_1, \ldots, a_n), g_m(b_1, \ldots, b_m)] &= \{a_1, \ldots, a_n, \{b_1, \ldots, b_m\}^g_m\}^f_{n+1}\\
&\quad- \pm_K\{b_1, \ldots, b_m, \{a_1, \ldots, a_n\}^f_n\}^g_{m+1},
\end{aligned}$$

where the Koszul sign relates the sequences

$$f_n, a_1, \ldots, a_n, g_m, b_1, \ldots, b_m \quad\text{and}\quad g_m, b_1, \ldots, b_m, f_n, a_1, \ldots, a_n$$

and then it is equal to

$$\pm_K = (-1)^{(\overline{f_n}+\overline{a_1}+\cdots+\overline{a_n})(\overline{g_m}+\overline{b_1}+\cdots+\overline{b_m})} = (-1)^{\overline{f_n(a_1,\ldots,a_n)}\;\overline{g_m(b_1,\ldots,b_m)}}.$$

We are now ready to prove the theorem, i.e., to prove the formula

$$P[f,g]_n = \sum_{k=1}^{n} [Pf_k, Pg_{n-k+1}]_{NR}.$$

By the previous computation we have:

$$\begin{aligned}
&P[f,g]_n(a_1,\ldots,a_n)\\
&= \sum_{k=0}^{n} \sum_{\sigma\in S(k,n-k)} \pm_K\ P[f_k(a_{\sigma(1)},\ldots,a_{\sigma(k)}), g_{n-k}(a_{\sigma(k+1)},\ldots,a_{\sigma(n)})],\\
&= \sum_{k=0}^{n} \sum_{\sigma\in S(k,n-k)} \pm_K\ Pf_{k+1}(a_{\sigma(1)},\ldots,a_{\sigma(k)}, Pg_{n-k}(a_{\sigma(k+1)},\ldots,a_{\sigma(n)}))\\
&\quad -\sum_{k=0}^{n} \sum_{\sigma\in S(k,n-k)} \pm_K\ Pg_{n-k+1}(a_{\sigma(1)},\ldots,a_{\sigma(n-k)}, Pf_k(a_{\sigma(n-k+1)},\ldots,a_{\sigma(n)}))\\
&= \sum_{k=1}^{n} Pf_k \barwedge Pg_{n-k+1}(a_1,\ldots,a_n) - (-1)^{\bar{f}\,\bar{g}} \sum_{k=1}^{n} Pg_{n-k+1} \barwedge Pf_k(a_1,\ldots,a_n)\\
&= \sum_{k=1}^{n} [Pf_k, Pg_{n-k+1}]_{NR}(a_1,\ldots,a_n).
\end{aligned}$$

□

Corollary 10.7.8 *In Setup 10.7, the linear map*

$$\ker P \to \mathrm{Hom}^*_{\mathbb{K}}(\oplus_{n=0}^{\infty} A^{\odot n}, A), \qquad m \mapsto \sum_{n=1}^{\infty} \{\cdots\}_n^m,$$

is a morphism of graded Lie algebras.

Proof Immediate from Theorem 10.7.7 since $\ker P$ is a graded Lie subalgebra implies that the adjoint derivation $[m,-]$ belongs to L. □

Corollary 10.7.9 *In the situation of Theorem 10.7.7, assume that $d \in \mathrm{Der}^1(M,M)$, $d^2 = 0$ and $d(\ker P) \subset \ker P$, i.e., $(M, d, [-,-])$ is a DG-Lie algebra and $\ker P$ is a DG-Lie subalgebra.*

Then the sequence $q_k = \{\cdots\}_n^d$ gives an $L_\infty[1]$ structure on A.

Notice that, according to Example 10.7.5, every $L_\infty[1]$ structure can be constructed as in the above corollary.

Example 10.7.10 Given a graded commutative unitary algebra A, according to Example 10.7.4, the Koszul braces of linear maps $f\colon A \to A$ such that $f(1) = 0$

are derived brackets of inner derivations. By Corollary 10.7.8, for every $f, g \in \operatorname{Hom}^*_{\mathbb{K}}(A, A)$ such that and $f(1) = g(1) = 0$ we have

$$\Phi^n_{[f,g]} = \sum_{a+b=n+1} [\Phi^a_f, \Phi^b_g]_{NR}. \tag{10.20}$$

In particular, if $f \in \operatorname{Hom}^1_{\mathbb{K}}(A, A)$, $f^2 = 0$ and $f(1) = 0$, then the sequence Φ^n_f gives an $L_\infty[1]$ structure on A.

10.8 Exercises

10.8.1 Let V be a graded vector space and consider the linear map

$$\eta \colon V[t] \to V[t] \otimes \mathbb{K}[t], \qquad \eta(vt^n) = \sum_{i=0}^{n} vt^{n-i} \otimes t^i.$$

Prove that a differential $D \in \operatorname{Hom}^1_{\mathbb{K}}(V[t], V[t])$ may be written as in (10.11) if and only if $\eta D = (D \otimes \operatorname{Id}_{\mathbb{K}[t]})\eta$.

10.8.2 In the notation of Sect. 10.3, prove that the functor $B \colon \mathbf{DG}_\infty \to \mathbf{DG}$ is a left adjoint of the functor $i \colon \mathbf{DG} \to \mathbf{DG}_\infty$. This means that there exists a natural transformation of functors $\eta \colon B \circ i \to \mathbf{1}_{\mathbf{DG}}$ inducing for every $V \in \mathbf{DG}$ and every $W \in \mathbf{DG}_\infty$ a bijective map

$$\operatorname{Hom}_{\mathbf{DG}_\infty}(W, i(V)) \xrightarrow{\simeq} \operatorname{Hom}_{\mathbf{DG}}(B(W), V), \qquad f \mapsto \eta_V \circ B(f).$$

10.8.3 Prove that there exists a faithful functor $\widehat{B} \colon \mathbf{DG}_\infty \to \mathbf{DG}$ such that

$$\widehat{B}(V, d_1, d_2, \ldots) = (V[[t]], \widehat{D}), \qquad \widehat{D}\left(\sum_n v_n t^n\right) = \sum_{n,i} d_{i+1}(v_n) t^{n+i}.$$

10.8.4 Let $0 < k < n$ be integers. Prove that:

1. a permutation σ is a $(k, n-k)$-shuffle if and only if $\sigma(i) < \sigma(i+1)$ for every $i \neq k$;
2. for a shuffle permutation $\sigma \in S(k, n-k)$ we have $\sigma(n) < n$ if and only if $\sigma(k) = n$;
3. for a shuffle permutation $\sigma \in S(k, n-k)$ either $\sigma(1) = 1$ or $\sigma(k+1) = 1$.

10.8.5 Let A be a unitary graded commutative algebra over a field $\mathbb{K}$ of characteristic 0. For every $d \in \operatorname{Hom}^*_{\mathbb{K}}(A, A)$ such that $d(1) = 0$ we shall denote by Φ^n_d, $n > 0$, the associated sequence of Koszul braces. Prove that if d has degree 1, $d^2 = 0$ and a is nilpotent, then a is a solution of the Maurer–Cartan equation of the corresponding $L_\infty[1]$ structure (Example 10.7.10) if and only if $d(e^a) = 0$.

10.8.6 Let A be a unitary graded commutative algebra over a field of characteristic 0 and let $f \in \operatorname{Hom}^*_{\mathbb{K}}(A, A)$ be a linear endomorphism such that $f(1) = 0$. Denote by

$$\mu_n \colon A^{\odot n+1} \to A, \qquad \mu_n(a_0, \ldots, a_n) = a_0 a_1 \cdots a_n, \quad n > 0,$$

the multiplication maps, by $[-,-] = [-,-]_{NR}$ the Nijenhuis–Richardson bracket on $\prod_{n>0} \operatorname{Hom}^*_{\mathbb{K}}(A^{\odot n}, A)$ and by Φ^n_f the Koszul braces of f. Prove that

$$\Phi^2_f = [f, \mu_1], \qquad \Phi^3_f = \frac{1}{2}([[f, \mu_1], \mu_1] - [f, \mu_2]). \tag{10.21}$$

Use the above equalities for proving that, if f has odd degree, then

$$[f, [f, \mu_n]] = [f^2, \mu_n], \qquad [f, \Phi^2_f] = \Phi^3_{f^2}, \qquad [f, \Phi^3_f] = \Phi^3_{f^2} - \frac{1}{2}[\Phi^2_f, \Phi^2_f].$$

Notice that if f is a differential operator of order ≤ 2, then $\Phi^3_f = 0$ and the last equation reduces to $\frac{1}{2}[\Phi^2_f, \Phi^2_f] = \Phi^3_{f^2}$ which, for $f = \Delta$, is the same equation as used in the proof of Lemma 9.4.2.

10.8.7 Let M be a DG-Lie algebra that, as a complex, is the direct sum $M = N \oplus I$ of a DG-Lie subalgebra N and an abelian differential graded Lie ideal I. Assume that for every surjective morphism $C \to B$ in $\mathbf{Art}_{\mathbb{K}}$ and for every $x \in \mathrm{MC}_M(C)$ the morphism $H^1(M \otimes C, d + [x, -]) \to H^1(M \otimes B, d + [x, -])$ is surjective. Prove that the morphism of deformation functor $\mathrm{Def}_M \to \mathrm{Def}_N$, induced by the projection $M \to N$, is smooth.

References

The notion of a DG_∞-vector space, and more generally of a DG_∞-module over a commutative ring, has been introduced by Lapin [166], although the same concept was already implicit in many works in homological perturbation theory, see e.g. [121].

Some authors, see e.g. [153], call the category of formal neighbourhood of graded vector spaces *pre-L_∞ algebras*. Thus, by Theorem 10.2.2, a pre-L_∞-morphism is nothing else than a morphism of graded vector spaces $\bigoplus_{n=1}^\infty V^{\odot n} \to W$.

In the literature there exists a discrepancy about the definition of L_∞-algebras and shuffles; our definition of shuffles corresponds to the notion of unshuffles given by some authors. Here we follow the terminology of [64, 170, 269], while, for instance, the papers [83, 163, 178] use the opposite convention. Similarly, there exist in the literature two different natural definitions of L_∞-algebras depending on different sign conventions in décalage isomorphisms. Here we follow [90, 151, 153], while in [163, 164] the maps l_k differ by the sign $(-1)^{k(k-1)/2}$. The name $L_\infty[1]$-algebra has been proposed by Schätz [234] and the name décalage isomorphism is taken from

[135, I.4.3.2]. As far as I know, the first appearance of L_∞-algebras in deformation theory was in the preprint [238]. Later L_∞-algebras appeared in some papers of string theory [164, 273] and in Kontsevich's proof of the formality theorem for Poisson manifolds [153], see also [67, 189].

Theorem 10.7.7 and Corollaries 10.7.8, 10.7.9 are proved by Voronov [265, 266], while the particular case of Koszul braces (Example 10.7.10) was proved earlier in [22, 161]. A generalization of the construction of Sect. 10.7 to the case where the graded Lie subalgebra A is not necessarily abelian has been given by Bandiera in [11]. We refer to [188] for the extension of formulas (10.21) to the whole family of Koszul braces.

Chapter 11
Coalgebras and Coderivations

The description of an $L_\infty[1]$ structure on a graded vector space V as an element $q \in \prod_{n>0} \operatorname{Hom}^1_{\mathbb{K}}(V^{\odot n}, V)$ satisfying the equation $[q, q]_{NR} = 0$ is somewhat difficult to handle and obscure to understand. One of the goals of this chapter is to reinterpret both the Nijenhuis–Richardson bracket and the category of formal neighbourhoods in the framework of graded coalgebras. This will allow us to give, in Chap. 12, a useful equivalent characterization of $L_\infty[1]$ structures which leads naturally to the notion of L_∞-morphisms.

As usual, we work over a fixed field $\mathbb{K}$ of characteristic 0 and the tensor product over $\mathbb{K}$ is denoted by $\otimes$.

11.1 Graded Coalgebras

Informally speaking, coalgebras are defined by dualizing the axioms defining algebras: for simplicity of exposition we consider here only coassociative coalgebras.

Definition 11.1.1 A (coassociative) **graded coalgebra** is a pair (C, Δ) consisting of a graded vector space C and of a morphism of graded vector spaces $\Delta\colon C \to C \otimes C$, called a **coproduct**, that satisfies the coassociativity equation:

$$(\Delta \otimes \mathrm{Id}_C)\Delta = (\mathrm{Id}_C \otimes \Delta)\Delta\colon C \to C \otimes C \otimes C.$$

Definition 11.1.2 Let (C, Δ) and (B, Γ) be graded coalgebras. A **morphism** of graded coalgebras $F\colon C \to B$ is a morphism of graded vector spaces that commutes with coproducts:

$$\Gamma F = (F \otimes F)\Delta\colon C \to B \otimes B.$$

M. Manetti, *Lie Methods in Deformation Theory*, Springer Monographs in Mathematics,
https://doi.org/10.1007/978-981-19-1185-9_11

Example 11.1.3 The base field $\mathbb{K}$ is a graded coalgebra (concentrated in degree 0) with the coproduct

$$\Delta\colon \mathbb{K} \to \mathbb{K} \otimes \mathbb{K}, \qquad \Delta(a) = a \otimes 1 = 1 \otimes a.$$

A linear map $f\colon \mathbb{K} \to \mathbb{K}$ is a coalgebra morphism if and only if either it is the identity ($f(1) = 1$) or it is trivial ($f(1) = 0$); in fact the relation $\Delta f = f^{\otimes 2}\Delta$ implies

$$f(1)\, 1 \otimes 1 = f(1) \otimes f(1) = f(1)^2\, 1 \otimes 1$$

and then $f(1) = f(1)^2$.

Example 11.1.4 Let $\mathbb{K}[t]$ be the polynomial ring in one indeterminate t of degree 0. The linear map

$$\Delta\colon \mathbb{K}[t] \to \mathbb{K}[t] \otimes \mathbb{K}[t], \qquad \Delta(t^n) = \sum_{i=0}^{n} t^i \otimes t^{n-i},$$

is the coproduct of a coalgebra structure. We leave to the reader the easy proof that, for every sequence $f_n \in \mathbb{K}$, $n > 0$, the linear map $F\colon C \to C$ defined by $F(1) = 1$ and

$$F(t^n) = \sum_{s=1}^{n} \sum_{i_1+\cdots+i_s=n} f_{i_1} f_{i_2} \cdots f_{i_s} t^s, \qquad n > 0, \tag{11.1}$$

is a morphism of graded coalgebras. Notice that the whole sequence f_n can be recovered from the coalgebra morphism F since $F(t^n) = f_n t + t^2(\cdots)$ for every $n > 0$.

Definition 11.1.5 A graded coalgebra (C, Δ) is called (graded) **cocommutative** if $\mathtt{tw} \circ \Delta = \Delta$, where

$$\mathtt{tw}\colon C \otimes C \to C \otimes C, \qquad \mathtt{tw}(x \otimes y) = (-1)^{\bar{x}\bar{y}} y \otimes x,$$

is the twist map.

Example 11.1.6 The polynomial coalgebra of Example 11.1.4 is cocommutative.

Example 11.1.7 Let C be a graded coalgebra with coproduct $\Delta\colon C \to C \otimes C$. The **convolution product** on the graded dual $\mathrm{Hom}^*_{\mathbb{K}}(C, \mathbb{K})$ is defined as:

$$\mathrm{Hom}^*_{\mathbb{K}}(C, \mathbb{K}) \times \mathrm{Hom}^*_{\mathbb{K}}(C, \mathbb{K}) \to \mathrm{Hom}^*_{\mathbb{K}}(C, \mathbb{K}), \qquad (f, g) \mapsto \mu(f \otimes g)\Delta,$$

where $\mu\colon \mathbb{K} \times \mathbb{K} \to \mathbb{K}$ is the usual product. We can immediately see that the convolution product is associative and then the dual of an associative graded coalgebra is an associative graded algebra.

In general, the dual of an algebra is not a coalgebra (with some exceptions, see e.g. Example 11.1.16). Heuristically, this asymmetry comes from the fact that, for an infinite-dimensional vector space V, there exists a natural map $V^\vee \otimes V^\vee \to (V \otimes V)^\vee$, while there does not exist any natural map $(V \otimes V)^\vee \to V^\vee \otimes V^\vee$.

Example 11.1.8 The dual of the polynomial coalgebra $\mathbb{K}[t]$ of Example 11.1.4 is exactly the algebra of formal power series $\mathbb{K}[[x]]$, where the duality pairing is $\langle x^n, t^m \rangle = \delta_{n,m}$. The coalgebra morphism $F\colon \mathbb{K}[t] \to \mathbb{K}[t]$ induced by a sequence $\{f_n\}$ as in (11.1) gives by transposition the morphism of local complete $\mathbb{K}$-algebras $F^T : \mathbb{K}[[x]] \to \mathbb{K}[[x]]$ such that $F^T(x) = \sum_{n>0} f_n x^n$.

Definition 11.1.9 Let (C, Δ) be a graded coalgebra. The **iterated coproducts** $\Delta^n : C \to C^{\otimes n+1}$ are defined by $\Delta^0 = \mathrm{Id}_C$, $\Delta^1 = \Delta$ and recursively for $n \geq 2$ by the formula $\Delta^n = (\mathrm{Id}_C \otimes \Delta^{n-1})\Delta$:

$$\Delta^n : C \xrightarrow{\Delta} C \otimes C \xrightarrow{\mathrm{Id}_C \otimes \Delta^{n-1}} C \otimes C^{\otimes n} = C^{\otimes n+1}.$$

For example, in the polynomial coalgebra $\mathbb{K}[t]$ of Example 11.1.4 we have

$$\Delta^n(t^p) = \sum_{i_0+\cdots+i_n=p} t^{i_0} \otimes t^{i_1} \otimes \cdots \otimes t^{i_n}.$$

It is plain that the morphisms of graded coalgebras $F\colon (C, \Delta) \to (B, \Gamma)$ commute with iterated coproducts; this means that for every $n \geq 0$ we have

$$\Gamma^n F = F^{\otimes n+1} \Delta^n : C \to B^{\otimes n+1}, \tag{11.2}$$

and therefore $F(\ker \Delta^n) \subset \ker \Gamma^n$ for every $n \geq 0$. In fact, by induction on n we have

$$\begin{aligned}\Gamma^n F &= (\mathrm{Id}_B \otimes \Gamma^{n-1})\Gamma F = (\mathrm{Id}_B \otimes \Gamma^{n-1})(F \otimes F)\Delta = (F \otimes \Gamma^{n-1} F)\Delta \\ &= (F \otimes F^{\otimes n} \Delta^{n-1})\Delta = F^{\otimes n+1}(\mathrm{Id}_C \otimes \Delta^{n-1})\Delta = F^{\otimes n+1}\Delta^n.\end{aligned}$$

Lemma 11.1.10 *Let (C, Δ) be a coassociative graded coalgebra. Then:*

1. *for every $a, b \geq 0$ we have*

$$\Delta^{a+b+1} = (\Delta^a \otimes \Delta^b)\Delta : C \to C^{\otimes a+b+2}; \tag{11.3}$$

2. *for every $s \geq 1$ and every $a_0, \ldots, a_s \geq 0$ we have*

$$(\Delta^{a_0} \otimes \Delta^{a_1} \otimes \cdots \otimes \Delta^{a_s})\Delta^s = \Delta^{s+\sum a_i}; \tag{11.4}$$

3. *for every $a, b \geq 0$ we have*

$$\Delta^a(\ker \Delta^{a+b}) \subset (\ker \Delta^b)^{\otimes a+1}. \tag{11.5}$$

In particular $\ker \Delta^n \subset \ker \Delta^{n+1}$ *and* $\Delta^n(\ker \Delta^{n+1}) \subset (\ker \Delta)^{\otimes n+1}$ *for every* $n \geq 0$.

Proof. We first prove (11.3) by induction on a; if $a = 0$ there is nothing to prove since by definition $\Delta^{b+1} = (\mathrm{Id}_C \otimes \Delta^b)\Delta$. If $a > 0$ then

$$\begin{aligned}(\Delta^a \otimes \Delta^b)\Delta &= ((\mathrm{Id}_C \otimes \Delta^{a-1})\Delta \otimes \Delta^b)\Delta = (\mathrm{Id}_C \otimes \Delta^{a-1} \otimes \Delta^b)(\Delta \otimes \mathrm{Id}_C)\Delta\\ &= (\mathrm{Id}_C \otimes \Delta^{a-1} \otimes \Delta^b)(\mathrm{Id}_C \otimes \Delta)\Delta = (\mathrm{Id}_C \otimes (\Delta^{a-1} \otimes \Delta^b)\Delta)\Delta\\ &= (\mathrm{Id}_C \otimes \Delta^{a+b})\Delta = \Delta^{a+b+1}.\end{aligned}$$

Next, we prove (11.4) by induction on s; the case $s = 1$ is exactly (11.3). If $s \geq 2$ we can write

$$\begin{aligned}(\Delta^{a_0} \otimes \Delta^{a_1} \otimes \cdots \otimes \Delta^{a_s})\Delta^s &= (\Delta^{a_0} \otimes \Delta^{a_1} \otimes \cdots \otimes \Delta^{a_s})(\mathrm{Id}_C \otimes \Delta^{s-1})\Delta\\ &= (\Delta^{a_0} \otimes (\Delta^{a_1} \otimes \cdots \otimes \Delta^{a_s})\Delta^{s-1})\Delta\\ &= (\Delta^{a_0} \otimes \Delta^{s-1+\sum_{i>0} a_i})\Delta = \Delta^{s+\sum a_i}.\end{aligned}$$

Notice that equation (11.4) gives in particular $\Delta^{n+1} = (\mathrm{Id}_{C^{\otimes n}} \otimes \Delta)\Delta^n$ and therefore $\ker \Delta^n \subset \ker \Delta^{n+1}$.

Finally, in order to show (11.5), we prove by induction on s that

$$\Delta^a(\ker \Delta^{a+b}) \subset (\ker \Delta^b)^{\otimes s} \otimes C^{\otimes a-s+1} \quad \text{for every} \quad 0 \leq s \leq a+1.$$

Since the above inclusion is trivially satisfied for $s = 0$, we can assume $0 < s \leq a+1$ and $\Delta^a(\ker \Delta^{a+b}) \subseteq (\ker \Delta^b)^{\otimes s-1} \otimes C^{\otimes a-s+2}$. Let $P \subset C$ be a graded vector subspace such that $C = P \oplus \ker \Delta^b$, then $\Delta^b \colon P \to C^{\otimes b+1}$ is split injective.

For every $x \in \ker \Delta^{a+b}$ we can therefore write $\Delta^a(x) = y_1 + y_2$ with

$$y_1 \in (\ker \Delta^b)^{\otimes s-1} \otimes \ker \Delta^b \otimes C^{\otimes a-s+1}, \quad y_2 \in (\ker \Delta^b)^{\otimes s-1} \otimes P \otimes C^{\otimes a-s+1}.$$

By (11.4) we have

$$0 = \Delta^{a+b}(x) = (\mathrm{Id}_C^{\otimes s-1} \otimes \Delta^b \otimes \mathrm{Id}_C^{\otimes a-s+1})\Delta^a(x) = (\mathrm{Id}_C^{\otimes s-1} \otimes \Delta^b \otimes \mathrm{Id}_C^{\otimes a-s+1})y_2,$$

and this implies $y_2 = 0$ since

$$(\ker \Delta)^{\otimes s-1} \otimes P \otimes C^{\otimes a-s+1} \xrightarrow{\mathrm{Id}_C^{\otimes s-1} \otimes \Delta^b \otimes \mathrm{Id}_C^{\otimes a-s+1}} (\ker \Delta)^{\otimes s-1} \otimes C^{\otimes a+b-s+2}$$

is split injective. □

Definition 11.1.11 Let (C, Δ) be a graded coalgebra. A morphism of graded vector spaces $p \colon C \to V$ is called a **cogenerator** of C if for every $x \in C$, $x \neq 0$, there exists

$n > 0$ such that $p^{\otimes n}\Delta^{n-1}(x) \neq 0$ in $V^{\otimes n}$. Equivalently, $p\colon C \to V$ is a cogenerator of C if the linear map

$$(p, p^{\otimes 2}\Delta, p^{\otimes 3}\Delta^2, \ldots)\colon C \longrightarrow \prod_{n>0} V^{\otimes n}$$

is injective. For a cogenerator $p\colon C \to V$ and a linear map $f\colon B \to C$, the composition $pf\colon B \to V$ is called the **corestriction** of f to p.

Example 11.1.12 In the notation of Example 11.1.4, the projection $\mathbb{K}[t] \to \mathbb{K}[t]/(t^2)$ is a cogenerator.

Proposition 11.1.13 *Let $p\colon B \to V$ be a cogenerator of a graded coalgebra (B, Γ). Then every morphism of graded coalgebras $F\colon (C, \Delta) \to (B, \Gamma)$ is uniquely determined by its corestriction $pF\colon C \to V$.*

Proof. Let $F, G\colon (C, \Delta) \to (B, \Gamma)$ be two morphisms of graded coalgebras such that $pF = pG$. In order to prove that $F = G$ it is sufficient to show that for every $x \in C$ and every $n \geq 0$ we have

$$p^{\otimes n+1}\Gamma^n(F(x)) = p^{\otimes n+1}\Gamma^n(G(x)).$$

By (11.2) we have $\Gamma^n F = F^{\otimes n+1}\Delta^n$ and $\Gamma^n G = G^{\otimes n+1}\Delta^n$. By assumption s $p^{\otimes n+1}F^{\otimes n+1} = (pF)^{\otimes n+1} = (pG)^{\otimes n+1} = p^{\otimes n+1}G^{\otimes n+1}$ for every $n \geq 0$ and therefore

$$p^{\otimes n+1}\Gamma^n F = p^{\otimes n+1}F^{\otimes n+1}\Delta^n = p^{\otimes n+1}G^{\otimes n+1}\Delta^n = p^{\otimes n+1}\Gamma^n G.$$

□

Definition 11.1.14 A graded coalgebra (C, Δ) is called **conilpotent** if $\Delta^n = 0$ for $n >> 0$. It is called **locally conilpotent** if $C = \cup_n \ker \Delta^n$.

Example 11.1.15 The vector space

$$\overline{T^c}(\mathbb{K}) := \{p(t) \in \mathbb{K}[t] \mid p(0) = 0\} = \bigoplus_{n>0} \mathbb{K}t^n$$

with the coproduct

$$\mathfrak{a}\colon \overline{T^c}(\mathbb{K}) \to \overline{T^c}(\mathbb{K}) \otimes \overline{T^c}(\mathbb{K}), \qquad \mathfrak{a}(t^n) = \sum_{i=1}^{n-1} t^i \otimes t^{n-i},$$

is a locally conilpotent coalgebra. If $\mathbb{K}[t]$ is the polynomial coalgebra of Example 11.1.4, then the projection $\mathbb{K}[t] \to \overline{T^c}(\mathbb{K})$, $p(t) \mapsto p(t) - p(0)$, is a morphism of coalgebras.

Example 11.1.16 Let $A = \oplus A_i$ be a finite-dimensional graded associative $\mathbb{K}$-algebra and let $C = A^\vee = \mathrm{Hom}^*_{\mathbb{K}}(A, \mathbb{K})$ be its graded dual. Since A and C are finite-dimensional, the pairing $\langle c_1 \otimes c_2, a_1 \otimes a_2\rangle = (-1)^{\overline{a_1}\,\overline{c_2}}\langle c_1, a_1\rangle\langle c_2, a_2\rangle$ gives a natural isomorphism $C \otimes C = (A \otimes A)^\vee$ and we may define Δ as the transpose of the multiplication map $\mu\colon A \otimes A \to A$. Then (C, Δ) is a graded coalgebra. Notice that C is conilpotent if and only if A is nilpotent.

Lemma 11.1.17 *Let (C, Δ) be a locally conilpotent graded coalgebra. Then every projection $p\colon C \to \ker \Delta$ is a cogenerator of C.*

Proof. According to (11.5) of Lemma 11.1.10 we have

$$\Delta^n(\ker \Delta^{n+1}) \subset (\ker \Delta)^{\otimes n+1}$$

for every $n \geq 0$. By assumption, for every $x \in C$, $x \neq 0$, there exists $n \geq 0$ such that $\Delta^n(x) \neq 0$ and $\Delta^{n+1}(x) = 0$. Thus $\Delta^n(x) \in \Delta^n(\ker \Delta^{n+1}) \subset (\ker \Delta)^{\otimes n+1}$ and then $p^{\otimes n+1}\Delta^n(x) = \Delta^n(x) \neq 0$. □

The proof of Lemma 11.1.17 shows that for a locally conilpotent graded coalgebra (C, Δ) over a field, a linear map $p\colon C \to V$ is a cogenerator if and only if its restriction $p\colon \ker \Delta \to V$ is injective.

It should be noticed that most of the the results of this section are also valid, mutatis mutandis, also when C is a graded module over a unitary commutative ring. The only possibly unvalid results over general commutative rings are (11.5) of Lemma 11.1.10 and therefore Lemma 11.1.17; however it is easy to see that Lemma 11.1.17 is valid whenever the injective map $\Delta\colon \ker p \to C \otimes C$ is either split injective or has a flat cokernel.

11.2 Comodules and Coderivations

As for the case of coalgebras, one can dualize the notion of a module and obtain the notion of a comodule. For simplicity of exposition we shall refer to as comodules only the bicomodules, i.e., comodules that are left and right in the obvious compatible way.

Definition 11.2.1 Let (C, Δ) be a graded coalgebra. A C-comodule is the data of a graded vector space M and two morphisms of graded vector spaces

$$r\colon M \to M \otimes C, \qquad l\colon M \to C \otimes M,$$

called respectively a right coaction and left coaction, such that:

1. $(\mathrm{Id}_M \otimes \Delta)r = (r \otimes \mathrm{Id}_C)r\colon M \to M \otimes C \otimes C$;
2. $(\Delta \otimes \mathrm{Id}_M)l = (\mathrm{Id}_C \otimes l)l\colon M \to C \otimes C \otimes M$;
3. $(l \otimes \mathrm{Id}_C)r = (\mathrm{Id}_C \otimes r)l\colon M \to C \otimes M \otimes C$.

If there is given only a right coaction r satisfying $(\mathrm{Id}_M \otimes \Delta)r = (r \otimes \mathrm{Id}_C)r$ we shall refer to M as a right comodule; left comodules are defined similarly.

Example 11.2.2 If $F: (D, \Gamma) \to (C, \Delta)$ is a morphism of graded coalgebras, then the maps

$$r = (\mathrm{Id}_D \otimes F)\Gamma: D \to D \otimes C, \qquad l = (F \otimes \mathrm{Id}_D)\Gamma: D \to C \otimes D,$$

give a structure of a C-comodule on D.

Example 11.2.3 This example is related with the description of DG_∞-morphism given in Sect. 10.3. Consider the polynomial coalgebra of Example 11.1.4:

$$\Delta: \mathbb{K}[t] \to \mathbb{K}[t] \otimes \mathbb{K}[t], \qquad \Delta(t^n) = \sum_{i=0}^{n} t^i \otimes t^{n-i}.$$

Then, for every graded vector space V, the space $V[t]$ of polynomials with coefficient in V is a $\mathbb{K}[t]$-comodule with right and left coactions defined by:

$$r(vt^n) = \sum_{i=0}^{n} vt^i \otimes t^{n-i}, \qquad l(vt^n) = \sum_{i=0}^{n} t^i \otimes vt^{n-i}.$$

Given any graded vector space W and any sequence $f_1, f_2, \ldots \in \mathrm{Hom}^0_{\mathbb{K}}(V, W)$ of morphisms of graded vector spaces, we can immediately check that the linear map

$$F: V[t] \to W[t], \qquad F(vt^n) = \sum_{i=0}^{n} f_{n+1}(v)t^{n-i},$$

is a morphism of comodules: this means that F commutes with coactions in the obvious ways $(F \otimes \mathrm{Id}_{\mathbb{K}[t]})r = rF$ and $(\mathrm{Id}_{\mathbb{K}[t]} \otimes F)l = lF$. It is not difficult to prove (Exercise 11.8.4) that every morphism of comodules $V[t] \to W[t]$ is obtained in this way.

Definition 11.2.4 Let (C, Δ) be a graded coalgebra and let M be a C-comodule with coactions

$$r: M \to M \otimes C, \qquad l: M \to C \otimes M.$$

A linear map $\alpha \in \mathrm{Hom}^n_{\mathbb{K}}(M, C)$ is called a **coderivation** of degree n if it satisfies the **coLeibniz rule**

$$\Delta\alpha = (\alpha \otimes \mathrm{Id}_C)r + (\mathrm{Id}_C \otimes \alpha)l: M \to C \otimes C.$$

Keep in mind that in the above definition we have adopted the Koszul sign convention; therefore for $x \in C$ and $m \in M$ we have:

$$(\alpha \otimes \mathrm{Id}_C)(m \otimes x) = \alpha(m) \otimes x, \qquad (\mathrm{Id}_C \otimes \alpha)(x \otimes m) = (-1)^{\overline{\alpha}\,\overline{x}} x \otimes \alpha(m).$$

In this book we are mainly interested in the C-comodule structure induced by a morphism of graded coalgebras $F: (D, \Gamma) \to (C, \Delta)$ as in Example 11.2.2. In this case, a morphism of graded vector spaces $\alpha \in \mathrm{Hom}^n_{\mathbb{K}}(D, C)$ is a coderivation of degree n if and only if

$$\Delta\alpha = (\alpha \otimes F + F \otimes \alpha)\Gamma.$$

The coderivations of degree n with respect to a coalgebra morphism $F: C \to D$ form a vector space denoted by $\mathrm{Coder}^n(C, D; F)$. For simplicity of notation we write $\mathrm{Coder}^n(C, C) = \mathrm{Coder}^n(C, C; \mathrm{Id}_C)$, viz.,

$$\mathrm{Coder}^n(C, C) = \{\alpha \in \mathrm{Hom}^n_{\mathbb{K}}(C, C) \mid \Delta\alpha = (\alpha \otimes \mathrm{Id}_C + \mathrm{Id}_C \otimes \alpha)\Delta\}.$$

Lemma 11.2.5 *Let C be a graded coalgebra. Then $\mathrm{Coder}^*(C, C) = \bigoplus_n \mathrm{Coder}^n(C, C)$ is a graded Lie subalgebra of $\mathrm{Hom}^*_{\mathbb{K}}(C, C)$.*

Proof. We only need to prove that $\mathrm{Coder}^*(C, C)$ is closed under the graded commutator. This is straightforward and it is left to the reader. □

Example 11.2.6 For every integer $k \geq -1$ denote by $f_k: \mathbb{K}[t] \to \mathbb{K}[t]$ the differential operator

$$f_k = t\left(\frac{d}{dt}\right)^{k+1}.$$

Then every f_k is a coderivation with respect to the coproduct

$$\Delta: \mathbb{K}[t] \to \mathbb{K}[t] \otimes \mathbb{K}[t], \qquad \Delta(t^n) = \sum_{i=0}^{n} \binom{n}{i} t^i \otimes t^{n-i}.$$

In fact, using the equalities

$$\binom{n}{k} = \frac{1}{k!}\prod_{i=0}^{k-1}(n-i), \qquad \binom{n}{0} = 1,$$

we have for every $n \geq 0$ and every $k \geq 0$

$$\frac{\Delta(f_{k-1}(t^n))}{k!} = \Delta\left(\binom{n}{k} t^{n-k+1}\right) = \sum_{i\geq 0}\binom{n}{k}\binom{n-k+1}{i} t^i \otimes t^{n-k-i+1},$$

$$\begin{aligned}\frac{(f_{k-1} \otimes \mathrm{Id})\Delta(t^n)}{k!} &= \sum_{j\geq k}\binom{n}{j}\binom{j}{k} t^{j-k+1} \otimes t^{n-j}\\ &= \sum_{i\geq 0}\binom{n}{i+k-1}\binom{i+k-1}{k} t^i \otimes t^{n-k-i+1},\end{aligned}$$

$$\frac{(\mathrm{Id}\otimes f_{k-1})\Delta(t^n)}{k!}=\sum_{i\geq 0}\binom{n}{i}\binom{n-i}{k}t^i\otimes t^{n-k-i+1},$$

and the conclusion follows from the straightforward equality

$$\binom{n}{k}\binom{n-k+1}{i}=\binom{n}{i+k-1}\binom{i+k-1}{k}+\binom{n}{i}\binom{n-i}{k}.$$

Notice that

$$[f_n, f_m]=f_n\circ f_m-f_m\circ f_n=(n-m)f_{n+m}.$$

and then the Lie subalgebra generated by the coderivations f_k is isomorphic to the Lie algebra generated by the derivations $g_k=-z^{k+1}\frac{d}{dz}$ of the polynomial algebra $\mathbb{K}[z]$.

Lemma 11.2.7 *Let $C\xrightarrow{\theta}D\xrightarrow{\rho}E$ be morphisms of graded coalgebras. The compositions with θ and ρ induce two properly defined linear maps*

$$\rho_*\colon Coder^n(C,D;\theta)\to Coder^n(C,E;\rho\theta),\qquad f\mapsto\rho f;$$

$$\theta^*\colon Coder^n(D,E;\rho)\to Coder^n(C,E;\rho\theta),\qquad f\mapsto f\theta.$$

Proof. Immediate consequence of the equalities

$$\Delta_E\rho=(\rho\otimes\rho)\Delta_D,\qquad \Delta_D\theta=(\theta\otimes\theta)\Delta_C.$$

□

Lemma 11.2.8 *Let $F\colon(C,\Delta)\to(D,\Gamma)$ be a morphism of graded coalgebras and let $d\in Coder^*(C,D;F)$ be a coderivation. Then:*

1. *for every $n\geq 0$*

$$\Gamma^n d=\left(\sum_{i=0}^{n}F^{\otimes i}\otimes d\otimes F^{\otimes n-i}\right)\Delta^n\colon C\to D^{\otimes n+1};$$

2. *if $p\colon D\to V$ is a cogenerator, then d is uniquely determined by its corestriction $pd\colon C\to V$.*

Proof. For the first part, we have:

$$\begin{aligned}\Gamma^n d&=(\mathrm{Id}_D\otimes\Gamma^{n-1})\Gamma d=(\mathrm{Id}_D\otimes\Gamma^{n-1})(F\otimes d+d\otimes F)\Delta\\&=(F\otimes\Gamma^{n-1}d+d\otimes\Gamma^{n-1}F)\Delta=(F\otimes\Gamma^{n-1}d)\Delta+(d\otimes F^{\otimes n}\Delta^{n-1})\Delta\end{aligned}$$

and the conclusion follows by induction on n, since

$$(F\otimes \Gamma^{n-1}d)\Delta=\left(F\otimes\left(\sum_{i=0}^{n-1}F^{\otimes i}\otimes d\otimes F^{\otimes n-i-1}\right)\Delta^{n-1}\right)\Delta$$
$$=\left(\sum_{i=1}^{n}F^{\otimes i}\otimes d\otimes F^{\otimes n-i}\right)(\mathrm{Id}_C\otimes\Delta^{n-1})\Delta,$$
$$(d\otimes F^{\otimes n}\Delta^{n-1})\Delta=(d\otimes F^{\otimes n})(\mathrm{Id}_C\otimes\Delta^{n-1})\Delta.$$

For the second item we need to prove that $pd=0$ implies $d=0$. Assume that $pd=0$, then for every $x\in C$ and every $n\geq 0$ we have

$$p^{\otimes n+1}\Gamma^n dx=\left(\sum_{i=0}^{n}(pF)^{\otimes i}\otimes pd\otimes(pF)^{\otimes n-i}\right)\Delta^n x=0$$

and then, since p is a cogenerator, we have $dx=0$. □

For later use we point out that if $\alpha\colon C\to C$ is a nilpotent coderivation of degree 0, then the map

$$e^{\alpha}=\sum_{n\geq 0}\frac{\alpha^n}{n!}\colon C\to C$$

is a morphism of graded coalgebras. The proof is essentially the same as in Lemma 2.1.2; in fact, in the associative algebra $\mathrm{Hom}^0_{\mathbb{K}}(C\otimes C, C\otimes C)$ we have

$$e^{\alpha}\otimes e^{\alpha}=\sum_{i,j\geq 0}\frac{1}{i!\,j!}\alpha^i\otimes\alpha^j=\sum_{n\geq 0}\frac{1}{n!}(\alpha\otimes\mathrm{Id}_C+\mathrm{Id}_C\otimes\alpha)^n$$

and then

$$(e^{\alpha}\otimes e^{\alpha})\Delta=\sum_{n\geq 0}\frac{1}{n!}(\alpha\otimes\mathrm{Id}_C+\mathrm{Id}_C\otimes\alpha)^n\Delta=\sum_{n\geq 0}\frac{1}{n!}\Delta\alpha^n=\Delta e^{\alpha}.$$

11.3 The Reduced Tensor Coalgebra

Let V be a graded vector space. The **reduced tensor coalgebra** generated by V is the graded vector space

$$\overline{T^c}(V)=\bigoplus_{n>0}V^{\otimes n}$$

equipped with the coassociative coproduct

$$\mathfrak{a}\colon \overline{T^c}(V) \to \overline{T^c}(V) \otimes \overline{T^c}(V)$$

$$\mathfrak{a}(v_1 \otimes \cdots \otimes v_n) = \sum_{r=1}^{n-1} (v_1 \otimes \cdots \otimes v_r) \otimes (v_{r+1} \otimes \cdots \otimes v_n).$$

The letter c at the exponent of $\overline{T^c}$ is only used to distinguish the reduced tensor coalgebra from the reduced tensor algebra $\overline{T}(V) = \bigoplus_{n>0} V^{\otimes n}$, equipped with the usual concatenation product.

Unless otherwise specified, we shall denote by $p_V \colon \overline{T^c}(V) \to V$ the projection with kernel $\bigoplus_{n\geq 2} V^{\otimes n}$. Then the graded coalgebra $\overline{T^c}(V)$ is locally conilpotent and p_V is a cogenerator; in fact, for every $s > 0$,

$$\mathfrak{a}^{s-1}(v_1 \otimes \cdots \otimes v_n) = \sum_{1\leq i_1<i_2<\cdots<i_s=n} (v_1 \otimes \cdots \otimes v_{i_1}) \otimes \cdots \otimes (v_{i_{s-1}+1} \otimes \cdots \otimes v_n)$$

and then

$$\ker \mathfrak{a}^{s-1} = \bigoplus_{i=1}^{s-1} V^{\otimes i}, \qquad p_V^{\otimes s}\mathfrak{a}^{s-1}(v_1 \otimes \cdots \otimes v_s) = v_1 \otimes \cdots \otimes v_s.$$

Morphisms

We can interpret $\overline{T^c}$ as a functor from the category of graded vector spaces to the category of locally conilpotent graded coalgebras; for every morphism of graded vector spaces $f\colon V \to W$, the linear map

$$\overline{T^c}(f)\colon \overline{T^c}(V) \to \overline{T^c}(W), \qquad \overline{T^c}(f)(v_1 \otimes \cdots \otimes v_n) = f(v_1) \otimes \cdots \otimes f(v_n),$$

is a morphism of graded coalgebras. The functor $\overline{T^c}$ is faithful but it is not full, see the Corollary 11.3.3 below.

If (C, Δ) is a locally conilpotent graded coalgebra then, for every $x \in C$, there exists $n > 0$ such that $\Delta^n(x) = 0$ and then there is a properly defined morphism of graded vector spaces

$$\sum_{n=0}^{\infty} \Delta^n \colon C \to \overline{T^c}(C), \qquad x \mapsto x + \Delta(x) + \Delta^2(x) + \cdots .$$

Definition 11.3.1 Let (C, Δ) be a locally conilpotent graded coalgebra and let V be a graded vector space. The (coassociative) **multiplicative expansion** of a morphism $f \in \mathrm{Hom}^0_{\mathbb{K}}(C, V)$ is the morphism $F \in \mathrm{Hom}^0_{\mathbb{K}}(C, \overline{T^c}(V))$ defined by the formula

$$F := \sum_{n=1}^{\infty} f^{\otimes n}\Delta^{n-1} \colon C \xrightarrow{\sum_{n=0}^{\infty}\Delta^n} \overline{T^c}(C) \xrightarrow{\overline{T^c}(f)} \overline{T^c}(V). \tag{11.6}$$

The importance of multiplicative expansions is clear from the following proposition.

Proposition 11.3.2 *Let (C, Δ) be a locally conilpotent graded coalgebra and let V be a graded vector space. Then every morphism of graded coalgebras $C \to \overline{T^c}(V)$ is equal to the multiplicative expansion of its corestriction. For every morphism of graded vector spaces $f\colon C \to V$, its multiplicative expansion*

$$F = \sum_{n=1}^{\infty} f^{\otimes n} \Delta^{n-1} \colon C \to \overline{T^c}(V)$$

is the unique morphism of graded coalgebras such that $p_V F = f$.

Proof. The equality $p_V F = f$ is clear and, since p_V is a cogenerator, by Proposition 11.1.13 the coalgebra morphism F is uniquely determined by $p_V F$. Thus, to conclude the proof it is sufficient to show that

$$\sum_{n=0}^{\infty} \Delta^n \colon (C, \Delta) \to (\overline{T^c}(C), \mathfrak{a})$$

is a morphism of graded coalgebras. It is useful to write $\mathfrak{a} = \sum_{n=2}^{+\infty} \sum_{a=1}^{n-1} \mathfrak{a}_{a,n-a}$, where:

$$\mathfrak{a}_{a,b} \colon V^{\otimes a+b} \to V^{\otimes a} \otimes V^{\otimes b}, \quad \mathfrak{a}_{a,b}(v_1 \otimes \cdots \otimes v_{a+b}) = (v_1 \otimes \cdots \otimes v_a) \otimes (v_{a+1} \otimes \cdots \otimes v_{a+b}).$$

We have

$$\begin{aligned} \left(\left(\sum_{n\geq 0} \Delta^n\right) \otimes \left(\sum_{n\geq 0} \Delta^n\right)\right) \Delta &= \sum_{n\geq 0} \sum_{a=0}^{n} (\Delta^a \otimes \Delta^{n-a}) \Delta \\ &= \sum_{n\geq 0} \sum_{a=0}^{n} \mathfrak{a}_{a+1,n+1-a} \Delta^{n+1} = \mathfrak{a}\left(\sum_{n\geq 0} \Delta^n\right), \end{aligned}$$

where in the last equality we have used that $\mathfrak{a}(C^{\otimes 1}) = 0$ and therefore $\mathfrak{a}\Delta^0 = 0$. □

Corollary 11.3.3 *Let U, V be graded vector spaces. Given any morphism of graded vector spaces $f\colon \overline{T^c}(U) \to V$, the linear map $F\colon \overline{T^c}(U) \to \overline{T^c}(V)$:*

$$F(v_1 \otimes \cdots \otimes v_n) = \sum_{s=1}^{n} \sum_{1\leq i_1 < i_2 < \cdots < i_s = n} f(v_1 \otimes \cdots \otimes v_{i_1}) \otimes \cdots \otimes f(v_{i_{s-1}+1} \otimes \cdots \otimes v_{i_s}),$$

is the unique morphism of graded coalgebras whose corestriction is f.

Proof. This is Proposition 11.3.2 for $C = \overline{T^c}(U)$. □

Corollary 11.3.4 *Let $F\colon \overline{T^c}(U) \to \overline{T^c}(V)$ be a morphism of reduced graded tensor coalgebras and denote by $F^i_j\colon U^{\otimes j} \to V^{\otimes i}$, $i, j > 0$, the components of F. Then:*

1. *$F^i_j = 0$ for every $i > j$;*
2. *$F^n_n = (F^1_1)^{\otimes n}$ for every $n > 0$;*
3. *F is an isomorphism if and only if F^1_1 is an isomorphism.*

Proof. The first two items follow immediately from Corollary 11.3.3. For every pair of morphisms of graded coalgebras $\overline{T^c}(U) \xrightarrow{F} \overline{T^c}(V) \xrightarrow{G} \overline{T^c}(W)$, since $F^i_1 = 0$ for every $i > 1$ we have $(GF)^1_1 = G^1_1 F^1_1$ and this implies that if F is an isomorphism, then also F^1_1 is an isomorphism. Conversely, if F^1_1 is an isomorphism then F^n_n is an isomorphism for every n; since $F^i_j = 0$ for every $i > j$, in order to prove that F is an isomorphism it is sufficient to prove that for every $n > 0$ the restriction $F\colon \bigoplus_{i=1}^n U^{\otimes i} \to \bigoplus_{i=1}^n V^{\otimes i}$ is an isomorphism. This follows easily from the snake lemma applied to the diagram

$$\begin{array}{ccccccccc}
0 & \longrightarrow & \oplus_{i<n} U^{\otimes i} & \longrightarrow & \oplus_{i\le n} U^{\otimes i} & \longrightarrow & U^{\otimes n} & \longrightarrow & 0 \\
 & & \downarrow{\scriptstyle F} & & \downarrow{\scriptstyle F} & & \downarrow{\scriptstyle F^n_n} & & \\
0 & \longrightarrow & \oplus_{i<n} V^{\otimes i} & \longrightarrow & \oplus_{i\le n} V^{\otimes i} & \longrightarrow & V^{\otimes n} & \longrightarrow & 0
\end{array}$$

and by induction on n. Notice that the same argument proves that F is injective if and only if F^1_1 is injective. If F^1_1 is surjective then F is surjective but the converse is generally false. □

Example 11.3.5 Let A be a graded associative algebra and consider the two maps $\mu, \mu^* \in \mathrm{Hom}^0_{\mathbb{K}}(\overline{T^c}(A), A)$:

$$\mu(a_1 \otimes \cdots \otimes a_n) = a_1 a_2 \cdots a_n, \qquad \mu^*(a_1 \otimes \cdots \otimes a_n) = (-1)^{n-1} a_1 a_2 \cdots a_n.$$

Then their multiplicative expansions, i.e., the morphisms of graded coalgebras

$$F, G\colon \overline{T^c}(A) \to \overline{T^c}(A) \quad \text{such that} \quad p_A F = \mu, \quad p_A G = \mu^*,$$

are isomorphisms and each one is the inverse of the other.

Since $F^1_1 = G^1_1 = \mathrm{Id}_A$, by Corollary 11.3.4 both F, G are isomorphisms and it is sufficient to prove that GF is the identity. Since GF is uniquely determined by its corestriction, this is equivalent to proving that $p_V GF = \mu^* F = p_V$. By Corollary 11.3.3

$$F(a_1 \otimes \cdots \otimes a_n) = \sum_{s=1}^{n} \sum_{1 \le i_1 < i_2 < \cdots < i_s = n} (a_1 a_2 \cdots a_{i_1}) \otimes \cdots \otimes (a_{i_{s-1}+1} \cdots a_{i_s})$$

and then $\mu^* F(a) = a$ for every $a \in A$. For every $n \ge 2$ we have

$$\begin{aligned}\mu^* F(a_1 \otimes \cdots \otimes a_n) &= \sum_{s=1}^{n} (-1)^{s-1} \sum_{1 \le i_1 < i_2 < \cdots < i_s = n} a_1 a_2 \cdots a_n \\ &= \sum_{s=1}^{n} (-1)^{s-1} \binom{n-1}{s-1} a_1 a_2 \cdots a_n \\ &= \left(\sum_{s=0}^{n-1} (-1)^s \binom{n-1}{s} \right) a_1 a_2 \cdots a_n = 0.\end{aligned}$$

Coderivations The above considerations about morphisms hold essentially also for coderivations.

Proposition 11.3.6 *Let (C, Δ) be a locally conilpotent graded coalgebra, V a graded vector space and $F\colon C \to \overline{T^c}(V)$ a morphism of graded coalgebras with corestriction $f = p_V F \in Hom^0_{\mathbb{K}}(C, V)$. Then, for every $q \in Hom^*_{\mathbb{K}}(C, V)$, the linear map*

$$Q = \sum_{n=0}^{\infty} (\sum_{i=0}^{n} (f^{\otimes i} \otimes q \otimes f^{\otimes n-i}) \Delta^n \colon C \to \overline{T^c}(V)$$

is the unique F-coderivation such that $q = p_V Q$. In particular, the corestriction map

$$Coder^*(C, \overline{T^c}(V); F) \to Hom^*_{\mathbb{K}}(C, V), \qquad Q \mapsto p_V Q,$$

is an isomorphism of graded vector spaces.

Proof. Recall from Proposition 11.3.2 that

$$F = \overline{T^c}(f) \circ \sum_{n=0}^{\infty} \Delta^n \colon C \to \overline{T^c}(V).$$

Since Q is the composition of the coalgebra morphism $\sum_{n=0}^{\infty} \Delta^n \colon C \to \overline{T^c}(C)$ with the map

$$R\colon \overline{T^c}(C) \to \overline{T^c}(V), \qquad R = \sum_{i,j \ge 0} f^{\otimes i} \otimes q \otimes f^{\otimes j},$$

it is sufficient to prove that R is a $\overline{T^c}(f)$-coderivation, or equivalently that satisfies the coLeibniz rule

$$(R \otimes \overline{T^c}(f) + \overline{T^c}(f) \otimes R)\mathfrak{a} = \mathfrak{a} R\,.$$

Writing $R_n = \sum_{i+j=n-1} f^{\otimes i} \otimes q \otimes f^{\otimes j}$, for every a, b we have

$$\mathfrak{a}_{a,b} R_{a+b} = (R_a \otimes f^{\otimes b} + f^{\otimes a} \otimes R_b)\mathfrak{a}_{a,b}$$

and the proof is obtained by taking the sum over all a, b. □

Corollary 11.3.7 *Let V be a graded vector space. Then every $q \in Hom^k_{\mathbb{K}}(\overline{T^c}(V), V)$ is the corestriction of a unique coderivation $Q \in Coder^k(\overline{T^c}(V), \overline{T^c}(V))$ given by the explicit formula:*

$$Q(v_1 \otimes \cdots \otimes v_n)$$
$$= \sum_{l=1}^{n} \sum_{i=0}^{n-l} (-1)^{k(\overline{v_1}+\cdots+\overline{v_i})} v_1 \otimes \cdots \otimes v_i \otimes q(v_{i+1} \otimes \cdots \otimes v_{i+l}) \otimes \cdots \otimes v_n.$$

Proof. This is precisely Proposition 11.3.6 in the case F equal the identity and hence $f = p_V : \overline{T^c}(V) \to V$. □

As a first consequence of the explicit formula of Corollary 11.3.7 we get that, for every $Q \in \text{Coder}^*(\overline{T^c}(V), \overline{T^c}(V))$ we have $Q(V^{\otimes n}) \subset \bigoplus_{k=1}^{n} V^{\otimes k}$ for every n. More generally, if $q = (q_1, q_2, q_3, \ldots)$ with $q_k : V^{\otimes k} \to V$ is the corestriction of Q and $q_k = 0$ for every $k \leq r$, then $Q(V^{\otimes n}) \subset \bigoplus_{k=1}^{n-r} V^{\otimes k}$ for every n.

The definition of the Gerstenhaber bracket (Example 5.8.2) extends naturally to a graded vector space V. We want to show that, via the corestriction isomorphism

$$\text{Coder}^*(\overline{T^c}(V), \overline{T^c}(V)) \xrightarrow{\simeq} \text{Hom}^*_{\mathbb{K}}(\overline{T^c}(V), V), \qquad Q \mapsto p_V Q,$$

the Gerstenhaber bracket to $\text{Hom}^*_{\mathbb{K}}(\overline{T^c}(V), V)$ corresponds to the commutator bracket in $\text{Coder}^*(\overline{T^c}(V), \overline{T^c}(V))$.

Recall that the Gerstenhaber bracket is defined as the graded commutator of the (nonassociative) Gerstenhaber product:

$$[f, g] = f \circ g - (-1)^{\overline{f}\,\overline{g}} g \circ f, \qquad f, g \in \text{Hom}^*_{\mathbb{K}}(\overline{T^c}(V), V),$$

where, for $f \in \text{Hom}^a_{\mathbb{K}}(V^{\otimes n+1}, V)$ and $g \in \text{Hom}^b_{\mathbb{K}}(V^{\otimes m+1}, V)$ we have $f \circ g \in \text{Hom}^{a+b}_{\mathbb{K}}(V^{\otimes n+m}, V)$:

$$f \circ g(v_0 \otimes \cdots \otimes v_{n+m})$$
$$= \sum_{i=0}^{n} (-1)^{b(\overline{v_0}+\cdots+\overline{v_{i-1}})} f(v_1 \otimes \cdots \otimes v_{i-1} \otimes g(v_i \otimes \cdots \otimes v_{i+m}) \otimes \cdots \otimes v_{n+m}).$$

According to Corollary 11.3.7, if $F, G \in \text{Coder}^*(\overline{T^c}(V), \overline{T^c}(V))$ corestrict to f, g respectively, then

$$p_V FG = fG = f \circ g, \qquad p_V GF = gF = g \circ f,$$

and therefore $p_V[F, G] = [f, g]$.

Keep in mind that, in general, composition of coderivations is not a coderivation and that the unique coderivation corestricting to $f \circ g$ is not FG.

11.4 Symmetrization

We have already seen in Sect. 10.1 that for every graded vector space V and every positive integer n, the twist map induces a right linear action of the symmetric groups Σ_n on $V^{\otimes n}$:

$$\mathrm{tw}\colon V^{\otimes n}\times\Sigma_n\to V^{\otimes n},\qquad \mathrm{tw}(v_1\otimes\cdots\otimes v_n,\sigma)=\epsilon(\sigma)(v_{\sigma(1)}\otimes\cdots\otimes v_{\sigma(n)}).$$

Recall that the symmetric powers of a graded vector space V are defined as the coinvariant spaces of the above actions. More precisely

$$V^{\odot n}=\frac{V^{\otimes n}}{I_n},$$

where I_n is the subspace generated by the vectors $v-\mathrm{tw}(v,\sigma)$, with $\sigma\in\Sigma_n$ and $v\in V^{\otimes n}$. Denoting by $\pi\colon V^{\otimes n}\to V^{\odot n}$ the natural projection we have

$$v_1\odot\cdots\odot v_n=\pi(v_1\otimes\cdots\otimes v_n). \tag{11.7}$$

Lemma 11.4.1 *Let V be a graded vector space. For every integer $n>0$ the morphism of graded vector spaces $N\colon V^{\odot n}\to V^{\otimes n}$,*

$$\begin{aligned}N(v_1\odot\cdots\odot v_n)&=\sum_{\sigma\in\Sigma_n}\epsilon(\sigma;v_1,\ldots,v_n)(v_{\sigma(1)}\otimes\cdots\otimes v_{\sigma(n)})\\&=\sum_{\sigma\in\Sigma_n}\mathrm{tw}(v_1\otimes\cdots\otimes v_n,\sigma).\end{aligned} \tag{11.8}$$

is properly defined and injective. Moreover $\frac{1}{n!}\pi N=Id_{V^{\odot n}}$ and its image is the subspace $(V^{\otimes n})^{\Sigma_n}$ of invariant tensors.

Proof. The unicity of N, the equality $\pi N=n!\,\mathrm{Id}_{V^{\odot n}}$ and the injectivity are clear. Consider the linear map

$$N'\colon V^{\otimes n}\to V^{\otimes n},\qquad N'(v_1\otimes\cdots\otimes v_n)=\sum_{\sigma\in\Sigma_n}\mathrm{tw}(v_1\otimes\cdots\otimes v_n,\sigma).$$

Then (11.8) is equivalent to $N'=N\pi$ and therefore N is properly defined if and only if the kernel of π is contained in the kernel of N'. By definition $\ker\pi=I_n$ is the subspace generated by all the vectors $v-\mathrm{tw}(v,\sigma)$, $\sigma\in\Sigma_n$, $v\in V^{\otimes n}$. Since $N'(v)=N'(\mathrm{tw}(v,\sigma))$ we have $\ker\pi\subset\ker N'$. Finally, the map

$$\frac{1}{n!}N'\colon V^{\otimes n}\to(V^{\otimes n})^{\Sigma_n}$$

is a projection and has the same image as N. It should be noted that the injectivity of N is equivalent to the equality $\ker \pi = \ker N'$ that is equivalent to the direct sum decomposition $V^{\otimes n} = I_n \oplus (V^{\otimes n})^{\Sigma_n}$. □

Lemma 11.4.2 *Let (C, Δ) be a graded cocommutative coalgebra. For every $n > 0$ the image of Δ^{n-1} is contained in the set of Σ_n-invariant elements of $C^{\otimes n}$ and therefore there exists a factorization*

$$\Delta^{n-1} = N\frac{\pi}{n!}\Delta^{n-1} \colon C \to C^{\otimes n},$$

where the maps N, π are defined as in (11.8) *and* (11.7).

Proof. The twist action of Σ_n on $C^{\otimes n}$ is generated by the operators

$$\mathtt{tw}_a = \mathrm{Id}_{C^{\otimes a}} \otimes \mathtt{tw} \otimes \mathrm{Id}_{C^{\otimes n-a-2}}, \qquad 0 \le a \le n-2,$$

and therefore it is sufficient to prove that $\mathtt{tw}_a\Delta^{n-1} = \Delta^{n-1}$ for every a, n. Since $\mathtt{tw}\Delta = \Delta$, according to Lemma 11.1.10 we have:

$$\begin{aligned}\mathtt{tw}_a\Delta^{n-1} &= \mathtt{tw}_a(\mathrm{Id}_{C^{\otimes a}} \otimes \Delta \otimes \mathrm{Id}_{C^{\otimes n-a-2}})\Delta^{n-2}\\ &= (\mathrm{Id}_{C^{\otimes a}} \otimes \Delta \otimes \mathrm{Id}_{C^{\otimes n-a-2}})\Delta^{n-2} = \Delta^{n-1}.\end{aligned}$$

□

We conclude the section with two technical lemmas that will be used in the study of symmetric coalgebras. As in the previous chapters we denote by $S(a, n-a) \subset \Sigma_n$ the subset of $(a, n-a)$-shuffles (Definition 10.4.1).

Lemma 11.4.3 *In the above notation, for every sequence of homogeneous vectors $v_1, v_2, \ldots, v_n \in V$ and every integer $a = 1, 2, \ldots, n$, via the natural identification $V^{\otimes a} \otimes V^{\otimes n-a} = V^{\otimes n}$ we have:*

$$N(v_1 \odot \cdots \odot v_n) = \sum_{\sigma \in S(a,n-a)} \epsilon(\sigma) N(v_{\sigma(1)} \odot \cdots \odot v_{\sigma(a)}) \otimes N(v_{\sigma(a+1)} \odot \cdots \odot v_{\sigma(n)}).$$

Proof. Since every $\eta \in \Sigma_n$ can be uniquely written as $\eta = \sigma\tau$ with $\sigma \in S(a, n-a)$ and $\tau \in \Sigma_a \times \Sigma_{n-a}$ we have:

$$\begin{aligned}N(v_1 \odot \cdots \odot v_n) &= \sum_{\eta \in \Sigma_n} \mathtt{tw}(v_1 \otimes \cdots \otimes v_n, \eta)\\ &= \sum_{\sigma \in S(a,n-a)} \sum_{\tau \in \Sigma_a \times \Sigma_{n-a}} \mathtt{tw}(\mathtt{tw}(v_1 \otimes \cdots \otimes v_n, \sigma), \tau)\\ &= \sum_{\sigma \in S(a,n-a)} \epsilon(\sigma) \sum_{\tau \in \Sigma_a \times \Sigma_{n-a}} \mathtt{tw}(v_{\sigma(1)} \otimes \cdots \otimes v_{\sigma(n)}, \tau)\end{aligned}$$

$$= \sum_{\sigma \in S(a,n-a)} \epsilon(\sigma) N(v_{\sigma(1)} \odot \cdots \odot v_{\sigma(a)}) \otimes N(v_{\sigma(a+1)} \odot \cdots \odot v_{\sigma(n)}).$$

□

Lemma 11.4.4 *Given two graded vector spaces V, M, a positive integer l and two linear maps $f \in Hom^0_{\mathbb{K}}(V, M)$ and $b \in Hom^k_{\mathbb{K}}(V^{\otimes l}, M)$ write $q = bN \in Hom^k_{\mathbb{K}}(V^{\odot l}, M)$ and for every integer $n \geq l$ define:*

1. *a map $B \in Hom^k_{\mathbb{K}}(V^{\otimes n}, M^{\otimes n-l+1})$ by the formula:*

$$B(v_1 \otimes \cdots \otimes v_n) = \sum_{i=0}^{n-l} (-1)^{k(\overline{v_1}+\cdots+\overline{v_i})} f(v_1) \otimes \cdots \otimes f(v_i)$$
$$\otimes\, b(v_{i+1} \otimes \cdots \otimes v_{i+l}) \otimes f(v_{i+l+1}) \otimes \cdots \otimes f(v_n);$$

2. *a map $Q \in Hom^k_{\mathbb{K}}(V^{\odot n}, M^{\odot n-l+1})$ by the formula:*

$$Q(v_1 \odot \cdots \odot v_n) = \sum_{\sigma \in S(l,n-l)} \epsilon(\sigma) q(v_{\sigma(1)} \odot \cdots \odot v_{\sigma(l)}) \odot f(v_{\sigma(l+1)}) \odot \cdots \odot f(v_{\sigma(n)}).$$

Then

$$BN = NQ \in Hom^k(V^{\odot n}, M^{\otimes n-l+1}).$$

Proof. This is completely straightforward and it is left to the reader. □

11.5 The Reduced Symmetric Coalgebra

Let V be a graded vector space. The **reduced symmetric coalgebra** generated by V is the graded vector space

$$\overline{S^c}(V) = \bigoplus_{n>0} V^{\odot n}$$

equipped with the cocommutative coproduct $\mathfrak{l}\colon \overline{S^c}(V) \to \overline{S^c}(V) \otimes \overline{S^c}(V)$:

$$\mathfrak{l}(v_1 \odot \cdots \odot v_n) = \sum_{a=1}^{n-1} \sum_{\sigma \in S(a,n-a)} \epsilon(\sigma)(v_{\sigma(1)} \odot \cdots \odot v_{\sigma(a)}) \otimes (v_{\sigma(a+1)} \odot \cdots \odot v_{\sigma(n)}).$$

As above the role of the letter c at the exponent is to distinguish the reduced symmetric coalgebra from the reduced symmetric algebra. The cocommutativity of $\mathfrak{l}$ is clear from the definition, while the coassociativity follows from the following lemma.

Lemma 11.5.1 *In the above notation, the map $\mathfrak{l}\colon \overline{S^c}(V) \to \overline{S^c}(V) \otimes \overline{S^c}(V)$ is a coassociative coproduct and the symmetrization morphism of Lemma 11.4.1:*

$$N\colon (\overline{S^c}(V), \mathfrak{l}) \to (\overline{T^c}(V), \mathfrak{a}),$$

is an injective morphism of graded coalgebras.

Proof. Since N is injective, we only need to prove that $\mathfrak{a}N = (N \otimes N)\mathfrak{l}$. According to Lemma 11.4.3, for every $0 < a < n$ we have

$$\begin{aligned}&\mathfrak{a}_{a,n-a}N(v_1 \odot \cdots \odot v_n)\\ &\quad = N \otimes N \sum_{\sigma \in S(a,n-a)} \epsilon(\sigma)(v_{\sigma(1)} \odot \cdots \odot v_{\sigma(a)}) \otimes (v_{\sigma(a+1)} \otimes \cdots \otimes v_{\sigma(n)})\end{aligned}$$

and then

$$\mathfrak{a}N(v_1 \odot \cdots \odot v_n) = \sum_{a=1}^{n-1} \mathfrak{a}_{a,n-a}N(v_1 \odot \cdots \odot v_n) = (N \otimes N)\mathfrak{l}(v_1 \odot \cdots \odot v_n).$$

In order to prove the coassociativity of $\mathfrak{l}$, since N is injective it is sufficient to prove that $N^{\otimes 3}(\mathrm{Id} \otimes \mathfrak{l})\mathfrak{l} = N^{\otimes 3}(\mathfrak{l} \otimes \mathrm{Id})\mathfrak{l}$, and this follows from the equality $\mathfrak{a}N = (N \otimes N)\mathfrak{l}$:

$$\begin{aligned}N^{\otimes 3}(\mathrm{Id} \otimes \mathfrak{l})\mathfrak{l} &= (N \otimes \mathfrak{a}N)\mathfrak{l} = (\mathrm{Id} \otimes \mathfrak{a})(N \otimes N)\mathfrak{l} = (\mathrm{Id} \otimes \mathfrak{a})\mathfrak{a}N\\ &= (\mathfrak{a} \otimes \mathrm{Id})\mathfrak{a}N = (\mathfrak{a} \otimes \mathrm{Id})(N \otimes N)\mathfrak{l} = N^{\otimes 3}(\mathfrak{l} \otimes \mathrm{Id})\mathfrak{l}.\end{aligned}$$

□

Although we defined $V^{\odot n}$ as a quotient of $V^{\otimes n}$, it is more convenient to think of the reduced symmetric coalgebra as a subcoalgebra of the reduced tensor coalgebra, via the identification provided by the map N. Keep in mind that for general V there does not exist any coproduct on $\overline{S^c}(V)$ such that the projection map $\pi : \overline{T^c}(V) \to \overline{S^c}(V)$ is a morphism of coalgebras; in fact, for every pair of linearly independent vectors $x, y \in V$, if $z = x \otimes y - (-1)^{\bar{x}\bar{y}} y \otimes x$, then $\pi(z) = 0$, while $\pi^{\otimes 2}\mathfrak{a}(z) \neq 0$.

For an element $x \in \overline{S^c}(V)$ we have $\mathfrak{l}^n(x) = 0$ if and only if $\mathfrak{a}^n(Nx) = 0$ and then $\ker \mathfrak{l}^n = \oplus_{i=1}^n V^{\odot i}$; in particular the graded coalgebra $\overline{S^c}(V)$ is locally conilpotent and the projection $p_V : \overline{S^c}(V) \to V$ with kernel $\oplus_{n>1} V^{\odot n}$ is a cogenerator.

Morphisms

We can interpret $\overline{S^c}$ as a functor from the category of graded vector spaces to the category of cocommutative locally conilpotent graded coalgebras; for every morphism of graded vector spaces $f\colon V \to W$ the induced map

$$\overline{S^c}(f)\colon \overline{S^c}(V) \to \overline{S^c}(W), \qquad \overline{S^c}(f) = \sum_{n>0} f^{\odot n},$$

is a morphism of graded coalgebras since

$$N \circ \overline{S^c}(f) = \overline{T^c}(f) \circ N\colon \overline{S^c}(V) \to \overline{T^c}(W),$$

and $\overline{T^c}(f)$, N are morphism of graded coalgebras.

Definition 11.5.2 Let (C, Δ) be a locally conilpotent graded cocommutative coalgebra and V a graded vector space. The (coassociative symmetric) **multiplicative expansion** of a morphism $f \in \mathrm{Hom}^0_{\mathbb{K}}(C, V)$ is the linear map $F\colon C \to \overline{S^c}(V)$ defined by the formula

$$F = \overline{S^c}(f) \sum_{n>0} \frac{\pi}{n!} \Delta^{n-1} = \sum_{n=1}^{\infty} f^{\odot n} \frac{\pi}{n!} \Delta^{n-1} = \sum_{n=1}^{\infty} \frac{\pi}{n!} f^{\otimes n} \Delta^{n-1}. \tag{11.9}$$

Proposition 11.5.3 *Let (C, Δ) be a locally conilpotent graded cocommutative coalgebra and V a graded vector space. Then for every morphism of graded vector spaces $f\colon C \to V$, its multiplicative expansion $F\colon C \to \overline{S^c}(V)$ is a morphism of graded coalgebras such that $p_V F = f$. Every morphism of graded coalgebras $C \to \overline{S^c}(V)$ is equal to the multiplicative expansion of its corestriction.*

Proof. For the first part of the proposition it is sufficient to prove that the map

$$\sum_{n>0} \frac{\pi}{n!} \Delta^{n-1} \colon (C, \Delta) \to (\overline{S^c}(C), \mathfrak{l})$$

is a morphism of graded coalgebras. We already know that $\sum_{n>0} \Delta^{n-1} \colon C \to \overline{T^c}(C)$ is a morphism of graded coalgebras. According to Lemma 11.4.2 we have

$$\sum_{n>0} \Delta^{n-1} = N\left(\sum_{n>0} \frac{\pi}{n!} \Delta^{n-1} \right)$$

and then the proof is an immediate consequence of the fact that N is an injective morphism of graded coalgebras. The second part follows from Proposition 11.1.13 because $p_V F = f$. □

Corollary 11.5.4 *For every pair of graded vector spaces V, W and every linear map $f \in \mathrm{Hom}^0_{\mathbb{K}}(\overline{S^c}(V), W)$, its multiplicative expansion $F\colon \overline{S^c}(V) \to \overline{S^c}(W)$ is the unique morphism of graded coalgebras such that $p_W F = f$. If $f = \sum_i f_i$ with $f_i\colon V^{\odot i} \to W$, then the components $F^i_j\colon V^{\odot j} \to W^{\odot i}$ of F satisfy the conditions:*

1. $F^1_n = f_n$;
2. *for every $1 < i < j$ the components F^i_j may be computed by the recursive formula*

$$F^i_j(v_1 \odot \cdots \odot v_j)$$
$$= \frac{1}{i} \sum_{a=1}^{j-i+1} \sum_{\sigma \in S(a, j-a)} \epsilon(\sigma) f_a(v_{\sigma(1)} \odot \cdots \odot v_{\sigma(a)}) \odot F^{i-1}_{j-a}(v_{\sigma(a+1)} \odot \cdots \odot v_{\sigma(j)}).$$

3. $F_j^j = f_1^{\odot j}$;
4. $F_j^i = 0$ *for every* $i > j$.

In particular, every element in the image of F_j^i *is a linear combination of vectors of the form*

$$f_{a_1}(u_1) \odot f_{a_2}(u_2) \odot \cdots \odot f_{a_i}(u_i), \qquad \sum_{k=1}^{i} a_k = j, \quad u_k \in V^{\odot a_k}.$$

Proof. The first part is an easy consequence of the formulas

$$\sum_j F_j^i = \frac{\pi}{i!} f^{\otimes i} \mathfrak{l}^{i-1} = \frac{\pi}{i!}(f \otimes f^{\otimes i-1}\mathfrak{l}^{i-2})\mathfrak{l}, \quad \sum_j F_j^{i-1} = \frac{\pi}{(i-1)!} f^{\otimes i-1}\mathfrak{l}^{i-2},$$

which are particular cases of Proposition 11.5.3. The statement about the image of F_j^i is an easy consequence of the formulas

$$F_j^i = \frac{\pi}{i!} f^{\otimes i} \mathfrak{l}^{i-1} = \frac{\pi}{i!}(f \otimes f^{\otimes i-1}\mathfrak{l}^{i-2})\mathfrak{l} \colon V^{\odot j} \to W^{\odot i}$$

and

$$\mathfrak{l}^{i-1}(V^{\odot j}) \subset \sum_{a_1+\cdots+a_i=j} V^{\odot a_1} \otimes \cdots \otimes V^{\odot a_i}.$$

□

The formulas of Corollary 11.5.4 give for instance

$$\begin{aligned} &F_3^2(x \odot y \odot z) \\ &= \frac{1}{2}\big(f_2(x \odot y) \odot f_1(z) + (-1)^{\overline{y}\,\overline{z}} f_2(x \odot z) \odot f_1(y) + f_1(x) \odot f_2(y \odot z)\big), \end{aligned}$$

while if either $f_1 = 0$ or $f_3 = 0$ we have:

$$F_4^2(v_1 \odot \cdots \odot v_4) = \frac{1}{2} \sum_{\sigma \in S(2,2)} f_2(v_{\sigma(1)} \odot v_{\sigma(2)}) \odot f_2(v_{\sigma(3)} \odot v_{\sigma(4)}).$$

Corollary 11.5.5 *Let* $F \colon \overline{S^c}(U) \to \overline{S^c}(V)$ *be a morphism of reduced symmetric coalgebras. Then* $F(U) \subset V$*. Moreover, if* $F_1^1 \colon U \to V$ *is injective (resp.: surjective, bijective), then also* F *is injective (resp.: surjective, bijective). Conversely, if* F *is an isomorphism, then also* F_1^1 *is an isomorphism.*

Proof. For every $n > 0$ there exists a morphism of short exact sequences:

$$\begin{array}{ccccccccc} 0 & \longrightarrow & \oplus_{i<n} U^{\odot i} & \longrightarrow & \oplus_{i\le n} U^{\odot i} & \longrightarrow & U^{\odot n} & \longrightarrow & 0 \\ & & \downarrow F & & \downarrow F & & \downarrow F^{\odot n}_{|U} & & \\ 0 & \longrightarrow & \oplus_{i<n} V^{\odot i} & \longrightarrow & \oplus_{i\le n} V^{\odot i} & \longrightarrow & V^{\odot n} & \longrightarrow & 0 \end{array}$$

and the first part is proved by using induction on n and the snake lemma. The last part is trivial since, if G is the inverse of F, then $G_1^1 F_1^1 = \mathrm{Id}_U$ and $F_1^1 G_1^1 = \mathrm{Id}_V$. □

Example 11.5.6 Let $U = V = \mathbb{K}$ be one-dimensional vector spaces, considered as graded vector spaces concentrated in degree 0. Given a positive integer p, consider a sequence of maps

$$f_n\colon U^{\odot n} \to V \text{ such that } f_i \neq 0 \text{ if and only if } i = p.$$

Denoting by $F\colon \overline{S^c}(U) \to \overline{S^c}(V)$ the associated morphism of reduced symmetric coalgebras, it is easy to see that $F_j^i \neq 0$ if and only if $j = ip$. In particular, the surjectivity of F does not imply the surjectivity of F_1^1.

Coderivations

We begin with the analogue of Proposition 11.3.6 in the cocommutative case.

Proposition 11.5.7 *Let V be a graded vector space and C a locally conilpotent cocommutative coalgebra. Then for every coalgebra morphism $F\colon C \to \overline{S^c}(V)$ the composition with $N\colon \overline{S^c}(V) \to \overline{T^c}(V)$ gives a bijection*

$$Coder^*(C, \overline{S^c}(V); F) \cong Coder^*(C, \overline{T^c}(V); NF).$$

In particular, the corestriction map

$$Coder^*(C, \overline{S^c}(V); F) \xrightarrow{\simeq} Hom^*_{\mathbb{K}}(C, V), \qquad Q \mapsto p_V Q,$$

is an isomorphism of graded vector spaces.

Proof. For the first part, we only need to prove that if $B\colon C \to \overline{T^c}(V)$ is a coderivation with respect to the coalgebra morphism NF, then $B = NP$ for some $P\colon C \to \overline{S^c}(V)$; the injectivity of N automatically implies that P is a coderivation with respect to F. If $f = p_V NF = p_V F$ and $b = p_V B$, then by Proposition 11.3.6 we have

$$B = \sum_{n=0}^{\infty}\sum_{i=0}^{n} (f^{\otimes i} \otimes b \otimes f^{\otimes n-i})\Delta^n\colon C \to \overline{T^c}(V),$$

where Δ is the coproduct of C. By Lemmas 11.4.2 and 11.4.4 we have

$$B = \sum_{n=0}^{\infty}\sum_{i=0}^{n}(f^{\otimes i}\otimes b\otimes f^{\otimes n-i})N\frac{\pi}{(n+1)!}\Delta^n = \sum_{n=0}^{\infty} NQ_{n+1}\frac{\pi}{(n+1)!}\Delta^n = NP,$$

where $Q_{n+1}\colon C^{\odot n+1}\to V^{\odot n+1}$ is the map

$$Q_{n+1}(c_0\odot\cdots\odot c_n) = \sum_{\sigma\in S(1,n)}\epsilon(\sigma)b(c_{\sigma(0)})\odot f(c_{\sigma(1)})\odot\cdots\odot f(c_{\sigma(n)}),$$

and $P = \sum_{n=0}^{\infty} NQ_{n+1}\dfrac{\pi}{(n+1)!}\Delta^n$. The second part follows from Proposition 11.3.6 and from the fact that for every map $P\colon C\to\overline{S^c}(V)$ we have $p_V P = p_V NP$. □

The proof of Proposition 11.5.7 also provides an explicit description of the inverse of the corestriction isomorphism $\mathrm{Coder}^*(C,\overline{S^c}(V);F)\xrightarrow{\simeq}\mathrm{Hom}^*_{\mathbb{K}}(C,V)$. For simplicity we study the inverse isomorphism only for F equal to the identity; this particular case is completely sufficient for our purposes.

Definition 11.5.8 Let V be a graded vector space. Define the (coassociative symmetric) **Leibniz expansion** of a linear map $q\in\mathrm{Hom}^*_{\mathbb{K}}(\overline{S^c}(V),V)$ as the linear map $Q\in\mathrm{Hom}^*_{\mathbb{K}}(\overline{S^c}(V),\overline{S^c}(V))$ defined by the formula

$$Q(v_1\odot\cdots\odot v_n) = \sum_{i=1}^{n}\sum_{\sigma\in S(i,n-i)}\epsilon(\sigma)q(v_{\sigma(1)}\odot\cdots\odot v_{\sigma(i)})\odot v_{\sigma(i+1)}\odot\cdots\odot v_{\sigma(n)}.$$

Corollary 11.5.9 *Let V be a graded vector space. Then the corestriction map*

$$Coder^*(\overline{S^c}(V),\overline{S^c}(V))\xrightarrow{\simeq} Hom^*(\overline{S^c}(V),V),\qquad Q\mapsto p_V Q,$$

is an isomorphism of graded vector spaces. For every map $q\in Hom^*_{\mathbb{K}}(\overline{S^c}(V),V)$, its Leibniz expansion $Q\in Hom^*_{\mathbb{K}}(\overline{S^c}(V),\overline{S^c}(V))$:

$$Q(v_1\odot\cdots\odot v_n) = \sum_{i=1}^{n}\sum_{\sigma\in S(i,n-i)}\epsilon(\sigma)q_i(v_{\sigma(1)}\odot\cdots\odot v_{\sigma(i)})\odot v_{\sigma(i+1)}\odot\cdots\odot v_{\sigma(n)},$$

is the unique coderivation of $\overline{S^c}(V)$ such that $p_V Q = q$.

Proof. We only need to prove that the map Q is a coderivation. If $q=\sum q_i$ with $q_i\colon V^{\odot i}\to V$, by linearity it is not restrictive to assume that $q_i = 0$ for every $i\neq l$. Let $b\in\mathrm{Hom}^*_{\mathbb{K}}(V^{\otimes l},V)$ be any linear map such that $bN = q_l = q$; for example one can take $b = \dfrac{\pi}{l!}q_l$. Let $B\in\mathrm{Coder}^*(\overline{T^c}(V),\overline{T^c}(V))$ be the unique coderivation such that $p_V B = b$; by Corollary 11.3.7 we have

$$B(v_1\otimes\cdots\otimes v_n) = \sum_i (-1)^{\overline{b}\,(\overline{v_1}+\cdots+\overline{v_i})} v_1\otimes\cdots\otimes v_i\otimes b(v_{i+1}\otimes\cdots\otimes v_{i+l})\otimes\cdots\otimes v_n.$$

According to Lemma 11.4.4 the equality $BN = NQ$ holds and then also Q is a coderivation. □

Remark 11.5.10 For later use it is important to point out that if $F\colon \overline{S^c}(V) \to \overline{S^c}(W)$ is a morphism of graded coalgebras, then $F(V^{\odot n}) \subset \bigoplus_{i\le n} W^{\odot i}$ for every $n > 0$. Similarly, if $Q\colon \overline{S^c}(V) \to \overline{S^c}(V)$ is a coderivation, then $Q(V^{\odot n}) \subset \bigoplus_{i\le n} V^{\odot i}$ for every $n > 0$. The first of the above claims follows from 11.5.4, the second from 11.5.9.

The corestriction isomorphism $\operatorname{Coder}^*(\overline{S^c}(V), \overline{S^c}(V)) \simeq \operatorname{Hom}^*_{\mathbb{K}}(\overline{S^c}(V), V)$ can be used for a different description of the Nijenhuis–Richardson bracket (Definition 10.6.5). Recall that, for $f \in \operatorname{Hom}^*_{\mathbb{K}}(V^{\odot n+1}, V)$ and $g \in \operatorname{Hom}^*_{\mathbb{K}}(V^{\odot m+1}, V)$ we have

$$[f, g] = f \barwedge g - (-1)^{\bar{f}\,\bar{g}} g \barwedge f \in \operatorname{Hom}^*_{\mathbb{K}}(V^{\odot n+m+1}, V),$$

where

$$f \barwedge g(v_0, \dots, v_{n+m}) = \sum_{\sigma\in S(m+1,n)} \epsilon(\sigma) f(g(v_{\sigma(0)}, \dots, v_{\sigma(m)}), v_{\sigma(m+1)}, \dots, v_{\sigma(m+n)}).$$

The following result is an immediate application of Corollary 11.5.9.

Lemma 11.5.11 *In the above notation, $f \barwedge g = fG$, where $G \in \operatorname{Coder}^*(\overline{S^c}(V), \overline{S^c}(V))$ is the unique coderivation corestricting to g. In particular the Nijenhuis-Richardson bracket corresponds to the commutator bracket in $\operatorname{Coder}^*(\overline{S^c}(V), \overline{S^c}(V))$ via the corestriction isomorphism.*

11.6 Scalar Extension and Restitution

Given two commutative graded algebras A, B, the usual tensor product of complexes $A \otimes B$ inherits a natural structure of commutative graded algebras with product

$$(a_1 \otimes b_1)(a_2 \otimes b_2) = (-1)^{\overline{b_1}\,\overline{a_2}} a_1a_2 \otimes b_1b_2.$$

It is plain that the twisting involution $\mathtt{tw}\colon A \otimes B \to B \otimes A$ is a morphism of algebras, if A, B are both unitary then also $A \otimes B$ has a unit, if either A or B is nilpotent then $A \otimes B$ is nilpotent.

The above construction applies in particular when $A = S(V)$ is the symmetric algebra generated by a graded vector space. Namely, for every graded commutative algebra B, the tensor product

$$S(V) \otimes B = \bigoplus_{n=0}^{\infty} V^{\odot n} \otimes B$$

is again a graded commutative algebra with product

$$(x \otimes a) \odot (y \otimes b) = (-1)^{\overline{a}\,\overline{y}}(x \odot y) \otimes ab, \qquad x, y \in S(V), \quad a, b \in B,$$

and there exists a natural morphism of graded algebras $\alpha \colon S(V \otimes B) \to S(V) \otimes B$

$$\alpha((x_1 \otimes a_1) \odot \cdots \odot (x_n \otimes a_n)) = (-1)^{\sum_{i<j} \overline{a_i}\,\overline{x_j}}(x_1 \odot \cdots \odot x_n) \otimes a_1 \cdots a_n.$$

Lemma 11.6.1 *Let A be a nilpotent graded commutative algebra, V a graded vector space and $x \in (V \otimes A)^0$. Consider x as an element of the nilpotent algebra $S(V) \otimes A$ and denote by $y = \sum_{n>0} x^{\odot n}/n! = e^{\odot x} - 1 \in \overline{S^c}(V) \otimes A$. Then*

$$(\mathfrak{l}^{n-1} \otimes Id_A)(y) = \mu(y^{\otimes n}) \in \overline{S^c}(V)^{\otimes n} \otimes A, \tag{11.10}$$

where $\mu \colon (\overline{S^c}(V) \otimes A)^{\otimes n} \to \overline{S^c}(V)^{\otimes n} \otimes A$ is the map induced by the multiplication in A:

$$\mu((x_1 \otimes a_1) \otimes \cdots \otimes (x_n \otimes a_n)) = (-1)^{\sum_{i<j} \overline{a_i}\,\overline{x_j}}(x_1 \otimes \cdots \otimes x_n) \otimes a_1 \cdots a_n.$$

Proof. By the definition of μ there exists a commutative diagram

$$\begin{array}{ccccc}
\overline{S^c}(V \otimes A) & & \xrightarrow{\qquad\qquad \alpha \qquad\qquad} & & \overline{S^c}(V) \otimes A \\
\Big\downarrow{\scriptstyle \mathfrak{l}^{n-1}} & & & & \Big\downarrow{\scriptstyle \mathfrak{l}^{n-1} \otimes \mathrm{Id}_A} \\
\overline{S^c}(V \otimes A)^{\otimes n} & \xrightarrow{\alpha^{\otimes n}} & (\overline{S^c}(V) \otimes A)^{\otimes n} & \xrightarrow{\mu} & \overline{S^c}(V)^{\otimes n} \otimes A.
\end{array}$$

Writing $x = \alpha(z)$ with $z \in (V \otimes A)^0 \subset \overline{S^c}(V \otimes A)$, a simple induction on n gives the formula

$$\frac{1}{m!}\mathfrak{l}^{n-1}(z^{\odot m}) = \sum_{\substack{i_1+\cdots+i_n=m, \\ i_1,\ldots,i_n>0}} \frac{z^{\odot i_1}}{i_1!} \otimes \cdots \otimes \frac{z^{\odot i_n}}{i_n!},$$

and therefore

$$\frac{1}{m!}(\mathfrak{l}^{n-1} \otimes \mathrm{Id}_A)(x^{\odot m}) = \sum_{\substack{i_1+\cdots+i_n=m, \\ i_1,\ldots,i_n>0}} \mu\left(\frac{x^{\odot i_1}}{i_1!} \otimes \cdots \otimes \frac{x^{\odot i_n}}{i_n!}\right).$$

The sum of the above equalities over all positive integers m gives the proof. □

Let V, W be graded vector spaces; for every graded commutative algebra A, in Sect. 10.6 we introduced the scalar extension operator

$$\mathrm{Hom}^*_{\mathbb{K}}(V^{\odot n}, W) \xrightarrow{(-)_A} \mathrm{Hom}^*_{\mathbb{K}}((V \otimes A)^{\odot n}, W \otimes A), \qquad f \mapsto f_A,$$

defined in the following way:

$$f_A(v_1 \otimes a_1, \ldots, v_n \otimes a_n) = (-1)^{\sum_{i<j} \overline{a_i}\,\overline{v_j}} f(v_1, \ldots, v_n) \otimes a_1 a_2 \cdots a_n.$$

Equivalently, the scalar extension f_A is the composition of $f \otimes \mathrm{Id}_A$ with the product in $S(V) \otimes A$:

$$(v_1 \otimes a_1) \odot \cdots \odot (v_n \otimes a_n) \mapsto (-1)^{\sum_{i<j} \overline{a_i}\,\overline{v_j}} (v_1 \odot \cdots \odot v_n) \otimes a_1 a_2 \cdots a_n\,.$$

The scalar extension is clearly functorial in the sense that it commutes with morphisms of commutative graded algebras.

Restitution maps are a standard tool in classical invariant theory. Let V, W be vector spaces over a field $\mathbb{K}$ of characteristic 0. The **restitution** of a symmetric multilinear map $f : V^{\odot n} \to W$ is the map of sets defined by the formula

$$\mathrm{MC}_f : V \to W, \qquad \mathrm{MC}_f(v) = \frac{1}{n!} f(v^{\odot n}).$$

The restitution operator is faithful, since the original map f can be recovered by MC_f by the well known **polarization formula**; namely, $f(v_1, v_2, \ldots, v_n) =$ *coefficient of* $t_1 \cdots t_n$ *in the polynomial* $\mathrm{MC}_f(t_1 v_1 + \cdots + t_n v_n)$.

If V, W are graded vector spaces, the restitution defined as above is no longer faithful since it vanishes, for every $n > 1$, on every homogeneous vector of odd degree and we need a more refined construction. Among the various possible solutions, we adopt here the one introduced in Sect. 10.2.

Recall that **NCGA** denotes the category of commutative graded algebras that are nilpotent and finite-dimensional as a vector space over $\mathbb{K}$, and the formal neighbourhood of a graded vector space V is defined as the functor

$$\widehat{V} : \mathbf{NCGA} \to \mathbf{Set}, \qquad \widehat{V}(A) = (V \otimes A)^0.$$

Now given any linear map $f \in \mathrm{Hom}^d_{\mathbb{K}}(V^{\odot n}, W) = \mathrm{Hom}^0_{\mathbb{K}}(V^{\odot n}, W[d])$ we define its restitution as the morphism of formal neighbourhoods $\mathrm{MC}_f : \widehat{V} \to \widehat{W[d]}$:

$$\mathrm{MC}_f(A) : (V \otimes A)^0 \to (W \otimes A)^d, \qquad A \in \mathbf{NCGA},$$

$$\mathrm{MC}_f(A)(x) = \frac{1}{n!} f_A(x^{\odot n}), \qquad x \in V \otimes A\,.$$

Finally, since we work with nilpotent algebras, it makes sense to extend the above definition to a map

$$\mathrm{MC} : \mathrm{Hom}^d_{\mathbb{K}}(\overline{S^c}(V), W) \to \mathrm{Mor}(\widehat{V}, \widehat{W[d]})$$

by setting

$$\mathrm{MC}_f(A)\colon (V\otimes A)^0 \to (W\otimes A)^d,\qquad \mathrm{MC}_f(A)(x)=\sum_{n=1}^{\infty}\frac{1}{n!}f_A(x^{\odot n}).$$

Notice that for $d=0$ we already defined the morphism MC_f (with the different symbol $\widehat{f}$) in Sect. 10.2. It is convenient to write

$$\mathrm{MC}_f(A)(x)=f\otimes \mathrm{Id}_A(e^{\odot x}-1),\qquad x\in V\otimes A,$$

where $e^{\odot x}-1=\sum_{n=1}^{\infty}\dfrac{x^{\odot n}}{n!}\in S(V)\otimes A$. For notational simplicity, when the algebra A is clear from the context, we simply write $\mathrm{MC}_f(x)=f(e^{\odot x}-1)$ instead of $\mathrm{MC}_f(A)(x)=f\otimes \mathrm{Id}_A(e^{\odot x}-1)$. More generally, whenever $\alpha\colon X\to Y$ is any linear map of graded vector spaces, we still denote by the same letter α the induced map $\alpha=$ "$\alpha\otimes \mathrm{Id}''_A\colon X\otimes A\to Y\otimes A$.

Lemma 11.6.2 *The restitution operator is faithful; given $f,g\in Hom^d_{\mathbb{K}}(\overline{S^c}(V),W)$ we have $f=g$ if and only if $MC_f=MC_g$.*

Proof. This is an immediate consequence of Theorem 10.2.2, in view of the canonical bijection $\mathrm{Hom}^d_{\mathbb{K}}(\overline{S^c}(V),W)=\mathrm{Hom}^0_{\mathbb{K}}(\overline{S^c}(V),W[d])$. □

Theorem 11.6.3 (Composition of restitutions) *Let U,V,W be graded vector spaces, $f\in Hom^0_{\mathbb{K}}(\overline{S^c}(U),V)$ and denote by $F\colon \overline{S^c}(U)\to\overline{S^c}(V)$ the multiplicative expansion of f. Then for every $g\in Hom^d_{\mathbb{K}}(\overline{S^c}(V),W)$ we have*

$$MC_g\circ MC_f=MC_{gF}\colon \widehat{U}\to\widehat{W[d]}.$$

Proof. By linearity it is not restrictive to assume that g has only one nonzero multilinear component $V^{\odot n}\to W$, i.e., that $g(V^{\odot m})=0$ for every $m\neq n$. Let $p=p_V\colon \overline{S^c}(V)\to V$ be the canonical projection and consider the map

$$\frac{\pi}{n!}\colon V^{\otimes n}\to V^{\odot n},\qquad v_1\otimes\cdots\otimes v_n\mapsto \frac{1}{n!}v_1\odot\cdots\odot v_n$$

as a left inverse of the symmetrization map $N\colon V^{\odot n}\to V^{\otimes n}$. Considering the composite map

$$g\frac{\pi}{n!}p^{\otimes n}\colon \overline{S^c}(V)^{\otimes n}\xrightarrow{p^{\otimes n}} V^{\otimes n}\xrightarrow{\pi/n!} V^{\odot n}\xrightarrow{g} W$$

our first step is to prove that:

$$g=g\frac{\pi}{n!}p^{\otimes n}\,\mathfrak{l}^{n-1}\colon \overline{S^c}(V)\to W\quad\text{whenever}\quad g(V^{\odot m})=0\ \text{for every}\ m\neq n. \tag{11.11}$$

If $x \in V^{\odot m}$ with $m \neq n$ it is immediate from the definition that $g(x) = g\frac{\pi}{n!}p^{\otimes n}\mathfrak{l}^{n-1}(x) = 0$; thus it is not restrictive to prove the equality (11.11) on the component $V^{\odot n}$.

Since $N\colon \overline{S^c}(V) \to \overline{T^c}(V)$ is a morphism of coalgebras, we have a commutative diagram

$$\begin{array}{ccc} \overline{S^c}(V) & \xrightarrow{\mathfrak{l}^{n-1}} & \overline{S^c}(V)^{\otimes n} \\ \downarrow{\scriptstyle N} & & \downarrow{\scriptstyle N^{\otimes n}} \\ \overline{T^c}(V) & \xrightarrow{\mathfrak{a}^{n-1}} & \overline{T^c}(V)^{\otimes n} \end{array}$$

that restricted to $V^{\odot n}$ gives

$$\begin{array}{ccc} V^{\odot n} & \xrightarrow{\mathfrak{l}^{n-1}} & V^{\otimes n} \\ \downarrow{\scriptstyle N} & & \downarrow{\scriptstyle \mathrm{Id}} \\ V^{\otimes n} & \xrightarrow{\mathrm{Id}} & V^{\otimes n} \end{array}$$

If $x \in V^{\odot n}$, then

$$g\frac{\pi}{n!}p^{\otimes n}\mathfrak{l}^{n-1}(x) = g\frac{\pi}{n!}N(x) = g(x).$$

Now let A be a fixed nilpotent graded commutative algebra and $x \in (U \otimes A)^0$. By the definition of restitution we have

$$\mathrm{MC}_f(x) = f(y), \qquad \mathrm{MC}_{gF}(x) = gF(y), \qquad y = e^{\odot x} - 1.$$

Since F is a morphism of coalgebras, by (11.11) and (11.10) we have

$$\begin{aligned} \mathrm{MC}_{gF}(x) &= gF(y) = \widetilde{g}\,\mathfrak{l}^{n-1}F(y) = g\frac{\pi}{n!}p^{\otimes n}F^{\otimes n}\mathfrak{l}^{n-1}(y) = g\frac{\pi}{n!}f^{\otimes n}\mu(y^n) \\ &= g\mu\left(\frac{\pi}{n!}f(y)^{\otimes n}\right) = \frac{1}{n!}g_A(f(y)^{\odot n}) = \mathrm{MC}_g(f(y)) \\ &= \mathrm{MC}_g(\mathrm{MC}_f(x)). \end{aligned}$$

□

As an immediate consequence of Theorems 10.2.2 and 11.6.3 we get that the category of reduced symmetric coalgebras is equivalent to the category of formal neighbourhoods of graded vector spaces.

Proposition 11.6.4 (Coderivation of exponential) *Let $q \in Hom^*_{\mathbb{K}}(\overline{S^c}(V), V)$ and denote by $Q \in Coder^*(\overline{S^c}(V), \overline{S^c}(V))$ its Leibniz expansion.*

*Then for every $A \in$ **NCGA** and every $x \in (V \otimes A)^0$ we have*

$$Q \otimes Id_A(e^{\odot x} - 1) = MC_q(A)(x) \odot e^{\odot x} \in \overline{S^c}(V) \otimes A.$$

Proof. Denoting by $p_n \colon \overline{S^c}(V) \to V^{\odot n}$ the natural projections, then $p_1 = p$ is the usual cogenerator and the proposition is equivalent to the sequence of equalities

$$p_n Q \otimes \mathrm{Id}_A(e^{\odot x} - 1) = \mathrm{MC}_q(A)(x) \odot \frac{x^{\odot n-1}}{(n-1)!}, \qquad n > 0.$$

Let $n > 0$ be a fixed integer. Equation (11.11) applied to $g = p_n$ implies that

$$p_n = \frac{\pi}{n!} p^{\otimes n} \mathfrak{l}^{n-1} \colon \overline{S^c}(V) \to V^{\odot n}$$

and then

$$p_n Q = \frac{\pi}{n!} p^{\otimes n} \mathfrak{l}^{n-1} Q = \frac{\pi}{n!} p^{\otimes n} \left(\sum_{i=0}^{n-1} \mathrm{Id}^{\otimes i} \otimes Q \otimes \mathrm{Id}^{\otimes n-i-1} \right) \mathfrak{l}^{n-1}.$$

Setting $y = e^{\odot x} - 1$ we have seen in (11.10) that $\mathfrak{l}^{n-1}(y) = \mu(y^{\otimes n})$. Therefore

$$\begin{aligned} p_n Q(e^{\odot x} - 1) &= \frac{\pi}{n!} p^{\otimes n} \mu \sum_{i=0}^{n-1} y^{\otimes i} \otimes Q(y) \otimes y^{\otimes n-i-1} \\ &= \frac{\pi}{n!} \mu \sum_{i=0}^{n-1} x^{\otimes i} \otimes q(y) \otimes x^{\otimes n-i-1} = \frac{1}{(n-1)!} q(y) \odot x^{\odot n-1} \\ &= \mathrm{MC}_q(A)(x) \odot \frac{x^{\odot n-1}}{(n-1)!}. \end{aligned}$$

□

11.7 Symmetric Coalgebras and Their Coderivations

Let $S(V)$ be the symmetric algebra generated by a graded vector space V and consider the (unique) morphism of graded algebras $\Delta \colon S(V) \to S(V) \otimes S(V)$ such that

$$\Delta(1) = 1 \otimes 1, \qquad \Delta(v) = v \otimes 1 + 1 \otimes v, \qquad v \in V.$$

We can immediately show that for every $v_1, \ldots, v_n \in V$ we have

$$\Delta(v_1 \odot \cdots \odot v_n) = \sum_{a=0}^{n} \sum_{\sigma \in S(a, n-a)} \epsilon(\sigma)(v_{\sigma(1)} \odot \cdots \odot v_{\sigma(a)}) \otimes (v_{\sigma(a+1)} \odot \cdots \odot v_{\sigma(n)}),$$

and therefore

$$\Delta(x) = 1 \otimes x + x \otimes 1 + \mathfrak{l}(x), \quad \text{for any } x \in V^{\odot n},\ n > 0.$$

By the universal properties of the tensor product, the two maps

$$\mathrm{Id} \otimes \Delta,\ \Delta \otimes \mathrm{Id}\colon S(V) \otimes S(V) \to S(V) \otimes S(V) \otimes S(V)$$

are morphisms of unitary graded commutative algebras. Since for every $v \in V$ we have

$$(\mathrm{Id} \otimes \Delta)\Delta(v) = (\Delta \otimes \mathrm{Id})\Delta(v) = v \otimes 1 \otimes 1 + 1 \otimes v \otimes 1 + 1 \otimes 1 \otimes v,$$

it follows that $(\mathrm{Id} \otimes \Delta)\Delta = (\Delta \otimes \mathrm{Id})\Delta$. In other words, we have proved that the pair $(S(V), \Delta)$ is a (coassociative) cocommmutative graded coalgebra, called a **symmetric coalgebra** generated by V.

Thus the same graded vector space $\oplus_{n\geq 0} V^{\odot n}$ carries both the structure of an algebra and a structure of a coalgebra; it is useful to denote the algebra structure by $S(V)$ and the coalgebra structure by $S^c(V)$. Notice that $S(\mathbb{K})$ is isomorphic to the polynomial ring $\mathbb{K}[t]$, while $S^c(\mathbb{K})$ is isomorphic to the coalgebra of Example 11.2.6.

It is important to point out that the coproduct $\Delta\colon S^c(V) \to S^c(V) \otimes S^c(V)$ is an injective map; in fact, denoting by $q\colon S^c(V) \to \mathbb{K}$ the projection we have $(\mathrm{Id} \otimes q)\Delta(v) = v \otimes 1$ for every $v \in S^c(V)$.

There exists a direct sum decomposition of graded vector spaces

$$S^c(V) = \bigoplus_{n\geq 0} V^{\odot n} = \mathbb{K} \oplus \overline{S^c}(V),$$

and it should be noted that $\overline{S^c}(V)$ is not a subcoalgebra of $S^c(V)$, while the natural projection $S^c(V) \to \overline{S^c}(V)$ is a morphism of graded coalgebras.

Lemma 11.7.1 *Let $F\colon S^c(V) \to S^c(W)$ be a nontrivial morphism of graded coalgebras. Then $F(1) = 1$ and $F\left(\overline{S^c}(V)\right) \subset \overline{S^c}(W)$.*

Proof. We first prove that $F(1) \neq 0$. Since F is nontrivial it makes sense to define the minimum integer n such that $F(v) \neq 0$ for some $v \in V^{\odot n}$; if $F(1) = 0$ then $F(V^{\odot 0}) = 0$ and therefore $n > 0$. If $v \in V^{\odot n}$ is a vector such that $F(v) \neq 0$, then the assumption $n > 0$ and $F(V^{\odot i}) = 0$ for every $i < n$ gives

$$\Delta(F(v)) = (F \otimes F)\Delta(v) = 0$$

in contradiction with the injectivity of Δ. Thus we have proved $F(1) \neq 0$.

Since $\Delta(F(1)) = (F \otimes F)\Delta(1) = F(1) \otimes F(1)$, denoting by $q\colon S^c(W) \to \mathbb{K}$ the projection, we have $(\mathrm{Id} \otimes q)\Delta(F(1)) = F(1) \otimes 1$,

$$F(1) \otimes q(F(1)) = (\mathrm{Id} \otimes q)\Delta(F(1)) = F(1) \otimes 1,$$

and therefore $F(1) = 1 + x$ for some $x \in \overline{S^c}(V)$. Finally, the relation

$$\Delta(1+x) = 1 \otimes 1 + x \otimes 1 + 1 \otimes x + \mathfrak{l}(x) = (1+x) \otimes (1+x)$$

implies $\mathfrak{l}(x) = x \otimes x$ and then $x = 0$, i.e., $F(1) = 1$.

Using the equality $F(1) = 1$ it is easy to prove that $F(V^{\odot n}) \subset \overline{S^c}(W)$ for every $n > 0$. Assume by induction that $F(V^{\odot i}) \subset \overline{S^c}(W)$ for every $i < 0 < n$ and let $v \in V^{\odot n}$. Then $(F \otimes F)\mathfrak{l}(v) \in \overline{S^c}(W) \otimes \overline{S^c}(W)$ and the equality

$$\begin{aligned} F(v) \otimes 1 &= (\mathrm{Id} \otimes q)\Delta(F(v)) = (\mathrm{Id} \otimes q)(F \otimes F)\Delta(v) \\ &= (F \otimes qF)(v \otimes 1 + 1 \otimes v + \mathfrak{l}(v)) = F(v) \otimes 1 + 1 \otimes qF(v) \end{aligned}$$

implies $qF(v) = 0$. □

Theorem 11.7.2 *Let V, W be graded vector spaces. Then the map*

$$Hom_{\mathbb{K}}(\overline{S^c}(V), \overline{S^c}(W)) \to Hom_{\mathbb{K}}(\mathbb{K} \oplus \overline{S^c}(V), \mathbb{K} \oplus \overline{S^c}(W)), \quad \overline{F} \mapsto F := Id_{\mathbb{K}} \oplus \overline{F},$$

induces a bijection between the set of morphisms of coalgebras $\overline{S^c}(V) \to \overline{S^c}(W)$ and the set of nontrivial morphisms of coalgebras $S^c(V) \to S^c(W)$.

Proof. Let $\overline{F}\colon \overline{S^c}(V) \to \overline{S^c}(W)$ be a morphism of graded vector spaces and write

$$F\colon S^c(V) \to S^c(W), \qquad F(1) = 1, \quad F(v) = \overline{F}(v), \quad v \in \overline{S^c}(V).$$

Then, for every $v \in \overline{S^c}(V)$ we have

$$(F \otimes F)\Delta(v) - \Delta(F(v)) = (\overline{F} \otimes \overline{F})\mathfrak{l}(v) - \mathfrak{l}(\overline{F}(v)),$$

and then F is a morphism of graded coalgebras if and only if $\overline{F}$ has the same property. It is now sufficient to apply Lemma 11.7.1. □

Lemma 11.7.3 *Given a graded vector space V and a coderivation of the symmetric coalgebra $h \in Coder^*(S^c(V), S^c(V))$ we have $h(S^c(V)) \subset \overline{S^c}(V)$ and $h(1) \in V$. Moreover, the restriction $h\colon \overline{S^c}(V) \to \overline{S^c}(V)$ is a coderivation of the reduced symmetric coalgebra if and only if $h(1) = 0$.*

Proof. Let h be a coderivation of $S^c(V)$ and write $h(1) = a + y$, with $a \in \mathbb{K}$ and $y \in \overline{S^c}(V)$. Then the equality $\Delta(h(1)) = (\mathrm{Id} \otimes h + h \otimes \mathrm{Id})\Delta(1)$ becomes

$$a \otimes 1 + 1 \otimes y + y \otimes 1 + \mathfrak{l}(y) = a \otimes 1 + 1 \otimes a + 1 \otimes y + y \otimes 1$$

and therefore $a = \mathfrak{l}(y) = 0$. Since $\ker \mathfrak{l} = V$ this implies $h(1) \in V$. Consider now any element $x \in \overline{S^c}(V)$ and write as above $h(x) = a + y$, with $a \in \mathbb{K}$ and $y \in \overline{S^c}(V)$. Then the equality $\Delta(h(x)) = (\mathrm{Id} \otimes h + h \otimes \mathrm{Id})\Delta(x)$ becomes

$$
\begin{aligned}
a\otimes 1+\Delta(y)=\Delta(h(x)) &= (h\otimes \mathrm{Id}+\mathrm{Id}\otimes h)\Delta(x)\\
&= (h\otimes \mathrm{Id}+\mathrm{Id}\otimes h)(x\otimes 1+1\otimes x+\mathfrak{l}(x))\\
&= h(x)\otimes 1+h(1)\otimes x+1\otimes h(x)+(-1)^{\overline{h}\,\overline{x}}x\otimes h(1)\\
&\quad+(h\otimes \mathrm{Id}+\mathrm{Id}\otimes h)\mathfrak{l}(x).
\end{aligned}
\tag{11.12}
$$

Applying the natural projection $S^c(V)\otimes S^c(V)\to \mathbb{K}\otimes\mathbb{K}$ to the above equality we obtain

$$a\otimes 1=1\otimes a+a\otimes 1$$

and therefore $a=0$, $h(\overline{S^c}(V))\subset \overline{S^c}(V)$. The equation (11.12) above becomes

$$
\begin{aligned}
h(x)\otimes 1+1\otimes h(x)+\mathfrak{l}(h(x))=\Delta(h(x)) &= (h\otimes \mathrm{Id}+\mathrm{Id}\otimes h)\Delta(x)\\
&= (h\otimes \mathrm{Id}+\mathrm{Id}\otimes h)(x\otimes 1+1\otimes x+\mathfrak{l}(x))\\
&= h(x)\otimes 1+h(1)\otimes x+1\otimes h(x)+(-1)^{\overline{h}\,\overline{x}}x\otimes h(1)\\
&\quad+(h\otimes \mathrm{Id}+\mathrm{Id}\otimes h)\mathfrak{l}(x),
\end{aligned}
\tag{11.13}
$$

which implies

$$\mathfrak{l}(h(x))=(h\otimes \mathrm{Id}+\mathrm{Id}\otimes h)\mathfrak{l}(x)\iff h(1)\otimes x+(-1)^{\overline{h}\,\overline{x}}x\otimes h(1)=0,$$

and this proves that h restricts to a coderivation of $\overline{S^c}(V)$ if and only if $h(1)=0$. □

Theorem 11.7.4 *Let V be a graded vector space. For every $x\in V$ the linear map*

$$\sigma_x\colon S^c(V)\to S^c(V),\qquad \sigma_x(v)=x\odot v,$$

is a coderivation. Every coderivation of $S^c(V)$ is written uniquely as the sum of a coderivation of $\overline{S^c}(V)$ and a coderivation of type σ_x.

Proof. For every $x\in V$ and $v\in S^c(V)$ we have

$$\Delta\sigma_x(v)=\Delta(x\odot v)=(1\otimes x+x\otimes 1)\odot\Delta(v)=(\mathrm{Id}\otimes\sigma_x+\sigma_x\otimes \mathrm{Id})\Delta(v)$$

and this proves that σ_x is a coderivation. Given $h\in \mathrm{Coder}^*(S^c(V),S^c(V))$, by Lemma 11.7.3 we have $h(1)\in V$ and

$$h-\sigma_{h(1)}\in\{\eta\in \mathrm{Coder}^*(S^c(V),S^c(V))\mid \eta(1)=0\}=\mathrm{Coder}^*(\overline{S^c}(V),\overline{S^c}(V)).$$

Conversely, if $\eta=h-\sigma_x\in \mathrm{Coder}^*(\overline{S^c}(V),\overline{S^c}(V))$ for some $x\in V$, again by Lemma 11.7.3 we have $\eta(1)=0$ and then $h(1)=\sigma_x(1)=x$. □

Corollary 11.7.5 *Given a graded vector space V and a coderivation of the symmetric coalgebra $h\in \mathrm{Coder}^*(S^c(V),S^c(V))$ we have*

$$h(V^{\odot n}) \subseteq \bigoplus_{i=1}^{n+1} V^{\odot i}, \quad \textit{for every } n \geq 0.$$

Moreover, the restriction $h\colon \overline{S^c}(V) \to \overline{S^c}(V)$ *is a coderivation of the reduced symmetric coalgebra if and only if* $h(1) = 0$.

Proof. Immediate consequence of Theorem 11.7.4. □

The Theorem 11.7.4 and Corollary 11.5.9 give an isomorphism of graded vector spaces $\mathrm{Coder}^*(S^c(V), S^c(V)) \simeq \mathrm{Hom}^*_{\mathbb{K}}(S^c(V), V)$ induced by corestriction. It is easy to check that, via this isomorphism, the Nijenhuis–Richardson bracket on $\mathrm{Hom}^*_{\mathbb{K}}(S^c(V), V)$ (Definition 10.6.5) corresponds to the graded commutator in $\mathrm{Coder}^*(S^c(V), S^c(V))$.

For instance, if $x \in V$ and $Q \in \mathrm{Coder}^*(\overline{S^c}(V), \overline{S^c}(V))$ is the Leibniz extension of a map $q \in \mathrm{Hom}^*_{\mathbb{K}}(V^{\odot n+1}, V)$, then since $p_V \sigma_x Q = 0$ we have

$$\begin{aligned} p_V[Q, \sigma_x](y_1 \odot \cdots \odot y_n) &= p_V Q(x \odot y_1 \odot \cdots \odot y_n) = q(x, y_1, \ldots, y_n) \\ &= [q, x]_{NR}(y_1, \ldots, y_n). \end{aligned} \tag{11.14}$$

We conclude this chapter by proving two technical results that we shall use in the homotopy classification of DG-Lie and L_∞-algebras.

Lemma 11.7.6 (Rectification of idempotents) *Let* $E\colon S^c(V) \to S^c(V)$ *be a non-trivial morphism of graded symmetric coalgebras such that* $EE = E$. *Then* $E = F^{-1}GF$, *where* F *is an isomorphism of coalgebras with linear part equal to the identity and* G *is induced by a linear projection on* V. *In other words, the homogeneous components of* F *and* G *satisfy the conditions* $F^1_1 = Id_V$, $G^1_1 G^1_1 = G^1_1$ *and* $G^1_n = 0$ *for every* $n > 1$.

Proof. Define $F, G\colon S^c(V) \to S^c(V)$ as the morphisms of graded coalgebras such that

$$F^1_1 = \mathrm{Id}, \quad G^1_1 = E^1_1, \qquad F^1_n = (2E^1_1 - \mathrm{Id})E^1_n, \quad G^1_n = 0 \quad n > 1\,.$$

The condition $EE = E$ is equivalent to $E^1_n = \sum_{i=1}^n E^1_i E^i_n$ and therefore for every $n \geq 2$ we have

$$\sum_{i=2}^n E^1_i E^i_n = E^1_n - E^1_1 E^1_n = (\mathrm{Id} - E^1_1)E^1_n\,.$$

Thus $(FE)^1_1 = E^1_1 = (GF)^1_1$, while for every $n > 1$ we have

$$(FE)^1_n = E^1_n + \sum_{i=2}^n (2E^1_1 - \mathrm{Id})E^1_i E^i_n = E^1_n + (2E^1_1 - \mathrm{Id})(\mathrm{Id} - E^1_1)E^1_n = E^1_1 E^1_n,$$

$$(GF)^1_n = E^1_1 F^1_n = \sum_{i=1}^n E^1_1(2E^1_1 - \mathrm{Id})E^1_n = E^1_1 E^1_n,$$

and therefore $GF = FE$. □

Proposition 11.7.7 *Let $E\colon S^c(V) \to S^c(V)$ be a nontrivial morphism of graded coalgebras such that $E = EE$. For a vector $v \in V$ the following conditions are equivalent:*

1. *there exists a coderivation $\alpha \in Coder^*(S^c(V), S^c(V))$ such that $\alpha E = E\alpha$ and $\alpha(1) = v$;*
2. *$E(v) = v$.*

Proof. If there exists α as above then $E(v) = E\alpha(1) = \alpha E(1) = \alpha(1) = v$. Conversely, if $E(v) = v$, by Lemma 11.7.6, it is not restrictive to assume the morphism E induced by a projection $e\colon V \to V$, i.e., $E(v_1 \odot \cdots \odot v_n) = e(v_1) \odot \cdots \odot e(v_n)$; in this case the usual formula

$$\alpha(x) = v \odot x, \qquad x \in S^c(V), \quad e(v) = v,$$

gives a coderivation with required properties. □

11.8 Exercises

11.8.1 A **counity** of a graded coalgebra (C, Δ) is a morphism of graded vector spaces $\epsilon\colon C \to \mathbb{K}$ such that $(\epsilon \otimes \mathrm{Id}_C)\Delta = (\mathrm{Id}_C \otimes \epsilon)\Delta = \mathrm{Id}_C$. Prove that if a counity exists, then it is unique.

11.8.2 Let (C, Δ) be a graded coalgebra. A graded subspace $I \subset C$ is called a **coideal** if $\Delta(I) \subset C \otimes I + I \otimes C$. Prove that a subspace is a coideal if and only if it is the kernel of a morphism of coalgebras.

11.8.3 A graded coalgebra (C, Δ) is said to be **connected** if there is an element $e \in C$ such that $\Delta(e) = e \otimes e$ (in particular $\deg(e) = 0$) and $C = \cup_{r=0}^{+\infty} F_r C$, where $F_r C$ is defined recursively in the following way:

$$F_0 C = \mathbb{K}e, \qquad F_{r+1}C = \{x \in C \mid \Delta(x) - e \otimes x - x \otimes e \in F_r C \otimes F_r C\}.$$

Prove that every locally conilpotent coalgebra is connected.

11.8.4 In the situation of Example 11.2.3, let $F\colon V[t] \to W[t]$ be a morphism of $\mathbb{K}[t]$-comodules and denote by $f\colon V[t] \to W$ the composition of F with the evaluation map $W[t] \to W$, $\sum_i w_i t^i \mapsto w_0$. Prove that

$$F(vt^n) = \sum_{i=0}^{n} f(vt^i)t^{n-i}$$

for every $v \in V, n \geq 0$. (Hint: writing $G = F - \tilde{F}$, where $\tilde{F}(vt^n) = \sum_{i=0}^{n} f(vt^i)t^{n-i}$, prove by induction on n that $G(vt^n) = 0$.)

11.8.5 Let (C, Δ) be a graded coalgebra, V a graded vector space and denote by $\mathtt{tw}\colon V \otimes C \to C \otimes V$ the usual twisting map. Prove that the graded vector space $M = V \otimes C$ has a natural structure of a C-comodule, where the right and left coactions are defined by:

$$r = \mathrm{Id}_V \otimes \Delta\colon V \otimes C \to V \otimes C \otimes C\,,$$

$$l = (\mathtt{tw} \otimes \mathrm{Id}_C)r\colon V \otimes C \to C \otimes V \otimes C\,.$$

11.8.6 Let A be a graded associative algebra over the field $\mathbb{K}$. For every local homomorphism of $\mathbb{K}$-algebras

$$\gamma\colon \mathbb{K}[[x]] \to \mathbb{K}[[x]], \qquad \gamma(x) = \sum_{n=1}^{\infty} \gamma_n x^n, \quad \gamma_n \in \mathbb{K},$$

let $F_\gamma\colon \overline{T^c}(A) \to \overline{T^c}(A)$ be the (unique) morphism of graded coalgebras with corestriction

$$f_\gamma = pF_\gamma\colon \overline{T^c}(A) \to A, \qquad f_\gamma(a_1 \otimes \cdots \otimes a_n) = \gamma_n a_1 \cdots a_n.$$

Prove the validity of the composition formula $F_{\gamma\delta} = F_\delta F_\gamma$. (Hint: Example 11.1.8.)

11.8.7 Let $F\colon \overline{S^c}(U) \to \overline{S^c}(V)$ be a surjective morphism of graded coalgebras. We have seen in Example 11.5.6 that its linear component $F_1^1\colon U \to V$ may not be surjective. Prove that if $U = \oplus_{i>0} U^i$ is concentrated in strictly positive degrees and F is surjective, then also F_1^1 is surjective.

11.8.8 Let V be a finite-dimensional vector space with basis $\partial_1, \ldots, \partial_m$. Prove that

$$\mathfrak{l}(\partial_1^{n_1} \cdots \partial_m^{n_m}) = \sum_{a_1,\ldots,a_m} \binom{n_1}{a_1} \cdots \binom{n_m}{a_m} \partial_1^{a_1} \cdots \partial_m^{a_m} \otimes \partial_1^{n_1-a_1} \cdots \partial_m^{n_m-a_m}$$

and deduce that the algebra $A = \mathrm{Hom}_{\mathbb{K}}(\overline{S^c}(V), \mathbb{K})$ is isomorphic to the maximal ideal of the power series ring $\mathbb{K}[[x_1, \ldots, x_m]]$, with pairing

$$\langle \partial_1^{n_1} \cdots \partial_m^{n_m}, f(x) \rangle = \frac{\partial^{n_1+\cdots+n_m} f}{\partial x_1^{n_1} \cdots \partial x_m^{n_m}}(0).$$

11.8.9 Let $F\colon \overline{S^c}(U) \to \overline{S^c}(V)$ be a morphism of reduced symmetric coalgebras and assume that there exists a morphism $p\colon V \to U$ of graded vector spaces such that $pF_1^1 = \mathrm{Id}_U$. Prove that there exists an isomorphism $G\colon \overline{S^c}(V) \to \overline{S^c}(V)$ of graded coalgebras such that $G_1^1 = \mathrm{Id}_V$ and $(GF)_n^1 = 0$ for every $n > 0$ (cf. Lemma 12.1.4).

Hint: use the relation $F^i_n\, p^{\odot n}\, F^n_n = F^i_n$ to define recursively a sequence of maps $G^1_n \colon V^{\odot n} \to V$, such that $G^1_1 = \mathrm{Id}_V$ and

$$\sum_{i=1}^{n} G^1_i F^i_n = 0 \qquad n \geq 2.$$

11.8.10 Let $F \colon \overline{S^c}(U) \to \overline{S^c}(V)$ be a morphism of reduced symmetric coalgebras and assume that there exists a morphism $p \colon V \to U$ of graded vector spaces such that $F^1_1 p = \mathrm{Id}_V$. Prove that there exists an isomorphism $G \colon \overline{S^c}(U) \to \overline{S^c}(U)$ of graded coalgebras such that $G^1_1 = \mathrm{Id}_U$ and $(FG)^1_n = 0$ for every $n > 0$ (cf. Lemma 12.1.4).

11.8.11 Prove that two graded vector spaces are isomorphic if and only if they have isomorphic formal neighbourhoods.

11.8.12 [The alternating coalgebra] Let W be a graded vector space and denote by s the suspension symbol of degree $+1$. For every $n \geq 0$ there exists a linear isomorphism of degree n:

$$s^{\otimes n} \colon W^{\odot n} \to (sW)^{\wedge n}, \qquad v_1 \odot \cdots \odot v_n \mapsto (-1)^{\sum (n-i)\overline{v_i}} s v_1 \wedge \cdots \wedge s v_n .$$

Define the map

$$\Gamma \colon (sW)^{\wedge n} \to \bigoplus_{i=1}^{n-1} (sW)^{\wedge i} \otimes (sW)^{\wedge n-i}, \qquad \Delta = \sum_{i=1}^{n-1} (s^{\otimes i} \otimes s^{\otimes n-i}) \lfloor (s^{\otimes n})^{-1},$$

and use Lemma 10.1.2 for proving the formula

$$\Gamma(sv_1 \wedge \cdots \wedge sv_n) = \sum_{i=1}^{n-1} \sum_{\sigma \in S(i,n-i)} \chi(\sigma)(sv_{\sigma(1)} \wedge \cdots \wedge sv_{\sigma(i)}) \otimes (sv_{\sigma(i+1)} \wedge \cdots \wedge sv_{\sigma(n)}).$$

References

We refer to [30, 72, 170, 252] for more results regardings coalgebras. The terminology related to locally conilpotent coalgebras is not well settled in the literature and here we follow [156]; locally conilpotent coalgebras are referred to in the literature also as conilpotent or connected coalgebras.

The equivalence between the Gerstenhaber bracket (Example 5.8.2) with the construction at the end of Sect. 11.3 is deeply discussed in [249].

For a detailed treatment of restitution and polarization in classical invariant theory we refer to [57, 216]. Notice that our definition of restitution is slightly different from that most used in the literature, namely, our definition of restitution of a symmetric n-linear map differs by a factor $1/n!$.

Most of the material of Sect. 11.7 is taken from [83].

Chapter 12
L_∞-Morphisms

In Chap. 10 we introduced the décalage isomorphisms, which can be interpreted as a canonical equivalence between the categories of L_∞ and $L_\infty[1]$-algebras, both endowed with linear morphisms. Here we want to enrich these categories by enhancing their sets of morphisms; more precisely, for every pair (V, q), (W, r) of $L_\infty[1]$-algebras we introduce the notion of L_∞-morphism $f\colon (V, q) \rightsquigarrow (W, r)$ together with the composition rule of two of them. The L_∞-morphisms between L_∞-algebras are then defined by imposing that the décalage isomorphisms give again an equivalence of categories. The key point in the above construction is the interpretation of every $L_\infty[1]$-algebra as formal pointed DG-manifolds equipped with a fixed cogenerator onto its tangential complex.

12.1 Formal Pointed DG-Manifolds

The notion of differential graded algebras has a natural analogue in the framework of coalgebras.

Definition 12.1.1 A **differential graded coalgebra** is the data of a graded coalgebra C together with a coderivation $d_C \in \operatorname{Coder}^1(C, C)$, called a differential, such that $d_C^2 = 0$. A morphism of differential graded coalgebras is a morphism of graded coalgebras commuting with differentials.

We shall denote by **DGC** the category of differential graded coalgebras. Together with the usual forgetful functor **DGC** $\to$ **DG**, $(C, \Delta_C, d_C) \mapsto (C, d_C)$, there is also properly defined the functor:

$$\mathbf{DGC} \to \mathbf{DG}, \qquad (C, \Delta_C, d_C) \mapsto (\ker \Delta_C, d_C).$$

M. Manetti, *Lie Methods in Deformation Theory*, Springer Monographs in Mathematics,
https://doi.org/10.1007/978-981-19-1185-9_12

In fact, if $x \in \ker \Delta_C$, then $\Delta_C d_C(x) = (\mathrm{Id} \otimes d_C + d_C \otimes \mathrm{Id})\Delta_C(x) = 0$ and therefore $d_C(x) \in \ker \Delta_C$, i.e., $(\ker \Delta_C, d_C) \in \mathbf{DG}$.

If $F\colon (C, \Delta_C, d_C) \to (D, \Delta_D, d_D)$ is a morphism of differential graded coalgebras and $x \in \ker \Delta_C$, then $\Delta_D F(x) = (F \otimes F)\Delta_C(x) = 0$ and therefore $F(x) \in \ker \Delta_D$. We shall refer to the DG-vector space $(\ker \Delta_C, d_C)$ as the **tangential complex** of the differential graded coalgebra (C, Δ_C, d_C).

Definition 12.1.2 A **formal pointed DG-manifold** is a differential graded coalgebra $\mathfrak{M}$ isomorphic to $(\overline{S^c}(V), Q)$ for some graded vector space V and some differential Q. A morphism of formal pointed DG-manifolds is a morphism of differential graded coalgebras.

Thus, by definition every formal pointed DG-manifold is a cocommutative locally conilpotent coalgebra. In the above definition a specific isomorphism of differential graded coalgebras $F\colon \mathfrak{M} \to (\overline{S^c}(V), Q)$ is not considered as a part of the data. Since V is precisely the kernel of the coproduct in $\overline{S^c}(V)$, the isomorphism F induces an isomorphism of DG-vector spaces $f\colon (\ker \Delta_{\mathfrak{M}}, d_{\mathfrak{M}}) \to (V, Q_{|V})$, and therefore F gives a projection map

$$q\colon \mathfrak{M} \xrightarrow{F} \overline{S^c}(V) \xrightarrow{p} V \xrightarrow{f^{-1}} \ker \Delta_{\mathfrak{M}}.$$

Since $\mathfrak{M}$ is locally conilpotent, the projection q lifts to the isomorphism of graded coalgebras $\overline{S}(f^{-1})F\colon \mathfrak{M} \to \overline{S^c}(\ker \Delta_{\mathfrak{M}})$.

Conversely, every projection $q\colon \mathfrak{M} \to \ker \Delta_{\mathfrak{M}}$ induces a morphism of graded coalgebras $G\colon \mathfrak{M} \to \overline{S^c}(\ker \Delta_{\mathfrak{M}})$; since the linear part of GF^{-1} is an isomorphism of graded vector spaces, also GF^{-1} and G are isomorphisms of graded coalgebras.

In conclusion, a formal pointed DG-manifold is a cocommutative locally conilpotent differential graded coalgebra $\mathfrak{M}$ such that every projection $q\colon \mathfrak{M} \to \ker \Delta_{\mathfrak{M}}$ is the corestriction of an isomorphism of graded coalgebras $G\colon \mathfrak{M} \to \overline{S^c}(\ker \Delta_{\mathfrak{M}})$, and therefore also of an isomorphism of differential graded coalgebras

$$G\colon \mathfrak{M} \to (\overline{S^c}(\ker \Delta_{\mathfrak{M}}), G\, d_{\mathfrak{M}}\, G^{-1}).$$

Definition 12.1.3 A morphism of formal pointed DG-manifolds $F\colon \mathfrak{M} \to \mathfrak{N}$ is called a **strict morphism** if $F(\ker \Delta_{\mathfrak{M}}) = F(\mathfrak{M}) \cap \ker \Delta_{\mathfrak{N}}$.

In other words, a morphism of formal pointed DG-manifolds $F\colon \mathfrak{M} \to \mathfrak{N}$ is strict if the morphism of pairs $(\mathfrak{M}, \ker \Delta_{\mathfrak{M}}) \to (\mathfrak{N}, \ker \Delta_{\mathfrak{N}})$ is strict as a morphism of filtered modules, according to the standard terminology.

Lemma 12.1.4 *For a morphism of formal pointed DG-manifolds $F\colon \mathfrak{M} \to \mathfrak{N}$ the following conditions are equivalent:*

1. *F is strict;*
2. *there exist two projections $p\colon \mathfrak{N} \to \ker \Delta_{\mathfrak{N}}$, $q\colon \mathfrak{M} \to \ker \Delta_{\mathfrak{M}}$ such that $pF = Fq$;*

3. *there exist two isomorphisms* $Q\colon \mathfrak{M} \to \overline{S^c}(U)$, $P\colon \mathfrak{N} \to \overline{S^c}(V)$ *of graded coalgebras such that* $PFQ^{-1}\colon \overline{S^c}(U) \to \overline{S^c}(V)$ *is the multiplicative expansion of a linear map* $U \to V$.

Proof The only nontrivial implications are (1) $\Rightarrow$ (2) $\Rightarrow$ (3). If F is strict we have an exact sequence

$$0 \to \ker F \cap \ker \Delta_{\mathfrak{M}} \to \ker \Delta_{\mathfrak{M}} \xrightarrow{F} F(\mathfrak{M}) \cap \ker \Delta_{\mathfrak{N}} \to 0.$$

Let $p\colon \mathfrak{N} \to \ker \Delta_{\mathfrak{N}}$ be any projection such that $p(F(\mathfrak{M})) = F(\mathfrak{M}) \cap \ker \Delta_{\mathfrak{N}}$. By basic linear algebra, every projection $q'\colon \ker F \to \ker \Delta_{\mathfrak{M}} \cap \ker F$ can be extended to a commutative diagram

$$\begin{array}{ccccccccc}
0 & \longrightarrow & \ker F & \longrightarrow & \mathfrak{M} & \xrightarrow{F} & F(\mathfrak{M}) & \longrightarrow & 0\\
 & & \downarrow{\scriptstyle q'} & & \downarrow{\scriptstyle q} & & \downarrow{\scriptstyle p} & & \\
0 & \longrightarrow & \ker F \cap \ker \Delta_{\mathfrak{M}} & \longrightarrow & \ker \Delta_{\mathfrak{M}} & \xrightarrow{F} & F(\mathfrak{M}) \cap \ker \Delta_{\mathfrak{N}} & \longrightarrow & 0
\end{array}$$

where q is a projection. Then $pF = Fq$ as required.

Given p, q as in the second item, we have already seen that p, q are the corestrictions of two isomorphisms of graded coalgebras

$$Q\colon \mathfrak{M} \to \overline{S^c}(\ker \Delta_{\mathfrak{M}}), \qquad P\colon \mathfrak{N} \to \overline{S^c}(\ker \Delta_{\mathfrak{N}}).$$

The corestriction of PF is equal to $pF = Fq$ and then it is equal to the corestriction of $\overline{S^c}(f)Q$, where $f\colon \ker \Delta_{\mathfrak{M}} \to \ker \Delta_{\mathfrak{N}}$ is the restriction of F. Thus $PF = \overline{S^c}(f)Q$ by Proposition 11.1.13. □

Given a morphism of formal pointed DG-manifolds $F\colon \mathfrak{M} \to \mathfrak{N}$ it is immediate from the definition that if the restriction $F\colon \ker \Delta_{\mathfrak{M}} \to \ker \Delta_{\mathfrak{N}}$ is surjective, then F is strict. The same conclusion holds if the restriction $F\colon \ker \Delta_{\mathfrak{M}} \to \ker \Delta_{\mathfrak{N}}$ is injective; in fact it is not restrictive to assume $\mathfrak{M} = \overline{S^c}(U)$, $\mathfrak{N} = \overline{S^c}(V)$ and by Corollary 11.5.4 the injectivity of the linear part $F_1^1\colon U \to V$ implies the injectivity of the components $F_n^n\colon U^{\odot n} \to V^{\odot n}$ for every n; therefore $u \in \overline{S^c}(U)$ and $F(u) \in V$ imply $u \in U$.

Definition 12.1.5 A morphism of formal pointed DG-manifolds $F\colon \mathfrak{M} \to \mathfrak{N}$ is called:

1. a **fibration** if $F\colon \ker \Delta_{\mathfrak{M}} \to \ker \Delta_{\mathfrak{N}}$ is surjective;
2. a **weak equivalence** if $F\colon \ker \Delta_{\mathfrak{M}} \to \ker \Delta_{\mathfrak{N}}$ is a quasi-isomorphism.

It follows immediately from the definitions and by Corollary 11.5.5 that a morphism of formal pointed DG-manifolds is a fibration if and only if is strict and surjective.

Lemma 12.1.6 *Consider a commutative diagram of locally conilpotent differential graded cocommutative coalgebra*

$$\begin{array}{ccc} B & \xrightarrow{\alpha} & \mathfrak{M} \\ \Big\downarrow{\scriptstyle i} & & \Big\downarrow{\scriptstyle F} \\ C & \xrightarrow{\beta} & \mathfrak{N}, \end{array} \tag{12.1}$$

where F is a fibration of formal pointed DG-manifolds, i is an injective morphism and $\Delta_C(C) \subset i(B) \otimes i(B)$. Then the obstruction to the lifting of (12.1) *belongs to the vector space $H^1(Hom^*_{\mathbb{K}}(coker(i), \ker \Delta_{\mathfrak{M}} \cap \ker F))$.*

More precisely, Lemma 12.1.6 means that there is naturally defined an "obstruction class"

$$e \in H^1\left(\mathrm{Hom}^*_{\mathbb{K}}(\mathrm{coker}(i), \ker \Delta_{\mathfrak{M}} \cap \ker F)\right)$$

that vanishes if and only if the diagram (12.1) has a lifting, i.e., if there exists a morphism of differential graded coalgebras $C \to \mathfrak{M}$ making the diagram

$$\begin{array}{ccc} B & \xrightarrow{\alpha} & \mathfrak{M} \\ \Big\downarrow{\scriptstyle i} & \nearrow & \Big\downarrow{\scriptstyle F} \\ C & \xrightarrow{\beta} & \mathfrak{N} \end{array}$$

commutative.

Proof It is not restrictive to assume $B \subset C$ and i the inclusion morphism, so that $\Delta_C(C) \subset B \otimes B$. Since every fibration is a strict morphism, by Lemma 12.1.4 there exist two projections $p\colon \mathfrak{N} \to \ker \Delta_{\mathfrak{N}}$, $q\colon \mathfrak{M} \to \ker \Delta_{\mathfrak{M}}$, such that $pF = Fq$.

Since F is a fibration we can lift the morphism $p\beta\colon C \to \ker \Delta_{\mathfrak{M}}$ to a commutative diagram of graded vector spaces

$$\begin{array}{ccc} B & \xrightarrow{q\alpha} & \ker \Delta_{\mathfrak{M}} \\ \Big\downarrow{\scriptstyle i} & \overset{\gamma_1}{\nearrow} & \Big\downarrow{\scriptstyle F} \\ C & \xrightarrow{p\beta} & \ker \Delta_{\mathfrak{N}}. \end{array}$$

According to Proposition 11.5.3 there exists a unique morphism of graded coalgebras $\gamma\colon C \to \mathfrak{M}$ such that $q\gamma = \gamma_1$ and $\alpha = \gamma i$. Since $Fq = pF$ the above diagram implies that

$$\begin{array}{ccc} B & \xrightarrow{\alpha} & \mathfrak{M} \\ \Big\downarrow{\scriptstyle i} & \overset{\gamma}{\nearrow} & \Big\downarrow{\scriptstyle F} \\ C & \xrightarrow{\beta} & \mathfrak{N} \end{array}$$

is a commutative diagram of graded coalgebras. Defining

$$\psi = d_{\mathfrak{M}}\gamma - \gamma d_C : C \to \mathfrak{M},$$

the equality $\psi = 0$ holds if and only if γ is a morphism of differential graded coalgebras. We have $\psi i = F\psi = \Delta_{\mathfrak{M}}\psi = 0$. In fact

$$\psi i = d_{\mathfrak{M}}\gamma i - \gamma d_C i = d_{\mathfrak{M}}\alpha - \alpha d_B = 0, \quad F\psi = F d_{\mathfrak{M}}\gamma - F\gamma d_C = d_{\mathfrak{N}}\beta - \beta d_C = 0.$$

$$\begin{aligned}
\Delta_{\mathfrak{M}} d_{\mathfrak{M}}\gamma &= (d_{\mathfrak{M}} \otimes \mathrm{Id} + \mathrm{Id} \otimes d_{\mathfrak{M}})\Delta_{\mathfrak{M}}\gamma = (d_{\mathfrak{M}} \otimes \mathrm{Id} + \mathrm{Id} \otimes d_{\mathfrak{M}})(\gamma \otimes \gamma)\Delta_C \\
&= (d_{\mathfrak{M}}\gamma \otimes \gamma + \gamma \otimes d_{\mathfrak{M}}\gamma)\Delta_C, \\
\Delta_{\mathfrak{M}}\gamma d_C &= (\gamma \otimes \gamma)\Delta_C d_C = (\gamma \otimes \gamma)(d_C \otimes \mathrm{Id} + \mathrm{Id} \otimes d_C)\Delta_C \\
&= (\gamma d_C \otimes \gamma + \gamma \otimes \gamma d_C)\Delta_C, \\
\Delta_{\mathfrak{M}}\psi &= ((d_{\mathfrak{M}}\gamma - \gamma d_C) \otimes \gamma + \gamma \otimes (d_{\mathfrak{M}}\gamma - \gamma d_C))\Delta_C \\
&= (\psi \otimes \gamma + \gamma \otimes \psi)\Delta_C = 0,
\end{aligned}$$

where the last equality follows from $\psi i = 0$ and the assumption $\Delta_C(C) \subset i(B) \otimes i(B)$. Thus ψ factors to a morphism $\psi \in \mathrm{Hom}^1_{\mathbb{K}}(C/B, \ker \Delta_{\mathfrak{M}} \cap \ker F)$ whose differential is

$$\delta\psi = d_{\mathfrak{M}}\psi + \psi d_C = 0\,.$$

In order to define the obstruction class e as the cohomology class

$$[\psi] \in H^1(\mathrm{Hom}^*_{\mathbb{K}}(C/B, \ker \Delta_{\mathfrak{M}} \cap \ker F))$$

and conclude the proof we need to show that $[\psi]$ does not depend on the choice of γ and that $[\psi] = 0$ if and only if γ can be chosen as a morphism of DG-coalgebras. Assume we have another commutative diagram of graded coalgebras

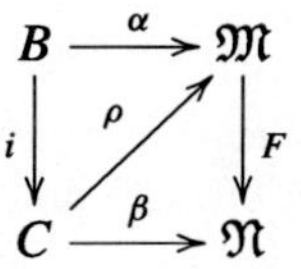

and consider the difference $h = \rho - \gamma : C \to \mathfrak{M}$. Then $h(B) = 0$, $Fh = 0$ and

$$\Delta_{\mathfrak{M}} h = \Delta_{\mathfrak{M}}\rho - \Delta_{\mathfrak{M}}\gamma = (\rho \otimes \rho - \gamma \otimes \gamma)\Delta_C = (h \otimes h + h \otimes \gamma + \gamma \otimes h)\Delta_C = 0.$$

This implies that $h \in \mathrm{Hom}^0_{\mathbb{K}}(C/B, \ker \Delta_{\mathfrak{M}} \cap \ker F)$,

$$d_{\mathfrak{M}}\rho - \rho d_C = \psi + d_{\mathfrak{M}} h - h d_C = \psi + \delta h$$

and therefore the cohomology class $[\psi]$ is well defined. Finally, if $\psi = \delta h$ for some $h \in \mathrm{Hom}^0_{\mathbb{K}}(C, \ker \Delta_{\mathfrak{M}} \cap \ker F)$ such that $h(B) = 0$, then it is straightforward to check that $\gamma - h$ is a morphism of differential graded coalgebras commuting with i, α, β and F. □

Theorem 12.1.7 (Lifting property) *Consider a commutative square of formal pointed DG-manifolds*

$$\begin{array}{ccc} \mathfrak{A} & \xrightarrow{\alpha} & \mathfrak{M} \\ {\scriptstyle f}\downarrow & & \downarrow{\scriptstyle g} \\ \mathfrak{B} & \xrightarrow{\beta} & \mathfrak{N} \end{array} \tag{12.2}$$

such that:

1. *the restriction* $f_1 \colon \ker \Delta_{\mathfrak{A}} \to \ker \Delta_{\mathfrak{B}}$ *is injective;*
2. *the restriction* $g_1 \colon \ker \Delta_{\mathfrak{M}} \to \ker \Delta_{\mathfrak{N}}$ *is surjective;*
3. *either* f_1 *or* g_1 *is a quasi-isomorphism.*

Then every commutative diagram

$$\begin{array}{ccc} \ker \Delta_{\mathfrak{A}} & \xrightarrow{\alpha_1} & \ker \Delta_{\mathfrak{M}} \\ {\scriptstyle f_1}\downarrow & \nearrow & \downarrow{\scriptstyle g_1} \\ \ker \Delta_{\mathfrak{B}} & \xrightarrow{\beta_1} & \ker \Delta_{\mathfrak{N}} \end{array} \tag{12.3}$$

of DG-vector spaces lifts to a commutative diagram

$$\begin{array}{ccc} \mathfrak{A} & \xrightarrow{\alpha} & \mathfrak{M} \\ {\scriptstyle f}\downarrow & \nearrow & \downarrow{\scriptstyle g} \\ \mathfrak{B} & \xrightarrow{\beta} & \mathfrak{N} \end{array}$$

of formal pointed DG-manifolds.

Proof By Corollary 11.5.5 the map f is injective; writing for simplicity $V = \ker \Delta_{\mathfrak{B}}$ and $U = f_1(\ker \Delta_{\mathfrak{A}}) \subset V$, we have already pointed out that the injectivity of f_1 implies that f is a strict morphism and by Lemma 12.1.4 there exists two isomorphisms of graded coalgebras $F \colon \mathfrak{A} \simeq (\overline{S^c}(U)\ G \colon \mathfrak{B} \simeq (\overline{S^c}(V), Q)$ such that f is induced by the inclusion $U \subset V$ and then the image of GfF^{-1} is precisely $\overline{S^c}(U)$. Writing $Q = Gd_{\mathfrak{B}}G^{-1}$ it is not restrictive to replace in the diagram (12.2) the formal pointed DG-manifold $\mathfrak{B}$ with $(\overline{S^c}(V), Q)$ and $\mathfrak{A}$ with $(\overline{S^c}(U), Q)$.

Consider now the increasing filtration of differential graded subcoalgebras of $(\overline{S^c}(V), Q)$:

$$C_0 = \overline{S^c}(U), \qquad C_n = \left(\oplus_{i=1}^n V^{\odot i}\right) \oplus \left(\oplus_{j>n} U^{\odot j}\right) = C_{n-1} + V^{\odot n}.$$

It is clear that $\Delta(C_{n+1}) \subset C_n \otimes C_n$ and there exist isomorphisms of DG-vector spaces

$$\frac{C_n}{C_{n-1}} = \frac{C_{n-1} + V^{\odot n}}{C_{n-1}} = \frac{V^{\odot n}}{V^{\odot n} \cap C_{n-1}} = \frac{V^{\odot n}}{U^{\odot n}}.$$

According to Proposition 10.1.7, if f_1 is a quasi-isomorphism, then every inclusion map $C_n \to C_{n+1}$ is a quasi-isomorphism.

The pasting of the diagrams (12.2), (12.3), gives a commutative diagram

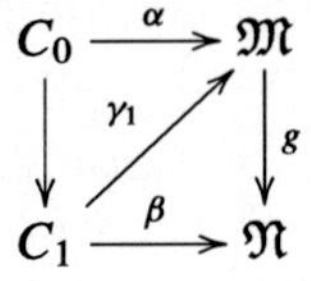

which, by Lemma 12.1.6 can be lifted to a sequence of diagrams

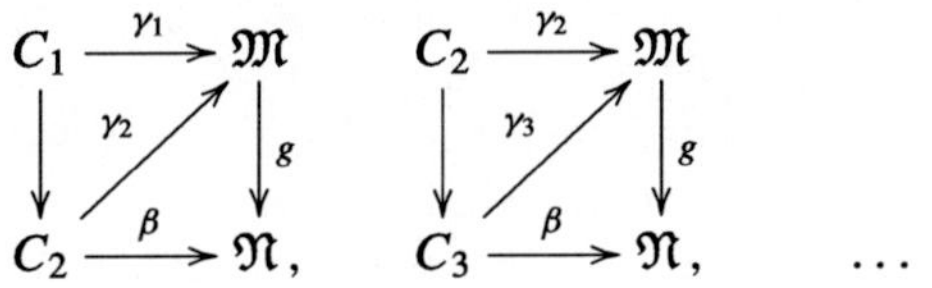

and the proof is done by taking the colimit of the sequence of morphisms γ_n. □

Definition 12.1.8 A differential graded cocommutative coalgebra Ω is said to be (formally) **fibrant** if it satisfies the following extension condition: given a locally conilpotent differential graded cocommutative coalgebra C and $B \subset C$ a differential graded subcoalgebra such that $\Delta_C(C) \subset B \otimes B$ and C/B is an acyclic complex, then the restriction map

$$\mathrm{Mor}_{\mathbf{DGC}}(C, \Omega) \to \mathrm{Mor}_{\mathbf{DGC}}(B, \Omega)$$

is surjective.

Theorem 12.1.9 *A locally conilpotent differential graded cocommutative coalgebra is fibrant if and only if it is a formal pointed DG-manifold.*

Proof Every formal pointed DG-manifold $\mathfrak{M}$ is fibrant by Lemma 12.1.6, applied with $\mathfrak{N} = 0$.

Conversely, let Ω be a locally conilpotent fibrant differential graded cocommutative coalgebra; choosing a projection $f : \Omega \to \ker \Delta_\Omega$, by Lemma 11.1.17 f is a cogenerator of Ω and there exists an injective morphism of graded coalgebras $F : \Omega \to \overline{S^c}(\ker \Delta_\Omega)$ corestricting to f.

In order to conclude the proof it is sufficient to prove that F is surjective. Choose a surjective morphism of complexes $h : V \to \ker \Delta_\Omega$ with V acyclic (e.g. $V = C(\mathrm{Id}_{\ker \Delta_\Omega})$ the mapping cocone of the identity on $\ker \Delta_\Omega$) and consider the filtration of acyclic conilpotent differential graded subcoalgebras of $\overline{S^c}(V)$:

$$C_1 = V \subset C_2 = V \oplus V^{\odot 2} \subset \cdots \subset C_n = \oplus_{i=1}^n V^{\odot i} \subset \cdots .$$

Since $\overline{S^c}(V) = \cup C_n$, the assumption that Ω is fibrant implies that the morphism $h\colon V \to \Omega$ extends to a morphism of differential graded coalgebras $H : \overline{S^c}(V) \to \Omega$. Since the linear part of FH is the surjective morphism h, by Corollary 11.5.5 the morphism FH is also surjective and a fortiori F is surjective too. □

12.2 L_∞-Morphisms of $L_\infty[1]$-Algebras

Given a graded vector space V and a linear map $q \in \operatorname{Hom}^1_{\mathbb{K}}(\overline{S^c}(V), V)$, we described the Leibniz expansion $Q \in \operatorname{Coder}^1(\overline{S^c}(V), \overline{S^c}(V))$ in Definition 11.5.8, and proved in Corollary 11.5.9 that Q is the unique coderivation that corestricts to q. The interpretation of the Nijenhuis–Richardson bracket as the graded commutator in the space of coderivations of a reduced symmetric coalgebra (Lemma 11.5.11) implies immediately that the pair (V, q) is an $L_\infty[1]$-algebras if and only if the pair $(\overline{S^c}(V), Q)$ is a differential graded coalgebra. Thus every $L_\infty[1]$-algebra gives canonically a formal pointed DG-manifold, and every formal pointed DG-manifold is isomorphic to one arising from an $L_\infty[1]$-algebra. It is therefore natural to consider, for every pair of $L_\infty[1]$-algebras the set of morphisms between them, considered as formal pointed DG-manifolds.

Definition 12.2.1 For every $L_\infty[1]$-algebras (V, q) let's denote by $\overline{S^c}(V, q)$ the differential graded coalgebra $(\overline{S^c}(V), Q)$, where Q is the Leibniz expansion of q.

Definition 12.2.2 (*L_∞-morphisms*) Given two $L_\infty[1]$-algebras (V, q) and (W, r), an **L_∞-morphism** $f\colon (V, q) \rightsquigarrow (W, r)$ is a linear map $f \in \operatorname{Hom}^0_{\mathbb{K}}(\overline{S^c}(V), W)$ such that its multiplicative expansion $F\colon \overline{S^c}(V, q) \to \overline{S^c}(W, r)$ is a morphism of differential graded coalgebras.

Equivalently, an L_∞-morphism $f\colon (V, q) \rightsquigarrow (W, r)$ is the corestriction of a morphism $F\colon \overline{S^c}(V, q) \to \overline{S^c}(W, r)$ of differential graded coalgebras.

Thus, by definition there exists a bijection

$$\{L_\infty\text{-morphisms}\colon (V, q) \rightsquigarrow (W, r)\} = \operatorname{Mor}_{\mathbf{DGC}}(\overline{S^c}(V, q), \overline{S^c}(W, r)).$$

Composition of L_∞-morphisms is defined in terms of the composition in the category **DGC**. More precisely, the composition of two L_∞-morphisms $f\colon (V, q) \rightsquigarrow (W, r)$ and $g\colon (W, r) \rightsquigarrow (U, s)$, is defined by the formula:

$$gf := gF \in \operatorname{Hom}^0_{\mathbb{K}}(\overline{S^c}(V), U), \qquad F = \text{multiplicative expansion of } f.$$

The use of the squiggly arrow $\rightsquigarrow$ distinguishes graphically L_∞-morphisms from linear morphisms of $L_\infty[1]$-algebras. Denoting by f_n, $n > 0$, the components of an L_∞-morphism f:

$$f = \sum_{n>0} f_n, \qquad f_n \in \operatorname{Hom}^0_{\mathbb{K}}(V^{\odot n}, W),$$

we shall call f_1 the *linear part* of f, f_2 the *quadratic part* and so on. According to Corollary 11.5.5, an L_∞-morphism is an isomorphism if and only if its linear part is an isomorphism. By Lemma 12.1.4 every linear morphism of $L_\infty[1]$-algebras is strict; conversely, every strict L_∞-morphism is linear up to left and right composition with L_∞ isomorphisms. The following proposition is a more explicit formulation of the notion of L_∞-morphisms.

Proposition 12.2.3 *Given two $L_\infty[1]$-algebras $(V, q_1, q_2, \ldots)$ and $(W, r_1, r_2, \ldots)$, a sequence of linear maps $f_n \in Hom^0_{\mathbb{K}}(V^{\odot n}, W)$, $n > 0$, gives an L_∞-morphism $f = \sum f_n$ if and only if for every $v_1, \ldots, v_n \in V$ we have*

$$\sum_{i=1}^{n} r_i F^i_n(v_1 \odot \cdots \odot v_n)$$
$$= \sum_{i=1}^{n} \sum_{\sigma \in S(i,n-i)} \epsilon(\sigma) f_{n-i+1}(q_i(v_{\sigma(1)} \odot \cdots \odot v_{\sigma(i)}) \odot \cdots \odot v_{\sigma(n)}), \qquad (12.4)$$

where the maps $F^i_n \colon V^{\odot n} \to W^{\odot i}$ are the components of the multiplicative expansion of f, defined recursively by the formulas $F^1_n = f_n$ and

$$F^i_n(v_1 \odot \cdots \odot v_n)$$
$$= \frac{1}{i} \sum_{a=1}^{n-i+1} \sum_{\sigma \in S(a,n-a)} \epsilon(\sigma) f_a(v_{\sigma(1)} \odot \cdots \odot v_{\sigma(a)}) \odot F^{i-1}_{n-a}(v_{\sigma(a+1)} \odot \cdots \odot v_{\sigma(n)}).$$

Proof Let's define $(\overline{S^c}(V), Q) = \overline{S^c}(V, q_1, q_2, \ldots)$, $(\overline{S^c}(W), R) = \overline{S^c}(V, r_1, r_2, \ldots)$ and let $F^i_n \colon V^{\odot n} \to W^{\odot i}$ be the components of the morphism of graded coalgebras $F \colon \overline{S^c}(V) \to \overline{S^c}(W)$ such that $f = p_W F$. Then $F^i_n = 0$ for $i > n$, while for $i \leq n$ the components F^i_n are determined by the recursive equations of Corollary 11.5.4. Thus the Eq. (12.4) are equivalent to the equality

$$rF = fQ \quad \iff \quad p_W(RF - FQ) = rF - fQ = 0.$$

Since $RF - FQ$ is an F-coderivation, by Lemma 11.2.8 the vanishing of its corestriction $p_W(RF - FQ)$ is equivalent to the vanishing of $RF - FQ$. □

For $n = 1$, the Eq. (12.4) reduces to $r_1 f_1 = f_1 q_1$ and then the linear component $f_1 \colon (V, q_1) \to (W, r_1)$ is a morphism of complexes. Notice that if $f_n = 0$ for every $n > 1$, the Eq. (12.4) becomes

$$r_n f_1^{\odot n} = r_n(f_1 \odot \cdots \odot f_1) = f_1 q_n$$

and therefore the linear morphisms of $L_\infty[1]$-algebras are exactly the L_∞-morphisms with trivial nonlinear components. Keep in mind that the notion of linear L_∞-morphism does not make sense in the framework of formal pointed DG-manifolds;

on the other hand, the notion of strict L_∞-morphism (=linear up to isomorphisms) also makes sense for DG-manifolds.

Remark 12.2.4 In the notation of Proposition 12.2.3, if $q_n = r_n = 0$ for every $n \geq 3$, then the sequence f_n gives an L_∞-morphism if and only if

$$\sum_{i=1}^{2} \sum_{\sigma\in S(i,n-i)} \epsilon(\sigma) f_{n-i+1}(q_i(v_{\sigma(1)}\odot\cdots\odot v_{\sigma(i)})\odot\cdots\odot v_{\sigma(n)})$$

$$= r_1(f_n(v_1\odot\cdots\odot v_n))$$

$$+\frac{1}{2}\sum_{\sigma\in S(a,n-a)} \epsilon(\sigma) r_2(f_a(v_{\sigma(1)}\odot\cdots\odot v_{\sigma(a)})\odot f_{n-a}(v_{\sigma(a+1)}\odot\cdots\odot v_{\sigma(n)})).$$

The L_∞-morphisms commute with scalar extensions; more precisely, let A be a commutative DG-algebra and $f\colon V \rightsquigarrow W$ an L_∞-morphism, then the scalar extension f_A of $f = \sum f_i$ is determined by the components

$$(f_A)_n\colon (V\otimes A)^{\odot n} \to V\otimes A,$$
$$(f_A)_n(v_1\otimes a_1,\ldots,v_n\otimes a_n) = (-1)^{\sum_{i<j}\overline{a_i}\,\overline{v_j}} f_n(v_1,\ldots,v_n)\otimes a_1a_2\cdots a_n,$$

and it is completely straightforward to check that $f_A\colon V\otimes A \rightsquigarrow W\otimes A$ remains an L_∞-morphism.

The product of $L_\infty[1]$-algebras defined in Example 10.4.6 is also a product in the category of $L_\infty[1]$-algebras with L_∞-morphisms; given two L_∞-morphisms

$$f\colon (U,q)\rightsquigarrow (V,r), \qquad g\colon (U,q)\rightsquigarrow (W,s),$$

we can immediately check that the linear map

$$(f,g)\colon \overline{S^c}(U)\to V\times W$$

gives an L_∞-morphism $(f,g)\colon (U,q)\rightsquigarrow (V,r)\times(W,s) = (V\times W, r\times s)$.

In general, an L_∞-morphism does not have any kernel in the category of $L_\infty[1]$-algebras. As an example, consider the vector spaces $V=\mathbb{K}^2$, $W=\mathbb{K}$, both of them equipped with the trivial $L_\infty[1]$ structure, and consider the L_∞-morphism

$$f\colon (V,0)\rightsquigarrow (W,0), \qquad f_2(x,y) = x_1y_2 + x_2y_1, \quad f_i = 0, \quad i\neq 2.$$

The inclusions $i\colon V_1 = \{x\in V \mid x_1 = 0\}\to V$ and $j\colon V_2 = \{x\in V\mid x_2 = 0\}\to V$ are linear L_∞-morphisms such that $fi = fj = 0$ and then, if $k\colon (H,h_1,h_2,\ldots)\rightsquigarrow (V,0)$ is a kernel of f then the image of the linear part k_1 should contain $V_1\cup V_2$.

12.3 L_∞-Morphisms of L_∞ and DG-Lie Algebras

We have already seen that to every L_∞-algebra $(L, l_1, l_2, \ldots)$ corresponds canonically an $L_\infty[1]$-algebra $(L[1], q_1, q_2, \ldots)$, where l_i and q_i are related by décalage isomorphisms as in Sect. 10.6. More explicitly, if $s\colon L[1] \to L$ is the tautological isomorphism of degree $+1$, then the maps l_i and q_i are related by the formulas

$$l_k(sv_1, \ldots, sv_k) = -(-1)^{\sum_i (k-i)\overline{v_i}} s q_k(v_1, \ldots, v_k)\,.$$

Definition 12.3.1 An L_∞-**morphism** of L_∞-algebras is an L_∞-morphism of the corresponding $L_\infty[1]$-algebras.

Given two L_∞-algebras $(H, h_1, h_2, \ldots)$ and $(L, l_1, l_2, \ldots)$, via the dècalage isomorphisms (10.16):

$$\text{déc}\colon \operatorname{Hom}^0_{\mathbb{K}}(H[1]^{\odot n}, L[1]) \xrightarrow{\simeq} \operatorname{Hom}^{1-n}_{\mathbb{K}}(H^{\wedge n}, L),$$

every L_∞-morphism $g\colon (H, h_1, h_2, \ldots) \rightsquigarrow (L, l_1, l_2, \ldots)$ is given by a sequence of maps

$$g_n \in \operatorname{Hom}^{1-n}_{\mathbb{K}}(H^{\wedge n}, L), \qquad n \geq 1,$$

such that the maps $f_n = \text{déc}^{-1}(g_n)$ satisfy the condition of Proposition 12.2.3. Although it is possible to write down the description of every L_∞-morphism $g\colon (H, h_1, h_2, \ldots) \rightsquigarrow (L, l_1, l_2, \ldots)$ of L_∞-algebras exclusively in terms of equations involving the maps g_n, h_n, l_n, it is very often convenient to work with the corresponding $L_\infty[1]$ counterpart; however, in the case of L_∞-morphisms between DG-Lie algebras it may be convenient to work directly with L_∞-structures.

Proposition 12.3.2 *Let L, M be two DG-Lie algebras and let $g_n \in Hom^{1-n}_{\mathbb{K}}(L^{\wedge n}, M)$, $n \geq 1$, be a sequence of maps. Define the Lie-cup products $g_p \cup g_q\colon L^{\wedge p+q} \to M$ by the formulas:*

$$g_p \cup g_q(a_1, \ldots, a_{p+q}) = \sum_{\sigma \in S(p,q)} \pm_\sigma \left[g_p(a_{\sigma(1)}, \ldots, a_{\sigma(p)}), g_q(a_{\sigma(p+1)}, \ldots, a_{\sigma(p+q)})\right],$$

where the sign $\pm_\sigma$ is equal to $\chi(\sigma)(-1)^{(1-q)(\overline{a_{\sigma(1)}}+\cdots+\overline{a_{\sigma(p)}})}$.

Then the sequence g_n gives an L_∞-morphism $g\colon L \rightsquigarrow M$ if and only if for every $n > 0$ and every $a_1, \ldots, a_n \in L$ we have:

$$\begin{aligned} dg_n(a_1,\ldots,a_n) &+ \frac{1}{2}\sum_{p=1}^{n-1}(-1)^{p(1-n+p)} g_p \cup g_{n-p}(a_1,\ldots,a_n) \\ &= (-1)^{n-1}\sum_{\sigma\in S(1,n-1)} \chi(\sigma) g_n(da_{\sigma(1)}, a_{\sigma(2)},\ldots,a_{\sigma(n)}) \\ &\quad + (-1)^{n-2}\sum_{\sigma\in S(2,n-2)} \chi(\sigma) g_{n-1}([a_{\sigma(1)},a_{\sigma(2)}],a_{\sigma(3)},\ldots,a_{\sigma(n)}), \end{aligned} \tag{12.5}$$

where $\chi(\sigma)=\chi(\sigma;a_1,\ldots,a_n)$ is the usual antisymmetric Koszul sign.

Proof For simplicity of notation we write $V=L[1]$ and $W=M[1]$ and we identify L, M with sV, sW respectively, where s is the usual suspension formal symbol of degree $+1$. The corresponding $L_\infty[1]$ structures $(V,q_1,q_2,0,\ldots)$ and $(W,r_1,r_2,0,\ldots)$ are therefore given by the maps:

$$q_1(v)=-dv, \qquad sq_2(v_1,v_2)=-(-1)^{\overline{v_1}}[sv_1,sv_2],$$

$$r_1(w)=-dw, \qquad sr_2(w_1,w_2)=-(-1)^{\overline{w_1}}[sw_1,sw_2].$$

By décalage isomorphisms the maps g_n correspond to the maps $f_n\in \mathrm{Hom}^0_{\mathbb{K}}(V^{\odot n},W)$ defined by

$$sf_n(v_1,\ldots,v_n)=(-1)^{\sum_i (n-i)\overline{v_i}} g_n(sv_1,\ldots,sv_n).$$

It is now a tedious but completely straightforward computation to verify that the via the above relations the Eq. (12.5) are the same described in Remark 12.2.4. □

It is useful and instructive to write the equations of Proposition 12.3.2 for $n=1,2$. For $n=1$ we simply have $dg_1=g_1d$ and then $g_1\colon L\to M$ is a morphism of DG-vector spaces.

For $n=2$, Eq. (12.5) becomes:

$$g_1([a,b])-[g_1(a),g_1(b)]=dg_2(a,b)+g_2(da,b)+(-1)^{\bar{a}}g_2(a,db),$$

and then g_1 is a Lie morphism up to a homotopy given by $g_2\in \mathrm{Hom}^{-1}_{\mathbb{K}}(L^2,M)$. In particular, every L_∞-morphism $g\colon L\rightsquigarrow M$ of DG-Lie algebras induces a morphism of graded Lie algebras $g_1\colon H^*(L)\to H^*(M)$.

An Example: Homotopy Transfer of Endomorphisms

Every couple of morphisms of DG-vector spaces

$$V\xrightarrow{\pi} H\xrightarrow{i} V,$$

defines by composition a morphism of DG-vector spaces

$$g_1=i^*\pi_*\colon \mathrm{Hom}^*_{\mathbb{K}}(V,V)\to \mathrm{Hom}^*_{\mathbb{K}}(H,H), \qquad g_1(a)=\pi a i.$$

If $i\pi = \mathrm{Id}_V$ then ϕ is also a morphism of differential graded Lie algebras. The next proposition, together with its constructive proof, may be considered as the L_∞ analogue of the ordinary perturbation lemma (Theorem 5.4.6).

Proposition 12.3.3 *In the above notation, if the morphism $i\pi\colon V \to V$ is homotopic to the identity, then g_1 is the linear component of an L_∞-morphism*

$$g\colon Hom^*_{\mathbb{K}}(V, V) \rightsquigarrow Hom^*_{\mathbb{K}}(H, H).$$

Proof We shall prove that an L_∞-morphism g as above can be explicitly contructed in terms of any homotopy operator $h \in \mathrm{Hom}^{-1}_{\mathbb{K}}(V, V)$ such that $hd_V + d_V h = i\pi - \mathrm{Id}_V$. The L_∞-morphism is given by the sequence

$$\begin{aligned} g_n &\colon \mathrm{Hom}^*_{\mathbb{K}}(V, V)^{\wedge n} \to \mathrm{Hom}^*_{\mathbb{K}}(H, H), \qquad n \geq 1, \\ g_n(a_1, \ldots, a_n) &= \sum_{\sigma\in\Sigma_n} \widehat{\chi}(\sigma)\, \pi\, a_{\sigma(1)}\, h\, a_{\sigma(2)}\, h \cdots h\, a_{\sigma(n)}\, i, \end{aligned}$$

where $\widehat{\chi}(\sigma) = \chi(\sigma)(-1)^{\sum_i (n-i)(1+\overline{a_{\sigma(i)}})}$ is the appropriate antisymmetric Koszul sign, determined by the algebraic relation:

$$h^{\wedge n-1} \wedge a_1 \wedge \cdots \wedge a_n = \widehat{\chi}(\sigma)\, a_{\sigma(1)} \wedge h \wedge a_{\sigma(2)} \wedge \cdots \wedge h \wedge a_{\sigma(n)} \in \mathrm{Hom}^*_{\mathbb{K}}(V, V)^{\wedge 2n-1}.$$

The proof that the above maps g_n satisfy the condition of Proposition 12.3.2 is rather straightforward, long and unexciting, although it can be simplified by using some techniques not developed in this book (graded polarization, symmetrization of A_∞-morphisms). Here we only prove the equality (12.5) in the particular, but nevertheless significant, situation where $a_1, \ldots, a_n$ are equal to a fixed endomorphism $a\colon V \to V$ of *odd degree* (cf. Theorem 5.4.6). In this case we have:

$$\begin{aligned} g_n(a, \ldots, a) &= n!\, \pi\, a\, h\, a\, h \cdots h\, a\, i = n!\, \pi (ah)^{n-1} ai, \\ g_p \cup g_{n-p}(a, \ldots, a) &= n!\, (-1)^{p(1-n+p)} [\pi(ah)^{p-1} ai, \pi(ah)^{n-p-1} ai], \\ \frac{1}{2}\sum_{p=1}^{n-1} (-1)^{p(1-n+p)} g_p \cup g_{n-p}(a, \ldots, a) &= \frac{n!}{2} \sum_{p=1}^{n-1} [\pi(ah)^{p-1} ai, \pi(ah)^{n-p-1} ai] \\ &= n! \sum_{p=1}^{n-1} \pi(ah)^{p-1} ai\pi(ah)^{n-p-1} ai. \end{aligned}$$

Denoting by the letter $\overline{\partial}$ the differentials of V, H and by the letter $d = [\overline{\partial}, -]$ the differentials of $\mathrm{Hom}^*_{\mathbb{K}}(V, V)$, $\mathrm{Hom}^*_{\mathbb{K}}(H, H)$, then

$$d\pi = di = 0, \qquad dh = \overline{\partial} h + h \overline{\partial} = i\pi - \mathrm{Id}_V,$$

and by the Leibniz formula

$$dg_n(a, \dots, a)$$
$$= n!\pi\left(\sum_{i=1}^{n}(ah)^{i-1}da(ha)^{n-i} - \sum_{i=1}^{n-1}\pi(ah)^{i-1}a(i\pi - \mathrm{Id}_V)a(ha)^{n-i-1}\right)i.$$

Since

$$h^{\wedge n-1} \wedge da \wedge a^{\wedge n-1} = (-1)^{n-1}da \wedge h \wedge a \wedge \cdots \wedge h \wedge a$$

we have

$$g_n(da, a, \dots, a) = (-1)^{n-1}(n-1)!\sum_{i=1}^{n}(ah)^{i-1}da(ha)^{n-i},$$

$$\begin{aligned} g_{n-1}([a,a], a, \dots, a) &= 2g_{n-1}(a^2, a, \dots, a) \\ &= 2(-1)^{n-2}(n-2)!\sum_{i=1}^{n-1}(ah)^{i-1}a^2(ha)^{n-i-1}, \end{aligned}$$

and therefore

$$(-1)^{n-1}\sum_{\sigma\in S(1,n-1)}\chi(\sigma)g_n(da, a, \dots, a) = n!\pi\sum_{i=1}^{n}(ah)^{i-1}da(ha)^{n-i}i,$$

$$(-1)^{n-2}\sum_{\sigma\in S(2,n-2)}\chi(\sigma)g_{n-1}([a,a], a, \dots, a) = n!\pi\sum_{i=1}^{n-1}(ah)^{i-1}a^2(ha)^{n-i-1}i.$$

The proof of (12.5) is now an immediate consequence of the above calculation. □

Another Example: The Hitchin–Martinengo L_∞-Morphism

In this subsection we sketch a simplified version of a nontrivial example of L_∞-morphism between DG-Lie algebras described by Martinengo in her study of Hitchin maps; as in the previous example, some details are left to the reader.

Given two finite-dimensional vector spaces H, E, consider the symmetric algebra $S(E) = \oplus_{n\geq 0}E^{\odot n}$ generated by E, the associative algebra

$$A = \mathrm{End}(H) \otimes S(E)$$

and the trace map

$$\mathrm{Tr}\colon A \to S(E), \qquad \mathrm{Tr}(f \otimes v) = \mathrm{Tr}(f)v.$$

Lemma 12.3.4 *In the notation above, for every $f, g \in A$ and every $k > 0$ we have:*

1. *Tr*$([f,g]) = $ *Tr*$(fg - gf) = 0$;
2. *Tr*$\left((f + t[f,g])^k\right) = $ *Tr*$(f^k) + t^2(\cdots)$.

Proof The first item follows immediately from the commutativity of the symmetric algebra $S(E)$. Using induction on k and the relation $[f^k, g] = f^{k-1}[f,g] + [f^{k-1}, g]f$ we can immediately see that $\sum_{i=0}^{k-1} f^i[f,g]f^{k-i-1} = [f^k, g]$ and therefore

$$(f + t[f,g])^k = f^k + t[f^k, g] + t^2(\cdots).$$

□

Given a fixed positive integer k and a fixed element $\theta \in \mathrm{End}(H) \otimes E$, we can construct the sequence of polarization maps:

$$g_n\colon (\mathrm{End}(H) \otimes E)^{\odot n} \to E^{\odot k}, \qquad n > 0,$$

$$g_n(f_1 \odot \cdots \odot f_n) = \text{coefficient of } t_1 t_2 \cdots t_n \text{ inTr } (\theta + t_1 f_1 + \cdots + t_n f_n)^k,$$

where the kth power is taken in the algebra A; notice that $g_1(f) = k\mathrm{Tr}(f\theta^{k-1})$.

Lemma 12.3.5 *In the above situation, for every $f_1, \ldots, f_n \in End(H) \otimes E$ and every $h \in End(H)$ we have*

$$g_{n+1}([h,\theta] \odot f_1 \odot \cdots \odot f_n) + \sum_{i=1}^{n} g_n(f_1 \odot \cdots \odot [h, f_i] \odot \cdots \odot f_n) = 0.$$

Proof By Lemma 12.3.4 the coefficient of $st_1 \cdots t_n$ in

$$\mathrm{Tr}(\theta + t_1 f_1 + \cdots + t_n f_n + s[h, \theta + t_1 f_1 + \cdots + t_n f_n])^k$$

vanishes. □

Assume now that our fixed element $\theta \in \mathrm{End}(H) \otimes E$ satisfies the integrability condition $[\theta,\theta] = 0$ in $\mathrm{End}(H) \otimes E^{\wedge 2}$. Then we have a differential graded Lie algebra $L = \mathrm{End}(H) \otimes \bigwedge^* E$, where the elements of $\mathrm{End}(H) \otimes \bigwedge^i E$ have degree i, the bracket is equal to

$$[f \otimes v, g \otimes w] = [f,g] \otimes (v \wedge w)$$

and the differential is equal to

$$d\colon L^i \to L^{i+1}, \qquad d(x) = [\theta, x].$$

Notice that for $h \in L^0 = \mathrm{End}(H)$ and $f \in L^1 = \mathrm{End}(H) \otimes E$, the brackets $[h, f]$ and $[f, h]$ *are the same* in the differential graded Lie algebra L and in the associative algebra A.

Let $(V, q_1, q_2, 0, \ldots)$ be the $L_\infty[1]$ structure associated to L via standard décalage isomorphisms; in particular, we have $V^i = L^{i+1} = \text{End}(H) \otimes E^{\wedge i+1}$ and:

1. $q_1(h) = [h, \theta]$ for every $h \in V^{-1}$;
2. $q_2(h \odot f) = [h, f]$ for every $h \in V^{-1}$, $f \in V^0$.

Extending the maps $g_n : (V^0)^{\odot n} \to E^{\odot k}$ to $V^{\odot n}$ by taking their compositions with the natural projections $V^{\odot n} \to (V^0)^{\odot n}$, using degree considerations and Lemma 12.3.5, it is easy to prove that

$$g : (V, q_1, q_2, 0, \ldots) \rightsquigarrow (E^{\odot k}, 0, 0, \ldots), \qquad g = g_1 + g_2 + \cdots + g_k \tag{12.6}$$

is an L_∞-morphism.

12.4 Transferring $L_\infty[1]$ Structures

The notion of differential graded Lie algebras is not stable under homotopy equivalence of the underlying complexes; this means in particular that, given a differential graded Lie algebra L, a DG-vector space V and an isomorphism $\phi : H^*(V) \to H^*(L)$, in general there does not exist a Lie bracket on V and a morphism of differential graded Lie algebras $f : V \to L$ inducing ϕ in cohomology. As an example one can take a non-formal differential graded Lie algebra L (see Example 6.7.2), $V = H^*(L)$ and ϕ as the identity.

However, in the above situation, the DG-Lie structure on L can be transferred to an L_∞ structure on V and ϕ can be lifted to an L_∞-morphism. More generally we shall see that every L_∞ structure on L can be transferred to an L_∞ structure on V. As usual, for simplicity of calculations, we work in the framework of $L_\infty[1]$-algebras.

Theorem 12.4.1 *Assume the following data are given:*

- *an $L_\infty[1]$-algebra $(V, q_1, q_2, \ldots)$ and a DG-vector space (W, r_1);*
- *two morphisms of DG-vector spaces $\pi : (V, q_1) \to (W, r_1)$, $f_1 : (W, r_1) \to (V, q_1)$ and a homotopy $h \in Hom_{\mathbb{K}}^{-1}(V, V)$ such that*

$$q_1 h + h q_1 = f_1 \pi - Id_V.$$

Defining

$$q_+ = \sum_{i \geq 2} q_i : \overline{S^c}(V) \to V, \qquad q = q_1 + q_+$$

and denoting by $p_V : \overline{S^c}(V) \to V$ the usual projection we have:

1. *there exists a unique morphism of graded coalgebras $F : \overline{S^c}(W) \to \overline{S^c}(V)$ such that its corestriction $f = p_V F$ satisfies the equation $f = p_V F = f_1 + h q_+ F$;*

2. *the map* $r = r_1 + \pi q_+ F\colon \overline{S^c}(W) \to W$ *is an* $L_\infty[1]$ *structure on* W *and*

$$f\colon (W, r) \rightsquigarrow (V, q)$$

is a strict L_∞*-morphism;*

3. *if in addition* $\pi h = 0$ *and* $\pi f_1 = Id_W$ *then* $\pi f_n = 0$ *for every* $n > 1$ *and* $r = \pi q F$.

Proof For a morphism of graded coalgebras $F\colon \overline{S^c}(W) \to \overline{S^c}(V)$, the equation $p_V F = f_1 + hq_+ F$ is equivalent to the sequence of equations

$$F_1^1 = f_1, \qquad F_n^1 = \sum_{i=2}^{n} hq_i F_n^i \qquad n > 1, \tag{12.7}$$

where the maps $F_j^i\colon W^{\odot j} \to V^{\odot i}$ are the components of F. Since by Corollary 11.5.4 the component F_n^i depends only on $F_1^1, \ldots, F_{n-i+1}^1$, it is clear that the above Eq. (12.7) gives a recursive proof of the existence and the unicity of the coalgebra morphism F. Notice that, since $F(W) \subset V$ we have $q_+ F(W) = 0$ and therefore

$$f = p_V F = f_1 + f_2 + \cdots, \qquad r = r_1 + \pi q_+ F = r_1 + r_2 + \cdots .$$

Next, let Q be the coderivation of $\overline{S^c}(V)$ such that $p_V Q = q$ and let R be the coderivation of $\overline{S^c}(W)$ such that $p_W R = r$. We want to prove that $R^2 = 0$ and that $QF = FR$; the first step is to prove that the F-coderivation $\delta = QF - FR$ satisfies the equation $p_V \delta = hq_+ \delta$. In fact:

$$\begin{aligned}
p_V(QF - FR) &= qF - p_V FR = q_1 p_V F + q_+ F - p_V FR\\
&= q_1 f_1 + q_1 hq_+ F + q_+ F - f_1 r - hq_+ FR\\
&= q_1 f_1 + (f_1 \pi - \mathrm{Id}_V - hq_1) q_+ F + q_+ F - f_1 r - hq_+ FR\\
&= q_1 f_1 + f_1 \pi q_+ F - hq_1 q_+ F - f_1 r - hq_+ FR\\
&= q_1 f_1 + f_1 \pi q_+ F - hq_1 q_+ F - f_1 r_1 - f_1 \pi q_+ F - hq_+ FR\\
&= q_1 f_1 - hq_1 q_+ F - f_1 r_1 - hq_+ FR\\
&= -hq_1 q_+ F - hq_+ FR.
\end{aligned}$$

Since $Q^2 = 0$ we have $0 = qQ = q_1 q + q_+ Q = q_1 q_+ + q_+ Q$, thus $q_1 q_+ = -q_+ Q$ and finally

$$p_V(QF - FR) = -hq_1 q_+ F - hq_+ FR = hq_+(QF - FR).$$

As a second step we prove by induction on n that the F-coderivation

$$\delta = QF - FR\colon \overline{S^c}(W) \to \overline{S^c}(V)$$

vanishes on $W^{\odot n}$. Let's assume $n > 0$ and $\delta(W^{\odot i}) = 0$ for every $i < n$; then by the coLeibniz rule, for every $w \in W^{\odot n}$ we have

$$\mathfrak{l}\delta(w) = (\delta \otimes F + F \otimes \delta)\mathfrak{l}(w) = 0$$

and this implies that $\delta(w) \in V$. Using the equality $p_V \delta = hq_+\delta$ we get

$$\delta(w) = p_V\delta(w) = hq_+\delta(w) = 0.$$

The last step is to prove that $R^2 = 0$. Since $r_1^2 = 0$ we have

$$p_W R^2 = rR = r_1 r + \pi q_+ FR = r_1 \pi q_+ F + \pi q_+ QF$$

and since $q_1 q_+ = -q_+ Q$ we have

$$p_W R^2 = r_1 \pi q_+ F - \pi q_1 q_+ F = (r_1 \pi - \pi q_1) q_+ F = 0.$$

By Proposition 12.2.3 and Eq. (12.7) we have $F(x_1 \odot \cdots \odot x_n) = 0$ whenever $f_1(x_i) = 0$ for some index i. This implies that f is a strict morphism, i.e., that $F(\overline{S^c}(W)) \cap V = f_1(W)$, since, writing $W = \ker(f_1) \oplus H$, the map $F_n^n = f_1^{\odot n}$ is injective on $H^{\odot n}$.

Finally, if $\pi h = 0$ and $\pi f_1 = \mathrm{Id}_W$, then $\pi f_n = \pi h(\cdots) = 0$ for every $n > 1$. Since by definition $r = r_1 + \pi q_+ F$, the equality $r = \pi q F$ is equivalent to $r_1 = \pi q_1 p_V F = \pi q_1 f$. We have $\pi q_1 f_1 = \pi f_1 r_1 = r_1$, while for every $n > 1$

$$\pi q_1 f_n = \pi q_1 h(\cdots) = \pi(f_1 \pi - \mathrm{Id}_V - hq_1)(\cdots) = \pi h q_1(\cdots) = 0.$$

□

Corollary 12.4.2 *Let $(V, q_1, q_2, \ldots)$ be an $L_\infty[1]$-algebra such that the complex (V, q_1) is acyclic. Then $(V, q_1, q_2, q_3, \ldots)$ is L_∞ isomorphic to $(V, q_1, 0, 0, \ldots)$.*

Proof We apply Theorem 12.4.1 with $W = V$, $r_1 = q_1$, $f_1 = \mathrm{Id}_V$, $\pi = 0$ and h any contracting homotopy of (V, q_1). □

Corollary 12.4.3 *Let $(V, q_1, \ldots)$ be an $L_\infty[1]$-algebra and let $f_1 \colon (W, r_1) \to (V, q_1)$ be a quasi-isomorphism of differential graded vector spaces. Then (W, r_1) can be extended to an $L_\infty[1]$-algebra $(W, r_1, r_2, \ldots)$ and f_1 can be lifted to an L_∞-morphism.*

Proof By the result of Sect. 5.1, every quasi-isomorphism of DG-vector spaces can be written as the composition of injective and surjective quasi-isomorphisms (see also Exercise 5.9.9). Thus it is sufficient to prove the corollary under the assumption that f_1 is either injective or surjective. If f_1 is injective, by Lemma 5.4.2 we can find a morphism of complexes $\pi \colon (V, q_1) \to (W, r_1)$ and a homotopy $h \in \mathrm{Hom}^{-1}(V, V)$ such that

$$q_1h + hq_1 = f_1\pi - \mathrm{Id}_V, \qquad \pi f_1 = \mathrm{Id}_W,$$

and the conclusion follows from Theorem 12.4.1. If f_1 is surjective, by Corollary 5.1.7 there exists a morphism of DG-vector spaces $\pi\colon (V, q_1) \to (W, r_1)$ such that $f_1\pi = \mathrm{Id}_V$ and the conclusion follows from Theorem 12.4.1 by considering the trivial homotopy $h = 0$. □

Corollary 12.4.4 *Let $f\colon (V, q_1, q_2, \ldots) \rightsquigarrow (W, r_1, r_2, \ldots)$ be an L_∞-morphism of $L_\infty[1]$-algebra such that its linear part $f_1\colon (V, q_1) \to (W, r_1)$ is a surjective quasi-isomorphism of DG-vector spaces. Then there exists an L_∞-morphism*

$$g\colon (W, r_1, r_2, \ldots) \rightsquigarrow (V, q_1, q_2, \ldots)$$

such that $fg = Id_W$.

Proof Since f_1 is a surjective quasi-isomorphism, by Corollary 5.1.7 there exists an injective quasi-isomorphism $h_1\colon (W, r_1) \to (V, q_1)$ such that $f_1h_1 = \mathrm{Id}_W$. By Corollary 12.4.3 we can extend h_1 to an L_∞-morphism

$$h\colon (W, r_1, \hat{r}_2, \hat{r}_3, \ldots) \rightsquigarrow (V, q_1, q_2, q_3, \ldots).$$

The L_∞-morphism

$$fh\colon (W, r_1, \hat{r}_2, \hat{r}_3, \ldots) \rightsquigarrow (W, r_1, r_2, r_3, \ldots)$$

has the linear part equal to the identity and thus it is an isomorphism. Then it is sufficient to define $g = h(fh)^{-1}$. □

Remark 12.4.5 In view of the multiplicative expansion formulas (Definition 11.5.2), the recursive formulas of Theorem 12.4.1 take a simple form when $q_i = 0$ for every $i \geq 3$, or equivalently if $q_+ = q_2$. Denoting by $\mathfrak{l}$ the coproduct in the reduced symmetric coalgebra, by $f = \sum_{i>0} f_i\colon \overline{S^c}(W) \to V$ and by $b\colon V \otimes V \to V$ the map $b(x, y) = q_2(x \odot y)$, the recursive formulas $p_V F = f_1 + hq_+F$ and $r = r_1 + \pi q_+ F$ becomes

$$f = f_1 + \frac{1}{2} hbf^{\otimes 2}\mathfrak{l}, \qquad r = r_1 + \frac{1}{2}\pi bf^{\otimes 2}\mathfrak{l}. \tag{12.8}$$

12.5 Homotopy Classification of L_∞ and $L_\infty[1]$-Algebras

One of the advantages of using L_∞ and $L_\infty[1]$-algebras is that for them there exists a homotopy classification very similar to Lemma 5.1.4.

As usual, for simplicity we consider only $L_\infty[1]$-algebras, keeping in mind that every result regarding $L_\infty[1]$-algebras has its analogue in the category of L_∞-algebras.

Definition 12.5.1 An $L_\infty[1]$-algebra $(V, q_1, q_2, \dots)$ is called:

1. **minimal** if $q_1 = 0$;
2. **contractible** if the complex (V, q_1) is acyclic;
3. **linear contractible** if (V, q_1) is an acyclic complex and $q_j = 0$ for every $j > 1$.

Corollary 12.4.2 implies that an $L_\infty[1]$-algebra is contractible if and only if it is L_∞-isomorphic to a linear contractible one. Notice that the properties to be contractible and minimal are invariant under L_∞-isomorphism.

Lemma 12.5.2 *Let $(U, d) = (U, d_1, 0, \dots)$ be a linear contractible $L_\infty[1]$-algebra and let*

$$f : (V, q) \rightsquigarrow (W, r) = (W, r_1, r_2, \dots)$$

be an L_∞-morphism of $L_\infty[1]$-algebras. Then for every morphism of complexes $j : (U, d_1) \to (W, r_1)$ there exists an L_∞-morphism

$$g : (V, q) \times (U, d) = (V \times U, q \times d) \rightsquigarrow (W, r)$$

such that $g_{|V} = f$ and $g_1(u) = j(u)$ for every $u \in U$.

Proof Immediate consequence of Theorem 12.1.7, applied to the commutative diagram of $L_\infty[1]$-algebras

$$\begin{array}{ccc} (V, q) & \overset{f}{\rightsquigarrow} & (W, r) \\ \downarrow & & \downarrow \\ (V, q) \times (U, d) & \longrightarrow & 0 \end{array}$$

and to the commutative diagram of DG-vector spaces

$$\begin{array}{ccc} (V, q_1) & \xrightarrow{f_1} & (W, r_1) \\ \downarrow & \nearrow^{f_1 + j} & \downarrow \\ (V, q_1) \times (U, d_1) & \longrightarrow & 0. \end{array}$$

□

Corollary 12.5.3 *Let $f : (H, r_1, r_2, \dots) \rightsquigarrow (V, q_1, q_2, \dots)$ be an L_∞-morphism such that its linear part $f_1 : (H, r_1) \to (V, q_1)$ is an injective quasi-isomorphism of complexes. Then there exists an L_∞-morphism*

$$p : (V, q_1, q_2, \dots) \rightsquigarrow (H, r_1, r_2, \dots)$$

such that $pf = Id_H$.

Proof By Lemma 5.4.2 we have a direct sum decomposition $V = f_1(H) \oplus U$, with U an acyclic subcomplex of (V, q_1). According to Lemma 12.5.2 the L_∞-morphism f extends to an L_∞-morphism

$$g\colon (H, r_1, r_2, \ldots) \times (U, q_1, 0, \ldots) \rightsquigarrow (V, q_1, q_2, \ldots),$$

with linear part g_1 an isomorphism of complexes. Hence g is an L_∞-isomorphism and we can take p as the composition of the inverse of g with the projection onto the first factor. □

Corollary 12.5.4 *Every $L_\infty[1]$-algebra is L_∞-isomorphic to the product of a minimal and a linear contractible $L_\infty[1]$-algebras.*

Proof Consider an $L_\infty[1]$-algebra $(V, q_1, q_2, \ldots)$; then we can write (V, q_1) as the direct sum of two subcomplexes $V = H \oplus A$ with $q_1(H) = 0$ and A acyclic; in particular, the embedding $f_1\colon (H, 0) \to (V, q_1)$ is an injective quasi-isomorphism. By Corollary 12.4.3 f_1 can be extended to an L_∞-morphism

$$f\colon (H, 0, r_2, \ldots) \rightsquigarrow (V, q_1, q_2, \ldots)$$

and Lemma 12.5.2, applied to the inclusion $j\colon A \to V$, gives an L_∞-isomorphism

$$g\colon (H, 0, r_2, \ldots) \times (A, q_1, 0, \ldots) \rightsquigarrow (V, q_1, q_2, \ldots).$$

□

Definition 12.5.5 An L_∞-morphism $f\colon (V, q_1, \ldots) \rightsquigarrow (W, r_1, \ldots)$ is called a **weak equivalence** if its linear component $f_1\colon (V, q_1) \to (W, r_1)$ is a quasi-isomorphism of DG-vector spaces.

Some authors call quasi-isomorphisms the weak equivalences; this terminology is potentially dangerous in the L_∞ setting. In fact, given a morphism of differential graded coalgebras $F\colon \overline{S^c}(V, q) \to \overline{S^c}(W, r)$ it is easy to prove that if its restriction $F\colon V \to W$ is a quasi-isomorphism, then F is a quasi-isomorphism, while the converse is generally false; for counterexamples see [151, p. 40] and [113, Ex. 9.1.2].

Notice that every weak equivalence between minimal $L_\infty[1]$-algebras is an isomorphism.

Definition 12.5.6 Two $L_\infty[1]$-algebras (V, q), (W, r) are called **homotopy equivalent** if there exists a weak equivalence $f\colon (V, q) \rightsquigarrow (W, r)$.

Theorem 12.5.7 (Minimal model theorem) *Homotopy equivalence of $L_\infty[1]$-algebras is an equivalence relation. Every $L_\infty[1]$-algebra is homotopy equivalent to a minimal $L_\infty[1]$-algebra, unique up to L_∞-isomorphism.*

Proof By Corollary 12.5.4, for every $L_\infty[1]$-algebra (V, q) there exist a minimal $L_\infty[1]$-algebra (H, r) and two weak equivalences

$$i: (H, r) \rightsquigarrow (V, q), \qquad p: (V, q) \rightsquigarrow (H, r).$$

If $f: (V, q) \rightsquigarrow (K, s)$ is a weak equivalence and (K, s) is minimal, then the composition fi is an isomorphism. Similarly, if $g: (K, s) \rightsquigarrow (V, q)$ is a weak equivalence and (K, s) is minimal, then the composition pg is an isomorphism.

If $f: (V, q) \rightsquigarrow (U, s)$ is a weak equivalence, taking the minimal models

$$i: (H, r) \rightsquigarrow (V, q), \qquad p: (U, s) \rightsquigarrow (K, p)$$

the composition $g = pfi$ is an isomorphism and then also $ig^{-1}p: (U, s) \rightsquigarrow (V, q)$ is a weak equivalence. □

The minimal model theorem implies in particular that the homotopy equivalence introduced in Definition 12.5.6 is the same as the homotopy equivalence in the category of $L_\infty[1]$-algebras and L_∞-morphisms, with respect to the homotopy theory defined by weak equivalences.

12.6 Homotopy Classification of DG-Lie Algebras

The main goal of this section is to prove that two differential graded Lie algebras are quasi-isomorphic if and only if they are homotopy equivalent as L_∞-algebras. Due to the algebraic background developed in this book it is more convenient to give a non-standard, and apparently new, proof of this fact, whereas the "standard" proof uses the general properties, not considered here, of bar and cobar resolutions.

Lemma 12.6.1 *Let $f: L \to M$ be a quasi-isomorphism of differential graded Lie algebras. Then there exists an L_∞-morphism $l: M \rightsquigarrow L$ such that its linear part $l_1: M \to L$ is a quasi-inverse of f.*

Proof By Theorem 6.1.1 there exist two surjective quasi-isomorphisms of DG-Lie algebras $p: N \to L$ and $q: N \to M$ such that $f = qi$, where i is a right inverse of p. By Corollary 12.4.4 there exists an L_∞-morphism $g: M \rightsquigarrow N$ such that its linear part $g_1: M \to N$ is a quasi-inverse of q. It is now sufficient to consider the composition $l = pg$. □

Corollary 12.6.2 *Let L, M be two quasi-isomorphic differential graded Lie algebras, then there exists an L_∞-morphism $L \rightsquigarrow M$ whose linear part is a quasi-isomorphism of complexes.*

Proof Immediate consequence of Lemma 12.6.1. □

Corollary 12.6.2 asserts in particular that quasi-isomorphic differential graded Lie algebras are homotopy equivalent as L_∞-algebras. Now we give a proof of the converse implication based on the existence of a canonical Cartan homotopy from any DG-Lie algebra to the corresponding Chevalley–Eilenberg DG-Lie algebra.

Following the notation introduced in Sect. 11.7, when $(L, d, [-,-])$ is a differential graded Lie algebra we shall denote by $(L[1], q_1, q_2, 0, \ldots)$ the associated $L_\infty[1]$-algebra:

$$L[1] = s^{-1}L, \qquad q_1 = -d, \qquad sq_2(s^{-1}x, s^{-1}y) = (-1)^{\overline{x}} s^{-1}[x, y].$$

Let $Q \in \operatorname{Coder}^1(\overline{S^c}(L[1]), \overline{S^c}(L[1]))$ be the Leibniz expansion of $q_1 + q_2$ and consider the differential $\delta = [Q, -]$ on the graded Lie algebra $\operatorname{Coder}^*(S^c(L[1]), S^c(L[1]))$; we shall call $(\operatorname{Coder}^*(S^c(L[1]), S^c(L[1])), \delta, [-,-])$ the **Chevalley–Eilenberg DG-Lie algebra** of L. Since $Q(1) = 0$, for every $h \in \operatorname{Coder}^*(S^c(L[1]), S^c(L[1]))$ we have

$$(\delta h)(1) = [Q, h](1) = Q(h(1)) = q_1(h(1)),$$

and this implies that the linear map of degree 0

$$ev_1 : \operatorname{Coder}^*(S^c(L[1]), S^c(L[1])) \to L[1], \qquad ev_1(h) = h(1),$$

is a morphism of DG-vector spaces. According to Theorem 11.7.4 the map ev_1 is properly defined and its kernel is the DG-Lie subalgebra $\operatorname{Coder}^*(\overline{S^c}(L[1])\overline{S^c}(L[1]))$. Moreover, ev_1 is surjective and, as a morphism of graded vector spaces, a right inverse of ev_1 is given by the map:

$$\sigma : L[1] \to \operatorname{Coder}^*(S^c(L[1]), S^c(L[1])), \qquad \sigma_{s^{-1}x}(v) = s^{-1}x \odot v, \quad x \in L.$$

Lemma 12.6.3 *In the above notation, the linear map*

$$\boldsymbol{i} : L \to Coder^*(S^c(L[1]), S^c(L[1])), \qquad \boldsymbol{i}_x = \sigma_{s^{-1}x},$$

is a Cartan homotopy with boundary

$$\boldsymbol{l} : L \to Coder^*(\overline{S^c}(L[1]), \overline{S^c}(L[1])), \quad \boldsymbol{l}_x = [Q, \boldsymbol{i}_x] + \boldsymbol{i}_{dx} = [Q, \sigma_{s^{-1}x}] - \sigma_{q_1(s^{-1}x)}.$$

Proof Since the image of σ is an abelian graded Lie subalgebra, in order to prove that $\boldsymbol{i}$ is a Cartan homotopy we only need to show that $\boldsymbol{i}_{[x,y]} = [\boldsymbol{i}_x, [Q, \boldsymbol{i}_y]]$. We have already proved in Eq. (11.14) that for every $u, v, y_1, \ldots, y_s \in L[1]$ we have

$$p_{L[1]}[[Q, \sigma_u], \sigma_v](y_1 \odot \cdots \odot y_s) = q_{s+2}(u, v, y_1, \ldots, y_s),$$

where $p_{L[1]} : S^c(L[1]) \to L[1]$ is the canonical projection. Since $q_n = 0$ for $n > 2$, we get $[[Q, \sigma_u], \sigma_v] = \sigma_{q_2(u,v)}$. Therefore, for $u = s^{-1}x$ and $v = s^{-1}y$ we obtain

$$[\boldsymbol{i}_x, [Q, \boldsymbol{i}_y]] = [\sigma_u, [Q, \sigma_v]] = -(-1)^{\overline{u}\,(\overline{v}+1)}[[Q, \sigma_v], \sigma_u,]$$
$$= -(-1)^{\overline{u}\,(\overline{v}+1)}\sigma_{q_2(v,u)} = -(-1)^{\overline{u}}\sigma_{q_2(u,v)} = \boldsymbol{i}_{[x,y]}.$$

Moreover,

$$\boldsymbol{l}_x(1) = [Q, \sigma_{s^{-1}x}](1) - \sigma_{q_1(s^{-1}x)}(1) = Q(s^{-1}x) - q_1(s^{-1}x) = 0,$$

while for every $v_1, v_2, \ldots \in L[1]$ we have

$$p_{L[1]}\boldsymbol{l}_x(v_1) = q_2(s^{-1}x, v), \qquad p_{L[1]}\boldsymbol{l}_x(v_1 \odot \cdots \odot v_h) = 0 \quad h > 1.$$

□

Proposition 12.6.4 *Every differential graded Lie algebra L is quasi-isomorphic to the homotopy fibre of the inclusion of DG-Lie algebras*

$$Coder^*(\overline{S^c}(L[1]), \overline{S^c}(L[1])) \xrightarrow{\chi} Coder^*(S^c(L[1]), S^c(L[1])).$$

Proof Since the Cartan homotopy of Lemma 12.6.3 induces an isomorphism of DG-vector spaces $L \simeq \operatorname{coker}(\chi)[-1]$, the proposition is a particular case of Lemma 8.8.10, with

$$M = \operatorname{Coder}^*(S^c(L[1]), S^c(L[1])), \quad H = \operatorname{Coder}^*(\overline{S^c}(L[1]), \overline{S^c}(L[1])).$$

□

The first consequence of Proposition 12.6.4 is that if the morphism χ is injective in cohomology, then by Theorem 6.1.6 the DG-Lie algebra L is quasi-isomorphic to an abelian DG-Lie algebra. It is easy to prove that χ is injective in cohomology if and only if $ev_1\colon \operatorname{Coder}^*(S^c(L[1]), S^c(L[1])) \to L[1]$ is surjective in cohomology.

Theorem 12.6.5 *Two DG-Lie algebras are weakly equivalent as L_∞-algebras if and only if are quasi-isomorphic as DG-Lie algebras.*

Proof We only need to prove the converse of Corollary 12.6.2, namely that if L, M are DG-Lie algebras and there exists an L_∞ weak equivalence $f\colon L \rightsquigarrow M$, then L is quasi-isomorphic to M. Consider first the case where the linear part $f_1\colon L \to M$ is surjective; under this assumption, by Corollary 12.4.4 there exists an L_∞-morphism $g\colon M \rightsquigarrow L$ such that $fg = \mathrm{Id}_M$.

Denoting by $F\colon S^c(L[1]) \to S^c(M[1])$ and $G\colon S^c(M[1]) \to S^c(L[1])$ the multiplicative expansions of f and g respectively we have $FG = \mathrm{Id}$ and $(GF)^2 = GF$. Notice that

$$f_1\colon \{x \in L[1] \mid x = g_1 f_1(x)\} = g_1(M[1]) \to M[1]$$

is an isomorphism. Let $\mathcal{L} \subset \operatorname{Coder}^*(S^c(L[1]), S^c(L[1]))$ be the DG-Lie subalgebra of coderivations η such that $GF\eta = \eta GF$. By Proposition 11.7.7 we have a short exact sequence

$$0 \to \{\eta \in \mathcal{L} \mid \eta(1) = 0\} \to \mathcal{L} \xrightarrow{\eta \mapsto \eta(1)} g_1(M[1]) \to 0.$$

On the other hand, every $\eta \in \mathcal{L}$ factors to a unique $\eta' \in \mathrm{Coder}^*(S^c(M[1]), S^c(M[1]))$ such that $F\eta = \eta' F$; in fact, whenever $F(x) = 0$ we have $GF\eta(x) = \eta GF(x) = 0$ and therefore also $F\eta(x) = 0$. In conclusion, we have a commutative diagram of DG-Lie algebras

$$\begin{array}{ccc}
\mathrm{Coder}^*(\overline{S^c}(M[1]), \overline{S^c}(M[1])) & \longrightarrow & \mathrm{Coder}^*(S^c(M[1]), S^c(M[1])) \\
\uparrow & & \uparrow \\
\{\eta \in \mathcal{L} \mid \eta(1) = 0\} & \longrightarrow & \mathcal{L} \\
\downarrow & & \downarrow \\
\mathrm{Coder}^*(\overline{S^c}(L[1]), \overline{S^c}(L[1])) & \longrightarrow & \mathrm{Coder}^*(S^c(L[1]), S^c(L[1]))
\end{array}$$

inducing quasi-isomorphisms between the homotopy fibres of the horizontal inclusions; the conclusion now follows from Proposition 12.6.4.

In general, we can always find an acyclic DG-vector space H and a surjective quasi-isomorphism $L \times H \to M$ extending f_1. Considering H as an abelian DG-Lie algebra, then L and $L \times H$ are quasi-isomorphic, while by Lemma 12.5.2 we can extend f to an L_∞-morphism $L \times H \rightsquigarrow M$ with linear part a surjective quasi-isomorphism. □

12.7 Homotopy Abelian DG-Lie and L_∞-Algebras

There exists the analogous notion of formality and homotopy abelianity also for L_∞-algebras.

Definition 12.7.1 An L_∞-algebra $L = (L, l_1, l_2, \ldots)$ is called **abelian** if $l_n = 0$ for every $n \geq 2$. It is called **homotopy abelian** if it is weakly equivalent to an abelian L_∞-algebra.

In particular, a minimal L_∞-algebra is abelian if and only if the L_∞-structure is trivial. For differential graded Lie algebras, by Theorem 12.6.5 the above notion of homotopy abelianity is the same as the one introduced in Definition 6.2.1.

Lemma 12.7.2 *For an L_∞-algebra $L = (L, l_1, l_2, \ldots)$ the following conditions are equivalent:*

1. *there exists an L_∞-isomorphism $f\colon (L, l_1, l_2, l_3, \ldots) \rightsquigarrow (L, l_1, 0, 0, \ldots)$ with linear component f_1 equal to the identity;*

2. *L is homotopy abelian;*
3. *the minimal model of L is abelian.*

Proof The implication (1) $\Rightarrow$ (2) is trivial. Since the isomorphism class of the minimal model is invariant under weak equivalences and every abelian L_∞-algebra is weakly equivalent to a minimal abelian L_∞-algebra we have (2)$\Rightarrow$(3).

Finally, if condition (3) holds, then according to Corollary 12.5.4 there exists an L_∞-isomorphism $f: (L, l_1, l_2, \ldots) \rightsquigarrow (M, d, 0, \ldots)$, where $(M, d, 0, \ldots)$ is the product of an abelian and a linear contractible L_∞-algebras. Taking the composition with the linear morphism of L_∞-algebras $f_1^{-1}: (M, d, 0, \ldots) \to (L, l_1, 0, \ldots)$ we get (1). □

By the homotopy classification of L_∞-algebras we get the following extension of Corollary 6.1.3.

Theorem 12.7.3 *Let $f: (L, l_1, l_2, \ldots) \rightsquigarrow (M, m_1, m_2, \ldots)$ be an L_∞-morphism of L_∞-algebras:*

1. *if $(M, m_1, m_2, \ldots)$ is homotopy abelian and $f_1: (L, l_1) \to (M, m_1)$ is injective in cohomology, then also $(L, l_1, l_2, \ldots)$ is homotopy abelian.*
2. *if $(L, l_1, l_2, \ldots)$ is homotopy abelian and $f_1: (L, l_1) \to (M, m_1)$ is surjective in cohomology, then also $(M, m_1, m_2, \ldots)$ is homotopy abelian.*

Proof Passing to minimal models, the proof follows trivially from Lemma 12.7.2. □

A necessary condition for a differential graded Lie algebra to be homotopy abelian is that the bracket is trivial in cohomology; therefore if $(L, d, [-, -])$ is a homotopy abelian DG-Lie algebra, then there exists a morphism $h: \bigwedge^2 L \to L$ of degree -1 such that $[a, b] = dh(a, b) + h(da, b) + (-1)^{\bar{a}} h(a, db)$ for every $a, b \in L$.

As we have already seen in Example 6.7.4 such a condition is not sufficient. However, it becomes sufficient under some additional assumptions on h and, in order to describe one of them in a convenient way, we introduce the following notation.

Definition 12.7.4 If V is a graded vector space and $f: V \times V \times V \to W$ is a trilinear map, we denote by $\oint f$ the composition of f with the sum over cyclic permutations, always taking care of Koszul signs:

$$\oint f(v_1, v_2, v_3) = f(v_1, v_2, v_3) + (-1)^{\overline{v_1}(\overline{v_2}+\overline{v_3})} f(v_2, v_3, v_1) + (-1)^{\overline{v_3}(\overline{v_1}+\overline{v_2})} f(v_3, v_1, v_2).$$

Notice that, using the above formalism, the Jacobi identity in a differential graded Lie algebra becomes

$$\oint [[a, b], c] = 0.$$

Proposition 12.7.5 *Let $L = (L, d, [-, -])$ be a differential graded Lie algebra. If there exists a bilinear map $h: L \times L \to L$ of degree -1 such that:*

1. $h(a,b) = -(-1)^{\overline{a}\,\overline{b}} h(b,a)$,
2. $[a,b] = dh(a,b) + h(da,b) + (-1)^{\overline{a}} h(a,db)$,
3. $\oint [h(a,b),c] + \oint h([a,b],c) = 0$.

Then L is homotopy abelian.

Proof It is convenient to consider the $L_\infty[1]$-algebra $(V, q_1, q_2, 0, \dots)$ corresponding to L via décalage isomorphisms: $V = L[1]$, $q_1 = -d$ and $q_2(a,b) = (-1)^{\overline{a}}[a,b]$. The "homotopy" h corresponds to the map

$$r \in \mathrm{Hom}^0_{\mathbb{K}}(V^{\odot 2}, V), \qquad r(a,b) = (-1)^{\overline{a}} h(a,b),$$

and it is straightforward to check that conditions (2) and (3) above are described in terms of the Nijenhuis–Richardson bracket by the equations:

$$[r, q_1]_{NR} = q_2, \qquad [r, q_2]_{NR} = 0.$$

Let's denote by $R, Q_1, Q_2 \in \mathrm{Coder}^*(\overline{S^c}(V), \overline{S^c}(V))$ the Leibniz expansions of r, q_1, q_2 respectively. We have that $R(V^{\odot n}) \subset V^{\odot n-1}$ and then it is well defined an isomorphism of graded coalgebras $e^R \colon \overline{S^c}(V) \to \overline{S^c}(V)$. To conclude the proof it is sufficient to prove that

$$Q_1 + Q_2 = e^R \circ Q_1 \circ e^{-R} = e^{[R,-]}(Q_1).$$

Taking corestrictions the above equality becomes an equation involving the Nijenhuis-Richardson bracket; namely, we get the equation

$$q_1 + q_2 = e^{[r,-]_{NR}}(q_1) = q_1 + [r, q_1]_{NR} + \frac{1}{2}[r, [r, q_1]_{NR}]_{NR} + \cdots,$$

which is clearly equivalent to $[r, q_1]_{NR} = q_2$ and $[r, q_2]_{NR} = 0$. □

12.8 Exercises

12.8.1 Let (C, Δ, d) be a differential graded coalgebra. Prove that the triple $(L, \delta, [\cdot,\cdot])$, where:

$$L = \oplus_{n\in\mathbb{Z}} \mathrm{Coder}^n(C, C), \quad [f, g] = fg - (-1)^{\overline{g}\,\overline{f}} gf, \quad \delta(f) = [d, f],$$

is a differential graded Lie algebra.

12.8.2 Give examples of two strict morphisms of formal pointed DG-manifolds such that their composition is not strict.

12.8.3 Show that, for L_∞-algebras, the properties of nilpotency, abelianity and linear contractability are not invariant under L_∞-isomorphisms.

12.8.4 Let $\mathbb{K}$ be any field of characteristic $\neq 2$ and denote by A, B, H the usual basis of the Lie algebra $\mathfrak{sl}_2(\mathbb{K})$, described in Example 2.2.5. Denote by $\mathfrak{l} \subset \mathfrak{sl}_2(\mathbb{K})$ the Lie subalgebra generated by A, H and by $\mathfrak{h} = \mathbb{K}H$, $\mathfrak{a} = \mathbb{K}A$. Prove that there exists a DG-Lie algebra structure of $\mathfrak{l} \oplus \mathfrak{h}[-1]$ such that the inclusion $i\colon \mathfrak{a} \to \mathfrak{l} \oplus \mathfrak{h}[-1]$ is a quasi-isomorphism of DG-Lie algebras and that there does not exist any quasi-isomorphism of DG-Lie algebras $\mathfrak{l} \oplus \mathfrak{h}[-1] \to \mathfrak{a}$.

12.8.5 Give an example of a differential graded Lie algebra L and two DG-Lie subalgebras $M, N \subset L$ such that $L = N \oplus M$ as differential graded vector spaces and such that there does not exist any L_∞-morphism $f\colon L \rightsquigarrow M$ with linear part the natural projection.

12.8.6 Prove that, if $\pi : M \to L$ is a surjective morphism of DG-Lie algebras, then L is quasi-isomorphic to the homotopy fibre of the inclusion of DG-Lie algebras $H \xrightarrow{\chi} \operatorname{Coder}^*(S^c(M[1]), S^c(M[1]))$, where

$$H = \{\eta \in \operatorname{Coder}^*(S^c(M[1]), S^c(M[1])) \mid \pi(\eta(1)) = 0\}.$$

12.8.7 Let $f : L \rightsquigarrow M$ be an L_∞-morphism of L_∞-algebras and assume that M is homotopy abelian. Denote by $\phi\colon H^*(L) \to H^*(M)$ the morphism induced by the linear component f_1. Prove that there exists a minimal model $(H^*(L), 0, l_2, l_3, \ldots)$ of L such that $\phi l_i = 0$ for every i. (Hint: prove the analogous statement for $L_\infty[1]$-algebras. It is not restrictive to assume L, M minimal, and hence that $\phi = f_1$. Consider the composition of f with a projection $p\colon M \to f_1(L)$ and apply Lemma 12.1.4.)

References

The definition of a formal pointed DG-manifold is taken from [156], although the same object was called a Q-manifold in [153]. I learned Theorem 12.1.7 from Schuhmacher's thesis [241], where it is also implicitly proved that, with the notion of fibration and weak equivalence of Definition 12.1.5, the category of formal pointed DG-manifolds is a pointed category of fibrant objects, cf. also [13].

The original unabridged version of the Hitchin–Martinengo L_∞-morphism was introduced in [191] as a tool for proving some deformation properties of the Hitchin map; see also [52].

The proof of Theorem 12.4.1 was privately communicated to the author by Fiorenza and has been recently used by Bandiera in [11] for an improvement of Theorem 10.7.7. The same proof was also proposed in the preprint version of [74] and then removed under explicit request by the referee.

The proof of Theorem 12.6.5 follows an idea by Bandiera. We refer to [151, 170] for the study of bar and cobar resolutions and for the "standard" proof that two differential graded Lie algebras are quasi-isomorphic if and only if they are homotopy equivalent as L_∞-algebras.

The criterion of Proposition 12.7.5 is taken from [76], while the notation introduced in Definition 12.7.4 is taken from [158]. We also mention another nice sufficient condition for homotopy abelianity proved in [134], namely that a DG-Lie algebra is homotopy abelian if the adjoint map $L \xrightarrow{\text{ad}} \operatorname{Der}^*(L, L)$ is trivial in cohomology.

Chapter 13
Formal Kuranishi Families and Period Maps

After the results regarding coalgebras and L_∞-morphisms proved in Chaps. 11 and 12 we are ready to give some applications of L_∞-algebras in deformation theory. The first two sections of this chapter are devoted to the proof that every L_∞-morphism induces natural transformations of both Maurer–Cartan and deformation functors, together with an interpretation of the formal Kuranishi family in terms of homotopy transfer of L_∞ structure.

In the remaining sections we go further into the study of Cartan homotopies and show that this notions is very useful in the framework of L_∞-algebras, and in particular in the description of the infinitesimal Griffiths' period map as a morphism of deformation theories.

13.1 L_∞-Morphisms and Deformation Functors

One of the most important features of L_∞-morphisms is that their restitutions (Sect. 11.6) commute with Maurer–Cartan and deformation functors. This means that for every L_∞-morphism $f\colon (V, q) \rightsquigarrow (W, r)$ of $L_\infty[1]$-algebras, every $A \in \mathbf{Art}_{\mathbb{K}}$ and every $x \in \mathrm{MC}_V(A) \subset V^0 \otimes \mathfrak{m}_A$, the element

$$\mathrm{MC}_f(x) := \sum_{n>0} \frac{1}{n!} f_n(x, \ldots, x) \in W^0 \otimes \mathfrak{m}_A \tag{13.1}$$

belongs to $\mathrm{MC}_W(A)$; this construction is functorial and factors to a natural transformation $\mathrm{Def}_V \to \mathrm{Def}_W$.

The proof of these facts is a consequence of Theorem 11.6.3 and Proposition 11.6.4. Denote by $Q \in \mathrm{Coder}^1(\overline{S^c}(V), \overline{S^c}(V))$ and $R \in \mathrm{Coder}^1(\overline{S^c}(W), (\overline{S^c}(W))$ the Leibniz expansions of q and r respectively and by $F\colon (\overline{S^c}(V), Q) \to$

M. Manetti, *Lie Methods in Deformation Theory*, Springer Monographs in Mathematics,
https://doi.org/10.1007/978-981-19-1185-9_13

$(\overline{S^c}(W), R)$ the multiplicative expansion of f; by definition, Q, R are the coderivations corestricting to q, r respectively and F is the morphism of differential graded coalgebras corestricting to f. The equality $FQ = RF$ is therefore equivalent to $fQ = rF$.

Since the maximal ideal of $A \in \mathbf{Art}_{\mathbb{K}}$ belongs to the category **NCGA** of nilpotent finite-dimensional graded commutative algebras, the restitutions of f, q, r give three natural transformations

$$\begin{array}{ccc} V^0 \otimes \mathfrak{m}_A & \xrightarrow{\mathrm{MC}_f} & W^0 \otimes \mathfrak{m}_A \\ \downarrow{\scriptstyle \mathrm{MC}_q} & & \downarrow{\scriptstyle \mathrm{MC}_r} \\ V^1 \otimes \mathfrak{m}_A & & W^1 \otimes \mathfrak{m}_A \end{array} \qquad \begin{array}{l} \mathrm{MC}_f(x) = \sum_{n>0} \frac{1}{n!} f_n(x^{\odot n}) \\ \mathrm{MC}_q(x) = \sum_{n>0} \frac{1}{n!} q_n(x^{\odot n}) \\ \mathrm{MC}_r(y) = \sum_{n>0} \frac{1}{n!} r_n(y^{\odot n}), \end{array}$$

where the $\odot$-powers of x and y are made in the nilpotent algebras $S(V) \otimes \mathfrak{m}_A$ and $S(W) \otimes \mathfrak{m}_A$ respectively. Recall that, by definition, the Maurer–Cartan functors associated to the $L_\infty[1]$-algebras (V, q) and (W, r) are:

$$\begin{aligned} \mathrm{MC}_V(A) &= \{x \in V^0 \otimes \mathfrak{m}_A \mid \mathrm{MC}_q(x) = 0\}, \\ \mathrm{MC}_W(A) &= \{y \in W^0 \otimes \mathfrak{m}_A \mid \mathrm{MC}_r(y) = 0\}. \end{aligned}$$

According to Theorem 11.6.3 we have $\mathrm{MC}_{rF} = \mathrm{MC}_r \circ \mathrm{MC}_f$, while by Proposition 11.6.4, for every $x \in V^0 \otimes \mathfrak{m}_A$ we have

$$\begin{aligned} \mathrm{MC}_r(\mathrm{MC}_f(x)) = \mathrm{MC}_{rF}(x) = \mathrm{MC}_{fQ}(x) &= f(Q(e^{\odot x} - 1)) \\ &= f(\mathrm{MC}_q(x) \odot e^{\odot x}). \end{aligned}$$

If $\mathrm{MC}_q(x) = 0$ then also $\mathrm{MC}_r(\mathrm{MC}_f(x)) = 0$ and this gives the following result.

Lemma 13.1.1 *In the above notation, the restitution morphism*

$$V^0 \otimes \mathfrak{m}_A \xrightarrow{MC_f} W^0 \otimes \mathfrak{m}_A, \qquad A \in \mathbf{Art}_{\mathbb{K}}, \quad MC_f(x) = \sum_{n>0} \frac{1}{n!} f_n(x^{\odot n}),$$

preserves the solutions of the Maurer–Cartan equations.

If $g\colon (W, r) \rightsquigarrow (U, s)$ is an L_∞-morphism of $L_\infty[1]$-algebras and $g = \sum g_i$, the map gF is the composite L_∞-morphism $gf\colon (V, q) \rightsquigarrow (U, s)$. Again by Theorem 11.6.3 we have $\mathrm{MC}_g \circ \mathrm{MC}_f = \mathrm{MC}_{gF}$ and therefore the restitution is functorial with respect to composition of L_∞-morphisms.

Remark 13.1.2 In the above situation, given a small extension

$$0 \to I \to A \xrightarrow{\beta} B \to 0$$

in $\mathbf{Art}_{\mathbb{K}}$ and an element $x \in V^0 \otimes \mathfrak{m}_A$ such that $\beta(x) \in \mathrm{MC}_V(B)$, then its image $\mathrm{MC}_q(x) \in H^1(V) \otimes I$ is the obstruction to the lifting of $\beta(x)$ described in Sect. 10.5, while $\mathrm{MC}_r(\mathrm{MC}_f(x))$ is the obstruction to the lifting of $\mathrm{MC}_f(\beta(x)) \in \mathrm{MC}_W(B)$. Since the ideal I is annihilated by $\mathfrak{m}_A$ we have $\mathrm{MC}_q(x) \odot e^{\odot x} = \mathrm{MC}_q(x)$, hence

$$\mathrm{MC}_r(\mathrm{MC}_f(x)) = f(\mathrm{MC}_q(x) \odot e^{\odot x}) = f_1(\mathrm{MC}_q(x)).$$

In other words, the linear component f_1 commutes with obstruction maps.

Recall that by Definition 10.5.4 the deformation functor

$$\mathrm{Def}_V : \mathbf{Art}_{\mathbb{K}} \to \mathbf{Set}$$

controlled by an $L_\infty[1]$-algebra is the quotient of MC_V under the equivalence relation generated by homotopy equivalence of Maurer–Cartan elements.

Lemma 13.1.3 *Let V be an $L_\infty[1]$-algebra. Then the homotopy equivalence in MC_V is an equivalence relation preserved by L_∞-morphisms.*

Proof The only nontrivial property of homotopy equivalence is transitivity. The commutative diagram of surjective quasi-isomorphisms of DG-algebras

$$\begin{array}{ccc} \mathbb{K}[t,s,dt,ds] & \xrightarrow{t\mapsto 0} & \mathbb{K}[s,ds] \\ \downarrow{\scriptstyle s\mapsto 0} & & \downarrow{\scriptstyle s\mapsto 0} \\ \mathbb{K}[t,dt] & \xrightarrow{t\mapsto 0} & \mathbb{K} \end{array}$$

induces a surjective quasi-isomorphism of DG-algebras

$$\mathbb{K}[t,s,dt,ds] \xrightarrow{p} \mathbb{K}[t,dt] \times_{\mathbb{K}} \mathbb{K}[s,ds]$$

and therefore a surjective linear L_∞-morphism

$$p\colon \mathbb{K}[t,s,dt,ds] \otimes V \to (\mathbb{K}[t,dt] \times_{\mathbb{K}} \mathbb{K}[s,ds]) \otimes V.$$

that is also a weak equivalence; the $L_\infty[1]$ structure in $\mathbb{K}[t,s,dt,ds] \otimes V$ is the one obtained by scalar extension, see Remark 10.4.7 and Eq. (10.18).

By Corollary 12.4.4 there exists an L_∞-morphism

$$g\colon (\mathbb{K}[t,dt] \times_{\mathbb{K}} \mathbb{K}[s,ds]) \otimes V \rightsquigarrow \mathbb{K}[t,s,dt,ds] \otimes V$$

such that pg is the identity. Suppose now $A \in \mathbf{Art}_{\mathbb{K}}$ and let $x, y, z \in \mathrm{MC}_V(A)$ be such that x is homotopy equivalent to y and y is homotopy equivalent to z; by definition there exists $\xi(t) \in \mathrm{MC}_{V[t,dt]}(A)$ and $\eta(s) \in \mathrm{MC}_{V[s,ds]}(A)$ such that $\xi(1) = x$, $\xi(0) = \eta(0) = y$ and $\eta(1) = z$. In particular, the pair $(\xi(t), \eta(s))$ is a Maurer–Cartan element in

$$(\mathbb{K}[t,dt] \times_{\mathbb{K}} \mathbb{K}[s,ds]) \otimes V \otimes \mathfrak{m}_A,$$

and then $\psi(s,t) = \mathrm{MC}_g(\xi(t), \eta(s))$ is a Maurer–Cartan element of the DG-Lie algebra $\mathbb{K}[t, s, dt, ds] \otimes V \otimes \mathfrak{m}_A$ such that $\psi(t, 0) = \xi(t)$ and $\psi(0, s) = \eta(s)$.

Now $\mu(t) = \psi(1 - t, t)$ is a Maurer–Cartan element of $\mathbb{K}[t, dt] \otimes V \otimes \mathfrak{m}_A$ and $\mu(0) = \xi(1) = x$, $\mu(1) = \eta(1) = z$.

Every L_∞-morphism $V \rightsquigarrow W$ induces a commutative diagram

$$\begin{array}{ccccc} V & \xleftarrow{t\mapsto 0} & V[t, dt] & \xrightarrow{t\mapsto 1} & V \\ \Big\downarrow\wr & & \Big\downarrow\wr & & \Big\downarrow\wr \\ W & \xleftarrow{t\mapsto 0} & W[t, dt] & \xrightarrow{t\mapsto 1} & W \end{array}$$

and taking the corresponding commutative diagram of Maurer-Cartan functors we get that homotopy equivalence is preserved by L_∞-morphisms. □

We are now ready for the main result of this section.

Theorem 13.1.4 *Every L_∞-morphism $f : V \rightsquigarrow W$ induces a natural transformation of functors*

$$MC_f : MC_V \to MC_W, \qquad x \mapsto \sum_n \frac{1}{n!} f_n(x^{\odot n}),$$

that factors to a natural transformation

$$Def_f : Def_V \to Def_W.$$

If f is a weak equivalence of L_∞-algebras, then $Def_f : Def_V \to Def_W$ is an isomorphism of functors.

Proof The first part is clear and already proved. For the second part notice that if the $L_\infty[1]$-algebra U is linearly contractible, then $\mathrm{Def}_U = 0$. Since the functor Def commutes with direct products, $\mathrm{Def}_{U\times V} = \mathrm{Def}_U \times \mathrm{Def}_V$, it is sufficient to apply Corollary 12.5.4 and the fact that a weak equivalence of minimal $L_\infty[1]$-algebras is an L_∞-isomorphism. □

13.2 Formal Kuranishi Families

According to Schlessinger's theorem (Theorem 3.5.10), if $F : \mathbf{Art}_{\mathbb{K}} \to \mathbf{Set}$ is a deformation functor with finite-dimensional tangent vector space, then there exist a local complete Noetherian $\mathbb{K}$-algebra R with residue field $\mathbb{K}$ and a smooth natural transformation $h_R \to F$ inducing an isomorphism on tangent spaces $T^1 h_R = T^1 F$. Moreover, R is unique up to non-canonical isomorphism. It is important to point out that this is purely an existence theorem and its proof, given in Sect. B.1, is usually very far to be constructive. A slight improvement of Schlessinger's theorem can be given by using obstruction theory.

Theorem 13.2.1 *Let $F\colon \mathbf{Art}_{\mathbb{K}} \to \mathbf{Set}$ be a deformation functor with finite-dimensional tangent vector space T^1F and with a complete obstruction space V. If $x_1, \dots, x_n$ is a coordinate system in T^1F there exists a linear map*

$$q\colon V^\vee \to \mathbb{K}[[x_1, \dots, x_n]]$$

such that the ideal $I = (q(V^\vee))$, generated by the image of q, is contained in $(x_1, \dots, x_n)^2$ and such that, writing $R = \mathbb{K}[[x_1, \dots, x_n]]/I$, there exists a smooth natural transformation $h_R \to F$ inducing an isomorphism on tangent spaces.

Proof Let $\eta\colon h_R \to F$ be the smooth natural transformation given by Schlessinger's theorem. Write $P = \mathbb{K}[[x_1, \dots, x_n]]$ and let $\mathfrak{m}_P \subset P$ be the maximal ideal. Since $T^1F = T^1h_R = (\mathfrak{m}_R/\mathfrak{m}_R^2)^\vee$, the coordinates $x_1, \dots, x_n$ give a basis of $\mathfrak{m}_R/\mathfrak{m}_R^2$; choosing a lifting of them to $\mathfrak{m}_R$ we get an isomorphism

$$R \cong \frac{P}{I}, \quad \text{for some ideal} \quad I \subset \mathfrak{m}_P^2.$$

According to Example 3.6.9 the universal obstruction space of the pro-representable functor h_R is isomorphic to $\operatorname{Hom}_P(I, \mathbb{K}) = \operatorname{Hom}_{\mathbb{K}}\left(\frac{I}{\mathfrak{m}_P I}, \mathbb{K}\right)$ and then the smooth natural transformation η induces an injective linear obstruction map

$$o_\eta\colon \operatorname{Hom}_P(I, \mathbb{K}) \to V.$$

Thus its dual map

$$V^\vee \xrightarrow{o_\eta^T} \operatorname{Hom}_P(I, \mathbb{K})^\vee = \frac{I}{\mathfrak{m}_P I}$$

is surjective and can be lifted to a linear map $q\colon V^\vee \to I$. By Nakayama's lemma the image of q generates the ideal I. □

In the remaining part of this section we give a direct and more constructive proof of Theorem 13.2.1 under the additional assumption that $F = \operatorname{Def}_L$ for a certain differential graded Lie algebra L, taking $H^2(L)$ as complete obstruction space.

First we reproduce the classical argument by Nijenhuis and Richardson, and then we give an interpretation in terms of L_∞ structures.

Let $(L, d, [-, -])$ be a differential graded Lie algebra and choose a splitting of the underlying cochain complex (L, d); this means that for every integer n, writing $Z^n = \{x \in L^n \mid dx = 0\}$ and $B^n = \{dx \mid x \in L^{n-1}\}$, we have the additional data of two direct sum decompositions of vector spaces:

$$Z^n = B^n \oplus \mathcal{H}^n, \qquad L^n = Z^n \oplus C^n\,. \tag{13.2}$$

Clearly, $\mathcal{H}^n$ is isomorphic to $H^n(L)$ and $d\colon C^n \to B^{n+1}$ is an isomorphism.

Let $H\colon L^n \to \mathcal{H}^n$ be the projection with kernel $B^n \oplus C^n$ and let $\delta\colon L^{n+1} \to L^n$ be the composition of:

1. the projection $L^{n+1} \to B^{n+1}$ with kernel $C^{n+1} \oplus \mathcal{H}^{n+1}$,
2. $d^{-1}\colon B^{n+1} \to C^n$,
3. the inclusion $C^n \to L^n$.

Notice that

$$B^n = d(L^{n-1}), \quad C^n = \delta(L^{n+1}), \quad \mathcal{H}^n = L^n \cap \ker d \cap \ker \delta,$$

and then the decomposition (13.2) can be recovered from the operator δ. Moreover, $d\delta d = d$, $\delta d\delta = \delta$, $\delta^2 = 0$ and $d\delta + \delta d = \mathrm{Id}_L - H$.

Recall, from Sect. 10.2, that the formal neighbourhood of a vector space V is, up to some canonical isomorphism, given by the functor of Artin rings:

$$\widehat{V}\colon \mathbf{Art}_{\mathbb{K}} \to \mathbf{Set}, \qquad \widehat{V}(A) = V \otimes \mathfrak{m}_A.$$

Lemma 13.2.2 *In the above notation, the natural transformation*

$$F\colon \widehat{L^1} \to \widehat{L^1}, \qquad L^1 \otimes \mathfrak{m}_A \ni x \mapsto F(x) = x + \frac{1}{2}\delta[x, x] \in L^1 \otimes \mathfrak{m}_A,$$

is an isomorphism of formal neighbourhoods.

Proof In view of the results of Chap. 3 it is convenient to consider $\widehat{L^1}$ as a homogeneous deformation functor. The morphism F is bijective on tangent spaces; moreover, $\widehat{L^1}$ is smooth and this implies that also F is smooth. Regarding the injectivity, if $x, y \in L^1 \otimes \mathfrak{m}_A$ and $F(x) = F(y)$, we have

$$y - x = \frac{1}{2}\delta([x, x] - [y, y]) = \frac{1}{2}\delta([x - y, x + y])$$

and then $x - y \in \mathfrak{m}_A^s$ implies $x - y \in \mathfrak{m}_A^{s+1}$. □

For later use we point out that the inverse natural transformation F^{-1} satisfies the equation:

$$F^{-1}(x) = x - \frac{1}{2}\delta[F^{-1}(x), F^{-1}(x)], \tag{13.3}$$

since, setting $x = F(y)$, the above equation becomes $y = F(y) - \frac{1}{2}\delta[y, y]$. Notice that (13.3) can be solved recursively as $F^{-1} = \sum_{n>0} f_n$ with

$$f_1(x) = x, \qquad f_n(x) = -\frac{1}{2}\sum_{i=1}^{n-1}\delta[f_i(x), f_{n-i}(x)], \qquad n > 1.$$

Definition 13.2.3 In the above notation, the formal **Kuranishi map** is the morphism of formal neighbourhoods $K: \widehat{\mathcal{H}^1} \to \widehat{\mathcal{H}^2}$:

$$K(x) = \frac{1}{2}H\left([F^{-1}(x), F^{-1}(x)]\right) = \frac{1}{2}H([x,x] - [x, \delta[x,x]] + \cdots).$$

The **Kuranishi functor** Kur: $\mathbf{Art}_{\mathbb{K}} \to \mathbf{Set}$ is the fibre of the Kuranishi map:

$$\begin{aligned}\mathrm{Kur}(A) &= \{x \in \mathcal{H}^1 \otimes \mathfrak{m}_A \mid H([F^{-1}(x), F^{-1}(x)]) = 0\}\\ &= \{x \in L^1 \otimes \mathfrak{m}_A \mid dx = \delta x = K(x) = 0\}.\end{aligned}$$

The Kuranishi map and functor are not canonically defined since they depend by the choice of the map δ; however, at the end of this section we shall be able to prove that different choices of δ, or equivalently different splittings of L, give isomorphic Kuranishi functors.

Assume now that $H^1(L)$ is a finite-dimensional vector space, then the functor $\mathrm{Kur} = h_R$ is pro-representable, where R is the local complete ring of the fibre of the Kuranishi map K. More precisely, fixing a basis $e = (e_1, \ldots, e_n)$ of $\mathcal{H}^1$ with associated coordinate system $x_1, \ldots, x_n$ and formal power series ring $P = \mathbb{K}[[x_1, \ldots, x_n]]$, we have an isomorphism of functors:

$$e: h_P \to \widehat{\mathcal{H}^1}, \quad (f: \mathbb{K}[[x_1, \ldots, x_n]] \to A) \mapsto e(f) = \sum_i e_i \otimes f(x_i) \in \mathcal{H}^1 \otimes \mathfrak{m}_A.$$

Similarly, if $\{\epsilon_j\}$ is an unordered basis of $\mathcal{H}^2$ we can write the Kuranishi map K in the form

$$K(\sum_i e_i \otimes t_i) = \sum_j \epsilon_j \otimes g_j(t_1, \ldots, t_n), \qquad t_i \in \mathfrak{m}_A,$$

where the g_j's are power series depending only on L, δ and the choice of the two bases e_i, ϵ_j. Since $K(x) = H([x,x]) + \textit{higher order terms}$, every power series g_i has multiplicity ≥ 2.

For any $f: P \to A$ we have $K(e(f)) = 0$ if and only if $f(g_j) = 0$ for every j and this implies that the Kuranishi functor Kur is pro-represented by the quotient of P by the ideal generated by the power series $g_j(t)$.

Proposition 13.2.4 *In the above situation, consider the differential graded Lie subalgebra $N \subset L$ defined as*

$$\begin{cases} N^i = 0 & \textit{for } i \leq 0,\\ N^1 = \ker(\delta: L^1 \to L^0), & \\ N^i = L^i & \textit{for } i \geq 2.\end{cases}$$

Then the isomorphism $F: \widehat{L^1} \to \widehat{L^1}$ of Lemma 13.2.2 induces an isomorphism of functors

$$F\colon Def_N = MC_N \xrightarrow{\simeq} \mathrm{Kur},$$

where both MC_N and Kur *are considered as subfunctors of $\widehat{L^1}$.*

Proof It is clear that $\mathrm{Def}_N = \mathrm{MC}_N$ since $N^0 = 0$. If $A \in \mathbf{Art}_{\mathbb{K}}$ and $x \in \mathrm{MC}_N(A)$, then $[x, x] = -2dx$, $\delta x = 0$ and therefore $F(x) \in \mathrm{Kur}(A)$, since:

$$dF(x) = dx + \frac{1}{2}d\delta[x, x] = dx - d\delta dx = 0,$$
$$\delta F(x) = \delta x + \frac{1}{2}\delta^2[x, x] = \delta x = 0,$$
$$K(F(x)) = \frac{1}{2}H([x, x]) = -H(dx) = 0.$$

Conversely, consider an element $x \in L^1 \otimes \mathfrak{m}_A$ such that $F(x) \in \mathrm{Kur}(A)$; we need to prove that $x \in \mathrm{MC}_N(A)$. Since $\delta^2 = 0$ we have $\delta x = \delta F(x) = 0$ and therefore $x \in N^1 \otimes \mathfrak{m}_A$.

Since $H([x, x]) = K(F(x)) = dF(x) = 0$ we have

$$dx + \frac{1}{2}[x, x] = dx + \frac{1}{2}(d\delta + \delta d)[x, x] = dF(x) + \frac{1}{2}\delta d[x, x] = \frac{1}{2}\delta d[x, x].$$

Since $H([x, x]) = dF(x) = [[x, x], x] = 0$ we have

$$\begin{aligned}\delta d[x, x] &= 2\delta[dx, x] = -\delta[d\delta[x, x], x] = -\delta\left[[x, x] - H([x, x]) - \delta d[x, x],\ x\right]\\ &= \delta[\delta d[x, x], x].\end{aligned}$$

By Bianchi identity we have

$$\delta d[x, x] = 2\delta[dx, x] = \delta[\delta d[x, x], x] - [[x, x], x] = \delta[\delta d[x, x], x]$$

and the same argument as used in the proof of the injectivity of F implies that if $z = \delta[z, x]$, $z \in L \otimes \mathfrak{m}_A$, then $z = 0$ and therefore

$$\delta d[x, x] = 0, \qquad dx + \frac{1}{2}[x, x] = \frac{1}{2}\delta d[x, x] = 0.$$

□

We have already seen in Example 6.4.7 that the inclusion $N \subset L$ induces a smooth morphism

$$\mathrm{MC}_N = \mathrm{Def}_N \to \mathrm{Def}_L,$$

which is an isomorphism on tangent spaces; clearly, the same holds to the composition

$$\mathrm{Kur} \xrightarrow{F^{-1}} \mathrm{MC}_N \to \mathrm{Def}_L$$

that, when $H^1(L)$ has finite dimension, gives a hull for the deformation functor Def_L.

The morphism F^{-1}, determined by Eq. (13.3), and the Kuranishi map K have a clear interpretation in the framework of homotopy transfer of $L_\infty[1]$ structures. To see this, let's denote by $(V, q_1, q_2, 0, \dots)$ the $L_\infty[1]$-algebra corresponding to the differential graded Lie algebra L:

$$V^i = L^{i+1}, \qquad q_1 = -d, \quad q_2(x, y) = -(-1)^{\deg_V(x)}[x, y].$$

Notice that $\delta q_1 + q_1\delta = H - \mathrm{Id}_V$. Denote as usual by $p\colon \overline{S^c}(V) \to V$ the projection and let $G\colon \overline{S^c}(V) \to \overline{S^c}(V)$ be the unique morphism of graded coalgebras such that $pG = p + \delta q_2 G$.

If $G^n_m\colon V^{\odot m} \to V^{\odot n}$ are the components of G, then by Corollary 11.5.4 for every $n \geq 2$ we have

$$\begin{aligned} &G^2_n(v_1, \dots, v_n)\\ &= \frac{1}{2}\sum_{a=1}^{n-1} \sum_{\sigma\in S(a,n-a)} \epsilon(\sigma) G^1_a(v_{\sigma(1)}, \dots, v_{\sigma(a)}) \odot G^1_{n-a}(v_{\sigma(a+1)}, \dots, v_{\sigma(n)}) \end{aligned}$$

and therefore

$$\begin{aligned} &G^1_n(v_1, \dots, v_n) = \delta q_2 G^2_n(v_1, \dots, v_n)\\ &= \frac{1}{2}\sum_{a=1}^{n-1} \sum_{\sigma\in S(a,n-a)} \epsilon(\sigma)\delta q_2(G^1_a(v_{\sigma(1)}, \dots, v_{\sigma(a)}), G^1_{n-a}(v_{\sigma(a+1)}, \dots, v_{\sigma(n)})). \end{aligned}$$

If $V_\#$ is the graded vector space V equipped with the trivial $L_\infty[1]$ structure, then the Maurer–Cartan functor $\mathrm{MC}_{V_\#}$ is exactly $\widehat{V^0} = \widehat{L^1}$ and the map $g = p_V G$ gives an L_∞-morphism $g\colon V_\# \rightsquigarrow V_\#$ whose action on the Maurer–Cartan functor is given by the restitution map

$$\begin{aligned} \mathrm{MC}_g(x) &= \sum_{n>0} \frac{1}{n!} G^1_n(x, \dots, x)\\ &= x + \sum_{n>1} \frac{1}{n!}\frac{1}{2}\sum_{a=1}^{n-1}\binom{n}{a}(-\delta[G^1_a(x, \dots, x), G^1_{n-a}(x, \dots, x)])\\ &= x - \frac{1}{2}\delta[\mathrm{MC}_g(x), \mathrm{MC}_g(x)]. \end{aligned}$$

Thus, according to (13.3), the morphism MC_g corresponds, at the level of Maurer–Cartan elements, to the natural transformation F^{-1}.

Let's now apply the machinery of homotopy transfer to the $L_\infty[1]$-algebra $(V, q_1, q_2, 0, \dots)$ and to the contraction

$$\mathcal{H}[1] \underset{\pi}{\overset{\imath}{\rightleftarrows}} V \circlearrowright \delta$$

This gives an $L_\infty[1]$ structure $(\mathcal{H}[1], 0, r_2, r_3, r_4, \ldots)$ and an L_∞-morphism $\imath\colon \mathcal{H}[1] \rightsquigarrow V$. The explicit recursive formulas given in Theorem 12.4.1 prove that the morphism $\imath$ is precisely the restriction of g to the graded subspace $\mathcal{H}[1]$.

The Maurer–Cartan elements of $\mathcal{H}[1]$ are the elements $x \in \mathcal{H}^1$ (for simplicity we omit the tensor products with maximal ideals of Artin rings) such that $\sum_{n>0} \frac{1}{n!} r_n(x, \ldots, x) = 0$. Again by Theorem 12.4.1 we have for every $n > 1$

$$r_n(x, \ldots, x) = Hq_2(G_n^2(x, \ldots, x)) = \frac{1}{2}\sum_{a=1}^{n-1}\binom{n}{a}(-H[G_a^1(x, \ldots, x), G_{n-a}^1(x, \ldots, x)])$$

and then

$$\begin{aligned}\sum_n \frac{1}{n!} r_n(x, \ldots, x) &= \frac{1}{2}\sum_n\sum_{a=1}^{n-1}\frac{1}{a!(n-a)!}(-H[G_a^1(x, \ldots, x), G_{n-a}^1(x, \ldots, x)])\\ &= -\frac{1}{2}H[\mathrm{MC}_g(x), \mathrm{MC}_g(x)]\end{aligned}$$

and we just recover, up to the correct décalage sign, the Kuranishi map. In particular, the Kuranishi functor Kur is canonically isomorphic to $\mathrm{MC}_{\mathcal{H}[1]}$.

Notice that the $L_\infty[1]$-algebra $\mathcal{H}[1]$ is a minimal model of V; in particular, the isomorphism class of Kur is independent of the choice of the splitting.

Example 13.2.5 (*Equivariant Kuranishi maps*) Let G be a group of automorphisms of a differential graded Lie algebra $(L, d, [-, -])$ and assume that there exists a splitting

$$Z^n = B^n \oplus \mathcal{H}^n, \qquad L^n = Z^n \oplus C^n$$

as in (13.2) such that $\mathcal{H}^n$ and C^n are G-invariant subspaces for every n; since we are in characteristic 0, this is always possible if G is a finite group.

Under this assumption, for every n, both the homotopy $\delta\colon L^{n+1} \to L^n$ and the harmonic projection $H\colon L^n \to \mathcal{H}^n$ commute with every element of G. It is clear from the above explicit formulas that both the construction of the Kuranishi map and the construction of the transferred L_∞ structure on $\mathcal{H}^*$ can be done in a G-equivariant way. In particular, there exists a G-action on the Kuranishi pro-representable functor which is compatible with the natural action of G on the deformation functor Def_L.

13.3 Cartan Homotopies and L_∞-Morphisms

We have introduced Cartan homotopies in Sect. 8.8 as a useful tool for proving obstruction vanishing theorems, for instance the Theorem 8.8.3. In this section we see that Cartan homotopies are a good source of (non-linear) explicit L_∞-morphisms between DG-Lie algebras.

Recall that, for L and M differential graded Lie algebras over the field $\mathbb{K}$, a linear map

$$\boldsymbol{i} \in \operatorname{Hom}_{\mathbb{K}}^{-1}(L, M)$$

is called a Cartan homotopy (Definition 8.8.1) if, for every $a, b \in L$, we have

$$[\boldsymbol{i}_a, \boldsymbol{i}_b] = 0, \qquad \boldsymbol{i}_{[a,b]} = [\boldsymbol{i}_a, d_M \boldsymbol{i}_b].$$

The boundary of a Cartan homotopy $\boldsymbol{i}$ as above is the morphism of DG-Lie algebras (see Lemma 8.8.6):

$$\boldsymbol{l} = d_M \boldsymbol{i} + \boldsymbol{i} d_L : L \to M.$$

Example 13.3.1 Let (A, d) be a (commutative) DG-algebra and consider the complex $A[-1]$ as an abelian DG-Lie algebra. Then the morphism

$$\boldsymbol{i} : A[-1] \to \operatorname{Hom}_{\mathbb{K}}^*(A, A), \qquad \boldsymbol{i}_a(b) = ab,$$

is a Cartan homotopy with a trivial boundary.

Example 13.3.2 Let X be a complex manifold with holomorphic tangent sheaf Θ_X and holomorphic de Rham complex (Ω_X^*, ∂). Then, for every open subset $U \subset X$, the contraction

$$\Theta_X(U) \otimes \Omega_X^k(U) \xrightarrow{\ \lrcorner\ } \Omega_X^{k-1}(U)$$

induces a linear map of degree -1:

$$\boldsymbol{i} : \Theta_X(U) \to \operatorname{Hom}_{\mathbb{K}}^*(\Omega_X^*(U), \Omega_X^*(U)), \qquad \boldsymbol{i}_\xi(\omega) = \xi \lrcorner \omega,$$

which is a Cartan homotopy in view of the holomorphic Cartan homotopy formulas (Proposition 8.5.4).

Lemma 13.3.3 *Let $\boldsymbol{i} : L \to M$ be a Cartan homotopy with boundary $\boldsymbol{l}$. Then*

$$\oint [\boldsymbol{i}_{[a,b]}, \boldsymbol{l}_c] = \oint [\boldsymbol{i}_a, \boldsymbol{l}_{[b,c]}] = 0$$

for every $a, b, c \in L$, where $\oint$ denotes the sum over cyclic permutations (Definition 12.7.4).

Proof Since $[\boldsymbol{i}_{[a,b]}, \boldsymbol{l}_c] = \boldsymbol{i}_{[[a,b],c]}$ and $[\boldsymbol{i}_a, \boldsymbol{l}_{[b,c]}] = \boldsymbol{i}_{[a,[b,c]]}$ it is sufficient to apply the operator $\boldsymbol{i}$ to the Jacobi identity on L. □

Theorem 13.3.4 *Let L, M be two DG-Lie algebras and let $\boldsymbol{i} : L \to M$ be a Cartan homotopy with boundary $\boldsymbol{l}$. Then, the sequence of linear maps $g_n : L^{\wedge n} \to M[t, dt]$:*

$$g_1(a) = t\,\boldsymbol{l}_a + dt\,\boldsymbol{i}_a, \qquad g_2(a \wedge b) = t(1-t)\,\boldsymbol{i}_{[a,b]}, \qquad g_n = 0 \ \ \textit{for} \ \ n \geq 3,$$

gives an L_∞*-morphism* $g\colon L \rightsquigarrow M[t, dt]$.

Proof We first note that, since $g_n = 0$ for $n \geq 3$, the condition (12.5) takes a simple form and it follows immediately from Proposition 12.3.2 that g is an L_∞-morphism if and only if:

1. g_1 is a morphism of complexes;
2. $dg_2(a \wedge b) + g_2(da \wedge b) + (-1)^{\bar{a}} g_2(a \wedge db) = g_1([a, b]) - [g_1(a), g_1(b)]$ for every $a, b \in L$;
3. $\oint [g_2(a \wedge b), g_1(c)] = \oint g_2([a, b] \wedge c)$ for every $a, b, c \in L$;
4. $g_2 \cup g_2 = 0$.

Since $\boldsymbol{l}$ is a morphism of differential graded Lie algebras and $d\boldsymbol{i} + \boldsymbol{i}d = \boldsymbol{l}$ for every $a \in L$ we have

$$\begin{aligned} d(g_1(a)) &= dt\boldsymbol{l}_a + td(\boldsymbol{l}_a) - dtd(\boldsymbol{i}_a) = dt\boldsymbol{l}_a + t\boldsymbol{l}_{da} - dt(\boldsymbol{l}_a - \boldsymbol{i}_{da}) = t\boldsymbol{l}_{da} + dt\boldsymbol{i}_{da} \\ &= g_1(da). \end{aligned}$$

As regards the second condition, for every $a, b \in L$:

$$\begin{aligned} dg_2(a \wedge b) &= d(t(1-t)\boldsymbol{i}_{[a,b]}) = (1-2t)dt\boldsymbol{i}_{[a,b]} + t(1-t)d(\boldsymbol{i}_{[a,b]}) \\ &= (1-2t)dt\boldsymbol{i}_{[a,b]} + (t-t^2)(\boldsymbol{l}_{[a,b]} - \boldsymbol{i}_{[da,b]} - (-1)^{\bar{a}}\boldsymbol{i}_{[a,db]}) \\ &= (1-2t)dt\boldsymbol{i}_{[a,b]} + (t-t^2)\boldsymbol{l}_{[a,b]} - g_2(da \wedge b) - (-1)^{\bar{a}} g_2(a \wedge db) \end{aligned}$$

and then

$$\begin{aligned} & g_1([a, b]) - [g_1(a), g_1(b)] = t\boldsymbol{l}_{[a,b]} + dt\boldsymbol{i}_{[a,b]} - [t\boldsymbol{l}_a + dt\boldsymbol{i}_a, t\boldsymbol{l}_b + dt\boldsymbol{i}_b] \\ &= t\boldsymbol{l}_{[a,b]} + dt\boldsymbol{i}_{[a,b]} - t^2[\boldsymbol{l}_a, \boldsymbol{l}_b] - tdt[\boldsymbol{i}_a, \boldsymbol{l}_b] - (-1)^{\bar{a}} tdt[\boldsymbol{l}_a, \boldsymbol{i}_b] \\ &= (t-t^2)\boldsymbol{l}_{[a,b]} + (1-2t)\boldsymbol{i}_{[a,b]} \\ &= dg_2(a \wedge b) + g_2(da \wedge b) + (-1)^{\bar{a}} g_2(a \wedge db). \end{aligned}$$

Condition (3) is satisfied in the stronger form that both sides of the equation are equal to 0. In fact, by the Jacobi identity we have

$$\oint g_2([a, b] \wedge c) = (t-t^2) \oint \boldsymbol{i}_{[[a,b],c]} = (t-t^2)\boldsymbol{i}_{\oint [[a,b],c]} = 0,$$

and Lemma 13.3.3 implies that

$$\oint [g_2(a \wedge b), g_1(c)] = (t-t^2) \oint [\boldsymbol{i}_{[a,b]}, t\boldsymbol{l}_c + dt\boldsymbol{i}_c] = (t^2-t^3) \oint [\boldsymbol{i}_{[a,b]}, \boldsymbol{l}_c] = 0.$$

The explicit definition of the Lie-cup product $g_2 \cup g_2 \colon L^{\wedge 4} \to M$ is given in Proposition 12.3.2; for our needs it is sufficient to observe that for every $a_1, a_2, a_3, a_4 \in L$ the element $g_2 \cup g_2(a_1 \wedge a_2 \wedge a_3 \wedge a_4)$ is a linear combination with rational coefficients of the 24 elements

$$t^2(1-t)^2[\boldsymbol{i}_{[a_{\sigma(1)},a_{\sigma(2)}]}, \boldsymbol{i}_{[a_{\sigma(3)},a_{\sigma(4)}]}], \qquad \sigma \in \Sigma_4,$$

and hence $g_2 \cup g_2 = 0$. □

Corollary 13.3.5 *In the same situation as Theorem 13.3.4, let $N \subset M$ be a differential graded Lie subalgebra such that $\boldsymbol{l}(L) \subset N$ and let*

$$K(\chi) = \{(x, y(t)) \in N \times M[t, dt] \mid y(0) = 0,\ y(1) = x\}$$

be the homotopy fibre of the inclusion $\chi \colon N \hookrightarrow M$. Then:

1. *the sequence of linear maps $f_n \colon L^{\wedge n} \to K(\chi)$:*

$$f_1(a) = (\boldsymbol{l}_a, t\boldsymbol{l}_a + dt\boldsymbol{i}_a), \quad f_2(a \wedge b) = (0, t(1-t)\,\boldsymbol{i}_{[a,b]}),$$

   $f_n = 0$ for $n \geq 3$, gives an L_∞-morphism $g \colon L \rightsquigarrow K(\chi)$;
2. *the composition of the linear part f_1 with the canonical quasi-isomorphism*

$$K(\chi) \to coker\chi[-1], \qquad (x, y(t)) \mapsto \int_0^1 y(t),$$

   is equal to $\boldsymbol{i} \colon L \to coker\chi[-1]$;
3. *via the isomorphism $Def_{K(\chi)} \cong Def_{0,\chi}$ of Corollary 6.6.4, the induced map $f \colon Def_L \to Def_{0,\chi}$ is described, at the level of Maurer-Cartan elements, by*

$$f(a) = e^{-\boldsymbol{i}_a}, \qquad a \in MC_L.$$

Proof By construction we have $f = (\boldsymbol{l}, g)$, where $g \colon L \rightsquigarrow M[t, dt]$ is the L_∞-morphism described in Theorem 13.3.4. Since the composition of g with the evaluation maps $e_0, e_1 \colon M[t, dt] \to M$ gives the morphisms of DG-Lie algebras 0 and $\boldsymbol{l}$ respectively, the first two items of the corollary follow immediately.

As regards the third item, recall from Definition 6.5.8 that, for $A \in \mathbf{Art}_{\mathbb{K}}$, the set $\mathrm{Def}_{0,\chi}(A)$ is a quotient of

$$\mathrm{MC}_{0,\chi}(A) = \left\{e^a \in \exp(M^0 \otimes \mathfrak{m}_A) \;\middle|\; e^a * 0 \in N^1\right\}$$

under a suitable equivalence. Let $(x, a(t)) \in \mathrm{MC}_{K(\chi)}(A)$, then $a(t) \in \mathrm{MC}_{M[t,dt]}(A)$, $a(0) = 0$, and then there exists a unique polynomial $p(t) \in M^0[t]$ such that $p(0) = 0$ and $a(t) = e^{p(t)} * 0$. According to Lemma 6.5.9 the isomorphism $\mathrm{Def}_{K(\chi)} \simeq \mathrm{Def}_{0,\chi}$ is induced by the map between Maurer–Cartan functors:

$$\mathrm{MC}_{K(X)}(A) \ni (x, a(t)) \mapsto e^{p(1)} \in \mathrm{MC}_{0,X}(A).$$

Therefore, for proving item (3) it is sufficient to show that for every $a \in L^1 \otimes \mathfrak{m}_A$ such that $da + \frac{1}{2}[a, a] = 0$ we have

$$\mathrm{MC}_f(a) = f_1(a) + \frac{1}{2} f_2(a, a) = (\boldsymbol{l}_a, e^{-t\boldsymbol{i}_a} * 0),$$

or equivalently that

$$e^{-t\boldsymbol{i}_a} * 0 = t\boldsymbol{l}_a + dt\boldsymbol{i}_a + \frac{1}{2}(t - t^2)\boldsymbol{i}_{[a,a]}. \tag{13.4}$$

Since

$$-d(-t\boldsymbol{i}_a) = d(t\boldsymbol{i}_a) = dt\boldsymbol{i}_a + t\boldsymbol{l}_a - t\boldsymbol{i}_{da}, \quad [-t\boldsymbol{i}_a, d(t\boldsymbol{i}_a)] = -t^2[\boldsymbol{i}_a, \boldsymbol{l}_a] = -t^2\boldsymbol{i}_{[a,a]},$$

we have

$$e^{-t\boldsymbol{i}_a} * 0 = \sum_{n \geq 0} \frac{[-t\boldsymbol{i}_a, -]^n}{(n+1)!}(-d(-t\boldsymbol{i}_a)) = dt\boldsymbol{i}_a + t\boldsymbol{l}_a - t\boldsymbol{i}_{da} - \frac{1}{2}t^2\boldsymbol{i}_{[a,a]}.$$

and the conclusion is obtained by replacing $\boldsymbol{i}_{da}$ with $-\frac{1}{2}\boldsymbol{i}_{[a,a]}$. □

Corollary 13.3.6 *Let L, M be two differential graded Lie algebras and let $\boldsymbol{i}\colon L \to M$ be a Cartan homotopy with boundary $\boldsymbol{l}$. Then, for $A \in \mathbf{Art}_{\mathbb{K}}$ and $a \in MC_L(A)$ we have $e^{-\boldsymbol{i}_a} * 0 = \boldsymbol{l}_a$.*

Proof It is sufficient to consider the evaluation at $t = 1$ of the formula (13.4) above. □

13.4 Formal Pointed Grassmann Functors

The standard construction of Quot and Hilbert schemes, both classical and derived, is based on the Grassmann functor: given a finite dimensional vector space V the Grassmann functor $\mathrm{G}_V\colon \mathrm{Schemes}_{\mathbb{K}}^{op} \to \mathbf{Set}$ is defined as

$$\mathrm{G}_V(S) = \{\text{locally free subsheaves of } V \otimes_{\mathbb{K}} \mathcal{O}_S \text{ on } S\}.$$

Therefore, given a vector subspace $F \subset V$ the formal Grassmannian at the point F is the functor of Artin rings $\mathrm{G}_{V,F}\colon \mathbf{Art}_{\mathbb{K}} \to \mathbf{Set}$:

$$\mathrm{G}_{V,F}(A) = \{\text{free } A\text{-submodules } \mathcal{F} \subset V \otimes_{\mathbb{K}} A \text{ such that } \mathcal{F} \otimes_A \mathbb{K} = F\}.$$

In view of future applications to deformation theory, we want to give a homotopy invariant definition of $\mathrm{G}_{V,F}$ when V and F are DG-vector spaces over a fixed field $\mathbb{K}$ of characteristic 0. As usual, unless otherwise specified, the symbol $\otimes$ denotes the tensor product $\otimes_{\mathbb{K}}$ over the field $\mathbb{K}$.

Definition 13.4.1 Let V be a DG-vector space over the field $\mathbb{K}$ and $F \subset V$ a DG-vector subspace. The (formal pointed) **Grassmann functor** of the pair (V, F) is the functor of Artin rings $\mathrm{G}_{V,F} \colon \mathbf{Art}_{\mathbb{K}} \to \mathbf{Set}$ defined as:

$$\mathrm{G}_{V,F}(A) = \frac{\left\{\begin{array}{l}\text{subcomplexes of flat} A\text{-modules}\\ \mathcal{F} \subset V \otimes A \text{ such that } \mathcal{F} \otimes_A \mathbb{K} = F\end{array}\right\}}{\left\{\begin{array}{l}A\text{-linear automorphisms of the complex} V \otimes A \text{lifting the}\\ \text{identity on} V \text{and inducing the identity in cohomology}\end{array}\right\}}.$$

In the above definition, the condition $\mathcal{F} \otimes_A \mathbb{K} = F$ for a subsheaf of A-modules $\mathcal{F} \subset V \otimes A$ means that the induced map $\mathcal{F} \otimes_A \mathbb{K} \to V \otimes A \otimes_A \mathbb{K}$ is injective and, via the canonical isomorphism $V \otimes A \otimes_A \mathbb{K} = V$, its image is equal to F.

The functoriality of $\mathrm{G}_{V,F}$ is an easy consequence of Proposition 4.1.5; for every morphism $\alpha \colon A \to B$ in $\mathbf{Art}_{\mathbb{K}}$ we have that $\mathcal{F} \otimes_A B$ is a complex of flat B-modules and, since

$$(\mathcal{F} \otimes_A B) \otimes_B \mathbb{K} = \mathcal{F} \otimes_A \mathbb{K},$$

the natural map $\mathcal{F} \otimes_A B \to V \otimes B$ is injective. Notice moreover that for every $\mathcal{F} \in \mathrm{G}_{V,F}(A)$ we have a short exact sequence

$$0 \to \mathcal{F} \to V \otimes A \to \frac{V \otimes A}{\mathcal{F}} \to 0,$$

and Corollary 4.1.6 implies that $V \otimes A/\mathcal{F}$ is a complex of flat A-modules. Clearly, if the differential of V is trivial, then

$$\mathrm{G}_{V,F}(A) = \{\text{graded flat } A\text{-modules } \mathcal{F} \subset V \otimes A \text{ such that } \mathcal{F} \otimes_A \mathbb{K} = F\}$$

and the we recover the usual notion of a Grassmann functor.

Definition 13.4.2 Let V be a DG-vector space and $F \subset V$ a DG-vector subspace. The **Grassmann DG-Lie algebra** $G(V, F)$ is defined as the homotopy fibre of the inclusion of differential graded Lie algebras

$$\{h \in \mathrm{Hom}^*_{\mathbb{K}}(V, V) \mid h(F) \subset F\} \hookrightarrow \mathrm{Hom}^*_{\mathbb{K}}(V, V)\,.$$

Notice that the cokernel of the above inclusion is $\mathrm{Hom}^*_{\mathbb{K}}(F, V/F)$ and then, according to Corollary 6.1.7

$$H^i(G(V, F)) = \mathrm{Hom}^{i-1}_{\mathbb{K}}(H^*(F), H^*(V/F)).$$

Our next result is to prove that the Grassamann functor is canonically isomorphic to the deformation functor associated to the Grassmann DG-Lie algebra. To this end it is convenient to introduce the following notation: for every DG-vector space V we denote by $\mathrm{End}^*_{\mathbb{K}}(V)$ the differential graded Lie algebra $\mathrm{Hom}^*_{\mathbb{K}}(V, V)$; given a sequence $V_0, V_1, \ldots$ of DG-vector subspaces of V we introduce the DG-Lie subalgebra

$$\mathrm{End}^*_{\mathbb{K}}(V; V_0, V_1, \ldots) = \{h \in \mathrm{End}^*_{\mathbb{K}}(V) \mid h(V_i) \subset V_i, \quad \text{for every } i\}.$$

For example, the DG-Lie algebra $G(V, F)$ is defined as the homotopy fibre of the inclusion $\mathrm{End}^*_{\mathbb{K}}(V; F) \to \mathrm{End}^*_{\mathbb{K}}(V)$.

Lemma 13.4.3 *In the above situation, denote by $d \in End^1_{\mathbb{K}}(V)$ the differential of V. Then there exist two natural isomorphisms of functors*

$$Def_{G(V,F)} \xrightarrow{\simeq} \mathcal{G}_{V,F} \xrightarrow{\simeq} G_{V,F},$$

where for every $A \in \mathbf{Art}_{\mathbb{K}}$ we have

$$\mathcal{G}_{V,F}(A) = \frac{\{e^f \in \exp(End^0_{\mathbb{K}}(V) \otimes \mathfrak{m}_A) \mid e^f * 0 \in End^1_{\mathbb{K}}(V; F) \otimes \mathfrak{m}_A\}}{\sim}$$

and $e^f \sim e^g$ if there exist $\alpha \in End^0_{\mathbb{K}}(V; F) \otimes \mathfrak{m}_A$ and $\beta \in End^{-1}_{\mathbb{K}}(V) \otimes \mathfrak{m}_A$ such that $e^\alpha e^f = e^g e^{d\beta+\beta d}$.

Proof The construction of the natural isomorphism $\mathrm{Def}_{G(V,F)} \xrightarrow{\simeq} \mathcal{G}_{V,F}$ is done in Corollary 6.6.4 for general homotopy fibres of injective morphisms of DG-Lie algebras, cf. the proof of Corollary 13.3.5.

Thus we only need to give a canonical isomorphism of functors of Artin rings $\psi\colon \mathcal{G}_{V,F} \to G_{V,F}$. Given $A \in \mathbf{Art}_{\mathbb{K}}$, write

$$U_{V,F}(A) = \{\text{graded flat } A\text{-submodules } \mathcal{F} \subset V \otimes A \text{ such that } \mathcal{F} \otimes_A \mathbb{K} = F\},$$

$$\mathrm{Aut}^0(V \otimes A) = \left\{\begin{array}{l}\text{automorphisms of the complex } V \otimes A \text{ lifting the identity}\\ \text{on } V \text{ and inducing the identity in cohomology}\end{array}\right\}.$$

Given $\mathcal{F} = \oplus \mathcal{F}^i \in U_{V,F}(A)$, the flatness of every $\mathcal{F}^i$ as A-module implies that every basis of F lifts to a basis of $\mathcal{F}$ and then the map

$$\exp(\mathrm{End}^0_{\mathbb{K}}(V) \otimes \mathfrak{m}_A) \to U_{V,F}(A), \qquad e^f \mapsto e^{-f}(F \otimes A), \tag{13.5}$$

is surjective. Notice that the graded subspace $e^{-f}(F \otimes A)$ is a subcomplex if and only if $e^f d e^{-f}(F \otimes A) \subset F \otimes A$, or equivalently if and only if

$$e^f * 0 = e^f d e^{-f} - d \in \mathrm{End}^1_{\mathbb{K}}(V; F) \otimes \mathfrak{m}_A.$$

Therefore the subset of elements of $U_{V,F}(A)$ that are also subcomplexes is canonically isomorphic to

$$\frac{\{e^f \in \exp(\operatorname{End}^0_{\mathbb{K}}(V) \otimes \mathfrak{m}_A) \mid e^f * 0 \in \operatorname{End}^1_{\mathbb{K}}(V; F) \otimes \mathfrak{m}_A\}}{\text{left action of } \exp(\operatorname{End}^0_{\mathbb{K}}(V; F) \otimes \mathfrak{m}_A)}.$$

In view of the definition of $\mathcal{G}_{V,F}$ it is sufficient to prove that $\operatorname{Aut}^0(V \otimes A)$ is the irrelevant stabilizer of 0 in the nilpotent DG-Lie algebra $\operatorname{End}^*_{\mathbb{K}}(V) \otimes \mathfrak{m}_A$:

$$\operatorname{Aut}^0(V \otimes A) = \{e^{d\beta+\beta d} \mid \beta \in \operatorname{End}^{-1}_{\mathbb{K}}(V) \otimes \mathfrak{m}_A\}.$$

This is easy and it is left as an exercise. □

Proposition 13.4.4 (Homotopy invariance of Grassmann DG-Lie algebras) *Let $\alpha\colon V \to W$ be a quasi-isomorphism of DG-vector spaces and let $V_0 \subset V$, $W_0 \subset W$ be two differential graded subspaces such that $\alpha(V_0) \subset W_0$ and $\alpha\colon V_0 \to W_0$ is a quasi-isomorphism. Then the differential graded Lie algebras $G(V, V_0)$ and $G(W, W_0)$ are quasi-isomorphic.*

Proof We first prove that if both $\alpha\colon V \to W$ and $\alpha\colon V_0 \to W_0$ are surjective quasi-isomorphisms and I is the kernel of $\alpha\colon V \to W$, then the inclusion

$$\imath\colon \operatorname{End}^*_{\mathbb{K}}(V; V_0, I) \to \operatorname{End}^*_{\mathbb{K}}(V; V_0)$$

and the projection

$$\pi\colon \operatorname{End}^*_{\mathbb{K}}(V; V_0, I) \to \operatorname{End}^*_{\mathbb{K}}(W; W_0)$$

are quasi-isomorphisms. This implies in particular that if $\alpha\colon V \to W$ is a surjective quasi-isomorphism then the inclusion $\imath\colon \operatorname{End}^*_{\mathbb{K}}(V; I) \to \operatorname{End}^*_{\mathbb{K}}(V)$ and the projection $\pi\colon \operatorname{End}^*_{\mathbb{K}}(V; I) \to \operatorname{End}^*_{\mathbb{K}}(W)$ are quasi-isomorphisms (take $V_0 = W_0 = 0$).

To see this, let $I_0 = I \cap V_0$ be the kernel of $\alpha\colon V_0 \to W_0$. Then the complexes

$$G_\alpha = \{f \in \operatorname{Hom}^*_{\mathbb{K}}(I, V/I) \mid f(I_0) \subset V_0/I_0 = (V_0 + I)/I\},$$

$$R_\alpha = \{f \in \operatorname{End}^*_{\mathbb{K}}(V; V_0) \mid f(V) \subset I\} = \{f \in \operatorname{Hom}^*_{\mathbb{K}}(V, I) \mid f(V_0) \subset V_0 \cap I\},$$

are acyclic. In fact, both I and I_0 are acyclic DG-vector spaces and the complex G_α is the kernel of the surjective morphism of acyclic complexes.

$$\operatorname{Hom}^*_{\mathbb{K}}(I, V/I) \to \operatorname{Hom}^*_{\mathbb{K}}(I_0, V/(V_0 + I)),$$

while R_α is the kernel of the surjective morphism of acyclic complexes

$$\operatorname{Hom}^*_{\mathbb{K}}(V, I) \to \operatorname{Hom}^*_{\mathbb{K}}(V_0, I/I_0).$$

Now we have exact sequences

$$0 \to \mathrm{End}^*_{\mathbb{K}}(V; V_0, I) \xrightarrow{\iota} \mathrm{End}^*(V; V_0) \to G_\alpha \to 0,$$

$$0 \to R_\alpha \to \mathrm{End}^*_{\mathbb{K}}(V; V_0, I) \xrightarrow{\pi} \mathrm{End}^*(W; W_0) \to 0,$$

and this proves the proposition when $\alpha(V) = W$ and $\alpha(V_0) = W_0$.

We are now ready to prove the proposition in general. Consider the two DG-vector spaces of semicosimplicial cochains (Definition 7.4.2):

$$H = C\left(V \oplus W \underset{\mathrm{Id}_W}{\overset{\alpha}{\rightrightarrows}} W \Rrightarrow 0 \right),$$

$$H_0 = C\left(V_0 \oplus W_0 \underset{Id}{\overset{\alpha}{\rightrightarrows}} W_0 \Rrightarrow 0 \right).$$

More explicitly:

$$H^i = V^i \oplus W^i \oplus W^{i-1}, \quad d(v_i, w_i, w_{i-1}) = (dv_i, dw_i, \alpha(v_i) - w_i - dw_{i-1}),$$

and H_0 is a subcomplex of H. A basic computation in homology shows that the four projection maps

$$\beta\colon H \to V, \quad \beta\colon H_0 \to V_0, \quad \gamma\colon H \to W, \quad \gamma\colon H_0 \to W_0,$$

are surjective quasi-isomorphisms and by the above computation we have two commutative diagrams of differential graded Lie algebras

$$\begin{array}{ccccc}
\mathrm{End}^*_{\mathbb{K}}(V; V_0) & \longleftarrow & \mathrm{End}^*_{\mathbb{K}}(H; H_0, \ker\beta) & \longrightarrow & \mathrm{End}^*_{\mathbb{K}}(H; H_0) \\
\downarrow & & \downarrow & & \downarrow \\
\mathrm{End}^*_{\mathbb{K}}(V) & \longleftarrow & \mathrm{End}^*_{\mathbb{K}}(H; \ker\beta) & \longrightarrow & \mathrm{End}^*_{\mathbb{K}}(H),
\end{array}$$

$$\begin{array}{ccccc}
\mathrm{End}^*_{\mathbb{K}}(H; H_0) & \longleftarrow & \mathrm{End}^*_{\mathbb{K}}(H; H_0, \ker\gamma) & \longrightarrow & \mathrm{End}^*_{\mathbb{K}}(W; W_0) \\
\downarrow & & \downarrow & & \downarrow \\
\mathrm{End}^*_{\mathbb{K}}(H) & \longleftarrow & \mathrm{End}^*_{\mathbb{K}}(H; \ker\gamma) & \longrightarrow & \mathrm{End}^*_{\mathbb{K}}(W),
\end{array}$$

where every horizontal map is a quasi-isomorphism. In particular, the homotopy fibres of the five vertical morphisms are quasi-isomorphic DG-Lie algebras. □

Theorem 13.4.5 *Let V be a DG-vector space and $F \subset V$ a DG-vector subspace. Then there exists a canonical isomorphism of functors of Artin rings*

$$\varphi\colon \mathit{Def}_{G(V,F)} \xrightarrow{\simeq} G_{V,F}.$$

In particular, the functor $G_{V,F}$ is invariant under quasi-isomorphism of the pair (V, F).

Proof Immediate consequence of Lemma 13.4.3 and Proposition 13.4.4. □

Corollary 13.4.6 *Let V be a DG-vector space and $F \subset V$ a DG-vector subspace such that the inclusion $F \to V$ is injective in cohomology. Then $G(V, F)$ is a homotopy abelian DG-Lie algebra, $H^i(G(V, F)) = Hom^{i-1}_{\mathbb{K}}(H^*(F), H^*(V)/H^*(F))$ and then the formal pointed Grassmann functor $G_{V,F}$ is smooth.*

Proof By Künneth's formula the morphism $\mathrm{End}^*_{\mathbb{K}}(V, V) \to \mathrm{Hom}^*_{\mathbb{K}}(F, V/F)$ is surjective in cohomology and therefore the inclusion $\mathrm{End}^*_{\mathbb{K}}(V; F) \to \mathrm{End}^*_{\mathbb{K}}(V, V)$ is injective in cohomology. By Corollary 6.1.7 the Grassmann DG-Lie algebra $G(V, F)$ is homotopy abelian. The smoothness of the Grassmann functor follows from Theorem 13.4.5. □

Example 13.4.7 In general the Grassmann functor $\mathrm{G}_{V,F}$ is not smooth. Consider for instance the complex

$$V\colon \quad \mathbb{K}\alpha \oplus \mathbb{K}\beta \xrightarrow{d} \mathbb{K}\gamma \oplus \mathbb{K}\delta, \qquad d\alpha = \gamma, \quad d\beta = 0,$$

and the subcomplex $F = \mathbb{K}\beta \xrightarrow{d=0} \mathbb{K}\gamma$. A deformation of F over $\mathbb{K}[\epsilon]$, $\epsilon^2 = 0$, is given by the subcomplex

$$\mathcal{F}\colon \quad \mathbb{K}[\epsilon](\beta + \epsilon\alpha) \xrightarrow{d} \mathbb{K}[\epsilon](\gamma + \epsilon\delta),$$

which does not extend to any subcomplex of $V \otimes \mathbb{K}[t]/(t^3)$.

In the setup of Definition 13.4.1, for every $A \in \mathbf{Art}_{\mathbb{K}}$ there is a well defined map

$$H_A\colon \mathrm{G}_{V,F}(A) \to \{\text{graded } A\text{-submodules of } H^*(V) \otimes A\},$$

sending an equivalence class $[\mathcal{F}] \in \mathrm{G}_{V,F}(A)$ into the image of the map

$$H^*(\mathcal{F}) \to H^*(V \otimes A) = H^*(V) \otimes A\,.$$

Keep in mind the fact that in general the maps H_A do not give a natural transformation onto the functor $\mathrm{G}_{H^*(V),K}$, where K is the image of $H^*(F) \to H^*(V)$.

Theorem 13.4.8 *In the above situation, assume that the inclusion $F \to V$ is injective in cohomology. Then the maps H_A, $A \in \mathbf{Art}_{\mathbb{K}}$, give an isomorphism of functors*

$$H\colon G_{V,F} \xrightarrow{\simeq} G_{H^*(V),H^*(F)}, \qquad V \otimes A \supset \mathcal{F} \mapsto H^*(\mathcal{F}) \subset H^*(V) \otimes A.$$

Proof It is possible to choose a "harmonic projection" $\alpha\colon V \to H^*(V)$ that is a quasi-isomorphism and such that $\alpha(F) = H^*(F)$. Let's denote by $I = \ker \alpha$ and by K the homotopy fibre of the inclusion $\operatorname{End}^*_{\mathbb{K}}(V; F, I) \to \operatorname{End}^*_{\mathbb{K}}(V; I)$. The computation used in the proof of Proposition 13.4.4 shows that the natural maps $p\colon K \to G(V, F)$ and $q\colon K \to G(H^*(V), H^*(F))$ are quasi-isomorphisms of differential graded Lie algebras and then there exists a commutative diagram of isomorphisms

$$\begin{array}{ccccc} \operatorname{Def}_K & \xrightarrow{p} & \operatorname{Def}_{G(V,F)} & \xrightarrow{\varphi} & \mathrm{G}_{V,F} \\ & \searrow^{q} & & & \downarrow{\psi} \\ & & \operatorname{Def}_{G(H^*(V),H^*(F))} & \xrightarrow{\varphi} & \mathrm{G}_{H^*(V),H^*(F)}. \end{array}$$

We have to prove that $\psi = H$. Using again Corollary 6.6.4, an element of $\operatorname{Def}_K(A)$ is represented by an endomorphism $e^f \in \exp(\operatorname{End}^0_{\mathbb{K}}(V; I) \otimes \mathfrak{m}_A)$ such that $e^f * 0 \in \operatorname{End}^1_{\mathbb{K}}(V; F, I) \otimes \mathfrak{m}_A$. Let $g \in \operatorname{End}^0_{\mathbb{K}}(H^*(V)) \otimes \mathfrak{m}_A$ be the endomorphism such that $\alpha f = g\alpha$, then

$$\begin{aligned} \alpha\varphi p(e^f) &= \alpha(e^{-f}(F \otimes A)) = e^{-g}(\alpha(F \otimes A)) = e^{-g}(H^*(F) \otimes A) = \varphi(e^g) \\ &= \varphi q(e^f). \end{aligned}$$

This proves that for $\mathcal{F} \in \mathrm{G}_{V,F}(A)$ we have that $\psi(\mathcal{F}) = \alpha(\mathcal{F})$ is A-flat and then

$$0 \to \mathcal{F} \cap (I \otimes A) \to \mathcal{F} \to \alpha(\mathcal{F}) \to 0$$

is a short exact sequence of complexes of flat A-modules. Applying the functor $\otimes_A \mathbb{K}$ we get that $(\mathcal{F} \cap (I \otimes A)) \otimes_A \mathbb{K} = I \cap F$ is acyclic and then, by Corollary 4.1.2 also $\mathcal{F} \cap I \otimes A$ is acyclic. In particular, $\alpha\colon \mathcal{F} \to \alpha(\mathcal{F})$ is a quasi-isomorphism and then $H^*(\mathcal{F}) \to \alpha(\mathcal{F})$ is an isomorphism. □

13.5 Formal Period Maps

Generally speaking, a formal period map is a natural transformation from a deformation problem to a Grassmann functor, sometimes subjected to some additional conditions. Although these additional conditions strongly depend on the particular deformation problem, there exists a general algebraic description of period maps as L_∞ morphisms associated to a rather simple quadruple of data.

Definition 13.5.1 A (formal) **period data** is a quadruple $(L, V, F, \boldsymbol{i})$ where:

1. L is a differential graded Lie algebra;
2. V is a DG-vector space and F is a differential graded subspace of V;
3. $\boldsymbol{i}\colon L \to \operatorname{End}^*_{\mathbb{K}}(V) = \operatorname{Hom}^*_{\mathbb{K}}(V, V)$ is a Cartan homotopy such that its boundary $\boldsymbol{l}$ satisfies the condition $\boldsymbol{l}_a(F) \subset F$ for every $a \in L$.

Using the notation introduced in Sect. 13.4, for all period data $(L, V, F, \boldsymbol{i})$ the boundary of $\boldsymbol{i}$ gives a morphism of DG-Lie algebras $\boldsymbol{l}\colon L \to \mathrm{End}^*_{\mathbb{K}}(V; F)$. Therefore, by Corollary 13.3.5 the period data $(L, V, F, \boldsymbol{i})$ gives a canonical L_∞-morphism

$$g\colon L \rightsquigarrow G(V, F) = \text{homotopy fibre of the inclusion } \mathrm{End}^*_{\mathbb{K}}(V; F) \subset \mathrm{End}^*_{\mathbb{K}}(V).$$

Definition 13.5.2 The (formal) **period map** of the period data $(L, V, F, \boldsymbol{i})$ is the natural transformation of functors

$$\mathfrak{p}\colon \mathrm{Def}_L \to \mathrm{G}_{V,F}$$

induced by the isomorphism $\varphi\colon \mathrm{Def}_{G(V,F)} \to \mathrm{G}_{V,F}$ of Theorem 13.4.5 and by the L_∞-morphism $g\colon L \rightsquigarrow G(V, F)$.

According to Corollary 13.3.5, Lemma 13.4.3 and (13.5), the formal period map $\mathfrak{p}$ is induced by the maps

$$\mathrm{MC}_L(A) \to \mathrm{G}_{V,F}(A), \qquad a \mapsto e^{\boldsymbol{i}_a}(F \otimes A), \quad A \in \mathbf{Art}_{\mathbb{K}}.$$

Moreover, the composition of the linear part $g_1\colon L \to G(V, F)$ of the L_∞-morphism g with the canonical quasi-isomorphism of complexes

$$G(V, F) \to \frac{\mathrm{End}^*_{\mathbb{K}}(V)}{\mathrm{End}^*_{\mathbb{K}}(V; F)}[-1] = \mathrm{Hom}^*_{\mathbb{K}}(F, V/F)[-1]$$

coincides with the composition of

$$L \xrightarrow{\boldsymbol{i}} \mathrm{End}^*_{\mathbb{K}}(V)[-1] \xrightarrow{\text{natural projection}} \mathrm{Hom}^*_{\mathbb{K}}(F, V/F)[-1].$$

In view of Theorem 8.8.3, when the inclusion $\mathrm{End}^*_{\mathbb{K}}(V; F) \to \mathrm{End}^*_{\mathbb{K}}(V)$ is injective in cohomology, the period map gives constraints on the functor Def_L. More precisely, the abstract BTT theorem applied to period maps gives the following corollary.

Corollary 13.5.3 *Let $(L, V, F, \boldsymbol{i})$ be period data such that the inclusion $F \hookrightarrow V$ is injective in cohomology. Then:*

1. *if $\boldsymbol{i}\colon H^*(L) \to \mathrm{Hom}^*_{\mathbb{K}}(H^*(F), H^*(V/F))[-1]$ is injective, then L is homotopy abelian;*
2. *the obstructions of the functor Def_L are contained in the kernel of*

$$\boldsymbol{i}\colon H^2(L) \to \mathrm{Hom}^1_{\mathbb{K}}(H^*(F), H^*(V/F)).$$

Proof According to Corollary 13.4.6 the differential graded Lie algebra $G(V, F)$ is homotopy abelian. Since $\boldsymbol{i}$ and the linear part of the L_∞-morphism inducing the period map are the same in cohomology, the first item follows from Corollary 6.1.3 and the second item follows from Theorem 8.8.3. □

We are now ready to provide Corollaries 8.8.4 and 8.8.5 with a new and geometrically insightful proof that is extendable to more general and purely algebraic situations. Let $A_X^{*,*}$ be the Rham complex of a complex manifold X and denote by $F_X^0 \supset F_X^1 \supset \cdots$ its Hodge filtration:

$$F_X^p = \oplus_{i\geq p,j} A_X^{i,j} \,.$$

Notice that $H^i(X, \Omega_X^p) = H^{i+p}(F_X^p/F_X^{p+1})$ and that $F_X^p \to A_X^{*,*}$ is injective in cohomology for every p if and only if the Hodge to de Rham spectral sequence of X degenerates at E_1 (Corollary C.6.7).

We have already seen in Example 8.8.2 that the convolution operator

$$\boldsymbol{i}\colon KS_X \to \mathrm{Hom}_{\mathbb{C}}^*(A_X^{*,*}, A_X^{*,*})$$

is a Cartan homotopy with boundary $\boldsymbol{l}_\eta = [\partial, \boldsymbol{i}_\eta]$. Thus $\boldsymbol{l}_a(F_X^p) \subset F_X^p$ for every $a \in KS_X$ and then for every p we have period data $(KS_X, A_X^{*,*}, F_X^p, \boldsymbol{i})$.

Theorem 13.5.4 (Kodaira principle) *Let X be a complex manifold and let $p > 0$ be an integer such that the inclusion $F_X^p \to F_X^0 = A_X^{*,*}$ is injective in cohomology. Then the obstructions of Def_X are contained in the kernel of the map*

$$H^2(X, \Theta_X) \xrightarrow{\ \boldsymbol{i}\ } \prod_i Hom_{\mathbb{C}}\left(H^i(F_X^p), H^{i+1}\left(\frac{F_X^0}{F_X^p}\right)\right). \tag{13.6}$$

If in addition the three inclusions $F_X^{p+1} \to F_X^p \to F_X^{p-1} \to F_X^0$ are injective in cohomology, then the obstructions of Def_X are contained in the kernel of the map

$$H^2(X, \Theta_X) \xrightarrow{\ \boldsymbol{i}\ } \prod_i Hom_{\mathbb{C}}\left(H^i(X, \Omega_X^p), H^{i+2}(X, \Omega_X^{p-1})\right).$$

Proof The first part follows immediately from Corollary 13.5.3 applied to the period data $(KS_X, A_X^{*,*}, F_X^p, \boldsymbol{i})$. The proof of the remaining part of the theorem is the same as the proof of Corollary 8.8.4. Since $\boldsymbol{i}_a(F_X^p) \subset F_X^{p-1}$ for every $a \in KS_X$, every component of the map (13.6) has the following factorization:

$$\begin{array}{ccc} H^2(X, \Theta_X) & & \\ \downarrow{\scriptstyle \boldsymbol{i}} & \searrow^{\boldsymbol{i}} & \\ \mathrm{Hom}_{\mathbb{C}}\left(H^i\left(\frac{F_X^p}{F_X^{p+1}}\right), H^{i+1}\left(\frac{F_X^{p-1}}{F_X^p}\right)\right) & \longrightarrow & \mathrm{Hom}_{\mathbb{C}}\left(H^i(F_X^p), H^{i+1}\left(\frac{F_X^0}{F_X^p}\right)\right). \end{array}$$

If the three inclusions $F_X^{p+1} \to F_X^p \to F_X^{p-1} \to F_X^0$ are injective in cohomology, then for every i the map $H^{i+p}(F_X^p) \to H^{i+p}(F_X^p/F_X^{p+1}) = H^i(X, \Omega_X^p)$ is surjective, the map $H^{i+2}(X, \Omega_X^{p-1}) = H^{i+p}(F_X^{p-1}/F_X^p) \to H^{i+1}(F_X^0/F_X^p)$ is injective and therefore the natural map

$$\mathrm{Hom}_{\mathbb{C}}\left(H^i(X, \Omega_X^p), H^{i+2}(X, \Omega_X^{p-1})\right) \to \mathrm{Hom}_{\mathbb{C}}\left(H^{i+p}(F_X^p), H^{i+p+1}\left(\frac{F_X^0}{F_X^p}\right)\right)$$

is injective. □

Theorem 13.5.5 *Let X be a compact Kähler manifold. Then the composition of the period map $\mathfrak{p}\colon Def_X \to G_{F_X^0, F_X^p}$ associated to the period data $(KS_X, A_X^{*,*}, F_X^p, \boldsymbol{i})$ with the isomorphism*

$$G_{F_X^0, F_X^p} \to G_{H^*(F_X^0), H^*(F_X^p)}$$

of Theorem 13.4.8 is the classical (formal) Griffiths period map.

Proof Let $A \in \mathbf{Art}_{\mathbb{K}}$ and $\eta \in \mathrm{MC}_{KS_X}(A)$. According to Remark 8.5.6 the deformation of X associated to η is given by the sheaf of locally free A-algebras

$$\mathcal{O}_\eta = \ker(\mathcal{A}_X^{0,0} \otimes A \xrightarrow{\overline{\partial}+\boldsymbol{l}_\eta} \mathcal{A}_X^{0,1} \otimes A) = \{f \in \mathcal{A}_X^{0,0} \mid \overline{\partial} f = \eta \lrcorner \partial f\},$$

while $\mathfrak{p}(\eta) \in \mathrm{G}_{F_X^0, F_X^p}$ is the subcomplex $e^{\boldsymbol{i}_\eta}(F_X^p \otimes A) \subset F_X^0 \otimes A$. Writing $\mathcal{F}_X^p = \oplus_{i,j\geq 0}\mathcal{A}_X^{i+p,j}$, in order to prove the theorem it is sufficient to show that $e^{\boldsymbol{i}_\eta}(\mathcal{F}_X^p \otimes A)$ is the ideal subsheaf of $\mathcal{A}_X^{*,*}$ generated by $df_1 \wedge \cdots \wedge df_p$, for $f_1, \ldots, f_p \in \mathcal{O}_\eta$; for standard rank considerations it is sufficient to prove $\wedge^p d\mathcal{O}_\eta \subset e^{\boldsymbol{i}_\eta}(\mathcal{F}_X^p \otimes A)$. Since $\boldsymbol{i}_\eta$ is a derivation, $e^{\boldsymbol{i}_\eta}$ is a morphism of graded algebras and then it is sufficient to prove that $d\mathcal{O}_\eta \subset e^{\boldsymbol{i}_\eta}(\mathcal{F}_X^1 \otimes A)$. By Corollary 13.3.6 we have

$$\boldsymbol{l}_\eta = e^{-\boldsymbol{i}_\eta} * 0 = e^{-\boldsymbol{i}_\eta} d e^{\boldsymbol{i}_\eta} - d.$$

Since $e^{\boldsymbol{i}_\eta}$ is the identity on $\mathcal{A}_X^{0,0}$ and $(\overline{\partial} + \boldsymbol{l}_\eta)\mathcal{O}_\eta = 0$ we have

$$e^{-\boldsymbol{i}_\eta}(d\mathcal{O}_\eta) = e^{-\boldsymbol{i}_\eta} d e^{\boldsymbol{i}_\eta}\mathcal{O}_\eta = (d + \boldsymbol{l}_\eta)\mathcal{O}_\eta = \partial\mathcal{O}_\eta \subseteq \partial\mathcal{A}_X^{0,0} \subseteq \mathcal{F}_X^1.$$

□

13.6 Period Data of Differential Graded BV-Algebras

In this section we show that every differential graded Batalin–Vilkovisky (BV) algebra gives canonically formal period data, and the application of Corollary 13.5.3 to this situation provides a different proof of a well known theorem by Terilla.

Let n be an odd integer and let (A, d, Δ) be a differential graded Batalin–Vilkovisky algebra of degree n (Definition 9.4.6). This means that (A, d) is a unitary commutative differential graded algebra and $\Delta \in \mathrm{Hom}_{\mathbb{K}}^{-n}(A, A)$ is a differential operator of order ≤ 2 such that

$$\Delta(1) = 0, \quad \Delta^2 = 0, \quad [d, \Delta] = d\Delta + \Delta d = 0.$$

According to Corollary 9.3.5, the conditions $\Delta \in \mathrm{Diff}^2_{A/\mathbb{K}}(A, A)$ and $\Delta(1) = 0$ are equivalent to the "seven-terms relation":

$$\begin{aligned}\Delta(abc) + \Delta(a)bc + (-1)^{\bar{a}\,\bar{b}}\Delta(b)ac + (-1)^{\bar{c}(\bar{a}+\bar{b})}\Delta(c)ab\\ = \Delta(ab)c + (-1)^{\bar{a}(\bar{b}+\bar{c})}\Delta(bc)a + (-1)^{\bar{b}\bar{c}}\Delta(ac)b.\end{aligned}$$

Recall, from Lemma 9.4.3, that the DG-Lie algebra underlying the DGBV-algebra (A, d, Δ) is $(A[n], -d, [-, -]_\Delta)$ where:

$$[-, -]_\Delta \colon A^p \times A^q \to A^{p+q-n}, \qquad [a, b]_\Delta = (-1)^p(\Delta(ab) - \Delta(a)b) - a\Delta(b)\,.$$

Let t be a central indeterminate of (even) degree $n + 1$ and denote by $A[[t]]$ the graded vector space of formal power series with coefficients in A: more precisely $A[[t]] = \oplus_i A[[t]]^i$, where

$$A[[t]]^i = \left\{ \sum_{h\geq 0} a_h t^h \;\middle|\; a_h \in A^{i-h(n+1)} \right\}.$$

Similarly, we denote by $A((t)) = \bigcup_{p\in\mathbb{Z}} t^p A[[t]]$ the graded vector space of formal meromorphic Laurent series with coefficients in A. On the space $A((t))$ we consider the differential of degree $+1$ given by $d - t\Delta$:

$$(d - t\Delta)\left(\sum_h a_h t^h\right) = \sum_h (da_h - \Delta a_{h-1})t^h.$$

Theorem 13.6.1 *The linear map of degree* $-n - 1$

$$\boldsymbol{i} \colon A \to Hom^*_{\mathbb{K}}(A((t)), A((t))), \qquad \boldsymbol{i}_a(\phi) = \frac{a\phi}{t},$$

gives a Cartan homotopy $\boldsymbol{i} \colon A[n] \to Hom^*_{\mathbb{K}}(A((t)), A((t)))$ *and also formal period data* $(A[n], A((t)), A[[t]], \boldsymbol{i})$.

Proof We need to prove that $\boldsymbol{i}$ is a Cartan homotopy and the subspace $A[[t]]$ is invariant under the coboundary $\boldsymbol{l}$ of $\boldsymbol{i}$. The identity $[\boldsymbol{i}_a, \boldsymbol{i}_b] = 0$ is a trivial consequence of the graded commutativity of A. Next, since n is odd, the differential on $A[n]$ is

$-d$ and therefore we have $\boldsymbol{l}_b = [d - t\Delta, \boldsymbol{i}_b] - \boldsymbol{i}_{db}$; more explicitly, for every $b \in A$ and $\phi \in A((t))$ we have

$$\begin{aligned}\boldsymbol{l}_b(\phi) &= [d - t\Delta, \boldsymbol{i}_b](\phi) - \frac{(db)\phi}{t} = (d - t\Delta)\frac{b\phi}{t} - (-1)^{\bar{b}}\boldsymbol{i}_b(d\phi - t\Delta(\phi)) - \frac{(db)\phi}{t}\\ &= \frac{1}{t}(d(b\phi) - (-1)^b b(d\phi) - (db)\phi) - \Delta(b\phi) + (-1)^{\bar{b}} b\Delta(\phi)\\ &= -\Delta(b\phi) + (-1)^{\bar{b}} b\Delta(\phi)\end{aligned}$$

showing $\boldsymbol{l}_b(A[[t]]) \subset A[[t]]$. Finally, we need to prove that $[\boldsymbol{i}_a, \boldsymbol{l}_b] = \boldsymbol{i}_{[a,b]_\Delta}$ for every $a, b \in A$. In fact,

$$\begin{aligned}&[\boldsymbol{i}_a, \boldsymbol{l}_b](\phi) - \boldsymbol{i}_{[a,b]_\Delta}(\phi)\\ &\quad = \boldsymbol{i}_a(-\Delta(b\phi) + (-1)^{\bar{b}} b\Delta(\phi)) - (-1)^{\bar{a}(\bar{b}+1)}\boldsymbol{l}_b\left(\frac{a\phi}{t}\right) - \frac{1}{t}[a, b]_\Delta\,\phi,\end{aligned}$$

$$\begin{aligned}&t([\boldsymbol{i}_a, \boldsymbol{l}_b](\phi) - \boldsymbol{i}_{[a,b]_\Delta}(\phi))\\ &\quad = -a\Delta(b\phi) + (-1)^{\bar{b}} ab\Delta(\phi) - (-1)^{\bar{a}(\bar{b}+1)}(-\Delta(ba\phi) + (-1)^{\bar{b}} b\Delta(a\phi))\\ &\qquad - (-1)^{\bar{a}}(\Delta(ab)\phi - \Delta(a)b\phi) + a\Delta(b)\phi\\ &\quad = 0,\end{aligned}$$

where the last equality follows from the seven-terms relation. □

Definition 13.6.2 A differential graded Batalin–Vilkovisky algebra (A, d, Δ) of (odd) degree n is said to have the **degeneration property** if for every $a_0 \in A$ such that $da_0 = 0$ there exists a sequence a_i, $i \geq 0$, such that

$$\deg(a_{i+1}) = \deg(a_i) - n - 1, \qquad da_{i+1} = \Delta a_i, \qquad i \geq 0.$$

It should be pointed out that it is enough to check the degeneration property for a set of representatives for the cohomology of (A, d); in fact if $a_0 = db$, then the sequence defined as $a_1 = -\Delta(b)$ and $a_i = 0$ for $i \geq 2$ satisfies the conditions $da_{i+1} = \Delta a_i$.

Corollary 13.6.3 *The differential graded Lie algebra associated to a differential graded Batalin–Vilkovisky algebra with the degeneration property is homotopy abelian.*

Proof By Corollary 13.5.3 it is sufficient to prove that the two morphisms of DG-vector spaces

$$(A[[t]], d - t\Delta) \hookrightarrow (A((t)), d - t\Delta), \quad \boldsymbol{i}\colon A[n+1] \to \operatorname{Hom}^*_{\mathbb{K}}\left(A[[t]], \frac{A((t))}{A[[t]]}\right)$$

are injective in cohomology. We can immediately see that the degeneration property is equivalent to the fact that the morphism of complexes

$$(A[[t]], d - t\Delta) \xrightarrow{t \mapsto 0} (A, d)$$

is surjective in cohomology. The DG-vector space $(A((t)), d - t\Delta)$ has a natural decreasing filtration of DG-subspaces

$$A((t)) = \bigcup_{p \in \mathbb{Z}} F^p, \qquad F^p = t^p A[[t]],$$

and the short exact sequence

$$0 \to tA[[t]] \hookrightarrow A[[t]] \xrightarrow{t \mapsto 0} A \to 0$$

implies that the degeneration property holds if and only if the inclusion $F^1 \subset F^0$ is injective in cohomology. Since the multiplication maps $t^p \colon F^q \to F^{p+q}$, $p, q \in \mathbb{Z}$, are (up to degree shifting) isomorphisms of DG-vector spaces, we have that the degeneration property holds if and only if for every $p < q$ the inclusion $F^q \subset F^p$ is injective in cohomology. Since $A((t)) = \cup F^p$, the degeneration property implies in particular that the inclusion $A[[t]] \subset A((t))$ is injective in cohomology.

Since $(d - t\Delta)1 = 0$, in order to prove that the map

$$\boldsymbol{i} \colon A[n+1] \to \mathrm{Hom}^*_{\mathbb{K}}\left(A[[t]], \frac{A((t))}{A[[t]]}\right)$$

is injective in cohomology it is sufficient to prove the same property for the composition with the evaluation at $1 \in A[[t]]$, i.e., it is sufficient to prove that the map

$$A[n+1] \to \frac{A((t))}{A[[t]]}, \qquad a \mapsto \frac{a}{t},$$

is injective in cohomology. Since $a \mapsto \dfrac{a}{t}$ gives an isomorphism of DG-vector spaces $A[n+1] \cong F^{-1}/F^0$, the above statement is equivalent to the fact that the inclusion

$$\frac{F^{-1}}{F^0} \hookrightarrow \frac{A((t))}{F^0}$$

is injective in cohomology; this follows easily from the fact that the inclusions $F^0 \subset F^{-1} \subset A((t))$ are injective in cohomology. □

Example 13.6.4 Let (A, d, Δ) be a DGBV-algebra of degree n and assume the validity of the (half) $d\Delta$-lemma:

$$\ker d \cap \Delta(A) = d\Delta(A).$$

Then (A, d, Δ) has the degeneration property. In fact if $a_0 \in A$ and $da_0 = 0$, then $\Delta a_0 \in d\Delta(A)$ and then there exists $b \in A$ such that $d\Delta(b) = \Delta a_0$. It is sufficient to take $a_1 = \Delta(b)$ and $a_i = 0$ for every $i \geq 2$.

Notice that the converse is generally false. For instance, if $d = \Delta$, then (A, d, Δ) has the degeneration property, while $\ker d \cap \Delta(A) = d\Delta(A)$ if and only if $d = \Delta = 0$.

Example 13.6.5 Let (A, d, Δ) be a DGBV-algebra of degree n and assume that there exists a linear map $f \in \mathrm{Hom}_{\mathbb{K}}^{-n-1}(A, A)$ such that

$$\Delta = [d, f], \qquad [f, \Delta] = -[f, [f, d]] = 0.$$

Then (A, d, Δ) has the degeneration property. In fact, if t is a central formal variable of (even) degree $n + 1$, then in the graded associative algebra $\mathrm{Hom}_{\mathbb{K}}^*(A, A)[[t]]$ we have

$$e^{tf} d e^{-tf} = e^{[tf,-]} d = d + t[f, d] + \cdots = d - t\Delta$$

and therefore

$$t\Delta e^{tf} = -e^{tf} d + d e^{tf} = [d, e^{tf}].$$

Let $a_0 \in A$ be such that $da_0 = 0$ and define the sequence a_i by the rule

$$a_i = \frac{f^i}{i!}(a_0) \iff \sum_{i\geq 0} a_i t^i = e^{tf}(a_0).$$

We have

$$\sum_{i\geq 0} t^{i+1} \Delta a_i = t\Delta e^{tf} a_0 = d e^{tf} a_0 = \sum_{i\geq 0} t^i da_i$$

and therefore $da_{i+1} = \Delta a_i$ for every i.

Example 13.6.6 (*The Milnor DGBV-algebra*) Consider the graded Batalin–Vilkovisky algebra $A = \mathbb{K}[x_1, \ldots, x_n, y_1, \ldots, y_n]$, where $\overline{x_i} = 0$ and $\overline{y_i} = -1$, equipped with the second order differential operator of degree 1 (cf. Example 9.4.5):

$$\Delta = -\sum_{i=1}^{n} \frac{\partial}{\partial x_i} \frac{\partial}{\partial y_i}.$$

Given a polynomial $f \in A^0 = \mathbb{K}[x_1, \ldots, x_n]$ with partial derivatives $f_i = \dfrac{\partial f}{\partial x_i}$ we can define a differential $d \in \mathrm{Der}^1(A, A)$ by setting $d(y_i) = f_i$. Then (A, d, Δ) is a differential graded Batalin–Vilkovisky algebra called the Milnor DGBV-algebra of the polynomial f. Notice that its underlying complex is the Koszul complex of the sequence $f_1, \dots, f_n \in A^0$.

Proposition 13.6.7 *In the above setup, the following are equivalent:*

1. *$H^p(A) = 0$ for every $p < 0$;*
2. *the Milnor DGBV-algebra (A, d, Δ) has the degeneration property;*
3. *the DG-Lie algebra $(A[-1], -d, [-,-]_\Delta)$ is homotopy abelian;*
4. *the graded Lie algebra $H^*(A[-1])$ is abelian.*

Proof According to Corollary 13.6.3 we only need to prove that:

1. if $H^p(A) = 0$ for every $p < 0$ then the Milnor DGBV-algebra (A, d, Δ) has the degeneration property;
2. if $H^p(A) \neq 0$ for some $p < 0$ then the graded Lie algebra $H^*(A[-1])$ is not abelian.

Assume first $H^p(A) = 0$ for every $p < 0$ and consider an element $a \in A$ such that $da = 0$; we need to prove that there exists a sequence $a_1, a_2, \dots$ such that

$$\Delta a = da_1, \quad \Delta a_1 = da_2, \quad \Delta a_2 = da_3, \quad \dots \quad \Delta a_n = da_{n+1}, \quad \dots . \tag{13.7}$$

If $a \in A^0$, then $\Delta(a) = 0$ and it is enough to take $a_i = 0$ for every $i > 0$. If $a \in A^p$ for some $p < 0$, then there exists $b \in A^{p-1}$ such that $a = db$, therefore $\Delta a = \Delta(db) = -d(\Delta b)$, $\Delta(-\Delta b) = 0$ and we can choose $a_1 = -\Delta b$, $a_i = 0$ for $i > 1$.

Conversely, if the A^0-module $H^p(A)$ is nontrivial for some $p < 0$, then there exists a maximal ideal $\mathfrak{m} \subset A^0$ such that $f_1, \dots, f_n$ is not a regular sequence in the local regular ring $A^0_{\mathfrak{m}}$. Therefore there exist elements $h_i \in A^0_{\mathfrak{m}}$ such that $\sum_i h_i f_i = 0$, and according to [193, p. 161] we can assume that there exists an index j such that $h_j \notin (f_1, \dots, f_n) \subset A^0_{\mathfrak{m}}$. After a possible multiplication by an element of $A^0 - \mathfrak{m}$ it is not restrictive to assume $h_i \in A^0$ for every i and $\sum_i h_i f_i = 0$, $h_j \notin (f_1, \dots, f_n)$ in the ring A^0. Writing $a = \sum h_i y_i \in A^{-1}$ we have

$$da = 0, \qquad [a, x_j]_\Delta = h_j,$$

and then the graded Lie algebra $H^*(A[-1])$ is not abelian. □

Notice that the degeneration property can be conveniently described in term of degeneration of spectral sequences; according to a standard and well known fact (Corollary C.6.7) the inclusion of DG-vector spaces $(A[[t]], d - t\Delta) \to (A((t)), d - t\Delta)$ is injective in cohomology if and only if the spectral sequence of the filtration $t^p A[[t]]$ degenerates at E_1.

As an application of Example 13.6.5 we can prove the Lie formality of Koszul bracket on the de Rham complex of a Poisson manifold (Definition 9.5.3).

Theorem 13.6.8 *Let (A^*_X, d) be the de Rham complex of a Poisson manifold (X, π) and consider the Koszul bracket $[-,-]_\pi$ on the shifted complex $A^*_X[1]$. Then the differential graded Lie algebra $(A^*_X[1], -d, [-,-]_\pi)$ is homotopy abelian.*

Proof Recall that the differential BV-algebra structure on (A_X^*, d) is given by the differential operator $\Delta = \mathcal{L}_\pi = [\boldsymbol{i}_\pi, d]$. Setting $f = -\boldsymbol{i}_\pi$ we have $\Delta = [d, f]$ and, by Cartan formulas,

$$[f, \Delta] = [\Delta, \boldsymbol{i}_\pi] = [\mathcal{L}_\pi, \boldsymbol{i}_\pi] = \boldsymbol{i}_{[\pi,\pi]_{SN}} = 0.$$

By Example 13.6.5 the DGBV-algebra (A_X^*, d, Δ) has the degeneration property and the conclusion follows from Corollary 13.6.3. □

It is clear from the proof that the above theorem extends without difficulty to holomorphic Poisson manifolds. This allows us to replace the Kähler assumption on Theorem 9.6.8 with the degeneration of the Hodge to de Rham spectral sequence.

Theorem 13.6.9 *Let (X, π) be a compact holomorphic Poisson manifold and assume that the Hodge to de Rham spectral sequence of X degenerates at E_1. Then for every cohomology class $\omega \in H^1(X, \Omega_X^1)$, the first order deformation $\pi^\#(\omega) \in H^1(X, \Theta_X)$ can be extended to a deformation over $\mathbb{C}[t]/(t^n)$, for every $n > 0$.*

Proof As in the proof of Theorem 9.6.8 it is sufficient to prove that the differential graded Lie algebra $(A_X^{0,*}(\Omega_X^1)[1], -\overline{\partial}, [-,-]_\pi)$ is homotopy abelian.

By the holomorphic version of Theorem 13.6.8 we have that the DG-Lie algebra $(A_X^{0,*}(\Omega_X^*)[1], -\overline{\partial}, [-,-]_\pi)$ is homotopy abelian. Since the Hodge to de Rham spectral sequence of X degenerates at E_1 the inclusion of DG-Lie algebras

$$(A_X^{0,*}(\Omega_X^{\geq 1})[1], -\overline{\partial}, [-,-]_\pi) \to (A_X^{0,*}(\Omega_X^*)[1], -\overline{\partial}, [-,-]_\pi)$$

is injective in cohomology, while the natural projection

$$(A_X^{0,*}(\Omega_X^{\geq 1})[1], -\overline{\partial}, [-,-]_\pi) \to (A_X^{0,*}(\Omega_X^1)[1], -\overline{\partial}, [-,-]_\pi)$$

is surjective in cohomology. It is now sufficient to apply Theorem 12.7.3. □

13.7 Toward an Algebraic Proof of the Kodaira Principle

One of the main ingredients for purely algebraic proofs of several theorems in deformation theory is the interplay between totalization and period data. Just to provide an example, in this section we shall give a sketch of an algebraic proof of Theorem 13.5.4, pointing out that the same argument can be easily adapted to other situations and exported in the setup of smooth algebraic varieties over any algebraically closed field of characteristic 0.

It is straightforward to check that Cartan homotopies are stable under scalar extension. More precisely, if $\boldsymbol{i}\colon L \to M$ is a Cartan homotopy with boundary $\boldsymbol{l}$ and A is a differential graded commutative algebra, then the maps

$$\boldsymbol{i}\otimes \mathrm{Id}_A\colon L\otimes A\to M\otimes A,\qquad (x\otimes a)\mapsto \boldsymbol{i}_x\otimes a,$$

$$\mathrm{Id}_A\otimes \boldsymbol{i}\colon A\otimes L\to A\otimes M,\qquad (a\otimes x)\mapsto (-1)^{\overline{a}}a\otimes \boldsymbol{i}_x,$$

are Cartan homotopies with boundaries $\boldsymbol{l}\otimes \mathrm{Id}_A$ and $\mathrm{Id}_A\otimes \boldsymbol{l}$ respectively.

Lemma 13.7.1 *Let A be a DG-algebra and denote by $\mu\colon A\to Hom^*_{\mathbb{K}}(A,A)$ the operator of left multiplication in A: $\mu_a(b)=ab$. If $(L,V,F,\boldsymbol{i})$ is period data, then $(A\otimes L, A\otimes V, A\otimes F, \mu\otimes \boldsymbol{i})$ is period data.*

Proof By definition the map $\mu\otimes\boldsymbol{i}\colon A\otimes L\to \mathrm{End}^*_{\mathbb{K}}(A\otimes V)$ is given by the formula

$$(\mu\otimes\boldsymbol{i})_{a\otimes x}(b\otimes v)=(-1)^{\overline{b}(\overline{x}-1)}ab\otimes \boldsymbol{i}_x(v),$$

and then $\mu\otimes\boldsymbol{i}=(\mu\otimes \mathrm{Id})\circ(\mathrm{Id}_A\otimes\boldsymbol{i})$. According to Remark 8.8.7 it is sufficient to prove that

$$\mu\otimes \mathrm{Id}\colon A\otimes \mathrm{Hom}^*(V,V)\to \mathrm{Hom}^*(A\otimes V, A\otimes V),$$

$$(\mu\otimes \mathrm{Id})_{a\otimes f}(b\otimes v)=(-1)^{\overline{b}\,\overline{f}}ab\otimes f(v),$$

is a morphism of differential graded Lie algebras. This is completely straightforward and it is left to the reader. □

Definition 13.7.2 A morphism of period data $(L,V,F,\boldsymbol{i})\to(M,W,H,\tilde{\boldsymbol{i}})$ is a pair (f,h) where:

1. $f\colon L\to M$ is a morphism of DG-Lie algebras,
2. $h\colon V\to W$ is a morphism of DG-vector spaces such that $h(F)\subset H$.

The pair (f,h) must be compatible with $\boldsymbol{i},\tilde{\boldsymbol{i}}$ in the obvious way: this means that

$$h(\boldsymbol{i}_a(v))=\tilde{\boldsymbol{i}}_{f(a)}(h(v)).\tag{13.8}$$

Thus it makes sense to consider the category of period data and therefore the notion of semicosimplicial period data.

Definition 13.7.3 The totalization of a semicosimplicial period data $(L,V,F,\boldsymbol{i})$ is the period data

$$(\mathrm{Tot}(L),\mathrm{Tot}(V),\mathrm{Tot}(F),\tilde{\boldsymbol{i}}),$$

where the components of

$$\tilde{\boldsymbol{i}}\colon \mathrm{Tot}(L)\to \mathrm{Hom}^*_{\mathbb{K}}(\mathrm{Tot}(V),\mathrm{Tot}(V))$$

are the scalar extensions of $\boldsymbol{i}$ as in Lemma 13.7.1:

$$\Omega_n\otimes L_n\xrightarrow{\boldsymbol{i}}\mathrm{Hom}^*_{\mathbb{K}}(\Omega_n\otimes V_n,\Omega_n\otimes V_n).$$

The compatibility condition (13.8) immediately implies that $\tilde{\boldsymbol{i}}$ is well defined and it is a Cartan homotopy. Thus for every semicosimplicial period data $(L, V, F, \boldsymbol{i})$ there exists a canonical L_∞-morphism

$$g\colon \operatorname{Tot}(L) \rightsquigarrow G(\operatorname{Tot}(V), \operatorname{Tot}(F))$$

and then a (formal) period map

$$\mathfrak{p}\colon \operatorname{Def}_{\operatorname{Tot}(L)} \to \mathrm{G}_{\operatorname{Tot}(V),\operatorname{Tot}(F)} \simeq \mathrm{G}_{C(V),C(F)},$$

where, as usual $C(-)$ denotes the cochain complex.

Consider now a smooth algebraic variety X; for every integer $p \geq 0$ and every open subset $U \subset X$ let's denote by $\Omega_X^{\geq p}(U)$ the complex

$$\Omega_X^p(U) \xrightarrow{d} \Omega_X^{p+1}(U) \xrightarrow{d} \cdots,$$

where every $\Omega_X^i(U)$ is in degree i. In particular, $\Omega_X^*(U) := \Omega_X^{\geq 0}(U)$ is the de Rham complex of U.

Given an affine open covering $\mathcal{U} = \{U_i\}$ of X and a fixed integer $p \geq 0$ we may consider the semicosimplicial period data $(\Theta_X(\mathcal{U}), \Omega_X^*(\mathcal{U}), \Omega_X^{\geq p}(\mathcal{U}), \boldsymbol{i})$, where $\Theta_X(\mathcal{U})$ is the semicosimplicial Lie algebra

$$\prod_i \Theta_X(U_i) \rightrightarrows \prod_{i,j} \Theta_X(U_{ij}) \substack{\longrightarrow\\ \longrightarrow\\ \longrightarrow} \prod_{i,j,k} \Theta_X(U_{ijk}) \substack{\longrightarrow\\ \longrightarrow\\ \longrightarrow\\ \longrightarrow} \cdots,$$

$\Omega_X^*(\mathcal{U})$ is the semicosimplicial DG-vector space

$$\prod_i \Omega_X^*(U_i) \rightrightarrows \prod_{i,j} \Omega_X^*(U_{ij}) \substack{\longrightarrow\\ \longrightarrow\\ \longrightarrow} \prod_{i,j,k} \Omega_X^*(U_{ijk}) \substack{\longrightarrow\\ \longrightarrow\\ \longrightarrow\\ \longrightarrow} \cdots,$$

$\Omega_X^{\geq p}(\mathcal{U})$ is defined in the same way and $\boldsymbol{i}$ is the semicosimplicial Cartan homotopy induced by the usual contraction as in Example 13.3.2. According to Corollary 13.3.5 the period data

$$(\operatorname{Tot}(\Theta_X(\mathcal{U})), \operatorname{Tot}(\Omega_X^*(\mathcal{U})), \operatorname{Tot}(\Omega_X^{\geq p}(\mathcal{U})), \tilde{\boldsymbol{i}})$$

induces an L_∞-morphism

$$g\colon \operatorname{Tot}(\Theta_X(\mathcal{U})) \rightsquigarrow G_{\operatorname{Tot}(\Omega_X^*(\mathcal{U})),\operatorname{Tot}(\Omega_X^{\geq p}(\mathcal{U}))}.$$

Now the cochain complex $C(\Omega_X^*(\mathcal{U}))$ computes the hypercohomology of the de Rham complex and $C(\Omega_X^{\geq i}(\mathcal{U}))$, $i \geq 0$, is the Hodge filtration with respect to the cover $\mathcal{U}$. In particular, if the Hodge to de Rham spectral sequence of X degenerates at E_1, then the inclusion $\operatorname{Tot}(\Omega_X^{\geq p}(\mathcal{U})) \to \operatorname{Tot}(\Omega_X^*(\mathcal{U}))$ is injective in cohomology and the DG-Lie algebra $G_{\operatorname{Tot}(\Omega_X^*(\mathcal{U})),\operatorname{Tot}(\Omega_X^{\geq p}(\mathcal{U}))}$ is homotopy abelian. Now the same argument of Theorem 13.5.4 tells us that the obstructions to deformation of X are contained in the kernel of the contraction map

$$H^2(X, \Theta_X) = H^2\left(\mathrm{Tot}(\Theta_X(\mathcal{U}))\right) \xrightarrow{\widetilde{i}} H^2\left(G_{\mathrm{Tot}(\Omega_X^*(\mathcal{U})),\mathrm{Tot}(\Omega_X^{\geq p}(\mathcal{U}))}\right).$$

Denoting by $\Omega_X^{<p}$ the cokernel of the inclusion $\Omega_X^{\geq p} \subset \Omega_X^*$ we have a short exact sequence

$$0 \to \mathrm{Tot}(\Omega_X^{\geq p}(\mathcal{U})) \to \mathrm{Tot}(\Omega_X^*(\mathcal{U})) \to \mathrm{Tot}(\Omega_X^{<p}(\mathcal{U})) \to 0$$

and then

$$\begin{aligned} H^2\left(G_{\mathrm{Tot}(\Omega_X^*(\mathcal{U})),\mathrm{Tot}(\Omega_X^{\geq p}(\mathcal{U}))}\right) &= H^1(\mathrm{Hom}_{\mathbb{K}}^*(\mathrm{Tot}(\Omega_X^{\geq p}(\mathcal{U})), \mathrm{Tot}(\Omega_X^{<p}(\mathcal{U})))) \\ &= H^1(\mathrm{Hom}_{\mathbb{K}}^*(C(\Omega_X^{\geq p}(\mathcal{U})), C(\Omega_X^{<p}(\mathcal{U})))) \\ &= \mathrm{Hom}_{\mathbb{K}}^1(H^*(C(\Omega_X^{\geq p}(\mathcal{U})), H^*(C(\Omega_X^{<p}(\mathcal{U}))) \\ &= \mathrm{Hom}_{\mathbb{K}}^1(\mathbb{H}^*(X, \Omega_X^{\geq p}), \mathbb{H}^*(X, \Omega_X^{<p})). \end{aligned}$$

On the other hand, $\widetilde{i}$ has a factorization

$$\begin{array}{ccc} H^2(X, \Theta_X) & & \\ \downarrow i & \searrow \widetilde{i} & \\ \mathrm{Hom}_{\mathbb{K}}^2(H^*(X, \Omega_X^p), H^*(X, \Omega_X^{p-1})) & \xrightarrow{\tau} & \mathrm{Hom}_{\mathbb{K}}^1(\mathbb{H}^*(X, \Omega_X^{\geq p}), \mathbb{H}^*(X, \Omega_X^{<p})), \end{array}$$

where τ is the truncation map in hypercohomology. If the Hodge to de Rham spectral sequence degenerates at E_1 then τ is injective and therefore the obstruction to deformations of X are contained in the kernel of

$$H^2(X, \Theta_X) \xrightarrow{i} \prod_{p,q} \mathrm{Hom}_{\mathbb{K}}(H^q(X, \Omega_X^p), H^{q+2}(X, \Omega_X^{p-1})).$$

According to the Lefschetz principle, the degeneration at E_1 of the Hodge to de Rham spectral sequence holds for every projective manifold over a field of characteristic 0. Moreover, Deligne and Illusie have given a purely algebraic proof of this fact, where Hodge theory is replaced by a suitable reduction in positive characteristic.

13.8 Exercises

13.8.1 Let (L, d) be a DG-vector space and let $\delta \in \mathrm{Hom}_{\mathbb{K}}^{-1}(L, L)$ be a linear map such that $\delta^2 = 0$, $\delta d \delta = \delta$ and $d = d\delta d$. Defining $Z^n = L^n \cap \ker d$ and $\mathcal{H}^n = Z^n \cap \ker \delta$, prove that for every n there exist two direct sum decompositions:

$$L^n = Z^n \oplus \delta(L^{n+1}), \qquad Z^n = d(L^{n-1}) \oplus \mathcal{H}^n.$$

13.8.2 In the notation of Proposition 13.2.4 prove that the homotopy type of the DG-Lie algebra N is independent of the choice of δ. (Hint: if V, W are the L_∞ minimal models of N, L respectively and $f\colon V \rightsquigarrow W$ is the L_∞ morphism induced by the inclusion $N \subset L$, then $W^{>0} \subset W$ is an L_∞-subalgebra and $f\colon V \rightsquigarrow W^{>0}$ is an L_∞ isomorphism.)

13.8.3 In the notation of Definition 13.2.3, prove that $F^{-1}(x) = x$ for every $x \in \widehat{\mathcal{H}^1}$ if and only if $[\mathcal{H}^1, \mathcal{H}^1] \subset \mathcal{H}^2$.

13.8.4 Let $(V, q_1, q_2, q_3, \ldots)$ be an $L_\infty[1]$-algebra and let $x \in V^0$ be a vector such that $q_n(y, x, x, \ldots, x) = 0$ for every $y \in V$ and every $n > n_0$. Assume moreover that x is a solution of the Maurer–Cartan equation

$$\sum_{n\geq 1} \frac{1}{n!} q_n(x, \ldots, x) = 0.$$

Prove that:

1. the operator

$$\delta_x\colon V \to V, \qquad \delta_x(y) = \sum_{n\geq 1} \frac{1}{(n-1)!} q_n(y, x, \ldots, x)$$

is a differential: $\delta_x^2 = 0$;
2. if $f\colon V \rightsquigarrow H$ is an L_∞-morphism onto an abelian $L_\infty[1]$-algebra H, then the linear map

$$g\colon V \to H, \qquad g(y) = \sum_{n\geq 1} \frac{f_n}{(n-1)!}(y, x, \ldots, x),$$

satisfies the equation $g\delta_x = 0$;
3. if V is a homotopy abelian $L_\infty[1]$-algebra, then for every $A \in \mathbf{Art}_{\mathbb{K}}$, every $x \in \mathrm{MC}_V(A)$ and every i, the A-module $H^i(V \otimes A, \delta_x)$ is free.

13.8.5 Let L be a nilpotent DG-Lie algebra. Prove that $x \in L^1$ is a Maurer–Cartan element if and only if the equality $e^{-s^{-1}x} * 0 \in L$ holds in the DG-Lie algebra $\mathrm{cone}(\mathrm{Id}_L) = s^{-1}L \oplus L$ of Example 5.6.8.

13.8.6 Prove that for a differential graded BV-algebra (A, d, Δ) the following conditions are equivalent:

1. (A, d, Δ) has the degeneration property;
2. the morphism of complexes $(A[[t]], d - t\Delta) \xrightarrow{t\mapsto 0} (A, d)$ is surjective in cohomology;

3. the morphism of complexes $(A[[t]], d - t\Delta) \xrightarrow{\text{multiplication by } t} (A[[t]], d - t\Delta)$ is injective in cohomology;
4. the inclusion of complexes $(A[[t]], d - t\Delta) \hookrightarrow (A((t)), d - t\Delta)$ is injective in cohomology;
5. the cohomology of $(A[[t]], d - t\Delta)$ is a flat module over the principal ideal domain $\mathbb{K}[[t]]$.

13.8.7 Let L, M be two differential graded Lie algebras, $\boldsymbol{i}\colon L \to M$ a Cartan homotopy and $r \in \mathbb{K}$. Prove that the sequence of maps $g_n\colon L^{\wedge n} \to M$,

$$g_1(a) = r\boldsymbol{l}_a, \qquad g_2(a \wedge b) = r(1-r)\boldsymbol{i}_{[a,b]}, \qquad g_n = 0 \quad \forall\, n \geq 3,$$

give an L_∞-morphism $L \rightsquigarrow M$.

13.8.8 Denote as usual by $K(N \to M)$ the Thom–Whitney homotopy fibre of a morphism $N \to M$ of DG-Lie algebras.

For every DG-Lie algebra L, describe an explicit L_∞ right inverse of the surjective quasi-isomorphism $K(L \xrightarrow{j} \operatorname{cone}(\mathrm{Id}_L)) \to L$.

13.8.9 A **Cartan calculus** is a commutative diagram of DG-Lie algebras

$$\begin{array}{ccc} L & \xrightarrow{h} & H \\ \downarrow{\scriptstyle j} & & \downarrow{\scriptstyle \chi} \\ \operatorname{cone}(\mathrm{Id}_L) & \xrightarrow{f} & M \end{array} \tag{13.9}$$

where $j\colon L \to \operatorname{cone}(\mathrm{Id}_L)$ denotes the natural inclusion introduced in Example 8.8.8. Prove that to give a Cartan calculus (13.9) as above is the same as giving two morphisms of DG-Lie algebras $L \xrightarrow{h} H \xrightarrow{\chi} M$ and a Cartan homotopy $\boldsymbol{i}\colon L \to M$ with boundary $\boldsymbol{l} = \chi h$ and deduce that (13.9) induces a canonical span of DG-Lie algebras

$$\begin{array}{ccccc} & & K(j) & & \\ & \swarrow{\scriptstyle \text{q-iso}} & & \searrow & \\ L & & & & K(\chi). \end{array}$$

References

The original Nijenhuis and Richardson proof of the formal Kuranishi theorem was published in [208, 209] and later reproduced, with minor modification, in [95, 176]. The involved terminology has changed in the last 50 years: for instance, in the old terminology the map F introduced in Lemma 13.2.2 was called the *Kuranishi map*, while the map K of Definition 13.2.3 was called the *obstruction map*; we already said that the Maurer–Cartan equation was called the *deformation equation*. The interpretation of the formal Kuranishi theorem in terms of the homotopy transfer of L_∞ structures comes essentially from [84].

Theorem 13.3.4 is taken from [132] and it is a simplification of an essentially equivalent result of [75]; we refer to [79] for a different interpretation of the L_∞-morphism associated to a Cartan homotopy.

The algebraic description of the Griffiths period map as an L_∞-morphism (Theorem 13.5.5) comes from [75]. A similar result holds also for formal Abel–Jacobi maps at smooth submanifolds [77].

Corollary 13.6.3 was first proved by Terilla in [256], and then by Katzarkov, Kontsevich and Pantev in [140] and by Iacono [128] with different approaches. The proof given here follows [128] and we refer to [11] for a generalization to commutative BV_∞-algebras.

Theorem 13.6.8 was first proved by Sharygin and Talalaev in [247] and then reobtained by Fiorenza and Manetti as a special case of a more general result regarding Poisson brackets [76]. The proof proposed here follows essentially [129], while the proof given in [76] uses the criterion of Proposition 12.7.5.

For a more complete treatment of the argument of Sect. 13.7 we refer to the papers [126, 130]. The paper [56] by Deligne and Illusie contains a purely algebraic proof of the degeneration of the Hodge to de Rham spectral sequence.

Chapter 14
Tree Summation Formulas

It is a common feature in mathematics that recursive formulas can often be expanded into summation formulas over a certain set of combinatorial data. In this chapter we consider the cases where the combinatorial data involved are rooted trees (Definition 14.1.2) possibly equipped with additional data (orientation, labelling etc.). In particular, we shall provide tree summation formulas for the BCH product (recursive formula of Definition 2.5.1) and for the homotopy transfer of $L_\infty[1]$-structures (recursive formula of Theorem 12.4.1).

14.1 Rooted Trees and Forests

Let's start by recalling the basic terminology about directed graphs.

Definition 14.1.1 A **directed graph** Γ is the data of two finite sets V_Γ, E_Γ and of two maps

$$i, t\colon E_\Gamma \to V_\Gamma.$$

Elements of V_Γ are called vertices, elements of E_Γ are called edges. A morphism of directed graphs $\Gamma \to \Delta$ is a pair of maps $V_\Gamma \to V_\Delta$, $E_\Gamma \to E_\Delta$ commuting with the maps i, t in the obvious way.

Every directed graph can geometrically represented by taking a small circle for every vertex and an arrow for every edge:

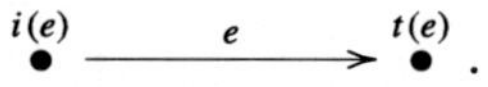

M. Manetti, *Lie Methods in Deformation Theory*, Springer Monographs in Mathematics,
https://doi.org/10.1007/978-981-19-1185-9_14

If e is an edge, then $i(e)$ and $t(e)$ are called respectively the initial and the terminal vertex of e. We shall also say that e is an outgoing edge of $i(e)$ and an incoming edge of $t(e)$.

The **arity** (also called valence in the literature) $|v|$ of a vertex v of a directed graph is the number of its incoming edges: $|v| = |t^{-1}(v)|$.

A (possibly empty) composable sequence of edges is called a directed path. More precisely, a directed path of length n from a vertex u to a vertex v is a sequence of n edges $e_1, \ldots, e_n$ such that

$$i(e_1) = u, \quad t(e_1) = i(e_2), \quad \ldots, \quad t(e_n) = v:$$

$$\overset{u}{\bullet} \xrightarrow{e_1} \bullet \xrightarrow{e_2} \cdots \xrightarrow{e_n} \overset{v}{\bullet}.$$

Definition 14.1.2 A directed graph is called a **rooted tree** if there exists a vertex, called a **root**, with the property that for every vertex x there exists a *unique* directed path from x to the root; clearly the root is unique. A rooted tree is called **planar** or **oriented** if for every vertex x it is given a total order in the set of incoming edges of x.

For example, the **singleton tree**, defined as the directed graph with one vertex and zero edges, is a planar rooted tree.

The above definition of a planar rooted tree is particularly useful from the combinatorial point of view. Every planar rooted tree can be visualized as a topological tree embedded in the oriented Euclidean half-plane $\{(x, y) \in \mathbb{R}^2 \mid y \leq 0\}$ with exactly one vertex (the root) lying in the border $\{y = 0\}$; for every vertex v, the set of incoming edges of v is ordered from left to the right.

Every rooted tree carries a natural partial ordering on the set of vertices. Given two vertices u, v in a rooted tree we define $u \leq v$ if the vertex v belongs to the (unique) directed path from u to the root.

A **leaf** is a vertex without incoming edges (arity 0); equivalently, a vertex is a leaf if it is a minimal element for the ordering $\leq$. We shall denote by $L_\Gamma \subseteq V_\Gamma$ the set of leaves of a rooted tree Γ.

A vertex is called **internal** if it is not a leaf (Fig. 14.1); notice that, if a rooted tree has at least two vertices, then the root is an internal vertex. A rooted tree is called **reduced** if every vertex is either a leaf or has arity ≥ 2.

A planar rooted tree Γ is a said to be a **binary tree** if every internal vertex has exactly two incoming edges.

The (finite) set of isomorphism classes of reduced rooted trees (resp.: planar reduced rooted trees, planar binary trees) with n leaves will be denoted by $\mathfrak{T}_n$ (resp.: $\mathcal{T}_n, \mathcal{B}_n$). We use the notation

$$\mathcal{T} = \bigcup_{n>0} \mathcal{T}_n, \qquad \mathcal{B} = \bigcup_{n>0} \mathcal{B}_n.$$

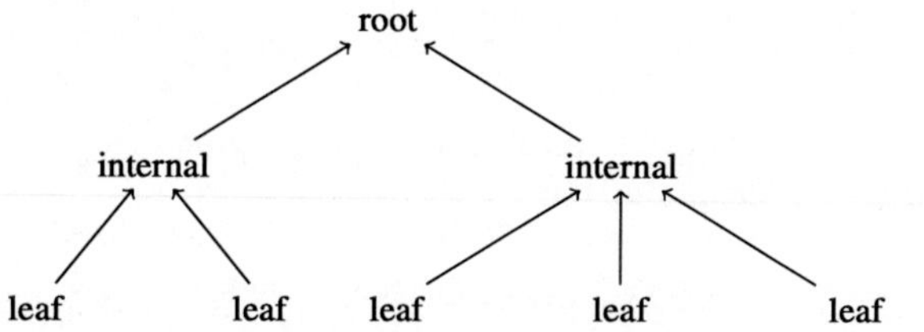

Fig. 14.1 Every finite planar rooted tree can be described graphically in the plane with the root at the top and the leaves at the bottom (every directed path moves upward), and with the anticlockwise order (from left to right) on every set of incoming edges on a vertex

Example 14.1.3 A rooted tree is called a **ladder** if every vertex has arity ≤ 1. Clearly, a ladder is reduced if and only if it is the singleton tree. A rooted tree is called a **corolla** if every vertex is either a leaf or the root.

There are (up to isomorphism) exactly three planar reduced rooted trees with three leaves:

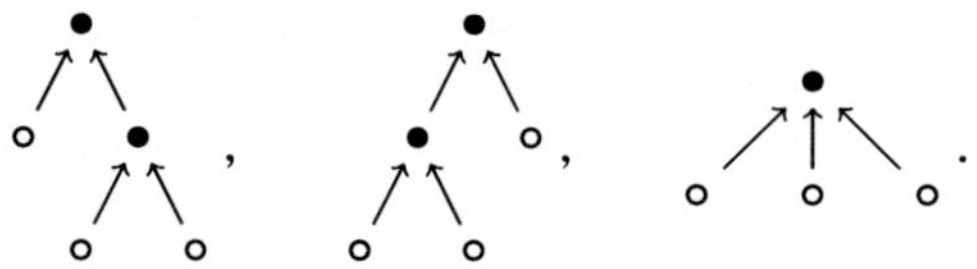

In the above pictures, the white circles are the leaves. The first two trees are binary trees (every vertex has arity either 0 or 2); the third picture is a corolla with root of arity 3. The first two pictures are isomorphic as rooted trees, hence $|\mathcal{B}_3| = 2$, $|\mathcal{T}_3| = 3$ and $|\mathfrak{T}_3| = 2$.

Example 14.1.4 (*Free magmatic algebras*) Planar rooted binary trees are strictly related to magmatic algebras; if R is a subset of a magmatic algebras $(M, \cdot)$ and $\Gamma \in \mathcal{B}$ is a planar rooted binary tree with subset of leaves $L_\Gamma \subset V_\Gamma$, then any application $f : L_\Gamma \to R$ is called a labelling of the leaves of Γ with elements of R. Given a labelling $f : L_\Gamma \to R$ we denote by $Z_\Gamma(f) \in M$ the element obtained by performing the product at every internal vertex, in the order arising from the planar structure of the directed tree. For example, the labelled planar rooted tree

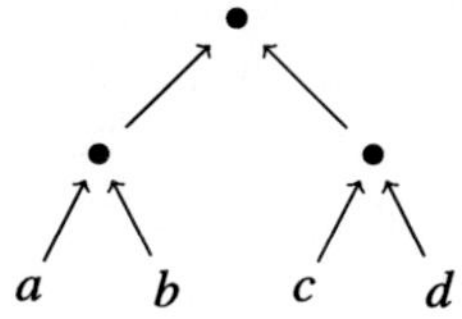

$$a, b, c, d \in R\,,$$

gives the product element $Z_\Gamma(f) = (a \cdot b) \cdot (c \cdot d) \in M$.

Consider now a fixed set S and denote by $M(S)$ the vector space with basis the set of (isomorphism classes of) pairs (Γ, f), with $\Gamma \in \mathcal{B}$ a binary planar rooted tree and $f : L_\Gamma \to S$ a map. Notice that there exists a bijection between S and the subset of pairs (Γ, f) as above such that $\Gamma = \bullet$ is the singleton graph (one vertex and zero edges).

The vector space $M(S)$ has a natural structure of magmatic algebras, with product equal to the bilinear extension of

$$(\Gamma, f) \cdot (\Delta, g) = \left(\widehat{\Gamma\Delta}, f \amalg g\right),$$

where the planar binary rooted tree $\widehat{\Gamma\Delta}$ is obtained by taking the disjoint union of Γ, Δ and a new vertex r, then adding an ordered pair of new edges v_1, v_2 such that

$$t(v_1) = t(v_2) = r, \qquad i(v_1) = \mathrm{root}(\Gamma), \qquad i(v_1) = \mathrm{root}(\Delta).$$

Since $L_{\widehat{\Gamma\Delta}}$ is the disjoint union of L_Γ and L_Δ, the labelling $f \amalg g$ is defined in the obvious way.

It is plain that for every magmatic algebra H and for every map $h \colon S \to H$, the linear map

$$M(S) \to H, \qquad (\Gamma, f) \mapsto Z_\Gamma(hf),$$

is the unique morphism of magmatic algebras extending h. In other words, $M(S)$ is the free magmatic algebra generated by S.

Definition 14.1.5 A **rooted forest** is a finite directed graph such that every connected component is a rooted tree. A rooted forest is called reduced if every component of it is a reduced rooted tree.

As in the case of rooted trees, if u, v are two vertices of a rooted forest we shall write $u \le v$ if there exists a directed path from u to v (such a path is necessarily unique); the relation $\le$ is a partial ordering in the set of vertices. As in the case of trees, a vertex u is a leaf if and only if $v \le u$ implies $u = v$, while it is a root if and only if $u \le v$ implies $u = v$.

Given a rooted forest Ω we denote by V_Ω the set of its vertices, by $R_\Omega \subset V_\Omega$ the subset of roots and by $L_\Omega \subset V_\Omega$ the subset of leaves.

Given two rooted forests Ω, Γ, a bijective map $\alpha \colon V_\Omega \to V_\Gamma$ extends to an isomorphism of directed graphs if and only if it is an isomorphism of partially ordered set, i.e., if and only if α and α^{-1} are order-preserving maps. Notice that every isomorphism of rooted forests is uniquely determined by its restriction to their leaves.

An **oriented** rooted forest is an ordered finite sequence of planar rooted trees. In other words, an oriented rooted forest is a rooted forest equipped with an **orientation**, defined as the data of:

1. a total ordering of the roots;
2. for every vertex, a total ordering of its incoming edges.

It is useful to give an equivalent definition of orientation of a rooted forest Ω as the data of a total ordering $\preceq$ of leaves such that if $i, j \in L_\Omega$, $i \preceq j$, $i \le z$ and $j \le z$ for some vertex $z \in V_\Omega$, then $h \le z$ for every $i \preceq h \preceq j$. The equivalence of the above two notions of orientation is easy to see and left to the reader.

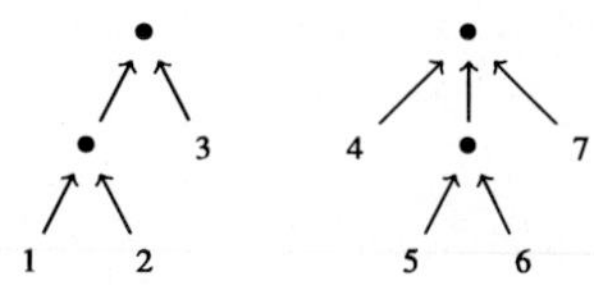

Fig. 14.2 An oriented rooted forest and its ordering on leaves

Every oriented rooted forest can be visualized as a disjoint union of topological trees immersed in the oriented Euclidean half-plane $\{(x, y) \in \mathbb{R}^2 \mid y \le 0\}$, where every tree component has exactly one vertex (the root) on the axis $y = 0$; the set of leaves (resp.: roots, incoming edges of a vertex) is conventionally ordered from left to right (Fig. 14.2).

Remark 14.1.6 Let Ω be a rooted forest with r roots and denote by $\text{Aut}(\Omega)$ the group of its automorphisms. Since every orientation of Ω is uniquely determined by a total ordering of the roots and, for every vertex, by a total ordering of incoming edges, it follows that the number of isomorphism classes of orientations on Ω is equal to

$$\frac{r!}{|\text{Aut}(\Omega)|}\prod_{v\in V_\Omega}|v|!\,.$$

Counting Planar Rooted Trees

We have already introduced the set $\mathcal{T}_n$ of isomorphism classes of planar reduced rooted trees with n leaves, as well as its subset $\mathcal{B}_n \subset \mathcal{T}_n$ of isomorphism classes of planar binary rooted trees. It is easy to describe explicitly the sets $\mathcal{T}_n$ and $\mathcal{B}_n$ for low values of n and compute their cardinality:

$$|\mathcal{B}_1| = 1,\quad |\mathcal{B}_2| = 1,\quad |\mathcal{B}_3| = 2,\quad |\mathcal{B}_4| = 5,\quad |\mathcal{B}_5| = 14,\quad |\mathcal{B}_6| = 42,\quad \dots;$$

$$|\mathcal{T}_1| = 1,\quad |\mathcal{T}_2| = 1,\quad |\mathcal{T}_3| = 3,\quad |\mathcal{T}_4| = 11,\quad |\mathcal{T}_5| = 45,\quad |\mathcal{T}_6| = 197,\quad \dots .$$

In order to compute the numbers $|\mathcal{B}_n|$, $|\mathcal{T}_n|$ for higher n it is useful to introduce the generating formal power series

$$\beta(t) = \sum_{n=1}^{\infty}|\mathcal{B}_n|t^n = t + t^2 + 2t^3 + \cdots,\qquad \tau(t) = \sum_{n=1}^{\infty}|\mathcal{T}_n|t^n = t + t^2 + 3t^3 + \cdots .$$

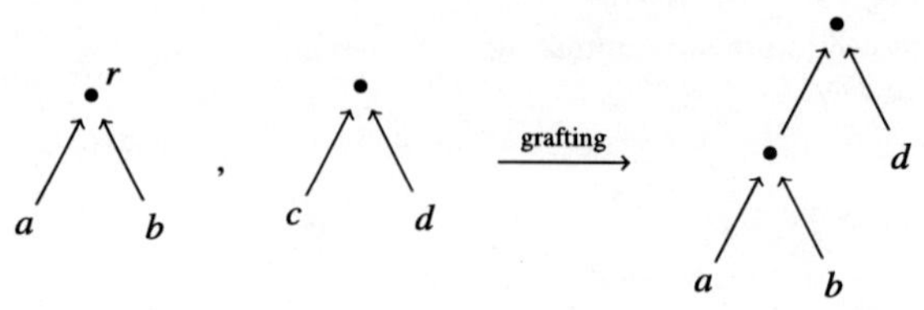

Fig. 14.3 Grafting the root r in the leaf c

Proposition 14.1.7 *In the above notation we have:*

$$\beta(t) = \sum_{n=1}^{\infty} |\mathcal{B}_n| t^n = \frac{1-\sqrt{1-4t}}{2},$$

$$\tau(t) = \sum_{n=1}^{\infty} |\mathcal{T}_n| t^n = \frac{t+1-\sqrt{1-6t+t^2}}{4}.$$

Proof We only prove the formula for $\tau(t)$; the (slightly simpler) proof for $\beta(t)$ is completely the same and it is left to the reader.

We denote by $\mathcal{F}_n^m$ the set of isomorphism classes of oriented reduced rooted forests with n leaves and m roots. Thus $\mathcal{F}_n^1 = \mathcal{T}_n$, $\mathcal{F}_n^m = \emptyset$ whenever $n < m$ and $\mathcal{F}_n^n$ contains only one element (n copies of the singleton). Since every element in $\mathcal{F}_n^m$ is an oriented sequence of m planar rooted trees, for every $k > 0$ we have

$$\sum_{n=1}^{\infty} |\mathcal{F}_n^m| t^n = \sum_{n=m}^{\infty} |\mathcal{F}_n^m| t^n = \tau(t)^m, \quad \sum_{n,m=k}^{\infty} |\mathcal{F}_n^m| t^n = \sum_{m=k}^{\infty} \tau(t)^m = \frac{\tau(t)^k}{1-\tau(t)}.$$

It is now useful to introduce the composition maps

$$\circ\colon \mathcal{F}_l^m \times \mathcal{F}_n^l \to \mathcal{F}_n^m, \qquad (\Gamma, \Omega) \mapsto \Gamma \circ \Omega,$$

where $\Gamma \circ \Omega$ is the grafting of the roots of Ω with the leaves of Γ via the unique bijection that is monotone with respect to the total ordering of the l roots of Ω (resp.: of the l leaves of Γ) induced by the orientation of Ω (resp.: the orientation of Γ), see Fig. 14.3. For example, we have:

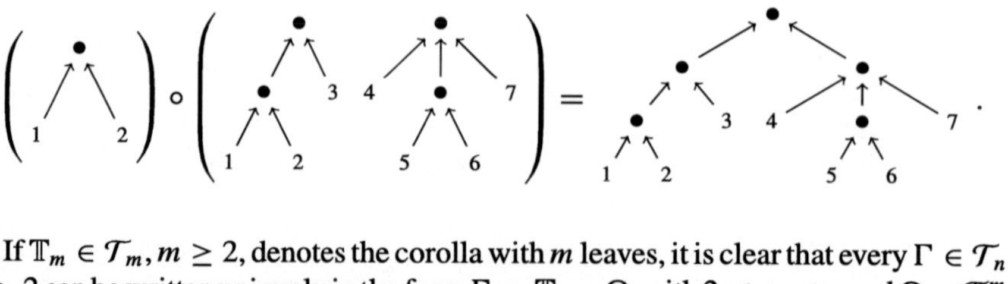

If $\mathbb{T}_m \in \mathcal{T}_m$, $m \geq 2$, denotes the corolla with m leaves, it is clear that every $\Gamma \in \mathcal{T}_n$, $n \geq 2$ can be written uniquely in the form $\Gamma = \mathbb{T}_m \circ \Omega$, with $2 \leq m \leq n$ and $\Omega \in \mathcal{F}_n^m$.

Hence for every $n \geq 2$ we have $|\mathcal{T}_n| = \sum_{m=2}^{n} |\mathcal{F}_n^m|$ and then, since $|\mathcal{T}_0| = 0, |\mathcal{T}_1| = 1$ we get the equation

$$\tau(0) = 0, \qquad \tau(t) = t + \sum_{n,m=2} |\mathcal{F}_n^m| t^n = t + \frac{\tau(t)^2}{1 - \tau(t)},$$

that is equivalent to $4\tau(t) = t + 1 - \sqrt{1 - 6t + t^2}$. □

For later use it is convenient to denote by

$$\amalg \colon \mathcal{F}_n^m \times \mathcal{F}_a^b \to \mathcal{F}_{n+a}^{m+b}$$

the obvious operation of ordered disjoint union:

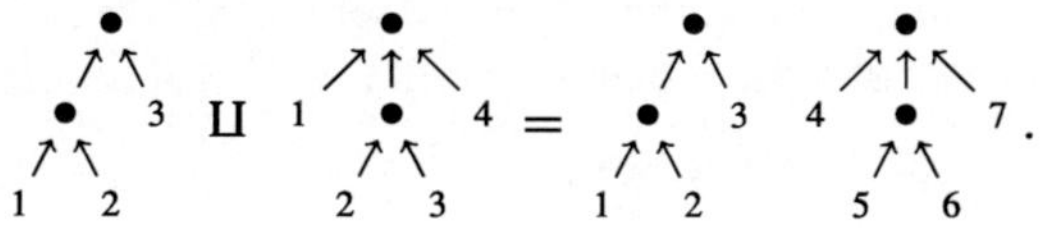

(14.1)

The operations $\circ$ and $\amalg$ are associative and satisfy the interchange law: if $V \circ A$ and $W \circ B$ are defined, then

$$(V \amalg W) \circ (A \amalg B) = (V \circ A) \amalg (W \circ B)\,.$$

14.2 A Tree Summation Formula for the BCH Product

In Example 14.1.4 we have seen that if S is a subset of a magmatic algebra R over a field $\mathbb{K}$, then the elements $Z_\Gamma(f)$, with $\Gamma \in \mathcal{B}$ and $f \colon L_\Gamma \to S$, are a set of generators, as a vector space, of the magmatic subalgebra generated by S. If the product in R is either symmetric or skew-symmetric (e.g. if R is a Lie algebra), then we may reduce the set of generators by a suitable choice of the labelling maps. For instance one can fix a total order on S and consider monotone labelling maps with respect to suitable order in the set of leaves.

In this section we are interested in Lie algebras and the Baker–Campbell–Hausdorff product; in this framework it is convenient to introduce the notion of subroot of planar rooted trees.

For every planar rooted tree Γ, with set of vertices V_Γ and set of leaves $L_\Gamma \subseteq V_\Gamma$ we define two maps

$$p \colon V_\Gamma \to L_\Gamma, \qquad d \colon V_\Gamma \to \mathbb{N}\,,$$

in the following recursive way: if v is a leaf, then $p(v) = v$ and $d(v) = 0$. If v is an internal vertex and $\overset{u}{\bullet} \xrightarrow{\;e\;} \overset{v}{\bullet}$ is the rightmost incoming edge on v, then $p(v) =$

$p(u)$ and $d(v) = d(u) + 1$. Recall that in our convention, where every directed edge moves upwards, rightmost means the maximum in the total ordering of incoming edges of a vertex.

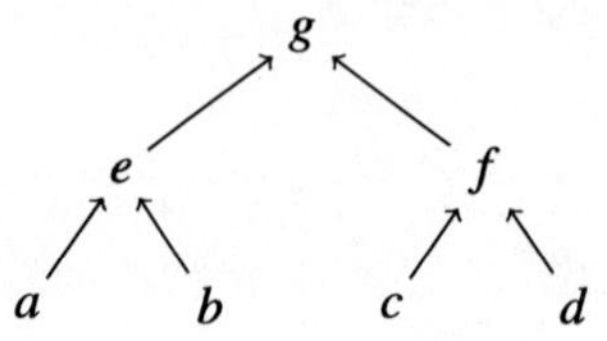

$$p(g) = p(f) = p(d) = d,$$
$$p(e) = p(b) = b,$$
$$p(a) = a,$$
$$p(c) = c.$$

In other words, if v is internal we move backwards throughout rightmost edges for $d(v)$ times until a leaf $p(v)$ is reached. Notice that for every vertex we have $p(v) \leq v$.

Definition 14.2.1 A **subroot** of a finite planar rooted tree is either the root or a vertex v such that its unique outgoing edge e is not the rightmost incoming edge of $t(e)$. The set of subroots of a finite planar rooted tree Γ will be denoted by S_Γ.

For example, in the planar rooted tree

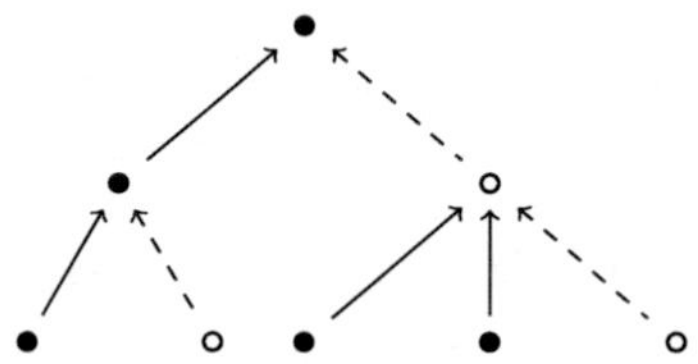

the rightmost edges are dashed and subroots are denoted by the black circles $\bullet$.

Notice that the restriction $p\colon S_\Gamma \to L_\Gamma$ is bijective, and a vertex v is a subroot if and only if $d(v) \geq d(u)$ for every vertex u such that $p(u) = p(v)$.

Definition 14.2.2 For every partially ordered set $(A, \leq)$, we denote by $\mathcal{B}(A)$ the set of pairs (Γ, f), where $\Gamma \in \mathcal{B}$ is a planar binary rooted tree and $f\colon L_\Gamma \to A$ is a map such that the composition

$$fp\colon (V_\Gamma, \leq) \to (A, \leq)$$

is an order-preserving map.

For example, if Γ is the planar rooted tree

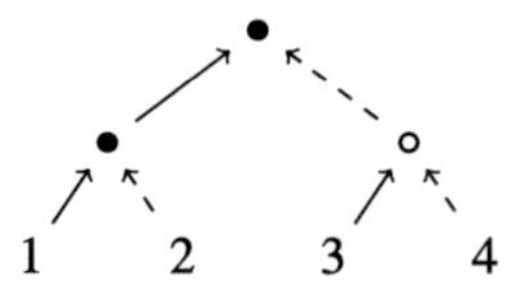

then a pair (Γ, f), $f\colon L_\Gamma = \{1, 2, 3, 4\} \to A$, belongs to $\mathcal{B}(A)$ if and only if $f(1) \leq f(2) \leq f(4)$ and $f(3) \leq f(4)$.

Example 14.2.3 The set $\mathcal{B}_3(y < x)$ contains exactly 9 labelled trees with 3 leaves, namely:

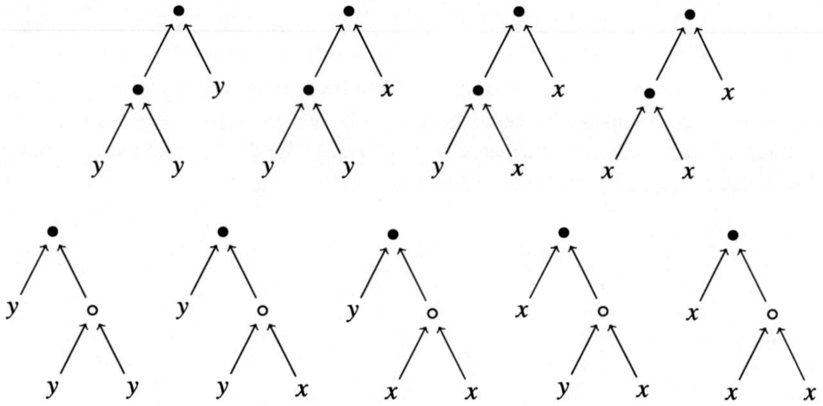

Example 14.2.4 For every integer $m \geq 0$ the set $\mathcal{B}_{m+1}(y < x)$ contains a unique labelled graph Ω_m with one leaf labelled x and m leaves labelled y:

$\Omega_0 = \overset{\bullet}{x}$, $\Omega_1 =$ 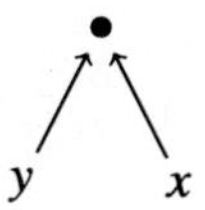, $\Omega_2 =$ 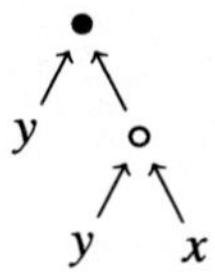, $\ldots$

We are now ready to write a sum over trees formula of the BCH product; in the next theorem we denote by b_n the Bernoulli number B_n multiplied by $n!$:

$$\sum_{n \geq 0} b_n t^n = \frac{t}{e^t - 1}.$$

Theorem 14.2.5 *Let L be a nilpotent Lie algebra. In the above notation, for every $x, y \in L$ their Baker–Campbell–Hausdorff product $x \bullet y$ is equal to*

$$x \bullet y = \sum_{(\Gamma, f) \in \mathcal{B}(y<x)} \left(\prod_{v \in S_\Gamma} \frac{b_{d(v)}}{o(v)} \right) Z_\Gamma(f), \tag{14.2}$$

where $S_\Gamma \subset V_\Gamma$ is the set of subroots, and for every subroot $v \in S_\Gamma$ the integer $o(v)$ is the number of leaves $u \in L_\Gamma$ such that $u \leq v$ and $f(u) = f(p(v))$.

Notice that if $v \in S(\Gamma) \cap L(\Gamma)$, then $d(v) = 0$, $o(v) = 1$ and then $b_{d(v)} o(v) = 1$; therefore, in the formula (14.2) it is sufficient to take into account only the internal subroots of Γ. Moreover, for most labelled trees (Γ, f), the element $Z_\Gamma(f)$ belongs either to the Lie ideal of $[x, x]$ or to the Lie ideal of $[y, y]$; in both cases $Z_\Gamma(f) = 0$.

Proof For every labelled planar binary tree (Γ, f) write

$$b_{(\Gamma,f)} = \prod_{v \in S_\Gamma} \frac{b_{d(v)}}{o(v)}.$$

Let $C(y < x) \subset \mathcal{B}(y < x)$ be the subset of labelled planar binary trees (Γ, f) such that $f(p(v)) = x$ for every internal vertex v. If $(\Gamma, f) \in \mathcal{B}(y < x) - C(y < x)$, then either $(\Gamma, f) = (\bullet, y)$, and hence $Z_\Gamma(f) = y$, or $Z_\Gamma(f)$ belongs to the Lie ideal generated by $[y, y]$, and then $Z_\Gamma(f) = 0$. Therefore:

$$\sum_{(\Gamma,f)\in\mathcal{B}(y<x)} b_{(\Gamma,f)} Z_\Gamma(f) = y + \sum_{(\Gamma,f)\in C(y<x)} b_{(\Gamma,f)} Z_\Gamma(f).$$

According to (2.5) we have $x \bullet y = \sum_{r\geq 0} Z_r$, where $Z_0 = y$ and

$$Z_{r+1} = \frac{1}{r+1} \sum_{m\geq 0} b_m \sum_{i_1+\cdots+i_m=r} (\operatorname{ad} Z_{i_1})(\operatorname{ad} Z_{i_2}) \cdots (\operatorname{ad} Z_{i_m})x, \quad \text{for } r \geq 0.$$

For every $r \geq 0$, let $C_r \subset C(y < x)$ be the subset of trees with exactly r leaves labelled x. Then for every $r > 0$ we have:

$$Z_r = \sum_{(\Gamma,f)\in C_r} b_{(\Gamma,f)} Z_\Gamma(f). \tag{14.3}$$

In fact, for $r = 1$ we have

$$Z_1 = \sum_{m\geq 0} b_m (\operatorname{ad} Z_0)^m x = \sum_{m\geq 0} b_m (\operatorname{ad} y)^m x,$$

whereas $C_1 = \{\Omega_m\}$, $m \geq 0$, and Ω_m is the unique element in C_1 with $m + 1$ leaves (Example 14.2.4). Therefore, the coefficient $b_{(\Omega_m)}$ is exactly b_m and so

$$Z_1 = \sum_{(\Gamma,f)\in C_1} b_{(\Gamma,f)} Z_\Gamma(f).$$

Finally, every element of C_{r+1} is obtained, in a unique way, by starting from a tree Ω_m and grafting, at each one of the m leaves labelled y, the roots of elements of $C_{i_1}, \ldots, C_{i_m}$, with $i_1 + \cdots + i_m = r$. Therefore, the proof of (14.3) follows easily from an induction on r. □

14.3 Automorphisms of $\overline{T^c}(V)$, $\overline{S^c}(V)$ and Inversion Formulas

Let V be a graded vector space. Given a linear map

$$f = \sum_{n=1}^{\infty} f_n \in \mathrm{Hom}^0_{\mathbb{K}}(\overline{T^c}(V), V), \quad \text{with} \quad f_1 = \mathrm{Id}_V, \quad f_n \colon V^{\otimes n} \to V\,,$$

and an oriented rooted forest $\Gamma \in \mathcal{F}^m_n$ with n leaves and m roots, we may interpret Γ as a flow chart where at every vertex of arity $k > 0$ the operation f_k is performed; this gives a well defined linear map $f_\Gamma \in \mathrm{Hom}^0_{\mathbb{K}}(V^{\otimes n}, V^{\otimes m})$.

For instance, the planar rooted trees

Γ: 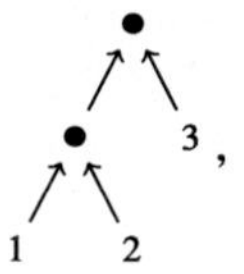

, Ω: 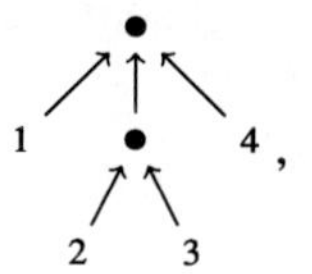

,

and their union $\Gamma \amalg \Omega$ (see (14.1)) give:

$$\begin{aligned} f_\Gamma(v_1 \otimes v_2 \otimes v_3) &= f_2(f_2(v_1 \otimes v_2) \otimes v_3),\\ f_\Omega(v_1 \otimes v_2 \otimes v_3 \otimes v_4) &= f_3(v_1 \otimes f_2(v_2 \otimes v_3) \otimes v_4),\\ f_{\Gamma\amalg\Omega}(v_1 \otimes \cdots \otimes v_7) &= f_\Gamma(v_1 \otimes v_2 \otimes v_3) \otimes f_\Omega(v_4 \otimes v_5 \otimes v_6 \otimes v_7). \end{aligned}$$

With a less informal approach we can also say that, if $\mathbb{T}_n \in \mathcal{F}^1_n$ denotes the corolla with n leaves (Fig. 14.4), then the morphisms $f_\Gamma \in \mathrm{Hom}^0_{\mathbb{K}}(V^{\otimes n}, V^{\otimes m})$ are uniquely determined by the conditions:

$$f_{\mathbb{T}_n} = f_n, \qquad f_{\Gamma\circ\Omega} = f_\Gamma \circ f_\Omega, \qquad f_{\Gamma\amalg\Omega} = f_\Gamma \otimes f_\Omega.$$

Definition 14.3.1 For every pair of positive integers n, m, let $\mathcal{S}^m_n \subseteq \mathcal{F}^m_n$ be the subset of (isomorphism classes of) oriented rooted forests such that every vertex is either a root or a leaf. Equivalently, $\Gamma \in \mathcal{S}^m_n$ if and only if $\Gamma = \mathbb{T}_{b_1} \amalg \cdots \amalg \mathbb{T}_{b_m}$ is a disjoint union of corollas. Moreover, define

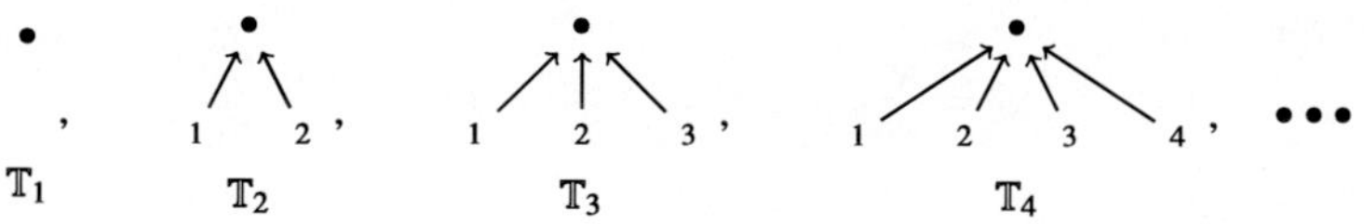

Fig. 14.4 The corollas

$$\mathcal{F} = \bigcup_{n,m} \mathcal{F}_n^m, \qquad \mathcal{S} = \bigcup_{n,m} \mathcal{S}_n^m.$$

Lemma 14.3.2 *Let V be a graded vector space and let $g \in \operatorname{Hom}^0_{\mathbb{K}}(\overline{T^c}(V), V)$ be a linear map such that $g_1 = \mathrm{Id}_V$. Then the maps*

$$G = \sum_{\Gamma \in \mathcal{S}} g_\Gamma : \overline{T^c}(V) \to \overline{T^c}(V), \qquad F = \sum_{\Gamma \in \mathcal{F}} g_\Gamma : \overline{T^c}(V) \to \overline{T^c}(V),$$

are isomorphisms of graded coalgebras and g is the corestriction of G.

Proof Since $\mathcal{F}_n^m = \mathcal{S}_n^m = \emptyset$ whenever $m > n$, the maps G and F are well defined morphisms of graded vector spaces; since $\mathcal{S}_n^1$ contains only the corolla $\mathbb{T}_n$ and $g_{\mathbb{T}_n} = g_n$, it is clear that g is the corestriction of G.

Since every $\Gamma \in \mathcal{S}_n^m$ admits a unique decomposition of type

$$\Gamma = \mathbb{T}_{b_1} \amalg \cdots \amalg \mathbb{T}_{b_m}, \qquad b_i \geq 1, \quad \sum b_i = n,$$

the fact that G is a morphism of coalgebras follows immediately from Corollary 11.3.3.

Similarly, every $\Gamma \in \mathcal{F}_n^m$ admits a unique decomposition of type

$$\Gamma = \Gamma_1 \amalg \cdots \amalg \Gamma_m, \qquad \Gamma_i \in \mathcal{F}_{b_i}^1, \quad \sum b_i = n,$$

and, as above, Corollary 11.3.3 implies that F is a morphism of coalgebras with corestriction

$$f = \sum_n f_n, \quad f_n = \sum_{\Gamma \in \mathcal{F}_n^1} g_\Gamma.$$

Finally, F and G are both isomorphisms of coalgebras since $g_1 = f_1 = \mathrm{Id}_V$. □

Theorem 14.3.3 (Inversion formula) *Let $H : \overline{T^c}(V) \to \overline{T^c}(V)$ be an isomorphism of reduced tensor graded coalgebras with corestriction $h = \sum h_n : \overline{T^c}(V) \to V$ and linear part $h_1 = \mathrm{Id}_V$.*

Setting $g_1 = h_1 = \mathrm{Id}_V$ and $g_n = -h_n$ for every $n \geq 2$, the inverse of H is equal to

$$H^{-1} = \sum_{\Gamma \in \mathcal{F}} g_\Gamma : \overline{T^c}(V) \to \overline{T^c}(V).$$

Proof Writing

$$F = \sum_{\Gamma \in \mathcal{F}} g_\Gamma : \overline{T^c}(V) \to \overline{T^c}(V),$$

we have already seen in Lemma 14.3.2 that

$$H = \sum_{\Gamma \in \mathcal{S}} h_\Gamma \colon \overline{T^c}(V) \to \overline{T^c}(V)$$

and that F is an isomorphism of graded coalgebras with corestriction $\sum_{n,\Gamma\in\mathcal{F}_n^1} g_\Gamma$ and the linear part equal to the identity.

It is therefore sufficient to prove that the corestriction of HF is the projection $\overline{T^c}(V) \to V$, or equivalently that for every $n > 1$ we have

$$\sum_{m=1}^{n} h_m \sum_{\Gamma\in\mathcal{F}_n^m} g_\Gamma = 0.$$

Since $h_1 = g_1 = \mathrm{Id}_V$ and $h_m = -g_m$ for $m > 1$ the above equality is equivalent to

$$\sum_{m=2}^{n} g_m \sum_{\Gamma\in\mathcal{F}_n^m} g_\Gamma = \sum_{\Gamma\in\mathcal{F}_n^1} g_\Gamma .$$

Every $\Gamma \in \mathcal{F}_n^1$ has a unique representation as $\Gamma = \mathbb{T}_m \circ \Gamma'$ with $2 \le m \le n$ and $\Gamma' \in \mathcal{F}_n^m$. Therefore

$$\sum_{\Gamma\in\mathcal{F}_n^1} g_\Gamma = \sum_{m=2}^{n} \sum_{\Gamma'\in\mathcal{F}_n^m} g_{\mathbb{T}_m\circ\Gamma'} = \sum_{m=2}^{n} g_m \sum_{\Gamma'\in\mathcal{F}_n^m} g_{\Gamma'} .$$

□

Our next goal is to exhibit the analogue of Theorem 14.3.3 for graded symmetric coalgebras. To this end we first need to study the behaviour of the operators g_Γ when composed of the operators

$$\pi\colon V^{\otimes n} \to V^{\odot n}, \qquad \pi(v_1 \otimes \cdots \otimes v_n) = v_1 \odot \cdots \odot v_n,$$

$$N\colon V^{\odot n} \to V^{\otimes n}, \qquad N(v_1 \odot \cdots \odot v_n) = \textstyle\sum_{\sigma\in\Sigma_n} \epsilon(\sigma) v_{\sigma(1)} \otimes \cdots \otimes v_{\sigma(n)}.$$

For every $n, m > 0$ let $\mathfrak{F}_n^m$ be the set of isomorphism classes of rooted forests with n leaves and m roots, and denote by $\rho\colon \mathcal{F}_n^m \to \mathfrak{F}_n^m$ the map forgetting the orientation.

Lemma 14.3.4 *In the above notation, for every* $g \in \mathrm{Hom}^0_{\mathbb{K}}(\overline{T^c}(V), V)$ *and every* $\Omega \in \mathfrak{F}_n^m$ *there exists a unique linear map* $\widetilde{g}_\Omega\colon V^{\odot n} \to V^{\odot m}$ *such that*

$$N\widetilde{g}_\Omega = \sum_{\rho(\Gamma)=\Omega} g_\Gamma N\colon V^{\odot n} \to V^{\otimes m} .$$

Proof Since N is injective and its image is the subspace of symmetric tensors, it is sufficient to prove that for every $1 \le i < m$ and every $v_1, \ldots, v_n \in V$ the element

$$w = \sum_{\rho(\Gamma)=\Omega} \sum_{\sigma\in\Sigma_n} \epsilon(\sigma) g_\Gamma(v_1 \otimes \cdots \otimes v_n)$$

is invariant under every transposition $\tau = (i, i+1) \in \Sigma_m$. Every $\Gamma \in \mathcal{F}_n^m$ admits a unique decomposition $\Gamma = \Gamma_1 \amalg \cdots \amalg \Gamma_m$ with every Γ_i a planar rooted tree. The transposition τ acts on Γ by exchanging Γ_i and Γ_{i+1}. It is easy to see that if $\Gamma_i = \Gamma_{i+1}$ then $\tau(\Gamma) = \Gamma$ and

$$\sum_{\sigma\in\Sigma_n} \epsilon(\sigma) g_\Gamma(v_1 \otimes \cdots \otimes v_n)$$

is τ-invariant, while if $\Gamma_i \neq \Gamma_{i+1}$ then $\tau(\Gamma) \neq \Gamma$ and in this case

$$\sum_{\sigma\in\Sigma_n} \epsilon(\sigma)(g_\Gamma + g_{\tau(\Gamma)})(v_1 \otimes \cdots \otimes v_n)$$

is τ-invariant. □

Lemma 14.3.5 *In the above notation, let $\Gamma, \Delta \in \mathcal{F}_n^m$ be reduced oriented rooted forests such that $\rho(\Gamma) = \rho(\Delta)$. Then for every $f = \sum f_n \in \mathrm{Hom}^0_{\mathbb{K}}(\overline{S^c}(V), V)$ we have*

$$\pi\, (f\pi)_\Gamma\, N = \pi\, (f\pi)_\Delta\, N \colon V^{\odot n} \to V^{\odot m}.$$

Proof It is sufficient to observe that π and N are invariant under the twist action of the symmetric group Σ_n on $V^{\otimes n}$ and that the change of orientation from Γ to Δ is described by a suitable permutation of leaves, inducing a permutation of the roots and a permutation of the incoming edges at every vertex. The permutation of leaves has no effect in view of the composition on the right with N, the permutation of roots has no effect in view of the composition on the left with π and the permutation of incoming edges has no effect because $f\pi$ is symmetric. □

Definition 14.3.6 Given a linear map $f \in \mathrm{Hom}^0_{\mathbb{K}}(\overline{S^c}(V), V)$ and a reduced rooted forest $\Omega \in \mathfrak{F}_n^m$ we define

$$f_\Omega := \pi\, (f\pi)_\Gamma\, N \colon V^{\odot n} \to V^{\odot m},$$

where Γ is any oriented rooted forest such that $\rho(\Gamma) = \Omega$.

The map f_Ω is properly defined in view of Lemma 14.3.5. For example, the rooted binary tree

Ω: 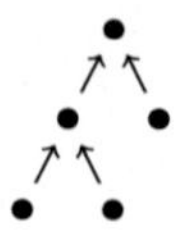

admits the following two orientations:

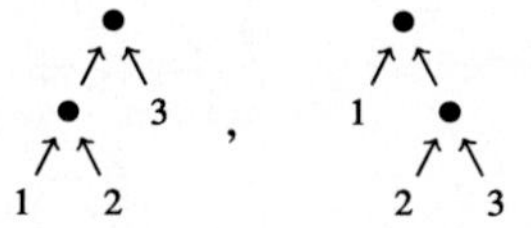

Using the first orientation we get

$$f_\Omega(v_1 \odot v_2 \odot v_3) = \sum_{\sigma\in\Sigma_3} \epsilon(\sigma)\, f_2(f_2(v_{\sigma(1)} \odot v_{\sigma(2)}) \odot v_{\sigma(3)}),$$

while the second orientation gives

$$f_\Omega(v_1 \odot v_2 \odot v_3) = \sum_{\sigma\in\Sigma_3} \epsilon(\sigma)\, f_2(v_{\sigma(1)} \odot f_2(v_{\sigma(2)} \odot v_{\sigma(3)})).$$

In practical terms, the formula for f_Ω is the same as the formula for f_Γ after replacing $\otimes$ with $\odot$ and taking the sum over all permutations. For example:

$\Omega =$ 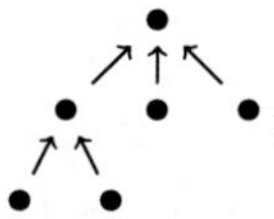 ; $\Omega' =$ 

give

$$f_\Omega(v_1 \odot v_2 \odot v_3 \odot v_4) = \sum_{\sigma\in\Sigma_4} \epsilon(\sigma)\, f_3(f_2(v_{\sigma(1)} \odot v_{\sigma(2)}) \odot v_{\sigma(3)} \odot v_{\sigma(4)}),$$

$$f_{\Omega'}(v_1 \odot v_2 \odot v_3 \odot v_4) = \sum_{\sigma\in\Sigma_4} \epsilon(\sigma)\, f_2(v_{\sigma(1)} \odot v_{\sigma(2)}) \odot v_{\sigma(3)} \odot v_{\sigma(4)}.$$

Lemma 14.3.7 *In the above situation, for a given map* $f = \sum f_n \in \operatorname{Hom}^0_{\mathbb{K}}(\overline{S^c}(V), V)$ *define*

$$g = \sum g_n \in \operatorname{Hom}^0_{\mathbb{K}}(\overline{T^c}(V), V), \qquad g_n = f_n \frac{\pi}{n!}.$$

Then for every $\Omega \in \mathfrak{F}^m_n$ *we have:*

$$\sum_{\rho(\Gamma)=\Omega} g_\Gamma\, N = N\, \frac{1}{|\mathrm{Aut}(\Omega)|} f_\Omega \colon V^{\odot n} \to V^{\otimes m}.$$

Proof We have already proved in Lemma 14.3.4 that the image of $\sum_{\rho(\Gamma)=\Omega} g_\Gamma\, N$ is contained in the space $N(V^{\odot m})$ of symmetric tensors and then it is sufficient to show that

$$\pi \sum_{\rho(\Gamma)=\Omega} g_\Gamma\, N = \pi N \frac{1}{|\mathrm{Aut}(\Omega)|} f_\Omega = \frac{m!}{|\mathrm{Aut}(\Omega)|} f_\Omega.$$

Writing $h_n = \dfrac{f_n}{n!}$ we have $g_n = h_n \pi$ and, in view of Definition 14.3.6, the above equation can be written as

$$\pi \sum_{\rho(\Gamma)=\Omega} (h\pi)_\Gamma\, N = \sum_{\rho(\Gamma)=\Omega} h_\Omega = \frac{m!}{|\mathrm{Aut}(\Omega)|} f_\Omega.$$

We have already observed in Remark 14.1.6 that the cardinality of the fibre $\rho^{-1}(\Omega)$ is equal to $\dfrac{m!}{|\mathrm{Aut}(\Omega)|} \prod_{v\in V_\Omega} |v|!$ and therefore the above equation follows from the obvious equality

$$f_\Omega = \left(\prod_{v\in V_\Omega} |v|!\right) h_\Omega.$$

□

Theorem 14.3.8 (Symmetric inversion formula) *Let V be a graded vector space and let $f = \sum f_n \in \mathrm{Hom}^0_{\mathbb{K}}(\overline{S^c}(V), V)$ be a map with linear component $f_1 = \mathrm{Id}_V$. Let $H^n_m, F^n_m \colon V^{\odot m} \to V^{\odot n}$ be the components of the isomorphisms of graded coalgebras defined by:*

$$H\colon \overline{S^c}(V) \to \overline{S^c}(V), \qquad H^1_1 = \mathrm{Id}_V, \quad H^1_n = -f_n, \quad n \ge 2,$$

$$F\colon \overline{S^c}(V) \to \overline{S^c}(V), \qquad F^1_1 = \mathrm{Id}_V, \quad F^1_n = \sum_{\Omega\in\mathfrak{F}^1_n} \frac{1}{|\mathrm{Aut}(\Omega)|} f_\Omega, \quad n \ge 2.$$

Then $F = H^{-1}$ and $F^m_n = \sum_{\Omega\in\mathfrak{F}^m_n} \dfrac{1}{|\mathrm{Aut}(\Omega)|} f_\Omega$ for every $n, m > 0$.

Proof Define $g_n = f_n \dfrac{\pi}{n!} \colon V^{\otimes n} \to V$, and denote by $K, G\colon \overline{T^c}(V) \to \overline{T^c}(V)$ the isomorphisms of graded coalgebras such that $K^1_1 = G^1_1 = \mathrm{Id}_V$ and

$$K^1_n = -g_n, \qquad G^1_n = \sum_{\Gamma\in\mathcal{F}^1_n} g_\Gamma, \qquad n \ge 2.$$

According to Theorem 14.3.3 we have $G = K^{-1}$. On the other hand,

$$(NH)^1_n = H^1_n = -f_n = -g_n N = (KN)^1_n,$$

and then $NH = KN\colon \overline{S^c}(V) \to \overline{T^c}(V)$. Similarly, Lemma 14.3.7 gives, for every $n \ge 2$,

$$(NF)^1_n = F^1_n = \sum_{\Omega \in \mathfrak{F}^1_n} \frac{1}{|\mathrm{Aut}(\Omega)|} f_\Omega = \sum_{\Gamma \in \mathcal{F}^1_n} g_\Gamma \; N = (GN)^1_n$$

and then $NF = GN$, $NFH = GNH = GKN = N$; the equality $F = H^{-1}$ follows from the injectivity of N. Finally, by Theorem 14.3.3 we have $G^m_n = \sum_{\Gamma \in \mathcal{F}^m_n} g_\Gamma$ and then

$$G^m_n \; N = \sum_{\Gamma \in \mathcal{F}^m_n} g_\Gamma \; N = N \sum_{\Omega \in \mathfrak{F}^m_n} \frac{1}{|\mathrm{Aut}(\Omega)|} f_\Omega .$$

□

14.4 Tree Summation Formula for Homotopy Transfer

As another application of rooted trees we shall a different description of homotopy transfer formulas of Theorem 12.4.1.

Definition 14.4.1 Let V, W be a graded vector space over the field $\mathbb{K}$. Given three linear maps

$$q \in \mathrm{Hom}^1_{\mathbb{K}}(\overline{S^c}(V), V), \qquad h \in \mathrm{Hom}^{-1}_{\mathbb{K}}(V, V), \qquad \imath \in \mathrm{Hom}^0_{\mathbb{K}}(W, V),$$

for every reduced rooted tree $\Omega \in \mathfrak{T}_n = \mathfrak{F}^1_n$ with $n \geq 2$ leaves we consider the operator

$$Z_\Omega(q, h, \imath) = q_k \circ (hq)_{\Omega'} \circ \imath^{\odot n} \colon W^{\odot n} \to V,$$

where k is the arity of the root of Ω and $\Omega' \in \mathfrak{F}^k_n$ is the rooted forest obtained by removing the root (and its incoming edges) from Ω, and $(hq)_{\Omega'}$ is defined as in Definition 14.3.6.

For later use we point out that Ω and Ω' have the same automorphisms group $\mathrm{Aut}(\Omega) = \mathrm{Aut}(\Omega')$ and $(hq)_\Omega = hq_k \circ (hq)_{\Omega'}$.

The above formal definition becomes more clear if we consider the following decoration of Ω:

1. every vertex of arity $k \geq 2$ is labelled q_k;
2. every edge containing a leaf is labelled $\imath$;
3. every edge not containing a leaf is labelled h.

Then, the operator $Z_\Omega(q, h, \imath)$ is defined by taking the composition of h, q_i and $\imath$ described by this decorated rooted tree. For instance, the decorated rooted tree

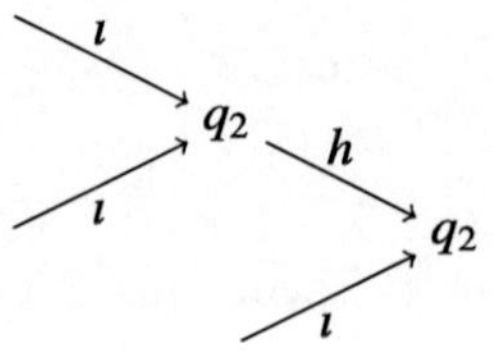

gives the operator

$$w_1 \odot w_2 \odot w_3 \mapsto \sum_{\sigma \in \Sigma_3} \epsilon(\sigma) q_2(hq_2(\iota(w_{\sigma(1)}) \odot \iota(w_{\sigma(2)})) \odot \iota(w_{\sigma(3)})).$$

Notice that every operator $Z_\Omega(q, h, \iota)$ has degree $+1$.

The above formalism can be used for a more explicit version of Theorem 12.4.1.

Theorem 14.4.2 *Assume the following data are given:*

- *an* $L_\infty[1]$*-algebra* $(V, q_1, q_2, \ldots)$ *and a DG-vector space* (W, r_1)*;*
- *two morphisms of DG-vector spaces* $\pi\colon (V, q_1) \to (W, r_1)$, $\iota\colon (W, r_1) \to (V, q_1)$ *and a homotopy* $h \in \mathrm{Hom}^{-1}_{\mathbb{K}}(V, V)$ *such that*

$$q_1 h + h q_1 = \iota\pi - \mathrm{Id}_V.$$

Write $\iota_1 = \iota$ *and for every* $n > 1$ *consider the operators:*

$$r_n = \sum_{\Omega \in \mathfrak{T}_n} \frac{1}{|\mathrm{Aut}(\Omega)|} \pi Z_\Omega(q, h, \iota) \in \mathrm{Hom}^1_{\mathbb{K}}(W^{\odot n}, W), \tag{14.4}$$

$$\iota_n = \sum_{\Omega \in \mathfrak{T}_n} \frac{1}{|\mathrm{Aut}(\Omega)|} h Z_\Omega(q, h, \iota) \in \mathrm{Hom}^0_{\mathbb{K}}(W^{\odot n}, V). \tag{14.5}$$

Then $(W, r_1, r_2, \ldots)$ *is an* $L_\infty[1]$*-algebra and* $\iota\colon (W, r_1, r_2, \ldots) \rightsquigarrow (V, q_1, q_2, \ldots)$ *is a strict* L_∞*-morphism.*

Proof Let $F\colon \overline{S^c}(W) \to \overline{S^c}(V)$ be the morphism of graded coalgebras such that

$$F^1_1 = \iota, \qquad F^1_n = \iota_n, \quad n \geq 2.$$

By Theorem 12.4.1 it is sufficient to prove that, for every $n \geq 2$:

1. $F^1_n = \sum_{m=2}^n hq_m F^m_n$,
2. $r_n = \sum_{m=2}^n \pi q_m F^m_n$.

To this end consider the morphism of graded coalgebras $E\colon \overline{S^c}(W) \to \overline{S^c}(V)$ such that $E^1_1 = \iota$ and $E^1_n = 0$ for every $n \geq$, together with the isomorphism of graded coalgebras

$$G \colon \overline{S^c}(V) \to \overline{S^c}(V)$$

with corestrictions

$$G_1^1 = \mathrm{Id}_V, \qquad G_n^1 = \sum_{\Omega \in \mathfrak{T}_n} \frac{1}{|\mathrm{Aut}(\Omega)|}(hq)_\Omega, \quad n > 1.$$

Since $E_n^n = \iota^{\odot n}$, $E_m^n = 0$ for $n \neq m$ and $hZ_\Omega(q, h, \iota) = (hq)_\Omega \iota^{\odot n}$ for every $\Omega \in \mathfrak{T}_n$, we have $F_n^1 = (GE)_n^1$ for every $n \geq 1$ and then $F = GE$. Notice that if $H \colon \overline{S^c}(V) \to \overline{S^c}(V)$ is the isomorphism of graded coalgebras with corestrictions $H_1^1 = \mathrm{Id}$ and $H_n^1 = -hq_n$ for every $n > 1$, then by Theorem 14.3.8 we have $H = G^{-1}$ and then $HF = E$.

Again for Theorem 14.3.8 for every $1 \leq m \leq n$ we have

$$G_n^m = \sum_{\Omega' \in \mathfrak{F}_n^m} \frac{1}{|\mathrm{Aut}(\Omega')|}(hq)_{\Omega'}$$

and then for every $n \geq 2$ we obtain

$$F_n^1 = G_n^1 \iota^{\odot n} = \sum_{m=2}^{n} hq_m\, G_n^m\, \iota^{\odot n} = \sum_{m=2}^{n} hq_m\, F_n^m,$$

$$r_n = \sum_{\Omega \in \mathfrak{F}_n^1} \frac{1}{|\mathrm{Aut}(\Omega)|} \pi Z_\Omega(q, h, \iota) = \sum_{m=2}^{n} \pi q_m \sum_{\Omega' \in \mathfrak{F}_n^m} \frac{1}{|\mathrm{Aut}(\Omega')|}(hq)_{\Omega'}\, \iota^{\odot n}$$

$$= \sum_{m=2}^{n} \pi q_m G_n^m \iota^{\odot n} = \sum_{m=2}^{n} \pi q_m F_n^m.$$

□

Example 14.4.3 Consider the Iwasawa DG-algebra $R = \mathbb{K}[\omega_1, \omega_2, \omega_3]$ (Example 6.7.2), where

$$\overline{\omega_1} = \overline{\omega_2} = \overline{\omega_3} = 1, \quad d\omega_1 = d\omega_2 = 0, \quad d\omega_3 = -\omega_1\omega_2.$$

Given a Lie algebra $\mathfrak{g}$ we want to prove that $R \otimes \mathfrak{g}$ is formal if and only if the functor $\mathrm{Def}_{\mathfrak{g}\otimes R}$ is smooth if and only if $\mathfrak{g}^{[3]} = [[\mathfrak{g}, \mathfrak{g}], \mathfrak{g}] = 0$. There exists a homotopy operator $h \colon R \to R$ defined by:

$$h(\omega_1\omega_2) = \omega_3, \qquad h(\omega_{i_1} \cdots \omega_{i_k}) = 0, \quad \forall\, (i_1, \ldots, i_k) \neq (1, 2),$$

that satisfy the formula $dh + hd = \pi - \mathrm{Id}$, where π is a projection on the graded vector subspace H spanned by $1, \omega_1, \omega_2, \omega_1\omega_3, \omega_2\omega_3, \omega_1\omega_2\omega_3$.

By homotopy transfer, for every Lie algebra $\mathfrak{g}$, we obtain a minimal L_∞-algebra $(\mathfrak{g}\otimes H, 0, l_2, l_3, \ldots)$ that is weakly equivalent to the differential graded Lie algebra $\mathfrak{g}\otimes R$.

Applying the tree summation formula of Theorem 14.4.2, the equalities

$$h(x)h(y) = h(h(x)y) = 0, \qquad x, y \in R,$$

implies that the only rooted trees giving a nontrivial contribution are:

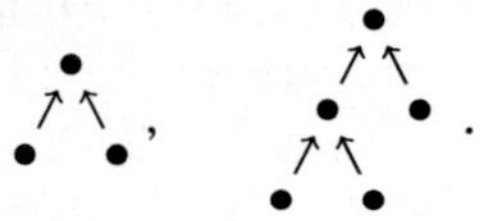

In particular, $l_n = 0$ for every $n \geq 4$. It is clear that if $\mathfrak{g}^{[3]} = 0$ then $l_3 = 0$ and then $R\otimes\mathfrak{g}$ is formal. Since $\pi(\omega_1\omega_2) = 0$ the map l_2 vanishes on the first cohomology subspace $H^1 \subset H$ and then the Maurer–Cartan equation in H is homogeneous of degree 3, viz.,

$$l_3(x^{\wedge 3}) = 0, \qquad x \in \mathfrak{g}\otimes H^1 = \mathfrak{g}\omega_1 \oplus \mathfrak{g}\omega_2\,.$$

Therefore if $R\otimes\mathfrak{g}$ is formal then l_3 vanishes in $H^1\times H^1\times H^1$. A straightforward computation shows that for every $a, b, c \in \mathfrak{g}$ we have

$$l_3(\omega_1\otimes a, \omega_2\otimes b, \omega_1\otimes c) = \pm\omega_1\omega_3\otimes[[a, b], c]$$

and then l_3 vanishes in $H^1\times H^1\times H^1$ if and only if $\mathfrak{g}^{[3]} = 0$ if and only if the functor $\mathrm{Def}_{\mathfrak{g}\otimes R}$ is smooth, cf. Proposition 6.7.3.

Example 14.4.4 We want to describe the $L_\infty[1]$ minimal model of the DG-Lie algebra $M = \oplus M^i$, which is defined in the following way:

1. $M^i = \mathbb{K}^2$ for $i = 1, 2$ and $M^i = 0$ otherwise;
2. if x, y is a basis of M^1 and u, v is a basis of M^2 we have

$$dx = 0, \quad dy = [x, x] = -u, \quad [x, y] = -v, \quad [y, y] = 0.$$

The $L_\infty[1]$-algebra (V, q) associated to M is

$$V^0 = \mathrm{Span}(x, y), \quad V^1 = \mathrm{Span}(u, v),$$

$$q_1(x) = 0,\ q_1(y) = u,\ q_2(x, x) = u,\ q_2(x, y) = v,\ q_2(y, y) = 0,$$

and $q_n = 0$ for every $n > 2$. The cohomology H of (V, q_1) is naturally isomorphic to the graded vector subspace generated by x, v and a contracting homotopy is $h(u) = -y, h(v) = 0, q_1h + hq_1 = \pi - \mathrm{Id}$ where $\pi\colon V \to H$ is the projection.

Since $q_2(h(-), h(-)) = 0$ and $hq_2(x, h(-)) = 0$ the only rooted trees giving a nontrivial contribution are the same as Example 14.4.3. Moreover, $\pi q_2(x, x) = 0$ and then in the minimal model (H, r) the unique nontrivial Taylor coefficient is r_3, namely

$$r_3(x^{\odot 3}) = \frac{3!}{2} q_2(hq_2(x, x), x) = -3v.$$

In particular, the DG-Lie algebra M is not formal, cf. the second case in Example 6.2.5.

14.5 An Example: $L_\infty[1]$ Structures on Mapping cones

Let $f\colon L \to M$ be a morphism of differential graded Lie algebras with mapping cocone

$$C(f) = L \times sM, \qquad d(l, sm) = (dl, s(f(l) - dm)),$$

and denote by

$$K(f) = \{(l, m(t)) \in L \times M[t, dt] \mid m(0) = 0,\ m(1) = f(l)\}$$

its (Thom–Whitney) homotopy fibre. We have proved in Corollary 6.6.4 that the deformation functor associated to $K(f)$ is isomorphic to

$$A \mapsto \frac{\left\{(x, a) \in (L^1 \otimes \mathfrak{m}_A) \times (M^0 \otimes \mathfrak{m}_A) \;\middle|\; e^a * 0 = f(x)\right\}}{\sim}, \qquad A \in \mathbf{Art}_{\mathbb{K}},$$

where $(x, a) \sim (y, b)$ if there exists $\alpha \in L^0 \otimes \mathfrak{m}_A$ such that $e^\alpha * x = y$ and the diagram

$$\begin{array}{ccc} 0 & \xrightarrow{\;e^a\;} & f(x) \\ & \underset{e^b}{\searrow} & \Big\downarrow{\scriptstyle e^{f(\alpha)}} \\ & & f(y) \end{array} \tag{14.6}$$

is commutative in the Deligne groupoid of $M \otimes \mathfrak{m}_A$. The aim of this section is to interpret the above description in terms of the Maurer–Cartan equation and homotopy equivalence associated to an L_∞ structure on $C(f)$, or equivalently to an $L_\infty[1]$ structure on the mapping cone of $-f$ (Definition 5.2.2); it is useful to recall that $\operatorname{cone}(-f)^n = L^{n+1} \oplus M^n$, while the differential $d\colon L^n \oplus M^{n-1} \to L^{n+1} \oplus M^n$ is defined by the formula

$$d(v, w) = (-dv, dw - f(v)), \quad v \in L^n,\ w \in M^{n-1}.$$

As usual, when there is no risk of confusion, we shall denote a generic element of $M[t, dt]$ by the symbol $m(t)$. Writing for simplicity $V = K(f)[1]$, by décalage isomorphisms, the $L_\infty[1]$ structure on V is $(q_1, q_2, 0, 0, \ldots)$, where

$$q_1(x, p(t)\,m + q(t)dt\,n) = (-dx, -p(t)\,dm + q(t)dt\,dn - p(t)'dt\,m)$$

and the nontrivial components of q_2 are

$$q_2 \colon L^i \times L^j \to L^{i+j}, \quad q_2(x, y) = (-1)^i[x, y],$$

$$q_2 \colon M[t, dt]^i \times M[t, dt]^j \to M[t, dt]^{i+j}, \quad q_2(m(t), n(t)) = (-1)^i[m(t), n(t)].$$

We have already observed in Sect. 6.1 that there exists a contraction

$$\operatorname{cone}(-f) \underset{\pi}{\overset{\imath}{\rightleftarrows}} V \circlearrowright h, \quad hq_1 + q_1h = \imath\pi - \mathrm{Id}_V,$$

where

$$\imath(x, m) = (x, tf(x) + dt\,m), \qquad \pi(x, m(t)) = \left(x, \int_0^1 m(t)\right),$$

$$h(x, m(t)) = \left(0, \int_0^t m(s) - t\int_0^1 m(s)\right).$$

Thus we are in the position to transfer the $L_\infty[1]$ structure from V to $\operatorname{cone}(-f)$ by using the above homotopy operator h and the tree summation formula of Theorem 14.4.2. Denote by $(\operatorname{cone}(-f), r_1, r_2, \ldots)$ the induced $L_\infty[1]$ structure (14.4) and by $\imath \colon \operatorname{cone}(-f) \rightsquigarrow V$ the induced L_∞-morphism (14.5).

The easily verified condition

$$q_2(h(V) \times h(V)) \subseteq \ker \pi \cap \ker h$$

implies that for every $n \geq 2$ there exists at most one isomorphism class of rooted trees with n leaves giving a nontrivial contribution to r_n and $\imath_n$. Namely:

- for $n = 2$, the decorated (with $h, \imath, q_2$) rooted tree

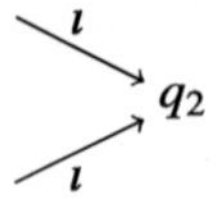

that gives the contributions

$$r_2(\gamma_1, \gamma_2) = \pi q_2(\imath(\gamma_1), \imath(\gamma_2)), \qquad \imath_2(\gamma_1, \gamma_2) = hq_2(\imath(\gamma_1), \imath(\gamma_2));$$

- for $n = 3$, the decorated rooted tree

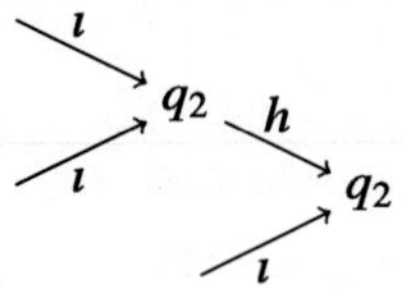

that gives the contributions

$$
\begin{aligned}
r_3(\gamma_1, \gamma_2, \gamma_3) &= \frac{1}{2} \sum_{\sigma \in \Sigma_3} \varepsilon(\sigma) \pi q_2(h q_2(\iota(\gamma_{\sigma(1)}), \iota(\gamma_{\sigma(2)})), \iota(\gamma_{\sigma(3)})) \\
&= \sum_{\substack{\sigma \in \Sigma_3 \\ \sigma(1)<\sigma(2)}} \varepsilon(\sigma) \pi q_2(h q_2(\iota(\gamma_{\sigma(1)}), \iota(\gamma_{\sigma(2)})), \iota(\gamma_{\sigma(3)}))\,;
\end{aligned}
$$

$$
\iota_n(\gamma_1, \ldots, \gamma_n) = \sum_{\substack{\sigma \in \Sigma_3 \\ \sigma(1)<\sigma(2)}} \varepsilon(\sigma) h q_2(h q_2(\iota(\gamma_{\sigma(1)}), \iota(\gamma_{\sigma(2)})), \iota(\gamma_{\sigma(3)}))\,;
$$

- for $n \geq 4$, the decorated rooted tree

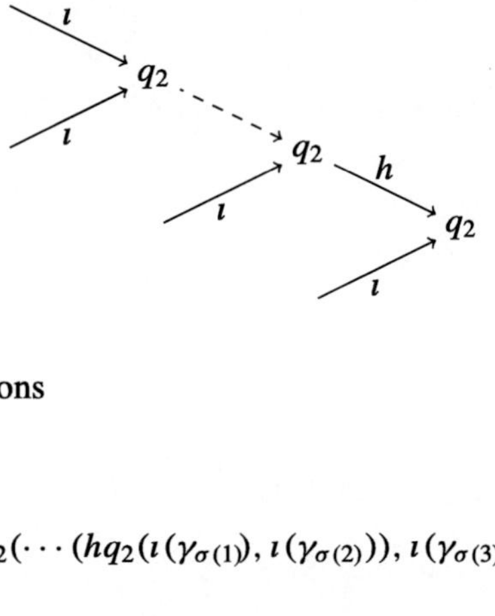

that gives the contributions

$$
\begin{aligned}
&r_n(\gamma_1, \ldots, \gamma_n) \\
&= \frac{1}{2} \sum_{\sigma \in \Sigma_n} \varepsilon(\sigma) \pi q_2(h q_2(\cdots (h q_2(\iota(\gamma_{\sigma(1)}), \iota(\gamma_{\sigma(2)})), \iota(\gamma_{\sigma(3)})), \ldots, \iota(\gamma_{\sigma(n)})) \\
&= \sum_{\substack{\sigma \in \Sigma_n \\ \sigma(1)<\sigma(2)}} \varepsilon(\sigma) \pi q_2(h q_2(\cdots (h q_2(\iota(\gamma_{\sigma(1)}), \iota(\gamma_{\sigma(2)})), \iota(\gamma_{\sigma(3)})), \ldots, \iota(\gamma_{\sigma(n)}));
\end{aligned}
$$

$$
\begin{aligned}
&\iota_n(\gamma_1, \ldots, \gamma_n) \\
&= \sum_{\substack{\sigma \in \Sigma_n \\ \sigma(1)<\sigma(2)}} \varepsilon(\sigma) h q_2(h q_2(\cdots (h q_2(\iota(\gamma_{\sigma(1)}), \iota(\gamma_{\sigma(2)})), \iota(\gamma_{\sigma(3)})), \ldots, \iota(\gamma_{\sigma(n)}))\,.
\end{aligned}
$$

A more refined description of the maps r_n, involving the original brackets in the differential graded Lie algebras L and M is obtained by decomposing the symmetric powers of cone$(-f)$ into types:

$$\text{cone}(-f)^{\odot n} = \bigoplus_{\lambda+\mu=n} L[1]^{\odot\lambda} \otimes M^{\odot\mu}.$$

The map r_2 decomposes as

$$\begin{aligned} r_2(x, y) &= (-1)^{\deg_L(x)}[x, y] \in L, \\ r_2(n, m) &= 0, \qquad\qquad\qquad\qquad x, y \in L, \quad n, m \in M. \\ r_2(x, m) &= \frac{1}{2}[f(x), m] \in M, \end{aligned}$$

Since $q_2(h(V) \times \imath(L[1])) = hq_2(\imath(L[1]) \times \imath(L[1])) = 0$, it follows that, for every $n \geq 2$, the restriction of r_{n+1} to $L[1]^{\odot a} \otimes M^{\odot n+1-a}$ can be nontrivial only for $a = 1$. For the same reason, since $(dt)^2 = 0$ for every $m_1, \dots, m_n \in M, n \geq 1$, and every $x \in L[1]$ we have

$$\begin{aligned} &r_{n+1}(x, m_1, \dots, m_n) \\ &= \sum_{\sigma\in S_n} \varepsilon(\sigma)\pi q_2(hq_2(\cdots(hq_2(tf(x), dt\, m_{\sigma(1)}), dt\, m_{\sigma(2)}), \dots, dt\, m_{\sigma(n)}). \end{aligned}$$

Define recursively a sequence of polynomials $\phi_n(t) \in \mathbb{Q}[t]$ and rational numbers I_n in the following way:

$$\phi_1(t) = t, \quad I_n = \int_0^1 \phi_h(t)dt, \quad \phi_{n+1}(t) = \int_0^t \phi_n(s)ds - tI_n, \quad n \geq 1. \tag{14.7}$$

According to the explicit description of the homotopy operator h, for every integer $n \geq 1$, and every $m \in M$ we have

$$h(0, \phi_n(t)dt\, m) = (0, \phi_{n+1}(t)\, m)\,,$$

and then, for every $m, n \in M$, $h \geq 0$ we have:

$$hq_2(\phi_h(t)\, n, dt\, m) = \phi_{h+1}(t)[n, m]\,.$$

Therefore, for $n \geq 1$ we may write:

$$
\begin{aligned}
&r_{n+1}(x, m_1, \ldots, m_n)\\
&\quad = \sum_{\sigma\in S_n} \varepsilon(\sigma)\pi q_2(hq_2(\cdots(hq_2(tf(x), dt\, m_{\sigma(1)}), dt\, m_{\sigma(2)}), \ldots)\\
&\quad = \sum_{\sigma\in S_n} \varepsilon(\sigma)\pi q_2(hq_2(\cdots(hq_2(\phi_2(t)[f(x), m_{\sigma(1)}], dt\, m_{\sigma(2)}), \ldots)\\
&\quad = I_n \sum_{\sigma\in S_n} \varepsilon(\sigma)[\cdots[[f(x), m_{\sigma(1)}], m_{\sigma(2)}], \ldots, m_{\sigma(n)}].
\end{aligned}
$$

A further simplification can be done by showing that $I_n = 0$ for every odd number $n \geq 3$. In fact, the following lemma relates the rational numbers I_n with the Bernoulli numbers.

Lemma 14.5.1 *Consider the sequence $I_n \in \mathbb{Q}$, $n \geq 1$, defined in Eq. (14.7). Then*

$$
\sum_{n=1}^{\infty} I_n x^n = 1 - \frac{x}{e^x - 1} \qquad \Longleftrightarrow \qquad I_n = -\frac{B_n}{n!},
$$

where the B_n are the Bernoulli numbers.

Proof In the notation of the recursive equation (14.7), consider the polynomials $\psi_0(t) = 1$ and $\psi_n(t) = \phi_n(t) - I_n$ for $n \geq 1$. Then, for any $n \geq 1$,

$$
\frac{d}{dt}\psi_n(t) = \psi_{n-1}(t), \qquad \int_0^1 \psi_n(t)dt = 0.
$$

Setting

$$
F(t, x) = \sum_{n=0}^{\infty} \psi_n(t)x^n,
$$

we have

$$
\frac{d}{dt}F(t, x) = \sum_{n=1}^{\infty} \psi_{n-1}(t)x^n = xF(t, x), \qquad \int_0^1 F(t, x)dt = 1.
$$

Therefore $F(t, x) = F(0, x)e^{tx}$,

$$
1 = \int_0^1 F(t, x)dt = F(0, x)\int_0^1 e^{tx}dt = F(0, x)\frac{e^x - 1}{x},
$$

and then

$$
F(0, x) = \frac{x}{e^x - 1}.
$$

Since $\psi_n(0) = -I_n$ for any $n \geq 1$ we get

$$\frac{x}{e^x-1} = F(0,x) = 1 - \sum_{n=1}^{\infty} I_n x^n.$$

□

In conclusion, we have proved that the only nontrivial components of the maps r_{n+1}, $n \geq 2$, are given by

$$r_{n+1}(x, m_1, \dots, m_n) = \frac{-B_n}{n!} \sum_{\sigma \in S_n} \varepsilon(\sigma)[\cdots[[f(x), m_{\sigma(1)}], m_{\sigma(2)}], \dots, m_{\sigma(n)}],$$
$$x \in L, \quad m_1, \dots, m_n \in M.$$

It is also easy to write the Maurer–Cartan equation for the $L_\infty[1]$-algebra $(\text{cone}(-f), r_1, r_2, \dots)$ in terms of the DG-Lie algebra structures of L and M. Assuming that M is nilpotent, then $x + m \in \text{cone}(-f)^1 = L^1 \oplus M^0$ satisfies the Maurer–Cartan equation

$$\sum_{n>0} \frac{1}{n!} r_n((x+m)^{\odot n}) = 0$$

if and only if x satisfies the Maurer–Cartan equation on L and $e^m * f(x) = 0$. In fact we have

$$\sum_{n>0} \frac{1}{n!} r_n((x+m)^{\odot n}) = -dx - \frac{1}{2}[x,x],\ dm - f(x) + \sum_{n>0} \frac{1}{n!} r_{n+1}(x, m, \dots, m)$$

and, since

$$\frac{1}{n!} r_{n+1}(x, m, \dots, m) = -\frac{B_n}{n!}[-, m]^n(f(x)) = -\frac{B_n}{n!}[-m, -]^n(f(x)),$$

we have

$$dm - f(x) + \sum_{n>0} \frac{1}{n!} r_{n+1}(x, m, \dots, m) = dm - \frac{[-m,-]}{e^{[-m,-]} - \mathrm{Id}}(f(x)).$$

Applying the invertible operator $\dfrac{e^{[-m,-]} - \mathrm{Id}}{[-m,-]}$ to both sides we obtain that

$$dm - f(x) + \sum_{n>0} \frac{1}{n!} r_{n+1}(x, m, \dots, m) = 0$$

if and only if

$$f(x) = \frac{e^{[-m,-]} - \mathrm{Id}}{[-m,-]}(dm) = e^{-m} * 0.$$

Example 14.5.2 (*Homotopy equivalence of derived brackets*) The above construction applies in particular in the setup of Sect. 10.7.

Assume there is given a differential graded Lie algebra $(M, d, [-,-])$ and a decomposition $M = L \oplus A$ as graded vector spaces, with L a differential graded Lie subalgebra and A an abelian graded Lie subalgebra. Denoting by $P \in \operatorname{Hom}^0_{\mathbb{K}}(M, M)$ the projection on A with kernel equal to L, the derived brackets (Definition 10.7.1) associated to this data are the graded symmetric maps

$$\{\cdots\}^d_n \colon A^{\odot n} \to A, \quad \{a_1, \ldots, a_n\}^d_n = P[\cdots[[d(a_1), a_2], a_3], \ldots, a_n], \quad n > 0,$$

and we have already proved (Corollary 10.7.9) that they give an $L_\infty[1]$ structure on A.

We sketch a proof that the homotopy class of this $L_\infty[1]$ structure depends only on the inclusion $\imath \colon L \to M$ and it is independent of the choice of A.

The map

$$f_1 \colon A \to \operatorname{cone}(-\imath), \qquad f_1(a) = (da - Pda, a),$$

is a quasi-isomorphism of DG-vector spaces, where the differential on A is equal to $\{-\}^d_1 = Pd$. Denoting by $P^\perp = \operatorname{Id}_M - P \colon M \to L$ and by

$$d_n \colon A^{\odot n} \to M, \quad d_n(a_1, \ldots, a_n) = \frac{1}{n!}\sum_{\sigma \in S_n} \varepsilon(\sigma)[\cdots[da_{\sigma(1)}, a_{\sigma(2)}], \ldots, a_{\sigma(n)}],$$

it is not difficult to prove that f_1 together with the sequence of maps

$$f_n \colon A^{\odot n} \to \operatorname{cone}(-\imath), \qquad f_n(a_1, \ldots, a_n) = (P^\perp d_n(a_1, \ldots, a_n), 0), \quad n \geq 2,$$

give an L_∞-morphism $f \colon A \rightsquigarrow \operatorname{cone}(-\imath)$ which is also a weak equivalence. In particular, the homotopy class of the $L_\infty[1]$ structure on A depends only on the inclusion of differential graded Lie algebras $\imath \colon L \to M$.

14.6 Exercises

14.6.1 Prove that a binary rooted tree with n leaves has exactly $2n - 1$ vertices and $2n - 2$ edges.

14.6.2 Consider the formal power series $f, g \in \mathbb{K}[[y]]$:

$$f(y) = y - y^2, \qquad g(y) = \frac{y(1-2y)}{(1-y)} = y - y^2 - y^3 - \cdots.$$

Use Example 11.1.4 and the inversion formula Theorem 14.3.3 in the case $V = \mathbb{K}$ for proving that

$$f\left(\sum_{n=1}^{\infty}|\mathcal{B}_n|x^n\right)=g\left(\sum_{n=1}^{\infty}|\mathcal{T}_n|x^n\right)=x$$

and deduce the counting formulas

$$\sum_{n=1}^{\infty}|\mathcal{B}_n|x^n=\frac{1-\sqrt{1-4x}}{2},\qquad \sum_{n=1}^{\infty}|\mathcal{T}_n|x^n=\frac{x+1-\sqrt{1-6x+x^2}}{4}.$$

14.6.3 (*Catalan numbers*) By definition, the nth Catalan number is equal to

$$C_n=\frac{1}{n+1}\binom{2n}{n},\qquad n=0,1,2,\ldots,$$

and the first 11 Catalan numbers are 1, 1, 2, 5, 14, 42, 132, 429, 1430, 4862, 16796.
Consider the formal power series

$$f(t)=\sum_{n=0}^{+\infty}\binom{2n}{n}t^n,\qquad g(t)=\sum_{n=0}^{\infty}C_nt^{n+1},$$

and prove that $f(0)=1$, $g(0)=0$, $g(t)'=f(t)$,

$$\binom{2n+2}{n+1}=\frac{4n+2}{n+1}\binom{2n}{n},\qquad (1-4t)f'(t)=2f(t).$$

Deduce that

$$f(t)=\frac{1}{\sqrt{1-4t}},\qquad g(t)=\frac{1-\sqrt{1-4t}}{2}$$

and then $|\mathcal{B}_n|=C_{n-1}$ for every $n>0$.

14.6.4 Prove that it is not possible to define canonically a bracket on the mapping cocone $C(f)$ of a morphism of Lie algebras $f\colon L\to M$ such that $C(f)$ becomes a differential graded Lie algebra and such that the projection $C(f)\to L$ is a morphism of differential graded Lie algebras. Here canonically means in particular that for every commutative diagram

$$\begin{array}{ccc} L & \xrightarrow{\alpha} & H \\ {\scriptstyle f}\downarrow & & \downarrow{\scriptstyle g} \\ M & \xrightarrow{\beta} & K \end{array}$$

of Lie algebras the induced map $C(f)\to C(g)$ is a morphism of differential graded Lie algebras.

(Hint: consider first the commutative square with $L=M=H=K$ an abelian Lie algebra, $f=g=\mathrm{Id}$, $\alpha=\beta=-\mathrm{Id}$ and prove that in this case the bracket in

$C(f)$ must be trivial. In general, for every $x \in L$ the span of $x, f(x)$ is a graded Lie subalgebra of $C(f)$ and this implies that for every $x, y \in L$ we have

$$[x + y, f(x) + f(y)] = [x, f(x)] = [y, f(y)] = 0 \iff [x, f(y)] = [f(x), y]$$

and by the Leibniz rule we get

$$[x, f(y)] = \frac{1}{2} f([x, y]).$$

It is now easy to see, taking for instance $L = M = \mathfrak{sl}_2(\mathbb{K})$ and $f = \mathrm{Id}$ that the Jacobi identity is not verified.)

References

Trees are an important combinatorial tool in enumerative geometry and moduli theory. A very small sample of very important papers in enumerative geometry and deformation theory where trees and graphs play an essential role is [20, 152, 154, 155].

The explicit formula of Theorem 14.2.5 has been proved in [131] and it is sufficient for our purposes, including the possibility for the nonexpert reader to practice with rooted trees in a simple case, in view of their less trivial application in homotopy transfer formulas (Sect. 14.4).

The L_∞ structure on mapping cocones is taken from [74]. An alternative proof of Lemma 14.5.1 can be done by observing that the polynomials $n!\psi_n(t)$ satisfy the recursive relations of the Bernoulli polynomials (see e.g. [223]).

For a detailed treatment of Example 14.5.2 we refer to [11].

Appendix A
Topics in the Theory of Analytic Algebras

This appendix can be considered as a short introductory course on analytic algebras and analytic singularities (Definition 1.6.2) where, among other things, we give an elementary proof of the Nullstellensatz for the ring $\mathbb{C}\{z_1, \dots, z_n\}$ of convergent complex power series. A certain importance in deformation theory is played by the smoothness criterion (Proposition A.3.4) and the two dimension bounds (Lemma A.5.2 and Theorem A.5.3).

As usual, unless otherwise specified, whenever A is a local commutative ring we denote by $\mathfrak{m}_A$ its maximal ideal.

A.1 Analytic Algebras

Let $\mathbb{C}\{z_1, \dots, z_n\}$ be the ring of convergent power series with complex coefficients. Equivalently, $\mathbb{C}\{z_1, \dots, z_n\}$ is the ring of Taylor expansions at 0 of holomorphic functions defined in an open neighbourhood of $0 \in \mathbb{C}^n$.

If f is a holomorphic function in a neighbourhood of 0 and $f(0) \neq 0$ then $1/f$ is holomorphic in a (possibly smaller) neighbourhood; this implies that a convergent power series f is invertible in $\mathbb{C}\{z_1, \dots, z_n\}$ if and only if $f(0) \neq 0$ and therefore $\mathbb{C}\{z_1, \dots, z_n\}$ is a local ring with maximal ideal

$$(z_1, \dots, z_m) = \{f \in \mathbb{C}\{z_1, \dots, z_n\} \mid f(0) = 0\}.$$

Moreover, it is known that:

- $\mathbb{C}\{z_1, \dots, z_n\}$ is a Noetherian ring ([97, 100], [107, II.B.9]);
- $\mathbb{C}\{z_1, \dots, z_n\}$ is a unique factorization domain ([97, 100], [107, II.B.7]);
- $\mathbb{C}\{z_1, \dots, z_n\}$ is a Henselian ring ([96, 97, 100]);
- $\mathbb{C}\{z_1, \dots, z_n\}$ is a regular local ring of dimension n [97].

M. Manetti, *Lie Methods in Deformation Theory*, Springer Monographs in Mathematics,
https://doi.org/10.1007/978-981-19-1185-9

Recall that the dimension of a Noetherian local ring A may be defined as the minimum integer d such that there exist $f_1, \ldots, f_d \in \mathfrak{m}_A$ with the property $\sqrt{(f_1, \ldots, f_d)} = \mathfrak{m}_A$. In particular, $\dim A = 0$ if and only if $\sqrt{0} = \mathfrak{m}_A$, i.e., $\dim A = 0$ if and only if $\mathfrak{m}_A$ is a nilpotent ideal.

A morphism of local rings $\phi \colon A \to B$ is called local if $\phi(\mathfrak{m}_A) \subset \mathfrak{m}_B$. Not every morphism of local rings is a local morphism; for instance, the inclusion of $\mathbb{C}\{z\}$ into the field $\mathbb{C}\{z\}[z^{-1}]$ of all meromorphic Laurent series, is not a local morphism.

Recall, from Definition 1.6.2, that an **analytic algebra** is a $\mathbb{C}$-algebra isomorphic to a quotient $\mathbb{C}\{z_1, \ldots, z_n\}/I$, for some $n \geq 0$ and some ideal $I \subset (z_1, \ldots, z_n)$.

Every analytic algebra is a Noetherian local ring. We denote by **An** the category of analytic algebras; the morphisms in **An** are by definition the local homomorphisms of local $\mathbb{C}$-algebras. It is useful to point out that if R, S are analytic algebras, then every morphism of $\mathbb{C}$-algebras $f \colon R \to S$ is local; in fact if $x \in \mathfrak{m}_R$ there exists a unique complex number $t \in \mathbb{C}$ such that $f(x) - t \in \mathfrak{m}_S$. If $t \neq 0$ then $x - t$ is invertible in R and then also $f(x - t) = f(x) - t$ is invertible in S; therefore $t = 0$ and $f(x) \in \mathfrak{m}_S$.

Example A.1.1 Every Artin, local and finitely generated $\mathbb{C}$-algebra A is an analytic algebra. In fact, by the Hilbert Nullstellensatz [10, Cor. 5.24] its residue field is $\mathbb{C}$ and, if $a_1, \ldots, a_n$ is a set of generators of the maximal ideal $\mathfrak{m}_A$, an easy induction on k shows that the maps

$$\frac{\mathbb{C}\{z_1, \ldots, z_n\}}{(z_1, \ldots, z_n)^k} = \frac{\mathbb{C}[z_1, \ldots, z_n]}{(z_1, \ldots, z_n)^k} \xrightarrow{z_i \mapsto a_i} \frac{A}{\mathfrak{m}_A^k}, \qquad k > 0,$$

are surjective. Since $\mathfrak{m}_A^k = 0$ for $k >> 0$ there exists a surjective local morphism

$$\mathbb{C}\{z_1, \ldots, z_n\} \xrightarrow{z_i \mapsto a_i} A.$$

Since $\bigcap_{s>0}(z_1, \ldots, z_n)^s = 0$ it makes sense to define the multiplicity of a nonzero power series $f \in \mathbb{C}\{z_1, \ldots, z_n\}$, $f \neq 0$, as the biggest integer s such that $f \in (z_1, \ldots, z_n)^s$.

The following theorem is well known, see e.g. [107, p. 15], with plenty of proofs in the existing literature.

Theorem A.1.2 (Implicit function theorem) *Let*

$$f_1, \ldots, f_m \in \mathbb{C}\{x_1, \ldots, x_n, y_1, \ldots, y_m\}$$

be convergent power series of positive multiplicity such that $\det\left(\frac{\partial f_i}{\partial y_j}(0)\right) \neq 0$. *Then the morphism of analytic algebras*

$$\mathbb{C}\{x_1, \ldots, x_n\} \xrightarrow{x_i \mapsto x_i} \frac{\mathbb{C}\{x_1, \ldots, x_n, y_1, \ldots, y_m\}}{(f_1, \ldots, f_m)}$$

is an isomorphism.

Definition A.1.3 The **embedding dimension** of an analytic algebra R is the dimension of the complex vector space $\dfrac{\mathfrak{m}_R}{\mathfrak{m}_R^2}$.

For instance, if $R = \dfrac{\mathbb{C}\{z_1, \ldots, z_k\}}{I}$ for some ideal $I \subset (z_1, \ldots, z_k)^2$, since

$$\mathfrak{m}_R = \frac{(z_1, \ldots, z_k)}{I}, \qquad \mathfrak{m}_R^2 = \frac{(z_1, \ldots, z_k)^2}{I},$$

the embedding dimension of R is k. The converse implication holds by virtue of the following result.

Corollary A.1.4 *Let R be an analytic algebra of embedding dimension k. Then R is isomorphic to $\dfrac{\mathbb{C}\{z_1, \ldots, z_k\}}{I}$ for some ideal $I \subset (z_1, \ldots, z_k)^2$.*

Proof Let $R = \mathbb{C}\{z_1, \ldots, z_n\}/J$ be an analytic algebra. If the ideal J is contained in $(z_1, \ldots, z_n)^2$ then $k = n$ and we have completed the proof. Otherwise there exists $f \in J$ and an index i such that $\dfrac{\partial f}{\partial z_i}(0) \neq 0$; up to permutation of indices we may suppose $i = n$. By Theorem A.1.2 the analytic algebra $\mathbb{C}\{z_1, \ldots, z_n\}/(f)$ is isomorphic to $\mathbb{C}\{z_1, \ldots, z_{n-1}\}$ and then R is isomorphic to $\mathbb{C}\{z_1, \ldots, z_{n-1}\}/I$, where I is the kernel of the surjective composite morphism

$$\mathbb{C}\{z_1, \ldots, z_{n-1}\} \xrightarrow{\simeq} \frac{\mathbb{C}\{z_1, \ldots, z_{n-1}, z_n\}}{(f)} \to R.$$

The conclusion follows by induction on n. □

Lemma A.1.5 *Let R be an analytic algebra with maximal ideal $\mathfrak{m}_R$. For every finite sequence $r_1, \ldots, r_n \in \mathfrak{m}_R$ there exists a unique morphism of analytic algebras*

$$f \colon \mathbb{C}\{z_1, \ldots, z_n\} \to R$$

such that $f(z_i) = r_i$.

Proof We first note that, by the Artin–Rees lemma (see e.g. [10, Cor. 10.19]), we have $\cap_n \mathfrak{m}_R^n = 0$ and then every local homomorphism $f \colon \mathbb{C}\{z_1, \ldots, z_n\} \to R$ is uniquely determined by its sequence of factorizations

$$f_s \colon \frac{\mathbb{C}\{z_1, \ldots, z_n\}}{(z_1, \ldots, z_n)^s} \to \frac{R}{\mathfrak{m}_R^s}, \qquad s > 0.$$

Since $\mathbb{C}\{z_1, \ldots, z_n\}/(z_1, \ldots, z_n)^s$ is a $\mathbb{C}$-algebra generated by $z_1, \ldots, z_n$, every f_s is uniquely determined by $f(z_i)$; this proves the unicity.

In order to prove the existence, since there exists an integer $m > 0$ and a surjective morphism of analytic algebras $\mathbb{C}\{u_1, \dots, u_m\} \to R$, it is not restrictive to assume $R = \mathbb{C}\{u_1, \dots, u_m\}$. The convergent power series r_i give a germ of holomorphic map

$$r = (r_1, \dots, r_n)\colon (\mathbb{C}^m, 0) \to (\mathbb{C}^n, 0),$$

and $r^*(z_i) = r_i$, where $r^*\colon \mathbb{C}\{z_1, \dots, z_n\} \to \mathbb{C}\{u_1, \dots, u_m\}$ is the induced morphism of analytic algebras. □

One of the consequences of Lemma A.1.5 is that the category of analytic algebras contains coproducts; given two analytic algebras R, S there exists a third analytic algebras $R \amalg S$ and two morphisms $i\colon R \to R \amalg S$, $j\colon S \to R \amalg S$, that satisfy the universal property of coproducts. This means that for every pair of morphisms of analytic algebras $f\colon R \to A$, $g\colon S \to A$ there exists a unique morphism of analytic algebras $h\colon R \amalg S \to A$ such that $hi = f$ and $hj = g$.

The construction of the coproduct can be done in the following way: if

$$R = \frac{\mathbb{C}\{z_1, \dots, z_n\}}{I}, \qquad S = \frac{\mathbb{C}\{y_1, \dots, y_m\}}{J},$$

denoting by

$$\mathbb{C}\{z_1, \dots, z_n\} \xrightarrow{i} \mathbb{C}\{z_1, \dots, z_n, y_1, \dots, y_m\} \xleftarrow{j} \mathbb{C}\{y_1, \dots, y_m\}$$

the natural inclusions, and by $K = (i(I)) + (j(J)) \subset \mathbb{C}\{z_1, \dots, z_n, y_1, \dots, y_m\}$ the ideal generated by the images of I and J, it is easy to see, as a consequence of Lemma A.1.5, that

$$R \amalg S = \frac{\mathbb{C}\{z_1, \dots, z_n, y_1, \dots, y_m\}}{K}.$$

is the coproduct of R and S.

Remark A.1.6 Given two analytic algebras R, S the maps $R \to R \amalg S$ and $S \to R \amalg S$ induce a morphism of $\mathbb{C}$-algebras $f\colon R \otimes S \to R \amalg S$. Then the morphism f is an isomorphism if and only if $R \otimes S$ is an analytic algebra. The "only if" part is obvious; conversely, if $R \otimes S$ is an analytic algebra, by the universal property of the coproduct there exists a morphism of analytic algebras $h\colon R \amalg S \to R \otimes S$ such that hf is the identity. In particular, h is surjective, $h(\mathfrak{m}_{R\amalg S}) = \mathfrak{m}_{R\otimes S}$ and then $f(\mathfrak{m}_{R\otimes S}) \subset \mathfrak{m}_{R\amalg S}$. We have therefore proved that f and fh are morphisms of analytic algebras; finally, fh is the identity since it commutes with the natural maps $R \to R \amalg S$ and $S \to R \amalg S$.

For notational simplicity, for every analytic algebra R and every integer $n \geq 0$, we write

$$R\{z_1, \dots, z_n\} = R \amalg \mathbb{C}\{z_1, \dots, z_n\}.$$

Notice that $(R\{z_1, \dots, z_n\})\{z_{n+1}, \dots, z_m\} = R\{z_1, \dots, z_m\}$.

A.2 Weierstrass Polynomials

In this section, after recalling the well known Weierstrass division and preparation theorems, we prove that under certain, and quite restrictive, additional conditions the usual (algebraic) tensor product of two analytic algebras remains an analytic algebra.

Definition A.2.1 A convergent power series $p \in \mathbb{C}\{z_1, \dots, z_n, t\}$ is called a **Weierstrass polynomial** in t of degree $d \geq 0$ if

$$p(z_1, \dots, z_n, t) = t^d + \sum_{i=0}^{d-1} p_i(z_1, \dots, z_n)t^i, \qquad p_i \in \mathbb{C}\{z_1, \dots, z_n\}, \ p_i(0) = 0.$$

For later use, it is useful to point out that if $p(z_1, \dots, z_n, t)$ is a Weierstrass polynomial in t of degree d then $p(0, \dots, 0, t) = t^d$ and therefore every monic factor of p in the ring $\mathbb{C}\{z_1, \dots, z_n\}[t]$ is a Weierstrass polynomial.

Theorem A.2.2 (Weierstrass preparation theorem) *Let $f \in \mathbb{C}\{z_1, \dots, z_n, t\}$ be a convergent power series such that $f(0, \dots, 0, t) \neq 0$, i.e., f does not belong to the ideal generated by $z_, \dots, z_n$. Then there exists a unique $e \in \mathbb{C}\{z_1, \dots, z_n, t\}$ such that $e(0) \neq 0$ and the product ef is a Weierstrass polynomial in t.*

Proof See e.g. [96, 97, 100, 101, 107, 145]. Notice that, since $e(0) \neq 0$, the degree in t of the Weierstrass polynomial ef is equal to the multiplicity of the power series $f(0, \dots, 0, t) \in \mathbb{C}\{t\}$. □

Corollary A.2.3 *Let $f \in \mathbb{C}\{z_1, \dots, z_n\}$ be a convergent power series of multiplicity d. Then, after a possible generic linear change of coordinates, there exists $e \in \mathbb{C}\{z_1, \dots, z_n\}$ such that $e(0) \neq 0$ and ef is a Weierstrass polynomial of degree d in z_n.*

Proof After a generic change of coordinates of the form $z_i \mapsto z_i + a_i z_n$, with $a_i \in \mathbb{C}$, the series $f(0, \dots, 0, z_n)$ has multiplicity d and the conclusion follows from Theorem A.2.2. □

Lemma A.2.4 *Let $f, g \in \mathbb{C}\{x_1, \dots, x_n\}[t]$ with g a Weierstrass polynomial in t. If $f = hg$ for some power series $h \in \mathbb{C}\{x_1, \dots, x_n, t\}$ then $h \in \mathbb{C}\{x_1, \dots, x_n\}[t]$.*

Proof Let s be the degree of g in t, then we can write $g = t^s + \sum_{i>0} g_i(x)t^{s-i}$, with $g_i(0) = 0$ for every $i > 0$. If

$$f = \sum_{i=0}^{r} f_i(x)t^{r-i}, \qquad h = \sum_{i=0}^{\infty} h_i(x)t^i,$$

we need to prove that $h_i = 0$ for every $i > r - s$. Assume the contrary and choose an index $j > r - s$ such that the multiplicity of h_j takes the minimum among all the multiplicities of the power series h_i, $i > r - s$. From the equality $0 = h_j + \sum_{i>0} g_i h_{j+i}$ we get a contradiction. □

Notice that if g is not a Weierstrass polynomial, then the above result is false; consider for instance the case $n = 0$, $f = t^3$, $g = t - t^2$ and $h = \sum_{i=2}^{\infty} t^i$.

Since $\mathbb{C}\{x_1, \dots, x_n\}$ is a unique factorization domain, by Gauss lemma also $\mathbb{C}\{x_1, \dots, x_n\}[t]$ is a unique factorization domain. The following lemma gives useful information about their irreducible elements.

Lemma A.2.5 *Let $f \in \mathbb{C}\{x_1, \dots, x_n\}[t]$ be an irreducible monic polynomial of degree $d > 0$ in t. Then $f(0, \dots, 0, t) = (t - c)^d$ for some $c \in \mathbb{C}$.*

Proof Let $c \in \mathbb{C}$ be a root of multiplicity $l > 0$ of the monic polynomial $f(0, \dots, 0, t) \in \mathbb{C}[t]$; since the polynomial $f(0, \dots, 0, t)$ has degree d it is sufficient to prove that $l \geq d$.

If $l < d$ then the multiplicity of the power series $f(0, \dots, 0, t + c) \in \mathbb{C}\{t\}$ is exactly l. By Theorem A.2.2 and Lemma A.2.4 the function $f(x_1, \dots, x_n, t + c)$ is divided in $\mathbb{C}\{x_1, \dots, x_n\}[t]$ by a Weierstrass polynomial of degree l in t, in contradiction with the irreducibility of f. □

Theorem A.2.6 (Weierstrass division theorem)
Let $p \in \mathbb{C}\{z_1, \dots, z_n, t\}$, $p \neq 0$, be a Weierstrass polynomial of degree $d \geq 0$ in t. Then for every $f \in \mathbb{C}\{z_1, \dots, z_n, t\}$ there exists a unique $h \in \mathbb{C}\{z_1, \dots, z_n, t\}$ such that $f - hp \in \mathbb{C}\{z_1, \dots, z_n\}[t]$ is a polynomial of degree $< d$ in t.

Proof For the proof we refer to [96, 97, 100, 101, 107, 145]. □

We note that an equivalent statement for the division theorem is the following:

Corollary A.2.7 *If $p \in \mathbb{C}\{z_1, \dots, z_n, t\}$, $p \neq 0$, is a Weierstrass polynomial of degree $d \geq 0$ in t, then $\mathbb{C}\{z_1, \dots, z_n, t\}/(p)$ is a free $\mathbb{C}\{z_1, \dots, z_n\}$-module with basis $1, t, \dots, t^{d-1}$.*

Proof Immediate consequence of Theorem A.2.6. □

Lemma A.2.8 *Let R, S be two analytic algebras. Then $R \otimes_{\mathbb{C}} S$ is a local algebra if and only if either R or S is an Artin algebra.*

Proof For the complete proof we refer to [253], cf. Exercise A.7.1. However, for our goals we only need the "if" part of the lemma, which is very easy to prove. In fact, if $\mathfrak{m}_R$ is nilpotent, then the natural projection $p\colon R \otimes_{\mathbb{C}} S \to \mathbb{C} \otimes_{\mathbb{C}} S = S$ is surjective and its kernel is a nilpotent ideal. In order to conclude the proof we only need to prove that $p^{-1}(\mathfrak{m}_S)$ is the unique maximal ideal of $R \otimes_{\mathbb{C}} S$, or equivalently that $x \in R \otimes_{\mathbb{C}} S$ is invertible if and only if $p(x)$ is invertible. One implication is clear; conversely, if $p(x)$ is invertible, there exists $y \in R \otimes_{\mathbb{C}} S$ such that $p(xy) = 1$ and then $xy = 1 + a$, with $p(a) = 0$. By assumption the kernel of p is a nilpotent ideal and therefore $1 + a$ is invertible. □

Theorem A.2.9 *Let $A \to B$ be a local morphism of local $\mathbb{C}$-algebras. Suppose that A is an analytic algebra and B is finitely generated as an A-module. Then B is an analytic algebra.*

Proof Let's first note that B is unitary and the residue field $B/\mathfrak{m}_B$ is a finite extension of $A/\mathfrak{m}_A = \mathbb{C}$; this implies that B, considered as a complex vector space, decomposes as $B = \mathbb{C} \oplus \mathfrak{m}_B$. Let $b_1, \ldots, b_k \in \mathfrak{m}_B$ be a set of generators of B as A-algebra. By induction on k it is sufficient to find an analytic algebra C and a local morphism $C \to B$ such that $b_2, \ldots, b_k$ generate B as a C-module.

It is not restrictive to assume $A = \mathbb{C}\{z_1, \ldots, z_n\}$; since B is Noetherian as an A-module there exists a monic polynomial $p(t) \in A[t]$ such that $p(b_1) = 0$ in B. According to Theorem A.2.2 and Lemma A.2.4 we may write $p(t) = q(t)h(t)$ with $q, h \in A[t]$, $q(t)$ a Weierstrass polynomial and $h(0)$ invertible in A. As a consequence $h(b_1)$ does not belong to the maximal ideal of B, hence it is invertible and therefore $q(b_1) = 0$. Thus we have

$$A \to C = \frac{A[t]}{(q(t))} \xrightarrow{t \mapsto b_1} B$$

and by Corollary A.2.7 the natural map

$$C = \frac{\mathbb{C}\{x_1, \ldots, x_n\}[t]}{(q(t))} \to \frac{\mathbb{C}\{x_1, \ldots, x_n, t\}}{(q(t))}$$

is an isomorphism and C is an analytic algebra. □

Corollary A.2.10 *Let A be a finitely generated Artin local $\mathbb{C}$-algebra. Then*

$$A\{z_1, \ldots, z_n\} = A \amalg \mathbb{C}\{z_1, \ldots, z_n\} = A \otimes \mathbb{C}\{z_1, \ldots, z_n\}.$$

Proof By Remark A.1.6 it is sufficient to show that $A \otimes \mathbb{C}\{z_1, \ldots, z_n\}$ is an analytic algebra, and this follows from Lemma A.2.8 and Theorem A.2.9. □

Corollary A.2.11 *Let $A \to R$, $A \to B$ be two morphisms of analytic algebras, with A, B local Artin algebras. Then $R \otimes_A B$ is an analytic algebra.*

Proof Since $R \otimes_A B$ is a quotient of $R \otimes_{\mathbb{C}} B$, by Lemma A.2.8 we have that $R \otimes_A B$ is a local A-algebra. By Theorem A.2.9 it is sufficient to prove that the natural morphism

$$R \to R \otimes_A B, \qquad r \mapsto r \otimes 1,$$

is a local morphism of local A-algebras. It is clear that the composition $R \to R \otimes_A B \to \mathbb{C} \otimes_A \mathbb{C} = \mathbb{C}$ is equal to the projection to the residue field and this proves that the map $R \to R \otimes_A B$ is a local morphism. □

A.3 Smooth Algebras

Analytic algebras have many common features with local complete algebras; one of them is that the surjectivity of a morphism can be detected at the level of Zariski cotangent spaces.

Lemma A.3.1 *A morphism of analytic algebras $f : A \to B$ is surjective if and only if the induced map*

$$\overline{f} : \frac{\mathfrak{m}_A}{\mathfrak{m}_A^2} \to \frac{\mathfrak{m}_B}{\mathfrak{m}_B^2}$$

is surjective.

Proof The only if part is clear and valid for every surjective local morphism. Assume conversely that $\overline{f}$ surjective and write

$$A = \frac{\mathbb{C}\{x_1, \ldots, x_n\}}{I}, \qquad B = \frac{\mathbb{C}\{y_1, \ldots, y_m\}}{J},$$

for some ideals $I \subset (x_1, \ldots, x_n)^2$ and $J \subset (y_1, \ldots, y_m)^2$. Then $x_1, \ldots, x_n$ is a basis of $\mathfrak{m}_A/\mathfrak{m}_A^2$, $y_1, \ldots, y_m$ is a basis of $\mathfrak{m}_B/\mathfrak{m}_B^2$ and up to a linear change of coordinates we may assume $m \le n$, $\overline{f}(x_i) = y_i$ for $i = 1, \ldots, m$ and $\overline{f}(x_i) = 0$ for $i > m$. According to Lemma A.1.5 the morphism f can be lifted to a morphism of analytic algebras $\Phi : \mathbb{C}\{x_1, \ldots, x_n\} \to \mathbb{C}\{y_1, \ldots, y_m\}$. Denoting by $\phi_i = \Phi(x_i) \in \mathbb{C}\{y_1, \ldots, y_m\}$, the restriction of Φ to the subalgebra $\mathbb{C}\{x_1, \ldots, x_m\}$ factors as

$$\mathbb{C}\{x_1, \ldots, x_m\} \to \frac{\mathbb{C}\{x_1, \ldots, x_m, y_1, \ldots, y_m\}}{(x_1 - \phi_1, \ldots, x_m - \phi_m)} \xrightarrow{x_i \mapsto \phi_i,\, y_j \mapsto y_j} \mathbb{C}\{y_1, \ldots, y_m\}.$$

Finally, both the above morphisms are isomorphisms by the implicit function theorem A.1.2, since the linear part of ϕ_i is equal to y_i for every $i = 1, \ldots, m$. In particular, Φ is surjective and therefore also f is surjective. □

Proposition A.3.2 *A morphism $f : A \to A$ from an analytic algebras into itself is an isomorphism if and only if the induced map*

$$\overline{f} : \frac{\mathfrak{m}_A}{\mathfrak{m}_A^2} \to \frac{\mathfrak{m}_A}{\mathfrak{m}_A^2}$$

is surjective.

Proof The only if part is obvious. Conversely, if $\overline{f}$ is surjective, then by Lemma A.3.1 the morphism f is surjective. Since $\ker f^n \subset \ker f^{n+1}$ and A is a Noetherian ring there exists a positive integer n such that $\ker f^{n+1} = \ker f^n$. Let $a \in \ker f$ and choose $b \in A$ such that $a = f^n(b)$. Then $f^{n+1}(b) = f(a) = 0$ and therefore $b \in \ker f^{n+1} = \ker f^n$, giving $a = 0$. We have thus proved that f is also injective. □

Definition A.3.3 An analytic algebra is called **smooth** if it is isomorphic to $\mathbb{C}\{z_1, \dots, z_k\}$ for some $k \geq 0$. A morphism of analytic algebras $f: R \to S$ is called **smooth** if there exists an isomorphism $\phi: S \to R\{z_1, \dots, z_k\}$ such that ϕf is the natural inclusion $R \hookrightarrow R\{z_1, \dots, z_k\}$.

Notice that an analytic algebra R is smooth if and only if the morphism $\mathbb{C} \to R$ is smooth. Composition of smooth morphisms is a smooth morphism.

Proposition A.3.4 *Let R be an analytic algebra. Then the following conditions are equivalent:*

1. *R is smooth;*
2. *for every surjective morphism of analytic algebras $B \to A$, the morphism*
$$\mathrm{Mor}_{\mathbf{An}}(R, B) \to \mathrm{Mor}_{\mathbf{An}}(R, A)$$
is surjective;
3. *for every $n \geq 2$ the morphism*
$$\mathrm{Mor}_{\mathbf{An}}\left(R, \frac{\mathbb{C}\{t\}}{(t^n)}\right) \to \mathrm{Mor}_{\mathbf{An}}\left(R, \frac{\mathbb{C}\{t\}}{(t^2)}\right)$$
is surjective.

Proof The proof that (1) implies (2) is an immediate consequence of Lemma A.1.5, while it is obvious that (2) implies (3). In order to prove that (3) implies (1), writing $R = \mathbb{C}\{z_1, \dots, z_k\}/I$, with $I \subset (z_1, \dots, z_k)^2$, we show that the lifting property (3) implies $I = 0$. Denote by $\pi: \mathbb{C}\{z_1, \dots, z_k\} \to R$ the projection to the quotient, assume $I = \ker \pi \neq 0$ and let $s \geq 2$ be the the greatest integer such that the ideal I is contained in $(z_1, \dots, z_k)^s$; we claim that
$$\mathrm{Mor}_{\mathbf{An}}\left(R, \frac{\mathbb{C}\{t\}}{(t^{s+1})}\right) \to \mathrm{Mor}_{\mathbf{An}}\left(R, \frac{\mathbb{C}\{t\}}{(t^2)}\right)$$
is not surjective. Choosing $f \in I - (z_1, \dots, z_k)^{s+1}$, after a possible generic linear change of coordinates of the form $z_i \mapsto z_i + a_i z_1$, $i \geq 2$, $a_2, \dots, a_k \in \mathbb{C}$, we may assume that f contains the monomial z_1^s with a nonzero coefficient, say $f = cz_1^s + \cdots, c \neq 0$. Let $\alpha: R \to \mathbb{C}\{t\}/(t^2)$ be the morphism defined by $\alpha(z_1) = t, \alpha(z_i) = 0$ for $i > 1$. If there exists $\beta: R \to \mathbb{C}\{t\}/(t^{s+1})$ that lifts α, then $\beta(z_1) - t, \beta(z_2), \dots, \beta(z_k) \in (t^2)$ and therefore $\beta\pi(f) \equiv ct^s \pmod{t^{s+1}}$, in contradiction with $\pi(f) = 0$. □

Proposition A.3.5 *Let $f: R \to S$ be a morphism of analytic algebras. Then the following conditions are equivalent:*

1. *f is smooth, and then $S \cong R\{z_1, \dots, z_n\}$;*

2. *for every surjective morphism of analytic algebras* $\alpha\colon B \to A$*, the morphism*

$$\mathbf{Mor_{An}}\,(S, B) \to \mathbf{Mor_{An}}\,(R, B) \times_{\mathbf{Mor_{An}}(R,A)} \mathbf{Mor_{An}}\,(S, A)$$

   is surjective;

3. *for any surjective morphism* $\alpha\colon B \to A$ *of zero-dimensional analytic algebras, the morphism*

$$\mathbf{Mor_{An}}\,(S, B) \to \mathbf{Mor_{An}}\,(R, B) \times_{\mathbf{Mor_{An}}(R,A)} \mathbf{Mor_{An}}\,(S, A)$$

   is surjective.

Proof The only nontrivial part of the proof is that condition (3) implies (1). Let $s_1, \dots, s_k \in \mathfrak{m}_S$ be elements such that their classes in $\mathfrak{m}_S/\mathfrak{m}_S^2$ form a basis of an algebraic complement of $f(\mathfrak{m}_R/\mathfrak{m}_R^2)$. Define $\phi\colon P = R\{z_1, \dots, z_k\} \to S$ as the unique extension of f such that $\phi(z_i) = s_i$; an easy application of Lemma A.3.1 shows that ϕ is surjective and our goal is to prove that condition (3) implies that ϕ is injective. To this end it is convenient to introduce the following general notation: for every analytic algebra Q and every positive integer n we denote by $Q_n = Q/\mathfrak{m}_Q^n$ and by $\pi_n\colon Q \to Q_n$ the natural projection.

In an informal language, condition (3) means that for every solid commutative diagram of analytic algebras

$$\begin{array}{ccc} R & \xrightarrow{\;f\;} & S \\ \big\downarrow & \swarrow & \big\downarrow \\ B & \xrightarrow{\;\alpha\;} & A\,, \end{array}$$

with α a surjective morphism of Artin local rings, there exists a dashed lifting. Looking at the diagram

$$\begin{array}{ccc} R & \xrightarrow{\;f\;} & S \\ {\scriptstyle \pi_2}\big\downarrow & \swarrow & \big\downarrow \\ R_2 & \xrightarrow{\;\alpha\;} & \mathbb{C} \end{array}$$

we obtain that $f\colon R_2 \to S_2$ is injective and then $\phi\colon P_2 \to S_2$ is an isomorphism. In particular, there exists a surjective morphism of analytic algebras $g_2\colon S \to P_2$ such that $g_2\phi = \pi_2\colon P \to P_2$ and, for every $k > 0$, we can lift g_2 to a surjective morphism $g_k\colon S \to P_k$ just by looking at the diagrams

$$\begin{array}{ccc} R & \xrightarrow{\;f\;} & S \\ {\scriptstyle i_k}\big\downarrow & \swarrow & \big\downarrow{\scriptstyle g_2} \\ P_k & \xrightarrow{\;\alpha\;} & P_2 \end{array} \qquad \text{where } \; i_k\colon R \hookrightarrow P \xrightarrow{\pi_k} P_k.$$

Notice now that $P_k \xrightarrow{\phi} S_k \xrightarrow{g_k} P_k$ is a surjective endomorphism of a finite-dimensional complex vector space and hence an isomorphism; thus $P_k \xrightarrow{\phi} S_k$ is injective for every $k > 0$ and then also $\phi\colon P \to S$ is injective. □

Example A.3.6 The analogue of condition (3) of Proposition A.3.4 for morphism does not imply smoothness. For example, the morphism of analytic algebras

$$R = \frac{\mathbb{C}\{x, y\}}{(x^3, y^3)} \to S = \frac{\mathbb{C}\{x, y\}}{(x^3, x^2y^2, y^3)}, \qquad x \mapsto x,\ y \mapsto y,$$

is not smooth, although for every $n \geq m$ the induced map

$$\mathrm{Mor}_{\mathbf{An}}\left(S, \frac{\mathbb{C}\{t\}}{(t^n)}\right) \to \mathrm{Mor}_{\mathbf{An}}\left(R, \frac{\mathbb{C}\{t\}}{(t^n)}\right) \times_{\mathrm{Mor}_{\mathbf{An}}\left(R, \frac{\mathbb{C}\{t\}}{(t^m)}\right)} \mathrm{Mor}_{\mathbf{An}}\left(S, \frac{\mathbb{C}\{t\}}{(t^m)}\right)$$

is surjective.

A.4 The Holomorphic Curve Selection Lemma

The aim of this section is to give an elementary proof of the following theorem.

Theorem A.4.1 (Holomorphic curve selection lemma) *Let $I \subset \mathbb{C}\{z_1, \ldots, z_n\}$ be a proper ideal and $h \notin \sqrt{I}$. Then there exists a morphism of analytic algebras $\phi\colon \mathbb{C}\{z_1, \ldots, z_n\} \to \mathbb{C}\{t\}$ such that $\phi(I) = 0$ and $\phi(h) \neq 0$.*

Before proving Theorem A.4.1 we need a series of results that we consider of independent interest. We assume that the reader has a basic knowledge of the resultant $R(p, q)$ of two polynomials $p, q \in A[t]$ with coefficients in an integral domain A, see e.g. [261].

Lemma A.4.2 *Let $p \in \mathbb{C}\{x\}[y]$ be a monic polynomial of positive degree d in y. Then there exists a ring homomorphism $\phi\colon \mathbb{C}\{x\}[y] \to \mathbb{C}\{t\}$ such that $\phi(p) = 0$ and $\phi(x) = t^s$ for some integer $s > 0$.*

Moreover, if p is an irreducible Weierstrass polynomial then ϕ extends to a morphism of analytic algebras $\phi\colon \mathbb{C}\{x, y\} \to \mathbb{C}\{t\}$ and $\ker \phi = p\mathbb{C}\{x, y\}$.

Proof If $d = 1$ then $p(x, y) = y - p_1(x)$ and we can consider the morphism ϕ given by $\phi(x) = t$, $\phi(y) = p_1(t)$. By induction we can assume that the theorem holds for monic polynomials of degree $< d$.

If p is reducible we have completed the proof; otherwise, writing $p = y^d + p_1(x)y^{d-1} + \cdots + p_d(x)$, after the coordinate change $x \mapsto x$, $y \mapsto y - p_1(x)/d$ we can assume $p_1 = 0$. For every $i \geq 2$ denote by α_i the multiplicity of p_i (we set $\alpha_i = +\infty$ if $p_i = 0$); by Lemma A.2.5 we have $\alpha_i > 0$ for every $i = 2, \ldots, d$. Let $j \geq 2$ be a fixed index such that $\dfrac{\alpha_j}{j} \leq \dfrac{\alpha_i}{i}$ for every i. Setting $m = \alpha_j$, the expression

$$q(\xi, y) = p(\xi^j, \xi^m y)\xi^{-dm} = y^d + \sum \frac{p_i(\xi^j)}{\xi^{mi}} y^{d-i} = y^d + \sum \eta_i(\xi) y^{d-i}$$

is a well defined element of $\mathbb{C}\{\xi\}[y]$ since for every i the multiplicity of $\eta_i(\xi)$ equal to $j\alpha_i - i\alpha_j \geq 0$.

Since $\eta_1 = 0$ and $\eta_j(0) \neq 0$ the polynomial q is not irreducible and then, by induction there exists a nontrivial morphism $\psi\colon \mathbb{C}\{\xi\}[y] \to \mathbb{C}\{t\}$ such that $\psi(q) = 0$, $\psi(\xi) = t^s$ and we can take $\phi(x) = \psi(\xi^j) = t^{js}$ and $\phi(y) = \psi(\xi^m y)$.

Assume now that p is an irreducible Weierstrass polynomial. The conditions $\phi(p) = 0$ and $\phi(x) \in (t)$ imply that $\phi(y) \in (t)$ and therefore ϕ extends to a morphism of analytic algebras $\phi\colon \mathbb{C}\{x, y\} \to \mathbb{C}\{t\}$.

For every $g \in \ker(\phi)$, by the division theorem there exists $r \in \mathbb{C}\{x\}[y]$ such that $g = hp + r$ and then $r \in \ker(\phi)$. Let $R(f, r) \in \mathbb{C}\{x\}$ be the resultant of the elimination of y on the polynomials p, r. By general properties of the resultant, $R(p, r)$ belongs to the ideal $(p, r) \subset \ker(\phi)$ and then $R(p, r) \in \ker(\phi) \cap \mathbb{C}\{x\} = 0$; since $\mathbb{C}\{x\}$ is a unique factorization domain, the vanishing of $R(f, r)$ implies that p divides r and then p divides g in $\mathbb{C}\{x, y\}$. □

Theorem A.4.3 (Newton–Puiseux) *Let $f \in \mathbb{C}\{x, y\}$ be a power series of positive multiplicity. Then there exists a nontrivial morphism of analytic algebras $\phi\colon \mathbb{C}\{x, y\} \to \mathbb{C}\{t\}$ such that $\phi(f) = 0$. Moreover, if f is irreducible then* $\ker \phi = (f)$.

In the above statement nontrivial means that either $\phi(x) \neq 0$ or $\phi(y) \neq 0$.

Proof After a linear change of coordinates we can assume $f(0, y)$ a nonzero power series of multiplicity $d > 0$; by the preparation theorem there exists an invertible power series e such that $p = ef$ is a Weierstrass polynomial of degree d in y. According to Lemma A.4.2 there exists a homomorphism $\phi\colon \mathbb{C}\{x\}[y] \to \mathbb{C}\{t\}$ such that $\phi(p) = 0$ and $0 \neq \phi(x) \in (t)$. Since p is a Weierstrass polynomial we have $\phi(y) \in (t)$ and then ϕ extends to a local morphism $\phi\colon \mathbb{C}\{x, y\} \to \mathbb{C}\{t\}$.

Since every irreducible factor of p in the ring $\mathbb{C}\{x\}[y]$ is a Weierstrass polynomial, if f is irreducible in $\mathbb{C}\{x, y\}$ then also p is irreducible in the ring $\mathbb{C}\{x\}[y]$ and then $\ker(\phi) = (p) = (f)$ by Lemma A.4.2. □

Lemma A.4.4 *Let A be an integral domain, $0 \neq \mathfrak{p} \subset A[t]$ a nontrivial prime ideal such that $\mathfrak{p} \cap A = 0$ and let $0 \neq p \in \mathfrak{p}$ be a polynomial of minimum degree in t. Then for every polynomial $q \notin \mathfrak{p}$, the resultant $R(p, q) \in A$ does not vanish.*

Proof Denote by $\mathbb{K}$ the fraction field of A and by $\mathfrak{p}^e \subset \mathbb{K}[t]$ the ideal generated by $\mathfrak{p}$. Then $\mathfrak{p}^e$ is a prime ideal of the ring $\mathbb{K}[t]$ and $\mathfrak{p}^e \cap A[t] = \mathfrak{p}$.

In fact, $\mathfrak{p}^e = \left\{ \frac{x}{a} \;\middle|\; x \in \mathfrak{p},\ a \in A - \{0\} \right\}$. If $\frac{x_1}{a_1}\frac{x_2}{a_2} \in \mathfrak{p}^e$ with $x_i \in A[t]$ and $a_i \in A$, then there exists $a \in A - \{0\}$ such that $ax_1x_2 \in \mathfrak{p}$. Since $\mathfrak{p} \cap A = 0$ it must be either $x_1 \in \mathfrak{p}$ or $x_2 \in \mathfrak{p}$. This shows that $\mathfrak{p}^e$ is prime. If $q \in \mathfrak{p}^e \cap A[t]$, then there exists $a \in A$, $a \neq 0$ such that $aq \in \mathfrak{p}$ and, since $\mathfrak{p} \cap A = 0$ we have $q \in \mathfrak{p}$.

Since $p \in \mathfrak{p} - \{0\}$ is of minimum degree in $\mathfrak{p}$, it also has minimum degree in $\mathfrak{p}^e$. Since $\mathbb{K}[t]$ is a Euclidean ring, $\mathfrak{p}^e = p\mathbb{K}[t]$ and, since $\mathfrak{p}^e$ is prime, the polynomial p is irreducible in $\mathbb{K}[t]$.

If $q \in A[t] - \mathfrak{p}$ then $q \notin \mathfrak{p}^e = p\mathbb{K}[t]$ and therefore $R(p, q) \neq 0$. □

Lemma A.4.5 *Let A be a unique factorization domain, $\mathfrak{p} \subset A[t]$ a prime ideal and denote $\mathfrak{q} = A \cap \mathfrak{p}$. If $\mathfrak{p} \neq \mathfrak{q}[t]$ (e.g. if $\mathfrak{p}$ is proper and contains a monic polynomial) then there exists an irreducible polynomial $p \in \mathfrak{p}$ such that for every polynomial $q \notin \mathfrak{p}$ we have $R(p, q) \notin \mathfrak{q}$.*

Proof The ideal $\mathfrak{q}$ is prime and $\mathfrak{q}[t] \subset \mathfrak{p}$, therefore the image of $\mathfrak{p}$ in the quotient ring $(A/\mathfrak{q})[t] = A[t]/\mathfrak{q}[t]$ is a prime ideal satisfying the hypothesis of Lemma A.4.4. It is therefore sufficient to take $\tilde{p}$ as any lifting of the element described in Lemma A.4.4 and use the functorial properties of the resultant in order to prove that $R(\tilde{p}, g) \notin \mathfrak{q}$ for every $q \notin \mathfrak{p}$. If $p \in \mathfrak{p}$ is an irreducible factor of $\tilde{p}$ and $h = \tilde{p}/p$, by the bilinearity relations of resultant we get $R(\tilde{p}, q) = R(p, q)R(h, q)$ and then we get $R(p, g) \notin \mathfrak{q}$ for every $q \notin \mathfrak{p}$. □

The Weierstrass division theorem and the general properties of the resultant allow us to extend (partially) the definition of the resultant to power series. More precisely, if $p \in \mathbb{C}\{z_1, \dots, z_n\}[t]$ is a Weierstrass polynomial in t of degree d, for every $f \in \mathbb{C}\{z_1, \dots, z_n, t\}$ we can define the resultant $R(p, f) \in \mathbb{C}\{z_1, \dots, z_n\}$ as the determinant of the morphism of free $\mathbb{C}\{z_1, \dots, z_n\}$-modules of rank d

$$f\colon \frac{\mathbb{C}\{z_1, \dots, z_n, t\}}{(p)} \to \frac{\mathbb{C}\{z_1, \dots, z_n, t\}}{(p)}$$

induced by the multiplication with f. It is easy to prove that this definition of the resultant coincides with the usual one whenever $f \in \mathbb{C}\{z_1, \dots, z_n\}[t]$. Moreover for every element $r \in \mathbb{C}\{z_1, \dots, z_n\}[t]$ such that $f - r \in (p)$ we have

$$R(p, f) = R(p, r) \in \mathbb{C}\{z_1, \dots, z_n\} \cap (p, r) = \mathbb{C}\{z_1, \dots, z_n\} \cap (p, f).$$

Given two Weierstrass polynomials p_1, p_2 and $f \in \mathbb{C}\{z_1, \dots, z_n, t\}$, choosing an element $r \in \mathbb{C}\{z_1, \dots, z_n\}[t]$ such that $f - r \in (p_1 p_2)$, the usual bilinearity relation give

$$R(p_1 p_2, f) = R(p_1 p_2, r) = R(p_1, r)R(p_2, r) = R(p_1, f)R(p_2, f)\,.$$

Lemma A.4.6 *Let $p \in \mathbb{C}\{z_1, \dots, z_n\}[t]$ be a Weierstrass polynomial of positive degree in t and let $V \subset \mathbb{C}\{z_1, \dots, z_n, t\}$ be a $\mathbb{C}$-vector subspace such that $R(p, f) = 0$ for every $f \in V$. Then there exists an irreducible factor $q \in \mathbb{C}\{z_1, \dots, z_n\}[t]$ of the polynomial p such that $V \subset q\mathbb{C}\{z_1, \dots, z_n, t\}$.*

Proof Let $p = p_1 p_2 \cdots p_s$ be the irreducible decomposition of p in $\mathbb{C}\{z_1, \dots, z_n\}[t]$ with every p_i a monic polynomial in t; then every p_i is a Weierstrass polynomial.

If $R(p, f) = 0$, the bilinearity relations implies that and $r = f - hp \in \mathbb{C}\{z_1, \ldots, z_n\}[t]$ is the remainder of the division then $R(p, r) = 0$. By the bilinearity relations there exists a factor p_i dividing r and therefore also dividing f. In particular, setting $V_i = V \cap (p_i)$, we have $V = \cup_i V_i$; therefore $V = V_i$ for at least one index i and we can take $q = p_i$. □

Lemma A.4.7 *Let $\mathfrak{p} \subset \mathbb{C}\{z_1, \ldots, z_n\}$ be a prime ideal and $h \notin \mathfrak{p}$. Then there exists a morphism of analytic algebras $\phi\colon \mathbb{C}\{z_1, \ldots, z_n\} \to \mathbb{C}\{t\}$ such that $\phi(\mathfrak{p}) = 0$ and $\phi(h) \neq 0$.*

Proof If h is an invertible element then the morphism $\phi\colon \mathbb{C}\{z_1, \ldots, z_n\} \to \mathbb{C}\{t\}$, $\phi(z_i) = 0$ for every i, satisfies the statement of the theorem. If $\mathfrak{p} = 0$ then, after a possible change of coordinates, we may assume $h(0, \ldots, 0, z_n) \neq 0$ and therefore we can take $\phi(z_i) = 0$ for $i = 1, \ldots, n-1$ and $\phi(z_n) = t$.

Thus we assume h of positive multiplicity, $\mathfrak{p} = 0$ and we prove the lemma by induction on n. If $n = 1$ the only prime nontrivial ideal is (z_1) and every $h \notin (z_1)$ is invertible.

Assume then $n > 1$, $\mathfrak{p} \neq 0$ and choose a nonzero element $g \in \mathfrak{p}$ with minimum multiplicity $d > 0$. After a possible linear change of coordinates and multiplication by invertible elements we may assume both h and g Weierstrass polynomials in the variable z_n. Writing

$$\mathfrak{r} = \mathfrak{p} \cap \mathbb{C}\{z_1, \ldots, z_{n-1}\}[z_n], \quad \mathfrak{q} = \mathfrak{p} \cap \mathbb{C}\{z_1, \ldots, z_{n-1}\} = \mathfrak{r} \cap \mathbb{C}\{z_1, \ldots, z_{n-1}\},$$

since $h \notin \mathfrak{r}$ we have $1 \notin \mathfrak{q}$ and hence $g \in \mathfrak{r} - \mathfrak{q}[t]$. By Lemma A.4.5 applied to the prime ideal $\mathfrak{r}$, there exists a polynomial $\varrho \in \mathfrak{r}$ such that $R(\varrho, h) \notin \mathfrak{q}$.

By induction on n there exists a morphism $\tilde{\psi}\colon \mathbb{C}\{z_1, \ldots, z_{n-1}\} \to \mathbb{C}\{x\}$ such that $\tilde{\psi}(\mathfrak{q}) = 0$ and $\tilde{\psi}(R(\varrho, h)) \neq 0$. Denoting by $\psi\colon \mathbb{C}\{z_1, \ldots, z_n\} \to \mathbb{C}\{x, z_n\}$ the natural extension of $\tilde{\psi}$, by functoriality of resultant we have $R(\psi(h), \psi(\varrho)) \neq 0$.

Since $R(g, f) \in \mathfrak{q}$ for every $f \in \mathfrak{p}$ we have $R(\psi(g), \psi(f)) = 0$ for every $f \in \mathfrak{p}$. Applying Lemma A.4.6 to the Weierstrass polynomial $\psi(g)$ and to the vector space $V = \psi(\mathfrak{p})$ we have an irreducible factor p of $\psi(g)$ in the ring $\mathbb{C}\{x\}[z_n]$ such that $\psi(\mathfrak{p}) \subset p\mathbb{C}\{x, z_n\}$. In particular, p divides $\psi(\rho)$, therefore $R(\psi(h), p) \neq 0$ and $\psi(h) \notin p\mathbb{C}\{x, z_n\}$.

By Lemma A.4.2 there exists $\eta\colon \mathbb{C}\{x, z_n\} \to \mathbb{C}\{t\}$ such that $\ker(\eta) = (p)$, and hence $\eta(\psi(h)) \neq 0$. It is therefore sufficient to take ϕ as the composition of ψ and η. □

Proof (*of Theorem* A.4.1) Since $\sqrt{I}$ is the intersection of all prime ideals containing I, there exists a prime ideal $\mathfrak{p}$ such that $I \subset \mathfrak{p}$ and $h \notin \mathfrak{p}$. The conclusion follows from Lemma A.4.7. □

A.5 Dimension Bounds

In this section we apply Theorem A.4.1 for proving two dimension bounds of analytic algebras that are very useful in deformation and moduli theory. The first bound (Lemma A.5.2) is completely standard and the proof is reproduced here for completeness; the second bound finds application in the T^1-lifting approach to deformation problems (see Sect. B.3).

Lemma A.5.1 (Krull's principal ideal theorem) *Let A be a Noetherian local ring and $f \in \mathfrak{m}_A$, then* $\dim A/(f) \geq \dim A - 1$. *Moreover:*

1. *if f is nilpotent, then* $\dim A/(f) = \dim A$;
2. *if f is not a divisor of* 0, *then* $\dim A/(f) = \dim A - 1$.

Proof See e.g. [10, Cor. 11.18]. □

Lemma A.5.2 *Let R be an analytic algebra with maximal ideal $\mathfrak{m}_R$. Then:*

1. $\dim R \leq \dim_{\mathbb{C}} \dfrac{\mathfrak{m}_R}{\mathfrak{m}_R^2}$ *and equality holds if and only if R is smooth;*
2. *for every proper ideal $I \subset R$ we have* $\dim \dfrac{R}{I} \geq \dim R - \dim_{\mathbb{C}} \dfrac{I}{\mathfrak{m}_R I}$.

In particular, if $R = \mathbb{C}\{z_1, \ldots, n\}/I$ then $n \geq \dim R \geq n - \dim_{\mathbb{C}} \dfrac{I}{\mathfrak{m}I}$, where $\mathfrak{m}$ is the maximal ideal of $\mathbb{C}\{z_1, \ldots, n\}$.

Proof Let $n = \dim_{\mathbb{C}} \dfrac{\mathfrak{m}_R}{\mathfrak{m}_R^2}$ be the embedding dimension of R; in order to prove that $\dim R \leq n$ it is sufficient to prove that the ideal $\mathfrak{m}_R$ may be generated by n elements. If $f_1, \ldots, f_n \in \mathfrak{m}_R$ induce a basis of $\dfrac{\mathfrak{m}_R}{\mathfrak{m}_R^2}$, then $f_1, \ldots, f_n$ is set of generators of the maximal ideal $\mathfrak{m}_R$; in fact $(f_1, \ldots, f_n) + \mathfrak{m}_R^2 = \mathfrak{m}_R$ and the conclusion follows from Nakayama's lemma.

According to Corollary A.1.4 we can write $R = \mathbb{C}\{z_1, \ldots, z_n\}/J$ for some ideal J contained in $(z_1, \ldots, z_n)^2$; in particular, $\dim R \leq \dim \mathbb{C}\{z_1, \ldots, z_n\} = n$. Since $\mathbb{C}\{z_1, \ldots, z_n\}$ is an integral domain, by Krull's principal ideal theorem [10, Cor. 11.18] we have $\dim R = n$ if and only if $J = 0$.

By Nakayama's lemma, the dimension of the complex vector space $I/\mathfrak{m}_R I$ is equal to the minimum number of generators of the ideal I and then the inequality $\dim R \geq n - \dim_{\mathbb{C}} I/\mathfrak{m}I$ is a direct consequence of Krull's principal ideal theorem. □

Theorem A.5.3 *Let $I \subset \mathbb{C}\{z_1, \ldots, z_n\}$ be an ideal. Denote by $\mathfrak{m} = (z_1, \ldots, z_n)$ the maximal ideal of $\mathbb{C}\{z_1, \ldots, z_n\}$ and by $J \subset I$ the ideal*

$$J = \left\{ f \in I \;\middle|\; \frac{\partial f}{\partial z_i} \in I,\ \forall\, i = 1, \ldots, n \right\}.$$

Then

$$\dim \frac{\mathbb{C}\{z_1, \ldots, z_n\}}{I} \geq n - \dim_{\mathbb{C}} \frac{I}{J + \mathfrak{m}I}.$$

Proof Let's denote for simplicity $P = \mathbb{C}\{z_1, \ldots, z_n\}$, $R = P/I$ and introduce the *curvilinear obstruction map*:

$$\gamma_I \colon \mathrm{Mor}_{\mathbf{An}}(P, \mathbb{C}\{t\}) \to \mathrm{Hom}_{\mathbb{C}}\left(\frac{I}{J + \mathfrak{m}I}, \mathbb{C}\right).$$

Given $\phi\colon P \to \mathbb{C}\{t\}$, if $\phi(I) = 0$ we define $\gamma_I(\phi) = 0$. If $\phi(I) \neq 0$ and s is the biggest integer such that $\phi(I) \subset (t^s)$, then we define, for every $f \in I$, $\gamma_I(\phi)f$ as the coefficient of t^s in the power series expansion of $\phi(f) = f(\phi(z_1), \ldots, \phi(z_n))$. It is clear that $\gamma_I(\phi)(\mathfrak{m}I) = 0$, while if $\phi(I) \subset (t^s)$ and $f \in J$ we have

$$\frac{d\phi(f)}{dt} = \sum_{i=1}^{n} \frac{\partial f}{\partial z_i}(\phi(z_1), \ldots, \phi(z_n)) \frac{d\phi(z_i)}{dt} \in (t^s)$$

and therefore $\phi(f) \in (t^{s+1})$ (this is the point where the characteristic of the field plays an essential role). The vectors in the image of γ_I are called *curvilinear obstructions* and we shall prove the theorem by showing that there exist at least $n - \dim R$ linearly independent curvilinear obstructions.

The ideal I is finitely generated, say $I = (f_1, \ldots, f_d)$; according to Nakayama's lemma we can assume $f_1, \ldots, f_d$ a basis of $I/\mathfrak{m}I$. By repeated applications of Theorem A.4.1 (and possibly reordering the f_i's) we can assume that there exists an $h \leq d$ such that the following holds:

1. $f_i \notin \sqrt{(f_1, \ldots, f_{i-1})}$ for $i \leq h$;
2. for every $i \leq h$ there exists a morphism of analytic algebras $\phi_i : P \to \mathbb{C}\{t\}$ such that $\phi_i(f_i) \neq 0$, $\phi_i(f_j) = 0$ for $j < i$, and the multiplicity of $\phi_i(f_j)$ is bigger than or equal to the multiplicity of $\phi_i(f_i)$ for every $j > i$;
3. $I \subset \sqrt{(f_1, \ldots, f_h)}$.

The last condition, together with Lemma A.5.1, implies that

$$\dim R = \dim P/(f_1, \ldots, f_h) \geq n - h.$$

Therefore it is enough to prove that $\gamma_I(\phi_1), \ldots, \gamma_I(\phi_h)$ are linearly independent in $\mathrm{Hom}_{\mathbb{C}}\left(\frac{I}{J + \mathfrak{m}I}, \mathbb{C}\right)$; this follows from the fact that the matrix $a_{ij} = \gamma_I(\phi_i)f_j$, $i, j = 1, \ldots, h$, is triangular with nonzero elements on the diagonal and therefore its rank is equal to h. □

Example A.5.4 Here is an example where the dimension bound of Theorem A.5.3 is better than the standard dimension bound of Lemma A.5.2.

Let $I = (g, g_x, g_y) \subset \mathbb{C}\{x, y\}$ be the ideal generated by the polynomial $g = x^5 + y^5 + x^3y^3$ together with its partial derivatives $g_x = 5x^4 + 3x^2y^3$, $g_y = 5y^4 + 3x^3y^2$.

In the notation of Theorem A.5.3, we want to prove that $J + \mathfrak{m}I \neq \mathfrak{m}I$ or equivalently that $J \not\subset \mathfrak{m}I$. Since $g \in J = \{f \in I \mid f_x, f_y \in I\}$ it is sufficient to show that $g \notin \mathfrak{m}I$. Assume $g = \alpha g_x + \beta g_y + \gamma g$ for some $\alpha, \beta, \gamma \in \mathfrak{m}$; since $1 - \gamma$ is invertible we have $g \in (g_x, g_y)$, say $g = ag_x + bg_y$. Looking at the homogeneous components of degree ≤ 5 we get

$$a \in \frac{x}{5} + \mathfrak{m}^2, \qquad b \in \frac{y}{5} + \mathfrak{m}^2,$$

and then

$$g - \frac{x}{5}g_x - \frac{y}{5}g_y = -\frac{1}{5}x^3y^3 \in \mathfrak{m}^2(g_x, g_y)$$

which is not possible.

A.6 Artin's Theorem on the Solution of Analytic Equations

It is not difficult to see that the algebraic results of this chapter that make sense also for the ring of formal power series $\mathbb{C}[[z_1, \ldots, z_n]]$ and its quotients, remain true; in particular, the Weierstrass preparation and division theorems hold with the same statement, see e.g. [165, IV.9]. In many cases, especially in deformation theory of complex manifolds, we seek for solutions of systems of analytic equations but we are able to solve these equations only formally; in this situation the following theorem, proved by Artin in 1968, is very helpful.

Theorem A.6.1 (Artin [5]) *Let n, m, N be nonnegative integers and let*

$$f_i(x, y) = f_i(x_1, \ldots, x_n, y_1, \ldots, y_N) \in \mathbb{C}\{x_1, \ldots, x_n, y_1, \ldots, y_N\},$$

$i = 1, \ldots, m$, be convergent power series. Assume that there exist N formal power series

$$\overline{y}_i(x) \in \mathbb{C}[[x_1, \ldots, x_n]], \qquad i = 1, \ldots, N,$$

without constant terms such that $f_i(x, \overline{y}(x)) = 0$ for every $i = 1, \ldots, m$. Then, for every positive integer c, there exists N convergent power series $y_i(x) \in \mathbb{C}\{x_1, \ldots, x_n\}$ such that $f_i(x, y(x)) = 0$ for every i and $y_i(x) \equiv \overline{y}_i(x) \pmod{\mathfrak{m}^c}$, where $\mathfrak{m}$ is the maximal ideal of $\mathbb{C}[[x_1, \ldots, x_n]]$.

In this book we give a simplified version of Artin's original proof under the additional assumption that the f_i's are polynomials; the proof contained in [5] uses the same ideas but it is technically more difficult, cf. [232]. More precisely, we give a complete proof of the following theorem.

Theorem A.6.2 *Let n, m, N be nonnegative integers and let*

$$f_i(x, y) = f_i(x_1, \ldots, x_n, y_1, \ldots, y_N) \in \mathbb{C}[x_1, \ldots, x_n, y_1, \ldots, y_N],$$

$i = 1, \ldots, m$, be m polynomials. Assume that there exists N formal power series

$$\overline{y}_i(x) \in \mathbb{C}[[x_1, \ldots, x_n]], \qquad i = 1, \ldots, N,$$

such that $f_i(x, \overline{y}(x)) = 0$ for every $i = 1, \ldots, m$. Then, for every positive integer c, there exist N convergent power series $y_i(x) \in \mathbb{C}\{x_1, \ldots, x_n\}$ such that $f_i(x, y(x)) = 0$ for every i and $y_i(x) \equiv \overline{y}_i(x) \pmod{\mathfrak{m}^c}$, where $\mathfrak{m}$ is the maximal ideal of $\mathbb{C}[[x_1, \ldots, x_n]]$.

The result is trivially true for $n = 0$ and we prove the theorem by induction on the integer n. The proof of inductive step is a consequence of the following two technical lemmas.

Lemma A.6.3 *Let $g, f_1, \ldots, f_m \in \mathbb{C}[x_1, \ldots, x_n, y_1, \ldots, y_N]$ and let c be a positive integer. Assume that there exist formal power series $\overline{y}_i(x) \in \mathbb{C}[[x_1, \ldots, x_n]]$, with $i = 1, \ldots, N$, such that:*

1. *$g(x, \overline{y}(x)) \neq 0$;*
2. *$g(x, \overline{y}(x))$ divides $f_i(x, \overline{y}(x))$ in the ring $\mathbb{C}[[x_1, \ldots, x_n]]$ for every $i = 1, \ldots, m$.*

Then there exist N convergent power series $y_i(x) \in \mathbb{C}\{x_1, \ldots, x_n\}$ such that $y_i(x) \equiv \overline{y}_i(x) \pmod{\mathfrak{m}^c}$ and:

3. *$g(x, y(x)) \neq 0$,*
4. *$g(x, y(x))$ divides $f_i(x, y(x))$ in the ring $\mathbb{C}\{x_1, \ldots, x_n\}$ for every $i = 1, \ldots, m$.*

Proof Let's denote by r the multiplicity of $g(x, \overline{y}(x))$; it is not restrictive to assume $c > r > 0$, so that the condition $g(x, y(x)) \neq 0$ is automatically satisfied. Up to a linear change of the coordinates x_i we have, by the Weierstrass preparation theorem,

$$g(x, \overline{y}(x)) = (x_n^r + \overline{a}_1(x)x_n^{r-1} + \cdots + \overline{a}_r(x))\overline{e}(x),$$

where $\overline{e}(x)$ is invertible and every $\overline{a}_i(x)$ is a formal power series of positive multiplicity in $x_1, \ldots, x_{n-1}$. By the Weierstrass division theorem for formal power series we have

$$\overline{y}_i(x) = (x_n^r + \overline{a}_1(x)x_n^{r-1} + \cdots + \overline{a}_r(x))u_i(x) + \sum_{j=0}^{r-1} \overline{y}_{ij}(x)x_n^j,$$

where the $\overline{y}_{ij}(x)$'s are formal power series in $x_1, \ldots, x_{n-1}$. Now, it follows easily from the Taylor formula that we may replace the power series $\overline{y}_i(x)$ with any power series of the form

$$\overline{y}_i(x) + h_i(x)(x_n^r + \overline{a}_1(x)x_n^{r-1} + \cdots + \overline{a}_r(x)), \qquad h_i \in \mathfrak{m}^{c-r},$$

and the assumption of the lemma remains satisfied. In particular, we may choose the h_i's in such a way that every $u_i(x)$ is a polynomial. Let's define the variables y_i by a set of new variables A_i, Y_{ij} via the formulas

$$y_i = (x_n^r + A_1 x_n^{r-1} + \cdots + A_r)u_i(x) + \sum_{j=0}^{r-1} Y_{ij} x_n^j.$$

We obtain a new set of polynomials

$$G(x, A, Y) = g(x, y), \qquad F_i(x, A, Y) = f_i(x, y),$$

and by Euclidean division with respect to the variable x_n in the ring $\mathbb{C}[x_i, A_j, Y_{ij}]$ we have:

1. $G(x, A, Y) = (x_n^r + A_1 x_n^{r-1} + \cdots + A_r)Q + \sum_{k=0}^{r-1} G_k x_n^k$,
2. $F_i(x, A, Y) = (x_n^r + A_1 x_n^{r-1} + \cdots + A_r)R_i + \sum_{k=0}^{r-1} F_{ik} x_n^k$.

Since $G(x, \overline{a}_i(x), \overline{y}_{ij}(x)) = g(x, \overline{y}(x))$, the unicity of the remainder in the Weierstrass division theorem implies $Q(0) \neq 0$ and $G_k(x, \overline{a}_i(x), \overline{y}_{ij}(x)) = 0$ for every k. Since $(x_n^r + \overline{a}_1(x)x_n^{r-1} + \cdots + \overline{a}_r(x))$ divides $f_i(x, \overline{y}(x))$, we also have $F_{hk}(x, \overline{a}_i(x), \overline{y}_{ij}(x)) = 0$ for every h, k.

By the inductive assumption we can find $a_i(x), y_{ij}(x) \in \mathbb{C}\{x_1, \ldots, x_{n-1}\}$, arbitrarily near to $\overline{a}_i(x), \overline{y}_{ij}(x)$ in the $\mathfrak{m}$-adic topology and such that

$$G_k(x, a_i(x), y_{ij}(x)) = F_{hk}(x, a_i(x), y_{ij}(x)) = 0, \quad \text{for every } h, k.$$

Therefore

1. $g(x, y(x)) = G(x, a(x), y_{ij}(x)) = (x_n^r + a_1(x)x_n^{r-1} + \cdots + a_r(x))Q$;
2. $f_i(x, y(x)) = F_i(x, a(x), y_{ij}(x)) = (x_n^r + a_1(x)x_n^{r-1} + \cdots + a_r(x))R_i$;

and then $g(x, y(x))$ divides $f_i(x, y(x))$ for every i. □

Lemma A.6.4 *Theorem A.6.2 holds under the following additional assumptions:*

1. $m \leq N$;
2. *if* $\delta(x, y) = \det\left(\dfrac{\partial f_i}{\partial y_j}\right)_{i,j=1,\ldots,m}$, *then* $\delta(x, \overline{y}(x)) \neq 0$.

Proof Since it is trivially true that $\delta^2(x, \overline{y}(x))$ divides $f_i(x, \overline{y}(x)) \equiv 0$, according to Lemma A.6.3, there exist convergent power series $u_i(x) \in \mathbb{C}\{x_1, \ldots, x_n\}$, $i = 1, \ldots, N$, such that $u_i(x) \equiv \overline{y}_i(x) \pmod{\mathfrak{m}^c}$ and:

1. $\delta(x, u(x)) \neq 0$;
2. $\delta^2(x, u(x))$ divides $f_i(x, u(x))$ for every $i = 1, \ldots, m$;

3. $f_i(x, u(x)) \in \mathfrak{m}^{c+r}$ for every $i = 1, \dots, m$, where r is the multiplicity of $\delta(x, u(x))$.

Let's write $f_i(x, u(x)) = \delta^2(x, u(x))F_i(x)$, $F_i(x) \in \mathbb{C}\{x_1, \dots, x_n\}$, and denote by $M_{ij}(x, y)$ the coefficients of the adjoint matrix of $\left(\dfrac{\partial f_i}{\partial y_j}\right)_{i,j=1,\dots,m}$. We have the Laplace identities:

$$\sum_j \frac{\partial f_i}{\partial y_j} M_{jk} = \begin{cases} 0 & \text{for } i \neq k, \\ \delta & \text{for } i = k. \end{cases}$$

We now prove that there exist $v_1(x), \dots, v_m(x) \in \mathbb{C}\{x_1, \dots, x_n\}$ such that the convergent power series

$$y_j(x) = u_j(x) + \delta(x, u(x)) \sum_k M_{jk}(x, u(x)) v_k(x), \qquad j = 1, \dots, m\ ,$$

$$y_j(x) = u_j(x) \qquad i = m+1, \dots, N\ ,$$

satisfy the requirement of Theorem A.6.2. For every index i, by the Taylor formula, we have:

$$\begin{aligned} f_i(x, y(x)) &= f_i(x, u(x)) + \delta(x, u(x)) \sum_{j=1}^m \frac{\partial f_i}{\partial y_j}(x, u(x)) \sum_k M_{jk}(x, u(x)) v_k(x) \\ &\quad + \delta^2(x, u(x)) Q_i \\ &= \delta^2(x, u(x))(F_i(x) + v_i(x) + Q_i), \end{aligned}$$

where every Q_i contains only monomials of degree ≥ 2 in the variables v_i. By the implicit function theorem A.1.2 we may solve the system of m equations

$$F_i(x) + v_i(x) + Q_i(x, v) = 0, \qquad i = 1, \dots, m,$$

in the ring $\mathbb{C}\{x_1, \dots, x_n\}$. Moreover, since $F_i(x) \in \mathfrak{m}^{c-r}$ we also have $v_i(x) \in \mathfrak{m}^{c-r}$ and then $y_i(x) \equiv u_i(x) \pmod{\mathfrak{m}^c}$. □

Proof (***of Theorem*** A.6.2) The power series $\overline{y_i}(x)$ give a morphism of rings

$$\phi\colon \mathbb{C}[x_1, \dots, x_n, y_1, \dots, y_N] \to \mathbb{C}[[x_1, \dots, x_n]], \qquad y_i \mapsto \overline{y_i}(x),$$

whose kernel $\mathfrak{p} = \ker(\phi)$ contains the polynomials f_i. Let $X = V(\mathfrak{p}) \subset \mathbb{C}^{n+N}$ be the affine irreducible variety defined by the prime ideal $\mathfrak{p}$ and denote by $r = n + N - \dim X$ its codimension. Possibly enlarging the set of equations it is not restrictive to assume that the polynomials f_i generate the prime ideal $\mathfrak{p}$, and therefore that $m \geq r$. Since $\mathfrak{p} \cap \mathbb{C}[x_1, \dots, x_n] = 0$, the projection $X \to \mathbb{C}^n$, $(x, y) \mapsto x$, is a dominant morphism of algebraic varieties.

According to the Bertini–Sard theorem, for a generic point $p \in X$, the Zariski tangent space T_pX has codimension r and the projection $T_pX \to \mathbb{C}^n$ is surjective. Up to a permutation of indices we may assume that T_pX is the annihilator of the r differentials $df_1, \ldots, df_r$. Then the determinant

$$\delta(x, y) = \det\left(\frac{\partial f_i}{\partial y_j}\right)_{i,j=1,\ldots,r}$$

does not vanish in p. In particular, $\delta(x, y) \notin \mathfrak{p}$ and then $\delta(x, \overline{y}(x)) \neq 0$.

According to Lemma A.6.4 there exist convergent power series $y_1(x), \ldots, y_N(x)$ such that $y_i(x) \equiv \overline{y}_i(x) \pmod{\mathfrak{m}^c}$ for every i and

$$f_j(x, y(x)) = 0 \qquad j = 1, \ldots, r.$$

Let $Y = V(f_1, \ldots, f_r) \subset \mathbb{C}^{n+N}$ be the zero locus of $f_1, \ldots, f_r$; clearly, $X \subset Y$ and, since X, Y have the same dimension in p, we have that X is an irreducible component of Y and then there exists a polynomial $h(x, y) \notin \mathfrak{p}$ such that $Y \subset X \cup V(h)$.

Since $h(x, \overline{y}(x)) \neq 0$, possibly increasing the integer c, we have $h(x, y(x)) \neq 0$. Let $\mathfrak{q}$ be the kernel of the morphism of rings

$$\mathbb{C}[x_1, \ldots, x_n, y_1, \ldots, y_N] \to \mathbb{C}\{x_1, \ldots, x_n\}, \quad y_i \mapsto y_i(x);$$

we want to prove that $\mathfrak{p} \subset \mathfrak{q}$. We have already proved that $f_1, \ldots, f_r \in \mathfrak{q}$, $h \notin \mathfrak{q}$, and therefore $V(\mathfrak{q}) \subset Y$ and $V(\mathfrak{q}) \not\subset V(h)$. Since $V(\mathfrak{q})$ is irreducible we have $V(\mathfrak{q}) \subset X = V(\mathfrak{p})$ and then $\mathfrak{p} \subset \mathfrak{q}$. □

A.7 Exercises

A.7.1 Prove that the power series

$$f_n(t) = \sum_{k \geq 0} \binom{n+k}{k} t^k = \frac{1}{n!} \frac{d^n}{dt^n}\left(\frac{t^n}{1-t}\right) \in \mathbb{C}\{t\}, \qquad n \geq 0,$$

are linearly independent over $\mathbb{C}$ and that

$$\sum_{n \geq 0} z^n f_n(t) = \frac{1}{1-t-z} = \sum_{n \geq 0} (z+t)^n \in \mathbb{C}[[z, t]].$$

Deduce that $1 - t - z$ is not invertible in $\mathbb{C}[[z]] \otimes \mathbb{C}[[t]]$ and that $\mathbb{C}\{z\} \otimes \mathbb{C}\{t\}$ is not a local ring.

A.7.2 Prove that the integral domain of entire holomorphic functions $f: \mathbb{C} \to \mathbb{C}$ is not Noetherian and that it is not a unique factorization domain. (Hint: consider the sine function $\sin(z)$ and its quotients $\sin(z)/\prod_{k=0}^{n}(z - \pi k)$.)

A.7.3 (*Cartan's Lemma*) Let R be an analytic algebra of embedding dimension n and let G be a finite group of automorphisms of R. Prove that there exist an injective homomorphism of groups $G \to GL(\mathbb{C}^n)$ and a G-equivariant isomorphism of analytic algebras $R \cong \mathbb{C}\{z_1, \ldots, z_n\}/I$ for some G-stable ideal I. (Hint: since G is finite there exists a direct sum decomposition $\mathfrak{m}_R = V \oplus \mathfrak{m}_R^2$ such that $gV \subset V$ for every $g \in G$.)

A.7.4 Prove that $f, g \in \mathbb{C}\{x, y\}$ have a common factor of positive multiplicity if and only if the $\mathbb{C}$-vector space $\mathbb{C}\{x, y\}/(f, g)$ is infinite-dimensional.

A.7.5 (*Rückert's Nullstellensatz*) Let $I, J \subset \mathbb{C}\{z_1, \ldots, z_n\}$ be proper ideals. Prove that $\sqrt{I} = \sqrt{J}$ if and only if

$$\mathrm{Mor}_{\mathbf{An}}\left(\frac{\mathbb{C}\{z_1, \ldots, z_n\}}{I}, \mathbb{C}\{t\}\right) = \mathrm{Mor}_{\mathbf{An}}\left(\frac{\mathbb{C}\{z_1, \ldots, z_n\}}{J}, \mathbb{C}\{t\}\right),$$

where the symbol "=" means equality of subsets of $\mathrm{Mor}_{\mathbf{An}}(\mathbb{C}\{z_1, \ldots, z_n\}, \mathbb{C}\{t\})$.

A.7.6 Prove that if $f: (X, x) \to (\mathbb{C}^d, 0)$ is a morphism of analytic singularities such that $f^{-1}(0) = \{x\}$, then $\dim(X, x) \le d$. (Hint: writing $f = (f_1, \ldots, f_d)$ then, by Rückert's Nullstellensatz we have $f^{-1}(0) = \{x\}$ if and only if $\sqrt{(f_1, \ldots, f_d)}$ is the maximal ideal of $\mathcal{O}_{X,x}$ and therefore $\dim \mathcal{O}_{X,x} \le d$.)

A.7.7 In the notation of Theorem A.5.3 prove that:

1. $I^2 \subset J$.
2. $I = J + \mathfrak{m}I$ if and only if $I = 0$.
3. If $I \subset \mathfrak{m}^2$ then
$$\mathrm{Hom}_{\mathbb{C}}\left(\frac{I}{J + \mathfrak{m}I}, \mathbb{C}\right) = \mathrm{Ext}^1_R(\Omega_R, \mathbb{C}),$$
where Ω_R is the R-module of separated differentials (see e.g. [171, p. 92]).
4. For every short exact sequence $0 \to E \to F \to G \to 0$ of R-modules of finite length, i.e., annihilated by some power of $\mathfrak{m}$, there is defined a map
$$ob\colon \mathrm{Der}_{\mathbb{C}}(R, G) \to \mathrm{Hom}_R\left(\frac{I}{J}, E\right)$$
with the property that $ob(\phi) = 0$ if and only if ϕ lifts to a derivation $R \to F$. Moreover, if $\mathfrak{m}_R E = 0$ then $\mathrm{Hom}_R\left(\frac{I}{J}, E\right) = \mathrm{Hom}_{\mathbb{C}}\left(\frac{I}{J + \mathfrak{m}I}, E\right)$.

A.7.8 By using Theorem A.6.1, prove that:

1. Every irreducible convergent power series $f \in \mathbb{C}\{z_1, \ldots, z_n\}$ is also irreducible in $\mathbb{C}[[z_1, \ldots, z_n]]$.
2. The ring $\mathbb{C}\{z_1, \ldots, z_n\}$ is integrally closed in $\mathbb{C}[[z_1, \ldots, z_n]]$.

References

For a shorter, but less elementary, proof of Theorem A.4.1 we refer to [69]; the proof given here is taken from [180].

To our knowledge, the Theorem A.5.3 was first proved in [236], cf. also [31]; the proof given here is essentially taken from [70].

Exercise A.7.3 is a slight generalization of the classical Cartan's lemma [35].

Appendix B
Special Obstructions and T^1-Lifting

For a deformation functor F, equipped with a complete obstruction theory (V, v_e) as defined in Chap. 3, the study of a fixed obstruction map v_e is strongly influenced by the algebraic properties of the small extension e. In particular, there exist several special classes of small extensions where the corresponding obstruction maps are easier to compute and, at the same time, give lots of useful information on the functor F.

In this chapter we consider three special classes of small extensions, namely *curvilinear*, *simple* and *semitrivial* extensions, together with their associated obstruction maps. The semitrivial obstructions are usually the easiest to compute and the main result of this chapter, the T^1-lifting theorem (Corollary B.3.8), asserts that in characteristic 0 a deformation functor is smooth if and only if every semitrivial obstruction vanishes.

We adopt the notation and terminology of Chap. 3; in particular $\mathbb{K}$ is a fixed field and $\mathbf{Art}_{\mathbb{K}}$ is the category of Artin local $\mathbb{K}$-algebra with residue field $\mathbb{K}$.

B.1 The Approximation Theorem

In Lemma 3.3.4 we have proved that the vanishing of a small class of obstruction maps is sufficient to ensure the smoothness of a pro-representable functor. In order to extend this result to general deformation functors we first prove that, in a certain sense, every deformation functor has the same obstruction theory of a colimit of pro-representable functors.

Lemma B.1.1 *Let* $0 \to J \to R \to A \to 0$ *be a small extension in* $\mathbf{Art}_{\mathbb{K}}$*, let* $F\colon \mathbf{Art}_{\mathbb{K}} \to \mathbf{Set}$ *be a deformation functor and* $a \in F(A)$ *a fixed element. Then there exists a* $\mathbb{K}$*-vector subspace* $I \subseteq J$*, depending on* a*, with the following property: given a vector subspace* $H \subseteq J$*, the element* a *lifts to* $F(R/H)$ *if and only if* $I \subseteq H$*.*

M. Manetti, *Lie Methods in Deformation Theory*, Springer Monographs in Mathematics,
https://doi.org/10.1007/978-981-19-1185-9

Proof The proof is a very easy consequence of Theorem 3.6.7 about the existence of a complete obstruction theory (V, v_e); in fact, if $\xi \in V \otimes J$ is the obstruction to the lifting of a to $F(R)$, then it is sufficient to take I as the smallest subspace of J such that $\xi \in V \otimes I$.

However, it is instructive to give also a direct and more elementary proof. Denote by $\mathcal{F}$ the family of vector subspaces $H \subseteq J$ such that a lifts to $F(R/H)$; if $H \in \mathcal{F}$ and $H \subseteq K \subseteq J$, then, according to the factorization

$$F(R/H) \to F(R/K) \to F(R/J) = F(A)$$

we also have $K \in \mathcal{F}$. Therefore it is sufficient to show that $\mathcal{F}$ has a minimum, and since J is finite-dimensional it is sufficient to prove that if $H, K \in \mathcal{F}$ then $H \cap K \in \mathcal{F}$. Assume $H, K \in \mathcal{F}$ and choose a vector subspace $L \subseteq J$ such that $K \subseteq L$, $H + L = J$ and $H \cap L = H \cap K$. The assumption $K \subseteq L$ implies that $L \in \mathcal{F}$; the assumption $H + L = J$ implies that the natural map

$$\frac{R}{H \cap L} \to \frac{R}{H} \times_A \frac{R}{L}$$

is an isomorphism and this implies that $H \cap L \in \mathcal{F}$ in view of Definition 3.2.5. □

Lemma B.1.2 *Let $F\colon \mathbf{Art}_{\mathbb{K}} \to \mathbf{Set}$ be a deformation functor, $A \in \mathbf{Art}_{\mathbb{K}}$ and $a \in F(A)$. Then there exists a small extension*

$$0 \to J \to B \xrightarrow{\alpha} A \to 0$$

and an element $b \in F(B)$ such that:

1. *$\alpha(b) = a$;*
2. *the cotangent map $\alpha\colon \dfrac{\mathfrak{m}_B}{\mathfrak{m}_B^2} \to \dfrac{\mathfrak{m}_A}{\mathfrak{m}_A^2}$ is an isomorphism;*
3. *given a small extension $0 \to M \to D \to C \to 0$ and a morphism $\gamma\colon A \to C$ in $\mathbf{Art}_{\mathbb{K}}$, the element $\gamma(a)$ lifts to $F(D)$ if and only if $\gamma\alpha$ lifts to a commutative diagram*

$$\begin{array}{ccc} B & \xrightarrow{\alpha} & A \\ \downarrow & & \downarrow{\scriptstyle\gamma} \\ D & \longrightarrow & C \end{array}$$

in $\mathbf{Art}_{\mathbb{K}}$.

Proof Write $A = P/\mathfrak{a}$, where $P = \mathbb{K}[[x_1, \ldots, x_k]]$ and $\mathfrak{a} \subset \mathfrak{m}_P^2$. Consider the small extension

$$0 \to J \to R \to A \to 0, \qquad \text{where } R = \frac{P}{\mathfrak{m}_P\mathfrak{a}}, \quad J = \frac{\mathfrak{a}}{\mathfrak{m}_P\mathfrak{a}},$$

and denote by $I \subseteq J$ the vector subspace with the property described in Lemma B.1.1. We claim that $B = R/I$ gives the required small extension; the only non trivial condition to verify is the "only if" part of (3). Let $0 \to M \to D \to C \to 0$ be a small extension and $\gamma\colon A \to C$ a morphism in $\mathbf{Art}_{\mathbb{K}}$ such that $\gamma(a)$ lifts to $F(D)$. Any lifting $P \to D$ of the composite map $P \to A \xrightarrow{\gamma} C$ gives a morphism of small extensions

$$\begin{array}{ccccccccc} 0 & \longrightarrow & J & \longrightarrow & R & \longrightarrow & A & \longrightarrow & 0 \\ & & \big\downarrow{\scriptstyle \gamma'} & & \big\downarrow & & \big\downarrow{\scriptstyle \gamma} & & \\ 0 & \longrightarrow & M & \longrightarrow & D & \longrightarrow & C & \longrightarrow & 0 \end{array}$$

with γ' independent of the choice of the lifting. We need to prove that $I \subseteq \ker \gamma'$ or equivalently, in view of the defining properties of I, that a lifts to $R/\ker \gamma'$. Let $H \subseteq M$ be a vector subspace such that $\gamma'(J) \oplus H = M$, then we have a commutative diagram

$$\begin{array}{ccccccccc} 0 & \longrightarrow & \dfrac{J}{\ker \gamma'} & \longrightarrow & \dfrac{R}{\ker \gamma'} & \longrightarrow & A & \longrightarrow & 0 \\ & & \big\downarrow{\scriptstyle \gamma'} & & \big\downarrow & & \big\downarrow{\scriptstyle \gamma} & & \\ 0 & \longrightarrow & M & \longrightarrow & D & \longrightarrow & C & \longrightarrow & 0 \\ & & \big\downarrow{\scriptstyle \pi} & & \big\downarrow & & \big\| & & \\ 0 & \longrightarrow & M/H & \longrightarrow & D/H & \longrightarrow & C & \longrightarrow & 0. \end{array}$$

Since $\pi\gamma'$ an isomorphism of vector spaces the natural map

$$\frac{R}{\ker \gamma'} \to \frac{D}{H} \times_C A$$

is an isomorphism and, since $\gamma(a)$ lifts to $F(D/H)$, we also have that a lifts to $F(R/\ker \gamma')$. □

Theorem B.1.3 (Approximation theorem) *Let F be a deformation functor, $A \in$ $\mathbf{Art}_{\mathbb{K}}$ and $a \in F(A)$. Then there exist a local complete Noetherian $\mathbb{K}$-algebra S with residue field $\mathbb{K}$, a natural transformation $f\colon h_S \to F$ and a local morphism of $\mathbb{K}$-algebras $\pi\colon S \to A$ with the following properties:*

1. *$a = f(\pi)$, where π is considered as an element of $h_S(A)$;*
2. *the cotangent map $\pi\colon \dfrac{\mathfrak{m}_S}{\mathfrak{m}_S^2} \to \dfrac{\mathfrak{m}_A}{\mathfrak{m}_A^2}$ is an isomorphism;*
3. *the induced obstruction map $f\colon O_{h_S} \to O_F$ is injective.*

Proof The pair $(S_0, a_0) = (A, a)$ can be extended to a sequence (S_n, a_n) where $a_n \in F(S_n)$, $\pi_n\colon S_{n+1} \to S_n$ is a small extension as in Lemma B.1.2 and $\pi_n(a_{n+1}) = a_n$. Taking S as the limit of S_n, the sequence $\{a_n\}$ gives a natural transformation $f\colon h_S \to F$, in fact every $\beta \in h_S(C)$ factors to a morphism $\beta\colon S_n \to C$ for $n >> 0$

and then $f(\beta) = \beta(a_n)$ defines f. In order to prove that the obstruction map is injective consider a small extension $e:\quad 0 \to M \to D \to C \to 0$ and an element $\beta \in h_S(C)$ such that $f(\beta)$ lifts to $F(D)$; we need to prove that β lifts to $h_S(D)$. Choosing an integer $n >> 0$ such that $\beta\colon S_n \to C$ we have $f(\beta) = \beta(a_n)$. By the properties of S_{n+1}, the morphism β can be lifted to a morphism $S \to S_{n+1} \to D$. □

Remark B.1.4 The proof of Theorem B.1.3 is based on the ideas of Schlessinger's original proof of Theorem 3.5.10. According to Corollary 3.6.10, the natural transformation f in Theorem B.1.3 is smooth if and only if it is surjective on tangent spaces.

We are now ready to prove Schlessinger's theorem as an immediate consequence of the approximation theorem and the standard smoothness criterion.

Proof (***of Schlessinger's theorem*** 3.5.10) One implication is trivial. Conversely, the finite dimensionality of the tangent space implies that there exist an $A \in \mathbf{Art}_{\mathbb{K}}$ and a morphism $h_A \to F$ inducing an isomorphism on tangent spaces; take for instance $A = \mathbb{K} \oplus \mathrm{Hom}_{\mathbb{K}}(T^1F, \mathbb{K})$.

Taking a natural transformation $f\colon h_S \to F$ as in Theorem B.1.3, then f is bijective on tangent spaces and then, according to Corollary 3.6.10, the map f is a hull. □

B.2 Curvilinear Obstructions

We have already introduced curvilinear obstructions for functors of Artin rings that are pro-represented by analytic algebras in the proof of Theorem A.5.3. The same notion extends naturally to every deformation functor and every base field.

Definition B.2.1 A **curvilinear extension** is a small extension in $\mathbf{Art}_{\mathbb{K}}$ that is isomorphic to

$$0 \to \mathbb{K} \xrightarrow{t^n} \frac{\mathbb{K}[t]}{(t^{n+1})} \to \frac{\mathbb{K}[t]}{(t^n)} \to 0$$

for some $n \geq 2$. The **curvilinear obstructions** of a deformation functor are the obstructions arising from the curvilinear extensions.

Theorem B.2.2 *Let $\mathbb{K}$ be an infinite field and $F\colon \mathbf{Art}_{\mathbb{K}} \to \mathbf{Set}$ a deformation functor. Then F is smooth if and only if for every integer $n \geq 2$ the natural map*

$$F\left(\frac{\mathbb{K}[t]}{(t^{n+1})}\right) \to F\left(\frac{\mathbb{K}[t]}{(t^n)}\right)$$

is surjective. Therefore a deformation functor is smooth if and only if every curvilinear obstruction vanishes.

Proof If F is pro-representable this is exactly Lemma 3.3.4. In the general case we need to prove that if $0 \to M \to B \to A \to 0$ is a small extension and $a \in F(A)$ does not belong to the image of $F(B)$, then F has at least one nontrivial curvilinear obstruction. By the approximation Theorem B.1.3 there exists a pro-representable functor h_S, a morphism $f: h_S \to F$ and an element $x \in h_S(A)$ such that $f(x) = a$ and $f: O_{h_S} \to O_F$ is injective. Clearly, x does not lift to $h_S(B)$, thus h_S is obstructed and then $f(O_{h_S})$ contains some nontrivial curvilinear obstructions. □

According to Remark 3.3.5, the second part of the above theorem, namely that a deformation functor is smooth if and only if every curvilinear obstruction vanishes, holds also for finite fields.

For a deformation functor F, let O^c_F denote the vector subspace of O_F generated by curvilinear obstructions. In order to avoid a common mistake we point out that the above theorem says that $O_F = 0$ if and only if $O^c_F = 0$, but **does not imply** that $O_F = O^c_F$. An example with $O_F \neq O^c_F$ is given by the pro-representable functor h_R, where $R = P/I$, $P = \mathbb{K}[[x, y]]$ and $I = (x^3, y^3, x^2y^2)$. In fact, the element x^2y^2 is not trivial in $I/\mathfrak{m}_P I$ and it is easy to prove that it is annihilated by every curvilinear obstruction; details are left as an exercise. A less elementary case is described in Example B.2.8 below.

Remark B.2.3 The proof of Theorem A.5.3 implies that, when $\mathbb{K} = \mathbb{C}$, R is an analytic algebra and $F = h_R$, then $\dim R \geq \dim T^1F - \dim O^c_F$, where O^c_F is the vector space generated by curvilinear obstructions. The same result holds over any algebraically closed field [70, 142], while it is false whenever $\mathbb{K}$ is not algebraically closed, see e.g. Exercise B.5.6.

Every small extension $0 \to J \to A \to B \to 0$ in $\mathbf{Art}_{\mathbb{K}}$ gives the second fundamental exact sequence of Kähler differentials [193, Thm. 25.2]:

$$J \xrightarrow{d} \Omega_{A/\mathbb{K}} \otimes_A B \to \Omega_{B/\mathbb{K}} \to 0,$$

where, for $x \in J$ we have $d(x) = d_{A/\mathbb{K}}(x) \otimes 1$.

Definition B.2.4 A small extension $0 \to J \to A \to B \to 0$ in $\mathbf{Art}_{\mathbb{K}}$ is said to be **simple** if the natural map $J \xrightarrow{d} \Omega_{A/\mathbb{K}} \otimes_A B$ is injective.

For instance, in characteristic 0 every curvilinear extension is simple.

Example B.2.5 Let $\mathbb{K}$ be a field of characteristic 0 and consider the ideal

$$I = (x^3, y^3, x^2y^2) \subset P = \mathbb{K}[[x, y]].$$

Then the small extension

$$0 \to J = \frac{I}{\mathfrak{m}_P I} \to S \to R \to 0, \quad \text{where} \quad R = \frac{P}{I}, \quad S = \frac{P}{\mathfrak{m}_P I}, \tag{B.1}$$

is simple. In fact $\mathfrak{m}_P I = (x^4, x^3y, xy^3, y^4)$ and $dx^3, dy^3, d(x^2y^2)$ are linearly independent in

$$\Omega_{S/\mathbb{K}} \otimes_S R = \frac{P}{(x^3, y^3, x^2y)}dx \oplus \frac{P}{(x^3, y^3, xy^2)}dy.$$

Example B.2.6 (*A non-simple extension*) Let $\mathbb{K}$ be a field of characteristic 0 and $f = x^3 + xy^5 + y^7 \in \mathbb{K}[[x, y]]$. It is an easy computation, left as an exercise, to show that f does not belong to the ideal generated by its partial derivatives $f_x = 3x^2 + y^5$ and $f_y = 5xy^4 + 7y^6$. Here we prove that the sequence

$$0 \to \mathbb{K} \xrightarrow{f} \frac{\mathbb{K}[[x, y]]}{(f_x, f_y, xf, yf)} \to \frac{\mathbb{K}[[x, y]]}{(f_x, f_y, f)} \to 0$$

is a non-simple small extension. The vanishing of df is clear and the only nontrivial condition to verify is that $f \notin (f_x, f_y, xf, yf)$. If $f \in (f_x, f_y, xf, yf)$ then there exists $a, b \in \mathbb{K}[[x, y]]$ such that $f + axf + byf \in (f_x, f_y)$ and, since $1 + ax + by$ is invertible, we also have $f \in (f_x, f_y)$ and we have observed that this is false. The same construction works also for $f = x^7 + x^4y^4 + y^7$.

Lemma B.2.7 *Let $0 \to J \to A \to B \to 0$ be a simple small extension. Then for every vector subspace $H \subseteq J$ the small extension*

$$0 \to \frac{J}{H} \to \frac{A}{H} \to B \to 0$$

is simple.

Proof Writing $C = \dfrac{A}{H}$, the right exact functor $\otimes_C B$ applied to the exact sequence

$$H \xrightarrow{d} \Omega_{A/\mathbb{K}} \otimes_A C \to \Omega_{C/\mathbb{K}} \to 0$$

gives a commutative diagram with exact rows

$$\begin{array}{ccccccc}
0 \longrightarrow & H & \longrightarrow & J & \longrightarrow & J/H & \longrightarrow 0 \\
 & \| & & \downarrow{\scriptstyle d} & & \downarrow{\scriptstyle d} & \\
 & H & \longrightarrow & \Omega_{A/\mathbb{K}} \otimes_A B & \longrightarrow & \Omega_{C/\mathbb{K}} \otimes_C B & \longrightarrow 0
\end{array}$$

and the conclusion follows by a straightforward diagram chasing. □

Let F be a deformation functor and let C be a class of small extensions in $\mathbf{Art}_{\mathbb{K}}$. The obstructions arising from C are defined as the obstructions $f(ob_e(x)) \in O_F = O_F \otimes \mathbb{K}$, with

$$e\colon \quad 0 \to M \to A \to B \to 0$$

is a small extension in C, $x \in F(B)$ and $f : M \to \mathbb{K}$ is a linear map. We have already noticed that in general the obstruction space O_F is not generated by curvilinear obstructions.

Example B.2.8 We describe an example of deformation functor F such that the obstruction space O_F is not generated by obstructions arising from simple extensions. Let $\mathbb{K}$ be a field of characteristic 0 and consider the pro-representable functor

$$F = h_R : \mathbf{Art}_{\mathbb{K}} \to \mathbf{Set}, \qquad R = \frac{\mathbb{K}[[x, y]]}{(f, f_x, f_y)}, \qquad f = x^3 + xy^5 + y^7.$$

Denoting by $\mathfrak{m}$ the maximal ideal of $\mathbb{K}[[x, y]]$, we have seen in Example B.2.6 that f, f_x, f_y is a minimal set of generators of the ideal $I = (f, f_x, f_y)$ and then f, f_x, f_y is a basis of $O_F^\vee = I/\mathfrak{m}I$. Our goal is to prove that for every obstruction $\xi \in O_F \otimes M$ arising from a simple extension $0 \to M \to A \to B \to 0$ we have $\xi(f) = 0 \in M$; this implies that simple obstructions are contained in the annihilator of f. Let $\alpha : R \to B$ be a local morphism of $\mathbb{K}$-algebras and consider a lifting of α to a commutative diagram

$$\begin{array}{ccccccccc}
0 & \longrightarrow & (f, f_x, f_y) & \longrightarrow & \mathbb{K}[[x, y]] & \longrightarrow & R & \longrightarrow & 0 \\
 & & \downarrow{\scriptstyle \xi} & & \downarrow & & \downarrow{\scriptstyle \alpha} & & \\
0 & \longrightarrow & M & \longrightarrow & A & \longrightarrow & B & \longrightarrow & 0.
\end{array}$$

Then the class of ξ in $O_F \otimes M = \mathrm{Hom}_{\mathbb{K}}(I/\mathfrak{m}I, M)$ is precisely the obstruction to lifting α to a morphism $R \to A$. Taking differentials we get a commutative diagram

$$\begin{array}{ccccc}
 & & (f, f_x, f_y) & \xrightarrow{d} & \Omega_{\mathbb{K}[[x,y]]/\mathbb{K}} \otimes_{\mathbb{K}[[x,y]]} R \\
 & & \downarrow{\scriptstyle \xi} & & \downarrow \\
0 & \longrightarrow & M & \xrightarrow{d} & \Omega_{A/\mathbb{K}} \otimes_A B.
\end{array}$$

Since $df = 0$ we have $d\xi(f) = 0$ and then $\xi(f) = 0$.

B.3 Primary and Semitrivial Obstructions

For every field $\mathbb{K}$ we have an isomorphism in $\mathbf{Art}_{\mathbb{K}}$

$$\frac{\mathbb{K}[x, y]}{(x^2, xy, y^2)} = \frac{\mathbb{K}[x]}{(x^2)} \times_{\mathbb{K}} \frac{\mathbb{K}[y]}{(y^2)}$$

inducing, for every deformation functor F, the bijection

$$F\left(\frac{\mathbb{K}[x,y]}{(x^2,xy,y^2)}\right) = T^1F \times T^1F.$$

Definition B.3.1 Let (V, v_e) be an obstruction theory for a deformation functor F; the **primary obstruction map** $[-,-]\colon T^1F \times T^1F \to V$ is the obstruction map arising from the small extension

$$0 \to \mathbb{K} \xrightarrow{xy} \frac{\mathbb{K}[x,y]}{(x^2,y^2)} \to \frac{\mathbb{K}[x,y]}{(x^2,xy,y^2)} \to 0.$$

Example B.3.2 If $\mathbb{K}$ has characteristic 0, $F = \mathrm{Def}_L$ for some differential graded Lie algebra L and we consider the natural obstruction theory described in Sect. 6.4, then the primary obstruction map

$$T^1F \times T^1F = H^1(L) \times H^1(L) \to H^2(L)$$

is equal to the induced bracket in cohomology. In particular, if L is a formal differential graded Lie algebra, then Def_L is smooth if and only if its primary obstruction map is trivial.

Lemma B.3.3 *The primary obstruction map is bilinear symmetric. In the setup of Definition B.3.1, if* $q\colon T^1F \to V$ *is the obstruction map arising from the curvilinear extension*

$$0 \to \mathbb{K} \xrightarrow{t^2} \mathbb{K}[t]/(t^3) \to \mathbb{K}[t]/(t^2) \to 0,$$

then $2q(u) = [u,u]$ *for every* $u \in T^1F$.

Proof The bilinearity and the symmetry are easy consequences of base change property for obstruction theories and their proof is left as an exercise. The morphism of algebras $\alpha\colon \mathbb{K}[t] \to \mathbb{K}[x,y]$, $\alpha(t) = x+y$, induces a morphism of small extensions

$$\begin{array}{ccccccccc}
0 & \longrightarrow & \mathbb{K} & \xrightarrow{t^2} & \dfrac{\mathbb{K}[t]}{(t^3)} & \longrightarrow & \dfrac{\mathbb{K}[t]}{(t^2)} & \longrightarrow & 0\\
 & & \Big\downarrow{\scriptstyle 2} & & \Big\downarrow{\scriptstyle \alpha} & & \Big\downarrow{\scriptstyle \alpha} & & \\
0 & \longrightarrow & \mathbb{K} & \xrightarrow{xy} & \dfrac{\mathbb{K}[x,y]}{(x^2,y^2)} & \longrightarrow & \dfrac{\mathbb{K}[x,y]}{(x^2,xy,y^2)} & \longrightarrow & 0
\end{array}$$

and by base change the obstruction of lifting $\alpha(u) = (u,u) \in T^1F \times T^1F$ is equal to $2q(u)$.

Therefore, in characteristic $\neq 2$, we have $[u,v] = q(u+v) - q(u) - q(v)$ and the vanishing of the primary obstruction map is equivalent to the surjectivity of $F(\mathbb{K}[t]/(t^3)) \to F(\mathbb{K}[t]/(t^2))$. □

We consider now another class of small extensions that is quite useful because in concrete cases the induced obstructions are usually much easier to understand; see for instance Theorem B.3.12.

For every $A \in \mathbf{Art}_{\mathbb{K}}$ and every A-module M we denote by $A \oplus M$ the trivial extension (with multiplication rule $(a, m)(b, n) = (ab, an + bm)$). Notice that M is finitely generated as an A-module if and only if $A \oplus M \in \mathbf{Art}_{\mathbb{K}}$.

Definition B.3.4 A small extension in $\mathbf{Art}_{\mathbb{K}}$ is called **semitrivial** if it is isomorphic to

$$0 \to K \to A \oplus M \xrightarrow{\alpha} A \oplus N \to 0$$

for some $A \in \mathbf{Art}_{\mathbb{K}}$ and some short exact sequence $0 \to K \to M \xrightarrow{\beta} N \to 0$ of finitely generated A-modules with $\mathfrak{m}_A K = 0$; the morphism α is the trivial extension of β, i.e., $\alpha(a, m) = (a, \beta(m))$. A **semitrivial obstruction** is an obstruction arising from a semitrivial extension.

Notice that in the above definition it is not required that the A-modules M, N are annihilated by the maximal ideal and therefore α may not have a right inverse morphism.

Example B.3.5 The primary obstruction is semitrivial; more generally, for every $n \geq 1$ the small extension

$$0 \to \mathbb{K} \xrightarrow{x^n y} \frac{\mathbb{K}[x, y]}{(x^{n+1}, y^2)} \to \frac{\mathbb{K}[x, y]}{(x^{n+1}, x^n y, y^2)} \to 0$$

is semitrivial and isomorphic to

$$0 \to \mathbb{K} \xrightarrow{x^n y} A \oplus Ay \to A \oplus \frac{A}{(x^n)} y \to 0, \qquad \text{where} \quad A = \frac{\mathbb{K}[x]}{(x^{n+1})}.$$

Lemma B.3.6 *Every semitrivial small extension is simple.*

Proof In the setup of Definition B.3.4, we want to prove that

$$0 \to K \to A \oplus M \xrightarrow{\alpha} A \oplus N \to 0$$

is a simple small extension, that is the map

$$K \xrightarrow{d} \Omega_{A \oplus M/\mathbb{K}} \otimes_{A \oplus M} (A \oplus N)$$

is injective. The projection $\pi : A \oplus M \to M$ is a derivation and then there exists a morphism of $A \oplus M$-modules $\phi : \Omega_{A \oplus M/\mathbb{K}} \to M$ such that $\phi(a\, db) = a\, \pi(b)$ for every $a, b \in A \oplus M$. Since $KM = 0$ we have $\phi(K\Omega_{A \oplus M/\mathbb{K}}) = 0$ and then ϕ factors through

$$\Omega_{A\oplus M/\mathbb{K}} \otimes_{A\oplus M} A \oplus N = \Omega_{A\oplus M/\mathbb{K}} \otimes_{A\oplus M} \frac{A \oplus M}{K} = \frac{\Omega_{A\oplus M/\mathbb{K}}}{K\Omega_{A\oplus M/\mathbb{K}}}.$$

We conclude the proof by observing that the composition $\phi d \colon K \to M$ is the inclusion of K into M. □

Theorem B.3.7 (T^1-lifting trick) *Let $\mathbb{K}$ be a field of characteristic 0 and let $F \colon$ **Art**$_{\mathbb{K}} \to$ **Set** be a deformation functor. Then F is smooth if and only if for every $n \geq 2$ the map*

$$F\left(\frac{\mathbb{K}[x,y]}{(x^n, y^2)}\right) \to F\left(\frac{\mathbb{K}[x,y]}{(x^n, x^{n-1}y, y^2)}\right)$$

is surjective.

Proof Apply Theorem B.2.2 and base change to the sequence of morphisms of small extensions

$$\begin{array}{ccccccccc}
0 & \longrightarrow & \mathbb{K} & \xrightarrow{t^n} & \dfrac{\mathbb{K}[t]}{(t^{n+1})} & \longrightarrow & \dfrac{\mathbb{K}[t]}{(t^n)} & \longrightarrow & 0 \\
 & & \downarrow{\scriptstyle n} & & \downarrow{\scriptstyle t\mapsto x+y} & & \downarrow{\scriptstyle t\mapsto x+y} & & \\
0 & \longrightarrow & \mathbb{K} & \xrightarrow{x^{n-1}y} & \dfrac{\mathbb{K}[x,y]}{(x^n, y^2)} & \longrightarrow & \dfrac{\mathbb{K}[x,y]}{(x^n, x^{n-1}y, y^2)} & \longrightarrow & 0,
\end{array}$$

for $n \geq 2$. □

Corollary B.3.8 (Abstract T^1-lifting theorem) *Over a field of characteristic 0, a deformation functor is smooth if and only if every semitrivial obstruction vanishes.*

Proof Immediate from Theorem B.3.7 and Example B.3.5. □

Example B.3.9 Let $(L, d, [-,-])$ be a differential graded Lie algebra. Assume that for every $n \geq 2$ and every $x \in \mathrm{MC}_L\left(\dfrac{\mathbb{K}[t]}{(t^n)}\right)$ the natural morphism

$$H^1\left(L \otimes \frac{\mathbb{K}[t]}{(t^n)}, d + [x, -]\right) \to H^1\left(L \otimes \frac{\mathbb{K}[t]}{(t^{n-1})}, d + [x, -]\right) \tag{B.2}$$

is surjective. Then Def_L is smooth.

The proof is an easy application of Theorem B.3.7; in fact, denoting by $A = \dfrac{\mathbb{K}[t]}{(t^n)}$ and by $B = \dfrac{\mathbb{K}[t]}{(t^{n-1})}$ we have that, for either $M = A$ or $M = B$:

$$\mathrm{MC}_L(A \oplus M) = \{(x, y) \mid x \in \mathrm{MC}_L(A),\ y \in L^1 \otimes M,\ dy + [x, y] = 0\}.$$

The surjectivity of (B.2) implies that for every $(x, y) \in \mathrm{MC}_L(A \oplus B)$ there exists $(x, z) \in \mathrm{MC}_L(A \oplus A)$ such that

$$y = z + du + [x, u] \in L^1 \otimes B, \qquad u \in L^0 \otimes B.$$

Choosing any lifting $u' \in L^0 \otimes A$ of u we have that (x, y) lifts to

$$(x, z + du' + [x, u']) \in \mathrm{MC}_L(A \oplus A).$$

Be careful of the fact that the flatness of $H^1\left(L \otimes \frac{\mathbb{K}[t]}{(t^n)}, d + [x, -]\right)$ as a $\frac{\mathbb{K}[t]}{(t^n)}$-module is not sufficient to ensure the surjectivity of the map in (B.2) (cf. Appendix A of [267]).

Proposition B.3.10 *Let F be a deformation functor. Then every obstruction of F arising from a simple small extension is semitrivial. Therefore over a field of characteristic 0 we have the inclusions:*

$$\mathrm{Span}(\textit{curvilinear obstructions}) \subset \mathrm{Span}(\textit{semitrivial } (= \textit{simple}) \textit{ obstructions}) \subset O_F.$$

Proof Given a simple small extension $0 \to J \to A \xrightarrow{f} B \to 0$ we have a morphism of small extensions

$$\begin{array}{ccccccccc}
0 & \longrightarrow & J & \longrightarrow & A & \longrightarrow & B & \longrightarrow & 0 \\
 & & \| & & \downarrow{\scriptstyle g} & & \downarrow{\scriptstyle h} & & \\
0 & \longrightarrow & J & \longrightarrow & B \oplus (\Omega_{A/\mathbb{K}} \otimes_A B) & \longrightarrow & B \oplus \Omega_{B/\mathbb{K}} & \longrightarrow & 0
\end{array}$$

where $g(a) = (f(a), da)$ and $h(b) = (b, db)$. The bottom row is a semitrivial extension and the conclusion follows by the base change property of obstructions maps. □

Lemma B.3.11 *Let* $F : \mathbf{Art}_{\mathbb{K}} \to \mathbf{Set}$ *be a deformation functor, let* $\alpha : B \to A$ *be a morphism in* $\mathbf{Art}_{\mathbb{K}}$ *and let*

$$0 \to K \to M \xrightarrow{\beta} N \to 0 \tag{B.3}$$

be an exact sequence of finitely generated A-modules. If (B.3) is a split exact sequence of B-modules, then the natural map

$$F(A \oplus M) \times_{F(A)} F(B) \to F(A \oplus N) \times_{F(A)} F(B)$$

is surjective.

Proof The morphism α induces two natural isomorphisms of local rings

$$(A \oplus M) \times_A B = B \oplus M, \qquad (A \oplus N) \times_A B = B \oplus N,$$

and therefore a commutative diagram

$$\begin{array}{ccc} F(B\oplus M) & \longrightarrow & F(B\oplus N) \\ \downarrow & & \downarrow \\ F(A\oplus M)\times_{F(A)}F(B) & \longrightarrow & F(A\oplus N)\times_{F(A)}F(B) \end{array}$$

with surjective vertical arrows. By assumption there exists a morphism of B-modules $\gamma\colon N\to M$ such that $\beta\gamma=\mathrm{Id}_N$ and then also $F(B\oplus M)\to F(B\oplus N)$ is surjective. □

Theorem B.3.12 *Every semitrivial obstruction of a deformation group functor* $G\colon \mathbf{Art}_{\mathbb{K}}\to\mathbf{Grp}$ *vanishes. Therefore, if the base field* $\mathbb{K}$ *has characteristic 0, then* G *is smooth.*

Proof Denote by $\cdot$ the multiplication on G and let $0\to K\to A\oplus M\xrightarrow{\alpha}A\oplus N\to 0$ be a semitrivial small extension. Denoting by $G(\mathbb{K})=\{1\}$ the trivial group, by $\pi\colon A\oplus N\to A$ the projection and by $i\colon A\to A\oplus M$ the inclusion we have $\pi\alpha i=\mathrm{Id}_A$ and then, for every $g\in G(A\oplus N)$,

$$(g\cdot(\alpha i\pi(g^{-1})),1)\in G(A\oplus N)\times_{G(A)}G(\mathbb{K}).$$

By Lemma B.3.11 there exists a lifting $(h,1)\in G(A\oplus M)\times_{G(A)}G(\mathbb{K})$ and then $h\cdot(i\pi(g))\in G(A\oplus M)$ is a lifting of g. □

The abstract T^1-lifting theorem Corollary B.3.8 is false in positive characteristic. Consider for instance a field $\mathbb{K}$ of characteristic $p>0$ and the algebra $R=\mathbb{K}[x]/(x^p)$. Then h_R is a group functor; given two algebra morphisms $\alpha,\beta\colon R\to A$, their product is defined by setting

$$\alpha\cdot\beta(x)=\alpha(x)+\beta(x).$$

We can immediately see that $\cdot$ is properly defined and associative. The unit element e is defined by $e(x)=0$ and the inverse α^{-1} by $\alpha^{-1}(x)=-\alpha(x)$.

According to Theorem B.3.12 every semitrivial obstruction of h_R vanishes. On the other hand, we already know that h_R is obstructed; for instance, the identity on R does not lift to a morphism $R\to\mathbb{K}[x]/(x^{p+1})$.

B.4 The Ran–Kawamata T^1-Lifting Theorem

The goal of this is to give a slight improvement of Theorem B.3.7, generally known as the *Ran–Kawamata* T^1*-lifting theorem*, that we present here in the enhanced version by Fantechi and Manetti. This is probably the most famous T^1-lifting-type theorem and it is largely used in deformation theory, very often also in cases where the simpler versions (Theorem B.3.7 and Corollary B.3.8) suffice.

Definition B.4.1 For every integer $n \geq 0$ let's consider the following local Artin rings:

$$A_n = \frac{\mathbb{K}[x]}{(x^{n+1})}, \qquad B_n = \frac{\mathbb{K}[x, y]}{(x^{n+1}, y^2)} = A_n \otimes \frac{\mathbb{K}[y]}{(y^2)},$$

together with the natural projection $B_n \to A_n$. A functor of Artin rings F is said to have the ***T^1–lifting property*** if, for every integer $n \geq 0$, the natural map

$$F(B_{n+1}) \to F(B_n) \times_{F(A_n)} F(A_{n+1})$$

is surjective.

Theorem B.4.2 (Ran–Kawamata T^1-lifting) *Let $\mathbb{K}$ be a field of characteristic 0. Then every deformation functor satisfying the T^1-lifting property is smooth.*

Proof By Theorem B.2.2 it is sufficient to prove that for every fixed integer $m \geq 1$ the map $F(A_{m+1}) \to F(A_m)$ is surjective. Fix such an m and an element $a \in F(A_m)$.

For every n consider Artin local $\mathbb{K}$-algebra

$$C_n = \frac{\mathbb{K}[x, y]}{(x^{n+1}, x^n y, y^2)} = \frac{B_n}{(x^n y)} \in \mathbf{Art}_{\mathbb{K}},$$

and notice that $B_{n-1} = C_n/(x^n)$. In order to reduce notation to a minimum, the following convention will be in force. First we rename the indeterminate x of A_n as t, giving $A_n = \mathbb{K}[t]/(t^{n+1})$; then all $\mathbb{K}$-algebras we consider are quotients of either $A = V = \mathbb{K}[t, s]$ or $B = C = \mathbb{K}[x, y]$, a morphism between these rings and the induced morphism between any two quotients will be denoted by the same letter. Denote by:

1. $i \colon A \to A$ and $j \colon B \to B$ the identity maps;
2. $f \colon A \to B$ defined by $f(t) = x + y$, $f(s) = x^m$;
3. $g \colon B \to A$ defined by $g(x) = t$, $g(y) = 0$;
4. $q \colon A \to A$ defined by $q = gf$, i.e., $q(t) = t$, $q(s) = t^m$.

□

Since $m + 1 \neq 0$ in $\mathbb{K}$, we have a cartesian diagrams with surjective horizontal maps

$$\begin{array}{ccc} A_{m+1} & \xrightarrow{i} & A_m \\ \downarrow f & & \downarrow f \\ B_m & \xrightarrow{j} & C_m \end{array}$$

and then an isomorphism $A_{m+1} \xrightarrow{\simeq} B_m \times_{C_m} A_m$ inducing a surjective map

$$F(A_{m+1}) \to F(B_m) \times_{F(C_m)} F(A_m).$$

The surjectivity of the above map allows us to prove that a lifts to $F(A_{m+1})$ by showing that $f(a) \in F(C_m)$ lifts to $F(B_m)$. The cartesian diagram

$$\begin{array}{ccc} C_m & \xrightarrow{g} & A_m \\ \Big\downarrow{\scriptstyle j} & & \Big\downarrow{\scriptstyle i} \\ B_{m-1} & \xrightarrow{g} & A_{m-1} \end{array}$$

gives an isomorphism $C_m \xrightarrow{\simeq} B_{m-1} \times_{A_{m-1}} A_m$ and therefore a surjective map

$$F(C_m) \xrightarrow{(j,g)} F(B_{m-1}) \times_{F(A_{m-1})} F(A_m). \tag{B.4}$$

Since $gf\colon A_m \to A_{m-1}$ is the natural projection we have

$$(f(a), a) \in F(B_{m-1}) \times_{F(A_{m-1})} F(A_m)$$

and the T^1-lifting condition implies that $(f(a), a)$ lifts to an element $b \in F(B_m)$. Writing $c = j(b) \in F(C_m)$ we have that c lifts $(f(a), a)$ under the map (B.4). On the other hand also $f(a) \in F(C_m)$ lifts $(f(a), a)$ under the map (B.4) and then if this map is injective, for instance if either $m = 1$ or F is homogeneous, then $f(a) = c$ and b is the required lifting of $f(a)$. If (B.4) is not injective, and then $m \geq 2$, we need a longer argument.

Assume $m \geq 2$ and consider the following auxiliary Artin local $\mathbb{K}$-algebras:

$$V_n = \frac{\mathbb{K}[t, s]}{(t^{n+1}, t^2 s, s^2)}, \qquad n \geq 2.$$

Again, since $m + 1 \neq 0$ we have two cartesian diagrams:

$$\begin{array}{ccc} V_{m+1} & \xrightarrow{i} & V_m \\ {\scriptstyle f}\Big\downarrow & & {\scriptstyle f}\Big\downarrow \\ B_m & \xrightarrow{j} & C_m\,, \end{array} \qquad \begin{array}{ccc} V_{m+1} & \xrightarrow{i} & V_m \\ {\scriptstyle q}\Big\downarrow & & {\scriptstyle q}\Big\downarrow \\ A_{m+1} & \xrightarrow{i} & A_m\,. \end{array} \tag{B.5}$$

Claim B.4.3 There exists $v \in F(V_m)$ such that $f(v) = c$ and $q(v) = a$.

Proof Defining by

$$A'_n = \frac{\mathbb{K}[t, s]}{(t^{n+1}, ts, s^2)} = A_n \times_{\mathbb{K}} \frac{\mathbb{K}[s]}{(s^2)}, \qquad C'_n = C_n \times_{\mathbb{K}} \frac{\mathbb{K}[s]}{(s^2)},$$

we have a factorization

$$\begin{array}{ccccc} & & V_m & & \\ & \swarrow^{f} & \downarrow{\scriptstyle i} & \searrow^{q} & \\ C_m & \xleftarrow[f]{} & A'_m & \xrightarrow[q]{} & A_m \end{array}$$

and there exists $w \in T^1F$ such that $\theta(f(a) + w) = (f(a), c)$, where

$$\theta\colon F(C_m) \times T^1F \to F(C_m \times_{B_{m-1}} C_m)$$

is induced by the canonical isomorphism $C'_m \cong C_m \times_{B_{m-1}} C_m$ described in the proof of Lemma 3.5.4. This implies that $a' = a + w \in F(A'_m)$ satisfies the equalities:

$$i(a') = a, \quad f(a') = c, \qquad q(a') = gf(a') = g(c) = a.$$

To conclude the proof, it remains to show that a' lifts to $F(V_m)$. This follows from the equality $F(A_m \times_{\mathbb{K}} A_m) = F(A_m) \times F(A_m)$ and the cartesian diagram

$$\begin{array}{ccc} V_m & \xrightarrow{\quad i \quad} & A'_m \\ \Big\downarrow{\scriptstyle (i,q)} & & \Big\downarrow{\scriptstyle (i,q)} \\ \dfrac{A_{m+1} \times_{\mathbb{K}} A_{m+1}}{I} & \longrightarrow & A_m \times_{\mathbb{K}} A_m \end{array}$$

where I is the ideal generated by (t^{n+1}, t^{n+1}); notice moreover that the diagonal map $A_m \to A_m \times_{\mathbb{K}} A_m$ lifts to a morphism of local $\mathbb{K}$-algebras

$$A_m \to \frac{A_{m+1} \times_{\mathbb{K}} A_{m+1}}{I}.$$

□

The above claim implies immediately the theorem; the first diagram of (B.5) implies that v lifts to $v' \in F(V_{m+1})$; considering the second diagram, we deduce that a lifts to $F(A_{m+1})$. □

B.5 Exercises

B.5.1 Let $F\colon \mathbf{Art}_{\mathbb{K}} \to \mathbf{Set}$ be a deformation functor and n_0 a positive integer. Prove that F is smooth if and only if for every integer $n \geq n_0$ the natural map

$$F\left(\frac{\mathbb{K}[t]}{(t^{n+1})}\right) \to F\left(\frac{\mathbb{K}[t]}{(t^n)}\right)$$

is surjective.

B.5.2 Prove that the primary obstruction map of a deformation functor is bilinear symmetric.

B.5.3 Use Example B.3.5 and the proof of Theorem B.3.7 in order to prove directly that, over a field of characteristic 0, every curvilinear obstruction is semitrivial.

B.5.4 Curvilinear obstructions cannot be used in order to prove that a morphism of deformation functors is smooth; this means that there exist morphisms $F \to G$ that are not smooth and such that for every positive integer n the natural map

$$F\left(\frac{\mathbb{K}[t]}{(t^{n+1})}\right) \to F\left(\frac{\mathbb{K}[t]}{(t^n)}\right) \times_{G\left(\frac{\mathbb{K}[t]}{(t^n)}\right)} G\left(\frac{\mathbb{K}[t]}{(t^{n+1})}\right)$$

is surjective (cf. Example A.3.6). Consider for instance the morphisms $h_R \to h_S$ induced by the following quotient morphisms of complete local algebras:

1. $S = \mathbb{K}[x, y]/(x^3, y^3) \to R = \mathbb{K}[x, y](x^3, y^3, x^2y^2)$;
2. $S = \mathbb{R}[[x, y]]/(x^2 + y^2) \to R = \mathbb{R}[[x, y]]/(x^2, y^2)$.

B.5.5 (*Kawamata*) Let F be a deformation functor such that for every $n \geq 1$ the map

$$F(B_n) = F\left(\frac{\mathbb{K}[[x, y]]}{(x^{n+1}, y^2)}\right) \to F\left(\frac{\mathbb{K}[[x, y]]}{(x^{n+1}, x^n y, y^2)}\right) = F(C_n) \qquad \text{(B.6)}$$

is surjective. Prove that F satisfies the T^1-lifting property.

Conversely, prove that if F is pro-representable and satisfies the T^1-lifting property, then (B.6) is surjective for every $n \geq 1$. Use this fact for an easier proof of Theorem B.4.2 in the case where F is a pro-representable functor.

B.5.6 Consider the $\mathbb{R}$-algebra $R = \mathbb{R}[x, y]/(x^3, y^3, x^2 + y^2)$. Prove that the curvilinear obstructions of the functor h_R generate a subspace of dimension 1.

References

The approximation Theorem B.1.3 has been proved in [69] under the name factorization theorem; the proof given here is slightly different and corrects a minor mistake contained in the original paper. Theorem B.2.2 is well known and widely used in the literature under the additional assumption that T^1F is finite-dimensional; the generalization given here was proved in [69].

Example B.2.6 is taken from [236] and Example B.2.8 from [182]. To my knowledge, the class of semitrivial extensions was first used in [99]; here we follow the presentation given in [182]. On the other side, simple extension are a standard tool in the interplay between deformation theory and variations of Hodge structures, see e.g. [24].

Theorem B.3.12 was proved in [69] and generalizes the well known fact that every group scheme over a field of characteristic 0 is formally smooth and reduced.

The statement of Theorem B.4.2 appeared first, in a particular case and with an incomplete proof, in [220]. Complete proofs were given later in [141] for pro-representable functors and in [70] for functors satisfying Schlessinger's conditions and admitting a linear complete obstruction theory. The proof presented here, valid for deformation functors, is a simplified version of the one given in [70], while the proof by Kawamata is proposed here in Exercise B.5.5. However, in most concrete geometric applications the use of the Ran–Kawamata's T^1-lifting theorem may be easily replaced by the use of either the T^1-lifting trick (Theorem B.3.7) or the Abstract T^1-lifting theorem (Corollary B.3.8).

Appendix C
Kähler Manifolds

The aim of this appendix is to give a basic introduction to Kähler manifolds. Our treatment is very far from being complete and exhaustive, and it is focused on the proof of two results largely used in this book: the $\partial\overline{\partial}$-lemma (Corollary C.5.7) and the degeneration at E_1 of the Hodge to de Rham spectral sequence for compact Kähler manifolds.

C.1 Alternating Forms on Hermitian Vector Spaces

For every finite dimensional real vector space V we shall denote by

$$\mathcal{A}_V^1 = \mathrm{Hom}_{\mathbb{R}}(V, \mathbb{C})$$

the complex vector space of complex valued linear 1-forms on V; for every non-negative integer p the space of $\mathbb{C}$-multilinear alternating p-forms on V is denoted by $\mathcal{A}_V^p = \bigwedge_{\mathbb{C}}^p \mathcal{A}_V^1$. The usual conjugation in $\mathbb{C}$ extends linearly and naturally to

$$\overline{\cdot}: \mathcal{A}_V^1 \to \mathcal{A}_V^1, \quad \overline{f}(v) = \overline{f(v)}, \quad f \in \mathcal{A}_V^1,\ v \in V,$$

and, for every p, also to

$$\overline{\cdot}: \mathcal{A}_V^p \to \mathcal{A}_V^p, \qquad \overline{f_1 \wedge \cdots \wedge f_p} = \overline{f_1} \wedge \cdots \wedge \overline{f_p}.$$

A p-form $f \in \mathcal{A}_V^p$ is said to be a **real form** if $\overline{f} = f$. If $H \subset \mathcal{A}_V^p$ is a complex vector subspace, its conjugate $\overline{H} = \{\overline{f} \mid f \in H\}$ is a complex vector subspace too.

If E is a finite dimensional complex vector space with complex-dual vector space $E^\vee = \mathrm{Hom}_{\mathbb{C}}(E, \mathbb{C})$, we have an obvious inclusion $E^\vee \subset \mathcal{A}_E^1$ and there exists a projection

M. Manetti, *Lie Methods in Deformation Theory*, Springer Monographs in Mathematics,
https://doi.org/10.1007/978-981-19-1185-9

$$\pi : \mathcal{A}^1_E \to E^\vee, \qquad \pi(f)(v) = \frac{f(v) - if(iv)}{2}, \qquad v \in E,\ i = \sqrt{-1}.$$

Notice moreover that the kernel of π is the subspace

$$\overline{E^\vee} = \{f \in \mathcal{A}^1_E \mid f(iv) = -if(v) \quad \forall\, v \in E\}$$

and then there exists a direct sum decomposition $E^\vee \oplus \overline{E^\vee} = \mathcal{A}^1_E$. Therefore, for every $p \geq 0$ we have a direct sum decomposition

$$\mathcal{A}^p_E = \oplus_{a,b} \mathcal{A}^{a,b}_E,$$

where $\mathcal{A}^{a,b}_E \cong \bigwedge^p E^\vee \otimes \bigwedge^q \overline{E^\vee}$ is the subspace generated by forms of type

$$f_1 \wedge \cdots \wedge f_a \wedge \overline{g_1} \wedge \cdots \wedge \overline{g_b}, \qquad f_j, g_j \in E^\vee.$$

Notice that

$$\overline{f_1 \wedge \cdots \wedge f_a \wedge \overline{g_1} \wedge \cdots \wedge \overline{g_b}} = (-1)^{ab}\, g_1 \wedge \cdots \wedge g_b \wedge \overline{f_1} \wedge \cdots \wedge \overline{f_a},$$

and therefore $\overline{\mathcal{A}^{a,b}_E} = \mathcal{A}^{b,a}_E$.

Definition C.1.1 For every pair of non-negative integers a, b, an element of $\mathcal{A}^{a,b}_E$ is called an **(a, b)-form** on the complex vector space E. More generally, a form on E is an element of the bigraded vector space $\mathcal{A}^{*,*}_E = \oplus_{p,q} \mathcal{A}^{p,q}_E$.

If $\dim_{\mathbb{C}} E = n$ and $z_1, \ldots, z_n$ is a system of coordinates, then $z_1, \ldots, z_n$ is a basis of $E^\vee$ and $\overline{z_1}, \ldots, \overline{z_n}$ is a basis of $\overline{E^\vee}$. Therefore a basis of $\mathcal{A}^{a,b}_E$ is given by the $\binom{n}{a}\binom{n}{b}$ elements

$$z_{i_1} \wedge \cdots \wedge z_{i_a} \wedge \overline{z_{j_1}} \wedge \cdots \wedge \overline{z_{j_b}}, \qquad i_1 < \cdots < i_a, \quad j_1 < \cdots < j_b,$$

and

$$\dim_{\mathbb{C}} \mathcal{A}^p_E = \binom{2n}{p} = \sum_{a=0}^{p} \binom{n}{a}\binom{n}{p-a} = \sum_{a=0}^{p} \dim_{\mathbb{C}} \mathcal{A}^{a,p-a}_E.$$

Definition C.1.2 A $\mathbb{C}$-linear map $F : \mathcal{A}^{*,*}_E \to \mathcal{A}^{*,*}_E$ is called a **real operator** if it commutes with the conjugation $\overline{\cdot} : \mathcal{A}^{*,*}_E \to \mathcal{A}^{*,*}_E$:

$$\overline{F(\omega)} = F(\overline{\omega}) \quad \text{for every} \quad \omega \in \mathcal{A}^{*,*}_E.$$

As an example, for every a, b denote by $P_{a,b} : \mathcal{A}^{*,*}_E \to \mathcal{A}^{a,b}_E$ the natural projection, then $P_{a,b}(\overline{\eta}) = \overline{P_{b,a}(\eta)}$ and therefore $P_{a,b} + P_{b,a}$ is a real operator.

For simplicity of exposition, from now on we shall restrict ourselves to finite-dimensional complex vector spaces. Recall that a Hermitian form on a finite complex vector space E is a $\mathbb{R}$-bilinear map $h\colon E \times E \to \mathbb{C}$ such that for every $v, w \in E$ and every $a \in \mathbb{C}$:

1. $h(av, w) = ah(v, w)$,
2. $h(v, aw) = \overline{a}\, h(v, w)$,
3. $h(w, v) = \overline{h(v, w)}$.

Notice that $h(v, v) \in \mathbb{R}$ for every $v \in E$. A Hermitian form h is called positive definite if $h(v, v) > 0$ for every $v \neq 0$. If $\dim_{\mathbb{C}} E = n$, it is well known that h is positive definite if and only if there exists a basis $e_1, \ldots, e_n$ of E such that $h(e_i, e_j) = \delta_{ij}$; such a basis is called unitary.

Definition C.1.3 A **Hermitian vector space** is a pair (E, h) where E is a finite-dimensional complex vector space and h is a positive definite Hermitian form on E.

Example C.1.4 Let (V, ρ) be an Euclidean vector space of dimension n: by definition V is an n-dimensional real vector space and $\rho\colon V \times V \to \mathbb{R}$ is a positive definite inner product. Then ρ induces a canonical structure of a Hermitian vector space on $\mathcal{A}^1_V = \mathrm{Hom}_{\mathbb{R}}(V, \mathbb{C})$. The Hermitian form is defined by the formula

$$(f, g)_\rho = \sum_{i=1}^{n} f(v_i)\overline{g(v_i)}, \qquad f, g\colon V \to \mathbb{C},$$

where $v_1, \ldots, v_n$ is an orthonormal basis of V. Notice that if $x_1, \ldots, x_n \in \mathcal{A}^1_V$ is the dual basis of $v_1, \ldots, v_n$, i.e., $x_i(v_j) = \delta_{ij}$ for every i, j, then $x_1, \ldots, x_n$ is a unitary basis for $(-, -)_\rho$.

Example C.1.5 If (E, h) is a Hermitian space and ρ is the real part of $h = \rho + i\omega$, then ρ is a positive definite inner product in the underlying real vector space.

According to Example C.1.4 the Hermitian form h induces, via ρ, the positive definite Hermitian form $(-, -)_h = (-, -)_\rho$ on $\mathrm{Hom}_{\mathbb{R}}(E, \mathbb{C})$. Since $\rho(iv, iw) = \rho(v, w)$ for every $v, w \in E$, if $e_1, \ldots, e_n \in E$ is a unitary complex basis for h, then $e_1, ie_2, \ldots, e_n, ie_n$ is an oriented orthonormal real basis for ρ and then

$$(f, g)_h = \sum_{j=1}^{n} f(e_j)\overline{g(e_j)} + f(ie_j)\overline{g(ie_j)}, \qquad f, g \in \mathrm{Hom}_{\mathbb{R}}(E, \mathbb{C}). \qquad \text{(C.1)}$$

Example C.1.6 Every Hermitian vector space (E, h) induces canonical Hermitian forms, denoted by the same letter h, on the exterior powers of E:

$$h\colon \bigwedge^p E \times \bigwedge^p E \to \mathbb{C}, \qquad h(v_1 \wedge \cdots \wedge v_p, w_1 \wedge \cdots \wedge w_p) = \det(h(v_i, w_j)).$$

If $e_1, \ldots, e_n$ is a unitary basis of E, then $e_{i_1} \wedge \cdots \wedge e_{i_p}$, $i_1 < \cdots < i_p$, is a unitary basis of $\bigwedge^p E$. We can extend the above Hermitian forms to $\bigwedge^* E$ by setting

$h(x, y) = x\overline{y}$ if $x, y \in \mathbb{C} = \bigwedge^0 E$, and $h(\eta, \mu) = 0$ if $\eta \in \bigwedge^p E$, $\mu \in \bigwedge^q E$ and $p \neq q$.

Definition C.1.7 Given a Hermitian space (E, h) denote by $(-, -)_h$ the Hermitian product on $\bigoplus_{p,q} \mathcal{A}_E^{p,q} = \bigwedge_{\mathbb{C}}^* \mathrm{Hom}_{\mathbb{R}}(E, \mathbb{C})$ defined as in Example C.1.5 for 1-forms, and extended to arbitrary forms as in Example C.1.6.

More concretely, if $e_1, \ldots, e_n$ is a fixed unitary basis of $(E, h = \rho + i\omega)$ then $e_1, \ldots, e_n, ie_1, \ldots, ie_n$ is a orthonormal basis of the Euclidean vector space (E, ρ). Hence if z_j is the associated coordinate system, according to (C.1) for every a, b we have

$$(z_a, z_b)_h = 2\delta_{a,b}, \qquad (z_a, \overline{z}_b)_h = 0, \qquad (\overline{z}_a, \overline{z}_b)_h = \overline{(z_b, z_a)_h} = 2\delta_{a,b},$$

and then $z_1/\sqrt{2}, \ldots, z_n/\sqrt{2}, \overline{z_1}/\sqrt{2}, \ldots, \overline{z_n}/\sqrt{2}$ is a unitary basis of $\mathcal{A}_E^1$ with respect to the Hermitian product $(-, -)_h$. According to Example C.1.6, a unitary basis of $\mathcal{A}_E^{*,*} = \bigoplus_{p,q} \mathcal{A}_E^{p,q}$ is given by the set of (p, q)-forms:

$$\frac{1}{\sqrt{2^{p+q}}} z_{a_1} \wedge \cdots \wedge z_{a_p} \wedge \overline{z}_{b_1} \wedge \cdots \wedge \overline{z}_{b_q}, \qquad a_1 < \cdots < a_p, \quad b_1 < \cdots < b_q.$$

Definition C.1.8 (*Kähler form*) The **Kähler form** of a finite-dimensional Hermitian vector space (E, h) is the $(1, 1)$-form $\omega \in \mathcal{A}_E^{1,1}$ defined in the following way. Choosing a basis $e_1, \ldots, e_n$ of E and denoting by $z_1, \ldots, z_n \in E^\vee$ its dual basis we have

$$\omega = \frac{i}{2} \sum_{r,s} h(e_r, e_s) z_r \wedge \overline{z_s}, \qquad i = \sqrt{-1}.$$

The Kähler form ω is real, i.e., $\overline{\omega} = \omega$, and properly defined. The first assertion is clear since

$$\overline{\omega} = \frac{-i}{2} \sum_{r,s} \overline{h(e_r, e_s)}\, \overline{z_r} \wedge z_s = \frac{i}{2} \sum_{r,s} h(e_s, e_r)\, z_s \wedge \overline{z_r} = \omega.$$

In order to see that ω does not depend from the basis we observe that $\mathcal{A}_E^2$ is isomorphic to the space of $\mathbb{R}$-bilinear skew-symmetric forms $E \times E \to \mathbb{C}$; this isomorphism is canonically defined, according to the various conventions existing in the literature, up to multiplication of an invertible rational number. Here, following Lemma 8.5.1, for every $f, g \in \mathcal{A}_E^1$ and every $v, w \in E$ we use the conventional pairing

$$(f \wedge g)(v, w) = \boldsymbol{i}_{f \wedge g}(v \wedge w) = \boldsymbol{i}_f \boldsymbol{i}_g(v \wedge w) = g(v) f(w) - f(v) g(w).$$

Thus if $v = \sum_r v_r e_r$ and $w = \sum_s w_s e_s$ we have $h(v, w) = \sum_{r,s} h(e_r, e_s) v_r \overline{w_s}$,

$$\omega(v,w) = \frac{i}{2}\sum_{r,s} h(e_r,e_s)(\overline{v_s}w_r - v_r\overline{w_s})$$
$$= -\frac{i}{2}\sum_{r,s}\left(h(e_r,e_s)v_r\overline{w_s} - \overline{h(e_r,e_s)v_r\overline{w_s}}\right)$$
$$= -\frac{i}{2}\left(h(v,w) - \overline{h(v,w)}\right),$$

and therefore ω is precisely the imaginary part of the Hermitian form h.

Definition C.1.9 (*The Hodge operator* $*$) Given a Hermitian vector space (E, h) of dimension $n = \dim_{\mathbb{C}} E$, we denote by

$$*\colon \mathcal{A}_E^{*,*} \to \mathcal{A}_E^{*,*}$$

the unique $\mathbb{C}$-linear operator such that:

1. $*\mathcal{A}_E^{p,q} \subset \mathcal{A}_E^{n-q,n-p}$ for every $0 \le p, q \le n$;
2. $*$ is a real operator: $*\overline{\eta} = \overline{*\eta}$ for every $\eta \in \mathcal{A}_E^{*,*}$;
3. if ω is the Kähler form, then for every $\eta, \mu \in \mathcal{A}_E^{*,*}$ we have

$$\eta \wedge *\overline{\mu} = \eta \wedge \overline{*\mu} = (\eta,\mu)_h \frac{\omega^{\wedge n}}{n!}. \tag{C.2}$$

We only need to prove the existence of the operator $*$ as above; the unicity is clear because the wedge product $\mathcal{A}_E^{p,q} \times \mathcal{A}_E^{n-p,n-q} \to \mathcal{A}_E^{n,n}$ is nondegenerate.

Consider the dual basis $z_1, \ldots, z_n$ of a unitary basis of (E, h); then we can write

$$\omega = \frac{i}{2}\sum_j z_j \wedge \overline{z_j} = \sum_j u_j, \qquad \frac{\omega^{\wedge n}}{n!} = u_1 \wedge \cdots \wedge u_n, \quad \text{where} \quad u_j = \frac{i}{2} z_j \wedge \overline{z_j},$$

and Eq. (C.2) becomes

$$\eta \wedge *\overline{\mu} = \eta \wedge \overline{*\mu} = (\eta,\mu)_h\, u_1 \wedge \cdots \wedge u_n.$$

If $A = \{a_1, \ldots, a_p\} \subset \{1, \ldots, n\}$ and $a_1 < a_2 < \cdots < a_p$, we denote $|A| = p$ and

$$z_A = z_{a_1} \wedge \cdots \wedge z_{a_p}, \qquad \overline{z}_A = \overline{z}_{a_1} \wedge \cdots \wedge \overline{z}_{a_p}, \qquad u_A = u_{a_1} \wedge \cdots \wedge u_{a_p}.$$

For every decomposition of $\{1, \ldots, n\} = A \cup B \cup M \cup N$ into four disjoint subsets, we write

$$z_{A,B,M,N} = \frac{1}{\sqrt{2^{|A|+|B|}}} z_A \wedge \overline{z}_B \wedge u_M \in \mathcal{A}_E^{|A|+|M|,|B|+|M|}. \tag{C.3}$$

These elements give a basis of $\mathcal{A}_E^{*,*}$, called *standard basis*, that is unitary for the Hermitian product $(-,-)_h$. Since

$$u_A = \frac{(-1)^{\frac{|A|(|A|+1)}{2}}}{(2i)^{|A|}} z_A \wedge \overline{z}_A = \frac{i^{|A|}(-1)^{\frac{|A|(|A|+1)}{2}}}{2^{|A|}} \overline{z_A} \wedge z_A,$$

we have

$$z_{A,B,M,N} \wedge z_{B,A,N,M} = i^{|A|-|B|}(-1)^{\frac{|A|(|A|+1)}{2}+\frac{|B|(|B|+1)}{2}} u_1 \wedge \cdots \wedge u_n.$$

We define the $\mathbb{C}$-linear operator $*\colon \mathcal{A}_E^{*,*} \to \mathcal{A}_E^{*,*}$ in the standard basis by the formula:

$$*z_{A,B,M,N} = i^{|A|-|B|}(-1)^{\frac{(|A|+|B|)(|A|+|B|+1)}{2}} z_{A,B,N,M}.$$

Since $\overline{z_{A,B,M,N}} = (-1)^{|A|\,|B|} z_{B,A,M,N}$ we have

$$\overline{*z_{A,B,M,N}} = *\overline{z_{A,B,M,N}} = i^{|B|-|A|}(-1)^{\frac{|A|(|A|+1)}{2}+\frac{|B|(|B|+1)}{2}} z_{B,A,N,M},$$

$$z_{A,B,M,N} \wedge *\overline{z_{A,B,M,N}} = u_1 \wedge \cdots \wedge u_n,$$

proving that $*$ is a real operator together with Eq. (C.2) and therefore that Definition C.1.9 makes sense.

Notice also that $*$ is an isometry for the Hermitian product $(-,-)_h$ and that

$$*^2 = \sum_{a,b}(-1)^{a-b} P_{a,b} = \sum_{a,b}(-1)^{a+b} P_{a,b}. \tag{C.4}$$

Remark C.1.10 It is important to point out that the form $\omega^{\wedge n}/n!$ appearing in Eq. (C.2) is the volume form of the real Euclidean space (E,ρ) underlying the Hermitian vector space $(E, h = \rho + i\omega)$.

Let $z_1,\ldots,z_n$ be the coordinate system of a unitary basis $e_1,\ldots,e_n$ for h. If $z_j = x_j + iy_j$ with $x_i, y_i \in \operatorname{Hom}_{\mathbb{R}}(E,\mathbb{R})$, then $u_j = \frac{i}{2} z_j \wedge \overline{z_j} = x_j \wedge y_j$ and therefore

$$\frac{\omega^{\wedge n}}{n!} = u_1 \wedge \cdots \wedge u_n = x_1 \wedge y_1 \wedge \cdots \wedge x_n \wedge y_n.$$

Notice that $x_1, y_1, \ldots, x_n, y_n$ is precisely the coordinates system of the basis $e_1, ie_1, \ldots, e_n, ie_n$ of the real oriented vector space E. Since $e_1, ie_1, \ldots, e_n, ie_n$ is a orthonormal basis of (E,ρ) we have that $x_1 \wedge y_1 \wedge \cdots \wedge x_n \wedge y_n$ is precisely the volume form of (E,ρ).

C.2 The Lefschetz Decomposition

Let (E, h) be a Hermitian vector space of dimension n. Then the wedge product with the Kähler form ω is a well defined nilpotent endomorphism of the space $\mathcal{A}_E^{*,*}$. The goal of this section is to give a detailed study of the corresponding Jordan canonical form.

Definition C.2.1 (*Lefschetz operators*) Let (E, h) be a Hermitian vector space of dimension n with projection operators $P_{a,b}\colon \mathcal{A}_E^{*,*} \to \mathcal{A}_E^{a,b}$, Kähler form $\omega \in \mathcal{A}_E^{1,1}$ and Hodge operator $*\colon \mathcal{A}_E^{p,q} \to \mathcal{A}_E^{n-q,n-p}$.

The Lefschetz operators $C, L, \Lambda\colon \mathcal{A}_E^{*,*} \to \mathcal{A}_E^{*,*}$ are defined in the following way:

1. $C = \sum_{a,b} i^{a-b} P_{a,b}$;
2. $L(\eta) = \eta \wedge \omega$, for every $\eta \in \mathcal{A}_E^{*,*}$;
3. $\Lambda = *^{-1} L * = * L *^{-1}$.

The equality $*^{-1} L * = * L *^{-1}$, equivalent to $*^2 L = L *^2$, follows immediately from (C.4). Notice that $L^{n+1} = \Lambda^{n+1} = 0$ for trivial reasons of degree.

Lemma C.2.2 *The operators $*$, C, L, Λ satisfy the relations:*

$$CL = LC, \qquad C\Lambda = \Lambda C, \qquad C* = *C, \qquad L* = *\Lambda, \qquad \Lambda * = *L.$$

The operators L, Λ are adjoint with respect to the Hermitian product $(-,-)_h$.

Proof The equalities $*\Lambda = L*$ and $\Lambda * = *L$ follow immediately from the definition of Λ. Fixing a standard basis $z_{A,B,M,N}$ as in (C.3), an easy computation gives:

$$C\, z_{A,B,M,N} = i^{|A|-|B|} z_{A,B,M,N},$$

$$L\, z_{A,B,M,N} = \sum_{i \in N} z_{A,B,M \cup \{i\}, N-\{i\}},$$

$$\Lambda\, z_{A,B,M,N} = \sum_{i \in M} z_{A,B,M-\{i\}, N \cup \{i\}}.$$

Since the standard basis is unitary for the product $(-,-)_h$ we have

$$(L\, z_{A,B,M,N}, z_{A',B',M',N'})_h = \begin{cases} 1 & \text{if } A = A',\ B = B',\ M \subset M' \\ 0 & \text{otherwise} \end{cases}$$
$$= (z_{A,B,M,N}, \Lambda\, z_{A',B',M',N'})_h$$

and therefore L, Λ are adjoint operators. The equalities $CL = LC$ and $C\Lambda = \Lambda C$ are clear. Since

$$*z_{A,B,M,N} = i^{|A|-|B|} (-1)^{\frac{(|A|+|B|)(|A|+|B|+1)}{2}} z_{A,B,N,M},$$

we obtain that $*$ and C are commuting operators. For later use we also observe that

$$C^{-1} * z_{A,B,M,N} = (-1)^{\frac{(|A|+|B|)(|A|+|B|+1)}{2}} z_{A,B,N,M}. \tag{C.5}$$

□

Lemma C.2.3 *Let (E, h) be a Hermitian vector space of dimension n. Then for every integer $r \geq 1$ we have*

$$[\Lambda, L^r] = \sum_{a,b} r(n-a-b-r+1)L^{r-1}P_{a,b}.$$

Proof Consider first the case $r = 1$. Fixing a standard basis $z_{A,B,M,N}$ we have:

$$Lz_{A,B,M,N} = \sum_{i \in N} z_{A,B,M\cup\{i\},N-\{i\}}$$

$$\Lambda z_{A,B,M,N} = \sum_{i \in M} z_{A,B,M-\{i\},N\cup\{i\}}$$

$$\Lambda L z_{A,B,M,N} = \sum_{i \in N} z_{A,B,M,N} + \sum_{j \in M}\sum_{i \in N} z_{A,B,M\cup\{i\}-\{j\},N\cup\{j\}-\{i\}},$$

$$L\Lambda z_{A,B,M,N} = \sum_{i \in M} z_{A,B,M,N} + \sum_{j \in M}\sum_{i \in N} z_{A,B,M\cup\{i\}-\{j\},N\cup\{j\}-\{i\}}.$$

Therefore

$$\begin{aligned}(\Lambda L - L\Lambda)z_{A,B,M,N} &= (|N| - |M|)z_{A,B,M,N} \\ &= (n - |A| - |B| - 2|M|)z_{A,B,M,N}\end{aligned}$$

and then

$$[\Lambda, L] = \sum_{a,b}(n-a-b)P_{a,b}.$$

If $r > 1$, by induction we have

$$\begin{aligned}&= [\Lambda, L^r]L + L^r[\Lambda, L] \\ &= \sum_{a,b} r(n-a-b-r+1)L^{r-1}P_{a,b}L + \sum_{a,b}(n-a-b)L^r P_{a,b}.\end{aligned}$$

Since $P_{a,b}L = LP_{a-1,b-1}$ we get

$$= \sum_{a,b} r(n-a-b-r+1)L^r P_{a-1,b-1} + \sum_{a,b}(n-a-b)L^r P_{a,b}$$
$$= \sum_{a,b} r(n-a-b-r-1)L^r P_{a,b} + \sum_{a,b}(n-a-b)L^r P_{a,b}$$
$$= \sum_{a,b}((r+1)(n-a-b)-r(r+1))L^r P_{a,b}$$
$$= \sum_{a,b}(r+1)(n-a-b-r)L^r P_{a,b}.$$

□

Remark C.2.4 Denoting by $H = \sum_{a,b}(n-a-b)P_{a,b}$ we have the equalities

$$[\Lambda, L] = H, \qquad [H, \Lambda] = 2\Lambda, \qquad [H, L] = -2L$$

and then a structure of a $\mathfrak{sl}_2(\mathbb{C})$-module on $\mathcal{A}_E^{a,b}$. In fact, most of the results of this section can be also proved by using the general theory of representations of the Lie algebra $\mathfrak{sl}_2(\mathbb{C})$.

Definition C.2.5 A form $v \in \mathcal{A}_E^{*,*}$ is called **primitive** if $\Lambda v = 0$.

Since $\Lambda(\mathcal{A}_E^{a,b}) \subset \mathcal{A}_E^{a-1,b-1}$, a form v is primitive if and only if every component $P_{a,b}(v)$ is primitive. If either $a = 0$ or $b = 0$ then every form in $\mathcal{A}_E^{a,b}$ is primitive.

For every integer $0 \le p \le 2n$, the space of p-forms on E is denoted by

$$\mathcal{A}_E^p = \bigoplus_{a+b=p} \mathcal{A}_E^{a,b}.$$

Proposition C.2.6 *Let (E, h) be a Hermitian vector space of dimension n, let $v \in \mathcal{A}_E^p$ be a primitive p-form and write $\alpha = n - p$. Then:*

1. *$L^b v = 0$ for every $b > \max(\alpha, -1)$. In particular, if $\alpha < 0$ then $v = L^0 v = 0$.*
2. *If $\alpha \ge 0$, then for every $\alpha \ge a > b \ge 0$*

$$\Lambda^{a-b}L^a v = \left(\prod_{r=b+1}^{a} r(\alpha - r + 1)\right) L^b v = \frac{a!}{b!}\frac{(\alpha-b)!}{(\alpha-a)!}L^b v;$$

in particular, $\Lambda^\alpha L^\alpha v = (\alpha!)^2 v$ and then $L^r v = 0$ if and only if $r > \alpha = n - p$.

Proof For $s, r \ge 1$, since v is primitive we have $\Lambda^s L^r v = \Lambda^{s-1}[\Lambda, L^r]v$ and therefore, by Lemma C.2.3,

$$\Lambda^s L^r v = \Lambda^{s-1}[\Lambda, L^r]v = r(\alpha - r + 1)\Lambda^{s-1}L^{r-1}v.$$

A recursive application of the above formula gives, for every $a > b \ge 0$,

$$\Lambda^{a-b}L^a v = \left(\prod_{r=b+1}^{a} r(\alpha - r + 1)\right) L^b v.$$

If $a > b > \max(\alpha, -1)$ then $r(\alpha - r + 1) \neq 0$ for every $r > b$ and therefore $L^b v = 0$ if and only if $\Lambda^{a-b}L^a v = 0$; taking $a >> 0$ we get the required vanishing of $L^b v$. Finally, if $\alpha \geq a > b \geq 0$ we can write the above formula in the form $\Lambda^{a-b}L^a v = \dfrac{a!}{b!}\dfrac{(\alpha - b)!}{(\alpha - a)!} L^b v$. □

Lemma C.2.7 *Let (E, h) be a Hermitian vector space of dimension n and let $v \in \mathcal{A}_E^p$ be a primitive p-form, with $p \leq n$. Then:*

$$C^{-1} * L^r v = \begin{cases} (-1)^{\frac{p(p+1)}{2}} \dfrac{r!}{(n-p-r)!} L^{n-p-r} v & \text{if } 0 \leq r \leq n-p, \\ 0 & \text{if } r > n-p. \end{cases}$$

The Lemma C.2.7, which plays a central role also in the proof of the Kähler identities (Theorem C.3.3), requires a quite long combinatorial proof that, for the clarity of exposition, is postponed to the end of this section.

Theorem C.2.8 (Lefschetz decomposition) *Let (E, h) be a Hermitian vector space of dimension n.*

1. *Every p-form $v \in \mathcal{A}_E^p$ can be written in a unique way as*

$$v = \sum_{r=\max(p-n,0)}^{n} L^r v_r$$

   with every $v_r \in \mathcal{A}_E^{p-2r}$ primitive (notice the inequality $p - 2r \leq n$ for every $r \geq \max(p - n, 0)$).
2. *In the situation of the above item, if $L^a v = 0$ for an integer $a \geq 0$, then $v_r = 0$ for every $r \geq a + p - n$.*
3. *There exist noncommutative polynomials $G^n_{p,r}(\Lambda, L)$ with rational coefficients depending only on n, p, r, such that in the above notation $v_r = G^n_{p,r}(\Lambda, L)v$ for every r.*

Proof Write $\alpha = n - p$ and assume first $\alpha \geq 0$. We prove the existence of the decomposition $v = \sum_{r\geq 0} L^r v_r$ as above by induction on the minimum q such that $\Lambda^q v = 0$; it is clear that $q \leq n + 1$. If $q \leq 1$ then the form v is already primitive. If $q \geq 1$ and $\Lambda^{q+1} v = 0$, then the form $w = \Lambda^q v \in \mathcal{A}_E^{p-2q}$ is primitive and therefore, setting $\gamma = \prod_{r=1}^{q} r(\alpha + 2q - r + 1)$, we have $\gamma > 0$ and

$$\Lambda^q \left(v - L^q \frac{w}{\gamma}\right) = w - \Lambda^q L^q \frac{w}{\gamma} = 0$$

by Proposition C.2.6. By the inductive assumption we have

$$v - L^q \frac{w}{\gamma} = \sum_{r=\max(p-n,0)}^{n} L^r v_r$$

with every v_r primitive, and this proves the existence of the Lefschetz decomposition when $\alpha \geq 0$.

If $\alpha < 0$ then $C^{-1} * v \in \mathcal{A}_E^{2n-p}$ and we can write: $C^{-1} * v = \sum_{r \geq 0} L^r v_r$ with $v_r \in \mathcal{A}_E^{2n-p-2r}$ primitive for every r. Since $(C^{-1}*)^2 = \mathrm{Id}$ we have

$$v = \sum_{r \geq 0} C^{-1} * L^r v_r$$

and by Lemma C.2.7 every element $C^{-1} * L^r v_r$ is a rational multiple of $L^{p-n+r} v_r$.

The unicity of the decomposition and the existence of the polynomials $G^n_{p,r}(\Lambda, L)$ are proved at the same time. If $q \leq n$ and

$$v = \sum_{r=\max(-\alpha,0)}^{q} L^r v_r$$

is a decomposition with every $v_r \in \mathcal{A}_E^{p-2r}$ primitive, by Proposition C.2.6 we have $L^{n-p+q} L^r v_r = 0$ for every $r < q$ and therefore $L^{n-p+q} v = L^{n-p+2q} v_q$. In view of the inequality $q \geq \max(p-n, 0)$ we have $n - p + 2q = 2q - (p-n) \geq 0$ and by Proposition C.2.6

$$\Lambda^{n-p+2q} L^{n-p+2q} v_q = ((n-p+2q)!)^2 v_q.$$

Equivalently $v_q = G^n_{p,q}(\Lambda, L) v$, where

$$G^n_{p,q}(\Lambda, L) = \frac{1}{((n-p+2q)!)^2} \Lambda^{n-p+2q} L^{n-p+2q},$$

and this implies that v_q is uniquely determined by v. Since

$$v - L^q v_q = (1 - L^q G^n_{p,q}(\Lambda, L)) v = \sum_{r=\max(-\alpha,0)}^{q-1} L^r v_r$$

we can repeat the above procedure and construct recursively the polynomials $G^n_{p,r}(\Lambda, L)$.

Finally, we prove the second item of the theorem. Let $v = \sum_{r \geq 0} L^r v_r$ be the Lefschetz decomposition of a p-form v such that $L^a v = 0$. Since v_r is a primitive $(p-2r)$-form, by Proposition C.2.6 we have $L^b v_r \neq 0$ if and only if $v_r \neq 0$ and

$b \leq n - p + 2r$. Since $\sum_{r>0} L^{a+r} v_r$ is the Lefschetz decomposition of $L^a v = 0$ we have $L^{a+r} v_r = 0$ for every r and therefore $v_r = 0$ whenever $a + r \leq n - p + 2r$. □

Corollary C.2.9 *If $n \geq p$, then a p-form $v \in \mathcal{A}_E^p$ is primitive if and only if $L^{n-p+1} v = 0$.*

Proof If v is a primitive form then $L^{n-p+1} v = 0$ by Proposition C.2.6. Conversely, if $v = \sum_{r \geq 0} L^r v_r$ is the Lefschetz decomposition and $L^{n-p+1} v = 0$ then $v_r = 0$ for every $r \geq (n - p + 1) + p - n = 1$ and therefore $v = v_0$ is primitive. □

Proof of Lemma C.2.7
Let $z_{A,B,M,N}$ be the standard basis associated to a fixed unitary basis of E and write

$$\mathcal{A}_E^{*,*} = \bigoplus_{A,B} V_{A,B}, \qquad A, B \subset \{1, \ldots, n\}, \quad A \cap B = \emptyset,$$

where $V_{A,B}$ is the subspace generated by all the forms $z_{A,B,M,N}$, where M varies among the subsets of $\{1, \ldots, n\}$ disjoint from $A \cup B$.

By Lemma C.2.2 and Eq. (C.5), every $V_{A,B}$ is an invariant subspace for the operators L, Λ and $C^{-1}*$. Thus it is sufficient to prove the equality

$$C^{-1} * L^r v = \begin{cases} (-1)^{\frac{p(p+1)}{2}} \dfrac{r!}{(n-p-r)!} L^{n-p-r} v & \text{if } 0 \leq r \leq n - p, \\ 0 & \text{if } r > n - p, \end{cases}$$

for a primitive p-form in $V_{A,B}$, $p \leq n$. For a given pair of disjoint subsets $A, B \subset \{1, \ldots, n\}$, denote by $K = \{1, \ldots, n\} - (A \cup B)$ the complement of their union and by

$$k = |K| = n - |A| - |B|, \qquad \tau = (-1)^{\frac{(|A|+|B|)(|A|+|B|+1)}{2}} = (-1)^{\frac{(n-k)(n-k+1)}{2}}.$$

Writing

$$M^c = K - M, \qquad u_M = z_{A,B,M,M^c},$$

for every subset $M \subset K$ we have $u_M \in \mathcal{A}_E^{n-k+2|M|}$,

$$L u_M = \sum_{i \in M^c} u_{M \cup \{i\}}, \qquad \Lambda u_M = \sum_{i \in M} u_{M - \{i\}}, \qquad C^{-1} * u_M = \tau u_{M^c},$$

and a simple combinatorial argument shows that for every $r \geq 0$ we have

$$L^r u_M = r! \sum_{M \subset N, |N| = |M| + r} u_N. \tag{C.6}$$

There exists a direct sum decomposition

$$V_{A,B} = \bigoplus_{\alpha \equiv k \pmod 2} V_\alpha, \qquad V_\alpha = V_{A,B} \cap \mathcal{A}_E^{n-\alpha},$$

where V_α is generated by the forms u_M with $|M^c| - |M| = k - 2|M| = \alpha$ and

$$L\colon V_\alpha \to V_{\alpha-2}, \qquad \Lambda\colon V_\alpha \to V_{\alpha+2}, \qquad C^{-1}*\colon V_\alpha \to V_{-\alpha}.$$

Denoting by $P_\alpha\colon V_{A,B} \to V_\alpha$ the projection, we have that a form $v \in V_{A,B}$ is primitive if and only if $P_\alpha(v)$ is primitive for every $\alpha \equiv k \pmod 2$. We have already seen that

$$[\Lambda, L] = \sum_\alpha \alpha P_\alpha, \qquad LC^{-1}* = C^{-1} * \Lambda, \qquad C^{-1} * L = \Lambda C^{-1} *.$$

Lemma C.2.10 *In the notation above, let $0 \le m \le k/2$ be an integer and let*

$$v = \sum_{M \subset K, |M|=m} a_M u_M \in V_{k-2m}, \qquad a_M \in \mathbb{C},$$

be a primitive form. Then for every indexing subset M in the above sum we have

$$a_M = (-1)^m \sum_{N \subset M^c, |N|=m} a_N.$$

Proof For $m = 0$ the above equality becomes $a_\emptyset = a_\emptyset$ and therefore we can assume $m > 0$. Let $M \subset K$ be a fixed subset of cardinality m; since

$$0 = \Lambda v = \sum_{|H|=m} a_H \sum_{i \in H} u_{H-\{i\}} = \sum_{|N|=m-1} u_N \sum_{i \in N^c} a_{N\cup\{i\}}$$

for every subset $N \subset K$ of cardinality $m-1$ the coefficient of u_N in the above sum is equal to $0 = \sum_{i \in N^c} a_{N\cup\{i\}}$, and writing $N^c = (M - N) \cup (M \cup N)^c$ we get

$$R_N: \qquad \sum_{i \in M-N} a_{N\cup\{i\}} = - \sum_{i \in (M\cup N)^c} a_{N\cup\{i\}}.$$

For every $0 \le r \le m$ write

$$S_r = \sum_{|H|=m, |H\cap M|=r} a_H.$$

Fixing an integer $1 \le r \le m$ and taking the sum of the equalities R_N, for all $N \subset K$ such that $|N \cap M| = r - 1$ we get

$$rS_r = -(m - r + 1)S_{r-1}$$

and then

$$a_M = S_m = -\frac{S_{m-1}}{m} = \frac{2S_{m-2}}{m(m-1)} = \cdots = (-1)^m \frac{m!}{m!} S_0 = (-1)^m \sum_{N \subset M^c, |N|=m} a_N.$$

□

Lemma C.2.11 *In the above notation, let $0 \le m \le k/2$ be an integer and let $v \in V_{k-2m}$ be a primitive form. Then for every $0 \le r \le k-2m$ we have*

$$C^{-1} * L^r v = \tau(-1)^m \frac{r!}{(k-2m-r)!} L^{k-2m-r} v.$$

Proof Write $\alpha = k - 2m$, so that $V_{k-2m} \subset \mathcal{A}_E^{n-\alpha}$, and consider first the case $r = 0$; writing $v = \sum a_N u_N$ with $|N| = m$, $a_N \in \mathbb{C}$, by (C.6) we have:

$$\frac{L^\alpha v}{\alpha!} = \sum_{|N|=m} a_N \sum_{\substack{N \subset M \\ |M|=m+\alpha}} u_M = \sum_{|N|=m} a_N \sum_{\substack{M \subset N^c \\ |M|=m}} u_{M^c} = \sum_{|M|=m} u_{M^c} \sum_{\substack{N \subset M^c \\ |N|=m}} a_N.$$

On the other hand $C^{-1} * v = \tau \sum_{|M|=m} a_M u_{M^c}$ and the equality

$$C^{-1} * v = \frac{\tau(-1)^m}{\alpha!} L^\alpha v$$

follows immediately from Lemma C.2.10. If $r \ge 1$ then

$$C^{-1} * L^r v = \Lambda^r C^{-1} * v = \frac{\tau(-1)^m}{\alpha!} \Lambda^r L^\alpha v$$

and by Proposition C.2.6 we get

$$C^{-1} * L^r v = \frac{\tau(-1)^m}{\alpha!} \frac{\alpha!}{(\alpha-r)!} \frac{r!}{0!} L^{\alpha-r} v = \tau(-1)^m \frac{r!}{(\alpha-r)!} L^{\alpha-r} v.$$

□

Now the proof of Lemma C.2.7 follows immediately from the above results and by the fact that if $v \in V_{k-2m}$ is primitive, then $v \in \mathcal{A}_E^p$ with $p = n - k + 2m$ and

$$\tau(-1)^m = (-1)^{\frac{(n-k)(n-k+1)+2m}{2}} = (-1)^{\frac{p(p+1)}{2}}.$$

C.3 Kähler Identities and Harmonic Forms

Let X be a complex manifold of dimension n and denote by $\mathcal{A}_X^{*,*}$ the sheaf of differential forms on X. By definition $\mathcal{A}_X^{*,*} = \oplus_{a,b}\mathcal{A}_X^{a,b}$, where $\mathcal{A}_X^{a,b}$ is the sheaf of sections of the complex vector bundle $\bigwedge^a T_X^\vee \otimes \bigwedge^b \overline{T_X^\vee}$. As in the previous chapters, we also denote by $A_X^{a,b} = \Gamma(X, \mathcal{A}_X^{a,b})$ the $\mathbb{C}$-vector space of differential forms of type (a, b) on X.

The projections $P_{a,b}$ and the operator $C = \sum_{a,b} i^{a-b} P_{a,b}$, defined on the fibres of the above bundles, extend in the obvious way to morphisms of sheaves $P_{a,b}, C\colon \mathcal{A}_X^{*,*} \to \mathcal{A}_X^{*,*}$. It is convenient, for every integer $0 \le r \le 2n$, to denote $\mathcal{A}_X^r = \oplus_{a+b=r}\mathcal{A}_X^{a,b}$ and $P_r = \sum_{a+b=r} P_{a,b}\colon \mathcal{A}_X^{*,*} \to \mathcal{A}_X^r$.

Denoting by $d\colon \mathcal{A}_X^r \to \mathcal{A}_X^{r+1}$ the de Rham differential we also introduce the operators:

$$d^C = C^{-1}dC, \qquad \partial = \frac{d + id^C}{2}, \qquad \overline{\partial} = \frac{d - id^C}{2}.$$

Then

$$d = Cd^C C^{-1}, \qquad d = \partial + \overline{\partial}, \qquad d^C = i(\overline{\partial} - \partial),$$

and it is easy to see that $d, d^C, \partial, \overline{\partial}$ are derivations of the sheaf of graded algebras $\mathcal{A}_X^{*,*}$, since C commutes with the wedge product. It is well known, and in any case easy to prove by using the existence of local holomorphic coordinates, that for every differential form $\eta \in \mathcal{A}_X^{p,q}$ we have

$$\partial\eta \in \mathcal{A}_X^{p+1,q}, \quad \overline{\partial}\eta \in \mathcal{A}_X^{p,q+1}, \quad d\eta = \partial\eta + \overline{\partial}\eta, \quad d^C(\eta) = i^{-1}\partial\eta + i\overline{\partial}\eta.$$

Since $0 = d^2 = \partial^2 + \partial\overline{\partial} + \overline{\partial}\partial + \overline{\partial}^2$ we get $0 = \partial^2 = \partial\overline{\partial} + \overline{\partial}\partial = \overline{\partial}^2$ and then $(d^C)^2 = 0$, $dd^C = 2i\partial\overline{\partial} = -d^C d$.

Using the structure of a graded Lie algebra on the space of $\mathbb{C}$-linear operators of the sheaf of graded algebras $\mathcal{A}_X^{*,*}$ (with the total degree $\overline{v} = a + b$ if $v \in \mathcal{A}_X^{a,b}$), the above relations can be written as

$$[d, d] = dd + dd = 2d^2 = 0, \quad [d^C, d^C] = [d, d^C] = [\partial, \partial]=[\overline{\partial}, \overline{\partial}] = [\partial, \overline{\partial}] = 0.$$

Note finally that d and C are real operators and then also d^C is; moreover, $\overline{\partial}\overline{\eta} = \overline{\partial\eta}$.

Recall that a Hermitian metric on X is a C^∞ positive definite Hermitian form h on the tangent vector bundle T_X. This means that if $z_1, \ldots, z_n$ are local holomorphic coordinates then $h_{ij} = h\left(\frac{\partial}{\partial z_i}, \frac{\partial}{\partial z_j}\right)$ is a smooth function for every i, j and the matrix (h_{ij}) is Hermitian and positive definite at every point. The associated Kähler form is

$$\omega = \frac{i}{2}\sum_{i,j} h_{ij}dz_i \wedge d\overline{z}_j \in A_X^{1,1} = \Gamma(X, \mathcal{A}_X^{1,1}).$$

Therefore, the Hermitian metric h induces, for every open subset $U \subset X$, the Hodge and Lefschetz operators:

$$L\colon \Gamma(U, \mathcal{A}_X^{a,b}) \to \Gamma(U, \mathcal{A}_X^{a+1,b+1}), \qquad Lv = v \wedge \omega,$$

$$*\colon \Gamma(U, \mathcal{A}_X^{a,b}) \to \Gamma(U, \mathcal{A}_X^{n-b,n-a}), \qquad \eta \wedge *\overline{\mu} = (\eta, \mu)_h \frac{\omega^{\wedge n}}{n!},$$

$$\Lambda\colon \Gamma(U, \mathcal{A}_X^{a,b}) \to \Gamma(U, \mathcal{A}_X^{a-1,b-1}), \qquad \Lambda = *^{-1}L* = (C^{-1}*)^{-1}LC^{-1}*.$$

The commuting relations between them

$$[L, C] = [\Lambda, C] = [*, C] = [L, *^2] = 0, \qquad [\Lambda, L^r] = \sum_p r(n-p-r+1)P_p$$

are still valid. A differential form v is primitive if $\Lambda v = 0$; the existence of the polynomials $G^n_{p,r}(\Lambda, L)$ described in Theorem C.2.8 gives the existence and unicity of Lefschetz decomposition for every differential p-form:

$$v = \sum_{r \geq \max(p-n,0)} L^r v_r, \qquad v \in \mathcal{A}_X^p, \quad \Lambda v_r = 0.$$

Definition C.3.1 A Hermitian metric with associated Kähler form ω is called a **Kähler metric** if $d\omega = 0$.

Example C.3.2 Probably the simplest example of a Kähler metric is the standard metric in $\mathbb{C}^n$. If $z_1, \ldots, z_n$ are the canonical coordinates, then the differentials $dz_1, \ldots, dz_n$ are a unitary basis of the holomorphic cotangent space at every point and the the Kähler form

$$\omega = \frac{i}{2} \sum_{i=1}^n dz_i \wedge d\overline{z_i}$$

is closed.

Kähler metrics have several nice and useful properties, and almost all of them follow by certain identities involving the operators L, Λ, ∂ and $\overline{\partial}$. For a nice description of these identities, in addition to d, d^C, ∂ and $\overline{\partial}$ we introduce the following operators:

$$\delta = -*d*, \qquad \delta^C = -*d^C* = C^{-1}\delta C,$$

$$\partial^* = -*\overline{\partial}* = \frac{\delta - i\delta^C}{2}, \qquad \overline{\partial}^* = -*\partial* = \frac{\delta + i\delta^C}{2}.$$

Theorem C.3.3 (Kähler identities) *Given a fixed Kähler metric on a complex manifold, the associated differentials and Lefschetz operators satisfy the following commuting relations:*

$[L, d] = 0$	$[L, d^C] = 0$	$[L, \partial] = 0$	$[L, \overline{\partial}] = 0$
$[\Lambda, d] = -\delta^C$	$[\Lambda, d^C] = \delta$	$[\Lambda, \partial] = i\overline{\partial}^*$	$[\Lambda, \overline{\partial}] = -i\partial^*$
$[L, \delta] = d^C$	$[L, \delta^C] = -d$	$[L, \partial^*] = i\overline{\partial}$	$[L, \overline{\partial}^*] = -i\partial$
$[\Lambda, \delta] = 0$	$[\Lambda, \delta^C] = 0$	$[\Lambda, \partial^*] = 0$	$[\Lambda, \overline{\partial}^*] = 0$

where the bracket $[-, -]$ *is the graded commutator in the space of endomorphisms of the sheaf of graded vector spaces* $\mathcal{A}_X^{*,*}$.

Proof We first prove that $[L, d] = Ld - dL = 0$. Denoting by $\omega \in A_X^{1,1}$ the Kähler form, since $d\omega = 0$, for every p-form v we have

$$[L, d]v = dv \wedge \omega - d(v \wedge \omega) = -(-1)^p v \wedge d\omega = 0.$$

Next, we prove that $[\Lambda, d] = \Lambda d - d\Lambda = -\delta^C$. According to the Lefschetz decomposition it is sufficient to prove that $[\Lambda, d]L^r u = -\delta^C L^r u$ for every $r \geq 0$ and every primitive p-form u ($p \leq n$).

Given a primitive form $u \in \mathcal{A}_X^p$, with $p \leq n$, since $L^{n-p+1}u = 0$ and $[L, d] = 0$, we have $L^{n-p+1}du = dL^{n-p+1}u = 0$ and by Theorem C.2.8 we have $du = u_0 + Lu_1$ with $u_0 \in \mathcal{A}_X^{p+1}$ and $\mathcal{A}_X^{p-1}$ primitive forms. Setting $\alpha = n - p$, we have

$$\begin{aligned} L^r u &= \Lambda dL^r u - d\Lambda L^r u = \Lambda L^r du - d\Lambda L^r u \\ &= \Lambda L^r u_0 + \Lambda L^{r+1} u_1 - r(\alpha - r + 1)dL^{r-1}u \\ &= r(\alpha - r)L^{r-1}u_0 + (r+1)(\alpha - r + 1)L^r u_1 \\ &\quad - r(\alpha - r + 1)L^{r-1}(u_0 + Lu_1) \\ &= -rL^{r-1}u_0 + (\alpha - r + 1)L^r u_1. \end{aligned}$$

On the other hand, we have by Lemma C.2.7

$$\begin{aligned} -\delta^C L^r u = C^{-1} * d * CL^r u &= C^{-1} * dC^2 C^{-1} * L^r u \\ &= C^{-1} * dC^2 (-1)^{p(p+1)/2} \frac{r!}{(\alpha - r)!} L^{\alpha - r} u \end{aligned}$$

and then

$$-\delta^C L^r u = (-1)^{p(p-1)/2} \frac{r!}{(\alpha - r)!} C^{-1} * L^{\alpha - r}(u_0 + Lu_1).$$

Again by Lemma C.2.7,

$$C^{-1} * L^{\alpha-r} u_0 = (-1)^{(p+1)(p+2)/2} \frac{(\alpha - r)!}{(r-1)!} L^{r-1} u_0,$$

$$C^{-1} * L^{\alpha-r+1} u_1 = (-1)^{(p-1)p/2} \frac{(\alpha - r + 1)!}{r!} L^r u_1.$$

Putting all the terms together, we obtain the equality $[\Lambda, d]L^r u = -\delta^C L^r u$.

The remaining 14 identities follow easily form $[L, d] = 0$ and $[\Lambda, d] = -\delta^C$. In fact, since $\Lambda = *^{-1} L* = *L*^{-1}$ we have

$$[\Lambda, \delta] = (*L*^{-1})(- * d*) - (- * d*)(*^{-1} L*) = - * [L, d]* = 0.$$

$$\begin{aligned}[L, \delta] &= -L * d * + * d * L = - * \Lambda d * + * d \Lambda * = - * [\Lambda, d]* \\ &= *\delta^C * = - *^2 d^C *^2 = d^C.\end{aligned}$$

The second column in the above table follows from the first and by the fact that C commutes with L and Λ. The last two columns are linear combinations of the first two. □

Corollary C.3.4 *Let ω be the Kähler form associated to a Kähler metric on a complex manifold. Then*

$$d\omega^{\wedge p} = d^C \omega^{\wedge p} = \delta \omega^{\wedge p} = \delta^C \omega^{\wedge p} = 0, \qquad \textit{for every} \quad p \geq 0.$$

Proof The equality $d\omega^{\wedge p} = 0$ follows immediately by the Leibniz rule. Since $\omega^{\wedge p}$ is a (p, p)-form, we have $C\omega^{\wedge p} = \omega^{\wedge p}$ and then also $d^C \omega^{\wedge p} = 0$; similarly, $\delta^C \omega^{\wedge p} = 0$ if and only if $\delta \omega^{\wedge p} = 0$. To conclude the proof we prove by induction of p that $\delta \omega^{\wedge p} = 0$. This is trivial for $p = 0$ since the operator δ has degree -1. Assuming $\delta \omega^{\wedge p-1} = 0$, the Kähler identity $[L, \delta] = d^C$ gives

$$0 = d^C \omega^{\wedge p-1} = L\delta \omega^{\wedge p-1} - \delta L \omega^{\wedge p-1} = -\delta \omega^{\wedge p}.$$

□

The Laplacian quartet is composed of the following operators:

1. $\Delta_d = \Delta = [d, \delta] = d\delta + \delta d,$
2. $\Delta_{d^C} = \Delta^C = C^{-1} \Delta C = [d^C, \delta^C] = d^C \delta^C + \delta^C d^C,$
3. $\Delta_\partial = \Box = [\partial, \partial^*] = \partial\partial^* + \partial^*\partial,$
4. $\Delta_{\overline{\partial}} = \overline{\Box} = [\overline{\partial}, \overline{\partial}^*] = \overline{\partial}\,\overline{\partial}^* + \overline{\partial}^*\,\overline{\partial}.$

In every one of the lines above the definition is given by the second and third equalities, while the first equality provides two different notations for the same operator. A straightforward computation shows that $\Delta + \Delta^C = 2\Box + 2\overline{\Box}$.

Corollary C.3.5 *In the above notation, if h is a Kähler metric then:*

$$[d, \delta^C] = [d^C, \delta] = [\partial, \overline{\partial}^*] = [\overline{\partial}, \partial^*] = 0, \qquad \frac{1}{2}\Delta = \frac{1}{2}\Delta^C = \Box = \overline{\Box}.$$

In particular, Δ is bihomogeneous of degree $(0, 0)$.

Proof According to Theorem C.3.3 and the Jacobi identity we have

$$[d, \delta^C] = [d, [d, \Lambda]] = \frac{1}{2}[[d, d], \Lambda] = 0.$$

The proof of $[d^C, \delta] = [\partial, \overline{\partial}^*] = [\overline{\partial}, \partial^*] = 0$ is the same and it is left as an exercise. For the equalities among Laplacians it is sufficient to show that $\Delta = \Delta^C$ and $\Box = \overline{\Box}$. According to the Kähler identities

$$\Delta = [d, \delta] = [d, [\Lambda, d^C]] = [[d, \Lambda], d^C] + [\Lambda, [d, d^C]].$$

Since $[d, d^C] = dd^C + d^C d = 0$ we have

$$\Delta = [d, \delta] = [[d, \Lambda], d^C] = [\delta^C, d^C] = \Delta^C.$$

The proof of $\Box = \overline{\Box}$ is the same. □

Corollary C.3.6 *In the above notation, if h is a Kähler metric, then Δ commutes with all the operators $P_{a,b}$, $*$, L, C, Λ, d, d^C, ∂, $\overline{\partial}$, δ, δ^C, ∂^*, $\overline{\partial}^*$.*

Proof Since Δ has bidegree $(0, 0)$ it is clear that it commutes with the projections $P_{a,b}$. Recalling that $\delta = - * d*$ and $*^2 d = -d *^2$ we get $d = *\delta*$ and then:

$$\begin{aligned} *\Delta &= *d\delta + *\delta d = - * d * d * + * \delta * \delta * = \delta d * + d\delta * = \Delta *, \\ [L, \Delta] &= [L, [d, \delta]] = [[L, d], \delta] + [[L, \delta], d] = [d^C, d] = 0, \\ [d, \Delta] &= [d, [d, \delta]] = \frac{1}{2}[[d, d], \delta] = 0. \end{aligned}$$

Now it is sufficient to observe that all the operators listed in the statement of the corollary belong to the $\mathbb{C}$-algebra generated by $P_{a,b}$, $*$, d and L. □

Definition C.3.7 A differential form $\eta \in \mathcal{A}_X^{*,*}$ is called **harmonic** if $\Delta\eta = 0$.

The harmonicity is preserved by Lefschetz decompositions, more precisely:

Corollary C.3.8 *Let h be a Kähler metric and let $v = \sum_r L^r v_r$ be the Lefschetz decomposition of a p-form. Then v is harmonic if and only if v_r is harmonic for every r.*

Proof Since Δ commutes with L, if $\Delta v_r = 0$ for every r then also $\Delta v = 0$. Conversely, if n is the dimension of the manifold since $v_r = G^n_{p,r}(\Lambda, L)v$ for suitable noncommutative polynomials with rational coefficients $G^n_{p,r}$, and Δ commutes with Λ, L, we obtain that $\Delta v = 0$ implies $\Delta v_r = 0$ for every r.

An alternative proof can be obtained by observing that, since $[\Delta, \Lambda] = 0$, if v is a primitive form, then Δv is primitive too. Therefore $\sum_r L^r \Delta v_r$ is the Lefschetz decomposition of Δv. In particular, $\Delta v = 0$ if and only if $\Delta v_r = 0$ for every r. □

Corollary C.3.9 *Every holomorphic differential p-form is harmonic with respect to any Kähler metric.*

Proof A holomorphic differential p-form is a $(p, 0)$-form ϕ such that $\overline{\partial}\phi = 0$. According to the equality $\Delta = 2\overline{\Box}$ it is sufficient to prove that $\overline{\partial}^* \phi = 0$, and this is clear since $\overline{\partial}^*$ is an operator of bidegree $(0, -1)$. □

Corollary C.3.10 *In the above notation, if h is a Kähler metric and v is a closed primitive (p, q)-form then v is harmonic.*

Notice that if either $p = 0$ or $q = 0$ then v is always primitive. The example of holomorphic forms on $\mathbb{C}^n$ shows that a primitive harmonic form is not necessarily closed; however, we shall prove in Lemma C.4.2 that if X is compact then every harmonic form is closed.

Proof Since v is closed and primitive we have $dv = \Lambda v = 0$, $[d, \Lambda]v = 0$, $\Delta v = d\delta v$, and then it is sufficient to prove that $\delta v = 0$. This follows from Kähler identities since

$$\delta v = C\delta^C C^{-1} v = i^{q-p} C\delta^C v = i^{q-p} C[d, \Lambda]v = 0.$$

□

C.4 Hodge Decomposition in Compact Hermitian Manifolds

Recall that a Hermitian manifold is a complex manifold equipped with a Hermitian metric on its tangent bundle. The goal of this section is to give a short review of the harmonic theory on compact Hermitian manifolds and in particular of the Hodge decomposition theorem C.4.3.

Let X be a compact Hermitian manifold of dimension n with Hermitian metric h and Kähler form ω. We have already seen in Remark C.1.10 that $\omega^{\wedge n}/n!$ is the volume form of the underlying Riemannian manifold; this implies in particular that

$$\mathrm{Vol}(X) = \frac{1}{n!}\int_X \omega^{\wedge n}.$$

The Hermitian metric induces a structure of a (Hermitian) pre-Hilbert space on $A_X^{*,*} = \Gamma(X, \mathcal{A}_X^{*,*})$ where the Hermitian product is given by

$$(\phi, \psi) = \int_X \phi \wedge *\overline{\psi} = \int_X (\phi, \psi)_h \frac{\omega^{\wedge n}}{n!}, \qquad \phi, \psi \in A_X^{*,*},$$

and $(-,-)_h$ is the Hermitian product of Definition C.1.7. Notice that there exists a orthogonal direct sum of pre-Hilbert spaces

$$A_X^{*,*} = \bigoplus_{a,b}^{\perp} A_X^{a,b}.$$

We have already observed that the operator $*\colon A_X^{*,*} \to A_X^{*,*}$ is a pointwise isometry for $(-,-)_h$ and then

$$(*\phi, *\psi) = (\phi, \psi), \qquad \phi, \psi \in A_X^{*,*}.$$

Similarly, by Lemma C.2.2 the operators L, Λ are mutually adjoint with respect to $(-,-)_h$ and then

$$(L\phi, \psi) = (\phi, \Lambda\psi), \qquad \phi, \psi \in A_X^{*,*}.$$

Proposition C.4.1 *With respect to the above pre-Hilbert space structures on $A_X^{*,*}$ we have the following pairs (written in columns) of mutually formally adjoint operators:*

operator	d	d^C	∂	$\overline{\partial}$	L	$*$
formal adjoint	δ	δ^C	∂^*	$\overline{\partial}^*$	Λ	$*^{-1}$

In particular, all the four Laplacians are formally self-adjoint operators.

Proof Recall that two operators f, g on $A_X^{*,*}$ are called formally adjoint if $(f\phi, \psi) = (\phi, g\psi)$ for every $\phi, \psi \in A_X^{*,*}$. We have already proved that L, Λ are adjoint and that $*$ is an isometry.

We show here only that δ is the formal adjoint of d. The proof for the operators d^C, ∂ and $\overline{\partial}$ is essentially the same and it is left as an exercise. Let ϕ be a p-form and ψ a $p+1$-form. By the Stokes theorem

$$0 = \int_X d(\phi \wedge \overline{*\psi}) = \int_X d\phi \wedge \overline{*\psi} + (-1)^p \int_X \phi \wedge \overline{d{*}\psi}.$$

Since $\overline{d{*}\psi} = \overline{d * \psi}$ and $d * \psi = (-1)^{2n-p} *^2 d * \psi = -(-1)^p * \delta\psi$ we get

$$0 = \int_X d\phi \wedge \overline{*\psi} - \int_X \phi \wedge \overline{*\delta\psi} = (d\phi, \psi) - (\phi, \delta\psi).$$

□

Let D be any of the operators $d, d^C, \partial, \overline{\partial}$; denote by D^* its formal adjoint and by

$$\Delta_D = DD^* + D^*D$$

its Laplacian (i.e., $\Delta_d = \Delta$, $\Delta_{\overline{\partial}} = \overline{\Box}$ etc.).

Lemma C.4.2 *In the above notation we have* $\ker \Delta_D = \ker D \cap \ker D^*$ *and* $\ker D$ *is orthogonal to* $\operatorname{Im} D^*$.

Proof The inclusion $\ker D \cap \ker D^* \subset \ker \Delta_D$ is immediate from the definitions of the Laplacian. Conversely, if $\Delta_D \phi = 0$ then

$$\begin{aligned} 0 &= (\Delta_D \phi, \phi) = (DD^*\phi, \phi) + (D^*D\phi, \phi) = (D^*\phi, D^*\phi) + (D\phi, D\phi) \\ &= \|D^*\phi\|^2 + \|D\phi\|^2 \end{aligned}$$

and this is possible only if $D^*\phi = D\phi = 0$. Finally, if $D\phi = 0$, then for every ψ we have $(\phi, D^*\psi) = (D\phi, \psi) = 0$. □

Theorem C.4.3 (Hodge decomposition on forms) *Let X be a compact Hermitian manifold. In the notation above, if D is one of the operators $d, d^C, \partial, \overline{\partial}$, then the spaces of D-harmonic forms $\mathcal{H}^p_D$ are finite dimensional and there exist orthogonal decompositions*

$$A^{*,*}_X = \ker \Delta_D \overset{\perp}{\bigoplus} \operatorname{Im} \Delta_D = \ker \Delta_D \overset{\perp}{\bigoplus} \operatorname{Im} D \overset{\perp}{\bigoplus} \operatorname{Im} D^*,$$

$$\ker D = \ker \Delta_D \overset{\perp}{\bigoplus} \operatorname{Im} D.$$

Proof This is a deep and nontrivial result on the theory of elliptic self-adjoint operators on compact manifolds; for a proof we refer to the existing literature, e.g. [101, 145, 264, 271].

Here we only notice that the equality $\operatorname{Im} \Delta_D = \operatorname{Im} D \oplus^{\perp} \operatorname{Im} D^*$ is a formal consequence of the decompositions $A^{*,*}_X = \ker \Delta_D \oplus \operatorname{Im} \Delta_D$. In fact we have already proved that the two subspaces $\operatorname{Im} D, \operatorname{Im} D^*$ are orthogonal and it is obvious that $\operatorname{Im} \Delta_D \subset \operatorname{Im} D \oplus \operatorname{Im} D^*$. Given any $Dx \in \operatorname{Im} D$, we can write $x = x_0 + \Delta_D x_1$, with $\Delta_D(x_0) = 0$. By Lemma C.4.2 we have $Dx_0 = 0$ and then

$$Dx = Dx_0 + D\Delta_D x_1 = \Delta_D D x_1 \in \operatorname{Im} \Delta_D.$$

A similar proof shows that $D^*y \in \operatorname{Im} \Delta_D$ for every y and then $\operatorname{Im} D \oplus \operatorname{Im} D^* \subset \operatorname{Im} \Delta_D$. Finally, the equality $\ker D = \ker \Delta_D \oplus \operatorname{Im} D$ follows immediately from $\ker \Delta_D \oplus \operatorname{Im} D \subset \ker D$ and $\ker D \perp \operatorname{Im} D^*$. □

A consequence of the Hodge decomposition $A^{*,*}_X = \ker \Delta_D \bigoplus^{\perp} \operatorname{Im} \Delta_D$ is that the map $\Delta_D \colon \operatorname{Im} \Delta_D \to \operatorname{Im} \Delta_D$ is bijective. This permits us to define the **Green operator** $G_D \colon A^{*,*}_X \to A^{*,*}_X$ as the composition

$$G_D \colon A^{*,*}_X \xrightarrow{\text{orthogonal projection}} \operatorname{Im} \Delta_D \xrightarrow{\Delta_D^{-1}} \operatorname{Im} \Delta_D \xrightarrow{\text{inclusion}} A^{*,*}_X,$$

and the **harmonic projection** $H_D\colon A_X^{*,*} \to A_X^{*,*}$:

$$H_D\colon A_X^{*,*} \xrightarrow{\text{orthogonal projection}} \ker \Delta_D \xrightarrow{\text{inclusion}} A_X^{*,*}.$$

It is immediate from the definition of G_D and H_D that

$$\Delta_D G_D = G_D \Delta_D = \mathrm{Id} - H_D, \qquad G_D H_D = H_D G_D = 0.$$

In particular the Green operator commutes with the laplacian Δ_D and the harmonic projection H_D.

Lemma C.4.4 *In the above notation, if a linear operator $K\colon A_X^{*,*} \to A_X^{*,*}$ commutes with Δ_D then also commutes with G_D; in particular, $D^*G_D = G_D D^*$. The operator $\rho = G_D D^*$ is called the* ***D*-propagator** *and satisfies the contraction*[1] *equalities:*

$$\rho^2 = \rho H_D = H_D \rho = 0, \qquad \rho D + D\rho = \mathrm{Id} - H_D.$$

Proof Since $K\Delta_D = \Delta_D K$ we have $K(\ker \Delta_D) \subset \ker \Delta_D$ and $K(\mathrm{Im}\, \Delta_D) \subset \mathrm{Im}\, \Delta_D$. If $x \in \ker \Delta_D$ then $KG_D(x) = 0 = G_D K(x)$, while if $x \in \mathrm{Im}\, \Delta_D$, denoting by $y = G_D x \in \mathrm{Im}\, \Delta_D$ we have $x = \Delta_D y$ and then $G_D K x = G_D \Delta_D K y = K y = K G_D x$.

Since $D^* D^* = 0$ we have $\rho^2 = 0$; since $D^* H_D = H_D D^* = 0$ we have $\rho H_D = H_D \rho = 0$ and finally, since $DG_D = G_D D$ we obtain

$$\rho D + D\rho = G_D D^* D + D G_D D^* = G_D \Delta_D = \mathrm{Id} - H_D.$$

□

Corollary C.4.5 *In the notation above, if $v \in A_X^p$ and $Dv = 0$ then there exists $w \in A_X^{p-1}$ such that $Dw = v - H_D v$.*

Proof Take $w = \rho(v)$, where ρ is the D-propagator. □

Definition C.4.6 Let X be a compact Hermitian manifold and D one of the operators $d, d^C, \partial, \overline{\partial}$. The space of ***D*-harmonic *p*-forms** on X is

$$\mathcal{H}_D^p = \ker \Delta_D \cap A_X^p.$$

When $D = d$ we recover the usual notion of harmonic form given in Definition C.3.7. Lemma C.4.2 implies in particular that on compact Hermitian manifolds every harmonic form is closed; notice that this is generally false for noncompact manifolds, cf. Corollary C.3.9.

Probably the most important property of d-harmonic forms is that they are a set of representatives for the de Rham cohomology of the compact Hermitian manifold X. In fact, for every p we have

---

[1] Up to a sign, cf. Definition 5.4.1.

$$\mathcal{H}_d^p = \ker \Delta_d \cap A_X^p \subset \ker(d\colon A_X^p \to A_X^{p+1})$$

and then there is a natural projection map

$$\mathcal{H}_d^p \to H^p(X, \mathbb{C}) = \frac{\ker(d\colon A_X^p \to A_X^{p+1})}{dA_X^{p-1}}.$$

Corollary C.4.7 *For every compact Hermitian manifold X, the natural projections $\mathcal{H}_d^p \to H^p(X, \mathbb{C})$ are isomorphisms. In particular, the dimension of $\mathcal{H}_d^p$ is independent of the choice of the Hermitian metric.*

Proof By the Hodge decomposition on forms and Lemma C.4.2 we have

$$A_X^p = \mathcal{H}_d^p \oplus d(A_X^{p-1}) \oplus \delta(A_X^{p+1}), \qquad \mathcal{H}_d^p \oplus d(A_X^{p-1}) \subset \ker d,$$

and then it is sufficient to prove that $\ker d \cap \operatorname{Im} \delta = 0$, i.e., that $d\delta x = 0$ if and only if $\delta x = 0$. This follows immediately from the adjunction formula $0 = (d\delta x, x) = (\delta x, \delta x) = \|\delta x\|^2$. □

Similar considerations hold for the space of $\overline{\partial}$-harmonic forms. The operators $\overline{\partial}, \overline{\partial}^*: A_X^{*,*} \to A_X^{*,*}$ are homogeneous of bidegree (0, 1) and (0, −1) respectively, hence $\overline{\Box}(A_X^{p,q}) \subset A_X^{p,q}$. Denoting by $\mathcal{H}_{\overline{\partial}}^{p,q} = \ker \overline{\Box} \cap A_X^{p,q}$ the Hodge decomposition on forms gives

$$A_X^{p,q} = \mathcal{H}_{\overline{\partial}}^{p,q} \oplus \overline{\partial}(A_X^{p,q-1}) \oplus \overline{\partial}^*(A_X^{p,q+1}), \qquad \mathcal{H}_{\overline{\partial}}^{p,q} \oplus \overline{\partial}(A_X^{p,q-1}) \subset \ker \overline{\partial}.$$

Corollary C.4.8 *For every compact Hermitian manifold X, the natural projection maps*

$$\mathcal{H}_{\overline{\partial}}^{p,q} \to H_{\overline{\partial}}^q(X, \Omega_X^p) = \frac{\ker(\overline{\partial}\colon A_X^{p,q} \to A_X^{p,q+1})}{\overline{\partial} A_X^{p,q-1}}$$

are isomorphisms.

Proof As in the proof of Corollary C.4.7 it is sufficient to prove that if $\overline{\partial}\,\overline{\partial}^* x = 0$ then also $\overline{\partial}^* x = 0$, and this follows from the adjunction $(\overline{\partial}\,\overline{\partial}^* x, x) = \|\overline{\partial}^* x\|^2$. □

Finally, it is worth observing that if the Kähler form is closed (Kähler metric) then, by Corollary C.3.5, $\Delta = \Delta^C = 2\Box = 2\overline{\Box}$ and therefore the four operators $d, d^C, \partial, \overline{\partial}$ have the same harmonic forms.

C.5 Compact Kähler Manifolds

In this section we prove that certain good properties concerning the topology and the complex structure of compact complex manifolds are true whenever we assume the existence of a Kähler metric. This motivates the following definition:

Definition C.5.1 A complex manifold equipped with a Kähler metric is called a **Kähler manifold.**

In the literature, a manifold admitting a Kähler metric, without actually fixing one, is either called a manifold of Kähler type or, with an abuse of notation, a Kähler manifold.

We note that, while every complex manifold admits a Hermitian metric (this is an easy application of partitions of unity), not every complex manifold is Kähler, see e.g. the Example C.5.5 below. On the positive side there is the following well known theorem:

Theorem C.5.2 *We have:*

1. $\mathbb{C}^n$, $\mathbb{P}^n$ *and complex tori are Kähler manifolds;*
2. *If* M *is a Kähler manifold and* $N \subset M$ *is a regular submanifold then also* N *is a Kähler manifold.*

In particular, every complex projective variety is a compact Kähler manifold.

From now on X is a fixed compact manifold of Kähler type of dimension n and we denote by $H^*(X, \mathbb{C})$ its de Rham cohomology. Since the de Rham differential commutes with conjugation of differential forms, the $\mathbb{R}$-linear operator $v \mapsto \overline{v}$ factors to cohomology.

Similarly, for every pair of integers (p, q) we denote by $H_{\partial}^{p,q}(X)$ the pth cohomology group of the complex $(A_X^{*,q}, \partial)$ and by $H_{\overline{\partial}}^{p,q}(X)$ the qth cohomology group of the complex $(A_X^{p,*}, \overline{\partial})$. Since $\overline{\partial v} = \overline{\partial}\overline{v}$, the complex conjugation of forms induces for every p, q a $\mathbb{C}$-linear isomorphism $H_{\overline{\partial}}^{p,q}(X) \simeq \overline{H_{\partial}^{q,p}(X)}$.

Theorem C.5.3 (Hodge decomposition on cohomology) *Let* X *be a compact manifold of Kähler type and denote by*

$$H_d^{p,q}(X) = \frac{\{\eta \in A_X^{p,q} \mid \partial\eta = \overline{\partial}\eta = 0\}}{A_X^{p,q} \cap d(A_X^{p+q-1})} \subset H^{p+q}(X, \mathbb{C})$$

the subspace of de Rham cohomology classes represented by d-closed differential forms of type (p, q). *Then*

$$H^m(X, \mathbb{C}) = \bigoplus_{p+q=m} H_d^{p,q}(X), \qquad H_d^{p,q}(X) = \overline{H_d^{q,p}(X)},$$

and the natural morphisms

$$H_{\partial}^{p,q}(X) \leftarrow \{\eta \in A_X^{p,q} \mid \partial\eta = \overline{\partial}\eta = 0\} \rightarrow H_{\overline{\partial}}^{p,q}(X)$$

induce two isomorphisms $H_{\partial}^{p,q}(X) \cong H_d^{p,q}(X) \cong H_{\overline{\partial}}^{p,q}(X)$.

Proof The equality $H_d^{p,q}(X) = \overline{H_d^{q,p}(X)}$ is clear; since $H_{\overline{\partial}}^{p,q}(X) \simeq \overline{H_{\partial}^{q,p}(X)}$ it is sufficient to prove that $H^m(X, \mathbb{C}) = \bigoplus_{p+q=m} H_d^{p,q}(X)$ and that the natural map $A_X^{p,q} \cap \ker \partial \cap \ker \overline{\partial} \to H_{\overline{\partial}}^{p,q}(X)$ induces an isomorphism $H_d^{p,q}(X) \cong H_{\overline{\partial}}^{p,q}(X)$.

Choose a Kähler metric on X and use it to define the Laplacians Δ and $\overline{\Box}$; according to Corollary C.3.5 $\Delta = 2\Box = 2\overline{\Box}$ and then Δ is bihomogeneous of bidegree $(0, 0)$, i.e., $\Delta(A_X^{p,q}) \subset A_X^{p,q}$ for every p, q. Therefore:

$$\ker \Delta \cap A_X^{p,q} = \ker \overline{\Box} \cap A_X^{p,q}, \qquad \ker \Delta \cap A_X^m = \bigoplus_{p+q=m} \ker \Delta \cap A_X^{p,q}.$$

By Corollaries C.4.7 and C.4.8 it is sufficient to prove that for every p, q the natural injective map

$$\ker \Delta \cap A_X^{p,q} = \ker \overline{\Box} \cap A_X^{p,q} \to H_d^{p,q}(X)$$

induced by the inclusion $\ker \Delta \cap A_X^{p,q} = \ker \overline{\Box} \cap A_X^{p,q} \subset A_X^{p,q} \cap \ker \partial \cap \ker \overline{\partial}$, is also surjective. Let η be a d-closed differential form of type (p, q), by Corollary C.4.5 the form η and its harmonic projection $H_d(\eta)$ have the same de Rham cohomology class. Again the equality $\Delta = 2\overline{\Box}$ implies that $H_d = H_{\overline{\partial}}$ is an operator of bidegree $(0, 0)$ and hence $H_d(\eta) \in \ker \Delta \cap A_X^{p,q}$. □

Corollary C.5.4 *Let X be a compact manifold of Kähler type; denote its Betti numbers by $b_i = \dim_{\mathbb{C}} H^i(X, \mathbb{C})$ and its Hodge numbers by $h^{p,q} = \dim_{\mathbb{C}} H^q(X, \Omega_X^p)$. Then:*

1. $b_i = \sum_{a+b=i} h^{a,b}$;
2. $h^{p,q} = h^{q,p}$, *in particular b_i is even if i is odd;*
3. $h^{p,p} > 0$, *in particular $b_i > 0$ if i is even;*
4. *every holomorphic form on X is d-closed;*
5. *every antiholomorphic form on X is d-closed.*

Proof Items 1 and 2 are immediate consequences of the Hodge decomposition on cohomology. Take a Kähler metric with Kähler form and use it to define the four Laplacians. According to Corollary C.3.4, $\omega^{\wedge p}$ is harmonic and then

$$\ker \overline{\Box} \cap A_X^{p,p} = \ker \Delta \cap A_X^{p,p} \neq 0$$

for every $0 \leq p \leq n$. Finally, by Corollary C.3.9, holomorphic forms are Δ-harmonic and therefore d-closed. Every antiholomorphic form is the conjugate of a holomorphic form. □

Example C.5.5 The Hopf surfaces (Example 1.1.5) have $b_1 = b_3 = 1$, $b_2 = 0$ and then they are not of Kähler type.

Finally we are ready to prove the $\partial\overline{\partial}$-lemma as a consequence of the existence of a contracting homotopy for the operator ∂ in the complex $(A_X^{*,*}, \overline{\partial})$.

Theorem C.5.6 *Let X be a compact manifold of Kähler type. Every choice of a Kähler metric on X determines a linear operator $\sigma\colon A_X^{*,*} \to A_X^{*,*}$ of bidegree $(0,-1)$ such that*

$$[\partial, \sigma] = 0, \qquad [\overline{\partial}, \sigma]\partial = [\overline{\partial}, \sigma\partial] = \partial.$$

Proof Let $G_{\overline{\partial}}$ be the Green operator associated to $\overline{\partial}$ and define $\sigma = G_{\overline{\partial}}\overline{\partial}^*$. According to Corollaries C.3.5, C.3.6 and Lemma C.4.4 we have

$$\sigma = \overline{\partial}^* G_{\overline{\partial}}, \quad [\partial, \sigma] = 0, \quad [\overline{\partial}, \sigma]\partial = G_{\overline{\partial}}\Delta_{\overline{\partial}}\partial = \partial,$$

where the last equality holds since the image of ∂ is contained in the image of $\Delta_{\overline{\partial}} = \Delta_{\partial}$. □

Corollary C.5.7 ($\partial\overline{\partial}$-Lemma) *Let X be a compact manifold of Kähler type. Then*

$$\operatorname{Im} \partial\overline{\partial} = \ker \partial \cap \operatorname{Im} \overline{\partial} = \ker \overline{\partial} \cap \operatorname{Im} \partial.$$

In particular, $\partial A_X^{*,*}$ and $\overline{\partial} A_X^{*,*}$ are acyclic subcomplexes of the de Rham complex $A_X^{*,*}$.

Proof It is sufficient to prove $\operatorname{Im} \partial\overline{\partial} = \ker \overline{\partial} \cap \operatorname{Im} \partial$, since the other equality is the conjugate of this one.

The inclusion $\subset$ is evident. Conversely, let $x = \partial y$ be a $\overline{\partial}$-closed and ∂-exact differential form; by Theorem C.5.6 there exists a linear operator $\sigma\colon A_X^{*,*} \to A_X^{*,*}$ of degree -1 such that $[\partial, \sigma] = 0$, $[\overline{\partial}, \sigma]\partial = \partial$ and therefore

$$x = \partial y = [\overline{\partial}, \sigma]\partial y = \overline{\partial}\sigma\partial y + \sigma\overline{\partial}\partial y = -\overline{\partial}\partial\sigma y - \sigma\overline{\partial}x = \partial\overline{\partial}\sigma y.$$

□

Corollary C.5.8 *Let X be a compact manifold of Kähler type. Then for every p, q the natural maps*

$$\frac{\ker \partial \cap \ker \overline{\partial} \cap A_X^{p,q}}{\partial\overline{\partial} A_X^{p-1,q-1}} \to \frac{\ker \partial \cap \ker \overline{\partial} \cap A_X^{p,q}}{\overline{\partial}(\ker \partial \cap A_X^{p,q-1})} \to \frac{\ker \overline{\partial} \cap A_X^{p,q}}{\overline{\partial} A_X^{p,q-1}} = H^q(X, \Omega_X^p),$$

$$\frac{\ker \partial \cap \ker \overline{\partial} \cap A_X^{q,p}}{\partial\overline{\partial} A_X^{q-1,p-1}} \to \frac{\ker \partial \cap \ker \overline{\partial} \cap A_X^{q,p}}{\partial(\ker \overline{\partial} \cap A_X^{q-1,p})} \to \frac{\ker \partial \cap A_X^{q,p}}{\partial A_X^{q-1,p}},$$

are isomorphisms.

Proof The two lines are conjugates each other and then it is sufficient to prove that the maps on the first row are isomorphisms. Since

$$\operatorname{Im} \partial\overline{\partial} = \operatorname{Im} \overline{\partial}\partial = \overline{\partial}(\operatorname{Im} \partial) \subset \overline{\partial}(\ker \partial) \subset \operatorname{Im} \overline{\partial} \cap \ker \partial$$

by the $\partial\,\overline{\partial}$-lemma we get $\operatorname{Im}\partial\overline{\partial} = \overline{\partial}(\ker\partial)$ and therefore

$$\frac{\ker\partial\cap\ker\overline{\partial}\cap A_X^{p,q}}{\partial\overline{\partial}A_X^{p-1,q-1}} = \frac{\ker\partial\cap\ker\overline{\partial}\cap A_X^{p,q}}{\overline{\partial}(\ker\partial\cap A_X^{p,q-1})}.$$

Consider now the short exact sequence of complexes

$$0\to(\ker\partial,\overline{\partial})\to(A_X^{*,*},\overline{\partial})\to(\operatorname{Im}\partial,\overline{\partial})\to 0.$$

By $\partial\overline{\partial}$-lemma the complex on the right is acyclic, hence $(\ker\partial,\overline{\partial})\to(A_X^{*,*},\overline{\partial})$ is a quasi-isomorphism and then

$$\frac{\ker\partial\cap\ker\overline{\partial}\cap A_X^{p,q}}{\overline{\partial}(\ker\partial\cap A_X^{p,q-1})} = \frac{\ker\overline{\partial}\cap A_X^{p,q}}{\overline{\partial}A_X^{p,q-1}}.$$

□

C.6 Degeneration of Spectral Sequences

Very frequently, certain properties of mathematical objects are expressed in terms of degeneration of spectral sequences. Since in deformation theory via DG-Lie algebras we often meet double complexes contained in two or more quadrants (e.g. the cochain complex of a semicosimplicial DG-Lie algebra), for completeness' sake we study here degenerations of (not necessarily regular) spectral sequences arising from complete and exhaustive filtrations; it is worth noticing that the majority of textbooks of homological algebra treat only degenerations of regular spectral sequences, for instance the ones arising from first-quadrant double complexes.

For simplicity of exposition we consider only the case of decreasing filtrations of DG-vector spaces, although everything can be easily extended to the complexes in a fixed abelian category.

Our basic setup is the data of a DG-vector space (V, d) equipped with a decreasing filtration

$$\cdots\ F^pV\subset F^{p-1}V\ \cdots,\qquad p\in\mathbb{Z}, \tag{C.7}$$

of DG-vector subspaces $F^pV\subset V$. Since by assumption every F^pV is a graded vector subspace of V, we have $F^pV=\oplus_n F^pV^n$, where $F^pV^n = F^pV\cap V^n$.

Definition C.6.1 (*cf. Sect.* 5.3) The decreasing filtration in (C.7) is called **exhaustive** if $\bigcup_p F^pV = V$; it is called **complete** if for every $n\in\mathbb{Z}$ the natural map

$$V^n\to\lim\frac{V^n}{F^pV^n}$$

is an isomorphism.

Example C.6.2 In the above setup, assume that that for every fixed degree n we have $F^pV^n = V^n$ for $p << 0$ and $F^qV^n = 0$ for $q >> 0$. Then the filtration F^pV is complete and exhaustive.

By definition, the spectral sequence:

$$(E_r^p, d_r), \quad r \geq 0, \quad p \in \mathbb{Z}, \quad E_r^p \in \mathbf{DG}, \quad d_r \in \operatorname{Hom}^1_{\mathbb{K}}(E_r^p, E_r^{p+r}),$$

of the decreasing filtration (C.7) is equal to:

$$Z_r^p = \{x \in F^pV \mid dx \in F^{p+r}V\}, \qquad E_r^p = \frac{Z_r^p}{Z_{r-1}^{p+1} + dZ_{r-1}^{p-r+1}},$$

and the maps $d_r \colon E_r^p \to E_r^{p+r}$ are induced by d in the obvious way. We can immediately verify that every d_r is properly defined and that $d_r^2 = 0$. Notice that

$$dZ_{r-1}^{p-r+1} = \{dx \mid x \in Z_{r-1}^{p-r+1}\} = d(F^{p-r+1}V) \cap F^pV,$$

$E_0^p = F^pV/F^{p+1}V$ and that the DG-vector space E_r^p is minimal (= trivial differential) for every $r > 0$ since $Z_r^p \subset Z_{r-1}^{p-r+1}$.

The following proposition is the basic property of spectral sequences.

Proposition C.6.3 *For every $p \in \mathbb{Z}$ and every $r \geq 0$ there exists a natural isomorphism of graded vector spaces*

$$E_{r+1}^p = \frac{\ker(d_r \colon E_r^p \to E_r^{p+r})}{d_r(E_r^{p-r})}.$$

Proof This a well known result and its proof is widely present in the literature. We don't need this result and the tedious but straightforward proof is omitted. □

Example C.6.4 The **Hodge filtration** on a complex manifold X is the decreasing bounded filtration on the de Rham complex $A_X^{*,*}$ defined as

$$F^pA_X^{*,*} = \bigoplus_{i \geq p} A_X^{i,*} = \bigoplus_{i \geq p} \bigoplus_j A_X^{i,j}.$$

The associated spectral sequence (E_r^p, d_r) is called the **Hodge to de Rham**[2] spectral sequence of X. Notice that $E_0^p = A_X^{p,*}$, $d_0 = \overline{\partial}$ and then $E_1^p = H_{\overline{\partial}}^{p,*}(X) = H^p(X, \Omega_X^p)$; moreover, it is easy to see that $d_1 \colon E_1^p \to E_1^{p+1}$ is equal to the map induced by ∂ in cohomology.

[2] In the literature the Hodge to de Rham spectral sequence is also called Fröhlicher or Hodge-Fröhlicher.

Definition C.6.5 A spectral sequence (E_r^p, d_r) **degenerates at** E_k if $d_r = 0$ for every $r \geq k$.

Equivalently, a spectral sequence (E_r^p, d_r) degenerates at E_k if $E_k^p = E_r^p$ for every $p \in \mathbb{Z}$ and every $r \geq k$.

Theorem C.6.6 *Let (E_r^p, d_r) be the spectral sequence of an exhaustive decreasing filtration $\{F^pV\}$ and let $k \geq 0$ be a fixed integer. Then the following four conditions are equivalent:*

1. $d_r = 0$ *for every* $r \geq k$*, i.e., the spectral sequence degenerates at* E_k*;*
2. $F^pV \cap d(V) \subset d(F^{p-k+1}V) + F^{p+1}V$ *for every* $p \in \mathbb{Z}$*;*
3. $F^pV \cap d(V) \subset d(F^{p-k+1}V) + F^qV$ *for every* $p, q \in \mathbb{Z}$*;*
4. $F^pV \cap d(F^{p-k}V) \subset d(F^{p-k+1}V) + F^qV$ *for every* $p, q \in \mathbb{Z}$.

Moreover, if the filtration is complete, then the above four conditions are equivalent to:

5. $F^pV \cap d(V) \subset d(F^{p-k+1}V)$ *for every* $p \in \mathbb{Z}$.

Proof We first prove the chain of implications $(5 \Rightarrow 3 \Rightarrow 2 \Rightarrow 1 \Rightarrow 4 \Rightarrow 3)$ and then, assuming completeness of the filtration, the implication $(2 \Rightarrow 5)$. Notice that (3) and (4) are trivially satisfied whenever $q \leq p$.

The implications $(5 \Rightarrow 3 \Rightarrow 2)$ are trivial. As regards $(2 \Rightarrow 1)$, assume

$$F^pV \cap d(V) \subset d(F^{p-k+1}V) + F^{p+1}V$$

for every p and let $r \geq k$ an integer. Given an element $x \in Z_r^p$ we have

$$dx \in F^{p+r}V \cap d(V) \subset d(F^{p+r-k+1}V) + F^{p+r+1}V \subset d(F^{p+1}V) + F^{p+r+1}V.$$

Thus $dx = dy + z$, with $y \in F^{p+1}V$ and $z \in F^{p+r+1}V$ and then $y \in Z_{r-1}^{p+1}$, $z \in Z_{r-1}^{p+r+1}$ since $dz = 0$ and $dy = dx - z \in F^{p+r}V$. Therefore $dx \in dZ_{r-1}^{p+1} + Z_{r-1}^{p+r+1}$ and this is implies that $d_r x = 0$ in E_r^{p+r}.

For the proof of $(1 \Rightarrow 4)$ we need to show that if $d_r = 0$ for every $r \geq k$, then

$$F^pV \cap d(F^{p-k}V) = dZ_k^{p-k} \subset d(F^{p-k+1}V) + F^qV$$

for every p, q. We may use induction on $q \geq p$, since the result is trivially true for $p \geq q$. Assume $q > p$ and, by induction, that $dZ_k^{p-k} \subset d(F^{p-k+1}V) + F^{q-1}V$. Given $x \in Z_k^{p-k}$ we can write

$$dx = dy + z, \qquad y \in F^{p-k+1}V, \quad z \in F^{q-1}V,$$

and then

$$x - y \in Z_r^{p-k}, \qquad \text{where } r = q - p + k - 1 \geq k.$$

Using $d_r = 0$ we get

$$z = d(x - y) \in Z_{r-1}^{p-k+r+1} + dZ_{r-1}^{p-k+1} = Z_{r-1}^{q} + dZ_{r-1}^{p-k+1} \subset d(F^{p-k+1}V) + F^q V.$$

We now prove (4 $\Rightarrow$ 3). As above it is not restrictive to assume $p \leq q$; since $d(V) = \cup d(F^p V)$, it is sufficient to show by induction on $r \geq k$ that

$$F^p V \cap d(F^{p-r}V) = dZ_r^{p-r} \subset d(F^{p-k+1}V) + F^q V \quad \text{for every} \quad q \geq p;\ r \geq k.$$

For $r = k$ this is exactly condition (4). Assume $r > k$ and by induction that

$$F^p V \cap d(F^{p-r+1}V) \subset d(F^{p-k+1}V) + F^q V, \quad \text{for every} \quad p, q.$$

We have

$$F^p V \cap d(F^{p-r}V) \subset F^{p+k-r}V \cap d(F^{p-r}V) \subset d(F^{p-r+1}V) + F^q V,$$

where the second inclusion follows by condition (4) with $p + k - r$ in place of p, and then every $dx \in F^p V \cap d(F^{p-r}V)$ may be written in the form $dx = dy + z$ with $z \in F^q V \subset F^p V$, $dy \in d(F^{p-r+1}V)$. By the inductive assumption

$$dy = dx - z \in F^p V \cap d(F^{p-r+1}V) \subset d(F^{p-k+1}V) + F^q V$$

and therefore

$$dx = dy + z \in d(F^{p-k+1}V) + F^q V\,.$$

Finally, we assume that the filtration is complete and prove (2 $\Rightarrow$ 5). Given $dx \in F^p V$, by condition (2) we can write

$$dx = dy_p + z_p, \qquad y_p \in F^{p-k+1}V, \quad z_p \in F^{p+1}V.$$

Since $z_p = d(x - y_p)$ we have $z_p \in F^{p+1}V \cap d(V)$ and then $z_p = dy_{p+1} + z_{p+1}$, with $y_{p+1} \in F^{p-k+2}$ and $z_{p+1} \in F^{p+2}V$. Repeating the procedure we get two sequences $y_q \in F^{q-k+1}V$, $z_q \in F^{q+1}V$, $q \geq p$ such that

$$\begin{aligned} dx &= dy_p + z_p = d(y_p + y_{p+1}) + z_{p+1} = d(y_p + y_{p+1} + y_{p+2}) + z_{p+2} = \cdots \\ &= d(y_p + y_{p+1} + \cdots + y_q) + z_q. \end{aligned}$$

Since the filtration is complete we have $dx = d(\sum_{q \geq p} y_q) \in d(F^{p-k+1}V)$. □

The following two corollaries are immediate consequences of Theorem C.6.6:

Corollary C.6.7 *Let* (E_r^p, d_r) *be the spectral sequence of a complete exhaustive decreasing filtration* $\{F^p V\}$ *of a DG-vector space* V*. Then the following conditions are equivalent:*

1. *$d_r = 0$ for every $r \geq 1$, i.e., the spectral sequence degenerates at E_1;*
2. *$F^pV \cap d(V) = d(F^pV)$ for every p;*
3. *the inclusion $F^pV \to V$ is injective in cohomology for every p.*

Corollary C.6.8 *Let (E_r^p, d_r) be the spectral sequence of a complete exhaustive decreasing filtration $\{F^pV\}$ of a DG-vector space V. For every integer $k > 0$ the following conditions are equivalent:*

1. *$d_r = 0$ for every $r \geq k$, i.e., the spectral sequence degenerates at E_k;*
2. *$F^pV \cap d(V) = d(F^{p-k+1}V)$ for every p;*
3. *the inclusions $F^pV \to V$ and $F^pV \to F^{p-k+1}V$ have the same kernel in cohomology for every p.*

Corollary C.6.9 *The Hodge to de Rham spectral sequence of a compact manifold of Kähler type degenerates at E_1.*

Proof Let X be a compact complex manifold. Since the Hodge filtration is complete and exhaustive, by Corollary C.6.7 the Hodge to de Rham spectral sequence degenerates at E_1 if and only if the inclusion $F^{p+1}A_X^{*,*} \subset F^pA_X^{*,*}$ is injective in cohomology for every p. Since the quotient complex $F^pA_X^{*,*}/F^{p+1}A_X^{*,*}$ is isomorphic to $(A_X^{p,*}, \overline{\partial})$, it is sufficient to prove that the projection $(F^pA_X^{*,*}, d) \to (A_X^{p,*}, \overline{\partial})$ is surjective in cohomology. Let $\eta \in A_X^{p,*}$ such that $\overline{\partial}\eta = 0$; if $d\eta = 0$ we have completed the proof, otherwise by the $\partial\overline{\partial}$-lemma there exists $\mu \in A_X^{p,*}$ such that $\partial\overline{\partial}\mu = \partial\eta$ and therefore $\eta - \overline{\partial}\mu$ is d-closed and in the same $\overline{\partial}$-cohomology class as η.

For a different proof, let $\mathcal{H}_X^{*,*} = \oplus_{a,b}\mathcal{H}_X^{a,b}$ be the space of harmonic forms with respect to a Kälher metric and consider the filtration $F^p\mathcal{H}_X^{*,*} = \oplus_{a \geq p,b}\mathcal{H}_X^{a,b}$; since the differential on $\mathcal{H}_X^{*,*}$ is trivial, its spectral sequence degenerates at E_0. The inclusion $\imath \colon \mathcal{H}_X^{*,*} \to \mathcal{A}_X^{*,*}$ is a morphism of filtered complexes, giving a morphism of spectral sequences that is an isomorphism at E_1 in view of the Hodge decomposition on cohomology. By Proposition C.6.3 the Hodge to de Rham spectral sequence degenerates at E_1. □

C.7 Exercises

C.7.1 Let $v \neq 0$ be a primitive differential (p, q)-form on a compact Kähler manifold X. If ω is the Kähler form of the metric, prove that

$$\int_M v \wedge \overline{v} \wedge \omega^{\wedge n-p-q} \neq 0.$$

C.7.2 In a first quadrant double complex $(L^{*,*}, d, \delta)$ of vector spaces, with d, δ differentials of respective bidegrees $(1, 0)$ and $(0, 1)$, consider the following conditions:

(A) there exists a linear operator $\sigma\colon L^{*,*} \to L^{*,*}$ of bidegree $(0, -1)$ such that

$$[d, \sigma] = 0, \qquad [\delta, \sigma]d = [\delta, \sigma d] = d;$$

(B) $\operatorname{Im} d\delta = \ker \delta \cap \operatorname{Im} d$;
(C) there exists a linear operator $\tau\colon L^{*,*} \to L^{*,*}$ of bidegree $(1, -1)$ such that

$$[d, \tau] = 0, \qquad [\delta, \tau] = d;$$

(D) for every $a_0 \in L^{p,q}$ such that $\delta a_0 = 0$ there exists a sequence $a_i \in L^{p+i,q-i}$ such that $da_i = \delta a_{i+1}$;
(E) the spectral sequence of the filtration $F^p = \oplus_{j\geq p} L^{j,*}$ degenerates at E_1.

Prove that $A \iff B \Rightarrow C \Rightarrow D \iff E$ and show by an example that $D \not\Rightarrow C$. (Hint: the implication $A \Rightarrow B$ is the same as in Corollary C.5.7. In order to prove $B \Rightarrow A$ write $L^{a,b} = F^{a,b} \oplus C^{a,b}$ with $F^{a,b} = dL^{a-1,b}$ and observe that the complexes $(F^{a,*}, \delta)$ are acyclic. Define first $\sigma\colon F^{a,b} \to F^{a,b-1}$ such that $[\delta, \sigma]d = d$ and then $\sigma\colon C^{a,b} \to C^{a,b-1}$ such that $[d, \sigma] = 0$. For the proof of $C \Rightarrow D$ see Example 13.6.5.)

References

General references for Kähler manifolds and are [55, 101, 264, 270, 271] and the approach used here follows essentially the classical book by Weil [270]. For a proof of Theorem C.5.2 and for more applications of Kähler metrics to algebraic geometry we refer to [101].

The standard proof about the existence of Hermitian metric on any complex manifold can be found for instance in [145, Thm. 3.14].

Good references for a general theory of spectral sequences are [91, 101, 264, 269]. The reader should be aware that in the literature there exists a certain confusion about the definition of degenerate spectral sequence, for instance the notion of degeneration of [91] is different from C.6.5. Here, following Deligne [53], we adopt the standard notion of degeneration used in algebraic geometry [101, 264].

It is worth pointing out that for non-Kähler compact complex manifolds, the Hodge to de Rham spectral sequence can be arbitrarily non-degenerate [23]; this means that for every $k \geq 2$ there exists a compact complex manifold such that this spectral sequence does not degenerate at the E_k.

References

1. Abraham, R., Marsden, J.: Foundations of Mechanics. Benjamin, New York (1967)
2. André, M.: Méthode simpliciale en algèbre homologique et algèbre commutative. Lect. Notes Math. **32** (1967)
3. Arapura, D.: Algebraic Geometry Over the Complex Numbers. Springer (2012)
4. Artamkin, I.V.: On deformations of sheaves. Math. USSR Izvestiya **32**(3), 663–668 (1989)
5. Artin, M.: On the solutions of analytic equations. Invent. Math. **5**, 277–291 (1968)
6. Artin, M.: Versal deformations and algebraic stacks. Invent. Math. **27**, 165–189 (1974)
7. Artin, M.: Lectures on Deformations of Singularities. Tata Institute of Fundamental Research, Bombay (1976)
8. Artin, M., Grothendieck, A., Verdier, J.L.: Théorie des topos et cohomologie étale des schémas (SGA4). Springer Lecture Notes in Mathematics, vol. 269, 270, 305. Springer (1972–73)
9. Atiyah, M.F.: Complex analytic connections in fibre bundles. Trans. Am. Math. Soc. **85**, 181–207 (1957)
10. Atiyah, M.F., Macdonald, I.G.: Introduction to Commutative Algebra. Addison-Wesley, Reading, Mass (1969)
11. Bandiera, R.: Nonabelian higher derived brackets. J. Pure Appl. Algebra **219**, 3292–3313 (2015)
12. Bandiera, R., Manetti, M.: On coisotropic deformations of holomorphic submanifolds. J. Math. Sci. Univ. Tokyo **22**, 1–37 (2015)
13. Bandiera, R., Manetti, M.: Algebraic models of local period maps and Yukawa algebras. Lett. Math. Phys. **108**, 2055–2097 (2018)
14. Bandiera, R., Manetti, M., Meazzini, F.: Formality conjecture for minimal surfaces of Kodaira dimension 0. Compositio Mathematica **157**, 215–235 (2021)
15. Banica, C., Stanasila, O.: Méthodes algébrique dans la théorie globale des espaces complexes. Gauthier-Villars (1977)
16. Barth, W., Peters, C., van de Ven, A.: Compact complex surfaces. Springer Ergebnisse Math. Grenz. **4** (1984)
17. Bauer, I., Catanese, F.: Burniat surfaces III: deformations of automorphisms and extended Burniat surfaces. Doc. Math. **18**, 1089–1136 (2013)
18. Bauer, I., Catanese, F.: On rigid compact complex surfaces and manifolds. Adv. Math. **333**, 620–669 (2018)
19. Bauer, I., Pignatelli, R.: Rigid but not infinitesimally rigid compact complex manifolds. Duke Math. J. Duke Math. J. **170**(8), 1757–1780 (2021)

M. Manetti, *Lie Methods in Deformation Theory*, Springer Monographs in Mathematics,
https://doi.org/10.1007/978-981-19-1185-9

20. Behrend, K., Manin, Yu.: Stacks of stable maps and Gromov–Witten invariants. Duke Math. J. **85**, 1–60 (1996)
21. Bering, K.: Non-commutative Batalin-Vilkovisky algebras, homotopy Lie algebras and the Courant bracket. Comm. Mat. Phys. **274**, 297–341 (2007)
22. Bering, K., Damgaard, P.H., Alfaro, J.: Algebra of higher antibrackets. Nucl. Phys. B **478**, 459–503 (1996)
23. Bigalke, L., Rollenske, S.: The Frölicher spectral sequence can be arbitrarily non-degenerate. Math. Ann. **341**, 623–628 (2008); Erratum: Math. Ann. **358**, 1119–1123 (2014)
24. Bloch, S.: Semi-regularity and de Rham cohomology. Invent. Math. **17**, 51–66 (1972)
25. Bogomolov, F.: Hamiltonian Kählerian manifolds. Dokl. Akad. Nauk SSSR **243**, 1101–1104 (1978). Soviet Math. Dokl. **19**, 1462–1465 (1979)
26. Böhning, C., von Bothmer, H.C.G., Pignatelli, R.: A rigid, not infinitesimally rigid surface with K ample. Boll. Unione Mat, Ital (2021)
27. Bousfield, A.K., Gugenheim, V.K.A.M.: On PL de Rham theory and rational homotopy type. Mem. Amer. Math. Soc. **8** (1976)
28. Bousfield, A.K., Kan, D.M.: Homotopy Limits, Completions and Localizations. Springer Lecture Notes in Mathematics, vol. 304. Springer (1972)
29. Brown, K.S.: Abstract homotopy theory and generalized sheaf cohomology. Trans. Amer. Math. Soc. **186**, 419–458 (1973)
30. Brzezinski, T., Wisbauer, R.: Corings and Comodules. Cambridge University Press (2003)
31. Buchweitz, R.-O., Flenner, H.: Deformation Theory. Draft version (2009)
32. Buijs, U., Félix, Y., Murillo, A., Tanré, D.: Lie Models in Topology. Progress in Mathematics, Birkhäuser (2020)
33. Burns, D., Wahl, J.: Local contribution to global deformations of surfaces. Invent. Math. **26**, 67–88 (1974)
34. Cartan, H.: Notion d'algèbre différentielle; applications aux groupes de Lie et aux variétés où opère un groupe de Lie. Colloque de Topologie, tenu à Bruxelles, 5–8 juin 1950, CBRM, Georges Thone, Liège, pp. 15–27 (1950)
35. Cartan, H.: Quotient d'un espace analytique par un groupe d'automorphismes. Algebraic geometry and topology: a symposium in honour of S. Lefschetz. Princeton Math. Series **12**, 90–102 (1957)
36. Cartan, H.: Elementary Theory of Analytic Functions of One or Several Complex Variables. Addison-Wesley (1963)
37. Cartan, H., Eilenberg, S.: Homological Algebra. Princeton University Press, Princeton (1956)
38. Catanese, F.: On the moduli spaces of surfaces of general type. J. Diff. Geom. **19**, 483–515 (1984)
39. Catanese, F.: Moduli of algebraic surfaces. Theory of Moduli (Montecatini Terme, 1985), 1–83, Lecture Notes in Mathematics, vol. 1337. Springer (1988)
40. Catanese, F.: Everywhere non reduced moduli space. Invent. Math. **98**, 293–310 (1989)
41. Catanese, F.: Moduli and classification of irregular Kaehler manifolds (and algebraic varieties) with Albanese general type fibrations. Invent. Math. **104**, 263–289 (1991)
42. Catanese, F.: Deformation in the large of some complex manifolds, I. Ann. Mat. Pura Appl. (4) **183**, Volume in memory of Fabio Bardelli, 261–289 (2004)
43. Catanese, F., Frediani, P.: Deformation in the large of some complex manifolds, II. Recent progress on some problems in several complex variables and partial differential equations, 21–41, Contemp. Math., **400**, Amer. Math. Soc., Providence, RI (2006)
44. Catanese, F.: A superficial working guide to deformations and moduli. Handbook of moduli. Vol. I, 161–215, Adv. Lect. Math. (ALM), 24, Int. Press, Somerville (2013)
45. Cattaneo, A., Fiorenza, D., Longoni, R.: Graded Poisson algebras. In: Naber, J.-P., Sheung Tsun Tsou, G. (eds.) Encyclopedia of Mathematical Physics, pp. 560–567. Academic Press, Elsevier (2006)
46. Cayley, A.: On the theory of the analytical forms called trees. Philos. Mag. **13**, 172–176 (1857)
47. Chevalley, C., Eilenberg, S.: Cohomology theory of Lie groups and Lie algebras. Trans. Am. Math. Soc. **63**, 85–124 (1948)

48. Clemens, H.: Geometry of formal Kuranishi theory. Adv. Math. **198**(1), 311–365 (2005)
49. Coutinho, S.C.: A Primer of Algebraic D-Modules. Cambridge University Press, Cambridge (1995)
50. Curtin, D.J.: Obstructions to deforming a space curve. Trans. A.M.S. **287**, 83–94 (1981)
51. Curtis, E.B.: Simplicial homotopy theory. Adv. Math. **6**, 107–209 (1971)
52. Dalakov, P.: On the L-infinity description of the Hitchin map. Rend. Ist. Mat. Univ. Trieste **46**, 1–17 (2014)
53. Deligne, P.: Théorème de Lefschetz et critères de dégénérescence de suites spectrales. Publ. Math. IHES **35**, 107–126 (1968)
54. Deligne, P.: Letter to J. Millson, April 24, 1986. Scanned copy available on J. Millson's webpage
55. Deligne, P., Griffiths, P., Morgan, J., Sullivan, D.: Real homotopy theory of Kähler manifolds. Invent. Math. **29**, 245–274 (1975)
56. Deligne, P., Illusie, L.: Relévements modulo p^2 et décomposition du complexe de de Rham. Invent. Math. **89**, 247–270 (1987)
57. Dolgachev, I.: Lectures on Invariant Theory. London Mathematical Society Lecture Note Series. Cambridge University Press, Cambridge (2003)
58. Dorfman, I.: Dirac Structures and Integrability of Nonlinear Evolution Equations. Wiley (2003)
59. Douady, A.: Obstruction primaire à la déformation. Sém. Cartan **13** (1960/61) Exp. 4
60. Drinfeld, V.: Letter to V. Schechtman, Sept. 18 (1988)
61. Dupont, J.L.: Curvature and Characteristic Classes. Lecture Notes in Mathematics, vol. 640. Springer (1978)
62. Dynkin, E.B.: Calculation of the coefficients of the Campbell-Hausdorff formula. Dokl. Akad. Nauk., **57**, 323–326 (1947). English translation in: Dynkin, E.B., Yushkevich, A.A., Seitz, G.M., Onishchik, A.L. (Eds.), Selected papers of E.B. Dynkin with commentary. American Mathematical Society/International Press, Providence, RI/Cambridge, MA (2000)
63. Ehresmann, C.: Sur les espaces fibrés differentiables. C. R. Acad. Sci., Paris **224**, 1611–1612 (1947)
64. Eilenberg, S., Mac Lane, S.: On the groups H(π, n). I. Ann. Math. **58**, 55–106 (1953)
65. Eilenberg, S., Steenrod, N.: Foundations of Algebraic Topology. Princeton University Press, Princeton (1952)
66. Eilenberg, S., Zilber, J.A.: Semi-simplicial complexes and singular homology. Ann. Math. **2**(51), 499–513 (1950)
67. Esposito, C.: Formality Theory. From Poisson Structures to Deformation Quantization. Springer Briefs in Mathematical Physics, vol. 2. Springer, Cham (2015)
68. Fantechi, B., Göttsche, L., Illusie, L., Kleiman, S.L., Nitsure, N., Vistoli, A.: Fundamental Algebraic Geometry: Grothendieck's FGA Explained. American Mathematical Society (2005)
69. Fantechi, B., Manetti, M.: Obstruction calculus for functors of Artin rings. I. J. Algebra **202**, 541–576 (1998)
70. Fantechi, B., Manetti, M.: On the T^1-lifting theorem. J. Algebr. Geom. **8**, 31–39 (1999)
71. Fantechi, B., Pardini, R.: Automorphism and moduli spaces of varieties with ample canonical class via deformations of abelian covers. Comm. Algebra **25**, 1413–1441 (1997)
72. Félix, Y., Halperin, S., Thomas, J.: Rational Homotopy Theory. Graduate Texts in Mathematics, vol. 205, Springer, New York-Berlin (2001)
73. Fiorenza, D., Iacono, D., Martinengo, E.: Differential graded Lie algebras controlling infinitesimal deformations of coherent sheaves. J. Eur. Math. Soc. (JEMS) **14**, 521–540 (2012)
74. Fiorenza, D., Manetti, M.: L_∞ structures on mapping cones. Algebra Number Theory **1**, 301–330 (2007)
75. Fiorenza, D., Manetti, M.: A period map for generalized deformations. J. Noncommut. Geom. **3**, 579–597 (2009)
76. Fiorenza, D., Manetti, M.: Formality of Koszul brackets and deformations of holomorphic Poisson manifolds. Homol. Homotopy Appl. **14**(2), 63–75 (2012)

77. Fiorenza, D., Manetti, M.: Formal Abel–Jacobi maps. Int. Math. Res. Not. **4**, 1035–1090 (2020)
78. Fiorenza, D., Manetti, M., Martinengo, E.: Cosimplicial DGLAs in deformation theory. Commun. Algebra **40**, 2243–2260 (2012)
79. Fiorenza, D., Martinengo, E.: A short note on infinity-groupoids and the period map for projective manifolds. Publ. nLab **2**(1), 1–13 (2012)
80. Fischer, G.: Complex Analytic Geometry. Lecture Notes in Mathematics, vol. 538. Springer (1976)
81. Flenner, H.: Über Deformationen holomorpher Abbildungen. Habilitationsschrift, Osnabrück (1978)
82. Flenner, H.: Ein Kriterium für die Offenheit der Versalität. Math. Z. **178**, 449–473 (1981)
83. Frégier, Y., Zambon, M.: Simultaneous deformations of algebras and morphisms via derived brackets. J. Pure Appl. Algebra **219**, 5344–5362 (2015)
84. Fukaya, K.: Deformation theory, homological algebra and mirror symmetry. Geometry and physics of branes (Como, 2001), Ser. High Energy Phys. Cosmol. Gravit., IOP Bristol, 121–209 (2003)
85. Gabriel, P., Zisman, M.: Calculus of Fractions and Homotopy Theory. Springer (1967)
86. Gelfand, S.I., Manin, Y.I.: Methods of homological algebra. Springer Monographs in Mathematics, Springer, Berlin (2003)
87. Gerstenhaber, M.: The cohomology structure of an associative ring. Ann. Math. **78**, 267–288 (1963)
88. Gerstenhaber, M.: On the deformation of rings and algebras. Ann. Math. **79**, 59–103 (1964)
89. Getzler, E.: Batalin–Vilkovisky algebras and two-dimensional topological field theories. Comm. Math. Phys. **159**, 265–285 (1994)
90. Getzler, E.: Lie theory for nilpotent L_∞-algebras. Ann. Math. **170**(1), 271–301 (2009)
91. Godement, R.: Topologie algébrique et théorie des faisceaux. Hermann, Paris (1958)
92. Goerss, P.G., Jardine, J.F.: Simplicial Homotopy Theory. Birkäuser (1999)
93. Goldman, W.M., Millson, J.J.: Deformations of flat bundles over Kähler manifolds. Geometry and topology (Athens, GA, 1985), 129–145, Lecture Notes in Pure and Appl. Math., **105**, Dekker, New York, (1987)
94. Goldman, W.M., Millson, J.J.: The deformation theory of representations of fundamental groups of compact Kähler manifolds. Publ. Math. I.H.E.S. **67**, 43–96 (1988)
95. Goldman, W.M., Millson, J.J.: The homotopy invariance of the Kuranishi space. Ill. J. Math. **34**, 337–367 (1990)
96. Grauert, H., Remmert, R.: Analytische Stellenalgebren, vol. 176. Spinger, Grundlehren (1971)
97. Grauert, H., Remmert, R.: Coherent Analytic Sheaves, vol. 265. Spinger, Grundlehren (1984)
98. Greub, W., Halperin, S., Vanstone, R.: Connections, Curvature and Cohomology. Academic Press, III (1976)
99. Greuel, G.M., Looijenga, E.: The dimension of smoothing components. Duke Math. J. **52**, 263–272 (1985)
100. Greuel, G.M., Lossen, C., Shustin, E.: Introduction to Singularities and Deformations. Springer (2007)
101. Griffiths, P., Harris, J.: Principles of Algebraic Geometry. Wiley-Interscience publication (1978)
102. Griffiths, P.H., Morgan, J.W.: Rational Homotopy Theory and Differential Forms. Birkhäuser Progress in Mathematics, vol. 16, 2nd edn. (2013)
103. Grothendieck, A.: A General Theory of Fiber Spaces. University of Kansas (1955)
104. Grothendieck, A.: Techniques des construction en géometrié analytique. VI. Étude locale des morphismes: germes d'espaces analytiques, platitude, morphismes simples. Sém. H. Cartan (1960–1961) exp. 13, p. 1–13
105. Grothendieck, A.: Éléments de géométrie algébrique. III. Étude cohomologique des faisceaux cohérents. II. Inst. Hautes Études Sci. Publ. Math. **17** (1963)
106. Grothendieck, A.: Éléments de géométrie algébrique IV, quatrième partie. Publ. Math. IHES **32**, 5–361 (1967)

107. Gunning, R., Rossi, H.: Analytic Functions of Several Complex Variables. Prenctice-Hall (1965)
108. Guralnick, R.M.: A note on commuting pairs of matrices. Linear Multilinear Alg. **31**, 71–75 (1992)
109. Hartshorne, R.: Algebraic Geometry, vol. 52. Springer GTM (1977)
110. Hartshorne, R.: Deformation Theory. Graduate Texts in Mathematics, vol. 257. Springer, New York (2010)
111. Helmstetter, J.: Radical d'une algèbre symétrique à gauche. Annales de l'Institut Fourier **29**(4), 17–35 (1979)
112. Hinich, V.: Descent of Deligne groupoids. Internat. Math. Res. Notices **5**, 223–239 (1997)
113. Hinich, V.: DG coalgebras as formal stacks. J. Pure Appl. Algebra **162**, 209–250 (2001)
114. Hirschhorn, P.S., Kan, D.M., Smith, J.H.: Homotopy limit functors on model categories and homotopical categories. American Mathematical Soc (2005)
115. Hirzebruch, F.: Topological Methods in Algebraic Geometry. Classics in Mathematics. Springer, Berlin Heidelberg New York (1978)
116. Hitchin, N.: Deformations of holomorphic Poisson manifolds. Mosc. Math. J. **12**(3), 567–591 (2012)
117. Horikawa, E.: Deformations of holomorphic maps I. J. Math. Soc. Jpn. **25**, 372–396 (1973)
118. Horikawa, E.: Deformations of holomorphic maps II. J. Math. Soc. Jpn. **26**, 647–667 (1974)
119. Horikawa, E.: Deformations of holomorphic maps III. Math. Ann. **222**, 275–282 (1976)
120. Huang, L.: On joint moduli spaces. Math. Ann. **302**, 61–79 (1995)
121. Huebschmann, J., Kadeishvili, T.: Small models for chain algebras. Math. Z. **207**, 245–280 (1991)
122. Humphreys, J.E.: Introduction to Lie Algebras and Representation Theory. Springer (1972)
123. Huybrechts, D.: Complex Geometry. Springer (2005)
124. Iacono, D.: Differential graded Lie algebras and deformations of holomorphic maps. Ph.d. Thesis, Università di Roma (2006)
125. Iacono, D.: L_∞-algebras and deformations of holomorphic maps. Int. Math. Res. Not. **8** 36 pp. (2008)
126. Iacono, D.: Deformations of algebraic subvarieties. Rend. Mat. Appl. **30**, 89–109 (2010)
127. Iacono, D.: A semiregularity map annihilating obstructions to deforming holomorphic maps. Canad. Math. Bull. **54**, 472–486 (2011)
128. Iacono, D.: Deformations and obstructions of pairs (X, D). Int. Math. Res. Not. **19**, 9660–9695 (2015)
129. Iacono, D.: On the abstract Bogomolov–Tian–Todorov theorem. Rend. Mat. Appl. **38**, 175–198 (2017)
130. Iacono, D., Manetti, M.: An algebraic proof of Bogomolov–Tian–Todorov theorem. In: Deformation Spaces, vol. 39, pp. 113–133. Vieweg (2010)
131. Iacono, D., Manetti, M.: Posetted trees and Baker–Campbell–Hausdorff product. Mediterr. J. Math. **10**(2), 611–623 (2013)
132. Iacono, D., Manetti, M.: Semiregularity and obstructions of complete intersections. Adv. Math. **235**, 92–125 (2013)
133. Iacono, D., Manetti, M.: On deformations of pairs (manifold, coherent sheaf). Canad. J. Math. **71**, 1209–1241 (2019)
134. Iacono, D., Manetti, M.: On the adjoint map of homotopy abelian DG-Lie algebras. Archivum Mathematicum (Brno) Tomus 55, 7–15 (2019)
135. Illusie, L.: Complexe cotangent et déformations I, II, vol. 239. Springer LNM (1971), vol. 283 (1972)
136. Jacobson, N.: Lie Algebras. Wiley, New York (1962)
137. Kaledin, D.: Some remarks on formality in families. Mosc. Math. J. **7**(4), 643–652 (2007)
138. Kas, A.: On obstructions to deformations of complex analytic surfaces. Proc. Nat. Acad. Sci. U.S.A. **58**, 402–404 (1967)
139. Kashiwara, M., Schapira, P.: Sheaves on manifolds. Grundlehren der Mathematischen Wissenschaften, vol. 292, Springer, Berlin (1990)

140. Katzarkov, L., Kontsevich, M., Pantev, T.: Hodge theoretic aspects of mirror symmetry. From Hodge theory to integrability and TQFT tt*-geometry, 87–174, Proc. Sympos. Pure Math., vol. 78. American Mathematical Society, Providence (2008)
141. Kawamata, Y.: Unobstructed deformations - a remark on a paper of Z. Ran. J. Algebraic Geom. **1**, 183–190 (1992)
142. Kawamata, Y.: Unobstructed deformations. II. J. Algebraic Geom. **4**, 277–279 (1995)
143. Kobayashi, S.: Differential Geometry of Complex Vector Bundles. Princeton University Press (1987)
144. Kodaira, K.: On stability of compact submanifolds of complex manifolds. Amer. J. Math. **85**, 79–94 (1963)
145. Kodaira, K.: Complex Manifolds and Deformation of Complex Structures. Springer (1986)
146. Kodaira, K., Nirenberg, L., Spencer, D.C.: On the existence of deformations of complex analytic structures. Ann. Math. **68**, 450–459 (1958)
147. Kodaira, K., Spencer, D.C.: On the variation of almost complex structures. In: Algebraic Geometry and Topology. Princeton University Press, pp. 139–150 (1957)
148. Kodaira, K., Spencer, D.C.: A theorem of completeness for complex analytic fibre spaces. Acta Math. **100**, 281–294 (1958)
149. Kodaira, K., Spencer, D.C.: A theorem of completeness of characteristic systems of complete continuous systems. Amer. J. Math. **81**, 477–500 (1959)
150. Kodaira, K., Spencer, D.C.: On deformations of complex analytic structures, I–II, III. Ann. Math. **67**, 328–466 (1958); **71**, 43–76 (1960)
151. Kontsevich, M.: Topics in algebra-deformation theory. Lecture notes of a course given at Berkeley (1994)
152. Kontsevich, M.: Enumeration of rational curves via torus actions. The moduli space of curves (Texel Island, 1994), 335–368, Progr. Math., vol. 129, Birkhäuser Boston, Boston, MA (1995)
153. Kontsevich, M.: Deformation quantization of Poisson manifolds. I. Lett. Math. Phys. **66**, 157–216 (2003)
154. Kontsevich, M., Manin, Yu.: Gromov–Witten classes, quantum cohomology, and enumerative geometry. Comm. Math. Phys. **164**, 525–562 (1994)
155. Kontsevich, M., Soibelman, Y.: Deformations of algebras over operads and the Deligne conjecture. Conférence Moshé Flato 1999, Vol. I (Dijon), 255–307, Math. Phys. Stud., 21, Kluwer Acad. Publ., Dordrecht (2000)
156. Kontsevich, M., Soibelman, Y.: Deformation theory, I. Book in progress. Available from Y. Soibelman's web page
157. Kosmann-Schwarzbach, Y.: Derived brackets. Lett. Math. Phys. **69**, 61–87 (2004)
158. Kosmann-Schwarzbach, Y., Magri, F.: Poisson–Nijenhuis structures. Ann. Inst. Henri Poincaré **53**, 35–81 (1990)
159. Koszul, J.-L.: Lectures on Fibre Bundles and Differential Geometry. Tata institute of fundamental research, Bombay (1960)
160. Koszul, J.-L.: Crochet de Schouten–Nijenhuis et cohomologie. Astérisque (Numero Hors Serie), pp. 257–271 (1985)
161. Kravchenko, O.: Deformations of Batalin–Vilkovisky algebras. In: Poisson geometry (Warsaw, 1998), vol. 51. Banach Center Publ., pages 131–139. Polish Acad. Sci., Warsaw (2000)
162. Kuranishi, M.: On the locally complete families of complex analytic structures. Ann. Math. **75**, 536–577 (1962)
163. Lada, T., Markl, M.: Strongly homotopy Lie algebras. Comm. Algebra **23**, 2147–2161 (1995)
164. Lada, T., Stasheff, J.D.: Introduction to sh Lie algebras for physicists. Int. J. Theor. Phys. **32**, 1087–1104 (1993)
165. Lang, S.: Algebra, 3rd edn. Springer (2002)
166. Lapin, S.V.: Differential perturbations and D_∞ differential modules. Sb. Math. **192**(11–12), 1639–1659 (2001)
167. Lazzeri, F.: Analytic singularities. In: Tognoli, A. (ed.) Singularities of Analytic Spaces. Bressanone 1974, Springer reprint (2010)

168. Lian, B.H., Zuckerman, G.J.: New perspectives on the BRST-algebraic structure of string theory. Commun. Math. Phys. **154**, 613–646 (1993)
169. Lichnerowicz, A.: Les variéts de Poisson et leurs algèbres de Lie associées. J. Differential Geom. **12**, 253–300 (1977)
170. Loday, J.-L., Vallette, B.: Algebraic operads. Grundlehren der mathematischen Wissenschaften, vol. 346. Springer (2012)
171. Looijenga, E.J.N.: Isolated singular points on complete intersections. London Math. Soc. L.N.S. **77** (1983)
172. Lunardon, L.: Some remarks on Dupont contraction. Rend. Mat. Appl. **39**, 79–96 (2018)
173. Lunts, V.A.: Formality of DG algebras (after Kaledin). J. Algebra **323**(4), 878–898 (2010)
174. Lurie, J.: Moduli problems for ring spectra. Proceedings of the International Congress of Mathematicians. Volume II, Hindustan Book Agency, New Delhi, pp. 1099–1125 (2010)
175. Mac Lane, S.: Categories for the Working Mathematician. Springer (1971)
176. Manetti, M.: Deformation theory via differential graded Lie algebras. In: Seminari di Geometria Algebrica 1998–1999. Scuola Normale Superiore, Pisa (1999)
177. Manetti, M.: On the moduli space of diffeomorphic algebraic surfaces. Invent. Math. **143**, 29–76 (2001)
178. Manetti, M.: Extended deformation functors. Int. Math. Res. Not. **14**, 719–756 (2002)
179. Manetti, M.: Cohomological constraint to deformations of compact Kähler manifolds. Adv. Math. **186**, 125–142 (2004)
180. Manetti, M.: Lectures on deformations of complex manifolds. Rend. Mat. Appl. **24**, 1–183 (2004)
181. Manetti, M.: Lie description of higher obstructions to deforming submanifolds. Ann. Sc. Norm. Super. Pisa Cl. Sci. **6**, 631–659 (2007)
182. Manetti, M.: Differential graded Lie algebras and formal deformation theory. In: Algebraic Geometry: Seattle 2005. Proc. Sympos. Pure Math. **80**, 785–810 (2009)
183. Manetti, M.: On some examples of obstructed irregular surfaces. Sci. China Math. **54**(8), 1713–1724 (2011)
184. Manetti, M.: Kodaira-Spencer formality of products of complex manifolds. In: Trends in Contemporary Mathematics. Springer INdAM Series, vol. 8, pp. 85–95 (2014)
185. Manetti, M.: On some formality criteria for DG-Lie algebras. J. Algebra **438**, 90–118 (2015)
186. Manetti, M.: Uniqueness and intrinsic properties of non-commutative Koszul brackets. J. Homotopy Relat. Struct. **12**, 487–509 (2017)
187. Manetti, M., Meazzini, F.: Deformations of algebraic schemes via Reedy-Palamodov cofibrant resolutions. Indagationes Mathematicae **31**, 7–32 (2020)
188. Manetti, M., Ricciardi, G.: Universal Lie formulas for higher antibrackets. SIGMA **12**, 053, 20 pp (2016)
189. Markl, M.: Deformation theory of algebras and their diagrams. Regional conference series in mathematics, vol. 116, American Mathematical Soc. (2012)
190. Markl, M.: On the origin of higher braces and higher-order derivations. J. Homotopy Relat. Struct. **10**(3), 637–667 (2015)
191. Martinengo, E.: Infinitesimal deformations of Hitchin pairs and Hitchin map. Internat. J. Math. **23**(7), 1250053, 30 pp (2012)
192. Matsumura, H.: On algebraic groups of birational transformations. Rend. Accad. Lincei Ser. **8**(34), 151–155 (1963)
193. Matsumura, H.: Commutative Ring Theory. Cambridge University Press (1986)
194. May, J.P.: Simplicial Objects in Algebraic Topology. Chicago Lectures in Mathematics (1967)
195. Merkulov, S.A.: Nijenhuis infinity and contractible dg manifolds. Compositio Mathematica **141**, 1238–1254 (2005)
196. Moore, J.C.: Homotopie des complexes monoïdaux, I. Séminaire Henri Cartan (1954–1955), exp. 18
197. Moore, J.C.: Algébre homologique et homologie des espaces classifiants. Séminaire Henri Cartan (1959–1960), exp. 7
198. Morrow, J., Kodaira, K.: Complex Manifolds. AMS Chelsea Publishing (1971)

199. Mukai, S.: An Introduction to Invariants and Moduli. Cambridge Studies in Advanced Mathematics, vol. 81. Cambridge University Press, Cambridge (2003)
200. Mumford, D.: Further pathologies in algebraic geometry. Amer. J. Math. **84**, 642–648 (1962)
201. Mumford, D.: Lectures on Curves on an Algebraic Surface. Annals of Mathematics Studies, vol. 59. Princeton University Press (1966)
202. Nakamura, I.: Complex parallelisable manifolds and their small deformations. J. Differ. Geom. **10**, 85–112 (1975)
203. Namba, M.: Families of Meromorphic Functions on Compact Riemann Surfaces. Lecture Notes in Mathematics, vol. 767. Springer, New York-Berlin (1979)
204. Narasimhan, R.: Introduction to the Theory of Analytic Spaces. Lecture Notes in Mathematics, vol. 25. Springer, Berlin-New York (1966)
205. Navarro Aznar, V.: Sur la théorie de Hodge–Deligne. Invent. Math. **90**, 11–76 (1987)
206. Newlander, A., Nirenberg, L.: Complex analytic coordinates in almost complex manifolds. Ann. Math., Second Series, **65**(3), 391–404 (1957)
207. Nijenhuis, A.: Jacobi-type identities for bilinear differential concomitants of certain tensor fields I. Indagationes Math. **17**, 390–403 (1955)
208. Nijenhuis, A., Richardson, R.W.: Cohomology and deformations of algebraic structures. Bull. Amer. Math. Soc. **70**(3), 406–411 (1964)
209. Nijenhuis, A., Richardson, R.W.: Cohomology and deformations in graded Lie algebras. Bull. Amer. Math. Soc. **72**, 1–29 (1966)
210. Nijenhuis, A., Richardson, R.W.: Deformation of Lie algebra structures. J. Math. Mech. **17**, 89–105 (1967)
211. Palamodov, V.P.: Deformations of complex spaces. Uspekhi Mat. Nauk. **31:3**, 129–194 (1976). Transl. Russian Math. Surveys **31:3**, 129–197 (1976)
212. Palamodov, V.P.: Deformations of complex spaces. In: Several complex variables IV. Encyclopaedia of Mathematical Sciences, vol. 10, pp. 105–194, Springer (1986)
213. Pardini, R.: Abelian covers of algebraic varieties. J. Reine Angew. Math. **417**, 191–213 (1991)
214. Pridham, J.: Unifying derived deformation theories. Adv. Math. **224**(3), 772–826 (2010)
215. Pridham, J.: Derived deformations of schemes. Commun. Anal. Geom. **20**, 529–563 (2012)
216. Procesi, C.: Lie Groups: An Approach Through Invariants and Representations. Springer (2007)
217. Quillen, D.: Rational homotopy theory. Ann. Math. **90**, 205–295 (1969)
218. Quillen, D.: On the (co)homology of commutative rings. Proc. Sympos. Pure Math. **17**, 65–87 (1970)
219. Ran, Z.: Deformations of maps. In: Algebraic Curves and Projective Geometry. Lecture Notes in Mathematics, vol. 1389 Springer, Berlin, Heidelberg (1989)
220. Ran, Ziv: Deformations of manifolds with torsion or negative canonical bundle. J. Algebraic Geom. **1**, 279–291 (1992)
221. Ran, Ziv: Semiregularity, obstructions and deformations of Hodge classes. Ann. Scuola Norm. Sup. Pisa Cl. Sci. **28**(4), 809–820 (1999)
222. Ran, Ziv: Canonical infinitesimal deformations. J. Algebraic Geom. **9**, 43–69 (2000)
223. Remmert, R.: Theory of Complex Functions, vol. 122. Springer GTM (1991)
224. Reutenauer, C.: Free Lie Algebras. Clarendon Press, Oxford (1993)
225. Riehl, E.: Categorical Homotopy Theory. Cambridge University Press (2014)
226. Rim, D.S.: Formal deformation theory. In: SGA 7 I, Exp. VI. Lecture Notes in Mathematics, vol. **288**, pp. 32–132. Springer (1972)
227. Rim, D.S.: Equivariant G-structure on versal deformations. Trans. Amer. Math. Soc. **257**, 217–226 (1980)
228. Roger, C.: Gerstenhaber and Batalin-Vilkovisky algebras; algebraic, geometric, and physical aspects. Archivum Mathematicum (Brno) **45**, 301–324 (2009)
229. Rollenske, S.: Lie-algebra Dolbeault-cohomology and small deformations of nilmanifolds. J. Lond. Math. Soc. **2**(79), 346–362 (2009)
230. Rollenske, S.: Geometry of nilmanifolds with left-invariant complex structure and deformations in the large. Proc. Lond. Math. Soc. (3) **99**(2), 425–460 (2009)

231. Rollenske, S.: The Kuranishi space of complex parallelisable nilmanifolds. J. Eur. Math. Soc. **13**, 513–531 (2011)
232. Ruiz, J.M.: The Basic Theory of Power Series. Vieweg Advanced Lectures in Mathematics (1993)
233. Shafarevich, I.R.: Basic Algebraic Geometry 1, Varieties in Projective Space, 3rd edn. Springer, Berlin (2013)
234. Schätz, F.: Coisotropic submanifolds and the BFV-complex. Dissertation zur Erlangung der naturwissenschaftlichen Doktorwürde, Zürich (2009)
235. Schechtman, V.: Local Structure of Moduli Spaces. Preprint (1997)
236. Scheja, G., Storch, U.: Über differentielle Abhängigkeit bei Idealen analytischer Algebren. Math. Z. **114**, 101–112 (1970)
237. Schlessinger, M.: Functors of Artin rings. Trans. Amer. Math. Soc. **130**, 208–222 (1968)
238. Schlessinger, M., Stasheff, J.D: Deformation Theory and Rational Homotopy Type. Preprint (1979)
239. Schlessinger, M., Stasheff, J.D: The Lie algebra structure of tangent cohomology and deformation theory. J. Pure Appl. Algebra **38**, 313–322 (1985)
240. Schouten, J.A.: Über Differentialkonkomitanten zweier kontravarianter Größen. Indagationes Math. **2**, 449–452 (1940)
241. Schuhmacher, F.: L_∞-déformations et cohomologie de Hochschild. Thése de Doctorat de Mathématiques de l'Université Joseph Fourier, Grenoble (2004)
242. Sernesi, E.: Deformations of algebraic schemes. Grundlehren der Mathematischen Wissenschaften, vol. 334. Springer, New York Berlin (2006)
243. Serre, J.P.: Faisceaux algébriques cohérents. Ann. Math. **61**, 197–278 (1955)
244. Serre, J.P.: Lie Algebras and Lie Groups, 2nd edn. Springer (1992)
245. Seshadri, C.S. Theory of moduli. Algebraic geometry (Proc. Sympos. Pure Math., Vol. 29, Humboldt State Univ., Arcata, Calif., 1974), pp. 263–304. Amer. Math. Soc., Providence, RI (1975)
246. Severi, F.: Sul teorema fondamentale dei sistemi continui di curve sopra una superficie algebrica. Ann. Mat. Pura Appl. **23**, 149–181 (1944)
247. Sharygin, G., Talalaev, D.: On the Lie-formality of Poisson manifolds. J. K-Theory **2**, no. 2, Special issue in memory of Yurii Petrovich Solovyev. Part 1, 361–384 (2008)
248. Shimizu, Y., Ueno, K.: Advances in moduli theory. Translations of Mathematical Monographs, vol. 206. American Mathematical Society, Providence (2002)
249. Stasheff, J.D.: The intrinsic bracket on the deformation complex of an associative algebra. J. Pure Appl. Algebra **89**, 231–235 (1993)
250. Stasheff, J.D.: The (secret?) homological algebra of the Batalin-Vilkovisky approach. Secondary calculus and cohomological physics (Moscow, 1997), Contemp. Math., vol. 219, pp. 195–210. Amer. Math. Soc. (1998)
251. Sullivan, D.: Infinitesimal computations in topology. Publications mathématiques de l'I.H.É.S. **47**, 269–331 (1977)
252. Sweedler, M.E.: Hopf Algebras. Benjamin, New York (1969)
253. Sweedler, M.E.: When is the tensor product of algebras local? Proc. Am. Math. Soc. **48**, 8–10 (1975)
254. Talpo, M., Vistoli, A.: Deformation theory from the point of view of fibered categories. In: Handbook of moduli, vol. III, 281–397, Adv. Lect. Math. (ALM), 26, Int. Press, Somerville (2013)
255. Tamarkin, D., Tsygan, B.: The ring of differential operators on forms in noncommutative calculus. In: Graphs and patterns in mathematics and theoretical physics, Proc. Sympos. Pure Math., vol. 73, pp. 105–131. Amer. Math. Soc., Providence, RI (2005)
256. Terilla, J.: Smoothness theorem for differential BV algebras. J. Topol. **1**, 693–702 (2008)
257. Tian, G.: Smoothness of the universal deformation space of compact Calabi–Yau manifolds and its Peterson–Weil metric. In: Yau, S.T. (ed.) Mathematical Aspects of String Theory, pp. 629–646. Singapore (1988)

258. Todorov, A.N.: The Weil–Peterson geometry of the moduli space of $SU(n \geq 3)$ (Calabi-Yau) manifolds I. Commun. Math. Phys. **126**, 325–346 (1989)
259. Vakil, R.: Murphy's law in algebraic geometry: badly-behaved deformation spaces. Invent. Math. **164**, 569–590 (2006)
260. Vaisman, I.: Lectures on the geometry of Poisson manifolds. Progress in Mathematics, vol. **118**. Birkhäuser (1994)
261. van der Waerden, B.L.: Algebra I, 7th edn. Springer (1991)
262. Venkata Balaji, T.E.: An introduction to families, deformations and moduli. Universitätsdrucke Göttingen, Göttingen (2010)
263. Vinberg, E.B.: The theory of homogeneous convex cones. Transl. Moscow Math. Soc. **12**, 340–403 (1963)
264. Voisin, C.: Théorie de Hodge et géométrie algébrique complexe. Société Mathématique de France, Paris (2002)
265. Voronov, T.: Higher derived brackets and homotopy algebras. J. Pure Appl. Algebra **202**(1–3), 133–153 (2005)
266. Voronov, T.: Higher derived brackets for arbitrary derivations. Travaux mathématiques. Fasc. XVI, 163–186, Univ. Luxemb., Luxembourg (2005)
267. Wahl, J.M.: Equisingular deformations of normal surface singularities. I. Ann. Math. **104**(2), 325–356 (1976)
268. Wawrik, J.J.: Obstruction to the existence of a space of moduli. In: Global Analysis. Princeton Univ. Press, 403–414 (1969)
269. Weibel, C.A.: An introduction to homological algebra. Cambridge Studies in Advanced Mathematics, vol. 38. Cambridge University Press, Cambridge (1994)
270. Weil, A.: Introduction à l'étude des variétés Kählériennes. Hermann, Paris (1958)
271. Wells, R.O.: Differential Analysis on Complex Manifolds. Springer (1980)
272. Whitney, H.: Geometric Integration Theory. Princeton University Press, Princeton, NJ (1957)
273. Witten, E., Zwiebach, B.: Algebraic structures and differential geometry in $2D$ string theory. Nucl. Phys. B **377**, 55–112 (1992)
274. Xu, Ping: Gerstenhaber algebras and BV-algebras in Poisson geometry. Comm. Math. Phys. **200**, 545–560 (1999)
275. Yekutieli, A.: MC elements in pronilpotent DG Lie algebras. J. Pure Appl. Algebra **216**, 2338–2360 (2012)
276. Zappa, G.: Sui sistemi continui di curve sopra una rigata algebrica. Gior. Mat. Battaglini **77** (1947)

Index

M. Manetti, *Lie Methods in Deformation Theory*, Springer Monographs in Mathematics, https://doi.org/10.1007/978-981-19-1185-9